Environmental
Bioassay Techniques and
their Application

Developments in Hydrobiology 54

Series editor

H. J. Dumont

Environmental Bioassay Techniques and their Application

Proceedings of the 1st International Conference held in Lancaster, England, 11–14 July 1988

Edited by
M. Munawar. G. Dixon, C. I. Mayfield,
T. Reynoldson and M. H. Sadar

Reprinted from Hydrobiologia, vols. 188/189 (1989)

Kluwer Academic Publishers

Dordrecht / Boston / London

Library of Congress Cataloging-in-Publication Data

Environmental bioassay techniques & their application / edited by M.
  Munawar ... [et al.].
      p.   cm. -- (Developments in hydrobiology ; 54)
    ISBN 0-7923-0498-5
    1. Biological assay. 2. Environmental impact analysis.
  I. Munawar, M. II. Title: Environmental bioassay techniques and
  their application. III. Series.
  QH541.15B54E58 1989
  574.5'222--dc20   89-27796

LIVERPOOL INSTITUTE
OF HIGHER EDUCATION

Order No./Invoice No. £176.15
L3328 / B704647

Accession No.
279178

Class No.
574.5222 MUN

Control No.
ISBN

Catal.
4 8/94

ISBN 0-7923-0498-5 ✔

Printed on acid-free paper

Kluwer Academic Publishers incorporates the publishing programmes of Dr W. Junk Publishers,
MTP Press, Martinus Nijhoff Publishers, and D. Reidel Publishing Company.

Distributors

for the United States and Canada: Kluwer Academic Publishers, 101 Philip Drive, Norwell,
MA 02061, U.S.A.
for all other countries: Kluwer Academic Publishers Group, P.O. Box 322, 3200 AH Dordrecht,
The Netherlands

Copyright

© 1989 Kluwer Academic Publishers, Dordrecht

All rights reserved. No part of this publication may be reproduced, stored in a retrieval system,
or transmitted in any form or by any means, mechanical, photocopying, recording, or otherwise,
without the prior written permission from the copyright owners.

PRINTED IN BELGIUM

Dedicated to
our common future through sustainable environment

Contents

VIII

Hydrobiologia **188/189**: XI–XII, 1989.
M. Munawar, G. Dixon, C. I. Mayfield, T. Reynoldson and M. H. Sadar (eds)
Environmental Bioassay Techniques and their Application.

Preface

The exponential expansion of human development, industrialization and exploitation of various resources has resulted in an undesirable nutrient enrichment and contamination of the aquatic environment. The loadings from municipal, industrial, urban, agricultural and atmospheric pollutants are a serious threat to the integrity and the conservation of the global aquatic ecosystem. It is, however, encouraging to see that concern for environmental and ecological sustainability has been echoed at the global level in the World Commission Report on Environment and Development (1987), the Canadian Environmental Protection Act (1988), the International Joint Commission (1987), the International Workshop on the Ecological Effects of *in situ* Sediment Contaminants (1987), the International Symposium on the Phycology of Large Lakes of the World (1987), the European Conference on Ecotoxicology (1988), and also in a review by Schindler (1987).

Traditionally, environmental assessment has been mainly geared towards chemical monitoring to develop preliminary guidelines for regulatory purposes. Needless to say, such a dogmatic approach obviously lacked the predictive and dynamic nature of the toxicological information presently available to managers based on laboratory and field bioassays results. The recognition of the need to base environmental management on experimental assays was an important step towards achieving ecosystem sustainability, and maintaining the integrity and health of the resident biota. It is now apparent that an 'ecosystem health' approach must inevitably adopt a holistic approach by employing a multi-trophic, multi-species, and multi-bioassay strategy from bacteria to fish. In order to obtain a state-of-the-art evaluation of this subject, an International Conference on Environmental Bioassay Techniques and Their Application was convened at the University of Lancaster, Lancaster, U.K., during the period July 11–14, 1988, which was sponsored by the International Association of Sediment Water Science. Approximately 100 papers were presented in various symposia of the conference including eight keynote and plenary papers.

A proposal for a refereed publication (not a proceedings) received overwhelming support from the contributors, which resulted in the submission of 90 manuscripts. Additional papers, though not presented at the conference, were included if they were relevant to the theme of the conference. The Kluwer Academic Publishers (The Netherlands), was chosen as the vehicle for publication. An Editorial Board (M. Munawar, G. Dixon, C. Mayfield, T. Reynoldson, M. H. Sadar) was established to undertake the massive processing and review of more than 90 manuscripts. The Board developed guidelines for the review and revision process and I was requested to serve as Editor-in-Chief in order to coordinate the entire editorial process and to maintain the consistency and quality of the publication. Every manuscript was evaluated by at least two referees and reviewed by the relevant Editor and the Editor-in-Chief. The massive effort of coordinating a large number of referees for the review necessitated the appointment of the following Associate Editors who contributed significantly to the review process: Dr. A. Niimi (Fisheries and Oceans Canada, Burlington, Ontario, Canada), and Dr. T. Edsall (U.S. Fish and Wildlife Service, Ann Arbor, Michigan, USA).

The entire effort has materialized with the acceptance of 67 manuscripts which have been carefully arranged, in sequence, into four parts:

Part One: Keynote, plenary and conceptual papers (11 papers),
Part Two: Bacteria, protozoa, phytoplankton and plants (26 papers),
Part Three: Invertebrates (15 papers) and
Part Four: Fish and General (15 papers).

The multi-trophic and multi-disciplinary aspect of the papers along with their international flavour is apparent. The volume provides a state-of-the-art coverage of bioassay technology from 15 countries including a few timely reviews and conceptual papers. Most of the contributions came from Canada (36%), United Kingdom (18%), U.S.A. (13%), The Netherlands (12%), France (3%), West Germany (3%), and Hong Kong (3%). Equal contributions came from Brazil, China, Hungary, Italy, New Zealand, Norway, Sweden and Switzerland.

I am grateful to the members of the Programme Committee and the Editorial Board for their hard work and excellent cooperation. I wish to express my gratitude to Dr. R. L. Thomas for his advice and assistance. I would like to commend Dr. John Cairns Jr. for his excellent effort in writing the foreword for this publication. I sincerely thank the numerous referees and Drs. A. Niimi and T. Edsall for their meticulous and high quality reviews which laid a solid foundation for this volume. I must thank all the contributors for their cooperation in responding patiently to the reviews and the recommended revisions which I believe has enhanced the standard of publication.

I wish to sincerely thank Dr. J. Cooley, Director of Great Lakes Laboratory for Fisheries and Aquatic Sciences, Fisheries and Oceans Canada, for his support and facilities. Thanks are also due to several colleagues and personnel who assisted in the conference programme and publication of the volume namely: J. Whittaker, I. F. Munawar, L. McCarthy, G. Leppard, W. Page, S. Daniels, W. Finn, H. F. Nicholson, J. Jones, and L. Keeler. Finally, I wish to express my gratitude to Wil Peters, publishers, and Dr. Henri Dumont, Editor-in-Chief of Hydrobiologia, for their continuing support, excellent cooperation and interest in this publication.

Environmentalists and toxicologists can now apply a multi-trophic battery of tests with a field-to-laboratory strategy instead of using only traditional lethality testing. These technologies have given toxicologists effective tools with which to probe, experimentally, the frontiers of ecosystem health. The emerging multi-trophic and multi-disciplinary approach has a tremendous potential for developing an early warning, predictive, preventive and remedial action oriented strategy for environmental management. This compendium of papers is a modest attempt to present the state-of-the-art in environmental bioassay technology, which we hope will be useful to researchers, managers and a variety of users who are striving to make our world environmentally sustainable.

'What is needed now is a new era of economic growth – growth that is forceful and at the same time socially and environmentally sustainable.'

Gro Harlem Brundtland, Chairman
The World Commission on Environment and Development

References

Canadian Environmental Protection Act (CEPA), 1988. Enforcement and compliance policy. Environment Canada. 58 pp.

European Conference on Ecotoxicology. 1988. Copenhagen.

International Joint Commission. 1987. Guidance on characterization of toxic substances problems in Areas of Concern in the Great Lakes basin. Report to the Great Lakes Water Quality Board, Windsor, Ontario. 179 pp.

International Symposium on *the Phycology of Large Lakes of the World*, 1987. M. Munawar (ed), Arch. Hydrobiol. Beih. Ergebn. Limnol. 25. 256 pp.

International Workshop on *the Ecological Effects of in situ Sediment Contaminants*, 1987. R. Thomas, R. Evans, A. Hamilton, M. Munawar, T. Reynoldson and M. H. Sadar (eds.). Hydrobiologia 149: 272 pp.

Schindler, D. W., 1987. Detecting ecosystem responses to anthropogenic stress. Can. J. Fish. Aquat. Sci. 44 (Suppl. 1): 6–25.

World Commission Report on Environment and Development. 1987. *Our Common Future*. Oxford University Press. 400 pp.

DR. M. MUNAWAR, CHAIRMAN
Programme Committee & Editorial Board
Environment Bioassay Techniques
and their Application
Burlington, Ontario, Canada.
September, 1989.

Hydrobiologia **188/189**: XIII–XIV, 1989.
M. Munawar, G. Dixon, C. I. Mayfield, T. Reynoldson and M. H. Sadar (eds)
Environmental Bioassay Techniques and their Application.

Foreword

This conference was held at a particularly crucial period in the continuing evolution of bioassay techniques and, particularly, the application of their results to the protection of complex natural systems. Most environmental assessment in the 1940s and 1950s was carried out by sanitary engineers and chemists on purely chemical/physical measurements. Bioassays were considered a luxury or novelty rather than a standard practice. When short-term tests were used, they were batch solutions of a test mixture as a rule, and continuous-flow or renewal of test solution was virtually unheard of except in experimental situations. There was even a period in the United States when technological solutions for environmental protection, such as best applicable technology (BAT) or best practicable technology (BPT), were thought adequate. In short, if the best treatment system possible were installed, the environment would be inevitably protected. Dependence on technology and chemical/physical measurements in the absence of biological measurements quickly proved inadequate, and bioassays were given more recognition. Ultimately, there was the realization that no instrument devised by man can measure toxicity – only living material can be used for this purpose. A period followed of developing more sensitive end points than lethality, extending the duration of the tests, utilizing more life history stages in the tests, and expanding the number of species routinely used for bioassay purposes. In 1978, the first publication of the series of books now known as the Pellston Conferences on Hazard Evaluation coupled the estimation of environmental fate of chemical compounds with environmental effects. Previously, these two essential components had not been routinely considered simultaneously or in an integrated fashion. More recently, attention has been given to the predictive quality of bioassays including extrapolation from laboratory to field conditions, extrapolation from one level of biological organization to another (e.g., single species to communities and ecosystems), extrapolation from short exposure times to generational exposure, and extrapolation from one set of environmental conditions to another. The coupling of environmental fate with effects information naturally led to testing in more complex laboratory systems (microcosms and mesocosms) or compartments of natural systems (field enclosures) so that both types of information could be generated by the same system. These conditions made correlations between events easier and also produced more cost-effective generation of bioassay information. The correspondence of tests in simple laboratory systems to responses in complex natural systems has been of considerable interest in recent years, as has the problem of validating predictions based on laboratory results or surrogates of natural systems such as microcosms or mesocosms. This problem has yet to be resolved, but remarkable progress has been made in a relatively short time. It is instructive to compare the titles in the world's leading environmental and ecotoxicological journals with titles in the past, especially before such journals existed.

Although I have been a critic (hopefully a constructive one) of present bioassay techniques (including my own), I have never waivered in my conviction that decisions made on the crudest bioassays are more defensible and produce far better results than those made without any bioassay information whatsoever! However, it must be recognized that extrapolation from any artificial system, including those used for bioassays, to a complex natural system is a challenging and difficult task. As a consequence, the rapid evolution of the field should be an inspiration because the truth is being uncovered much more rapidly than anyone could have predicted in the past. This, of course, does not mean that complacency should set in and that present methods are satisfactory. However, scientists should be gratified that the degree of environmental protection available through bioassays is far better than it was when I entered the field under the direction of Ruth Patrick in 1948. My students look at pictures of the apparatus used in those days and exclaim, 'You actually used that!' I am confident that their students will do the same, which means that the rapidity of development of this field is phenomenal. Not every development leads to improvement, but those of us professionally involved in carrying out bioassays realize how much better off we are today than we were a few years ago.

Conferences such as this one pull together present and past developments of the field and are catalysts for the next leap to further improvement in the scientific and technical foundations of bioassay methodology and application.

JOHN CAIRNS, JR.
University Distinguished Professor and Director
University Center for Environmental and Hazardous Materials Studies
Virginia Polytechnic Institute and State University
Blacksburg, VA 24061
USA

Hydrobiologia **188/189**: 1–3, 1989.
M. Munawar, G. Dixon, C. I. Mayfield, T. Reynoldson and M. H. Sadar (eds)
Environmental Bioassay Techniques and their Application.
© 1989 *Kluwer Academic Publishers. Printed in Belgium.*

International Conference on Environmental Bioassay Techniques and their Application

An environmental research and management view

Sir Hugh Fish CBE
Chairman, Natural Environment Research Council of the United Kingdom, Polaris House, North Star Avenue Swindon, Wiltshire SN2 1EU, U.K.

Key words: objectives, standards, needs, priorities

Abstract

A short overview is given of the primary objectives of environmental research and management and of the importance of bioassay to these objectives. Current deficiencies in the availability of quantifiable bioassays and in the formulation of research needs and priorities are highlighted.

Mr. Chairman, Ladies and Gentlemen

My objective this morning is to start off this important International Conference on the correct track, and the correct track is of course that which matches the Conference's objectives. These have been clearly stated in the Conference documents under six headings. Five of these can be divided into two categories – namely those relating to the bioassay techniques themselves, and those relating to the application of these techniques in environmental management.

However there is one objective – to assess the research needs in the field of environmental bioassay – which must take into account on the one hand the present and future state of research in bioassay, and on the other hand the needs of environmental management. I am quite sure that your undoubted skills and knowledge will be able to take full and proper account of the state of, and trends in, the science and technology of environmental bioassay without any guidance from me. One of the reasons why I am sure of this is that several of the scientists employed in NERC's laboratories and grant-aided associations are presenting papers at this Conference. So it should help the proceedings if I include in my contribution some observations on the needs of environmental management derived from my experience in that activity.

First of all, I would remind you of the objectives of environmental research. These are:
1. To understand environmental processes, whether these are on the global, regional, local, and so on right down to the molecular, scale.
2. To understand the impact of man's activities on these processes.
3. To apply these understandings to achieve the optimal beneficial use of environmental resources consistent with proper sustainment of the environment.

Now of these objectives the first and second mainly involve basic research, which has important cultural value as well as its ultimate utilitarian value. The execution of this basic research depends, of course, on the technological consequences of other basic research. The third objective embraces mainly strategic and applied environmental research, again with the indispensable input of other applied sciences. Thus, the proceedings of this Conference should, and will, embrace the entire spectrum of basic, strategic and applied science.

Now as regards the environmental management scene, I will not bore you with lengthy quotes of environmental history. Yet this history, judged with the advantage of hindsight which advancing knowledge provides, shows to me that most endeavours in environmental management in the past were at best only partially right. It was of course considerations of expediency and lack of knowledge that brought this about. A striking example of this occurred in the great drive to improve public health in the United Kingdom in the mid-nineteenth century. Then to reduce urban filth and squalor and associated disease, water-borne sanitation was developed. Unfortunately this caused greatly increased pollution of rivers used as untreated sources of drinking water, giving rise to massive outbreaks of waterborne disease. Another example was the relief of local gross air pollution in the United Kingdom in the 1950's and 1960's, which has apparently contributed to the regional acid-rain problem of northern Europe. From this kind of occurrence we have learned that we simply cannot kick pollution around until we lose it – it will always turn up again in some consequence, somewhere.

Fortunately we are now in the age of pollution abatement characterised by standards for water quality. Also we are moving into environmental impact assessments in which biological considerations are predominant. Water quality standards are woefully short of biological standards based on bioassay either in the specific-test context or in the general indicator context. NERC is vitally interested in these matters, at our Plymouth Marine Laboratory and at the Windermere Laboratory of the Freshwater Biological Association. One of the great problems involved in the development of bioassay for environmental purposes is to separate the effects of chemical pollution stress from other environmental stresses. One new way of untangling this difficulty, in the case of freshwaters, has been developed at the River Laboratory of the Freshwater Biological Association. A similar approach to estuarial waters would yield

dividends. In the case of open coastal waters, where virtually the only man-made stress lies in chemical pollution, the problems of interpretation of bioassay are different.

It is important to note that good progress is being made in the development of quantifiable bioassay techniques for fresh and estuarine waters, and in using bioassay as the basis of predicting the consequences of man-made development. This importance springs from the fact that we shall never succeed in replacing, or powerfully augmenting, chemical standards with biological standards without the availability of bioassay techniques which produce quantified results. Chemical standards certainly have serious shortcomings, but because they are quantifiable they serve the purpose of the administrator and of some environmental managers.

Turning to the water management scene, what the modern water manager, including the proposed National River Authority in the United Kingdom, needs are sound bioassay techniques to indicate:
1. The specific effects of existing point and non-point sources of environmental stress on biota.
2. The effects of new environmental stress arising from proposed new pollution, or new water abstraction or new land drainage, and of how adverse effects of these may be reduced to acceptable levels.
3. How the techniques may be applied in a surveillance role, particularly in maintaining conformity with regulatory requirements and in giving warning of the occurrence of gross pollution.

This Conference should be able to assess how far, to date, these management needs have been, or are about to be, delivered as a result of bioassay research. In this context the acid test of delivery or non-delivery will be the answers to the following questions. First, can the instrumentation and know-how be transferred to the water managers? Second, and alternatively, can a bioassay service be provided to water managers where technology transfer is not yet feasible? At this point of course, the current fact that in many cases water managers will not, until obliged to, meet the cost of technology transfer or a new service, really confounds us all.

On the question of research needs, posed as one of the objectives of this Conference, I have given you a very brief outline of the water manager's needs for the products of research. It will be for you to put flesh on that skeleton. But often when I hear scientists talking of research needs, I get the feeling that the needs of researchers, as distinct from the needs of research users, seem to be the major concern. I could deliver a powerful homily on this subject, but this is not the occasion for pyrotechnics on my part – above all you will not be able to question me hard on my views. However I can tell you that unless researchers in the strategic and applied aspects of science really 'sell' the excellence, timeliness and economic and social relevance of their proposed research, to the users of the research results or to Government as the proxy user of research results for the public, sources of funds to finance the proposed research will be very difficult to find. This is now the position in the United Kingdom, and it increasingly applies in most of the indus-trialised countries. Furthermore, in the UK and elsewhere, on-going strategic and applied research carried out on public funds will have to be given priorities against the claims of new research, and low priority work, current or new, curtailed. We do not like these realities, but we shall have to cope with them now and for some time to come.

I hope therefore that this Conference will not conclude, in the research context, that all ideas for new research in bioassay should be equal contenders for funding. I would advise the Conference to endeavour to attach some priority, if only in the broadest terms, to stated research needs. If that can be done the conference will more than fulfil its objectives.

Two things are certain. The importance of and need for effective environmental bioassay techniques are already great and growing fast. Secondly, there is a bright, and seemingly endless future in this field of activity, particularly when viewed against the background of the advances that molecular biosciences will bring. On that note I will conclude, and wish you all an enjoyable, instructive and productive Conference.

Hydrobiologia **188/189**: 5–20, 1989.
M. Munawar, G. Dixon, C. I. Mayfield, T. Reynoldson and M. H. Sadar (eds)
Environmental Bioassay Techniques and their Application.
© 1989 *Kluwer Academic Publishers. Printed in Belgium.*

The scientific basis of bioassays

John Cairns, Jr.[1] & James R. Pratt[2]
[1]*University Center for Environmental and Hazardous Materials Studies and Department of Biology, Virginia Polytechnic Institute and State University, Blacksburg, Virginia 24061, U.S.A.;* [2]*School of Forest Resources, The Pennsylvania State University, University Park, Pennsylvania 16802, U.S.A.*

Key words: bioassay, ecotoxicology, stress, microcosm, mesocosm

Abstract

The ultimate goal of ecotoxicological testing is to predict ecological effects of chemicals and other stressors. Since damage should be avoided rather than corrected after it occurs, the predictive value of such tests is crucial. A modest base of evidence shows that, in some cases, extrapolations from bioassays on one species to another species are reasonably accurate and, in other cases, misleading. Extrapolations from laboratory bioassays to response in natural systems at the population level are effective if the environmental realism of the bioassay is sufficiently high. When laboratory systems are poor simulations of natural systems, gross extrapolation errors may result. The problem of extrapolating among levels of biological organization has not been given the serious attention it deserves, and currently used methodologies have been chosen for reasons other than scientific validity. As the level of biological organization increases, new properties are added (e.g., nutrient cycling, energy transfer) that are not readily apparent at the lower levels. The measured responses (or end points) will not be the same at all levels of biological organization, making the validation of predictions difficult. Evidence indicates that responses of ecologically complex laboratory systems correspond to predicted and documented patterns in stressed ecosystems. The difficulties of improving the ecological evidence used to predict adverse effects are not insurmountable since the essence of predictive capability is the determination of effects thresholds at all levels of organization. The dilemma between basing predictive schemes on either traditional or holistic methods can only be solved by facing scientific and ethical questions regarding the adequacy of evidence used to make decisions of environmental protection.

'When very little is known about an important subject, the questions people raise are almost invariably ethical. Then as knowledge grows, they become more concerned with information and amoral, in other words more narrowly intellectual. Finally, as understanding becomes sufficiently complete, the questions turn ethical again. Environmentalism is now passing from the first to the second phase, and there is reason to hope that it will proceed directly on to the third.'

> E. O. Wilson,
> 'The Conservative Ethic', 1984
> Reprinted by permission
> of Harvard University
> Press, *Biophilia*, p. 119.

Introduction

The agricultural and industrial revolutions emerged from the need to manage the human environment; the environmental revolution now seeks to manage ecosystems for their necessary services. Ecosystem services include the 'free' provision of biomass (food and fiber) and the assimilation of wastes. Earlier in this century, the wholesale disposal of wastes and alteration of ecosystems made ecosystems undependable in providing expected services: too few fish, undesirable species, poor drinking water quality. We acted to prevent further degradation, and gross pollution is now much less common in the industrialized world. Now we are interested in improving our risk assessment techniques to minimize hazards. We have, indeed, passed into a new information age. Important questions remain about how we obtain information to estimate and manage environmental hazards.

Both biological and chemical evidence are needed to assess risk to ecosystems effectively. The ability to detect a compound does not ensure that biological effects can be predicted, and the failure to detect a released chemical does not preclude its effects. Where several chemicals interact, the integration of effects by biological material is the *only* reliable evidence for predicting or detecting adverse impacts.

A bioassay is a procedure that uses living material to estimate chemical effects. In ecotoxicology, we use bioassays to predict the levels of chemicals that produce no observable adverse effects on populations, communities, and ecosystems and to identify biological resources at risk. This discussion gives a brief overview of the use of standard bioassays, examines possible ecosystem effects of stress and means for detecting these, and summarizes some of the ethical questions motivating needed current and future research.

Bioassays: an overview

Newcomers to the field of ecotoxicology are startled to learn that non-biological methods have been considered superior to biological methods for estimating the hazard to the environment of chemicals and other stressors (e.g., heated wastewater, suspended solids). When one of us (Cairns) left graduate school in 1948 to work with Ruth Patrick, one of her primary tasks was to convince regulatory personnel that biological evidence should be used in addition to chemical and physical evidence in protecting the environment. The chemical and physical determinations that were then the primary basis for regulatory decisions were quantitative, used generally understood methodology often endorsed as standard methods, and were familiar to the sanitary engineers and chemists who then dominated the water pollution control field.

Only a few biologists made any effort to provide methodology suitable for pollution assessment. The single species fish bioassay was introduced by Hart *et al.* (1945). Following its endorsement and publication (Doudoroff *et al.*, 1951), biological evidence became more common and eventually was broadened to include not only fish but invertebrates and algae. Stream surveys above and below industrial or municipal discharge were also popular, but these were not predictive. Although they recorded biological degradation after it occurred (the damage could frequently be reversed), they did not provide as good a management tool as the more predictive bioassays.

Aquatic toxicology is an outgrowth of mammalian toxicology (Sloof, 1983) which had as its primary focus the protection of a single species, *Homo sapiens*. Because of this, there was unfortunate but general acceptance of single species bioassays as appropriate for environmental toxicology. The difficulties of extrapolating from one species to another are well recognized in mammalian toxicology and have even been adequately documented in aquatic toxicology (e.g., Kenaga, 1978; Doherty, 1983; Mayer & Ellersieck, 1986). The scientific basis for bioassays currently used is limited simply because most of the basic scientific questions about extrapolation from one level of biological organization (e.g., species to ecosystem) remain unanswered.

It was quite understandable for early toxi-cologists to determine the responses of individual species to toxicants under conditions that would permit replication by others. This inevitably meant low environmental realism (i.e., relatively simple test systems), focusing on a few well-understood test species, and developing tests that could be used by modestly trained people. Practi-cally all of the funding for research in the United States came from the regulatory agencies and, as far as we can determine, there were similar cir-cumstances elsewhere.

Regulatory agencies, understandably, estab-lished funding priorities to address regulatory, not scientific problems. Macek (1982) describes the attitude of the regulatory agencies beautifully. 'I think the reason that the regulators tend routinely to ask for the same kinds of answers (data) and can't assimilate other types of data into the regula-tory process is that to do otherwise would require some kind of scientific judgment on the part of the people implementing the process. Making scienti-fic judgments about a science in which few funda-mental principles and underlying concepts exist can be a very risky business. Thus, those in regu-latory positions shy away from making such judg-ments and accepting the responsibility for them. Why should they risk it? It's much safer for job security to mechanize the process, make it objec-tive, and avoid having to make scientific judg-ments and taking the attendant risks.'

We examine problems further in this dis-cussion. The regulatory stance in the United States has been to develop a few tests and test species for aquatic ecosystems and provide in-struction for performing these tests. Those wish-ing a relatively recent view of the U.S. Environ-mental Protection Agency position would do well to read Wall & Hanmer (1987). Unfortunately, the scientific basis and/or fundamental questions underlying the problem of prediction of hazard are not examined in any depth in that article. This is distressing in a 'feature article' in one of the world's leading pollution journals.

In the 1970s in the United States, a technologi-cal solution to the pollution problem was attempt-ed that generally occurred under the umbrella of

BAT (best applicable technology) or BPT (best practical technology). The assumption was that, if the best technology presently available or the best practical technology was installed for waste treatment, nothing more could be done to protect natural systems. Since the best available tech-nology was being used, environmental measure-ments were not necessary.

This is an over-simplification of the situation that is discussed in somewhat greater detail in Cairns (1983). However, some of the fallacies of this assumption are quite evident: (1) the size and assimilative capacity of the ecosystem into which the waste is discharged is not factored into the regulatory decision, (2) each discharge is treated as if it were the only one in existence whereas in many areas there are multiple discharges located quite close together, (3) there could easily be 'over-treatment' (i.e., treatment that provided no additional biological benefits) that would go un-noticed without biomonitoring, (4) since industry would be required to install the latest technology, there would be no incentive for new technologies to be developed because personnel would have to be constantly retrained in the use of this tech-nology. This would be very expensive and would not necessarily increase environmental protec-tion. The technology-based standards were eventually modified, and the use of bioassays returned.

The purpose of bioassays

Bioassays are usually carried out as a means of determining the no-adverse-biological effects con-centration of a chemical in the environment. In relatively few cases, bioassays may be carried out to determine if a particular waste treatment pro-cess has reduced the toxicity of an effluent or to determine why chemicals in mixtures are acting differently than one would expect from their individual behavior.

Although these exercises are ultimately related to the environment, they do not require direct extrapolation of effects to natural systems. How-ever, the majority of the bioassays are carried out with the assumption that the test organisms are surrogates for the larger body of organisms comprising natural ecosystems. Basically, bioassays are intended to predict harm or no harm after exposure of living organisms to certain concentrations of a chemical (or mixture of chemicals) for certain periods of time. They are not reactive in the sense of documenting harm after it is done, which is much better accomplished by *in situ* surveys (e.g., Cairns, 1982). Therefore, the scientific basis for using bioassays must ultimately depend upon the degree to which the accuracy of the predictions made with bioassays can be validated or confirmed in natural systems. To be scientifically justifiable, indirect evidence is not enough – that is, failure to observe adverse effects in ecosystems at concentrations predicted to have no harm must consist of direct evidence, not absence of evidence due to inattention, lack of sufficiently detailed study, or failure to study the right end points or parameters.

There is, unquestionably, scientific justification for using living material to detect toxicity since no instrument devised by man will do so. Therefore, bioassays are superior to predictions made on the basis of chemical/physical measurements alone or assumptions of no harm based on the quality of the technology of the waste treatment system. However, the confidence we can place in predictions based on bioassays would be vastly improved if more attention were given to validating these predictions in natural systems or surrogates thereof.

The standard single species bioassay (both acute and chronic) has often been used for purposes not directly associated with ecosystem protection or impact prediction. For example, Tebo (1985) summarized the objectives of bioassay tests used by the U.S. Environmental Protection Agency for establishing chemical limits and we quote them as follows from pages 20 and 21* of *Multispecies Toxicity Testing* (Cairns, 1985):

---

* Reprint with permission, Pergamon Press.

"Screening

Tests should be rapid and inexpensive and should have wide applicability. Response should have high sensitivity to stress so that there will be low possibility of false-negatives.
Establishing Limitations

Tests should be of known precision with exposures that simulate environmental exposures and should be applicable to a wide range of site-specific situations.

The response should be directly related to environmental hazard and should be easy to interpret and meaningful to the public and courts.

Outputs should be directly translatable into specific decision criteria.

To avoid possibilities of varying interpretation, it is preferable that the end point be a discrete variable. If the end point is not a discrete variable (such as death), justifiable decision criteria should be provided.
Monitoring

Tests should be rapid, inexpensive, and of known precision.

Response should be sensitive and preferably related to the type of limitation imposed.

The desirable attributes of tests used for regulatory purposes are a function of these objectives."

These requirements, however, fail to note the limitations of bioassays. The most well-developed test methods (in terms of standardization) are acute bioassays. Inspection of the known precision of such tests shows that coefficients of variation are commonly 50–100% of the LC50/EC50 estimators (c.f., Mayer & Ellersieck, 1986). The measured response (death) is only environmentally meaningful in the grossest sense (dead animals are a problem), hence their public meaningfulness and utility in courts of law. Biologically, acute toxicity is uninteresting in the sense that rapid death and destruction, even when relatively widespread, results in short-term disruption of ecosystem services during the recovery period (Cairns *et al.*, 1971) unless the acute concentration approaches the expected environmental concentration over a long period. Long-term effects are more serious in their disruption of ecosystem processes and typically occur at concentrations near chronic toxicity test end points which are often near expected environmental concentrations of trace contaminants.

Experimental results have been translated into decision criteria using the aging application factor approach. Methods using bioassay data to estimate impacts on the larger biota require comparatively large data sets upon which interpretation of dose-response is made and only a few taxa (est. 5%) are allowed to be affected.

The result of these limitations has been a truncation of normal, healthy scientific debate over the ecological meaning of the measured response variables (typically mortality, reproductive output, and growth). In this paper we provide evidence useful for developing decision criteria based on direct measurement of ecosystem structures and functions. Interestingly, USEPA has supplemented discrete variable measures in recent short-term chronic fish growth tests (Horning & Weber, 1985). The relationship between the statistical analysis of these tests and any environmental impact is only implicit in the expression of results. Typically, chronic tests are used to estimate a no-effect level which is used in establishing water quality criteria.

In arguing against the use of new or more robust methods, scientists often argue either that the methods are not well developed or are too complicated. Agencies responsible for managing the environment control many of the research and development funds that might be directed toward developing and testing ecologically appropriate methods. When funding is inadequate to fully develop methods, reliance must again be placed on older methods. Sources funding basic ecological research often see research with practical application as 'too applied' for funding under their aegis even though useful ecological information results from 'applied' experiments. So poor is the link between 'basic' and 'applied' ecosystem scientists that the National Academy of Sciences (which includes few applied scientists) commonly goes outside its membership to assemble the appropriate blue-ribbon panel for important deliberations on the environment (e.g., NRC, 1981, 1986).

Tebo (1985, pp. 21–22) suggests the following as essentials for the predictive utility of bioassays.

'To be useful for regulation, the results of toxicity tests must be evaluated based on:
– *Interpretation*. What does the laboratory response mean in terms of environmental hazard?
– *Extrapolation*. Do responses in the laboratory simulate responses in the receiving system?
– *Sensitivity*. Is the response sufficiently sensitive to avoid excessive false-negatives? The sensitivity desired is a function of the objective of this study.
– *Variability*. Is the precision known (or can it be determined) and is it sufficiently high that impacts can be detected?
– *Replicability*. For regulatory purposes, toxicity tests must be sufficiently simple and standardized that they can be carried out by governmental, academic, and private laboratories of widely varying capabilities. There must be sufficient quality control to allow consistent interlaboratory and intralaboratory levels of precision.'

Attention to each of these points is essential for scientists developing new or more robust bioassay methods. While each of the questions posed seems reasonable and appropriate, procedures currently used to estimate risk fail on one or more criteria. Many test methods, including those using nontraditional methodologies, could meet these criteria if properly developed and validated. Each of the criteria mentioned by Tebo is discussed below.

Interpretation. Death, growth, and reproductive success are usually easily understood measures used in single species toxicity tests. Effects of toxicants on organisms under laboratory conditions might reasonably be expected to correspond to observed effects in real ecosystems, assuming that the realized environmental concentration of a toxicant in the compartment of interest was comparable to that tested in the laboratory and that the route of exposure was the same (typically in the water column). Where processes such as bioconcentration of toxicant in food, adsorption by sediments, or microbial degradation, are important, then single species laboratory tests may be poor predictors of ecosystem effects because they lack necessary environmental realism and biotic interactions.

Extrapolation. Correspondence between test results (and criteria based on test results) and environmental hazard is poorly understood. In general, chronic discharge criteria for particular chemicals are based on protection of sensitive or economically important species, usually fish. In combination with assumptions about minimal stream flows (see Biswas & Bell, 1984), discharges may infrequently reach or exceed criterion levels for a particular toxicant. Whole effluent bioassays are probably more representative of the total toxic effect of the complex effluent mixture, but often fail to simulate in-stream conditions including the upstream contaminant background or modification on the toxicant or mixture in the actual receiving system. Additionally, since ecosystems probably differ widely in their resistance to toxicants, and factors other than water hardness probably add to uncertainty about the action of a particular toxicant in a particular ecosystem, extrapolation of test results on individual compounds is not only problematic but unrealistic.

Sensitivity. Response sensitivity is a resolution problem. That is, the level of biological organization tested, the duration of the test, and the inherent variability of the measured response affect the outcome of the test procedures. Biological systems sensitive to toxicant action occur in all ecosystems, but these may be the standard test species (Cairns, 1986). Rather, they may be enzyme systems or microbial consortia or fish reproductive systems. Sound scientific judgement is required to determine which responses are the most valid and practical measures of adverse responses. However, the sensitivity of a response may have little meaning unless it is highly correlated with observable adverse ecological effects.

Variability. Determination of acceptable variability is a methodological and statistical problem (Giesy & Allred, 1985). Depending on the inherent variability of the measured response, the number of experimental units required to detect a given change or difference can be estimated (Sokal & Rohlf, 1981; Green, 1979). The size of the experimental array may conflict with the availability of resources, but this is a practical problem separate from the underlying science.

Replicability. Replicability of tests may be a statistical problem. Here, Tebo means repeatability: repeating the test gives the same results time after time. In addition to standard methods for single species tests, procedures for conducting standardized toxicological assessments of many types are reviewed and published by several organizations and professional journals (e.g., ASTM, *Toxicity Assessment*). The level of expertise required to carry out standard methods varies widely. However, we should not tolerate substandard performance from laboratories or establish test methodologies based on some least common denominator of performance. Not every laboratory will be capable of carrying out every applicable assessment method, as has been shown by efforts to assess interlaboratory variability of results from traditional test methods (Buikema, 1983).

Tebo (1985, p. 23) also summarizes his perceptions of decision criteria in setting limits based on bioassay results:

'Decision criteria
- *Social Relevance.* Is the response meaningful to the public and the courts?
- *Technical Relevance.* Does the response provide a realistic measure of population-, community-, or ecosystem-level impact? Is it possible to provide margins of safety based on objective criteria?
- *Legal Relevance.* Is the response (end point) usable for establishing limitations on the discharge of a substance? If the response is a continuous variable, is there an objective means of establishing a limiting exposure

concentration to avoid hazard? The 'no-effect' level in terms of mortality, as determined in single species tests, is an example of perhaps the only discrete variable resulting from toxicity tests.
- *Cost and Timing.* Is the cost reasonable in terms of the objectives of the test? Cost is largely a function of the time necessary to conduct the tests, the space required, and the level of expertise necessary both to conduct the test and to evaluate results. Decisions as to cost are largely a function of the degree of certainty required.'

Each of Tebo's points are discussed below.

Social relevance. Results of biological testing need to be communicated to the public and the courts. Obviously, a well-informed public (and its decision makers) will only understand bioassay results and the predicted risks of environmental stressors if they have a basic understanding of the underlying environmental science. The public and courts will understand the importance of environmental protection and the measures needed to assure it if they are provided with accurate and understandable information by responsible environmental scientists. It is incorrect to assume that the public, the courts, and non-scientist decision makers can only understand the grossest effects (death) on the most familiar organisms (fish and other vertebrates).

Technical relevance. To date, testing procedures and criteria have focused on the population level of biological organization. Methods to examine community- and ecosystem-level effects are being developed and need to be incorporated into risk assessment schemes.

Legal relevance. 'No-effect' levels reported for standard tests are statistically based and have been used to establish water quality criteria. Appropriate experimental designs and statistical tests are the only objective means of determining at which concentrations a chemical might be hazardous. The nature of the measured response (growth, reproduction, productivity, respiration, enzyme activity) is comparatively unimportant in the outcomes of statistical procedures except where the method of data collection or the nature of the collected data causes violation of the assumptions underlying statistical models. Discrete data (such as mortality) presents special problems in statistical analysis of experimental results. Other discrete variables measured in toxicity tests include numbers of offspring and numbers of species (in the case of microcosm experiments), but these can often be assumed to be continuous for purposes of statistical analysis.

Cost and timing. Costs of traditional and non-traditional tests are surprisingly similar for many types of procedures. Most tests last from 1 to 6 weeks, although life cycle tests may require several months to a year or more to complete. The costs of risk assessments of differing kinds have been summarized by Perez & Morrison (1985), and several authors have provided cost estimates and comparisons for single and multispecies tests (see discussions in Cairns, 1986). Most investigators have developed test methods that require common laboratory space and equipment. Additionally, the costs (time, space) of culturing standard test species are not always considered in comparisons of novel methods and traditional test methods.

With an estimated 63000 chemicals in daily use (Maugh, 1978) and literally millions on the American Chemical Society Computer Registry of Chemicals (personal communication), it is clear that society cannot wait for the perfection of a totally scientifically justifiable bioassay method. We know that chemical information alone can be misleading for the following reasons: (1) environmental quality mediates chemical toxicity (e.g., water hardness, pH, dissolved organics); (2) chemicals may act differently individually and in mixture; (3) chemicals may produce toxicological effects at concentrations below analytical capability; (4) various transformations may occur making the chemical more or less toxic; (5) chemicals undergo partitioning in the environment, and making the measurements in the wrong compartments or in only one compartment may produce misleading information; (6) concentrations of a chemical may vary, but living organisms integrate the toxicological effects of this continuous but varied exposure much better than can be done with nonliving models. Despite the weaknesses of chemical information alone, the correct interpretation of bioassay results would be impossible without accompanying chemical information. Neither biological nor chemical information should be examined alone when determining toxicological response, and impacts at different levels of biological organization need to be estimated.

Predicting and detecting ecosystem changes

Many concepts but few general principles for system level responses to human-induced stress exist in basic ecological theory or in environmental science as a separate discipline. The burgeoning of concepts is illustrated by a survey of the British Ecological Society (Cherrett, 1988) in which 236 concepts were identified by fewer than 1000 ecologists. Recent accounts of the application of basic theory to environmental problems (National Research Council, 1986) are helpful case studies of stress and summarize applicable theoretical paradigms that may apply to particular situations. However, predictable patterns of stress effects are elusive.

Ecosystems are hierarchically structured (Webster, 1979). New properties emerge at increasing levels of biological complexity that are not simply the sum of structures and activities at lower levels. Properties of communities and eco-

systems results from the simultaneous presence and functioning of many species. Many of these properties may only have meaning in the community or ecosystem context and are not predictable from properties of lower levels of organization. For example, predator-prey interactions only emerge from the concurrent activity of two species. The properties of such an interaction are not predictable from knowledge of the individual populations. Similar statements might be made for successional events and other complex interactions. Additionally, collective properties such as diversity, biomass allocation, or production summarize net effects in the whole community or ecosystem. However, these properties have meaning at more than one level of organization. Using population data from single species experiments (typically survival, growth, and reproduction) to imply effects on collective and emergent properties of systems is a conceptual leap that many ecologists are unwilling to make.

Odum (1985) has summarized predictions of system-level responses to stress that are drawn primarily from earlier work on successional trends in ecosystems (Odum, 1969). After many years of work on stress effects in freshwater lakes, Schindler (1987) has recently summarized the most useful indicators of anthropogenic stress. We summarize below (Tables 1–3) Odum's predictions for ecosystem change, Schindler's analysis of the detectability of these changes in the Experimental Lakes Area (ELA) of northern On-

Table 1. Predicted effects of stress on ecosystem energetics (Odum, 1969, 1985). Arrows indicate increases ($\uparrow$) or decreases ($\downarrow$) in system states or rates. Detectability refers to whether predicted changes are detectable (+) or not detectable (−) in microcosm experiments and ecosystem evaluations (Schindler, 1987).

Variable	Young → Mature Succession (Odum)	Stressed (Odum)	Detectability	
			Microcosm	Ecosystem (Schindler)
Community respiration	$\uparrow$	$\uparrow$	+	−
P/R	approaches 1	unbalanced	+	−
P/B or R/B	$\downarrow$	$\downarrow$	+ (not as predicted)	−

Table 2. Predicted effects of stress on ecosystem nutrient cycling. See Table 1 legend.

Variable	Young → Mature Succession (Odum)	Stressed (Odum)	Detectability	
			Microcosm	Ecosystem (Schindler)
Nutrient turnover	↓	↑	+	−
Nutrient loss	↓	↑	+	−

Table 3. Predicted effects of stress on community structure. See Table 1 legend.

Variable	Young → Mature Succession (Odum)	Stressed (Odum)	Detectability	
			Microcosm	Ecosystem (Schindler)
Proportion of r-stategists	↓	↑	?	probably
Organism size (life-span)	↑	↓	+	+ (but opposite to predictions)
Food chains	lengthen	shorten	difficult	probably
Diversity	↑	↓ (usually)	+ (not always as predicted)	+
Symbiosis	mutualistic	parasitic	?	+

tario (Canada), and examples from experiments on laboratory ecosystems.

Several caveats accompany this summary, but it provides an indication of our ability to measure holistic, ecosystem-level changes. First, Odum's work is primary on terrestrial ecosystems, although much microcosm work has been carried out under his direction. Second, the ELA manipulations for nutrient addition and acidification mimic actual mechanisms in nature such as pulse dosing during discrete pollution events. Little ELA work has been carried out on toxics such as those found in industrial discharges to surface waters, expecially streams. However, Schindler has also reviewed the work of scientists conducting large experimental ecosystem manipulations.

Energetics. Patterns of energy flow in ecosystems often define the peculiar nature of the organisms and processes of that system. For example, many ecosystems (but not all) are photosynthetically dominated: most of the energy available in the ecosystem is fixed in reduced carbon compounds using light. Primary production is quickly passed to several groups of consumers. In such systems, primary producers dominate. However, small streams are often driven by dead organic matter on a much slower photosynthetic cycle: dead plant parts (chiefly leaves) fall into streams. The differential decay rates of these leaves power the stream for much of the year. Aging (succession) in the former ecosystems is accompanied by increases in dead material (in the soil or on lake bottoms, for example). In these systems respiration increases through time. Stress serves also to initiate increases in community respiration through shifts to maintenance (repair), although sometimes this shift can be caused by differential sensitivity of the primary producers.

Measuring such energetic shifts can be accomplished by one of several methods available for monitoring primary production or community respiration. Primary production is usually monitored by some variant of the light-dark bottle method using either oxygen evolution or uptake of inorganic radioactive carbon by primary producers. Whole system can be monitored for diurnal cycles of primary production and respiration (e.g., Beyers, 1963; Giddings *et al.*, 1984; Stay *et al.*, 1985). Respiration is estimated by uptake of heterotrophically convertible substances, usually labeled, such as glucose or by the diurnal method noted above.

Similar methods are used for monitoring microcosms and larger ecosystems, but the timing of sampling to coincide with the presence of a stressor is important. In aquatic systems, primary production is dominated by rapidly reproducing (and rapidly recovering) populations of microorganisms.

Like many other researchers, we have observed shifts in energetic balance in microcosms using several stressors. For example, in experiments with atrazine, we noted elevation of primary production relative to biomass followed by nearly total collapse of photosynthesis (Pratt *et al.*, 1988). After 7 d, P/B ratios decreased dramatically as microecosystems shifted toward heterotrophy (Fig. 1). After 28 d, microcosms developed resistant floras that re-established primary production. Similar effects were seen by deNoyelles & Kettle (1985) in pond experiments and by Stay *et al.* (1985) in flask microcosms.

Schindler writes that energetic measures are poor indices of stress in large ecosystems, but we have observed altered production in large outdoor streams (USEPA Monticello Ecological Research Station) stressed with selenium. Low concentrations of selenium are known to stimulate algal growth. However, other researchers working in these streams with other stressors have not detected such functional shifts (Eaton *et al.*, 1985). While Odum predicted that P/B or R/B ratios would increase under stress (possibly due to loss of biomass), we have observed that P/B decreases with stress, probably indicating either direct effects on primary production or elevation in community respiration with stress.

Nutrient cycling. The behavior of nutrients in ecosystems is often difficult to monitor. Major nutrients are often in very low concentration; analytical methods are often inadequate to detect very low levels of biologically important nutrients, especially forms of nitrogen and phosphorus. Because nutrients are tightly cycled, even disruptions in the natural order of nutrient processing may be difficult to detect unless stress is severe. In flowing systems, disruption of nutrient cycles allows nutrients to be lost from the local system. Dramatic demonstration of losses of macronutrients is well known for small watersheds (Bormann & Likens, 1970). In lakes, nutrient cycle disruptions are more difficult to detect because of basin flushing times.

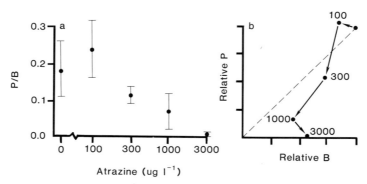

Fig. 1. A. Effect of atrazine on the ratio of primary production (P) to total biomass (B) in laboratory microcosms. B. Phase plane plot of relative primary production (P) and total biomass (B) in atrazine treated microcosms. Control values for P and B were set to 1 (or 100%) and values for other treatments adjusted to this scale. Plotted points are values of nominal atrazine treatments (μg l^{-1}).

We observed altered nutrient cycling as changes in the recovery rate of macronutrients such as major cations and organic phosphorus by the alkaline phosphatase enzyme systems of microbes (Sayler *et al.*, 1979). Similarly, we have quantified changes in community nutrient pools in response to stress. In response to added zinc, microcosm communities lost cations (Table 4). Similarly, phosphorus was lost and alkaline phosphatase activity increased (Fig. 2), clear demonstration of stress effects on nutrient cycles. Similar effects have been reported for nitrate assimilation in synthesized microcosms (Taub *et al.*, 1986); in these experiments nitrate assimilation was inhibited by copper toxicity.

Community structure. Changes in the biotic composition of sampled communities has been considered to be direct evidence of adverse impacts on ecosystems. The most important changes reflect simplification of communities measured as the loss of species or as reduction in species diversity indices of several sorts. See Green (1979) for a discussion of the relative merits

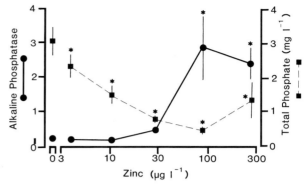

Fig. 2. Effect of zinc on phosphate retention and cycling in laboratory microcosms. Alkaline phosphatase activity is nmole p-NP mg protein^{-1} h^{-1}. Asterisks (*) identify values significantly different from controls ($p < 0.05$). Based on Pratt *et al.* (1987).

of species numbers versus mathematical diversity indices.

Many investigators have considered the changes that occur in stressed communities to represent a reversal of normal successional processes: reversion to smaller organisms with shorter life spans, higher reproductive rates

Table 4. Responses of microbial communities to zinc treatment in laboratory microcosms. Data are means (SD). Asterisks (*) denote values significantly different from controls ($p < 0.05$, Dunnet's test). Units for alkaline phosphatase activity are nmole p-NP mg protein^{-1} h^{-1}. Data from Pratt *et al.* (1987).

Average zinc dose μg l^{-1}	Species Number	Dry weight mg l^{-1}	In vivo fluor. FU	Total phosphate mg l^{-1}	Alkaline Phosphatase Activity
Control (<2.0)	33.7 (4.2)	179.7 (39.5)	507 (38)	3.06 (0.42)	180.0 (20.9)
4.2	30.3 (4.0)	139.9* (12.3)	472 (49)	2.27* (0.30)	181.1 (37.1)
10.7	28.7 (2.5)	89.7* (10.7)	290* (46)	1.46* (0.26)	176.3 (27.3)
29.8	26.7 (6.4)	36.7* (2.3)	88* (12)	0.75* (0.12)	461.3 (69.3)
89.2	22.7* (3.2)	25.3* (2.3)	18* (4)	0.54* (0.08)	2860* (936)
279.8	14.3* (0.6)	39.0* (8.2)	12* (2)	1.36* (0.58)	2389* (530)

among extant taxa, shorter food chains. However, nutrient enrichment in oligotrophic waters may be accompanied by increases in species diversity. Similarly, toxic stress that eliminates important controls in communities such as keystone predators or dominant taxa may result in increases in species richness (e.g., Fig. 3). In microcosms, atrazine addition resulted in reduction in producer species and a net increase in protozoa, most of which were heterotrophic. Perhaps the most general statement that can be made about stress effects on diversity is that diversity shifts from the nominal state. Usually, but not always, this shift is toward lower species richness (and diversity). It is unclear if species are actually eliminated from communities or if they are simply reduced to population sizes too small to detect. In the former case, the resilience of the ecosystem would be more seriously impaired.

Schindler reports that changes in the species richness of small, rapidly reproducing organisms such as algae were reliable indicators of ecosystem stress. He suggests that these are useful warning signals of stress effects. Unfortunately, such populations are rarely monitored in the field concomitant with fish and macroinvertebrates. In microcosm experiments, we routinely monitor protozoa since this group includes both photosynthetic (phytoflagellate algae) and heterotrophic forms. Stress typically reduces species number, and severe stress causes a return to heterotrophically dominated communities of small flagellates, similar to early successional communities (Henebry & Cairns, 1984).

Stress affects obvious symbiotic relationships, often assisting in the development of opportunistic diseases. Schindler reports that predictions of shifts in symbiosis toward parasitism are borne out by greater incidence of fish disease. Similarly, reservoir fish populations in Kentucky Lake (Tennessee River) during severe oxygen stress in 1986, there were many reported incidents of unexplained fish disease including both internal and external lesions (John Condor, Tennessee Wildlife Resources Agency, personal communication). Thermal stress is suspected of selecting for pathogenic strains of free-living amoebae that produce human meningitis.

Systematists tell us that there are between 3 and 30 million species, and every ecosystem is composed of at least several thousand of these. In the United States, decisions about pesticide licensing and chemical discharge limits are often based on the responses of three to six species. Typically, these species represent fish and invertebrates (often called fish food). Rarely are any primary producers tested (except in pesticide studies), and even more rarely are benthic, detritivorous species examined. This seems curious since it is well known that nearly 90% of all the carbon in most aquatic ecosystems is nonliving and much of this is in the detrital pool (Fenchel & Blackburn, 1979). While testing fish and planktonic invertebrates may ease the problem of communicating to the public and decision makers, it cannot be said to be scientifically justified, despite the claims of defenders of standard testing regimes.

The U.S. Federal Water Pollution Control Act prohibits the introduction of toxic substances in toxic amounts into U.S. waters. The introduction of a toxic chemical in any amount probably induces some local change in biological structure resulting in altered function. The question becomes one of determining if changes can be detected, if important processes are repeatedly or perpetually disrupted, and if species vital to the maintenance of system function are compromised. The protection of recreationally or

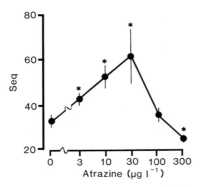

Fig. 3. Effect of atrazine on estimated equilibrium species number (Seq) for protozoa in laboratory microcosms. Asterisks (*) identify values significantly different from controls ($p < 0.05$). Based on Pratt et al. (1987).

economically important species is a secondary biological question, but is very important economically and socially. Maintenance of water quality that is protective of human uses (e.g., drinking water) is similarly important.

Developing a scientific basis for bioassay methodology

The core of science is an explicit statement of a hypothesis and the design of an experiment that will validate or falsify it. To meet the criteria for scientific acceptance, direct rigorously controlled experimental evidence is highly desirable and is mandatory in most cases. Scientists are required to analyze how compatible their evidence and hypotheses are with evidence generated by others. Possibly because aquatic toxicology is an outgrowth of mammalian toxicology, the basic hypotheses are sometimes only implicit. This has held back development of the scientific basis for bioassays immeasurably because mammalian toxicology and aquatic toxicology or environmental toxicology do not share key hypotheses.

In mammalian toxicology, a hypothesis might state that one can extrapolate toxicological responses from a surrogate organism (e.g., rats, mice, *Rhesus* monkeys, etc.) to humans. While this hypothesis has not been validated as well as practitioners would like, the limitations of extrapolations based on certain species of experimental organisms to humans are now beginning to be fairly well understood so that corrections can be made or, more importantly, appropriate test species for a particular chemical can be selected on the basis of experimental evidence already generated. If there is an explicit statement of the hypothesis underlying ecotoxicological bioassays, it is not widely known to practitioners and is certainly difficult to find in the literature.

The annual review of the literature in the 'Journal of the Water Pollution Control Federation' (e.g., June 1987, vol. 59) has a variety of groupings based primarily on the kinds of organisms used, but an examination of the many years of annual reviews does not emphasize anywhere in the discussions that fundamental hypotheses

are being tested. However, the key hypothesis is implicit when one examines the practices of the major U.S. regulatory agency (Wall & Hanmer, 1987). Clearly, it is the responsibility of these agencies to protect the environment, which most people would agree includes natural ecosystems. The practices presently endorsed are three short-term single species toxicity tests involving an alga, an invertebrate, and a fish. The implicit hypothesis is that, on the basis of bioassays carried out individually with three freshwater aquatic organisms (an alga, an invertebrate, and a fish), extrapolations to the response of an entire aquatic ecosystem can be effectively made so that standards for environmental concentrations of potentially toxic chemicals can be stated at or below which no significant adverse biological effects, in either structural or functional attributes, will occur at any level of biological organization (i.e., ecosystem to community to population and below). This hypothesis is difficult to find explicitly stated in the scientific literature; there is little evidence to support it.

Unfortunately, publications on single species toxicity tests (or even multispecies toxicity tests) rarely address a fundamental question: How will this information be used? If it is used primarily to compare relative toxicity, any species will do. If it is intended to be used to protect entire ecosystems, then some evidence must be provided about the reliability with which extrapolations can be made from the test organism to larger, more complex biological systems. The problem is that the scientific basis for applying information generated from bioassays has not expanded markedly in recent years and is unlikely to do so in the situation where funding is governed by regulatory agencies unlikely to ask fundamental questions and industries which must meet regulatory requirements. As Macek (1982) notes, 'There has been an incredible increase in data but virtually no increase in knowledge.'

The above hypothesis is so simply stated and seemingly so easily validated or falsified that it would be reasonable to wonder why it has not been done. To a certain degree, some checks have been made (e.g., Mount *et al.*, 1984, Crossland &

Wolff, 1985, Geckler *et al.*, 1976). Cairns & Cherry (1983) examined the correspondence of field and laboratory evidence at the same level of biological organization (single species) and found it to be relatively good. But as Cairns & Smith (1989) have noted, determining correspondence is not the same thing as validating a prediction. Furthermore, to ensure environmental protection, the validation should include higher levels of biological organization, such as community and ecosystem and both structural and functional attributes at each of the higher levels. But this would require the use of parameters or end points at the ecosystem or community level; ecologists as a profession have not formally endorsed such end points useful in determining ecosystem condition.

With regard to the scientific basis for the development of a predictive capability, ecologists are not on particularly solid ground either. As Harper (1982) notes ecology has tended to be highly descriptive in nature and has made little progress toward reaching maturity as a vigorous experimental and predictive science. Both ecologists and those developing bioassays may be so involved with the minutia of their methodology, such as solving highly site-specific problems, that they ignore the larger questions such as the scientific basis for their actions. Alternatively, since both fields are relatively young and still trying to acquire scientific respectability, investigators may fear to ask the larger questions because the answers may throw the spotlight on glaring deficiencies in the scientific justification for their present actions.

Ethical considerations

In the beginning of this discussion, we introduced the idea that ethical considerations motivated the human desire to manage, conserve, and protect environmental resources. Over the past 20 yr, ethical considerations were superceded by information needs (Wilson's amoral phase) leading to our present data rich/principle poor state (Macek, 1982). Recent syntheses of ethical-philosophical studies suggest that a significant debate on environmental ethics is now in progress (Lemons, 1985), Wilson's third phase. For ecotoxicologists, the timing and structure of this philosophical debate are particulary important. Important questions are being asked by ethicists about the way we assess environmental stresses and about the relation of these assessments to the important factors that influence ecological systems.

Two concepts drive much of ecological science (Golley, 1986). First, ecosystems and the interactions the organism has with its environment are hierarchically organized. Second, ecosystems are dynamic systems with flows and stores of energy, matter, and information. Individuals and whole systems are characterized by interactions among components and dynamic flows. Traditional ecotoxicological science has failed to account for ecosystem dynamics in its methods for predicting effects of stressors such as toxic chemicals on system health.

Debates about the adequacy and appropriateness of methods for estimating chemical effects in ecosystems are necessarily ethical. Where toxicological methods are incongruent with our world view and our concept of ecological functioning, decisions will necessarily be hotly debated. The time for these debates is now, the third phase of our developing awareness of ecological ethics.

Recent evidence indicates that our collective environmental ethic has shifted from theistic (God-centered) to humanistic (man-centered) – toward a reverence for life. Whether continuing development will bring us toward a deep ecological ethic (*sensu* Naess, Lemons, 1985; Golley, 1986) is uncertain. Clearly, protection of rare species, landscapes, and environmental aesthetics (e.g., scenic vistas) transcends ecological science. A greater discussion of environmental ethics among environmental scientists is needed to clarify competing world views as propounded by environmental laws and the missions of regulatory agencies. The information that we choose to use to predict environmental consequences of human activities will be determined from such discussions.

Acknowledgements

This work was supported in part by the U.S. Army Medical Research and Development Command under Contract No. DAMD17-88-C-8068. Opinions, interpretations, conclusions, and recommendations are those of the authors and are not necessarily endorsed by the U.S. Army.

References

Beyers, R. J., 1963. The metabolism of twelve aquatic laboratory microecosystems. Ecol. Monogr. 33: 281–306.

Biswas, H. & B. A. Bell, 1984. A method for establishing site-specific stream design flows for wasteload allocation. J. Wat. Pollut. Cont. Fed. 56: 1123–1130.

Bormann, F. H. & G. E. Likens, 1970. Nutrient cycles of an ecosystem. Sci. Am. 220: 92–101.

Buikema, A. L., Jr., 1983. Inter- and intralaboratory variation in conducting static acute toxicity tests with *Daphnia magna* exposed to effluents and reference toxicants. American Petroleum Institute Publication 4362, Washington, D.C. 32 pp.

Cairns, J., Jr., 1982. Predictive and reactive systems for ecosystem quality control. In M. B. Fiering (ed.), Scientific Basis of Water-Resource Management. Geophysics Research Board, National Academy Press, Washington, D.C.: 72–84.

Cairns, J., Jr. (ed.), 1985. Multispecies Toxicity Testing. Pergamon Press, N.Y., 261 pp.

Cairns, J., Jr. (ed.), 1986. Community Toxicity Testing, STP 920. American Society for Testing and Materials, Philadelphia, Pa., 350 pp.

Cairns, J., Jr. (ed.), 1989. Functional Testing of Aquatic Biota for Estimating Hazards of Chemicals, STP 988. American Society for Testing and Materials, Philadelphia, Pa. 242 pp.

Cairns, J., Jr. & D. S. Cherry, 1983. A site-specific field and laboratory evaluation of fish and Asiatic clam population responses to coal fired power plant discharges. Water Sci. Tech. 15: 10–37.

Cairns, J., Jr. & E. P. Smith, 1989. Developing a statistic support system for environmental hazard assessment. Hydrobiologia. In press.

Cairns, J., Jr., J. S. Crossman, K. L. Dickson & E. E. Herricks, 1971. The recovery of damaged streams. Assoc. Southeast. Biol. Bull. 18: 79–106.

Cherrett, J. M., 1988. Ecological concepts – a survey of the views of the members of the British Ecological Society. Biologist 35: 64–66.

Crossland, N. O. & C. J. M. Wolff, 1985. Fate and biological effects of pentachlorophenol in outdoor ponds. Envir. Toxicol. Chem. 4: 73–86.

deNoyelles, F., Jr. & W. D. Kettle, 1985. Experimental ponds for evaluating bioassay predictions. In T. P. Boyle (ed.), Validation and Predictability of Laboratory Methods for Assessing the Fate and Effects of Contaminants in the Environment, STP 865. American Society for Testing and Materials, Philadelphia, Pa.: 91–103.

Doherty, F. G., 1983. Interspecies correlations of acute aquatic median lethal concentrations for four standard testing species. Envir. Sci. Technol. 17: 661–665.

Doudoroff, P., B. G. Anderson, G. E. Burdick, P. S. Galtsoff, R. Patrick, E. R. Strong, E. W. Surber & W. M. van Horn, 1951. Bio-assay for the evaluation of acute toxicity of industrial wastes to fish. Sewage Ind. Wastes 23: 1380–1397.

Eaton, J., J. Arthur, R. Hermanutz, R. Kiefer, L. Mueller, R. Anderson, R. Erikson, B. Nording, J. Rogers & H. Pritchard, 1985. Biological effects of continuous and intermittent dosing of outdoor experimental streams with chlorpyrifos. In R. C. Bahner & D. J. Hansen (eds.), Aquatic Toxicology and Hazard Assessment: Eighth Symposium, STP 891. American Society for Testing and Materials, Philadelphia, Pa.: 85–118.

Fenchel, T. & T. H. Blackburn, 1979. Bacteria and Mineral Cycling. Academic Press, N.Y., 225 pp.

Geckler, J. R., W. B. Horning, T. M. Neiheisel, Q. H. Pickering, E. L. Robinson & C. E. Stephan, 1976. Validity of laboratory tests for predicting copper toxicity in streams, EPA 600/3-76-116. National Technical Information Service, Springfield, Va.

Giddings, J. M., P. J. Franco, R. M. Cushman, L. A. Hook, G. R. Southworth & A. J. Stewart, 1984. Effects of chronic exposure to coal-derived oil on freshwater ecosystems: II. experimental ponds. Environ. Toxicol. Chem. 3: 465–488.

Giesy, J. P. & P. M. Allred, 1985. Replicability of aquatic multispecies test systems. In J. Cairns, Jr. (ed.), Multispecies Toxicity Testing, Pergamon Press, N.Y.: 187–247.

Golley, F. B., 1986. Deep ecology from the perspective of environmental science. Envir. Ethics 9: 45–55.

Green, R. H., 1979. Sampling Design and Statistical Methods for Environmental Biologists. J. Wiley & Sons, N.Y., 257 pp.

Harper, J. L., 1982. Beyond description. In E. J. Newman (ed.), The Plant Community as a Working Mechanism. Blackwell Scientific Publishers, London: 11–25.

Hart, W. B., P. Doudoroff & J. Greenbank, 1945. The Evaluation of the Toxicity of Industrial Wastes, Chemicals and Other Substances to Freshwater Fishes. Waste Control Laboratory, Atlantic Refining Co., Philadelphia, Pa. 376 pp.

Henebry, M. S. & J. Cairns, Jr., 1984. Protozoan colonization rates and trophic status of some freshwater wetland lakes. J. Protozool. 31: 456–467.

Horning, W. B. & C. I. Weber, 1985. Short-term Methods for Estimating the Chronic Toxicity of Effluents and Receiving Waters to Freshwater Organisms. EPA/600/4-85-014. United States Environmental Protection Agency, Office of Research and Development, Cincinnati, Oh., 162 pp.

Kenaga, E. E., 1978. Test organisms and methods useful for

early assessment if acute toxicity of chemicals. Envir. Sci. Technol. 12: 1322–1329.

Lemons, J., 1985. Ecological stress phenomena and holistic environmental ethics – a viewpoint. Int. J. envir. Stud. 27: 9–30.

Macek, K. J., 1982. Aquatic toxicology: anarchy or democracy? In J. G. Pearson, R. B. Foster & W. E. Bishop (eds.), Aquatic Toxicology and Hazard Assessment. 5th Conference. American Society for Testing and Materials, Philadelphia, Pa.: 3–8.

Maugh, T. H., 1978. Chemicals: how many are there? Science 199: 162.

Mayer, F. L. & M. R. Ellersieck, 1986. Manual for Acute Toxicity: Interpretation and Data Base for 410 Chemicals and 66 Species of Freshwater Animals. U.S. Department of the Interior, Fish, and Wildlife Service, Resource Publication 160, Washington, D.C. 505 pp.

Mount, D. I., N. A. Thomas, T. J. Norberg, M. T. Barbour, T. H. Rousch & W. F. Brandes. 1984. Effluent and ambient toxicity testing and instream community response on the Ottowa River, Lima, Ohio, EPA 600/3–84–080. National Technical Information Service, Springfield, Va. 85 pp.

National Research Council, 1981. Testing for the Effects of Chemicals on Ecosystems. National Academy Press, Washington, D.C. 103 pp.

National Research Council, 1986. Ecological Knowledge and Environmental Problem-Solving: Concepts and Case Studies. National Academy Press, Washington, D.C., 338 pp.

Odum, E. P., 1969. The strategy of ecosystem development. Science 164: 262–270.

Odum, E. P., 1985. Trends expected in stressed ecosystems. BioScience 35: 419–422.

Perez, K. T. & G. E. Morrison, 1985. Environmental assessment from simple test systems and a microcosm: comparisons of monetary costs. In J. Cairns, Jr. (ed.), Multispecies Toxicity Testing. Pergamon Press, N.Y.: 89–95.

Pratt, J. R., B. R. Niederlehner, N. J. Bowers & J. Cairns, Jr., 1987. Effects of zinc on freshwater microbial communities.

In S. E. Lindberg & T. C. Hutchinson (eds), International Conference on Heavy Metals in the Environment, Vol. 2. CEP Consultants Ltd., Edinburgh: 324–326.

Pratt, J. R., N. J. Bowers, B. R. Niederlehner & J. Cairns, Jr., 1988. Effects of atrazine on freshwater microbial communities. Arch. Envir. Contam. Toxicol. 17: 449–457.

Sayler, G. S., M. Puziss & M. Silver, 1979. Alkaline phosphatase assay for freshwater sediments: application to perturbed sediment systems. Appl. Envir. Microbiol. 38: 922–927.

Schindler, D. W., 1987. Detecting ecosystem responses to anthropogenic stress. Can. J. Fish. Aquat. Sci. 44 (Suppl. 1): 6–25.

Slooff, W., 1983. Biological Effects of Chemical Pollutants in the Aquatic Environment and their Indicative Value. Dissertation, University of Utrecht, The Netherlands, 191 pp.

Sokal, R. & F. Rohlf, 1981. Biometry. W. H. Freeman & Co., N.Y., 859 pp.

Stay, F. S., D. P. Larsen, A. Kato & C. M. Rohm, 1985. Effects of atrazine on community level responses in Taub microcosms. In T. P. Boyle (ed.), Validation and Prediction of Laboratory Methods for Assessing the Fate and Effect of Contaminants in Aquatic Ecosystems, STP 865. American Society for Testing and Materials, Philadelphia, Pa.: 75–90.

Taub, F. B., A. C. Kindig & L. L. Conquest, 1986. Preliminary results of interlaboratory testing of a standardized aquatic microcosm. In J. Cairns, Jr. (ed.), Community Toxicity Testing, STP 920. American Society for Testing and Materials, Philadelphia, Pa.: 93–120.

Tebo, L. B., Jr., 1985. Technical considerations related to the regulatory use of multispecies toxicity tests. In J. Cairns, Jr. (ed.), Multispecies Toxicity Testing. Pergamon Press, N.Y.: 19–26.

Wall, T. M. & R. W. Hanmer, 1987. Biological testing to control toxic water pollutants. J. Wat. Pollut. Cont. Fed. 59: 7–12.

Webster, J. R., 1979. Hierarchical organization of ecosystems. In A. Halfon (ed)., Theoretical Systems Ecology. Academic Press, N.Y.: 119–121.

Hydrobiologia **188/189**: 21–60, 1989.
M. Munawar, G. Dixon, C. I. Mayfield, T. Reynoldson and M. H. Sadar (eds)
Environmental Bioassay Techniques and their Application.
© *1989 Kluwer Academic Publishers. Printed in Belgium.*

L. I. H. E.
THE MA... AND LIBRAR...
STAND PARK ..., ...ERPOOL, L1...

Recent developments in and intercomparisons of acute and chronic bioassays and bioindicators

John P. Giesy[1] & Robert L. Graney[2]
[1] *Department of Fisheries and Wildlife, Pesticide Research Center and Center for Environmental Toxicology, Michigan State University, E. Lansing, MI 48824-1222 USA;* [2] *Mobay Chemical Co. Agricultural Chemicals Division, 17745 S. Metcalf, Stilwell, KS 66085 USA*

Key words: toxicity, acute to chronic ratio, energetics

Abstract

The ultimate goal of toxicity testing is to monitor or predict the effects of single compounds, elements or mixtures on the long-term health of individual organisms, populations, communities and ecosystems. Unfortunately, one does not always have all of the information required to determine the long-term or 'chronic' effects of toxicants on the survival, growth or reproduction of aquatic organisms. For this reason, the chronic effects of toxicants are often inferred or estimated from observations made during short-term or 'acute' field or laboratory studies, which may be conducted at greater concentrations of toxicant. The observations made in the short-term studies are then related to the chronic effects by some statistical relationship. There are basically two approaches: 1) The long-term effects on a parameter, such as survival (lethality) are predicted from observations on the same parameter, during short-term exposures; 2) Alternatively, the response of one parameter to long-term exposures of a toxicant can be predicted from the short-term responses of a different parameter. In this report we present several different examples of both types of methods for estimating chronic responses from information on more short-term responses and discuss the rationale, advantages and disadvantages of each. We also report on two biochemical indicators; energetic substrates and RNA/DNA ratio. These indicators both act as sensitive, integrative measures of sublethal effects of contaminants during both acute and chronic exposures.

Introduction

It has been estimated that there are approximately 63,000 chemicals in common use (Hunter, 1987). Over 8.5 million chemicals have been documented in the American Chemical Society's Chemical Abstract Service (CAS) as of April 1988. Additional compounds have been described in papers published before 1967. Many new chemicals are synthesized or identified each year and enter into commercial use in quantities large enough to be of evironmental concern to humans and/or animals. Exotoxicologists need sufficient information on the fate and potential effects of these commercially important chemicals so that they can be manufactured, transported and used in a beneficial and environmentally safe manner. There are so many chemicals that are routinely used in large quantities that it is impossible to test the effects of all of them on entire ecosystems or even many of the important species. Thus we are relegated to testing a compound of interest under limited conditions. The best situation is where one acquires both acute and chronic toxicity information about the compound for organisms with a range of sensitivities, however, this goal is seldom realized. Alternatively, we hope to have some information on the acute and chronic toxicity toward a few key species. Often, even this amount of information is lacking. Sometimes only acute toxicity information is available. In such a case we need to estimate the chronic or safe level of exposure from acute toxicity information alone. In the worst case, we have no information on the acute or chronic toxicity of the compound to any organisms nor any information on the chemodynamics in or effects on ecosystems. Unfortunately, few compounds other than pesticides have been tested for their potential fate in the environment or toxicity to animals. It is estimated that only 5 to 10% of the known chemicals have been tested for toxicity, and fewer than 1% of the 50 000 or so compounds manufactured in the United States have been tested for their toxicity to aquatic organisms (Martell *et al.*, 1988). Methods for estimating responses of populations of aquatic organisms from single-species toxicity testing are beginning to be developed (Barnthouse *et al.*, 1987) and hopefully, we will eventually be able to use some short-term measurements to make such predictions. However, until these methodologies are integrated into the hazard assessment scheme and more data on classes of compounds is collected, we will need to rely on less suitable measures.

Much of the effort in the field of ecotoxicology has focused on an attempt to identify the one level of organization, type of test or parameter to measure as the most sensitive, accurate, reproducible measure of potential effects of chemicals in the aquatic environment. Furthermore, because of the great complexity of the systems of interest and the number of potential toxic agents, the ecotoxicologist is forced to make such predictions from little or even no measured information about the chemical of interest (Kaiser, 1987). Populations, communities and ecosystems represent complex interactions among individuals, species and their environment (Giesy & Odum, 1980). Therefore, there are no perfect short-term tests or measurements which will allow one to predict with great

certainty the effect of a toxicant on ecosystems. There are many reasons such as the expense, why the true effects of pollutants on ecosystems cannot be known. However, it is also impossible to collect all of the relevant information. Even if this could be accomplished, the generality in space and time would be small and the ecosystem or some functioning part of it would need to be exposed to the toxicant (Perry & Troelstrup, 1988). Even if all of the information about a chemical and the ecosystem with which it will interact were known, due to the vast amount of information and the ecological uncertainty principal, it would be impossible to predict the effect that chemicals would have on the ecosystem (Barnthouse *et al.*, 1986). With these limitations, what is possible and of what utility are simple, short-term toxicity tests? Can such simple tests be used to predict more complex or long-term effects? Here we will explore the relationships between acute and chronic bioassays.

Limitations of resources put a premium on predicting long-term (chronic), lethal or sublethal effects which cause adverse effects on aquatic ecosystems, from 'acute' toxicity tests (Sheehan *et al.*, 1984; Vouk *et al.*, 1985). Since perfect prediction is not attainable here we will discuss the utility of predicting chronic effects of chemicals on individuals, populations, communities and ecosystems from information obtained in short-term (acute) laboratory exposures and we discuss the feasibility of assays which can be used to adequately protect aquatic communities from the adverse effects of chronic exposures to toxicants (Stephan, 1986).

We realize that we cannot expect to be able to predict the effects of many toxicants on ecosystem function from simple, single-species bioassays (Cairns, 1988). The mode of toxic action can be different during acute and chronic exposures. However, before abandoning completely what we consider to be valuable screening tests, which may have utility to protect resources without the need for complete, ecosystem-level or even multi-species toxicity testing, let us assess some of the possible predictive relationships. As we have discussed before (Giesy, 1985), more complex tests are not necessarily more predictive. Because of the urgent need to screen for the effects of thousands, even hundreds of thousands of compounds, we feel that we will need to rely, at least for the near-term, on information obtained by some of our traditional assays of acute toxicity.

While our search for new and better approaches continues, we should not neglect those techniques which have been developed and found to be useful in the past. It should be remembered that single-species toxicity tests, multi-species (microcosms) tests and ecosystem-level simulations all have advantages and limitations (Neuhold, 1986). There is no one perfect, short-term test which will allow one to predict the ecosystem-level effects of chemicals with certainty. We will discuss the relationship between acute and chronic exposures and ways to predict the no observable adverse effect level (concentration; NOAEL; also denoted as NOAEL or NOEL), which is that concentration to which organisms can be exposed indefinitely and suffer no observable adverse effects. We will not consider monitoring environments, communities or populations for adverse effects from chronic exposures to chemicals. Rather, we will restrict our comments to the prediction of potential chronic effects on survival, growth and reproduction from measurments made during acute exposures.

Levels of organization

There are many levels of organization which can be affected by contaminants and processes at each level of organization can be used to monitor or predict the effects of toxic materials on aquatic organisms (Giesy & Odum, 1980; Giesy & Allred, 1985; Cairns, 1986b; Giesy et al., 1981). Toxic materials exist as atoms or molecules and as such elicit their toxic effects by interacting with biological systems at the cellular and subcellular level of organization (Giesy et al., 1988a; Versteeg et al., 1988). Therefore, one would expect that biochemical and physiological measures of toxicant-induced stress would be useful as sensitive, specific predictors of effects at the level of re-

sponse of whole animals to the effects of toxicants. Furthermore, because changes must occur at the biochemcial, cellular and tissue levels of organization before effects will be observed at the organism level of organization it has been suggested that such measures may prove to be useful, acute measures of more chronic effects at higher levels of organization (McKee & Knowles, 1986). For this reason, biochemical measures of toxic effects, such as protein content, the RNA/DNA ratio, energy content or concentrations of specific enzymes or substrates have been suggested as potential short-term, functional measures of effects which can be used to predict the effects of chronic exposures to xenobiotics on the survival and fecundity of organisms (Giesy et al., 1988a; McKee & Knowles, 1986; Versteeg et al., 1988; Mitz & Giesy, 1985a; 1985b; Graney & Giesy, 1986; 1987; 1988).

Before using changes in biochemcial or a physiological parameters of organisms which have been exposed to xenobiotics for a short time to predict chronic effects one must understand the relationship between short-term biochemical effects and whole-animal or population-level effects (Woltering, 1985). We will not discuss the prediction of effects on communities or ecosystems here. This has been discussed elsewhere by Giesy and co-workers (Giesy & Odum, 1980; Giesy, 1985; Giesy & Allred, 1985). There are interactions which occur within and among individuals that make it difficult to predict effects at one level of organization from observations at another level of organization (Cairns, 1986a; b; Giesy et al., 1988a; Giesy & Odum, 1980). For instance, the partitioning behavior and metabolism of compounds in whole animals may be such that the dose to a particular biochemcial system is much different under in vivo, field conditions than in laboratory studies of tissues, cells or subcellular preparations. Furthermore, while studies are conducted on experimental organisms under laboratory conditions to measure the survival, growth and reproduction of organisms, the goal of the ecotoxicologist is to predict the responses of organisms to xenobiotics under ecologically relevant conditions where organisms are responding to dynamic conditions of their environment, population, and community. While we know the ultimate goal we will restrict our comments here to the prediction of chronic effects on individuals and populations from acute toxicity tests.

Predicting population-level responses from acute toxicity testing

Because we ultimately want to predict effects of contaminants on population dynamics, we should be interested in survivorship and fecundity. In animal populations, we, as managers, are not interested in the fate of individual organisms. This means that we want to protect the resource at a different level of organization than we do for human populations. Thus, since we want to make predictions of effects at a higher (population or community) level of organization, we will need different sorts of information than that obtained in simple assays. When possible, we would like to have information on parameters which can be related most directly to population dynamics. In addition to direct measures of effects of contaminants on survival and reproduction, information on the effects of contaminants on size (growth rate) and energetic state will be useful because these properties can be related to fecundity. While chronic biochemical effects measured in laboratory studies are sensitive and respond rapidly in ways that are specific to the toxicant and can be related to population-level effects, many of these types of sublethal effects will be more useful in monitoring populations under field conditions than for predicting effects under field conditions from the results of laboratory situations.

Acute toxicity tests are useful (Stephan, 1982) because they establish initial benchmark or relative toxic potential. They are ecologically relevant because they can be compared to or calibrated with observations in the real world. They are repeatable, rapid and are easily interpretable (Sprague, 1970). As such, they are a logical place to begin to understand the hazards of toxic materials to aquatic ecosystems. Ideally, such tests would only be the first step in a comprehensive assessment process. Since it is unlikely that we will obtain all of the required information, we must make the best decisions that we can based on the information that we can obtain. Acute toxicity tests provide information on the relative toxicity among toxicants and species and were not designed to and cannot substitute for definitive chronic tests. However, since one is not able to conduct all of the chronic tests, we must ask what utility short-term tests have to predict more chronic effects since most of the information known about the effects of toxicants on aquatic animals is related to acute lethality from whole animal studies. Presently almost all of the standards and criteria are based on this type of information.

The dose-response toxicity relationship relates the responding proportion of populations to the exposure concentrations. This relationship provides useful comparative information, but provides little information which is useful in predicting long-term, chronic effects from acute effects, other than by correlation. The use of toxicity curves has been suggested as a mechanism to extrapolate from the effect of a toxicant to some portion of the population under acute exposure to a more chronic exposure (Sprague, 1969; Stephan, 1982). A toxicity curve relates the duration of exposure to the intensity of exposure (Fig. 1). When death is the endpoint this methodology can be used to estimate the asymptotic (insipient lethal level) LC_{50}. When other endpoints are used one could estimate the asymptotic effective concentration (sometimes called infinite EC_{50}). We feel that the use of toxicity curves would greatly enhance the utility of acute, single-species toxicity tests. Unfortunately most acute toxicity tests have not been designed to provide the information required to generate toxicity curves, but it would be rather easy to do so.

Dose-response relationships express the response of a population of individuals as a function of intensity (concentration of toxicant) and duration of exposure (Fig. 1). This type of analysis assumes that there is a relationship between the intensity and duration of exposure such that there is some minimum time to effect and some thresh-

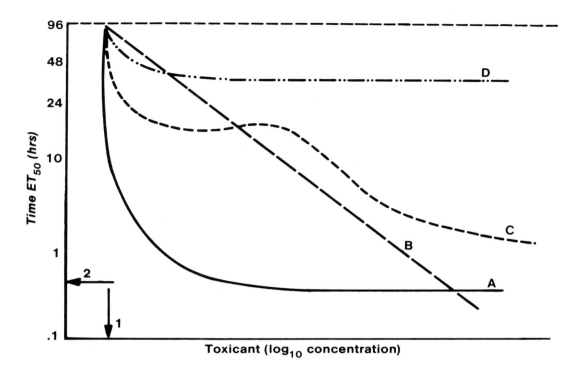

1 Incipient Lethal Concentration (Level)

2 Minimum Exposure Time

Fig. 1. Toxicity curves, which relate the duration of exposure, represented by the effective time to affect 50% of the population (ET-50) and the intensity of dose, represented by the concentration of contaminant to which the population was exposed. Redrawn from Warren, 1971.

old concentration to cause an effect. This is called the incipient lethal concentration or time-independent LC_{50}. For most classes of toxicants there seems to be a characteristic relationship between duration and intensity of dose. Several shapes of toxicity curves have been identified (Warren, 1971).

In some cases the duration and intensity are related linearly (Curve B, Fig. 1). In these cases one could substitute intensity for duration of exposure. This is essentially what is done when one conducts a short-term acute assay and then uses this information to predict more chronic effects. For other toxicants there is not a linear relationship between the intensity and duration of exposure. Many contaminants are represented by a curvilinear, hyperbolic relationship (Curve A, Fig. 1). This is due to differences in the rate of

damage and rate of repair and is the most common shape of toxicity curves. This shape of curve means that there is a complex interaction between duration and intensity of exposure and that it will be more difficult to define the relationship and thus to predict chronic effects from acute exposures. However, one can determine the region of effect which is a function of both duration and intensity of dose. That is, above the line, one would expect that a minimum of 50% of the population would respond to any combination of duration (time) of exposure and intensity (concentration). Below the isopleth one would expect to observe a smaller response. Some compounds, such as 2,3,7,8-tetrachlorodibenzo(p)dioxin have a mode of toxic action which takes a longer minimum period of time to be expressed than most other contaminants. This is thought to be due to the need for translocation of the chemical to the nucleus, interaction with DNA and subsequent pleothrophic effects. This results in a toxicity curve of type D. Curves of type C can be observed for toxicants which are actively metabolized.

The fact that several types of toxicity curves are observed indicates that it may be difficult to predict long-term, chronic effects of all toxicants from short-term exposures. In fact, the response is a function, not only of the relative rates of damage and repair, but is also a function of the rates of uptake, depuration, metabolism and detoxification. While this relationship may not be linear, it is often possible to develop useful working relationships between the duration and intensity of exposure. Toxicants which produce nonasymptotic toxicity curves are generally considered to be cumulative poisons and these toxicants would be expected to cause chronic toxicity at concentrations which are much less than those required to cause acute lethality. Thus, even if one were unable to predict chronic effects from acute exposures, one would be able to use toxicity curves to predict that the chronic toxicity would be different among compounds with similar acute toxicities, depending on the shapes of the toxicity curves (Stephan, 1982).

In addition to the biochemical and physiological reasons that different responses are observed during acute and chronic exposures, there are effects due to differences in sensitivity of individuals within a population (Fig. 2). While one can use the function which relates median time of exposure and concentrations (dose) to extrapolate from median acute to median chronic effects, such extrapolation ignores the variability of responses to toxicants among individuals in a population. For instance, when one substitutes exposure intensity for duration to conduct an acute test (96 hr.), the dose-dose rate relationship is much different than under chronic conditions. What is often observed is a change in the slope of the log dose-probability relationship (Attar & Maly, 1982). Since the slope of the dose-response relationship represents the heterogeneity of response among individuals at lower dose

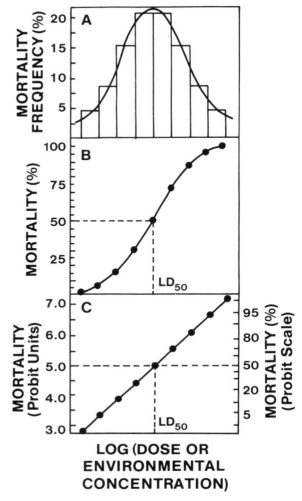

Fig. 2. A. Frequency distribution of the number (proportion) of a population responding to different concentrations of a toxicant; B. Cumulative frequency distribution as a function of concentration. C. Log-probit linearized relationship for estimating the dose-response relationship.

(Finney, 1971), the interpretation is that the probability of sensitive individuals surviving is greater and the distribution of sensitivity in the population can change. As long as the dose is sufficiently great, such a mechanism cannot be involved. However, at smaller doses, this is the mechanism which allows tolerance to develop. This is the situation where pesticide resistance has developed in insect populations. If, as we have discussed above, we are interested in protecting population and community integrity, then the use of short-term, acute toxicity testing is questionable. Not only are there physiological and biochemical reasons to conduct tests for longer periods of time but probabilistic population dynamic reasons as well. The longer that the genetic variations among individuals in a population have to express themselves, the greater the probability of making a wrong prediction.

Dose-dose rate

The responses of organisms to toxicants can change during the exposure such that chronic exposure will result in responses which could not be predicted from acute exposures. These effects are generally due to induction of enzymes, changes in substrates or exhaustion of a supply of enzyme or substrate. Examples of the former would be the induction of the mixed function oxygenase (MFO) system by compounds such as polychlorinated biphenyls (PcBs) or polycyclic aromatic hydrocarbons (PAH), or the induction of increased concentrations of metallothionene by exposure to metals.

Organisms have a finite capacity to repair damage caused by toxicants. For this reason one does not generally observe a linear toxicity curve. In addition to duration and intensity of exposure, one must consider the dose-rate. A continuous exposure to 50 units for 4 days may not be equivalent to exposure to 100 units on every other day. If one expects that the dose rate will have a significant effect on predicting chronic effects from acute effects a dose-dose rate response should be developed (Fig. 3). The effects of dose rate have been demonstrated for a metal, cadmium (Pascoe & Shazili, 1986) and an insecticide, fenvalerate (Curtis et al., 1985).

Because observed effects are a function of dose-rate as well as dose extrapolating from the results of acute toxicity tests to more chronic effects is often difficult. Thus, the reporting of the median post exposure lethal time (pe LT_{50}) has been suggested. This is a method of assessing and

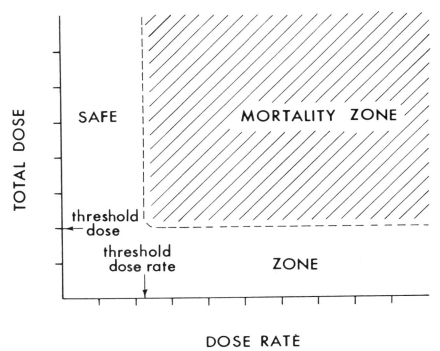

Fig. 3. Schematic representation of the relationship between total, cumulative dose and the dose rate. The hatched area above the dashed line is the region where the combination of effects would be expected to cause an effect. The effect plotted here is lethality.

comparing the results of a brief exposure to longer term exposures and is defined as the time measured from the end of the exposure period until 50% of the test population is dead (Pascoe & Shazili, 1986). If this information were reported for acute toxicity tests, it would enhance the ability to predict chronic responses.

Acute to chronic ratios

One method of predicting the threshold concentration which causes no observable effects during chronic exposures has been to establish the acute-to-chronic ratio (ACR) between the concentration which causes lethality during acute exposures (96 hr, LC_{50}) and the NOAEL (Stephan, 1982; equation 1):

$$ACR = \frac{96 \text{ hr } LC_{50}}{NOAEL} \qquad (1)$$

Although simple, the ACR has been found to be an effective tool to predict chronic effects when little information is available. While the acute-to-chronic ratio can be calculated for the interaction of a specific test organism with a single chemical the acute-to-chronic ratio for the species/chemical combination has no utility for predicting the chronic effects of that chemical to that species because the NOAEL is already known. The utility of establishing the acute-to-chronic ratio lies in using the relationship to predict the probable chronic responses of similar species or toxicants from known ACR values for other species or toxicants. The ACR approach has been used because almost 90% of the chemicals which have been assessed have had an acute LC_{50} value which was less than two orders of magnitude different from the chronic, no-effect concentration (Kenaga, 1982a). When the acute (LC_{50}) toxicity was correlated with the chronic (NOAEL) toxicity for 11 freshwater species and 126 chemicals the overall correlation coefficient was observed to be 0.88 (Slooff & Canton, 1983). Thus, it can be concluded that those compounds which tend to be acutely toxic are also chronically toxic. As a result, except

for a few classes of compounds which have particular modes of toxic action one can often predict chronic toxicity of a toxicant to a species from acute observations of the same chemical on the same species. For a large number of compounds which do not express toxicity through genotoxic mechanisms, but do affect membrane integrity or energy metabolism, the ACR may be an effective predictor of the effects of a compound on similar species or the response of a species to similar compounds. These predictions could be improved by correcting for effects of metabolism or bioconcentration if quantification were available. There is not always a definable relationship between the results of acute and chronic studies (Stephan, 1982). For instance, when the insect *Tanytarsus dissimilis* (Chironomidae) was exposed to copper, cadmium, zinc or lead, no effects on growth were observed when organisms were exposed to concentrations less than the LC_{50} (Anderson *et al.*, 1980).

The sensitivity to a compound varies greatly among species (Slooff, 1985). The NOAEL (concentration) can vary by almost 10 000 fold for the same species exposed to various chemicals and more than 100 fold among different species exposed to the same chemical (Slooff *et al.*, 1983; Sloof & Canton, 1983; Kenaga, 1977). Kenaga (1982b) surveyed the literature on acute-to-chronic ratios and found that the ACR ranged from 1 to over 18 000, but that an ACR of approximately 25 was appropriate for 93% of the chemicals studied. When metals and pesticides were excluded, the ACR for other industrial chemicals was approximately 12. Kenaga (1982b) sorted the ACR values into classes of compounds. This grouping by class increased the probability of accurately predicting chronic, no-effect concentrations from acute median lethality values, and suggested that ACR was a useful tool for predicting chronic, no-effect concentrations. While most species and the integrity of communities could be protected by this approach, the uncertainty of predicting exposure concentrations could be great even though the chronic no-effect levels of chemicals could be accurately predicted from acute exposures. Thus, the greatest utility of the ACR would be to predict the NOAEL from information obtained during acute exposures for other chemicals or species, for which one has no information on chronic effects.

The application factor (AF) is the inverse of the acute to chronic ratio and has been used to predict the maximum acceptable toxicant concentration (MATC). The MATC is set equal to the product of the AF and the 96 hr LC_{50} value. Values for application factors have been found to often vary by as much as a factor of 10 000 and sometimes as much as 100 000 among compounds (Kenaga, 1977). Furthermore, no relationship could be found between the acute lethality (LC_{50}) of compounds and the AF. However, when the relationships between the AFs for pairs of species were examined, it was found that the AF of one species could be predicted from that of another species. In a review of the use of application factors, Kenaga (1977) made some conclusions which are useful to reconsider here. These include:

(1) The LC_{50} is not useful to predict the NOAEL for other chemicals or species from the AF determined for another chemical-species combination.

(2) The AF may be used to predict the responses of similar species of organisms to a chemical.

In addition to these conclusions, the AF has some utility to predict the chronic effects of compounds within a homologous series, providing the modes of toxic action are similar. To this end a procedure has been developed to allow one to protect a specified proportion of a population or community which will respond to a specified proportion of the potential toxicants in a mixture. In this procedure, the final acute toxicity value is derived by the method proposed by Stephan & Erickson (1982). The final acute value (FAV) is then multiplied by an application factor to derive the final chronic value which is designed, theoretically, to protect 95% of the fish and invertebrate species from any chronic exposures to a chemical for which no information about the chronic effects for any species are known. The critical step in the application of this technique is the derivation of an appropriate application factor. The state of Michigan, USA (under Rule 57), has established an application factor from a probabilistic approach designed to protect organisms from 80% of the chemicals to which they would be exposed (Anon, 1982).

From our previous discussion we concluded that because of the large range in the acute-to-chronic ratios, it would be difficult to select a single, appropriate factor to predict chronic effects from acute toxicity information when the information consisted of only one toxicant or one species. In the procedure used by the state of Michigan to set water quality criteria, the AF was established by plotting the log of the chronic-to-acute ratios of 50 toxic chemicals , which have very different modes of toxic action as a function of the probability of inclusion (percentile rank) (Table 1; Fig. 4).

As we have discussed, there is a great amount of variation in sensitivity to toxicants among species. For this reason the greatest predictive power for protection of ecosystem function comes from using the most sensitive species. Because of different physiologies and biochemistries some toxicants can affect one species while not affecting another (Giesy et al., 1988a; 1988b; Giesy & Hoke, 1988). For instance the insecticide Lindane is approximately 300 times more toxic to guppies than it is to the bacterium *Photobacterium phosphoreum* (Hermans et al., 1985). Therefore, a battery of tests is generally used to maximize predictability (Schaeffer & Janardan, 1987; Dutka & Kwan, 1988). The Michigan procedure calculates final acute values (FAV) by using several species (Anon, 1982). When possible a number of species from several different families of aquatic organisms are used to calculate the FAV (Stephan & Erickson, 1982). For the Michigan procedure acute toxicity information for either *Daphnia magna*, rainbow trout (*Salmo gairdneri*) or fathead minnows (*Pimephales promelas*) is required. The acute toxicity information is considered in the couplets *D. magna*/fathead minnow (Fig. 5) or *D. magna*/rainbow trout (Fig. 6). The ratios of the LC_{50} of the most sensitive of the two species in each couplet, *D. magna*/fathead minnow or

Table 1. Information used to calculate the aquatic life chronic-acute application factor used in the Michigan waste water permitting regulations (Anon, 1982).

Chemical	Chronic Value	Acute Value	C/A Ratio	Rank
Hexachlorocyclopentadiene	5.2	7	0.7429	1.96
Acrolein	24	57	0.4211	3.92
Captan	25.2	64	0.3938	5.88
Endrin	0.22	0.60	0.3667	7.84
Hexachloroethane	540	1530	0.3529	9.80
Pentachlorophenol	116	364	0.3214	11.76
1,2,3,4-Tetrachlorobenzene	318	1070	0.2972	13.72
1,2,4-Trichlorobenzene	705	2870	0.2456	15.68
Zinc	138	671	0.2057	15.68
LAS	870	4350	0.2000	19.60
1,3-Dichlorobenzene	1510	7790	0.1938	21.56
1,4-Dichlorobenzene	763	4000	0.1908	23.52
Arsenic	912	5278	0.1728	25.48
1,2-Dichloroethane	20000	118000	0.1695	27.44
Pentachloroethane	1100	7300	0.1507	29.40
2,4-Dimethylphenol	2475	16750	0.1478	31.36
Antimony	2939	20291	0.1448	33.32
1,1,2,2-Tetrachloroethane	2400	20300	0.1182	35.28
1,1,2-Trichloroethane	9400	81700	0.1151	37.24
Copper	14.5	131	0.1107	39.20
Naphthalene	620	6600	0.0939	41.16
Hexachlorobutadiene	9.3	102	0.0912	43.12
Butylbenzylphthalate	311	3494	0.0890	45.08
Dieldrin	0.22	2.5	0.0880	47.04
2,4,6-Trichlorophenol	720	9040	0.0796	49.00
Lindane	8.1	104	0.0779	50.96
Phenol	2560	36000	0.0711	52.92
Cyanide	16	233	0.0688	54.88
Thallium	86	1280	0.0612	56.84
Tetrachloroethylene	840	13460	0.0624	58.80
Selenium	159	2600	0.0612	60.76
1,2-Dichloropropane	8100	139300	0.0581	62.72
2,4-Dichloropropane	365	8230	0.0443	64.68
1,3-Dichloropropane	5700	131100	0.0435	66.64
Carbaryl	0.38	9	0.0422	68.60
Malathion	26	738	0.0352	70.56
Nickel	130	4355	0.0299	75.52
Lead	39	1398	0.0279	74.48
Atrazine	187	6900	0.0271	76.44
Endosulfan	4.3	166	0.0259	78.40
Trifluralin	4.2	193	0.0218	80.36
Silver	0.12	6.4	0.0188	82.32
DDT	0.74	48	0.0154	84.28
Chlordane	0.8	59	0.0136	86.24
Chromium^{+3}	260	21700	0.0120	88.20
Chromium^{+6}	519	52970	0.0098	90.16
Mercury	0.52	74	0.0070	92.12
Toxaphene	0.14	22	0.0064	94.08
Cadmium	2.4	414	0.0058	96.04
Beryllium	5.3	2500	0.0021	98.00

Chronic value = NOAEL

Acute value = LC_{50}

C/A Ratio = Ratio of chronic-to-acute values.

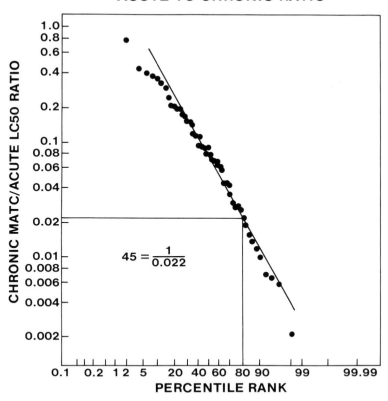

ACUTE TO CHRONIC RATIO

Fig. 4. Probability plot of the ratio of the MATC/final acute LC-50 for 50 toxicants. The chronic-to-acute ratio which includes 80% of the observations is 0.022, which is an application factor of 45.

D. magna/rainbow trout, were plotted on a log-probability diagram (Figs. 5 and 6). The relationship was nearly linear, which indicates that the distribution of the ratio is approximately log-normal and a value can be defined to include a defined proportion of the observations. The FAVs for chemicals not included in the test data set (Table 1) are calculated by dividing the LC_{50} of the most sensitive species (rainbow trout or daphnia) by seven if rainbow trout is used in the data set or nine if fathead minnow is included. These values are established to include 80% of the compounds tested for these two data sets (Figs. 5 and 6). This methodology is deemed to be effective because *D. magna* have been observed to be as or more sensitive to most toxicants as other test species (Giesy *et al.*, 1988a; Dutka & Kwan, 1988; Dutka *et al.*, 1988). Within a taxonomic

group of organisms, the variation in sensitivities among species is generally small relative to the variation among tests conducted for the same toxicant on the same species (Kenaga, 1978). Therefore, one should not expend precious time measuring the effects of a toxicant on a large number of organisms. Previously, we have proposed the use of a small battery of tests to screen for the effects of chemicals which have dissimilar modes of action. These chemicals may have differential effects on procaryotic or eucaryotic cells; plant or animal cells and unicellular or metazoan organisms (Giesy & Hoke, 1988). Because of their different biochemistries, physiologies and abilities to transform some xenobiotics and thus different sensitivity to some toxicants it has been deemed important to include fish in the predictive relationship (Giesy & Hoke, 1988).

FATHEAD MINNOW/<u>DAPHNIA</u>

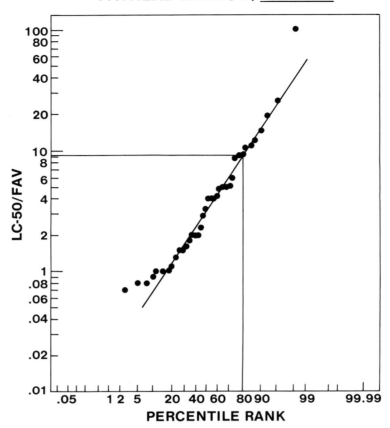

Fig. 5. Probability plot of the ratio of the acute toxicity (LC-50) of the more sensitive species, either *D. magna* or *P. promelas*, to the final acute value (FAV) as a function of the percentile rank.

The ACRs observed when 50 toxicants were investigated (Table 1) ranged over three orders of magnitude (Fig. 4). This probability plot resulted in a straight line which allowed the prediction of the ACR corresponding to the 80th percentile. The ACR that includes at least 80% of the compounds studied was found to be 45 which corresponds to an application factor (AF) of 0.022. The AF that includes 99% of the compounds in the test data set is approximately 0.002. In the case of the Michigan Rule 57 guidelines, the appropriate AF and correction for final acute values is set to include 80% of the compounds tested. That is, the critical value determined would be sufficient to assure that the final chronic value would not be exceeded for 80% of the compounds of the classes included in the analysis as well as 80% of the species sensitivities. Since most of the primary modes of toxic action likely to be encountered were included in the compounds used in the analysis, this application factor should be adequately protective without being overly protective. The advantage of this approach is that one can define the degree of certainty and probability of protection desired when making estimates of safe concentrations or chronic, no-effect concentrations from acute toxicity data. This is more desirable than the alternative which would most likely overprotect most of the time or an inappropriately large AF which would not adequately protect resources.

The proposed method does not reduce the

RAINBOW TROUT/<u>DAPHNIA</u> SP.

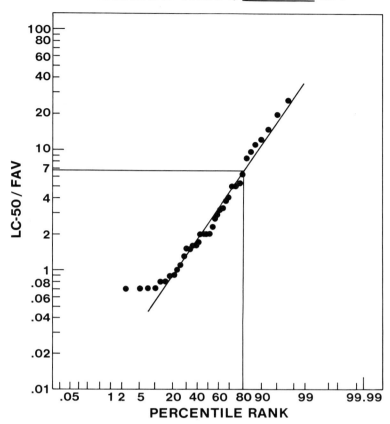

Fig. 6. Probability plot of the ratio of the acute toxicity (LC-50) of the more sensitive species, either *D. magna* or rainbow trout, to the final acute value (FAV) as a function of the percentile rank.

chances that a compound with a much larger ACR may occur and cause adverse effects. What it does is give some rational for the estimation of a NOAEL from very little information. As always, when one makes predictions with little information there is a chance for error. When possible it would be better to have measured information on the possible chronic effects of chemicals. This would be particularly true for compounds in classes, which are known to have a very small acute-to-chronic ratio. These classes include the co-planar polyhalogenated hydrocarbons such as PCBs, PBB (polybrominated biphenyls), and chlorinated-dibenzo-dioxins and chlorinated dibenzo-furans. Effects of chronic exposure should be measured for compounds, which are known to

bioaccumulate, are environmentally stable, are metabolically activated, are released in very large quantities and especially if they are subject to some combination of these factors. Also, the relationship between the acute and chronic effects of metals is by no means constant. Some metals cause chronic reproductive effects while others do not, and some metals cause chronic lethality while others do not.

The AFs of metals range from 0.37 to 0.003 for *Daphnia magna* (Biesinger & Christensen, 1972). An analysis of the AF seems to suggest that such factors have no utility in predicting the chronic effects of toxicants from acute toxicity information because of the great amount of variation in application factors among chemicals and species.

34

However, ecotoxicologists will be required, at least for the foreseeable future to use some justifiable application factor. Therefore, it seems reasonable that a value should be selected which will protect most organisms, from most chemicals most of the time (Kenaga, 1978). Finally, it will be impossible to conduct acute or chronic toxicity tests on all species, especially rare and endangered species. Therefore some rational method such as that proposed here will be needed to predict allowable concentrations to protect sensitive species.

Extrapolated NOAEL (LC$_0$)

An alternative to the use of acute to chronic ratios or application factors is the extrapolation to the LC$_0$ or infinite no-effect level (concentration) (Mayer et al., 1984). The application factor technique requires information obtained from both acute and chronic testing, and the extrapolation from one species to another or among toxicants within a homologous series. Extrapolation from information obtained during acute exposures to the chronic LC$_0$ predicts the concentration which would be expected to kill no organisms (probability of zero) during a chronic or infinitely long period of exposure. The estimated LC$_0$ avoids a number of problems which are encountered when application factors are calculated, and the application factor or acute to chronic ratio techniques are used. The endpoints observed and the degrees of response are generally not comparable between the acute and chronic data (Mayer et al., 1984). The endpoint in the chronic study may be something other than lethality, such as growth or fecundity. The degree of effect in the acute lethality test is generally the median or 50% level, while that in the chronic exposure is zero or a no-observable effect on some other parameter. Therefore, these two responses can only be compared directly if the slopes of the dose-response curves are the same, which they generally are not.

When an infinite LC$_0$ is estimated the dose, degree of response and time-to-effect are considered simultaneously. The endpoint (lethality) is

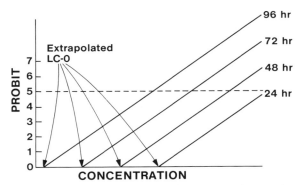

Fig. 7. Schematic representation of the methodology used to calculate the LC-0 by extrapolation of the probit, dose-response curve for four durations.

consistent between the two tests, and the degree of response (0%) is used to compare the results of the two tests. The proposed methodology assumes that the dose-response is a continuum in time and that the modes of toxic action during acute and chronic exposures are similar.

The apparent chronic LC$_0$ is estimated by first conducting acute lethality tests and extrapolating the dose-response lines to a probit value of zero (Fig. 7). This determines the threshold concentration to cause no lethality (incipient lethal concentration) during each duration of exposure. The durations of exposure are generally 24, 48, 72 and 96 hr. The goal is to estimate the concentration which will cause 0% mortality (probability of zero), although a probability of zero is undefined on a probability scale. On the probit scale, however, the value of zero is defined as -5 standard deviation units from the mean. This corresponds to a probability of less than 0.000001, which we will define as zero. The apparent LC$_0$ concentrations observed during each exposure period are plotted as a function of the reciprocal of time (1/T). The ultimate LC$_0$, or LC$_0$ as T (generally hours or days) approaches infinity, is determined as the intercept on the concentration axis (Fig. 8).

As discussed above, one would not expect these assumptions to be always met. However, when the methodology has been tested by comparing predicted values to definitive chronic tests, they have been very similar (Mayer et al., 1984). When the predicted LC$_0$ has been compared to the results of definitive whole-life or partial-life

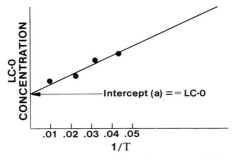

Fig. 8. Schematic representation of the extrapolation of LC-0 values at specified times to determine the infinite LC-0.

cycle tests, good agreement has been observed (Mayer *et al.*, 1984). The estimated LC_0 values were always found to be within a factor of two of the observed MATC and the no effect concentration was always within a factor of five of the LC_0. The greatest error in this analysis would be observed when predicting infinite LC_0s for compounds which are cumulative toxins. Thus, it is advisable that one have some knowledge of the mode of action of toxicants before this method is applied. The greatest utility of this method would be to estimate no effect concentrations for short-lived organisms for which no chronic observations can be made.

While the use of the LC_0 is not a perfect solution to determining the effect of chronic exposures of chemicals on organisms, we feel that it is a viable alternative to the use of application factors and acute-to-chronic ratios in preliminary assessments of single toxicants, especially when no information on the effects of chronic exposures are available, or complex mixtures.

Endpoints

Survival has been the endpoint most frequently monitored in both acute and chronic studies of the effects of chemicals on aquatic organisms. This endpoint has been the most reliable and least expensive to measure (Mayer *et al.*, 1986). There are a number of sublethal responses that can be measured (Sprague, 1988). These include whole-organism responses and within-organism re-

sponses. Recently, there has been pressure to measure other endpoints, such as growth and reproduction. They are perceived to be more sensitive and ecologically relevant than survival (Suter *et al.*, 1987). Mayer *et al.*, (1986) reports that the most sensitive endpoints all correlate well with survival. The correlation coefficients (R) range from 0.95 to 0.97 and the ratios among the endpoints were less than 5 approximately 95% of the time. It was also found that a correction factor of approximately 0.2 was sufficient approximately 95% of the time to correct no effect concentration (NOAEC) from measures of survival during chronic exposures (Mayer *et al.*, 1986).

Critical life stage tests

One way of reducing the duration of toxicity tests is to study only the most sensitive life stage of test organisms (Petrocelli, 1985; McKim, 1985; Kulshrestha *et al.*, 1986). While this technique does not take into account all of the possible differences between acute and chronic exposures, it has been found that the use of critical life stage tests has resulted in estimates of safe concentrations (MATRC) which are very similar to those determined in more long-term exposures (Macek & Sleight, 1977; Macek *et al.*, 1976; Kenaga, 1977). Chronic exposures to toxicants which affect fertility or fecundity will not be well predicted from acute exposures, (Kulshrestha *et al.*, 1986). For instance, in a survey of a number of endpoints for responses of fish to chronic toxic exposures Suter *et al.* (1987) found that effects on fecundity were generally a more sensitive measure of adverse effects than survival of adult organisms and that MATC values calculated from early life stage studies were not substitutes for life-cycle tests because they underestimate the potential for effects. Also, growth and reproduction of *D. magna* (Gersich *et al.*, 1985; Paulauski & Winner, 1988) and *Lymnaea palustris* (Borgman *et al.*, 1978) have been found to be more sensitive to the effects of contaminants than survival.

We have found that we can use a 10 day growth assay instead of the entire life-cycle test with

Chironomus tentans Gisey *et al.*, 1988a; 1989) because the second instar larvae of *C. tentans* is particularly sensitive to several metals and pesticides (Nebeker *et al.*, 1984; Hatakeyama, 1988). Many invertebrates are more sensitive to toxicants during their molt cycle (McCahon & Pascoe, 1988). Therefore, if one conducts an acute test which includes this sensitive stage, the chronic effects on survival can be accurately predicted.

Cell culture assays

The use of cell lines has been proposed as a rapid alternative to toxicity testing with whole organisms (Bols *et al.*, 1985). For a number of compounds there is a good general agreement between the results of whole animal assays and that with cell cultures. The tests are rapid and do not require the maintenance or treatment of whole animals. There are, however, a number of disadvantages of this technique. Few cell lines are available for study, especially for invertebrates. Because of the differences between exposure of whole organisms and *in vitro* tissue cultures such as hepatic cells the loss of specialized function increases the likelihood of missing potential adverse effects on whole animals. As we have mentioned, it is always difficult to predict effects at one level of organization from effects at another. The responses of cells are further removed from ecosystems than are whole organisms. Thus, the prediction of effects and subsequent protection of ecosystem-level function are less accurate.

Because cell and tissue systems can be isolated one from the other allowing one to test hypotheses, we feel that the greatest utility of tests which measure the effects on cell lines will be in elucidating the mechanisms of toxic action. Also tests of this type will be useful in developing better models to predict effects from quantitative-structure-activity-relationships (QSAR) but will not be useful as routine screening tests.

Alternative dosing regimes

Part of the difficulty and expense of conducting toxicity tests is the maintenance of the dosing or exposure system. Furthermore, when chronic effects are studied, the concentrations in target tissues are seldom known. To relate laboratory-derived results to field conditions it would be useful to have information on the dose received at a receptor, rather than the total concentration of toxicant in the water. For this reason oral intubation and intraperitoneal injections have been suggested as alternatives to exposure of whole organisms (Hodson *et al.*, 1984). We do not advocate these methods of exposure even though, for some compounds, the results of these tests can be calibrated to whole animal exposures. We do advocate the measurement of concentrations of toxicants in the tissues of exposed organisms.

Biochemical measures

Most of the information available on the effects of toxic materials on aquatic organisms is from whole organism tests. Sprague (1976) and Mount (1988) suggest that responses of whole organisms provide the most useful information for setting standards to protect aquatic organisms. We agree but also advocate the use of biochemical or physiological measures for monitoring adverse effects in aquatic systems under certain conditions, and for elucidating modes of toxic action. Here we will describe some of the advocated measures but do not endorse them for screening studies. The utility of measuring biochemical and physiological parameters, which are often classified as 'clinical' measures, is in determining the mode of toxic action of chemicals under laboratory conditions, determining the probable causes of lethal or sublethal effects under field conditions (Wedemeyer *et al.*, 1984; Versteeg & Giesy, 1984, 1985; Versteeg *et al.*, 1988) or monitoring for the combined adverse effects of long-term exposures to mixtures of toxicants. Also, because surface waters are seldom acutely toxic to aquatic organisms, clinical measures have been useful in monitoring for effects before they reach population- or community-levels (Versteeg *et al.*, 1988). In addition to their utility as measures of interactions between specific chemicals and receptors there are general clinical measures, that are useful in monitoring and integrating the effects of chemicals, with diverse modes of action, on important processes, such as biomass conversion.

There are several biochemical measures which can be useful in predicting chronic responses from acute exposures. These include: 1) concentration of energy-yielding substrates (lipid and/or carbohydrate content) in tissues and 2) RNA/DNA ratio. In the following sections we present the rationale for the use of such indicators and explain their utility in predicting chronic effects from acute exposures.

Energetics

Toxicants can have effects on the feeding rate, biomass conversion efficiency and energy requirements of aquatic organisms (Borgmann & Ralph, 1986; Borgmann *et al.*, 1978). The measurement of the effects of toxicants on biomass conversion, generally recorded as weight gain of individuals, integrates biochemical, physiological and behavioral effects into one easily interpreted parameter. These measurements are reported as the scope for growth and scope for reproduction, respectively (Warren, 1971; Giesy *et al.*, 1988d). Theoretically, reduced food consumption due to direct biochemical effects or decreased feeding efficiency can result in reduced calorie intake (Borgmann & Ralph, 1986) or increased energetic costs to resist a stressor (Versteeg *et al.*, 1988; Giesy *et al.*, 1988). Such nutritional effects can reduce the overall fitness of the organisms to resist the effect of other stressors or, due to effects on the scope for growth and thus size, reduce the efficiency of predation at critical life stages. The fecundity of individuals can be reduced by a reduced proportion of energy being channeled into gamete production. Also, indirect effects of nutrition on fecundity can be observed because smaller organisms contain fewer eggs. Egg pro-

duction is a function of body size. Alterations in bioenergetics, along with measures of effects on reproduction in laboratory studies, give a strong empirical basis for the examination of population responses (Capuzzo *et al.*, 1988). Finally, and most importantly, effects on biomass conversion at one level of organization can be used to predict effects at other levels of organization (Borgmann, 1985). Thus, such measures of chronic effects are ecologically relevant in a quantifiable way.

Energetic effects can be measured in several ways. The simplest way to quantify energetic effects is to measure weight gain over some period of time. A more detailed measure would be to determine the kinetic growth rate parameters as a function of maximum growth rate. This can easily be done by the use of the Monod (Michaelis-Menten), first order growth model. Using such a model allows one to extrapolate the effects of particular concentrations of a contaminant to effects on biomass conversion under different nutritional conditions. The extrapolation makes the measurement more ecologically relevant.

There are a multitude of energetic responses which organisms exhibit in response to toxicant-induced stress, and in many ways, these responses can be directly related to the general adaptation syndrome (GAS) (Selye, 1956; 1976). A review of the GAS and its implications for measuring the degree of stress under which an organism exists, is beyond the scope of this presentation, but is discussed at length elsewhere (Giesy *et al.*, 1988). In very simple terms the theory of the GAS states that there are a number of quantifiable, general responses of organisms under stress, and that organisms can resist the stressful effects of exposure to toxicants for a finite period of time. The basis for the use of energetics is that energy is required to resist the effects of toxicants and to maintain homeostasis. Eventually the organism expends all of its energy reserves and it passes into the exhaustion stage and ultimately dies. This energy can be used for movement, to transport ions or biomolecules, to replace damaged or biotransformed, structural or functional proteins and to metabolize toxicants (Table 2).

Acute responses, which are generally hormo-

Table 2. Energetic Considerations as Indicators of Toxicant-Induced Stress in Aquatic Organisms

Rationale:	Resisting a stressor requires energy. This will result in reduced energy reserves, altered energy pathways or changes in scope for activity, growth or reproduction.
Measures:	Lipids Adenylates Adenylate Energy Charge Glycogen/Glucose Amino/Acids - Proteins Condition Factors Plasma Lactate, Pyruvate
Example:	Chronic exposure of clams to cadmium results in decreases in energy reserves.
Advantage:	Energy is a universal, interchangeable commodity which allows comparison and integration of disparate stressor responses and integrates primary and secondary effects on a target organism and its support environment. Effects on behavior, are biochemically and physiologically integrated by energy.
Disadvantage:	Because it is non-specific, the mechanism of stressor-induced effects are unknown.
Conclusions:	Energetics are a useful integrative measure of multiple chemical and physical stressors in aquatic organisms in both field monitoring and laboratory bioassays.
Recommendations:	Energetics can be used in laboratory studies to decrease time required for tests. There is a need to establish critical ranges for target organisms and relate these to ecologically significant functions.

nally controlled, are considered to be the initial response to a stressor and often involve the increase of energy-related substrates in the plasma/hemolymph. These changes are transient, however, and will only occur under fairly severe stress (Mazeaud *et al.*, 1977). During gradual sublethal stress, more subtle changes are likely to occur and it is more difficult to separate these effects from normal background variability.

Chronic stress will initiate compensatory physiological adjustments such that changes in energy metabolism may be required to maintain homeostasis and/or homeokinesis. For instance, when Fishes such as *Tilapia mossambica* (Peters) and *Cyprinus carpio* were exposed to pentachlorophenol (PCP) the metabolic rate of the fish increased considerably while the activity of the organisms decreased (Peer *et al.*, 1983). Growth and maintenance metabolism demand the majority of the energy expenditure of an organism so any increased maintenance requirement will result in reduced growth. This concept has been developed into the scope for growth and the scope for reproduction (Warren, 1971). For instance, when perch (*Perca fluviatilis* L.) were exposed to chronic concentrations of copper, the growth rate was significantly reduced (Collvin, 1984). Also, since the reproductive potential of many species is directly related to the size of the organism, this can have direct effects on fecundity and potentially on the population. This being the case, an assessment of the energy status of an organism can be indicative of its overall condition. This has traditionally been done for fish by calculating the condition factor (Lagler, 1952). There is a strong negative correlation between condition factor and amount of body fat (Gibbons *et al.*, 1978), thus fish with smaller body condition factor (K) have smaller energy reserves.

Concentrations of a variety of energy-related biomolecules in plasma or hemolymph and tissue change predictably in response to stress. Under acute stress, these changes such as increased plasma lactate concentrations, result either from increased locomotive activity, or hormonal changes designed to mobilize energy stores in order to respond immediately to stress-induced energy demands. During chronic exposures to stress, altered regulatory enzyme activities and cellular energetic requirements cause changes in biochemical energy stores. Quantification of the energy stores and transport molecules have been used to assess the effects of a variety of acute and chronic stressors (Table 2).

The increased energy demand associated with stress may reduce energy stores such as glycogen and lipids and/or alter their distribution among tissues. Reduction or redistribution of energy stores are generally considered to be a non-specific indicator which is responsive to most any kind of stressor, natural or man-made, and can be applied as a biochemical indicator of stress in both field monitoring and laboratory hazard assessment schemes. Use in the field requires a large data base to enable one to separate toxic effects from normal variations or 'noise'. Noise is caused by accessory factors such as temperature, diet, salinity and reproductive status. Under field conditions, the requirement for static measurements of the energy content of whole organisms or specific tissues has limited usefulness of this approach. Separation of toxic effects from natural variability is extremely difficult and limits our ability to interpret the observed changes. The most successful approach has involved examining the entire energy budget for an organism.

As with some of the other biochemical indicators of stress, the most useful application of energetic measures may be as a short-term indicator of chronic toxicity. Under laboratory conditions, much of the natural variability can be eliminated, or at least controlled. Since for many of the standard bioassay organisms such as *Daphnia sp.*, fathead minnows and trout, reproduction is directly related to the size and/or energy content of the organism, short-term measurements of energy reserves may be predictive of both growth status and reproductive potential.

As with all clinical measures of toxicant-induced stress, our ability to interpret and understand the significance of changes in the energetic status of an organism is limited by the knowledge of the energetics and biochemistry of that specific organism. This is especially true for studies of energetics in the field since the primary energy substrate can vary under a variety of environmental and physiological conditions. For example, in the bivalve *Mytilus edulis*, the primary energy reserve during starvation stress varies seasonally, with carbohydrates being utilized during summer and fall and protein representing the predominate energy reserve in the winter and spring (Bayne, 1973). When salmon migrate lipids are their primary energy substrate, although for most fish, carbohydrates are most readily mobilized during stress (Love, 1970).

Both physical and chemical stressors cause changes in the concentrations of energy storage and transport biomolecules in fish and invertebrates. For example, blood sugar concentrations of freshwater fish do not vary with age (Kiermeir, 1939) but can be affected by nutrition. Anaerobiosis or oxygen stress can result in the mobilization of glycogen reserves (Wilps & Zebe, 1976). Assuming that $NADH^+$ can be regenerated, glycolysis can continue under anaerobic conditions and provide the energy required to maintain homeostasis. Therefore, when choosing an appropriate organism for use in bioassays which measure energetic responses, background information on its physiology and biochemistry is essential. In utilizing these biochemicals to assess the effects of stressors, it must be remembered that many factors including species, sex, season, temperature and hormonal status can affect the basal concentrations of energy substrates (Chavin & Young, 1970; Thorpe & Ince, 1974; Plisetskaya, 1980; Fletcher, 1984; Lehtinen et al., 1984).

Glucose

The onset of stress in vertebrates causes an increased concentration of cortisol in plasma which in turn increases the concentration of glucose in blood. Glucose is the major carbohydrate reservoir utilized during cortisol elevation (Leach & Taylor, 1982). However, the hyperglycemic response of rainbow trout seems to not be under strict control of cortisol. When fish were exposed to phenol after having been injected with betamethasone to inhibit cortisol release by suppressing ACTH release, hyperglycemia was still observed (Swift, 1982). When mullet were exposed acutely and chronically to cadmium, a dose-related increase in glucose concentration in plasma was observed. However, cadmium failed to elicit a corticorsteroid stress response. GAS is not always applicable in fish exposed to pollutants (Thomas, 1976). Cadmium exposure caused a reduced capacity of mullet to produce and secrete insulin (Thomas, 1976). Decreased serum insulin concentrations in fish exposed to zinc were concomitant with increased concentrations of glucose in plasma and decreased concentrations of glycogen in the liver (Wagner & McKeown, 1982). These effects may also be related to the suppression of insulin release and the hyperglycemic action of epinephrine. Blood glucose concentrations of fish can be decreased by starvation or increased in response to acute and chronic stressors. Change in blood glucose concentration is the most widely recognized and consistent response to stressors.

Short-term exposures of only a few minutes or several hours to a physical stressor can cause an increase in concentrations of glucose in plasma of fishes (Hattingh, 1976; Morata et al., 1982b; Carmichael et al., 1983; Nieminen et al., 1983). Glucose concentrations in plasma usually return rapidly to pre-exposure concentrations within approximately 8 h, and sometimes a temporary hypoglycemia may be observed (Fletcher, 1984). The decrease of glucose concentrations in plasma is affected by glycogen reserves in the liver (Nakano & Tomlinson, 1967) and the genetic strain of fish (Casillas & Smith, 1977). Glucose concentrations in blood can vary considerably among individuals of the same species. McCarthy et al., (1973) conducted a survey to determine the range of haematological parameters in rainbow trout blood. The fish were from different locations but were the same age and had a similar nutritional status. The mean glucose concentration was 127 mg/100 ml but the range was from 50 to 221 mg/100 ml. Hunt (1972) reported glucose concentrations which varied from 15.7 to 152.5 mg/100 ml for 9 species of fish. Larson et al. (1975) studied marine teleosts and reported a range of blood glucose concentrations from 1.1 to 66.9 mg/100 ml. The blood glucose concentration of *Limanda limanda* did not vary seasonally (Fletcher, 1984). This type of variation due to strain and edaphic factor differences make blood glucose concentrations a difficult biochemical indicator to interpret.

Glycogen

Glycogen is a branched polymer of glucose residues and represents the readily mobilizable storage form of glucose for most organisms. Glycogenolysis is catalyzed by the enzyme glycogen phosphorylase which is primarily under hormonal control *via* the secondary messenger, cAMP. Both increases and decreases in glycogenolysis can occur due to toxicant-induced stress. In the majority of cases, the increased energy demand associated with stress results in depletion of glycogen reserves (Bhagyalakashmi et al., 1984a; Thomas et al., 1981a; 1981b). Since glycogen storage and mobilization is restricted to certain tissues, alterations in glycogen concentrations observed are often tissue specific. In mussels, the hepatopancreas and mantle are the primary storage organs for glycogen and generally the first to be utilized (Bayne, 1973), although under stressful conditions mobilization from other tissues such as foot muscle can also occur (Moorthy et al., 1983).

In fish, the liver is the primary glycogen storage location (Grizzle & Rogers, 1976). Glycogen is stored in vacuoles which may appear large and open or small and grainy or both. Histological, examination may not be sufficient to discern the differences between lytic necrosis and glycogen stores in hepatocytes. Therefore, staining after amylase treatment is suggested.

Glycogen concentrations in liver, skeletal muscle and heart tissue decrease concomitantly with increases of glucose concentrations in plasma (Nakano & Tomlinson, 1967; Miles et al., 1974; Nieminen et al., 1983). Increased concentrations of the catecholamine and corticosteroid hormones during response to stressors mobilize these energy stores (Nakano & Tomlinson, 1967; Pickering et al., 1982) Increases in glycogen reserves have also been observed during toxicant exposure (Bhagyalakshmi et al., 1984a, b). This response was observed after an initial decrease in hepatopancreas glycogen concentrations and was considered to be an adaptation to the stressor. Glycogen is rapidly utilized at the site of energy requirement but is only slowly replaced by mobilization from liver stores.

As with blood glucose concentrations, concentrations of glycogen vary considerably among tissues and organisms (Prosser, 1973), with diet

(Bayne, 1973), and with reproductive conditions and season (Newell & Bayne, 1980; Gabbott *et al.*, 1979). Changes are not as transitory or sensitive to non-toxicant stress as glucose concentrations in plasma and, therefore, are more promising as a monitor of stress. However, accessory factors can interfere with interpretation of responses. A decrease of the glycogen concentration in liver tissue does not necessarily indicate a general decrease in whole-body energy reserves, since lipid and cholesterol concentrations in liver both increase during exposure to metals (Dubale & Shah, 1981; Katti & Sathyanesan, 1984).

Lactic acid

Lactic acid is the final product of anaerobic carbohydrate metabolism and is the primary source of energy in the white muscle of fishes (Miles *et al.*, 1974; Pickering *et al.*, 1982). Therefore, short-term stressors cause a rapid depletion of blood glucose and glycogen and an increase in lactic acid in plasma and white muscle tissue (Wedemeyer *et al.*, 1984). Increases in lactic acid concentrations are generally associated with swimming fatigue in fish (Wedemeyer & Yasutake, 1977). Capture stress can influence lactic acid concentration in the blood. This is especially true if fish are allowed to stay in nets or traps for long periods of time. The increase in blood lactic acid concentration is rapid (Morata *et al.*, 1982a, b) but usually reaches a maximum approximately four hours after the onset of a stressor and returns to normal within 12 h. Fish can return from short-term stress in as little as 30 min (Sholander, 1957). Therefore, to effectively use lactic acid concentrations in blood as an indicator of stressful conditions, one must be mindful of capture techniques and timing of sample collection in both laboratory and field studies.

Creatine phosphate

Creatine phosphate is an 'energy-rich' compound which is rapidly converted to useable energy.

During stress or exercise creatine phosphate concentrations in fish blood decrease significantly.

Pyruvate

Pyruvate is the final product of glycolysis and is formed by the removal of a phosphoryl group from phosphoenolpyruvate. This reaction is catalyzed by pyruvate kinase and results in the formation of one molecule of ATP. Depending upon the organism and its specific metabolic requirements, pyruvate can be converted to ethanol, lactate or acetyl coenzyme A. The accumulation of pyruvate as an intermediate compound has been suggested as an indicator of energy utilization and/or glycolytic activity and, therefore, has potential as a short-term indicator of stress.

Lipids

A relatively large body of literature exists on the lipid composition of fishes (Miller *et al.*, 1976; Glass *et al.*, 1975; Patton & Thomas, 1971). The total lipid content and tissue distribution varies greatly among fish and invertebrate species (Holland, 1978; Love, 1970). The lipid content of dark muscle in fish is much greater than that of white muscle. The relative concentrations of lipids in fish have been shown to change in response to environmental parameters (Plisetskaya, 1980) such as salinity (Daikoku *et al.*, 1982; Hansen & Abraham, 1983) and temperature (Hansen & Abraham, 1983; Zwingelstein *et al.*, 1978) as well as nutritional state (Fletcher, 1984; Yu & Sinnuhuber, 1975). The lipid content of fish is also dependent on diet (Phillips *et al.*, 1956). In fact, laboratory-reared fish generally have a greater lipid content due to the artificial diet they are fed. The total lipid content of fish varies with age and reproductive condition (Love, 1970). Lipid content is not affected by capture stress (Love, 1970). The fatty acid compositions of marine and freshwater fish are different (Reiser *et al.*, 1963). The C_{18} unsaturated acid group is greater and C_{20} and C_{22} groups less in freshwater fish. Palmitoleic and linoleic acid are more important components of freshwater fish. Beyond this, the degree of unsaturation is comparable between marine and freshwater fish.

Alterations in the lipid and associated fatty acid composition of organisms exposed to a variety of stressors has been studied extensively in both vertebrates and invertebrates. Reasons for observed changes are not always apparent and a variety of mechanisms exist by which the lipid composition of an organism can be affected by toxicant-induced stress (Leatherland & Sonstegard, 1984). The following discussion will be divided into three subsections based on the different types of lipid alterations commonly observed and the proposed rationale for those changes. The first section deals primarily with serum/hemolymph lipids. The second section discusses tissue-specific energy storage forms of lipids and their mobilization during stress. The final section involves membrane lipids and their alteration during toxicant exposure. As will become more apparent, the discussion of lipids not only involves energy metabolism but also includes effects on lipids and associated biochemical pathways which are not involved with energetics, such as transport lipids, cholesterol and membrane lipids. Because the lipid contents of tissues are so heterogeneous, the effects of stressors on lipid content can only be meaningful if specific tissues are sampled. Furthermore, the use of lipid profiles or total lipid content of fishes as a measure of toxicant-induced stress will be species-specific and effects of season, sex, reproductive and nutritional status will have to be considered.

Both marine and freshwater fishes can synthesize some polyunsaturated fatty acids but in general, the lipid contents of fishes are typical of their environments and changes in whole-body lipid profiles will probably not be particularly useful as biochemical indicators under field conditions but may prove to be useful under controlled laboratory exposure to single compounds.

Serum lipids

Interpretation of the toxicological significance of changes in hemolymph lipids can be extremely difficult. Increased energy demand may result in the mobilization of lipids from storage depots, such that transport lipoproteins and assocated triglycerides are elevated. Direct toxicant interference with lipid synthesis and degradation, and/or lipid transport processes could also alter the lipid profiles of the hemolymph.

The hemolymph/serum lipid profile varies tremendously among different species and can be influenced by a variety of exogenous factors. For example, phospholipids in crustaceans are the primary circulating lipid, whereas in insects diglycerides are the major transport lipid (Chang & O'Connor, 1983). Often a large proportion of hemolymph lipids are associated with complex lipoprotein moieties. In crustaceans, up to 95% of the lipids are classified as high density lipoproteins (HDL) and function in the transport of pigments, lipids and proteins (Lee & Pupprone, 1978). Depending upon the developmental stage, significant differences can be observed between males and females (Allen, 1972). Seasonal differences exist (Morris *et al.*, 1982) as do cyclic fluctuations during the molt cycle of the organism (O'Connor & Gilbert, 1969). Knowledge of these factors is required before using hemolymph lipid alterations in toxicological evaluations. Considering the wide array of toxicants of interest and the numerous potential target sites for interference with lipid metabolism, the monitoring of hemolymph lipid profiles in serum would provide, at best, a general, non-specific indicator of toxicant exposure, but may be useful in a particular species. Given the large array of functions in which hemolymph lipids are involved, it is not surprising that they fluctuate considerably.

Lipids and energetics

Lipids provide an essential, readily available energy source for a large number of aquatic invertebrates (Holland, 1978; Voogt, 1983). Their importance as a primary vs. secondary energy source varies among species and throughout seasons (Morris, 1971; Gardner & Riley, 1972). As the organism develops, it may alter its primary mechanism of storage. Benthic larvae often store neutral lipids, however, as they develop, their primary energy storage form may shift to glycogen (Holland, 1978). Similarly, during molting and/or reproductive cycles, the mobilization and utilization of different energy substrates may shift (Barclay *et al.*, 1983; O'Connor & Gilbert, 1969; Frank *et al.*, 1975). A variety of explanations have been proposed to account for these different strategies, however, it is not the purpose of this paper to discuss the extremely complex topic of invertebrate energy metabolisms. Suffice it to say that the energy pool which may be mobilized for the purpose of maintaining homeostatis during toxicant-induced stress may vary depending upon developmental stage, reproductive condition and molt status of the organism.

Another major factor which can influence an organism's lipid composition, or more specifically, the component fatty acid composition of the various lipid classes, is the quantitative and qualitative characteristics of an organism's diet (Voogt, 1983; Clarke, 1977). The importance of diet seems to be species-specific, however, and of equal importance to diet are the metabolic requirements of the organism (Hinchcliffe & Riley, 1972).

Monitoring changes in lipid content represents a non-specific indicator of the relative health of an organism but only with respect to appropriate control organisms. Since lipid deposition in eggs can represent a major energy source for the developing embryo, any interference with lipid metabolism in the adult can directly influence the survival of the next generation. In controlled laboratory experiments, the lipid content of larvae can be indicative of the relative health of their parents. Also, lipid contents, monitored over time, can be used as an indicator of growth and potential reproductive performance.

Membrane alterations

The ability of an organism to modify the fatty acid composition of membrane lipids in response to changing environmental conditions has become increasingly evident. Acclimation of invertebrates to lower temperatures increases the proportion of unsaturated fatty acids of membrane lipids relative to saturated fatty acids (Chapelle, 1978; Morris, 1971). This is thought to maintain membrane fluidity so that important functional properties such as water permeability and enzyme activity are retained (Hazel & Prosser, 1974). Lipophilic materials tend to partition into membranes and can reach extremely high concentrations (Conway, 1982). Interactions between the foreign material and unsaturated fatty acids may reduce membrane fluidity, potentially causing an increase in the relative concentration of unsaturated fatty acids. This situation was reported by Morris *et al.* (1982a, b), where *Gammarus duebeni* collected from polluted sites had a greater relative concentration of unsaturated fatty acids. Similarly, the membrane composition of bacteria exposed to insecticides was significantly different from that of the controls (Rosas *et al.*, 1985). Considerably more research is needed before this approach could be effectively utilized in the field as a monitoring tool, however, it does have potential as an indicator of ecosystem contamination by lipophilic materials.

Proteins

Serum and whole body protein concentrations have been shown to be influenced by a variety of environmental factors (Claybrook, 1983). Obviously, a large percentage of an organism's body is composed of structural proteins. However, it is not only the structural protein which can potentially change during toxicant-induced stress but also the soluble proteins. The rationale for changes in soluble or structural proteins are often quite different. Protein associated with the hemolymph can have a wide variety of functions such as enzymatic, transport or hormonal and can therefore change for a multitude of reasons. Serum enzyme activities can change in toxicant stressed organisms. Similarly, osmotic stress caused by changes in salinity or toxicant exposure can result in changes in hemolymph protein concentrations (Gilles, 1977; Boone & Schoffeniels, 1979). Oxygen stress can induce the production of transport pigments such as hemoglobin in certain species of crustaceans (Prosser, 1973), however, a decrease in hemoglobin concentrations have also been reported for *Daphnia* exposed to naphthalenes (Crider *et al.*, 1982). Given the wide range of natural events which can influence hemolymph protein concentrations, it will be difficult for one to distinguish toxic effects from background variability.

Crustaceans from the same population can have protein concentrations in hemolymph which vary threefold (Horn & Kerr, 1969). This indicates rather poor regulatory control over hemolymph protein concentrations. However, one needs to distinguish between measuring total soluble protein and measuring specific protein fractions. Changes in total protein may be difficult to interpret, however, alterations in specific enzymes or transport pigment may offer clues concerning potential sources or mode of toxicity.

The majority of the changes in structural protein which have been reported can be directly related to the oxidation of amino acids for energy. During starvation of invertebrates, both hemolymph and structural proteins are utilized as energy sources (Claybrook, 1983; Florkin & Sheer, 1970; Gilles, 1970) and have been shown to be mobilized under stressful conditions (Bayne, 1973). Glycogen and/or lipids are generally utilized first, however depending upon the season, reproductive status of the organisms and length of stressful conditions, proteins can become an important energy source (Bayne, 1973; Heath & Barnes, 1970). During toxicant-induced stress, protein catabolism was increased in oysters exposed to naphthalene (Riley & Mix, 1981) and in freshwater crabs exposed to sumithion (Bhagyalakshmi, *et al.*, 1983). Increased proteolysis to meet the energy demands of stress were assumed to cause the decline in structural proteins. Since protein content is generally less variable than either lipid or glycogen, separation of toxic effects from background 'noise' may be easier, although the secondary nature of protein catabolism for most organisms makes the sensitivity of the measurement questionable. Under laboratory conditions, however, McKee & Knowles (1985) have found protein to be a sensitive indicator of *Daphnia* growth and feel that total protein content, measured during acute exposures is an effective predictor of chronic survival, growth and reproductive potential.

Whole body calorimetry

Whole-body calorimetry has been suggested as a technique for assessing the energetic status of an organism during chronic toxicity (Krueger *et al.*, 1968). Using this technique, the conversion of energy equivalents among biomolecules would not mask effects on whole-body energy status. When juvenile pink salmon were exposed to oil, weight-specific caloric content decreased but weight-specific fat content was not affected (Moles & Rice, 1983). Presumably carbohydrate and protein content were decreased by the exposure. Further research in this area is needed to determine the utility of this method for predicting toxicity but the authors reported that caloric content was not as sensitive a measure of toxicity as growth. Exposure of *Cichlosoma bimaculactum* to pentachlorophenol (0.2 ppm) caused a decreased rate of growth as a result of an increase in the energetic cost of specific dynamic action and a greater energy loss which was observed to cause a decreased rate of growth (Krueger *et al.*, 1968).

The whole-body caloric content is similar to the condition factor technique which has long been used by fisheries biologists to determine the nutritional status of fish. Specific caloric measures may be more sensitive to short-term changes in the overall energy status of the organism than would be the condition factor. However, if biochemical or caloric indicators are to be successfully used, they will need to be correlated with the more traditional condition factor measures.

Toxicant effects on energetics
Fish
Glucose

A number of metals have been observed to cause hyperglycemia in fish, but chronic (30–60 days) effects of metals on energetic biomolecules of fish have not been extensively studied. Cadmium has been reported to both increase (Sastry & Subhadra, 1985) and decrease (Christensen *et al.*, 1977; Gill & Pant, 1983; Haux & Larsson, 1984) glucose concentrations in plasma during chronic exposure. Cadmium and zinc caused increased concentrations of glucose in plasma and a concomitant decrease in glycogen concentrations in tissue during acute exposures (Watson & McKeown, 1976; Shaffi, Gill & Pant, 1983). Cadmium-exposed fish were unable to remove the glucose load from their blood. Exposure of the freshwater teleost *Colisa fasciatus* to 80% of the 96 h LC_{50} for Ni caused hyperglycemia, which was attributed to pathological effects on the gills (Chaudhry, 1984). Zinc exposure caused hyperglycemia in rainbow trout (Watson & McKeown, 1976). Chronic exposure of *Channa punctatus* to chromium caused hypoglycemia, while glycogen concentrations increased in liver but decreased in muscle tissue of *C. punctatus* (Sastry & Sunita, 1983). Rainbow trout exposed to an effluent containing 260 mg Ti/l in an effluent had elevated plasma glucose and lactate concentrations when exposed at 13–15 °C, however, these effects were not observed when the fish were exposed at 7–8 °C (Lehtinen *et al.*, 1984). The results of that study indicate some of the problems that may be encountered in interpreting the results of energetic and metabolic indicators of toxicant-induced stress in fish.

Numerous organic compounds cause a rapid increase of glucose concentrations in plasma, when fish are exposed to acutely toxic concentrations (McLeay & Brown, 1975; DiMichele & Taylor, 1978; Casillas *et al.*, 1983; Thomas *et al.*, 1981b; Gluth & Hanke, 1984; Soivio *et al.*, 1983; Pant & Singh, 1983). The magnitude of the increase can be dependent on glycogen concentration in the liver (McLeay, 1977).

Most of the research into effects of organic toxicants on carbohydrate metabolism has been conducted with pesticides (Verma *et al.*, 1983). Pesticides cause predictable alterations in energy biomolecules in blood and other tisues of fish during acute exposures (Verma *et al.*, 1983). Most pesticides cause a rapid increase in concentrations of glucose in blood (Silbergeld, 1974; Singh & Srivastava, 1981; Murty & Devi, 1982; Gluth & Hanke, 1984; Srivastava & Mishra, 1983). Pesticide exposure causes glycogen concentrations in muscle tissue to increase within hours of the onset of exposure and then to decrease with continued exposure.

The effects on glucose of chronic exposure of fish to pesticides are similar to those acute effects mediated by the endocrine system (Singh & Singh, 1980). Snakehead (*C. punctatus*) exposed to the organophosphorus insecticide, sevin (carbaryl), for 30 and 60 days exhibited hyperglycemia and decreased glycogen concentrations in muscle and liver (Sastry & Siddiqui, 1982). Concentrations of glucose in the blood of the Indian catfish (*Heteropneustes fossilis*) were significantly increased by exposure for three hours to 5.6 mg/l parathion (Singh & Srivastava, 1981). Blood glucose of *Tilapia mossambicus* (Peters) decreased when fish were exposed to acutely toxic concentrations of the organophosphate insecticide sumithion (fenitrothion) (Koundinya & Ramamurthi, 1979). Exposure to an acutely toxic concentration of the herbicide 2,4-D caused hyperglycemia in the Indian catfish (Srivastava & Gupta, 1981). Alternatively, when Indian catfish were exposed to 80% of the 96 h LC_{50} of a mixture of aldrin and formithion, the concentration of glucose in the blood was significantly decreased after 3, 6, 12, 48 and 96 h of exposure (Singh & Srivastava, 1981).

Not all toxicants cause hyperglycemia in fish. Treatment with sublethal concentrations of naphthalene did not affect the glucose concentration in plasma of the cichlid fish *Sarotherodon mossambicus* (Peters) but did cause a decrease in glycogen and pyruvic acid and an increase in lactic acid. This change indicates a shift to anaerobic metabolism in naphthalene-intoxicated fish (Dange & Masurekar, 1982).

Anesthetics have been used to try to minimize the effects of capture and handling stress. However, anesthetization with MS-222 caused greater concentration of glucose in plasma relative to untreated controls (Nieminen *et al.*, 1983). Fish anesthetized with etomidate did not exhibit increased plasma glucose concentrations, however, netting and confining of these fish caused increased glucose concentrations in plasma relative to unanesthetized fish (Limsuwan *et al.*, 1983).

Glycogen

The glycogen content of fish tissues is affected by both acute and chronic *in vivo* exposure to metals. Long-term exposure to a single metal or mixture of metals generally causes a decrease in glycogen concentrations in brain, liver and skeletal muscle tissue (Dubale & Shah, 1981; Arillo *et al.*, 1982; Gill & Pant, 1983; Sastry & Subhadra, 1985). When *C. punctatus* were exposed to 2.6 mg Cr/l (as K_2CrO_4) for 60 and 120 days, the glycogen concentration of liver was greater than that of control fish, while that of muscle tissue was less than in untreated fish (Sastry & Sunita, 1983). Acute, *in vivo* exposures (0.8 of LC_{50} for 96 h) of the freshwater fish *C. fasciatus* to Ni caused a decrease of glycogen concentrations in muscle tissue (Chaudry, 1984). Chronic exposure of rainbow trout to Cd caused a decrease in glycogen concentration of muscle. This condition persisted 25 weeks after Cd exposure ceased (Haux & Larsson, 1984). Furthermore, after 57 wk of recovery there was a dose-dependent increase in glycogen content of liver tissue. These effects were thought to be due to selective uptake of Cd by the pancreas. The uptake caused inhibition of insulin secretion and subsequent insulin-mediated changes in carbohydrate metabolism. Therefore, unless the source term for exposure is known, it is difficult to interpret the responses of glycogen to chemical stressors.

Exposure of fish to organophosphate insecticides causes changes in the glycogen content of fish tissues. *In vivo* exposure of the Indian catfish (*H. fossilis*) to 5.6 mg/l parathion caused a significant decrease in glycogen concentrations in both liver and muscle after 3, 6, 12 and 96 h of exposure (Srivastava & Singh, 1981). Glycogen concentrations in liver tissue were also decreased after 3 and 6 h but there was a compensatory resynthesis of glycogen in the liver after 12 h of exposure. When the Indian catfish (*C. punctatus*) was exposed to endosulfan for 96 h, the glycogen content of liver and muscle were significantly decreased while that of kidney tissue was increased (Murty & Devi, 1982).

When Indian catfish were exposed to 80% of the 96 h LC_{50} of a mixture of aldrin and formathion glycogen concentrations in muscle decreased significantly after 3 h of exposure and stayed depressed until the end of the experiment at 96 h (Singh & Srivastava, 1981). Long-term exposure of a freshwater cyprinid to sublethal concentrations of carbaryl (0.194 mg/l) or dimethoate (0.683 mg/l) caused a significant depletion of glycogen in liver after 15 day (Pant & Singh, 1983). The glycogen content of brain tissue was decreased by exposure to carbaryl but increased by exposure to dimethoate. When the freshwater fish *S. mossambicus* was exposed to an acutely toxic concentration (6 mg/l) of the organophosphate insecticide sumithion, a decrease in hepatic glycogen concentrations occurred (Koundinya & Raramurthi, 1979). Fish exposed chronically to endrin or quinalpos exhibited greater concentrations of glycogen in muscle and liver. This observation indicates a reduction in basal metabolism or an effect of glycogenesis regulation (Grant & Mehrle, 1973; Sastry & Siddiqui, 1984). Endosulfan, on the other hand, causes a decrease in glycogen concentrations in liver and muscle (Sastry & Siddiqui, 1983). Acute, *in vivo* exposures of the freshwater fish *S. mossambicus* to acidic water caused tissue-specific changes. Glycogen concentration of red muscle was increased while that of white muscle was decreased due to acute exposures.

Lactic acid

Few studies have investigated toxicant-induced effects on lactate concentrations in fish blood.

Exposure to 2.6 mg K_2CrO_4/l for 120 d caused an increase in lactic acid concentration in plasma and muscle of the teleost fish *C. punctatus*, while the lactic acid concentration in liver tissue was less than that of untreated fish (Sastry & Sunita, 1983). Similarly, *in vivo* exposure of the freshwater fish *C. fasciatus* to 80% of the 96 h LC_{50} for Ni caused a significant increase in lactic acid content of the blood (Chaudhry, 1984). An increased tissue lactate:pyruvate ratio in fish exposed to naphthalene indicates a greater anaerobic energy production (Dange & Masurekar, 1982), which is believed to be caused by reduced tissue O_2 due to gill damage (McLeay & Brown, 1974).

The concentrations of lactate in the blood of fish has been studied in response to only a few organic toxicants. Concentrations of lactate in the blood of Indian catfish were increased by exposure to 80% of the 96 h LC_{50} of a mixture of aldrin and formithion for 3, 6, 12 and 96 h (Singh & Srivastava, 1981). The insecticides thiotox and dichlorfos caused increases in lactate concentration in blood of three species of the freshwater fishes, *Clarias batrachus*, *Saccobranchus fossilis* and *Mystus vittatus* (Verma *et al.*, 1983).

Pyruvic acid

The pyruvate concentration in blood of the Indian catfish (*H. fossilis*) was decreased from 0.35 to 0.29 mg/100 ml after fish were exposed to a mixture of two insecticides, aldrin and formithion for 6 h to 80% of the 96 h LC_{50} (Singh & Srivastava, 1981). Alternatively, the pyruvate concentration in blood of the Indian catfish was elevated due to exposure to a lethal concentration of 2,4-D for 3, 6, 12 and 48 h (Srivastava & Gupta, 1981).

Lipids

Relatively few studies have investigated the effects of toxicant-induced stress on the lipid content of fish. The effects of stressors on fatty acid contents are equivocal. The plasma, free, fatty acid concentrations can increase or decrease in response to stress, according to species (Mazeaud *et al.*, 1977).

When rainbow trout were exposed *in vivo* to chronic concentrations of copper (0.075–0.225 mg/l), the growth rate was depressed initially, but at the end of the exposure there was no significant effect on lipid content of the fish (Lett *et al.*, 1976). Cadmium (Cd) caused a decrease in total lipid content of brain, liver and gonadal tissues in catfish (*Clarias batrachus*) (L.) (Katti & Sathyanesan, 1984). The cholesterol response to Cd in this species was more complex. Cholesterol concentration increased in liver tissue but decreased in both brain and gonadal tissues.

Lipid metabolism is affected by acute exposures to pesticides. Total lipids in general and phospholipids in particular in muscle, gill and liver, while free fatty acid concentrations in muscle and liver tissues increased (Murty & Devi, 1982; Rao & Rao, 1984a, b). Lipid content of brain tissue was significantly decreased in *Cirrhinus mrigala* which had been exposed to the carbamate insecticide carbaryl (Rao *et al.*, 1984b). Cholesterol concentrations in serum were increased by exposure to parathion (Rao & Rao, 1984b) and decreased during exposure to a variety of chlorinated hydrocarbons (Gluth & Hanke, 1984). Because there is no consistent trend, it is difficult to interpret pesticide effects on cholesterol concentrations. Cholesterol is not a useful bioindicator of exposure to pesticides.

When rainbow trout were exposed to 0.01, 0.02 and 0.03 mg/l cyanide (HCN) for 18 d, there was a significant reduction in growth rate, especially during the first 9 days. Lipid content was also significantly less than that of unexposed fish (Dixon, 1975). This change in lipid content was positively correlated with growth rate.

Organic xenobiotics can alter blood and tissue lipids of fish both quantitatively and qualitatively during chronic exposures (Chen & Sonstegard, 1984; Verma *et al.*, 1984). Chronic exposure of the cod (*Gadus morhua*) to crude oil caused increases of fatty acids and phospholipids in fatty tissue while causing decreases in triglyceride concentrations. These effects indicate enhanced mobilization and utilization of stored triglycerides to meet increased energy demands (Dey *et al.*, 1983). Changes in lipid content due to exposure to xenobiotics does not affect profiles of fatty acids in all fish. Total lipid concentrations in both liver and adipose tissue decreased in rainbow trout which were fed diets containing phthalate esters (Sargent & Henderson, 1983). However, the profiles of fatty acids in liver, muscle and adipose tissue remained unchanged. Since phospholipids are the most actively degradable lipids, it is suggested that they are rapidly utilized to provide energy. Therefore, a decrease in total phospholipids as observed when *Oreochromis mossambicus* were exposed to methyl parathion (Rao & Rao, 1984a, b) could be attributed to their use of energy to resist a stressor. Therefore, in addition to total lipid content and tissue-specific lipid content, we suggest that phospholipid profiling may be a useful monitoring tool to determine intermediate term toxicant-induced stress effects in fish.

Invertebrates
Glucose

Compared to fish, few studies have been conducted on toxicant-induced effects on glucose in invertebrates. Hyperglycemic responses have been observed in freshwater crabs exposed to DDT (Fingerman *et al.*, 1981), sumithion (Bhagyalakshmi *et al.*, 1983) and BHC (Sreenivasula *et al.*, 1983), and in blue crabs and polychaete worms exposed to pentachlorophenol (Coglianese & Neff, 1982, Thomas *et al.*, 1981a, b). These responses have been attributed to the release of hyperglycemic hormones and subsequent mobilization of stored glycogen reserves. However, hyperglycemic responses are not always observed. The concentration of glucose in the coelomic fluid of *Neanthes virens*, decreased during acute PCP exposure (Thomas *et al.*, 1981a, b). However, hypoglycemia generally preceded hyperglycemia which is the primary response. Any hyperglycemia which did occur was only transitory and glucose concentrations eventually returned to normal. Significant differences were observed in glucose concentrations in coelomic fluid in polychaetes collected from polluted and unpolluted stations, however this was a limited study. The importance of the observed alterations was difficult to determine (Carr & Neff, 1984). The transitory nature of changes in glucose concentrations in hemolymph limits its usefulness as an indicator of stress (Plisetskaya *et al.*, 1978). In addition, the influence of accessory factors on glucose concentrations in hemolymph will impede ones ability to separate toxic effects from background variability. Factors such as sex (Dean & Vernberg, 1965); reproductive, seasonal and diurnal cycles (Telford, 1974); and molt status (Telford, 1974) will influence the glucose levels and need to be considered before using a particular organism.

Lipids

Specific concentrations of lipids in serum have been altered in aquatic invertebrates exposed to toxicants. Hypocholesterolemia has been reported in blue crabs exposed to pentachlorophenol (Coglianese & Neff, 1982), although no differences in fatty acid profile were observed in similar experiments (Bose & Fujiwara, 1978). Alterations in the cholesterol level may be an indirect result of toxicant effects on enzyme metabolism systems. O'Hara *et al.* (1985) have shown that polynuclear aromatic hydrocarbons can interfere with the cytochrome P-450 enzyme system, such that cholesterol metabolism is impaired.

The greater energy demand caused by the stress of toxicant exposure can result in a mobilization of lipid energy reserves and subsequent decline in the total lipid content of the organism. A decrease in lipid content was noted in amphipods exposed to fuel oil (Lee *et al.*, 1981), while increased lipid catabolism was measured in oysters exposed to naphthalene (Riley & Mix, 1981). Capuzzo *et al.*, (1984) found that petroleum hydrocarbons significantly interfere with lipid metabolisms in the American lobster. This interference results in consistently lesser triglycerol concentrations in serum of exposed larvae.

RNA, DNA and protein

Rationale

Exposure of aquatic organisms to stress frequently causes reduced growth rates. This effect is caused by a number of factors including interference with biochemical pathways and reduced scope for growth due to diversion of resources to resist stressors. Sensitive measurement of growth rates or a parameter related to growth rate, would enable determination of an adverse effect on growth without conducting a long-term growth study (Table 3).

The quantification of ribonucleic acid (RNA), deoxyribonucleic acid (DNA), and protein have been used to assess growth rates in aquatic organisms (Munro & Fleck, 1966; Buckley, 1979; Hartree, 1972). RNA content is directly related to the synthesis of structural protein. The ratio of RNA to DNA is generally believed to more accurately reflect growth since this ratio is not affected by cell number or size. The RNA/DNA ratio has been proposed as an integrative measure of stress effects, based on the assumption that exposure to stressors directly inhibits growth or that the energetic expenditure reduces scope for growth. Quantification of protein on the other hand, is a measure of actual tissue (protein) accumulation and represents a 'static' indicator of an organism's growth status, and gives no information concerning its growth rate.

Fish

RNA/DNA ratios have been used successfully to assess short- and long-term growth rates in fish (Bulow, 1970; 1971; Haines, 1973; Sower et al., 1983). This ratio varies rapidly and is directly related to nutrition, age and season (Haines, 1973; Bulow et al., 1981). The rapidity with which the RNA/DNA ratio can change in response to feeding is a potential limitation for use in both laboratory and field experiments.

A number of toxicants have been shown to affect RNA/DNA under laboratory conditions.

Table 3. Protein Content and Synthesis as an Indicator of Toxicant-Induced Stress in Aquatic Organisms

Rationale:	Stress causes a reduction in the growth rate; RNA, DNA, protein concentrations and the RNA/DNA ratio are biochemical measures of growth rate.
Measures:	RNA, DNA and protein are directly quantifiable in tissues and homogenates of organisms. Amino acid accumulation is measured by utilizing radio-labelled amino acids.
Example:	Xenobiotics cause increased energy expenditure. The biochemistry is shifted away from protein formation. Protein and RNA content decrease. This effect is quantifiable long before effects on body weight can be measured.
Applicability:	Useful for a wide variety of organisms. Most useful under laboratory conditions. Rapid measure of chronic effects.
Advantages:	A wide variety of stressors have been shown to decrease growth rates by decreasing food intake or increasing energy usage. This technique is a very sensitive measure of growth rate.
Disadvantages:	RNA content has been shown to change very rapidly. The techniques which are traditionally employed are subject to variability due to nutritional status. Growth may not always be affected. Growth may not be the most sensitive endpoint.
Conclusions:	Stressor-induced changes in RNA, DNA, protein concentration and amino acid assimilation can be used to determine toxicant effects in laboratory exposures. Interpretation of field monitoring would be more difficult and is not recommended.
Recommendations:	Research on the effects of long-term stressors on RNA, DNA and protein content of fish and invertebrates should be conducted to determine if these techniques can predict ecologically relevant effects on populations of organisms.

Protein, RNA-P, DNA-P and RNA/DNA ratio in whole fathead minnow larvae have been shown to be affected by toxicants (Barron & Adelman, 1984). Exposure of fish larvae to toxicants under laboratory conditions caused reduced concentrations of both RNA and protein. They reported good agreement between concentrations established as safe, based on 28 to 32d early life stage toxicity tests, and nucleotide and protein measurements after 96 h. RNA/DNA ratios were variable and ranged from 1.63 to 3.69 in control fish among the experiments. RNA, DNA and protein content per larvae and RNA/DNA ratio were sensitive to toxicant exposure, with RNA concentrations being the most sensitive measure. The synthetic detergent Idet-20[R] caused a statistically significant decrease in the RNA concentration in the liver of the fish C. batrachus (Verma et al., 1984).

Toxicants have also been observed to affect RNA/DNA ratios under field conditions. Kearns & Atchinson (1979) correlated RNA/DNA ratios and growth rates in yellow perch (Perca flavescens) with metal pollution (cadmium and zinc) in 2 areas of a lake. Regression analysis revealed that whole fish cadmium concentrations were significantly negatively correlated with RNA/DNA ratios. This correlation indicated that increased cadmium body burden resulted in decreased growth rate. However, the RNA/DNA ratios of fish exposed to carbaryl drift in a stream (0.79 mg/l) resulted in significantly elevated RNA/DNA ratios (Wilder & Stanley, 1983). This result was attributed to increased food availability due to toxicity to benthic invertebrates. This unexpected result underscores the type of assessory factors which will complicate interpretation of toxicant-induced effects on RNA/DNA ratios difficult under field conditions.

Invertebrates

Estimation of growth rates by measurement of the RNA, DNA and protein content of invertebrates has been successful. Correlations between RNA content and growth have been established for mosquito larvae (Lang et al., 1965), boll weevils (Vickers & Mitlin, 1966), blowfly (Price, 1969), amphipods (Sutcliffe, 1965) and brine shrimp (Dagg & Littlepage, 1972).

Studies have shown variability in RNA concentrations due to differences in nutritional history, molt cycle and developmental stage (Wigglesworth, 1963; Vickers & Mitlin, 1966; Dagg and Littlepage, 1972). Fluctuations in the RNA content of Rhodinus prolixus have been attributed to the stimulation of RNA synthesis by the molting hormone ecdysone (Wigglesworth, 1963). The influence of accessory factors and the inconsistent growth patterns of many invertebrate species may severely limit the use of RNA/DNA ratios as estimators of growth rate under field conditions.

To our knowledge, only one study has been conducted on the effect of toxicant-induced stress on the nucleic acid content of aquatic invertebrates. McKee & Knowles (1985) used the nucleic acid and protein content of the freshwater cladoceran D. magna as a short-term indicator of chronic reproductive effects. All of the biochemical parameters were significantly reduced by exposure to 60 µg/l chlordecone, however, protein content was found to be the most sensitive growth indicator. The no-effect level based on day 7 protein determinations was the same as that based on day 21 reproduction values (young/adult/day). The use of such biochemical measurements as short-term end points in chronic reproduction studies, which are routinely used in hazard assessment procedures, has potential and should be further investigated.

Amino acid incorporation growth index

The ^{14}C-amino acid incorporation index has been developed as a relative measure of the rate at which an organism is growing (Ottaway & Simkiss, 1977). The theoretical basis for this measure of growth is that free amino acids are incorporated into structural proteins. Therefore, the rate of incorporation of a labelled amino acid should be indicative of growth rate. The original method used scales removed from live fish but was useful only for large fish. The method has been adopted to measure ^{14}C incorporation into axial muscle of fish larvae and the gill and digestive gland of mussels (Viarengo et al., 1982).

An assay has been proposed by Adelman and Buscacker (1982) in which sections of larval fish are incubated in physiological saline which contains ^{14}C-labelled glycine. The procedure seems to work, however, to date there have been no studies to determine the effects of toxicants on ^{14}C-glycine incorporation. Toxicants have, however, been observed to affect assimilation of other amino acids. Assimilation of ^{14}C-glutamic acid, leucine and arginine were determined for muscle, gill and liver tissues of *T. mossambica* which had been exposed to 2 mg malathion/l for 48 h (Sahib *et al.*, 1984). The malathion-treated fish assimilated more glutamic acid than the untreated fish. An examination of the different accumulation patterns among amino acids indicated that this was caused by an elevated rate of synthesis of basic proteins. This technique would be most useful in shortening the time required to conduct chronic toxicity studies under laboratory conditions or to determine the growth rates of fry from different locations. ^{14}C-glycine incorporation into scales has been shown to be closely correlated with stressors, which are known to affect fish growth (Adelman, 1980) and may be an effective field monitoring tool if effects of toxicants are not obscured by variations due to season, sex, nutritional status, or handling stress.

Using ^{14}C-labelled leucine, Viarengo *et al.*, (1980; 1982) successfully developed a sublethal stress indicator based on protein synthesis in gill and digestive gland tissue of the marine mussel, *Mytilus galloprovincialies*. Under controlled laboratory conditions, they were able to correlate copper body burdens with reduced protein synthesis (Viarengo *et al.*, 1980), and then successfully applied the technique to organisms collected from polluted and unpolluted environments (Viarengo *et al.*, 1982). A similar assay has been developed to use incorporation of ^{14}C-labelled thymidine by embryos of marine invertebrates as a rapid toxicity assay (Jackim & Nacci, 1984). This assay has been used to assess the responses of embryos of the sea urchin (*Arbacia punctulata*) to a wide range of toxicants and has been found to be a useful predictor of reduced survival and growth in longer-term tests. This assay will be most useful in laboratory toxicity testing of pure compounds, effluents or contaminated sediments.

Protein

Tissue protein content has been suggested as an indicator of toxicant-induced stress. Endosulfan, an organochlorine insecticide, caused a significant increase in protein content of both kidney and brain tissue of *C. punctatus* exposed *in vivo* to concentrations ranging from 3.5 to 6.5 μg/l for 96 h (Murty & Devi, 1982). Alternatively, when the fish *Cirrhinus mrigala* was exposed to 2 mg/l the carbamate insecticide carbaryl *in vivo* for 96 h, the total protein concentrations in brain, gill and kidney tissue all increased significantly (Rao *et al.*, 1984a, b). Both of these studies indicate a disturbance in protein metabolism, however, one caused an increase, while the other caused a decrease.

The serum protein of fish has also been found to be affected by toxicants. For instance, when the brown bullhead (*Ictalurus nebulosus*, Lesueur) was exposed to 27 μg Cu-II/l for 6 d, serum total protein concentration decreased slightly relative to controls, while 49 and 107 μg Cu-II/l caused significant elevations. After 30 days exposure, total serum protein was greatly reduced (Christensen *et al.*, 1972). When brook trout (*Salvelinus fontinalis*) were exposed *in vivo* for 6 d to 24 and 39 μg Cu (II)/l, total serum protein (TSP) was increased significantly, relative to controls, whereas 67.5 μg Cu/l had no effect on TSP. After 30 d exposure, the TSP of fish exposed to the least concentration of copper had returned to normal, while that of *S. fontinalis* exposed to 67.5 μg Cu/l was significantly less than that of controls (McKim *et al.*, 1970). When brook trout alevins were exposed to methyl mercury (CH$_3$Hg(II), Cd (II) and lead (Pb II), small concentrations of CH$_3$Hg caused a decrease in TSP while a similar concentration of Cd caused an increase in TSP. Greater concentrations of Cd had no effect on TSP. These results indicate the responses of TSP to toxicants are very complex and the direction of response is dependent upon duration of exposure as well as concentration of toxicant. Therefore, TSP does not appear to be a viable biochemical indicator.

RNA, DNA and protein content, the

RNA/DNA ratio and amino acid incorporation can be used successfully to determine growth and growth rates in fish and invertebrates. The large number of factors which alter RNA, DNA and protein content in organisms (Gallis *et al.*, 1980; Bulow *et al.*, 1981), indicates that the best use of these measures are in young, rapidly growing organisms, exposed to stressors in a controlled laboratory experiment. Further use of these techniques to determine growth rates in the laboratory are recommended. In the field, determination of growth rates can be more easily and economically accomplished by other methods. In addition, the factors which alter RNA, DNA and protein content cannot be rigorously controlled. The best use of this technology is in short-term experiments where effects on growth rate need to be determined rapidly. Use of radiolabelled nucleoside, and amino acid incorporation into RNA, DNA and proteins, may increase the accuracy and amount of information obtained in assessing the turnover of these macromolecules.

Even though we know that biochemical changes, observed during acute exposures can be correlated with chronic effects on survival, growth or reproduction, so few studies have been conducted to calibrate the acute and chronic observations that presently it would be difficult to use any of these measures to routinely replace definitive chronic studies. We suggest that such calibrations be made for several classes of compounds on several bioassay species. In general we feel that the biochemical measures will have more utility in either understanding modes of action under laboratory conditions or serving as early warning measures for determining probable causes of effects on populations in the field.

Acknowledgements

Preparation of this manuscript was supported, in part, by the Michigan Agricultural Experiment Station, from which it is contribution number 12776. Portions of this manuscript were written while Professor Giesy was studying as a Fulbright fellow at the Chair of Ecological and Geological Chemistry of the University of Bayreuth, West Germany. The support and assistance of the Fulbright Commission, Bonn, West Germany and Professor O. Hutzinger and his faculty and staff, particularly H. Fiedler and G. Pusch is appreciated. A draft of the manuscript was reviewed by W. Gala, D. Tillitt, R. Hoke, L. Williams and K. Grzyb. Jane Thompson prepared the manuscript. We wish to thank the organizers of the conference, especially Drs. R. Thomas, T. Reynoldson and M. Munawar for inviting us to participate in and supporting travel to such an interesting and informative conference.

References

Adelman, I. R., 1980. Uptake of ^{14}C-glycine by scales as an index of fish growth: Effect of fish acclimation temperature. Trans. Am. Fish. Soc. 109: 187–194.

Adelman, I. R. & G. P. Busacker, 1982. Indicators of current growth rate as rapid methods for toxicity tests with fish larvae. In: R. A. Archibald (ed.), Environmental Biology State-of-the-Art Seminar, EPA Report 60019-82-007.

Allen, W. V., 1972. Lipid transport in the Dungeness crab, *Cancer magister* Dana. Comp. Biochem. Physiol. 43B: 193–207.

Anderson, R. L., C. T. Walbridge & J. T. Fiandt, 1980. Survival and Growth of *Tanytarsus dissimilis* (Chironomidae) exposed to copper, cadmium, zinc and lead. Arch. Envir. Contam. Toxicol. 9: 329–335.

Anon, 1982. Rule 57 Advisory Committee Report on Proposed Surface Water Quality Standard Derivation Procedures for Chemical Substances, Michigan Department of Natural Resources, Lansing, MI, 16 p.

Arillo, A., C. Margiocco, F. Melodia & P. Mensi, 1982. Biochemical effects on long term exposure to Cr, Cd, Ni on rainbow trout (*Salmo gairdneri* Rich.): Influence of sex and season. Chemosphere. 11: 47–57.

Attar, E. N. & E. J. Maly, 1982. Acute toxicity of cadmium, zinc and cadmium-zinc mixtures to *Daphnia magna*. Arch. Envir. Contam. Toxicol. 11: 291–296.

Barclay, M. C., W. Dall & D. M. Smith, 1983. Changes in lipid and protein during starvation and the moulting cycle in the tiger prawn, *Penaeus esculentus* Haswell. J. Mar. Biol. Ecol. 68: 229–244.

Barnthouse, L. W., G. W. Suter, S. M. Bartell, J. J. Beauchamp, R. H. Gardner, E. Linder, R. V. O'Neill & A. E. Rosen, 1986. Users Manual for Ecological Risk Assessment. ORNL-6251 Oak Ridge National Laboratory, pp. 227.

Barnthouse, L. W., G. W. Suter II, A. E. Rosen & J. J. Beauchamp, 1987. Estimating responses of fish populations to toxic contaminants. Envir. Toxicol. Chem. 6: 811–824.

Barron, M. G. & I. R. Adelman, 1984. Nucleic acid, protein content and growth of larval fish sublethally exposed to various toxicants. Can. J. Fish. Aquat. Sci. 41: 141–150.

Bayne, B., 1973. Aspects of the metabolism of *Mytilus edulis* during starvation. Nether. J. Sea Res. 7: 399–410.

Bhagyalakshmi, A., P. Sreenivasula Reddy & R. Ramamurthi, 1983. Muscle nitrogen metabolism of freshwater crab, *Oziotelphusa senex senex*, Fabricuis, during acute and chronic sumithion intoxication. Toxicol. Lett. 17: 89–93.

Bhagyalakshmi, A., R. S. Reddy & R. Raramurthi, 1984. *In-vivo* sub-acute physiological stress induced by sumithion on some aspects of oxidative metabolism in the freshwater crab. *Wat. Air Soil Pollut.* 23: 257–262.

Bhagyalakshmi, A., P. Sreenivasula Reddy & R. Ramamurthi, 1984. Subacute stress induced by sumithion on certain biochemical parameters in *Oziotelphusa senex senex*, the freshwater rice-field crab. Toxicol. Lett. 21: 127–134.

Biesinger, K. E. & G. M. Christensen, 1972. Effects of various metals on survival, growth, reproduction and metabolism of *Daphnia magna*. J. Fish. Res. Bd. Can. 29: 1691–1700.

Bols, N. C., S. A. Boliska, D. G. Dixon, P. V. Hodson & K. L. E. Kaiser, 1985. The use of fish cell cultures as an indication of contaminant toxicity to fish. Aquat. Toxicol. 6: 147–155.

Boone, W. R. & E. Schoffeniels, 1979. Hemocyanin synthesis during hypoosmotic stress in the shore crab *Carnius maenas*. Comp. Biochem. Physiol. 63B: 207–214.

Borgmann, U., 1985. Predicting the effect of toxic substances on pelagic ecosystems. Sci. Tot. Environ. 44: 111–121.

Borgman, U., O. Kramer & C. Loveridge, 1978. Rates of mortality, growth, and biomass production of *Lymnaea palustris* during chronic lead exposure. J. Fish. Res. Bd. Can. 35: 1109–1115.

Borgman, U. & K. M. Ralph, 1986. Effects of cadmium, 2,4-dichlorophenol and pentachlorophenol on feeding, growth, and particle-size-conversion efficiency of white sucker larvae and young common shiners. Arch. Envir. Contam. Toxicol. 15: 473–480.

Bose, A. K. & H. Fujiwara, 1978. Fate of pentachlorophenol in the blue crab, *Callinectes sapidus*. In: K. R. Rao (ed.), Pentachlorophenol: Chemistry, Pharmacology and Environmental Toxicology, Plenum Press, New York. 402 pp.

Buckley, L. J., 1979. Relationships between RNA-DNA ratio, prey density and growth rate in Atlantic cod (*Gadus morhua*) larvae. J. Fish. Res. Bd. Can. 26: 1497–1502.

Bulow, F. J., 1970. RNA-DNA ratios as indicators of recent growth rates of a fish. J. Fish Res. Bd. Can. 27: 2343–2349.

Bulow, F. J., 1971. Selection of suitable tissues for use in the RNA-DNA ratios technique of assessing recent growth rate of a fish. Iowa State J. Sci. 45: 71–78.

Bulow, F. J., M. E. Zeman, J. R. Winningham and W. F. Hudson, 1981. Seasonal variations in RNA-DNA ratios and in indicators of feeding, reproduction, energy storage, and condition in a population of bluegill sunfish, *Lepomis macrochirus* Rafinesque. J. Fish. Biol. 18: 237–244.

Cairns, J. Jr., 1968a. Predicting response from one level of biological organization to another. In: Cumulative Environmental Effects: A Binational Perspective. Canadian Environmental Research Council, Ottawa, 175 p.

Cairns, J. Jr. (Ed.), 1986b. Community Toxicity Testing. ASTM-STP 920. American Society for Testing and Materials, Philadelphia, PA, 350 p.

Cairns, J. Jr., 1988. Should regulatory criteria and standards be based on multispecies evidence? Envir. Prof. 10: 157–165.

Capuzzo, J. M., B. A. Lancaster & G. C. Sasaki, 1984. The effects of petroleum hydrocarbons on lipid metabolism and energetics of larval development and metamorphosis in the American lobster (*Homarus americanus* Milne Edwards). Mar. Envir. Res. 14: 201–228.

Capuzzo, J. M., M. N. Moore & J. Widdows, 1988. Effects of toxic chemicals in the marine environment: predictions of impacts from laboratory studies. Aquat. Toxicol. 11: 303–311.

Carmichael, G. J., A. Wedemeyer, J. P. McCraren & J. L. Millard, 1983. Physiological effects of handling and hauling stress on smallmouth bass. Prog. Fish Cult. 45: 110–113.

Carr, R. S. & J. M. Neff, 1984. Field assessment of biochemical stress indices for the sandworm *Neanthes virens* (Sars). Mar. Envir. Res. 14: 267–279.

Casillas, E., M. Myers & W. E. Ames, 1983. Relationship of serum chemistry values to liver and kidney histopathology in English sole (*Parophrys vetulus*) after acute exposure to carbon tetrachloride. Aquat. Toxicol. 3: 61–78.

Casillas, E. & L. S. Smith, 1977. Effect of stress on blood coagulation and hematology in rainbow trout (*Salmo gairdneri*). J. Fish Biol. 10: 481–491.

Chang, E. S. & J. D. O'Connor, 1983. Metabolism and transport of carbohydrates and lipids. In: L. H. Mantel (ed.), The Biology of Crustacea, Vol. 5, Internal Anatomy and Physiological Regulation, pp. 263–287. Academic Press, New York.

Chapelle, S., 1978. The influence of acclimation temperature on the fatty acid composition of an aquatic crustacean (*Carcinus maenas*). J. Exp. Zool. 204: 337–346.

Chaudhry, H. S. 1984. Nickel toxicity on carbohydrate metabolism of a freshwater fish, *Colista fasciatus*. Toxicol. Lett. 20: 115–121.

Chavin, W. & J. E. Young, 1970. Factors in the determination of normal serum glucose levels of goldfish, *Carassius auratus* L. Comp. Biochem. Physiol. 33: 629–653.

Chen, T. T. & R. A. Sonstegard, 1984. Development of a rapid, sensitive and quantitative test for the assessment of the effects of xenobiotics on reproduction in fish. *Mar. Envir. Res.* 14: 429–430.

Christensen, G. M., S. M. McKim, W. A. Brungs & E. P. Hunt, 1972. Changes in the blood of the brown bullhead (*Ictalurus nebulosus* (Lesueur)) following short- and long-term exposure to copper. Toxicol. Appl. Pharm. 23: 417–427.

Clarke, A., 1977. Lipid class and fatty acid composition of

54

Chorismus antarcticus (Pfeffer) (Crustacea: Decapoda) at south Georgia. J. Exp. Mar. Biol. Ecol. 28: 297–314.

Claybrook, D. L., 1983. Nitrogen metabolism. In: L. H. Mantel (ed.), The Biology of Crustacea, Vol. 5, Internal Anatomy and Physiological Regulation, pp. 163–213. Academic Press, New York.

Coglianese, M. P. & J. M. Neff, 1982. Biochemical responses of the blue crab, *Callinectes sapidus* to pentachlorophenol. In: W. B. Vernberg, A. Calabruse, F. P. Thurnberg and E. J. Vernberg (eds.), Physiological Mechanisms of Marine Pollution Toxicity, Academic Press, New York.

Collvin, L., 1984. The effects of copper on maximum respiration rate and growth rate of perch, *Perca fluviatilis* L. Wat. Res. 18: 139–144.

Conway, R. A., 1982. Environmental Risk Analysis for Chemicals. Van Nostrand Reinhold Company, New York, 558 pp.

Crider, J. Y., J. Wilhm & H. J. Harmon, 1982. Effects of naphthalene on the hemoglobin concentration and oxygen uptake of *Daphnia magna*. Bull. Envir. Contam. Toxicol. 28: 52–57.

Curtis, L. R., W. K. Seim & G. A. Chapman, 1985. Toxicity of fenvalerate to developing steelhead trout following continuous or intermittent exposure. J. Toxicol. Env. Health. 15: 445–457.

Dagg, M. J. & J. L. Littlepage, 1972. Relationships between growth rate and RNA, DNA, protein and dry weight in *Artemia salina* and *Euchaeta elongata*. Mar. Biol. 17: 162–170.

Daikoku, T., I. Yano & M. Musui, 1982. Lipid and fatty acid compositions and their changes in the different organs and tissues of guppy, *Poecilia reticulata* on sea water adaptation. Comp. Biochem. Physiol. 73A: 167–174.

Dange, A. D. & V. B. Masurekar, 1982. Naphthalene-induced changes in carbohydrate metabolism in *Sarotherodon mossambicus* Peters (Pisces: Aichlidae). Hydrobiologia. 94: 163–172.

Dean, J. M. & F. J. Vernberg, 1965. Variation in the blood glucose level of crustaceans. Comp. Biochem. Physiol. 14: 29–34.

Dey, A. C., J. W. Kiceniuk, U. P. Williams, R. A. Khan & J. F. Payne, 1983. Long term exposure of marine fish to crude petroleum I. Studies on liver lipids and fatty acids in cod (*Gadus morhua*) and winter flounder (*Pseudopleuronectes americanus*). Comp. Biochem. Physiol. 75C: 93–101.

DiMichele, L. & M. H. Taylor, 1978. Histopathological and physiological responses of *Fundulus heteroclitus* to naphthalene exposure. J. Fish. Res. Bd. Can. 35: 1060–1066.

Dixon, D. G., 1975. Some effects of chronic cyanide poisoning on the growth, respiration and liver tissue of rainbow trout. M. S. Thesis, Concordia University, Montreal Quebec, Canada, 77 pp.

Dubale, M. S. & P. Shah, 1981. Biochemical alterations induced by cadmium in the liver of *Channa punctatus*. Envir. Res. 26: 110–118.

Dutka, B. J. & K. K. Kwan, 1988. Battery of screening tests approach applied to sediment extracts. Toxicol. Assess. 3: 303–314.

Dutka, B. J., K. Jones, K. K. Swan, H. Bailey & R. McInnis, 1988. Use of microbial and toxicant screening tests for priority site selection of degraded areas in water bodies. Wat. Res. 22: 503–510.

Fingerman, M., M. H. Hanumate, V. P. Deshpunde & R. Nagabhushan, 1981. Increase in the total reducing substances in the hemolymph of the fresh water crab, *Barytelphusa guerini*, produced by a pesticide (DDT) and an indolealkylamine (serotonin). Experientia. 37: 178–179.

Finney, D. J., 1971. Porbit Analysis. 3rd edition, Cambridge University Press, Cambridge, England, 333 p.

Fletcher, D. J., 1984. Plasma glucose and plasma fatty acid levels of *Limanda limanda* (L.) in relation to season, stress, glucose loads and nutritional state. J. Fish Biol. 25: 629–648.

Florkin, M. & B. T. Scheer, 1970. Chemical Zoology, Arthropoda Vol. VI. Academic Press, New York, 460 pp.

Frank, J. R., S. D. Sulkin & R. P. Morgan, 1975. Biochemical changes during larval development of the xanthid crab *Rhethropanopeus harrisii* I. protein, total lipid, alkaline phosphatase and glutamic oxaloacetic transaminase. Mar. Biol. 32: 105–111.

Gabbott, P. A., P. A. Cook & M. A. Whittle, 1979. Seasonal changes in glycogen synthase activity in the mantle tissue of the mussel *Mytilus edulis* L.: regulation by tissue glucose. Biochem. Soc. Trans. 7: 895–896.

Gallis, J. L., F. Belloc & C. Beauvie, 1980. Freshwater adaptation in the euryhaline teleost, *Chelon labrosus* – IV. Changes of DNA and protein contents, RNA/DNA ratio and acid phosphatase activity in the branchial tissue. Comp. Biochem. Physiol. 67A: 69–76.

Gardner, D. & J. P. Riley, 1972. The component fatty acids of the lipids of some species of marine and freshwater molluscs. J. Mar. Biol. Ass. U.K. 52: 827–838.

Gersich, F. M., D. L. Hopkins, S. L. Applegath, C. G. Mendoza & D. P. Milazzo, 1985. The sensitivity of chronic endpoints used in *Daphnia magna* Straus life-cycle tests. In: R.C. Bahner & J. D. Hansen (eds.), Aquatic Toxicology and Hazard Assessment: Eighth Symposium ASTM-STP 891. pp. 245–252. American Society for Testing and Materials, Philadelphia, PA.

Gibbons, J. W., D. H. Bennett, G. W. Esch & T.C. Hazen, 1978. Effects of thermal effluent on body condition of largemouth bass. Nature. 274: 470–471.

Giesy, J. P. & E. P. Odum, 1980. Microcosmology: Introductory Comments. In: J. P. Giesy (ed.), Microcosms in Ecological Research. pp. 1–13. U.S. Technical Information Center, U.S. Department of Energy, Symposium Series 52 (Conf.-781101).

Giesy, J. P., C. S. Duke, G. W. Dickson, S. Denzer & G. J. Leversee, 1981. Effect of chronic cadmium exposure on the phosphoadenylate concentrations and energy charge in freshwater shrimp and crayfish. Verh. Int. Ver. Limnol. 21: 173–188.

Giesy, J. P., 1985. Multispecies tests: Research needs to assess the effects of chemicals on aquatic life. In: R. C. Bahner & D. J. Hansen (eds.), Aquatic Toxicology and Hazard Assessment: Eighth Symposium. ASTM STP 891, pp. 67–77. American Society for Testing and Materials, Philadelphia, PA.

Giesy, J. P. & P. M. Allred, 1985. Replicability of aquatic multispecies test systems. In: John Cairns (ed.), Multispecies Toxicity Testing, pp. 187–247, Pergamon Press, New York.

Giesy, J. P. & R. A. Hoke, 1989. The future of sediment toxicity assessments: Relevant answers require appropriate questions. J. Great Lakes Res. (In press).

Giesy, J. P., R. L. Graney, J. L. Newsted, C. Rosiu, A. Benda, R. G. Kreis & F. J. Horvath, 1988a. A comparison of three sediment bioassay methods for Detroit River sediment. Envir. Toxicol. Chem. 7: 483–498.

Giesy, J. P., C. Rosiu, R. L. Graney, J. C. Newsted, A. Benda, R. G. Kreis Jr. & F. J. Horvath, 1988b. Detroit River Sediment Toxicity. J. Great Lakes Res. 14: 502–513.

Giesy, J. P., C. J. Rosiu, R. L. Graney & M. G. Henry, 1989. Benthic invertebrate bioassays with toxic sediment and pore water. Envir. Toxicol. Chem. (in press).

Giesy, J. P., D. J. Versteeg & R. L. Graney, 1988. A review of selected clinical indicators of stress-induced changes in aquatic organisms. In: M. S. Evans (ed.), Toxic Contaminants and Ecosystem Health; A Great Lakes Focus. pp. 169–200, John Wiley and Sons, New York.

Gill, T. S. & J. C. Pant, 1983. Cadmium toxicity: Inducement of changes in blood and tissue metabolites in fish. Toxicol. Lett. 18: 195–200.

Gilles, R., 1970. Intermediate metabolism and energy production in some invertebrates. Arch. Inter. de Physiol. de Biochem. 78: 313–326.

Gilles, R., 1977. Effects of osmotic stresses on the protein concentrations and patterns of Eriocheir sineses blood. Comp. Biochem. Physiol. 56A: 109–114.

Glass, R. L., T. P. Krick, D. M. Sand, C. H. Rahn & H. Schlenk, 1975. Ruranoid fatty acids from fish lipids. Lipids. 10: 695–702.

Gluth, G. & W. Hanke, 1984. A comparison of physiology changes in carp, Cyprinus carpio, induced by several pollutants at sublethal concentrations II. The dependency on the temperature. Comp. Biochem. Physiol. 79C: 39–45.

Graney, R. L. & J. P. Giesy, 1986. Effects of long-term exposure to pentachlorophenol on the free amino acid pool and energy reserves on the freshwater Amphipod Gammarus pseudolimneus Bousfield (Crustacea, Amphipoda). Ecotoxicol. Environ. Safety 12: 233–251.

Graney, R. L. & J. P. Giesy, 1987. The effect of short-term exposure to pentachlorophenol and osmotic stress on the free amino acid pool of the freshwater amphipod (Gammarus pseudolimneas) Bousfield. Arch. Envir. Contam. Toxicol. 16: 167–176.

Graney, R. L. & J. P. Giesy, 1988. Alterations in the oxygen consumption, condition index and concentration of free

amino acids in Corbicula fluminea (Mollusca: Plecypoda) exposed to sodium dodecyl sulfate. Envir. Toxicol. Chem. 7: 301–315.

Grant, B. F. & P. M. Mehrle, 1973. Endrin toxicosis in rainbow trout (Salmo gairdneri). J. Fish. Res. Bd. Can. 30: 31–40.

Grizzle, J. A. & W. A. Rogers, 1976. Anatomy and histology of the channel catfish. Auburn University Agric. Expt. Station, Auburn, AL.

Haines, T. A., 1973. An evaluation of RNA-DNA ratio as a measure of long-term growth in fish populations. J. Fish Res. Bd. Can. 30: 195–199.

Hansen, H. J. M. & S. Abraham, 1983. Influence of temperature, environmental salinity and fasting on the patterns of fatty acids synthesized by gills and liver of the European eel (Anguilla anguilla). Comp. Biochem. Physiol. 75B: 581–587.

Hartree, E. F., 1972. Determination of protein: A modification of the Lowry method that gives a linear photometric response. Anal. Biochem. 48: 422–427.

Hatakeyama, S., 1988. Chronic effects of Cu on reproduction of Polypedium nubifer (Chironomidae) through water and food. Ecotox. Envir. Safety. 16: 1–10.

Hattingh, J., 1976. Blood sugar as an indicator of stress in the freshwater fish, Labeo capensis (Smith). J. Fish Biol. 10: 191–195.

Haux, C. & A. Larsson, 1984. Long-term sublethal physiological effects on rainbow trout, Salmo gairdneri, during exposure to cadmium and after subsequent recovery. Aquat. Toxicol. 5: 129–142.

Hazel, J. & C. L. Prosser, 1974. Molecular mechanisms of temperature compensations in poikilotherms. Physiol. Rev. 54: 620–677.

Heath, J. R. & H. Barnes, 1970. Some changes in biochemical composition with season and during the molting cycle of the common shore crab, Carcinus maenas. J. Exp. Mar. Biol. Ecol. 5: 199–233.

Henderson, R. J. & J. R. Sargent, 1983. Studies on the effects of di-2-ethylhexyl phthalate on lipid metabolism in rainbow trout (Salmo gairdneri) fed zooplankton rich in wax esters. Comp. Biochem. Physiol. 74C: 325–330.

Hermans, J., F. Busser, P. Leevwangh & A. Musch, 1985. Quantitative structure-activity relationships and mixture toxicity of organic chemicals in Photobacterium phosphoreum: the Microtox[R] test. Ecotoxicol. Envir. Safety. 9: 17–25.

Hinchcliffe, P. R. & J. P. Riley, 1972. The effect of diet on the component fatty acid composition of Artemia salina. J. Mar. Biol. Ass. U.K. 52: 203–211.

Hodson, P. V., D. G. Dixon & K. L. E. Kaiser, 1984. Measurement of median lethal dose as a rapid indication of contaminant toxicity to fish. Envir. Toxicol. Chem. 3: 243–254.

Holland, D. L., 1978. Lipid reserves and energy metabolism in the larvae of benthic marine invertebrates. In: D. C. Malins and J. R. Sargent (eds.), Biochemical and Biophysi-

56

cal Perspectives in Marine Biology, Vol. 4, Academic Press, New York, 222 pp.

Horn, E. C. & M. S. Kerr, 1969. The hemolymph proteins of the blue crab, *Callinectes sapidus* I. Hemocyanins and other major proteins. Comp. Biochem. Physiol. 29: 493–508.

Hunter, R. S., F. D. Culver, J. R. Hill & A. Fitzgerald, 1987. QSAR System User Manual: A Structure-Activity Based Chemical Modeling and Information System. Institute for Biological and Chemical Process Analysis, Montana State University, Bozeman, MT.

Jackim, E. & D. Nacci, 1984. A rapid aquatic toxicity assay utilizing labelled thymidine incorporation in sea urchin embryos. (Unpublished manuscript).

Kaiser, K. L. (ed.), 1987. QSAR in Environmental Toxicology – II. Reidel Publishing Co., Dordrecht, 465 pp.

Katti, S. R. & A. G. Sathyanesan, 1984. Changes in tissue lipid and cholesterol content in the catfish *Clarias batrachus* (L.) exposed to cadmium chloride. Bull. Envir. Contam. Toxicol. 32: 486–490.

Kearns, P. K. & G. J. Atchison, 1979. Effects of trace metals on growth of yellow perch (*Perca flavescens*) as measured by RNA-DNA ratios. Envir. Biol. Fishes 4: 383–387.

Kenaga, E. E., 1977. Aquatic test organisms and methods useful for assessment of chronic toxicity of chemicals. In: K. L. Dickson, A. W. Maki and J. Cairns (eds.), Analyzing the Hazard Evaluation Process. pp. 101–111, American Fisheries Society, Bethesda, MD.

Kenaga, E. E. 1978. Test organisms and methods useful for early assessment of acute toxicity of chemicals. Envir. Sci. Technol. 1322–1329.

Kenaga, E. E., 1982a. Review: the use of environmental toxicology and chemistry data in hazard assessment: Progress, needs, challenges. Envir. Toxicol. Chem. 1: 69–79.

Kenaga, E. E., 1982b. Predictability of chronic toxicity from acute toxicity of chemicals in fish and aquatic invertebrates. Envir. Toxicol. Chem. 1: 347–358.

Kiermeir, A., 1939. On the blood sugar of freshwater fish. Z. Vergl. Physiol. 27: 460–490.

Koundinya, P. R. & R. Ramamurthi, 1979. Effect of organophosphate pesticide sumithion (fenitrothion) on some aspects of carbohydrate metabolism in a freshwater fish, *Sarotherodon mossambicus* (Peters). Experimentia. 35: 1632–1633.

Krueger, H. M., J. B. Saddler, G. A. Chapman, I. J. Tinsley & R. R. Lowry, 1968. Bioenergetics, exercise and fatty acids of fish. Am. Zool. 8: 119–129.

Kulshrestha, S. K., N. Arora & S. Sharma, 1986. Toxicity of four pesticides on the fingerlings of Indian major carps *Labeo rohita*, *Catla catla* and *Cirrhinus mgigala*. Ecotoxicol. Environ. Safety. 12: 114–119.

Lang, C. A., H. Y. Lau & D. J. Jefferson, 1965. Protein and nucleic acid changes during growth and aging in the mosquito. Biochem. J. 95: 372–377.

Lagler, K. F., 1952. Freshwater Fishery Biology. Wm. C. Brown Co., Dubuque, IA, 421 pp.

Larsson, A., B. Bengtsson & O. Suanberg, 1976. Some haematological and biochemical effects of cadmium on fish. In: P. M. Lockwood (ed.), Effects of Pollutants on Aquatic Organisms. Vol. 2, pp. 35–45, Cambridge University Press, Cambridge, UK.

Leach, G. J. & M. H. Taylor, 1982. The effects of cortisol treatment on carbohydrate and protein metabolism in *Fundulus heteroclitus*. Gen. Comp. Endocrinol. 48: 76–83.

Leatherland, J. F. & R. A. Sonstegard, 1984. Pathobiological responses of feral teleosts to environmental stressors: Interlake studies of the physiology of Great Lakes salmon. In: V. W. Cairns, P. V. Hodson and J. O. Nriagu (eds.), Contaminant Effects on Fisheries, pp. 116–139, John Wiley and Sons, New York.

Lee, W. Y., S. A. Macko & J. A. C. Nicol, 1981. Changes in nesting behavior and lipid content of a marine amphipod (*Amphithoe valida*) to the toxicity of a No. 2 fuel oil. Wat. Air Soil Pollut. 15: 185–195.

Lee, R. F. & D. L. Pupprone, 1978. Serum lipoproteins in the spiny lobster, *Panularus interruptus*. Comp. Biochem. Physiol. 59B: 239–243.

Lehtinen, K., A. Larsson & G. Klingstedt, 1984. Physiological disturbances in rainbow trout, *Salmo gairdneri* (R.) exposed at two temperatures to effluents from a titanium dioxide industry. Aquat. Toxicol. 5: 155–166.

Lett, P., G. J. Farmer & F. W. H. Beamish, 1976. Effect of copper on some aspects of the bioenergetics of rainbow trout (*Salmo gairdneri*). J. Fish. Res. Bd. Can. 33: 1335–1342.

Limsuwan, C., T. Limsuwan, J. M. Grizzle & A. Plumb, 1983. Stress response and blood characteristics of channel catfish (*Ictalurus punctatus*) after anesthesia with etomidate. Can. J. Fish. Aquat. Sci. 40: 2105–2112.

Love, R. M., 1970. The Chemical Biology of Fishes. Academic Press, London, 547 pp.

Macek, K. J., K. S. Buxton, S. K. Derr, J. W. Dean & S. Sauter, 1976. Chronic toxicity of Lindane to selected aquatic invertebrates and fishes. U.S. Environmental Protection Agency EPA-600/3-76-046.

Macek, K. J. & B. H. Sleight, 1977. Utility of toxicity tests with embryos and fry of fish in evaluating hazards associated with the chronic toxicity of chemicals to fishes. In: F. L. Mayer & J. L. Hamelick (eds.), Aquatic Toxicology and Hazard Assessment, pp. 137–146, ASTM STP 634 American Society for Testing and Materials, Philadelphia, PA.

Martell, F. L., R. T. Motekaitis & R. M. Smith, 1988. Structure-stability relationships of metal complexes and metal speciation in environmental aqueous solutions. Envir. Toxicol. Chem. 7: 417–434.

Mayer, F. L., D. Buckler, M. Ellersieck & G. Krause, 1984. Estimating chronic toxicity of chemicals to fishes from acute toxicity test data: An alternative to the application factor. Presented to the annual meeting of the Society of Environmental Toxicology and Chemistry.

Mayer, F. L., K. S. Mayer & M. R. Ellersiech, 1986. Relation of survival to other endpoints in chronic toxicity tests with fish. Envir. Toxicol. Chem. 5: 737–748.

Mazeaud, M., F. Mazeaud & E. M. Donaldson, 1977. Primary and secondary effects of stress in fish: Some new data with a general review. Trans. Amer. Fish. Soc. 106: 201–212.

McCahon, C. P. & D. Pascoe, 1988. Cadmium toxicity to the freshwater amphipod *Gammarus pulex* (L.) during the molt cycle. Freshwat. Biol. 19: 197–203.

McCarthy, D. H., J. P. Stevenson & M. S. Roberts, 1973. Some blood parameters of the rainbow trout (*Salmo gairdneri* Richardson) I. The Kamloops variety. J. Fish Biol. 5: 1–8.

McKee, M. J. & C. O. Knowles, 1985. Relationships between biochemical parameters, survival, and reproduction in *Daphnia magna* exposed to chlordecone. Presented at 6th Annual Meeting of The Society of Environmental Toxicology and Chemistry. St. Louis, MO.

McKee, M. J. & C. O. Knowles, 1986. Protein, nucleic acid and adenylate levels in *Daphnia magna* during chronic exposure to chlordecone. Envir. Pollut. 42A: 335–351.

McKim, J. M., 1985. Early life stage toxicity tests. In: G. M. Rand & S. R. Petrocelli (eds.), Fundamentals of Aquatic Toxicology, pp. 58–95, Hemisphere, Publishers, Washington, DC.

McKim, J. M., G. M. Christensen & E. P. Hunt, 1970. Changes in the blood of brook trout (*Salvelinus fontinalis*) after short-term and long-term exposure to copper. J. Fish. Res. Bd. Can. 27: 1883–1889.

McLeay, D. J., 1977. Development of a blood sugar bioassay for rapidly measuring stressful levels of pulpmill effluent to salmonid fish. J. Fish. Res. Bd. Can. 34: 477–485.

McLeay, D. J. & D. A. Brown, 1974. Growth stimulation and biochemical changes in juvenile coho salmo (*Oncorhynchus kisutch*) exposed to bleached kraft pulpmill effluent for 200 days. J. Fish. Res. Bd. Can. 31: 1043–1049.

McLeay, D. J. & D. A. Brown, 1975. Effects of acute exposure to bleached kraft pulpmill effluent on carbohydrate metabolism of juvenile coho salmon (*Oncorhynchus kisutch*) during rest and exercise. J. Fish. Res. Bd. Can. 32: 753–760.

Miles, H. M., S. M. Loehner, D. T. Michard & S. L. Salivar, 1974. Physiological responses of hatchery reared muskellunge (*Esox masquinongy*) to handling. Trans. Am. Fish. Soc. 103: 336–342.

Miller, N. G. A., M. W. Hill & M. W. Smith, 1976. Positional and species analyses of membrane phospholipids extracted from goldfish adapted to different environmental temperatures. Biochem. Biophys. Acta. 455: 644–654.

Mitz, S. V. & J. P. Giesy, 1985a. Sewage Effluent Biomonitoring: I. Survival, Growth and Histopathological Effects in Channel Catfish. Ecotoxicol. Environ. Safety 10: 22–39.

Mitz, S. V. & J. P. Giesy, 1985b. Sewage Effluent Biomonitoring: II. Biochemical Indicators of Ammonia Exposure in Channel Catfish. Ecotoxicol. Environ. Safety 10: 40–52.

Moles, A. & S. D. Rice, 1983. Effects of crude oil and naphthalene on growth, caloric content, and fat content of pink salmon juveniles in seawater. Trans. Am. Fish. Soc. 112: 205–211.

Moorthy, L. S., M. D. Naidu, C. S. Chetty & K. S. Swami, 1983. Changes in carbohydrate metabolism in tissues of freshwater mussel (*Lamellidens marginalis*) exposed to Phosphamidon. Bull. Envir. Contam. Toxicol. 30: 219–222.

Morata, P., M. J. Faus, M. Perez-Palomo & F. Sanchez-Medina, 1982b. Effect of stress on liver and muscle glycogen phosphorylase in rainbow trout (*Salmo gairdneri*). Comp. Biochem. Physiol. 72B: 421–425.

Morata, P., A. M. Vargus, F. Sanchez-Medina, M. Garcia, G. Cardenete & S. Zamora. 1982a. Evolution of gluconeogenic enzyme activities during starvation in liver and kidney of the rainbow trout (*Salmo gairdneri*). Comp. Biochem. Physiol. 71B: 65–70.

Morris, R. J., 1971. Seasonal and environmental effects in the lipid composition of *Neomyses integer*. J. Mar. Biol. Ass. U.K. 51: 21–31.

Morris, R. J., A. P. M. Lockwood & M. E. Dawson, 1982a. An effect of acclimation salinity on the fatty acid composition of the gill phospholipids and water flux of the amphipod crustacean *Gammarus duebini*. Comp. Biochem. Physiol. 72A: 497–503.

Morris, R. J., A. P. M. Lockwood & M. E. Dawson, 1982b. Changes in the fatty acid composition of the gill phospholipids in *Gammarus duebeni* with degree of gill contamination. Mar. Pollut. Bull. 13: 345–348.

Mount, D. I., 1988. Ambient toxicity to assess biological impact. In: M. S. Evans (ed.), Toxic Contaminants and Ecosystem Health; A Great Lakes Focus, pp. 237–245, John Wiley and Sons, New York.

Munro, H. N. & A. Fleck, 1966. Recent developments in the measurement of nucleic acids in biological materials: A supplementary note. Analyst. 91: 78–88.

Murty, A. S. & A. P. Devi, 1982. The effect of endosulfan and its isomers on tissue protein, glycogen, and lipids in the fish (*Channa punctatus*). Pest. Biochem. Physiol. 17: 280–286.

Nakano, T. & N. Tomlinson, 1967. Catecholamine and carbohydrate concentrations in raibow trout (*Salmo gairdneri*) in relation to physical disturbance. J. Fish. Res. Bd. Can. 24: 1701–1715.

Nebeker, A. V., M. A. Cairns & C. M. Wise, 1984. Relative sensitivity of *Chironomus tentans* life stages to copper. Environ. Toxicol. Chem. 3: 151–158.

Neuhold, J. M., 1986. Toward a meaningful interaction between ecology and aquatic toxicology. In: T. M. Poston and R. Purdy (eds.), Aquatic Toxicology and Environmental Fate: Volume 9. ASTM STP 921, pp. 11–21, American Society for Testing and Materials, Philadelphia, PA.

Newell, R. I.E. & B. L. Bayne, 1980. Seasonal changes in the physiology, reproductive condition and carbohydrate content of the cockle *Cardium edule* (Cerastoderma, Bivalvia: Cardiidae). Mar. Biol. 56: 11–19.

Nieminen, M., P. Pasanen, M. Laitinen, 1983. Effects of formalin treatment on the blood composition of salmon (*Salmo salar*) and rainbow trout (*Salmo gairdneri*). Comp. Biochem. Physiol. 76C: 265–269.

O'Connor, J. D. & L. I. Gilbert, 1969. Alterations in lipid

metabolism associated with premolt activity in a land crab and freshwater crayfish. Comp. Biochem. Physiol. 29: 889–904.

O'Hara, S. C. M., A. C. Neal, E. D. S. Corner & A. L. Pulsford, 1985. Interrelationships of cholesterol and hydrocarbon metabolism in the shore crab, *Carcinus*. J. Mar. Biol. Ass. U.K. 65: 113–131.

Ottaway, E. M. & K. Simkiss, 1977. 'Instantaneous' growth rates of fish scales and their use in studies of fish populations. J. Zool. (London). 181: 407–419.

Pant, J. C. & T. Singh, 1983. Inducement of metabolic dysfunction by carbamate and organophosphorus compounds in a fish, *Puntius conchonius*. Pest. Biochem. Physiol. 20: 294–298.

Pascoe, D. & N. A. M. Shazili, 1986. Episodic pollution – A comparison of brief and continuous exposure of rainbow trout to cadmium. Ecotoxicol. Environ. Safety. 12: 189–198.

Patton, S. & A. J. Thomas, 1971. Composition of lipid foams from swim bladders of two deep ocean fish species. J. Lipid. Res. 12: 331–335.

Paulauski, J. D. & R. W. Winner, 1988. Effects of water hardness and humic acid on zinc toxicity to *Daphnia magna* Straus. Aquat. Toxicol. 12: 273–290.

Peer, M. M., J. Nirmala & M. N. Kutty, 1983. Effects of pentachlorophenol (NaPCP) on survival, activity and metabolism in *Rhionmugil corsula* (Hamilton), *Cyprinus carpio* (L) and *Tilapia mossambica* (Peters). Hydrobiologia. 107: 19–24.

Perry, J. A. & N. H. Troelstrup, Jr., 1988. Whole ecosystem manipulation: A productive avenue for test system research. Envir. Toxicol. Chem. 7: 941–951.

Petrocelli, S. R., 1985. Chronic toxicity tests. In: G. M. Rand and S. R. Petraocelli (eds.), Fundamental of Aquatic Toxicology, pp. 96–109, Hemisphere Publishers, Washington, DC.

Phillips, A. M., F. E. Lovelace, H. A. Podoliak, D. R. Brockway & G. C. Balzer, 1956. The nutrition of trout. Fish. Res. Bull. No. 19, New York, 56 pp.

Pickering, A. D., T. G. Pickering & P. Christie, 1982. Recovery of the brown trout, *Salmo trutta* L., from acute handling stress: a time course study. J. Fish Biol. 20: 229–244.

Plisetskaya, E., 1980. Fatty acid levels in blood of cyclostomes and fish. Envir. Biol. Fish. 5: 273–290.

Plisetskaya, E., V. K. Kazakoo, Soltitskaya & L. G. Leibson, 1978. Insulin-producing cells in the gut of freshwater bivalve molluscs *Anodonata cygnea* and *Unio pictorum* and the role of insulin in the regulation of their carbohydrate metabolism. Gen. Comp. Endocrin. 35: 133–145.

Price, G. M., 1969. Protein synthesis and nucleic acid metabolism in the fat body of the larva of the blowfly, *Calliphora erythrocephala*. J. Insect Physiol. 15: 931–944.

Prosser, C. L., 1973. Comparative Animal Physiology. W. B. Saunders Co. Philadelphia, 966 pp.

Rao, P. V. V., M. K. C. Sridhar & A. B. O. Desalu, 1984a.

Effect of acute oral cadmium on mitochondrial enzymes in rat tissue. Arch. Envir. Contam. Toxicol. 12: 293–297.

Rao, K. S. P. & K. V. R. Rao, 1984a. Changes in the tissue lipid profiles of fish (*Oreochromis mossambicus*) during methyl parathion toxicity – A time course study. Toxicol. Lett. 21: 147–153.

Rao, D. M., A. S. Murty & P. A. Swarup, 1984b. Relative toxicity of technical grade and formulated carbaryl and 1-naphthol to, and carbaryl-induced biochemical changes in the fish *Cirrhinus mrigala*. Envir. Pollut. 34: 47–54.

Riley, A. R. T. & M. C. Mix, 1981. The effects of naphthalene on glucose metabolism in the European flat oyster *Ostrea edulis*. Comp. Biochem. Physiol. 70C: 13–20.

Rosas, S. B., M. M. Carranza de Storani & N. E. Ghittoni, 1985. Fatty acids and phospholipids of membranes isolated from *Escherichia coli* growing in a medium with parathion. Bull. Envir. Contam. Toxicol. 34: 265–270.

Sahib, I. K. A., K. R. S. Rao & K. V. Rao, 1984. Effect of malathion on protein synthetic potentiality of the tissues of the teleost, *Tilapia mossambica* (Peters), as measured through incorporation of 14C amino acids. Toxicol. Lett. 20: 63–67.

Sastry, K. V. & A. A. Siddiqui. 1983. Metabolic changes in the snake head fish *Channa punctatus* chronically exposed to endosulfan. Wat. Air Soil Pollut. 19: 133–141.

Sastry, K. V. & A. A. Siddiqui, 1984. Some hematological, biochemical, and enzymological parameters of a freshwater teleost fish, *Channa punctatus*, exposed to sublethal concentrations of quinalphos. Pest. Biochem. Physiol. 22: 8–13.

Sastry, K. V. & K. Subhadra, 1985. *In vivo* effects of cadmium on some enzyme activities in tissues of the freshwater catfish, *Heteropneustes fossilis*. Envir. Res. 36: 32–45.

Sastry, K. V. & K. Sunita, 1983. Enzymological and biochemical changes produced by chronic chromium exposure in a teleost fish, *Channa punctatus*. Toxicol. Lett. 16: 9–15.

Schaeffer, D. J. & K. G. Janardan, 1987. Designing batteries of short-term tests with largest inter-tier correlation. Ecotoxicol. Environ. Safety 13: 316–323.

Selye, H., 1956. The Stress of Life. McGraw-Hill Book Company, New York.

Selye, H., 1976. Stress in Health and Disease. Butterworth Publishers, Boston, MA., 497 pp.

Shaffi, S. A., 1980. Zinc intoxication in some freshwater fishes. I. Variations in tissue energy reserves. Ann. Limnol. 16: 91–97.

Sheehan, P. J., D. R. Miller, G. C. Butler, P. Bourdeau & J. M. Ridgeway, 1984. Effects of pollutants at the ecosystem level. SCOPE 22, John Wiley and Sons, Chichester, 443 pp.

Silbergeld, E. K., 1974. Blood glucose: A sensitive indicator of environmental stress in fish. Bull. Envir. Contam. Toxicol. 11: 20–25.

Singh, H. & T. P. Singh, 1980. Effects of two pesticides on testicular 32 P uptake, gonadotrophic potency, lipid and cholesterol content of testis, liver and blood serum during spawning phase in *Heteropneustes fossilis* (Bloch). Endokrin. Band. 76: 228–296.

Singh, N. & A. K. Srivastava, 1981. Effect of a paired mixture of aldrin and formithion on carbohydrate metabolism in a fish, *Heteropneustes fossilis*. Pest. Biochem. Physiol. 15: 257–261.

Slooff, W. & J. H. Canton, 1983. Comparison of the susceptibility of 11 freshwater species to 8 chemical compounds. II. Semi-chronic toxicity tests. Aquat. Toxicol. 4: 271–282.

Slooff, W., 1985. The role of multispecies testing in aquatic toxicology: In: J. Cairns (ed.), Multispecies Toxicity Testing. pp. 45–60. Pergamon Press, Oxford, UK.

Slooff, W. & J. H. Canton, J. L. M. Hermens, 1983. Comparisons of the susceptability of 22 freshwater species to 15 chemical compounds. I. (Sub)Acute toxicity tests. Aquat. Toxicol. 4: 113–128.

Soivio, A., S. Lindgren & A. Oikari. 1983. Seasonal changes in certain metabolic and haematologic responses of *Salmo gairdneri* acutely exposed to dehydroabietic acid (DHAA). Comp. Biochem. Physiol. 75C: 281–284.

Sower, S. A., C. B. Schreck, M. Evenson, 1983. Effects of steroids and steroid antagonists on growth, gonadal development, and RNA/DNA ratios in juvenile steelhead trout. Aquaculture. 32: 243–254.

Sprague, J. B., 1969. Measurement of pollutant toxicity to fish. I. Bioassay methods for acute toxicity. Wat. Res. 3: 793–821.

Sprague, J. B., 1970. Measurement of pollutant toxicity to fish. II. Utilizing and applying bioassay results. Wat. Res. 4: 3–32.

Sprague, J. B., 1976. Current status of sublethal tests of pollutants on aquatic organisms. J. Fish. Res. Bd. Can. 33: 1888–1992.

Sprague, J. B., 1988. Fish tests that give useful information. In: M. S. Evans (ed.), Toxic Contaminants and Ecosystem Health; A Great Lakes Focus. pp. 247–256. John Wiley and Sons, New York.

Sreenivasula R. P., Bhagyalakshmi & R. Ramamurthi, 1983. In-vivo acute physiological stress induced by BHC in hemolymph biochemistry of *Oziotelphusa senex senex*, the Indian rice field crab. Toxicol. Lett. 18: 35–38.

Srivastava, A. K. & A. B. Gupta, 1981. The effect of sodium salt of 2,4D on carbohydrate metabolism in the Indian catfish *Heteropneustes fossilis* (Bloch). Acta. Hydrobiol. 23: 259–268.

Srivastava, A. K. & J. Mishra, 1983. Effects of fenthion on the blood and tissue chemistry of a teleost fish (*Heteropneustes fossilis*). J. Comp. Path. 93: 27–31.

Stephan, C. E., 1982. Increasing the usefulness of acute toxicity tests. In: J. G. Pearson, R. B. Foster & W. E. Bishop (eds.), Aquatic Toxicology and Hazard Assessment: Fifth Conference. ASTM STP 766, pp. 69–81, American Society for Testing and Materials, Philadelphia, PA.

Stephan, C. E., 1986. Proposed goal of applied aquatic toxicology. In: T. M. Poston & R. Purdy (eds.), Aquatic Toxicology and Environmental Fate: Ninth Volume. pp. 3–10. ASTM STP 921, American Society for Testing and Materials, Philadelphia, PA.

Stephan, C. E. & R. J. Erickson, 1982. Calculations of the final acute value for water quality criteria for aquatic life. Environ. Research Laboratory-Duluth (unpublished document).

Sutcliffe, W. H., 1965. Growth estimates from ribonucleic acid content in some small organisms. Limnol. Oceanogr. 10: 253–258.

Suter, G. W., A. E. Rosen, E. Linder & D. F. Parkhurst, 1987. Endpoints for responses of fish to chronic exposures. Environ. Toxicol. Chem. 6: 793–809.

Swift, D. J., 1982. Changes in selected blood component values of rainbow trout, *Salmo gairdneri* Richardson, following blocking of the cortisol stress response with betamethasone and subsequent exposure to phenol or hypoxia. J. Fish Biol. 21: 269–277.

Telford, M., 1974. Blood glucose in crayfish. I. Variations associated with molting. Comp. Biochem. Physiol. 47A: 461–468.

Thomas, P., 1982. Effect of cadmium exposure on plasma cortisol levels and carbohydrate metabolism in mullet (*Mugil cephalus*). J. Endocrinol. Supp. 94, 35 p.

Thomas, P., R. S. Carr & J. M. Neff, 1981b. Biochemical stress responses of mullett *Mugil cephalus* and polychaete worms *Neanthes virens* to pentachlorophenol. In: J. Vernberg, A. Calabrese, F. P. Thurberg & W. B. Vernberg, Biological Monitoring of Marine Pollutants, pp. 73–103, Academic Press, New York.

Thomas, P., H. W. Wofford & J. M. Neff, 1981a. Biochemical stress responses of striped mullet (*Mugil cephalus* L.) to fluorene analogs. Aquat. Toxicol. 1: 329–342.

Thorp, J. H., J. P. Giesy & S. A. Wineriter, 1979. Effects of chronic cadmium exposure on crayfish survival, growth and tolerance to elevated temperature. Arch. Envir. Contam. Toxicol. 8: 449–456.

Versteeg, D. J. & J. P. Giesy, 1985. Lysosomal enzyme release in the bluegill sunfish (*Lepomis machrochirus* Rafinesque) exposed to cadmium. Arch. Envir. Contam. Toxicol. 14: 631–640.

Versteeg, D. J. & J. P. Giesy, 1986. The histological and biochemical effects of cadmium exposure in the bluegill sunfish (*Lepomis machrochirus*). Ecotoxicol. Environ. Safety 11: 31–43.

Versteeg, D. J., R. L. Graney & J. P. Giesy, 1988. Field utilization of clinical measures for the assessment of xenobiotic stress in aquatic organisms. In: W. J. Adams, G. A. Chapman & W. G. Landis (eds.), Aquatic Toxicology and Hazard Assessment 10th Volume, pp. 289–306. ASTM STP 971, American Society for Testing and Materials, Philadelphia, PA.

Verma, S. R., S. Rani, I. P. Tonk & R. C. Dalela, 1983. Pesticide-induced dysfunction in carbohydrate metabolism in three freshwater fishes. Envir. Res. 32: 127–133

Verma, S. R., M. Saxena & I. P. Tonk, 1984. The influence of Idet 20 on the biochemical composition and enzymes in the liver of *Clarias batrachus*. Environ. Pollut. 33A: 245–255.

Viarengo, A., M. Pertica, G. Mancinelli, S. Palmero, G. Zanicchi & M. Oranesu, 1982. Evaluation of general and specific stress indices in mussels collected from populations subjected to different levels of heavy metal pollution. Mar. Envir. Res. 6: 235–243.

Viarengo, A., M. Pertica, G. Mancinelli, R. Capelli & M. Orunesu, 1980. Effects of copper on the uptake of amino acids, on protein synthesis and on ATP content in different tissues of *Mytilus galloprovincialis* L. Mar. Envir. Res. 4: 145–152.

Vickers, D. H. & N. Mitlin, 1966. Changes in nucleic acid content of the Boll Weevil *Anthonomus grandis* Boheman during its development. Physiol. Zool. 39: 70–76.

Voogt, P. A., 1983. Lipids: Their distribution and metabolism. In: P. W. Hochachka (ed.), The Mollusca, Vol. 1, Metabolic Biochemistry and Molecular Biomechanics. Academic Press, London. 510 pp.

Vos, J. G., 1981. Screening and function tests to detect immune suppression in toxicity studies. In: R. P. Sharma (ed.), Immunologic Considerations in Toxicology, Vol. 1, pp. 109–120, CRC Press, Boca Raton, FL.

Vouk, V. B., G. C. Butler, D. G. Hoel, D. B. Peakall, 1985. Methods for estimating risk of chemicals injury: Human and non-human biota and ecosystems, SCOPE 26, SCOMSEC2, John Wiley and Sons, Chichester, 680 pp.

Wagner, G. F. & B. A. McKeown, 1982. Changes in plasma insulin and carbohydrate metabolism of zinc-stressed rainbow trout, *Salmo gairdneri*. Can. J. Zool. 60: 2079–2084.

Warren, C. E., 1971. Biology and Water Pollution Control. W. B. Saunders Co, Philadelphia, 434 pp.

Watson, T. A. & B. A. McKeown, 1976. The effect of sub-lethal concentrations of zinc on growth and plasma glucose levels in rainbow trout, *Salmo gairdneri* (Richardson), J. Wildl. Diseases 12: 263–270.

Wedemeyer, G. A. & W. T. Yasutake, 1977. Clinical measures for the assessment of the effects of environmental stress on fish health. Technical paper No. 89, U.S. Fish and Wildlife Service, US Department of Interior, Washington, D.C., pp. 1–18.

Wedemeyer, G. A., D. J. McLeay & C. P. Goodyear, 1984. Assessing the tolerance of fish populations to environmental stress: the problems and methods of monitoring. In: V. W. Cairns, P. V. Hodson and J. P. Nriagu (eds.), Contaminant Effects on Fisheries, pp. 164–186, John Wiley and Sons, New York.

Wigglesworth, V. B., 1963. The action of moulting hormone and juvenile hormone at the cellular level in *Rhodnius prolixus*. J. Exp. Biol. 40: 231–245.

Wilder, I. B. & J. G. Stanley, 1983. RNA-DNA ratio as an index to growth in salmonid fishes in the laboratory and in streams contaminated by carbaryl. J. Fish Biol. 22: 165–172.

Wilps, H. & E. Zebe, 1976. The end-products of anaerobic carbohydrate metabolism in the larvae of *Chironomus thummi thummi*. Comp. Physiol. 112: 263–272.

Woltering, D. M., 1985. Population responses to chemical exposure in aquatic multispecies systems. In: J. Cairns, Jr. (ed.), Multispecies Toxicity Testing, pp. 61–75, Pergamon Press, New York.

Yu, T. C. & R. O. Sinnuhuber, 1975. Effect of dietary linoleic and linoleic acids upon growth and lipid metabolism of rainbow trout (*Salmo gairdneri*). Lipids. 10: 63–66.

Zwingelstein, G., N. A. Malak & G. Barichon, 1978. Effect of environmental temperature on biosynthesis of liver phosphatidylcholine in the trout (*Salmo gairdneri*). J. Therm. Biol. 3: 229–233.

Hydrobiologia **188/189**: 61–64, 1989.
M. Munawar, G. Dixon, C. I. Mayfield, T. Reynoldson and M. H. Sadar (eds)
Environmental Bioassay Techniques and their Application.
© 1989 *Kluwer Academic Publishers. Printed in Belgium.*

The choice and implementation of environmental bioassays

P. Calow
Department of Animal and Plant Sciences, University of Sheffield, Sheffield, S10 2TN, U.K.

Abstract

Bioassays play a crucial role in assessing the actual or potential impacts of anthropogenic agents on the natural environment. They can be used to probe the extent to which an ecosystem is being or has been polluted, and to predict the ecological impact of agents before release. Problems associated with the choice and implementation of bioassays in both these modes are discussed. A major general conclusion is that the experimental method should play a prominent part in addressing the problems.

Introduction

Environmental bioassays use biological systems to assess the actual or potential impact of substances derived from the activity of Man on the natural environment. As such they can be used predictively, to forecast the impact of a substance prior to its release, or as a monitor of actual effects in nature. These differences are illustrated in Fig. 1. In the predictive mode, a perturbation is applied to the system and a response noted that should be interpreted relative to its environmental implications. In the monitoring mode, characteristics of the system are used to estimate the

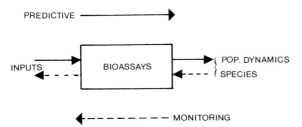

Fig. 1. The test system in bioassays is often treated as a black box that maps environmental inputs to outputs in terms of population dynamics or species composition. Predictive bioassays operate from left to right; the potential impact of a toxin on natural environment is judged from the response to a known input. Monitoring bioassays operate from right to left; from an assumed output, an attempt is made to characterize the input.

level of perturbation that is being or has been applied to it.

Problems associated with both these kinds of bioassays can be divided into those concerned with choice of assay and those concerned with implementation. This is a convenient, though not necessarily sharp distinction. I deal with the categories in reverse order.

Problems of implementation

These are largely concerned with the effective design of experimental and sampling programmes, such that they give decisive and statistically-reliable conclusions about the potential or actual impact of substances. The problems of design in the predictive mode are those usually associated with experimental work and present no special problems. On the other hand, the problems associated with design in the monitoring mode are more profound. The reason for this assymetry is captured in a fundamental theorem of systems analysis (Calow, 1976); in a deterministic system, a particular input will always produce the same output, but for complex systems, particular outputs may arise from one of several inputs. Thus, to find (usually through sampling and survey work; Hellawell, 1986) biological systems in the field in particular states (e.g. stunted individuals, low-density populations, low-diversity communities) does not guarantee that these have been caused by particular environmental perturbations (i.e. pollution) just because similar correlations have been noted previously. To have more confidence in these kinds of assays, relevant reference sites have to be used (e.g. before/after or upstream/downstream comparisons). However, even if this is done the monitor and reference sites have to be separated either in time or space, and so there can be no guarantee that the reference represents what the monitor would have been like without any disturbance (this is sometimes referred to as pseudoreplication; Hurlbert, 1984).

An alternative to using actual systems in nature as monitors of their own states, is to use 'planted' systems. Examples are systems in which the behavioural states of caged fishes or the physiological states of caged invertebrates (see respectively Rand, 1985 and Naylor et al., 1989) are used as monitoring bioassays. The advantages of these kinds of bioassays are that there can be full knowledge of their states prior to exposure, and standardization of upstream/downstream systems. This therefore facilitates statistical comparisons and gives a better chance of associating a particular change in a system (relative to 'before' or 'upstream' controls), causally, with particular environmental variables. The major problem associated with planted systems concerns the relevance of the responses within them to the state of the 'natural' ecosystems that surround them. This will be addressed in more detail below.

Problems of choice

There are at least 2 criteria that will influence the choice of bioassays; one depends on the fundamental scientific basis for using assays and the other on the needs of regulators.

Scientific criteria

An important role for scientific research is to investigate how one bioassay relates to others in sensitivity and reliability of responses through intercalibration exercises (e.g. Bayne et al., 1988). However, these are not in themselves sufficient. A more important role for scientific research is to investigate how the results of a bioassay relate to ecological impact; i.e. what there ecological relevance is. The kinds of questions that have to be addressed here are: Do 'key' species exist that can be used in assays? How is 'key' to be defined: in terms of contribution to the functional integrity of a community; or in terms of sensitivity; or in terms of being representative of a particular taxonomic or trophic group; or, finally, in terms of being present in a particular 'compartment' of the ecosystem under consideration (e.g. in the water column or sediment)? At what level should observations be made: molecular, or cellular, or physiological, or organismic or population or community? Can the effects at one level be related to the effects at higher levels in this hierarchy or are there so-called emergent properties (e.g. interaction between individuals at the population level, or between populations at the community level) that are important? Are multi-species tests more important than single-species tests, or are such species combinations too unique to allow repeatable and generalisable results? What kinds of measurements should be made on these systems; structural or functional?

These questions raise a wide range of fundamental issues for both ecotoxicology in particular and ecology in general, that can only be properly considered experimentally. For example, the identification of 'key' species in the functional sense (above), could not be achieved simply by observing community structures. Instead, the artificial exclusion of species, followed by observations on changes in the attributes of communities will have to be attempted (e.g. as in Paine, 1977). We also know too little about the relative sensitivities of species (and even genotypes within species; e.g. Baird et al., 1989) to toxins and whether some species are generally sensitive to and others generally tolerant of toxins ('indicator species philosophy' presumes some generality; Hellawell, 1986). What is required is an experimental programme that rigorously compares sensitivities between species and genotypes, and identifies mechanisms of toxic action and resistance processes.

An *a priori* expectation is that since all organisms are fundamentally similar at molecular and cellular levels, responses to toxins here will be very general. Moreover, it is at these levels that chemical toxins probably have their primary effects. On the other hand, some of the molecular and cellular responses may simply be homeostatic and, by definition, will not impinge on the survival, development and fecundity of individuals in which they occur. Hence, they will not influence the dynamics of populations and ecosystems and in this sense are not ecologically relevant. Discovering the relationship between changes at one biological level (e.g. physiological impact of a pollutant) and those at another (e.g. population dynamics), will only be possible by establishing mechanistic links experimentally between levels (in models; e.g. Metz & Diekmann, 1986; Maltby & Calow, 1989), and testing predictions based on these about changes at one level from observations of impact at the lower level.

It might be argued, on the other hand, that it must always be more relevant to work 'top down' (Cairns, 1986); i.e. by using assays that involve multi-species groupings (simulated ecosystems, microcosms, mesocosms). There are a number of problems with this approach, however. Firstly, knowledge on how to package species into stable, artificial test systems is rudimentary and deserves much more attention. An alternative is to use multi-species samples from nature in tests, but these are often not reproducible and need not be stable. Secondly, multi-species groups, being complex systems, may be unique in their response. Hence, there is a potential trade-off between generality and uniqueness, and this again deserves further consideration by careful comparison of the responses of different multi-species groups to the same toxicants. This, of course, will involve a considerable amount of experimental

effort and raises the final problem, that will be treated further below, of the time, effort and expense needed to carry out multi-species tests.

The needs of regulators

Whilst requiring bioassays to be relevant in the sense outlined above (i.e. to the ecosystems they are designed to assess), in general regulators also require that results from these should be decisive and easy to comprehend. Ideally, they should not only make the intensity of actual or potential impact obvious to experts but also to the general public, so that public opinion might be influenced by them. Finally, it is important that assays should be economic in time, effort and the funding needed to carry them out. These features are likely to constrain suggestions for assays that might arise from the scientific research programme (above).

Conclusions

There are three major conclusions that can be drawn from this note.

Firstly, the results of monitoring bioassays that derive from surveys of 'natural' systems, have to be treated with some caution because of the difficulties of ascribing causation and of pseudoreplication. For a more detailed review of these problems and a potential solution see Underwood & Peterson (1988). The use of planted monitors lessens these problems, but here questions about the relevance of responses can legitimately be raised.

Secondly, from the point of view of exploring and understanding the ecological relevance of results from bioassays, the experimental method (involving the testing of properly framed models) ought to play a prominent role. There is likely to be considerable overlap here between what is generally regarded as classical ecology and ecotoxicology.

Thirdly, and finally, pragmatism also has to

L. I. H. E.
THE MARKLAND LIBRARY
STAND PARK RD., LIVERPOOL, L16 9JD

play an important part in the choice and implementation of bioassays. And from this point of view there has to be effective dialogue between those who use bioassays, the research community and the regulators.

References

Baird, D. J., I. Barber, M. Bradley, P. Calow & V. M. Soares, 1989. The *Daphnia* bioassay: a critique. In: M. Munawar, G. Dixon, C. I. Mayfield, T. Reynoldson & M. H. Sadar (eds.), Environmental Bioassay Techniques and Their Application. Hydrobiologia. This volume.

Bayne, B. L., R. F. Addison, J. M. Capuzzo, K. R. Clarke, J. S. Gray, M. N. Moore & R. M. Warwick, 1988. An overview of the GEEP Workshop. Marine Ecology – Progress Series, 46: 235–243.

Cairns, J., 1986. The myth of the most sensitive species. Bioscience, 36: 670–672.

Calow, P., 1976. Biological Machines. A Cybernetic Approach to Life. Edward Arnold, London. 134 pp.

Hellawell, J. M., 1986. Biological Indicators of Freshwater Pollution and Environmental Management. Elsevier Applied Science Publishers, London. 546 pp.

Hurlbert, S. H., 1984. Pseudoreplication and the design of ecological field experiments. Ecological Monographs, 54: 187–211.

Maltby, L. & P. Calow, 1989. The application of bioassays in the resolution of environmental problems; past, present and future. In: M. Munawar, G. Dixon, C. I. Mayfield, T. Reynoldson & M. H. Sadar (eds.) Environmental Bioassay Techniques and their Application. Hydrobiologia. This volume.

Metz, J. A. J. & O. Diekmann, 1986. The Dynamics of Physiologically Structured Populations. Springer Verlag, Heidelberg. 511 pp.

Naylor, C., L. Maltby & P. Calow, 1989. Scope for growth in *Gammarus pulex*, a freshwater benthic detritivore. In: M. Munawar, G. Dixon, C. I. Mayfield, T. Reynoldson & M. H. Sadar (eds.) Environmental Bioassay Techniques and their Application. Hydrobiologia. This volume.

Paine, R. T., 1977. Controlled manipulations in the marine intertidal zone and their contribution to ecological theory. The changing Scene in Natural Sciences, 1776–1976. Academy of National Science Special Publication, 12: 245–270.

Rand, G. M., 1985. Behaviour. In: G. M. Rand & S. R. Petrocelli (Eds.), Fundamentals of Aquatic Toxicology. Hemisphere Publishing Corporation, Washington: pp. 221–263.

Underwood, A. J. & C. H. Peterson, 1988. Towards an ecological framework for investigating pollution. Marine Ecology Progress Series, 46: 227–234.

Hydrobiologia **188/189**: 65–76, 1989.
M. Munawar, G. Dixon, C. I. Mayfield, T. Reynoldson and M. H. Sadar (eds)
Environmental Bioassay Techniques and their Application.
© 1989 *Kluwer Academic Publishers. Printed in Belgium.*

The application of bioassays in the resolution of environmental problems; past, present and future

L. Maltby & P. Calow
Department of Animal and Plant Sciences, University of Sheffield, Sheffield S10 2TN, U.K.

Key words: freshwater, water quality, assessment, prediction

Abstract

Literature on bioassays for freshwater systems has been reviewed (between 1979 and 1987) and classified into studies concerned with prediction and assessment and, within these categories, into studies concerned with single- and multi-species bioassays. Changing trends in the response criteria and types of organisms used in the predictive tests are judged against results from a similar review carried out in 1979. This leads to the conclusion that though there may have been changes in detail, bioassays have remained surprisingly unchanged in general features over this time. The relative merits of, and relationship between single- and multi-species studies for both predicting and assessing the biological impact of toxicants are discussed. The conclusion is that some bioassays have more severe problems than others, but a concern with all of them is that responses observed in particular systems may not be relevant in general. The possibility of developing a general theoretical infrastructure for bioassays that addresses this problem is considered.

Introduction

Environmental bioassays involve the use of biological systems of varying levels of complexity to predict or assess the impact of pollutants on ecosystems. That biological systems should be capable of assaying ecological disturbance is self-evident, since without biological effects there is no disturbance. But this is a superficial view, for in developing bioassays there is often a need to simplify and generalize and this raises difficult questions such as: To what extent can one species or species group act as proxy for another? And, to what extent can bioassays be simplified, e.g. from multi- to single-species systems and from ecosystem to individual and even cellular and molecular systems? With these issues in mind, we here critically review bioassays that have been

developed for both the prediction and assessment of pollution impacts in freshwater systems. It is not surprising that practicality and convenience have been dominant in determining which bioassays have been used, but this often involves selecting particular test systems or assays for general use; i.e. to predict general environmental impact effects from particular test results, or to assess impact from correlations between presence/absence of species and species groups and particular stressors that have been observed in previous instances. These are dubious practices philosophically (Saarinen, 1980; Mentis, 1988), and an important need is to develop a mechanistic understanding of particular effects so that a general theoretical framework can be developed for bioassays.

66

Framework for this review

In carrying out this review we distinguished between those bioassays concerned with prediction (anticipating environmental impacts) and assessment (monitoring actual impacts) and, within each of these categories, between bioassays employing single and many species. A comprehensive review of the literature was carried out up to 1979 by Murphy (1979) for predictive studies, so we concentrated on papers published post-1979 to 1987. A computer search was carried out using 3 databases (BIOSIS – Biological Abstracts; ASFA – Aquatic Sciences & Fisheries Abstracts; AQUALINE – WRc). The search concentrated on 2 subject areas (environmental/ industrial toxicology & limnology) and key words included: indicator species, toxicity test, biotic indices, hazard assessment and evaluation, impact assessment and evaluation. The complete review is documented in WRc Report PRS 1715-M* and here we only draw attention to general features and trends.

Predictive Studies

More than 90% of all the studies that we classified as predictive were single-species laboratory tests. The kinds of criteria used in these are summarised in Fig. 1 together with the frequencies of papers that fell into each category. By far the

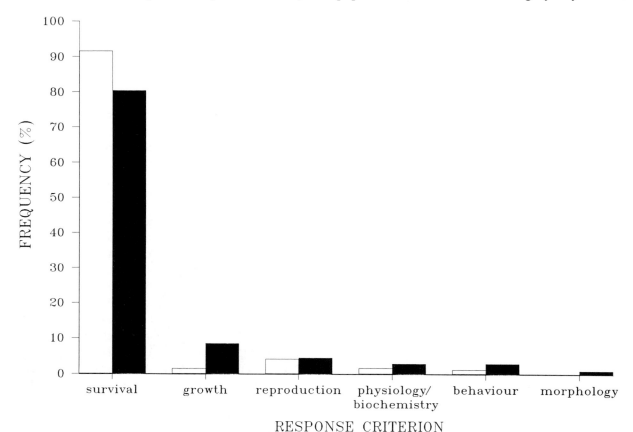

Fig. 1. Frequency (%) of criteria used in predictive, single-species laboratory tests in papers surveyed pre 1979 (n = 992) (open blocks) and those surveyed between 1979 and 1987 (n = 1175) (black blocks). See text for further information.

* Can be obtained from Water Research Centre, Medmenham Laboratory, PO Box 16, Henley Road, Medmenham, Marlow, Bucks. SL7 2HD, UK.

majority of tests in our review used death as a criterion of toxicity and were carried out over short time intervals; i.e. were acute tests. Invertebrates were the most commonly used test organisms (used in 74.8% of papers) followed by fish (23.9%). A detailed analysis of the invertebrates used in the studies is given in Fig. 2 and shows that cladocerans have been the most widely used animals. The most frequently used fish species were the fathead minnow (*Pimephales promelas*) and the rainbow trout (*Salmo gairdneri*).

Chronic and sublethal assays represented less than 20% of all the single-species studies and involved a number of response criteria (Fig. 1).

Also given in Figs. 1 & 2 are the distributions of papers from the 1979 study. There have been

few changes in terms of the general response criteria that have been employed, and although the detailed features of taxonomic distributions differ between surveys the general features remain remarkably similar.

Multi-species tests have been in the minority. Some of these have been laboratory studies involving mixed-flask cultures (e.g. Heath, 1979; Taub *et al.*, 1983; Stay *et al.*, 1985; Biesinger & Stokes, 1986; Larsen *et al.*, 1986) periphyton growing on artificial substrates (e.g. Buikema *et al.*, 1983; Niederlehner *et al.*, 1985; Cairns *et al.*, 1986; Yun-Fen *et al.*, 1986), pelagic microcosms (e.g. Harte *et al.*, 1980), pond microcosms (e.g. Harris, 1980; Giddings & Franco, 1985; Portier, 1985) and model streams (e.g. Shriner &

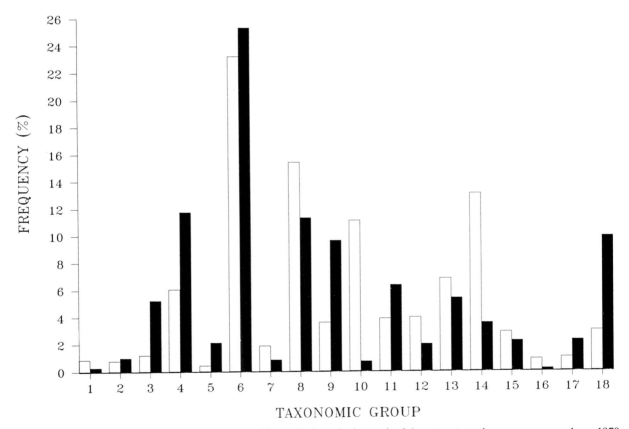

Fig. 2. Frequency (%) of types of invertebrates used in predictive, single-species laboratory tests in papers surveyed pre 1979 (n = 909) (open blocks) and those surveyed between 1979 and 1987 (n = 707) (black blocks). 1 = Protozoa; 2 = Nematoda; 3 = Platyhelminthes; 4 = Oligochaeta; 5 = Hirudinea; 6 = Cladocera; 7 = Copepoda; 8 = Amphipoda; 9 = Isopoda; 10 = Decapoda; 11 = Diptera; 12 = Trichoptera; 13 = Ephemeroptera; 14 = Plecoptera; 15 = Odonata; 16 = Coleoptera; 17 = Heteroptera; 18 = Mollusca.

Gregory, 1984; Muirhead-Thomson, 1987). Other studies have been carried out in the open and involved artificial streams (e.g. Arthur *et al.*, 1983; Kownacki *et al.*, 1985; Yasuno *et al.*, 1985; Zischke & Arthur, 1987) and ponds (e.g. Crossland, 1982, 1984), various enclosures in 'natural systems' (e.g. Solomon *et al.*, 1980; Dejoux, 1982; Lewis, 1986; Munawar & Munawar, 1987) and, finally, the manipulation of natural systems themselves (e.g. Schindler & Turner, 1982; Hall *et al.*, 1980, 1982, 1985; Ormerod *et al.*, 1987). Criteria used in these studies have involved both structural attributes (e.g. population density, species composition and diversity) and functional ones (e.g. primary productivity and P/R ratios).

It is no coincidence that single-species, acute tests have been dominant both pre- and post-1979. These are the most conveniently carried out and easily-controlled tests. Needless to say, the ability of single-species, laboratory tests to accurately predict the effects of toxins on complex communities has been widely debated (e.g. Cairns, 1986). *A priori* it is clear that whereas such tests can assay the direct effects of toxins on individuals, they cannot provide information on the impact of toxins on between-species interactions, the so-called indirect effects. Several comparative studies have been performed to assess the predictive ability of single- versus multi-species studies, and these are summarised in Table 1. The results from these studies are far from decisive, however. Sometimes, single-species tests are less sensitive than multi-species ones (e.g. Niederlehner *et al.*, 1985) and sometimes more sensitive (e.g. Chapman *et al.*, 1982a & b). Sometimes laboratory results suggest greater toxicity than field results (e.g. Lewis, 1986) and sometimes the reverse (e.g. Sherman *et al.*, 1987).

Assessment studies

In these studies ecological features of communities or their parts are used, retrospectively, to make judgements about how disturbed natural systems are. This is done by: (a) indicator groups

Table 1. Comparative studies assessing the predictive ability of single-species and multi-species tests.

SINGLE-SPECIES	Chapman *et al.* (1982a & b)
v	Giddings & Franco (1985)
LAB. MULTI-SPECIES	Hansen & Garton (1982)
	Larsen *et al.* (1986)
	Niederlehner *et al.* (1985)
SINGLE-SPECIES	Adams *et al.* (1983)
v	Crossland (1984)
FIELD MULTI-SPECIES	Crossland & Hillaby (1985)
	Crossland & Wolff (1985)
	Giddings & Franco (1985)
	Larsen *et al.* (1986)
	Lewis (1986)
	Sherman *et al.* (1987)
LAB. MULTI-SPECIES	deNoyelles & Kettle (1985)
v	Giddings *et al.* (1984)
FIELD MULTI-SPECIES	Giddings & Franco (1985)
	Larsen *et al.* (1986)
	Portier (1985)

(biotic indices); (b) general measures of community structure without reference to the particular species present (diversity indices); (c) measurements of community function. The ability of organisms to accumulate chemicals (bioaccumulation) can also be used to assay the presence of toxins (Hellawell, 1986) but will not be treated further here.

The presence of tolerant species or the absence of sensitive ones is often taken as an indicator of pollution (e.g. James & Evison, 1979). A list of some of the types of organisms used in this way is given in Table 2. Indicators are selected because experience has shown them to be particularly resistant or sensitive in certain situations. Unfortunately, though, an organism can be sensitive to one pollutant but tolerant of another (Slooff, 1983; Slooff & De Zwart, 1983; Cairns, 1986) so extrapolation seems dangerous. Most indicators have been selected to assay organic pollution. Much more work seems necessary on the basic biology of indicator species to gauge their relative sensitivity to a range of toxins and to understand the mechanisms that might be involved in allowing more or less tolerance to toxins.

Table 2. Organisms used as indicator species.

ALGAE	Gotoh & Negoro (1986)
	Reddy & Venkateswarlu (1986)
INVERTEBRATES	
Oligochaetes	Giani (1984)
	Krieger (1984)
Chironomids	Krieger (1984)
	Waterhouse & Farrell (1985)
Ephemeropterans	Howells *et al.* (1983)
	Simpson *et al.* (1985)
Molluscs	Krieger (1984)
	Mouthon (1981)
FISH	Howells *et al.* (1983)

Table 3. Biotic indices.

RATIOS	
Chironomids: insects	Winner *et al.* (1980)
Asellus: *Gammarus*	Watton & Hawkes (1984)
Trophic Condition Index	Mason *et al.* (1985)
QUALITATIVE INDICES	
Trent Biotic Index	Balloch *et al.* (1976)
	Ellis *et al.* (1986)
	Hamer & Soulsby (1980)
	Murphy (1978)
	Pinder *et al.* (1987)
	Pshenitsyna (1986)
	Vandelannoote *et al.* (1981)
	Watton & Hawkes (1984)
Grahams Biotic Index	Balloch *et al.* (1976)
Belgian BI	dePauw & Vanhooren (1983)
Tuffery – Verneux	Fontoura & Moura (1984)
	Vandelannoote *et al.* (1981)
BMWP Score	Armitage *et al.* (1983)
	Brooker (1984)
	Ellis *et al.* (1986)
	Extence *et al.* (1987)
	Pinder & Farr (1987a & b)
	Pinder *et al.* (1987)
	Watton & Hawkes (1984)
SEMI-QUANTITATIVE INDICES	
Chandler Biotic Score (+ Ave. CBS)	Balloch *et al.* (1976)
	Brooker (1984)
	Cook (1976)
	Hamer & Soulsby (1980)
	Murphy (1978)
	Pinder & Farr (1987a & b)
	Pinder *et al.* (1987)
	Scullion & Edwards (1980)
	Watton & Hawkes (1984)
Empirical biotic index (+ Hilsenhoff biotic index)	Chutter (1972)
	Hilsenhoff (1977, 1982)
	Jones *et al.* (1981)
	Mason *et al.* (1985)
	Narf *et al.* (1984)
	Seeley & Zimmerman (1985)
	Zimmerman & Skinner (1985)

An extension of the indicator-organism concept is the biotic index assay that uses combinations of indicator species to assay water quality. Those that have been employed are listed in Table 3. They range from simple ratios of tolerant to intolerant species to qualitative indices that assign scores to species present on the basis of their sensitivity, to semi-quantitative indices that not only assign scores but weight them by the abundance of each indicator species. In one sense, the culminations of this trend are diversity indices. However, these neither take note of the types of species present nor their presumed sensitivities, only their number and the distribution of individuals between them. Types of diversity indices that have been used in this context are listed in Table 4. There are two major problems with these diversity indices; firstly, in terms of how well they measure diversity (Taylor *et al.*, 1976; May, 1975) and, secondly, in terms of how they should be expected to respond to pollution (Pinder & Farr, 1987a & b).

All these techniques are fundamentally comparative; i.e. comparing the biological composition of communities at a site with that found/ expected in either other polluted or non-polluted situations. It is possible to make the comparative base explicit by using statistical techniques of various levels of complexity to compare community structures at putatively unpolluted and polluted sites – a list of suitable techniques and studies is given in Table 5.

Finally, functional responses of individuals within communities (e.g. scope for growth assays,

Table 4. Diversity indices.

SHANNON-WIENER	Balloch *et al.* (1976)
	Chadwick & Canton (1983)
	Cook (1976)
	Couture *et al.* (1987)
	Dills & Rogers (1974)
	Ellis *et al.* (1986)
	Fontoura & Moura (1984)
	Hamer & Soulsby (1980)
	Hodgkiss & Law (1985)
	Maret & Christiansen (1981)
	Mason *et al.* (1985)
	Matthews *et al.* (1980)
	Milbrink (1983)
	Moore (1979)
	Murphy (1978)
	Pinder & Farr (1987a & b)
	Pinder *et al.* (1987)
	Scullion & Edwards (1980)
	Simpson *et al.* (1985)
	Titmus (1981)
	Watton & Hawkes (1984)
MARGALEF	Cook (1976)
	Fontoura & Moura (1984)
	Mason *et al.* (1985)
	Murphy (1978)
	Pinder *et al.* (1987)
	Scullion & Edwards (1980)
	Vandelannoote *et al.* (1981)
SIMPSON	Hodgkiss & Law (1985)
	Milbrink (1983)
	Pinder & Farr (1987a & b)
	Pinder *et al.* (1987)
MENHINICK	Murphy (1978)
	Pinder *et al.* (1987)
	Vandelannoote *et al.* (1981)
FISHER'S ALPHA	Viaud & Lavandier (1982)
	Watton & Hawkes (1984)
KEMPTON-TAYLOR	Watton & Hawkes (1984)
SEQUENTIAL COMPARISON INDEX	Hodgkiss & Law (1985)

Table 5. Methods of comparing communities.

(1) Techniques for computing degree of similarity/dissimilarity

Jaccard coefficient	Scullion & Edwards (1980)
Product-moment correlation	Learner *et al.* (1983)
Spearman-rank correlation	Brooker & Morris (1980)
Kendall's rank correlation	Scullion & Edwards (1980)
	Learner *et al.* (1983)
Squared Euclidean-Distance	Learner *et al.* (1983)
Principle components analysis	Hamer & Soulsby (1980)
Detrended correspondence analysis	Furse *et al.* (1984)
	Wright *et al.* (1984)
	Armitage *et al.* (1987)
	Moss *et al.* (1987)
	Ormerod & Edwards (1987)
	Weatherley & Ormerod (1987)

(2) Techniques for clustering (1)

Nearest neighbour	Learner *et al.* (1983)
Average linkage	Brooker & Morris (1980)
	Learner *et al.* (1983)
Two-way indicator species analysis (TWINSPAN)	Furse *et al.* (1984)
	Wright *et al.* (1984)
	Armitage *et al.* (1987)
	Moss *et al.* (1987)
	Ormerod & Edwards (1987)
	Weatherley & Ormerod (1987)

pollution, how function relates to structure (and *vice versa*) and what such responses mean for the 'well being' of the ecosystem.

The distribution of papers between the various kinds of assessment techniques is illustrated in Fig. 3. Structural indices have been used most, and of these the Shannon-Wiener diversity function has been dominant.

Bayne *et al.*, 1985) and of whole communities (e.g. P/R ratios, Odum, 1975; the autotrophy index of Weber, 1973 and Matthews *et al.*, 1982) and total community respiration (Osborne & Davies, 1981; Grimm & Fisher, 1984) have also been used for purposes of assessment. Again, though, not enough is known about how functional attributes would be expected to respond to

Quo vadis?

Predictive bioassays are carried out to anticipate possible future impacts of chemicals on the basis of simplified tests, and assessment bioassays are concerned with interpreting present observations in the light of past experience. In both cases, generalisations are attempted from particular

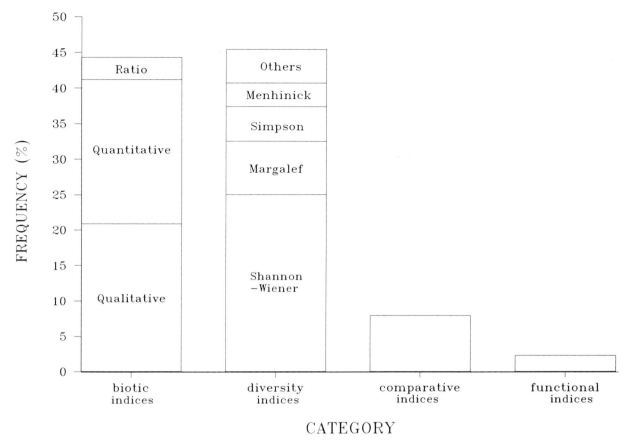

Fig. 3. Frequency (%) of community indices used to assess impact of pollutants, in papers surveyed post 1979. See text for further explanation.

observations; a particular test result is used to forecast general consequences, and particular links between species occurrences and absences and pollution are used to formulate general rules about the ecological impacts of pollution. Moreover, test organisms in predictive tests and indicators in assessments are often chosen because they are supposed to be 'typical', 'sensitive' or both, which is another form of generalisation (Cairns, 1986). Yet the logico-scientific problems with this approach, known as induction, are well known (Saarinen, 1980; Mentis, 1988); particular observations do not prove rules because, no matter how many confirmatory observations are made, there is no guarantee that counter-examples will not ultimately be discovered. A particular effect caused by a particular chemical on a particular organism or group of organisms does not

guarantee that other chemicals will have the same effect on the same system or that the same chemical will have the same effects on different organisms. Furthermore, there is no guarantee that the effects of pollutants on complex systems will be the same as those on simple ones. Finally, there is no guarantee that correlations between ecological variables and community composition in a set of habitats will apply across all habitats.

These are familiar kinds of criticisms that have been levelled at bioassays, in various ways, numerous times in the past. Yet despite these feelings of unease, that appear to be well founded, the foregoing review suggests that bioassays have changed little over the past ten years. This is probably due to a combination of legislative constraints (once tests are written into legislation, they have to be carried out routinely) and uncertainty about how to advance in a field that involves complex problems associated with complex systems.

A possible way forward, from a predictive point of view, might be to design and build more complex systems that approximate more and more to reality. And as suggested in the review, there is a trend in this direction from microcosm to mesocosm to perturbations of natural systems. There are two problems with this kind of approach, however. First, tests involving more and more complex systems are more and more expensive to run in time, effort and resources and hence can rarely be replicated very adequately. Second, this does not overcome the problem of induction, for there is no guarantee that complex systems, even natural ones, will behave in the same way as other complex systems, particularly in nature. So the trend towards complex test systems could involve a considerable waste of time and effort.

It would seem to us that the only dependable way forward is to move away from the inductive approach to one that seeks understanding of how perturbations impact biological systems at various levels and how the impact at one level influences systems at higher levels. For example, how do chemicals impact the molecular, cellular and physiological systems within organisms? Knowing something about the general features of these suborganismic systems, to what extent can these effects be generalized – from one organism to another and from one chemical to another? To what extent do these suborganismic effects influence the survival, developmental rates and fecundities of the organisms? To what extent do these organismic effects influence population dynamics, and to what extent do population effects influence community processes and structure? Some examples of models that begin to take account of this approach are those of Kooijman & Metz (1984). Indirect effects (p. 4) can impact between-organisms and between-species interactions, but how important are they relative to the direct causal chain just described? All these are challenging questions that can only be approached by framing hypotheses about mechanisms operating within and between levels, and testing these by critical experiment. This, of course, is the hypothetico-deductive approach (Mentis, 1988). The ultimate aim will be to develop complex general models that represent the mechanism of toxicological effects and which, by suitable modification of parameter values, can identify impacts and make predictions about them. This is a long-term programme. The crucial first stage, though, must be a shift away from induction (look-see) to the hypothetico-deductive (try to identify and understand mechanism) approaches in designing bioassays.

Acknowledgments

This paper is based on a literature review carried out under contract from the Water Research Centre, U.K. We would like to thank Susan Frank (University of Sheffield) and Maxine Forshaw (WRc) for running computer searches and Professor R. W. Edwards for useful discussion.

References

Adams, W, J., R. A. Kimerle, B. B. Heidolph & P. R. Michael, 1983. Field comparison of laboratory-derived acute and chronic toxicity data. In W. E. Bishop, R. D. Cardwell & B. B. Heidolph (eds), Aquatic Toxicology and Hazard Assessment, Sixth Symposium. ASTM STP 802: 367–385.

Armitage, P. D., R. J. M. Gunn, M. T. Furse, J. F. Wright & D. Moss, 1987. The use of prediction to assess macroinvertebrate response to river regulation. Hydrobiologia 144: 25–32.

Armitage, P. D., D. Moss, J. F. Wright & M. T. Furse, 1983. The performance of a new biological water quality score system based on macroinvertebrates over a wide range of unpolluted running water sites. Wat. Res. 17: 333–347.

Arthur, J. W., J. A. Zischke, K. A. Allen & R. O. Hermanutz, 1983. Effects of Diazinon on macroinvertebrates and insect emergence in outdoor experimental channels. Aquat. Toxicol. 4: 283–301.

Bayne, B. L., D. A. Brown, K. Burns, D. R. Dixon, A. Ivanovici, D. R. Livingstone, D. M. Lowe, M. N. Moore, A. R. D. Stebbing & J. Widdows, 1985. The effects of stress and pollution on marine animals. Praeger Special Studies, N.Y., 384 pp.

Balloch, D., C. E. Davies & F. H. Jones, 1976. Biological

assessment of water quality in three British rivers: The North Esk, the Ivel and the Taff. Wat. Poll. Contr. 75: 92–110.

Biesinger, K. E. & G. N. Stokes, 1986. Effects of synthetic polyelectrolytes on selected aquatic organisms. J. Water Pollut. Cont. Fed. 55: 207–213.

Brooker, M. P., 1984. Biological surveillance in Welsh rivers for water quality and conservation assessment. In D. Pascoe & R. W. Edwards (eds), Freshwater Biological Monitoring. Pergamon Press, Oxford: 25–33.

Brooker, M. P. & D. L. Morris, 1980. A survey of the macro-invertebrate riffle fauna of the River Wye. Freshwat. Biol., 10: 437–458.

Buikema, A. L., J. Cairns & W. H. Yongue, 1983. Correlation between the autotrophic index and protozoan colonization rates as indicators of pollution stress. In W. E. Bishop, R. D. Cardwell & B. B. Heidolph (eds), Aquatic Toxicology and Hazard Assessment. ASTM STP 802: 204–215.

Cairns, J., 1986. The myth of the most sensitive species. Bioscience 36: 670–672.

Cairns, J., J. R. Pratt, B. R. Niederlehner & P. V. McCormick, 1986. A simple cost-effective multispecies toxicity test using organisms with a cosmopolitan distribution. Envir. Mon. Assess. 6: 207–220.

Chadwick, J. W. & S. P. Canton, 1983. Coal mine drainage effects on a lotic ecosystem in Northwest Colorado, U.S.A. Hydrobiologia 107: 25–33.

Chapman, P. M., M. A. Farrell & R. O. Brinkhurst, 1982a. Relative tolerances of selected aquatic oligochaetes to individual pollutants and environmental factors. Aquat. Toxicol. 2: 47–68.

Chapman, P. M., M. A. Farrell & R. O. Brinkhurst, 1982b. Relative tolerances of selected aquatic oligochaetes to combination of pollutants and environmental factors. Aquat. Toxicol. 2: 69–78.

Chutter, F. M., 1972. An empirical biotic index of the quality of water of South African streams and rivers. Wat. Res. 6: 19–30.

Cook, S. E. K., 1976. Quest for an index of community structure sensitive to water pollution. Envir. Pollut. 11: 269–288.

Couture, P., G. Thellen, P. A. Thompson & J. C. Auclair, 1987. Structure and function of phytoplanktonic and microbial communities in relation to industrial waste water discharge: An ecotoxicological approach in a lotic system. Can. J. Fish. Aquat. Sci. 44: 167–175.

Crossland, N. O., 1982. Aquatic toxicology of cypermethrin. II. Fate and biological effects of pond experiments. Aquat. Toxicol. 2: 205–222.

Crossland, N. O., 1984. Fate and biological affects of methyl parathion in outdoor ponds and laboratory aquaria. Ecotox. Environ. Safety 8: 482–495.

Crossland, N. O. & J. M. Hillaby, 1985. Fate and effects of 3,4 dichloroaniline in the laboratory and in outdoor ponds. 2. Chronic toxicity to *Daphnia* and other invertebrates. Envir. Toxicol. Chem. 4: 489–500.

Crossland, N. O. & C. J. M. Wolff, 1985. Fate and biological effects of pentachlorophenol in outdoor ponds. Envir. Toxicol. Chem. 4: 73–86.

Dejoux, C., 1982. Recherche sur le devenir invertebrés derivant dans un cours d'eau tropical à la suite traitements antisimulidiens au temephos. Rev. Français Sci. de l'eau 1: 267–283.

deNoyelles, F. & W. D. Kettle, 1985. Experimental ponds for evaluating bioassay predictions. In T. P. Boyle (ed), Validation and predictability of laboratory methods for assessing the fate and effects of contaminants in aquatic ecosystems. ASTM STP 865: 91–103.

dePauw, N. & G. Vanhooren, 1983. Method for biological quality assessment of watercourses in Belgium. Hydrobiologia 100: 153–168.

Dills, G. & D. T. Rogers, 1974. Macroinvertebrate community structure as an indicator of acid pollution. Envir. Pollut. 6: 239–264.

Ellis, A. C., H. A. Hawkes, W. Pope, B. J. Owen & J. D. Cargill, 1986. Environmental impact assessment – A case study. Wat. Pollut. Contr. 85: 340–355.

Extence, C. A., A. J. Bates, W. J. Forbes & P. J. Barham, 1987. Biologically based water quality management. Envir. Pollut. 45: 221–236.

Fontoura, A. P. & A. M. G. Moura, 1984. Effects of some industrial effluents on the biological quality of the water of the River Lima, Portugal. Publ. Inst. Zool. 'Dr. Augusto Nobre' Fac. Cience, Porto 0: 1–21.

Furse, M. T., D. Moss, J. F. Wright & P. D. Armitage, 1984. The influence of seasonal and taxonomic factors in the ordination and classification of running water sites in Great Britain and the prediction of their macroinvertebrate communities. Freshwat. Biol. 14: 257–280.

Giani, N., 1984. The Riou-mort a tributary of the River Lot, France polluted by heavy metals. 4. A study of the oligochaetes. Ann. Limnol. 20: 167–182.

Giddings, J. M. & Franco, P. J., 1985. Calibration of laboratory bioassays with results from microcosms and ponds. In T. P. Boyle (ed), Validation and Predictability of Laboratory Methods for Assessing the Fate and Effects of Contaminants in Aquatic Ecosystems. ASTM STP 865: 104–119.

Giddings, J. M., P. J. Franco, R. M. Cushman, L. A. Hook, G. R. Southworth & A. J. Stewart, 1984. Effects of chronic exposure to coal-derived oil on freshwater ecosystems. 2. Experimental ponds. Envir. Toxicol. Chem. 3: 465–488.

Gotoh, T. & K. I. Negoro, 1986. Diatom vegetation of the less polluted river the U-Kawa River Kyoto Prefecture, Japan. Jap. J. Limnol. 47: 77–86.

Grimm, N. B. & S. G. Fisher, 1984. Exchange between interstitial and surface water: implication for stream metabolism and nutrient cycling. Hydrobiologia 111: 219–228.

Hall, R. J., C. T. Driscoll, G. E. Likens & J. M. Pratt, 1985. Physical, chemical, and biological consequences of episodic aluminium additions to a stream. Limnol. Oceanogr. 30: 212–220.

Hall, R. J., G. E. Likens, S. B. Fiance & G. R. Hendrey, 1980.

Experimental acidification of a stream in the Hubbard Brook experimental forest, New ampshire. Ecology 61: 976–989.

Hall, R. J., J. M. Pratt & G. E. Likens, 1982. Effects of experimental acidification on macroinvertebrate drift diversity in a mountain stream. Wat. Air Soil Pollut. 18: 273–288.

Hamer, A. D. & B. G. Soulsby, 1980. An approach to chemical and biological river monitoring systems. Wat. Pollut. Contr. 79: 56–69.

Hansen, S. R. & R. R. Garton, 1982. Ability of standard toxicity tests to predict the effects of the insecticide diflubenzuron on laboratory stream communities. Can. J. Fish. Aquat. Sci. 39: 1273–1288.

Harris, W. F., 1980. Microcosms as potential screening tests for evaluating transport and effects of toxic substances. Oak Ridge National Laboratories. Report No. EPA 600/3–80–042.

Harte, J., D. Levy, J. Rees & E. Saegebarth, 1980. Assessment of optimum aquatic microcosm design for pollution impact studies. Lawrence Berkeley Laboratory, Palo Alto, California, Final Report. Electric Power Research Institute, 170 pp.

Heath, R. T., 1979. Holistic study of an aquatic microcosm: theoretical and practical implications. Int. J. Env. Stud. 13: 87–93.

Hellawell, J. M., 1986. Biological Indicators of Freshwater Pollution and Environmental Management. Elsevier Applied Science Publishers, London, 546 pp.

Hilsenhoff, W. L., 1977. Use of arthropods to evaluate water quality in streams. Wisconsin Dept. Natural Resources Technical Bulletin, No. 100, 15 pp.

Hilsenhoff, W. L., 1982. Using a biotic index to evaluate water quality in streams. Wisconsin Dept. Natural Resources Technical Bulletin, No. 132, 23 pp.

Hodgkiss, I. J. & C. Y. Law, 1985. Relating diatom community structure and stream water quality using species diversity indices. Wat. Pollut. Contr. 84: 134–139.

Howells, E. J., M. E. Howells & J. S. Alabaster, 1983. A field investigation of water quality, fish and invertebrates in the Mawddach river system, Wales. J. Fish Biol. 22: 447–469.

James, A. & L. Evison, 1979 (Eds). Biological indicators of water quality. John Wiley & Sons, NY.

Jones, J. R., B. H. Tracy, J. L. Sebaugh, D. H. Hazelwood & M. M. Smart, 1981. Biotic index testing for ability to asess water quality of Missouri Ozark, USA, streams. Trans. Am. Fish. Soc. 110: 627–637.

Kooijman, S. A. L. M. & A. J. Metz, 1984. On the dynamics of chemically stressed populations: the deduction of population consequences from the effects of individuals. Ecotox. Environ. Safety, 8: 254–274.

Kownacki, A., E. Dumnicka, E. Grabacka, B. Kawecka & A. Starzecka, 1985. Stream ecosystems in mountain grassland (West Carpathians). 14. The use of experimental stream method in evaluating the effect of agricultural pollution. Acta Hydrobiology 27: 381–400.

Krieger, K. A., 1984. Benthic macroinvertebrates as indicators of environmental degradation in the southern near-shore zone of the central basin of Lake Erie, Canada. J. Great Lakes Res. 10: 197–209.

Larsen, D. P., F. deNoyelles, F. Stay & T. Shiroyama, 1986. Comparison of single-species microcosm and experimental pond responses to atrazine exposure. Envir. Toxicol. Chem. 5: 179–190.

Learner, M. A., J. W. Densem & T. C. Iles, 1983. A comparison of some classification methods used to determine benthic macro-invertebrate species associations in the river survey work based on data obtained from the River Ely, South Wales. Freshwat. Biol. 13: 13–36.

Lewis, M. A., 1986. Comparison of the effects of surfactants on freshwater phytoplankton communities in experimental enclosures and on algal population growth in the laboratory. Envir. Toxicol. Chem. 5: 319–332.

Maret, T. R. & C. C. Christiansen, 1981. A water quality survey of the Big Blue River, Nebraska, U.S.A. Trans. Nebr. Acad. Sci. 9: 35–48.

Mason, W. T. Jr., P. A. Lewis & C. I. Weber, 1985. An evaluation of benthic macroinvertebrate biomass methodology. Part 2. Field assessment and data evaluation. Envir. Monit. Assess. 5: 399–422.

Matthews, R. A., P. F. Kondratieff & A. L. Buikema, 1980. A field verification of the use of the autotrophic index in monitoring stress effects. Bull. Envir. Contam. Toxicol. 25: 226–233.

Matthews, R. A., P. F. Kondratieff & A. L. Buikema, 1982. A seasonal analysis of stress in a stream ecosystem using a nontaxonomic approach. In J. G. Pearson, R. B. Fosler & W. E. Bishop (eds), Aquatic Toxicology and Hazard Assessment. ASTM STP 766: 326–340.

May, R. M., 1975. Patterns of species abundance and diversity. In M. L. Cody & J. M. Diamond (eds), Ecology, Evolution and Communities. Harvard University Press, Mass, pp. 81–120.

Mentis, M., 1988. Hypothetico-deductive and inductive approaches in ecology. Functional Ecology 2: 5–14.

Milbrink, G., 1983. An improved environmental index based on the relative abundance of oligochaete species. Hydrobiologia 102: 89–97.

Moore, J. W., 1979. Diversity and indicator species as measures of water pollution in a subarctic lake. Hydrobiologia 66: 73–80.

Moss, D., M. T. Furse, J. F. Wright & P. D. Armitage, 1987. The prediction of the macro-invertebrate fauna of unpolluted running-water sites in Great Britain using environmental data. Freshwat. Biol. 17: 41–52.

Mouthon, J., 1981. Mollusks and the pollution of freshwaters; an outline of the sensitivity of the species to pollution. Bijdr Dierkd 51: 250–258.

Muirhead-Thomson, R. C., 1987. Pesticide impact on stream fauna with special reference to macroinvertebrates. Cambridge University Press, Cambridge: 275 pp.

Munawar, M. & I. F. Munawar, 1987. Phytoplankton bioassays for evaluating toxicity of *in situ* sediment contaminants. Hydrobiologia 149: 87–105.

Murphy, P. M., 1978. The temporal variability in biotic indices. Envir. Pollut. 17: 227–236.

Murphy, P. M., 1979. A manual for toxicity tests with freshwater macroinvertebrates and a review of the effects of specific toxicants. University of Wales Institute of Science and Technology, Cardiff, 249 pp.

Narf, R. P., E. L. Lange & R. C. Wildman, 1984. Statistical procedures for applying Hilsenhoffs biotic index. J. Freshwat. Ecol. 2: 441–448.

Niederlehner, B. R., J. R. Pratt, A. L. Buikema & J. Cairns, 1985. Laboratory tests evaluating the effects of cadmium on protozoan communities. Envir. Toxicol. Chem. 4: 155–166.

Odum, E. P., 1975. Ecology (2nd Edn.). Holt, Rinehart & Winston, London, N.Y, 244 pp.

Ormerod, S. J. & R. W. Edwards, 1987. The ordination and classification of macro-invertebrate assemblages in the catchment of the River Wye in relation to environmental factors. Freshwat. Biol. 17: 533–546.

Ormerod, S. J., P. Boole, C. P. McCahon & N. S. Weatherly, 1987. Short-term experimental acidification of a Welsh stream; comparing the biological effects of hydrogen ions and aluminium. Freshwat. Biol. 17: 341–356.

Osborne, L. L. & R. W. Davies, 1981. A device to measure in situ lotic benthic metabolism. Hydrobiologia 79: 261–264.

Pinder, L. C. V. & I. S. Farr, 1987a. Biological surveillance of water quality. 2. Temporal and spatial variation in macroinvertebrate fauna of the River Frome, a Dorset chalk stream. Arch. Hydrobiol. 109: 321–331.

Pinder, L. C. V. & I. S. Farr, 1987b. Biological surveillance of water quality. 3. The influence of organic enrichment on the macro-invertebrate fauna of small chalk streams. Arch. Hydrobiol. 109: 619–637.

Pinder, L. C. V., M. Ladle, T. Gledhill, J. A. B. Bass & A. M. Matthews, 1987. Biological surveillance of water quality. 1. A comparison of macroinvertebrate surveillance methods in relation to assessment of water quality, in a chalk stream. Arch. Hydrobiol. 109: 207–226.

Portier, R. J., 1985. Comparison of environmental effect and biotransformation of toxicants on laboratory microcosm and field microbial communities. In T. P. Boyle (ed), Validation and Predictability of Laboratory Methods for Assessing the Fate and Effects in Aquatic Ecosystems. ASTM STP 865: 14–30.

Pshenitsyna, V. N., 1986. Effectiveness of the Woodiwiss scale as a bioindicator of water quality. Hydrobiol. J. 22: 40–43.

Reddy, P. & U. V. Venkateswarlu, 1986. Ecology of algae in the paper mill effluents and their impact on the River Tungabhadra. J. Envir. Biol. 7: 215–223.

Saarinen, E., 1980. Conceptual Issues in Ecology. D. Reidel, Dordrecht, 374 pp.

Schindler, D. W. & M. A. Turner, 1982. Biological chemical and physical responses of lakes to experimental acidification. Wat. Air Soil Pollut. 18: 259–271.

Scullion, J. & R. W. Edwards, 1980. The effects of coal industry pollutants on the macro-invertebrate fauna of a small river in the South Wales coalfield. Freshwat. Biol. 10: 141–162.

Seeley, D. E. & M. C. Zimmerman, 1985. Use of the Hilsenhoff biotic index to determine the water quality of the Lycoming and Pine Creeks, Lycocing County, Pennsylvania, USA. Proc. PA. Acad. Sci. 59: 85.

Sherman, R. E., S. P. Gloss & L. W. Lion, 1987. A comparison of toxicity tests conducted in the laboratory and in experimental ponds using cadmium and the fathead minnow (Pimephales promelas). Wat. Res. 21: 317–323.

Shriner, C. & T. Gregory, 1984. Use of artificial streams for toxicological research. CRC Critical Reviews in Toxicology 13: 253–281.

Simpson, K. W., R. W. Bode & J. R. Colquhoun, 1985. The macroinvertebrate fauna of an acid-stressed head water stream system in the Adirondack Mountains, New York. Freshwat. Biol. 15: 671–681.

Slooff, W., 1983. Benthic macroinvertebrates and water quality assessment some toxicological considerations. Aquat. Toxicol. 4: 73–82.

Slooff, W. & D. De Zwart, 1983. Bioindicators and chemical pollution of surface waters. Envir. Mon. Assess. 3: 237–245.

Solomon, K. R., K. Smith, G. Guest, J. V. Yoo & N. K. Kaushik, 1980. Use of limnocorrals in studying the effects of pesticides in the aquatic ecosystem. Canadian Technical Report of Fisheries and Aquatic Sciences 975: 1–9.

Stay, F. S., P. Larsen, A. Katko & C. M. Rohm, 1985. Effects of atrazine on community level responses in Taub microcosms. In T. P. Boyle (ed), Validation and Predictability of Laboratory Methods for Assessing the Fate and Effects of Contaminants in Aquatic Ecosystems. ASTM STP 866: 75–90.

Taub, F. B., P. L. Read, A. C. Kindig, M. C. Harrass, H. J. Hartman, L. L. Conquest, F. J. Hardy & P. T. Munro, 1983. Demonstration of the ecological effects of streptomycin and malathion on synthetic aquatic microcosms. In W. E. Bishop, R. D. Cardwell & B. B. Heidolph (eds), Aquatic toxicology and hazard assessment: Sixth symposium. ASTM STP 802: 5–25.

Taylor, L. R., R. A. Kempton & J. P. Wolwood, 1976. Diversity statistics and the log-series model. J. Anim. Ecol. 45: 255–271.

Titmus, G., 1981. The use of Whittakers components of diversity in summarizing river survey data. Wat. Res. 15: 863–866.

Vandelannoote, A., G. De Gueldre & B. Bruylants, 1981. Ecological assessment of water quality: comparison of biological-ecological procedures in a rainfed lowland waterway Kleine Nete, Northern Belgium. Hydrobiol. Bull. 15: 161–164.

Viaud, M. & P. Lavandier, 1982. Comparison of some common indices for the evaluation of the water quality when applied to the detection of weak pollution. Bull. Soc. Hist. Nat. Toulouse 117: 221–230.

Waterhouse, J. C. & M. P. Farrell, 1985. Identifying pollution related changes in chironomid communities as a function of taxonomic rank. Can. J. Fish. Aquat. Sci. 42: 406–413.

Watton, A. J. & H. A. Hawkes, 1984. The performance of an invertebrate colonisation sampler in biological surveillance of lowland rivers. In D. Pascoe & R. W. Edwards (eds), Freshwater Biological Monitoring. Pergamon Press, Oxford: 15–24.

Weatherley, N. S. & S. J. Ormerod, 1987. The impact of acidification on macroinvertebrate assemblages in Welsh streams: towards an empirical model. Envir. Pollut. 46: 223–240.

Weber, C. I., 1973. Recent developments in the measurement of the response of phytoplankton and periphyton to changes in their environment. In G. E. Glass (ed), Bioassay Techniques and Environmental Chemistry. Ann Arbor Science, Michigan: 119–138.

Winner, R. W., M. W. Boesez & M. P. Farrell, 1980. Insect community structure as an index of heavy-metal pollution in lotic systems. Can. J. Fish. Aquat. Sci. 37: 647–655.

Wright, J. F., D. Moss, P. D. Armitage & M. T. Furse, 1984. A preliminary classification of running-water sites in Great Britain based on macro-invertebrate species and the prediction of community type using environmental data. Freshwat. Biol. 14: 221–256.

Yun-Fen, S., A. L. Buikema, W. H. Yongue, J. R. Pratt & J. Cairns, 1986. Use of protozoan communities to predict environmental effects of pollutants. J. Protozool. 33: 146–151.

Yasuno, M., V. Sugaya & T. Iwakuma, 1985. Effects of insecticides on the benthic communities in a model stream. Envir. Pollut. 38: 31–43.

Zimmerman, M. C. & W. F. Skinner, 1985. Application of the Hilsenhoff biotic index to determine the quality of a stream receiving fly ash effluent. Proc. PA. Acad. Sci. 59: 87.

Zischke, J. A. & J. W. Arthur, 1987. Effects of elevated ammonia levels on the fingernail clam, *Musculium transversum*, in outdoor experimental stream. Arch. Envir. Contam. Toxicol. 16: 225–231.

Hydrobiologia **188/189**: 77–86, 1989.
M. Munawar, G. Dixon, C. I. Mayfield, T. Reynoldson and M. H. Sadar (eds)
Environmental Bioassay Techniques and their Application.
© 1989 *Kluwer Academic Publishers. Printed in Belgium.*

The application of bioassay techniques to water pollution problems – The United Kingdom experience

D. W. Mackay, P. J. Holmes & C. J. Redshaw
Clyde River Purification Board, Rivers House, Murray Road, East Kilbride, Glasgow, Scotland, United Kingdom

Key words: bioassay, water, pollution, effluent, consent, toxicity

Abstract

The quality of the aquatic environment has long been assessed by chemical analyses and by biological surveillance of plant and animal communities. More recently, the biological response of living organisms has been used to evaluate the environmental impact of aqueous wastes.

Laboratory tests on single species have been used widely to evaluate the acute effects of potential pollutants. However, the value of such tests, often conducted on exotic species, is receiving increasing criticism. Measurements of more subtle chronic sub-lethal effects are now showing increasing promise as regulatory tools in environmental assessment and pollution control. The paper reviews the techniques being used, and those currently under development for, the water pollution control authorities in the UK. Practical examples of applications are provided and the future value of bioassays is discussed.

Introduction

The responsibility for controlling land-based discharges to fresh and saline waters in the United Kingdom is vested in catchment-based pollution control authorities. In England and Wales there are presently ten Water Authorities responsible for all aspects of river basin management, though major changes to their structure are planned. In Scotland there are seven River Purification Boards, which deal solely with water pollution control, and three Islands Councils which, like the Water Authorities, are multi-functional. All these authorities operate under the umbrella of the 1974 Control of Pollution Act (HM, Government, 1974). Control is exercised by the issuing of consents or licenses to discharge, the terms of which may include such reasonable conditions as the appropriate authority decides. Although a bioassay condition could be implemented in many consents this has, in practice, seldom been done. This situation is in contrast to the more prominent use of bioassays in national legislation (Lloyd, 1984) and international conventions (Haward, 1984) covering effluents discharged or dumped at sea from ships, oil platforms, aircraft or undersea pipelines.

This paper explores the role of bioassays in pollution control as administered by the United Kingdom water industry.

Historical Perspective

The quality of rivers and streams in the UK has for many years been assessed in terms of the community of macrobenthic fauna. The underlying principle that some members of the macroinvertebrate community are more sensitive to organic pollution than others, and will be progressively eliminated as levels of pollution rise, is well understood. A number of biotic indices have been produced and widely applied, all related to the range of species present and their relative abundance. More recently, the Freshwater Biological Association has refined the studies to the point of predicting what community of benthic macrofauna should be present in a stream of known geographical, physical, and hydrological characteristics. (Moss *et al.*, 1987). Similar techniques have been applied with more limited, but considerable, success to the estuarine and marine environments especially in relation to organically polluted areas such as sewage sludge dumping (Pearson *et al.*, 1986).

The complexity of biological monitoring, however, ensures that it is difficult, if not impossible, to include a biological standard, based on ecological community structure, in any consent. Even a simple 'catch all' phrase, as used in all Clyde River Purification Board (CRPB) consents (viz. 'shall not contain any substances in sufficient concentration either separately or in combination to be harmful to the flora and/or fauna') would be difficult to defend in a Court of Law; there are just too many variables.

In order to provide the basis for a more objective assessment of the biological significance of pollutants, the early Trent River Board pioneered some static lethal fish toxicity tests. A standardised static 48 hour LC50 test (lethal concentration required to kill 50% of test animals during, in this case, 48 hours) using the rainbow trout *Salmo gairdneri* Richardson was put forward by the renamed Trent River Authority to be considered, alongside an alternative procedure suggested by the Ministry of Agriculture, Fisheries and Food (MAFF), by the Ministry of Housing and Local Government (MHLG) (HMSO 1969). MAFF's test was a more sophisticated 48 hour LC50 flow-through test using the exotic harlequin fish *Rasbora heteromorpha* Dunker. Ultimately the more straightforward but less sensitive Trent test was adopted as a standard to be carried out with diluent water at one of two defined levels of hardness. Recommendations included the suggestion that a toxicity test should be included in consent conditions where suitable.

Further Government-recommended standard toxicity tests were put forward by the Department of the Environment in the 1980s (HMSO, 1983) after consultation with the Standing Committee of Analysts. Protocols were laid down for various acute, largely lethal, toxicity tests using rainbow trout, brown shrimp, *Crangon crangon* (Linnaeus), and the crustacean *Daphnia magna* (Straus). The *D. magna* test became enshrined as the first British standard (British Standard, 1983) when it was adopted by the Environment and Pollution Standards Committee. Morris & Buckley (1984) further reinforced recommendations for the use of *D. magna* tests in consent conditions after testing the potential of the organism on a discharge from a waste disposal landfill site for the Yorkshire Water Authority.

Despite recommendations to the contrary, toxicity tests have been little used in consents, rather their use has been to monitor suspect discharges or to attempt to predict the likely impact of new discharges, especially where 'novel' compounds or mixtures of compounds are present. Bioassay work is almost entirely confined to lethal tests and investigations involving more subtle and revealing sub-lethal tests have lagged behind.

The 1970s saw an upsurge in research orientated sub-lethal tests reflected in the publication of many novel bioassays such as reported in Cox (1974) and Cole (1979). Lloyd (1984), discussing ecotoxicological testing in Great Britain, suggests that environmental legislation covering freshwater is slanted towards the well-being of fish with the presumption that some modification of the invertebrate fauna is acceptable. In contrast, a greater range of biological responses has been examined for the marine environment, thereby providing increased scope for the implementation of bioassays in the statutory regulations controlling the various routes by which pollutants enter the sea.

The present decade has seen little change in the use of bioassays for statutory monitoring and control, rather there has been a consolidation of various tests introduced in the 1970s (Pearce, 1984). Fresh ideas continue to flow from those involved in research and there appears to be a feeling among the pollution control bodies that a review of practical current techniques would be timely.

Current Practice

In recent years the Water Research centre (WRc) has devoted considerable effort to the development of early warning systems to protect, in particular, the quality of drinking water at the 150 or so sites where it is abstracted directly from rivers (Pearce, 1984). The WRc has developed several automated fish monitors based on subtle changes in physiology. The Mark III Fish Monitor (Evans *et al.*, 1986) uses changes in the gill ventilation frequency of rainbow trout and is capable of raising an alarm within 15 minutes of the arrival of certain pollutants. However, although fish monitors have been installed at a number of sensitive sites throughout England and Wales, operators frequently complain of 'false alarms'.

Yorkshire Water Authority has recently been evaluating the efficacy of microbiological toxicity testing for assessing the quality of effluents discharging to sewage treatment works. It has developed a tiered system of effluent hazard assessment for evaluating the performance of sewage treatment works (R. Robinson, pers. commun.)

Severn Trent Water Authority is using the Ames test and other similar bioassays for screening raw and potable waters for mutagenicity (Tye, 1986). It is particularly interested in the impact of advanced water treatment processes (e.g. ozonation, filtration through activated carbon) on the mutagenicity of potable waters (J. Leahy, pers. commun.).

There are 3 discharges in the UK which qualify for registration under the European Community Directive on titanium dioxide (Council of the European Communities, 1978), viz two discharges to the Humber Estuary (controlled by the Anglian Water Authority) and one discharge to the Tees Estuary (controlled by the Northumbrian

Water Authority). The Directive specifies that acute toxicity tests shall be conducted on the brine shrimp *Artemia salina* Linnaeus and on species found commonly in the discharge area. It further specifies that over a period of 36 hours, and at an effluent dilution of 1/5000, these tests must not reveal more than 20% mortality for adult forms of the species tested and, for larval forms, mortality exceeding that of the control group.

A brief review of all UK water pollution control authorities revealed that only two out of ten water authorities, and two out of seven Scottish river purification boards, have ever used a toxicity test as part of the consent conditions. Some authorities have never used even a simple toxicity test for examining effluent toxicity.

Case studies from the Clyde River Purification Board

The Clyde River Purification Board (CRPB) is the water pollution control authority for a substantial area of Scotland with a population of 2.5 millions and a wide range of industrial discharges. The CRPB has used bioassays in a number of different situations but, before illustrating some of these, it is necessary to recap on the UK approach to pollution control based on the 'consent-to-discharge' system.

Historically, if a company wished to discharge an effluent to the aquatic environment, it would provide a simple list of the constituents contained therein – substances may include, for example, suspended solids, ammonia, certain heavy metals and so on. The controlling authority would assess the effects of this group of substances on the receiving environment and set consent conditions to provide the required degree of protection. There are two important assumptions in this procedure, namely:

(1) It is assumed that we know and understand the toxicity of each of the listed substances in the context of the environment into which they are being released.
(2) We assume that the listed substances include

all important constituents of the effluent under consideration. This is becoming rapidly more unlikely as discharges become increasingly complex. It is therefore very important that a rational, enforceable, bioassay should be included in the consent to discharge and CRPB has tried to achieve this.

Laboratory bioassays conducted by the Board are generally of short duration (one to four days), with death as the endpoint, and the rate of renewal of the test solution is dependent upon the stability of the substance under study. Acute toxicity tests have been conducted on a range of freshwater and marine species, using standard techniques wherever possible (e.g. HMSO, 1983), for a number of reasons, including:

(1) To screen effluents for potential toxicity – the first step in the hazard assessment of effluents is to identify potential problems and establish priorities for further study – the 96h LC50 of the effluent to 'ecologically-relevant' species is determined.
(2) To test compliance – in connection with a discharge consent, the lethality of an effluent is tested for compliance with a pre-determined toxicity consent condition which usually takes the form of the MHLG Test (HMSO, 1969).
(3) To screen receiving waters for potential toxicity – a test is conducted, often in conjunction with a field bioassay, to establish the toxicity of a water body suspected of being polluted.
(4) To evaluate the toxicity of a chemical – where insufficient data exist, tests may be conducted to establish the acute toxicity (e.g. 96h LC50) of a 'candidate' substance likely to be discharged to controlled waters within the Board's area – such information may be used in setting a receiving water standard for that substance.

The Board has issued a toxicity condition in the consent for certain complex discharges, particularly where effluent toxicity cannot be readily explained in terms of consented physical and chemical determinands. Toxicity consents have been issued for discharges to both freshwaters and tidal waters.

The toxicity consent condition specified for discharges to freshwater is based on the MHLG test (HMSO, 1969) i.e. a static test on rainbow trout, in which no more than 50% of the test fish shall be allowed to die following 48h exposure to effluent diluted by x volumes. A similar consent condition is specified for discharges to tidal waters, except that marine species, such as brown shrimp, turbot *Scophthalmus maximus* (Linnaeus) and plaice *Pleuronectes platessa* Linnaeus are exposed to pre-diluted effluent for 96h in a semi-static test. As a regulatory authority it is essential that these tests are relatively simple, robust and unambiguous so that they stand up in a Court of Law.

Control of complex effluents discharging to sea

(a) Derivation of a toxicity test consent condition
Whereas full details of this derivation are presented elsewhere (Haig *et al.*, 1989), relevant aspects are summarised below. In 1972, a pharmaceuticals factory was set up within the Board's area to manufacture the antibiotic, penicillin. The resulting complex effluent is discharged to Irvine Bay via a long sea pipeline.

The discharge was consented in terms of various physical and chemical parameters, including solvents and metals, known to be present in the effluent, limitations being set on both concentrations and monthly loadings (Table 1). Certain consent conditions were derived, *inter alia*, from laboratory tests of acute toxicity performed on mussels *Mytilus edulis* Linnaeus, brown shrimps and juvenile turbot, using effluent spiked with known constituent toxicants. Toxicity, determined for several effluent samples, agreed well with that predicted from chemical analyses, and there was judged to be no risk of toxic effects following initial dilution of the effluent upon discharge. Thus, effluent quality control was maintained by frequent analyses of the numerous physical and chemical parameters

Table 1. Summary of the main conditions governing a pharmaceutical company's consent to discharge to Irvine Bay.

1. The pH value shall not be less than 4 or greater than 12.

2. *The following constituents shall not exceed the maximum levels stated*

5 day biochemical oxygen demand (20 °C)	7000 mg l^{-1}	Suspended solids	5000 mg l^{-1}
Methyl isobutyl ketone	800 mg l^{-1}	Tetrahydrofuran	20 mg l^{-1}
Isopropyl alcohol	450 mg l^{-1}	Pyridine	5 mg l^{-1}
Acetone	300 mg l^{-1}	Zinc	2 mg l^{-1}
Methanol	500 mg l^{-1}	Copper	1.5 mg l^{-1}
Methylene dichloride	100 mg l^{-1}	Chromium	2 mg l^{-1}
Phenol	6 mg l^{-1}	Ammoniacal N	200 mg l^{-1}
Ethanol	300 mg l^{-1}	Toluene	60 mg l^{-1}
Ethylene dichloride	100 mg l^{-1}	Butanol	100 mg l^{-1}
Triethylamine	5 mg l^{-1}	Tertiary butyl amine	40 mg l^{-1}
Carbon tetrachloride	15 mg l^{-1}	Dimethyl formamide	20 mg l^{-1}

3. The maximum total input of individual constituents to the Firth of Clyde (Irvine Bay) in any period of 30 days shall not exceed:
   Suspended solids   350 tonnes
   Biochemical oxygen demand 600 tonnes

4. The volume of the discharge in any one day shall not exceed 6000 m^3 and the rate of discharge shall not exceed 400 m^3/h.

5. The effluent shall be conclusively deemed to comply with the terms of the toxicity Consent Condition when a sample thereof taken at the sampling point and diluted 125 times with sea water and tested according to the procedure set out in the document, headed 'Toxicity Test for Effluent Discharges to Saline Waters' attached to this consent, exhibits a cumulative percentage mortality as hereinafter defined of not greater than 50 per cent.

listed in the consent with verification of these standards by periodic tests of acute toxicity.

However, the Board recently detected a marked inexplicable increase in the toxicity of the effluent although, so far as was known, the chemical composition had not changed and the terms of the consent were being met. Faced with an elusive toxicant, the Board had to develop a means of achieving effective and lasting control over the effluent which, already variable in composition, was likely to increase in complexity as the factory continued to develop. Following lengthy negotiations with the Company, the Board derived a consent condition for effluent toxicity and implemented a test procedure in the consent as outlined below:

From engineering and hydrographic data the minimum initial design dilution of the present outfall is estimated at 250 × and this has subsequently been verified by dye release studies. In defining the 'mixing zone', in which the receiving water standard may be exceeded (Water Authorities Association, 1986), the Board allowed for secondary dilution of 5 × (achieved within 100 m of the outfall), and thereby a total dilution of 1250 ×.

For regulatory purposes, effluent toxicity is conventionally monitored in terms of acute effects such as lethality and immobilization. Since the toxicity of this effluent was known to be relatively non-persistent a standard of one-tenth of the mean 96h LC50 value (for ecologically relevant species) should protect the quality of the receiving water. Thus, for consent purposes, provided no more than 50% of the test animals are killed after 96h exposure to effluent at 125 × dilution, the effluent should not exert adverse effects on water quality outwith the mixing zone.

The toxicity test specified in the consent is a 'hybrid' of the procedure recommended by the Ministry of Housing and Local Government (HMSO, 1969), and that currently employed by MAFF for testing industrial wastes (Franklin, 1980), in that the consent requires that no more than 50% of the test animals shall die following 96h exposure to effluent at a pre-determined dilution.

The CRPB specifies that any one or all of four species must be used, including queen scallop *Chlamys opercularis* (Linnaeus), turbot, plaice and brown shrimp, thereby reflecting a range of relevant taxa present in Irvine Bay.

This toxicity test consent condition has subsequently been verified by field bioassays using caged shellfish deployed around the outfall [see Section (b)]. Subsequent investigations at the site identified the unknown toxicant as a detergent. This substance was previously cleared for use on the basis of its negligible toxicity to shrimp (although it was later found to be of high toxicity to fish) and, as such, had not been included in the discharge consent. Following substitution of the detergent with a less toxic formulation, the effluent now generally complies with the toxicity consent condition. This case clearly demonstrates the value of toxicity testing for complex effluent control and such testing will remain an integral part of the discharge consent.

(b) Biological monitoring of water quality using caged shellfish

The biological response of caged shellfish, deployed in surface waters, has been used to evaluate the impact on water quality of several coastal discharges. Typically, mussels and Pacific oysters *Crassostrea gigas* (Thunberg) have been deployed for periods of up to 5 months in specially designed cages (Curran *et al.*, 1986) located in near-surface waters about 100 m from the diffuser sections of the study outfalls and at a clean water 'control' site nearby. Parameters monitored include survival, growth (measured as increase in length), scope-for-growth (Bayne, 1989), tainting and the bioaccumulation of selected pollutants. Scope-for-growth was measured by the Water Research

centre under contract from the Government's Department of the Environment. Such studies have proved to be particularly useful in validating in the field the predictions of impact based on chemical analyses and laboratory tests on the toxicity of the effluent.

(1) Chemicals Plant

The Board first used caged shellfish to monitor the impact of a complex industrial discharge on water quality in Irvine Bay. The bioassay was used to verify, under field conditions, the consent conditions for this copper-rich discharge (Mackay *et al.*, 1986). Impact was evaluated in terms of survival of queen scallops, and by growth and bioaccumulation of copper in mussels. Shellfish were deployed in cages located at a number of stations moored close to the outfall and at a 'control' site some 8 km distant.

Scallops proved unsatisfactory as a test species, being unable to withstand physical buffeting by wave action. In contrast, the mussels survived and grew well, both around the outfall and at the 'control' site, and there was no evidence of copper bioaccumulation. Similar results were obtained for bioassays conducted over three consecutive years, suggesting that the outfall from this chemical manufacturing plant had little effect on receiving water quality.

(2) Other discharges in Irvine Bay

The bioassay was expanded in subsequent years to evaluate the impact on water quality of two adjacent discharges, one deriving from a pharmaceuticals company and the other from a major sewage outfall. Parameters studied were extended to include scope-for-growth in mussels, and tainting and bioaccumulation (of trace metals and organics) in oysters. The results of bioassays conducted over two consecutive years may be summarized as follows:

(i) shellfish survived at all sites
(ii) mussels grew well, although growth rate was depressed at certain stations located

around the pharmaceuticals and sewage outfalls

(iii) relative to the control site, scope for growth was reduced in mussels deployed around the sewage and pharmaceuticals outfalls, but not in mussels deployed around the Chemicals Plant outfall

(iv) oysters deployed around the 3 outfalls were tainted (rubbery and painty flavour) and were considered unfit for human consumption; in contrast, control site oysters were of satisfactory quality

(v) both copper and the moth-proofing agent, permethrin, are discharged via the sewer, and both contaminants were bioaccumulated in oysters deployed around the sewage outfall. (Cu levels of $155-320 \mu g \, g^{-1}$ dry wt. and permethrin levels of $200-270 \, ng \, g^{-1}$ wet wt. were detected). There was no evidence of significant bioaccumulation at other sites.

In view of the adverse effects detected in caged shellfish, and the high toxicity of the effluent, the pharmaceuticals factory outfall diffusers were subsequently modified to increase the initial dilution of the effluent upon discharge. The impact of the discharge was then re-examined and, in contrast to the results of previous studies, there was no evidence of a depressed growth rate in mussels deployed around the outfall. However, growth rate was still reduced in mussels deployed around the adjacent sewage outfall. Thus, the modifications to the pharmaceuticals outfall appear to have reduced the impact of the discharge.

Control of explosives factory effluents

The River Gryfe is 24 km long and rises in high ground in Renfrewshire, west of Glasgow. It is a clean river and the headwaters serve as an important spawning ground for sea trout *Salmo trutta* Linnaeus and salmon *Salmo salar* Linnaeus.

Close to its confluence with the Black Cart Water, the River Gryfe is joined by the Dargavel Burn. This burn is badly polluted by a variety of wastes generated in the nitration of glycerine, cellulose and guanadine at an explosives factory in Bishopton – the stretch of the burn downstream of the factory is devoid of aquatic life.

For many years the factory was operated by the Ministry of Defence and, as such, had Crown Exemption from pollution control legislation. Although the Board was allowed access to the site to sample the various discharges, it had no powers to control the quality of effluents. Consequently, the effluents and the receiving stream (Dargavel Burn) were of extremely poor quality. However, the factory has been privatized recently and, as such, the Board now has legal control of pollution emanating from the site. The Company is expending great effort to improve the quality of the discharges and various treatment plants are under construction.

The Board has recently set discharge consents for the seven outfalls at the site. A fish toxicity test is included in the consent conditions for six of these discharges in view of their complex chemical composition – pollutants include ammonia, lead, nitroglycerine, nitroguanadine and its derivatives. The toxicity test procedure and compliance criterion are based on the Ministry of Housing and Local Government Test (HMSO, 1969). The recommended test species is rainbow trout (*Salmo gairdneri*) and the recommended exposure period to effluent is 48 h. The dilution of effluent specified in each consent is based on a minimum flow of the receiving water which is exceeded 95% of the time (Q95) and the consent maximum flow for that effluent.

The quality of the receiving water body (Dargavel Burn) is too poor to support a fishery at present. However, water quality is expected to improve markedly following construction of the effluent treatment facilities and, once these processes are operating, caged rainbow trout will be used to monitor the quality of selected discharges and the receiving water course.

84

Future trends in the development and application of bioassays

It is evident that bioassays provide a rapid, sensitive and cost-effective tool for the monitoring and control of effluent discharges and receiving waters. The majority of present bioassays are concerned with the determination of survival as related to effluent or single aqueous toxicant concentrations – such tests have proven to be particularly useful for regulatory authorities in that they provide a rapid, inexpensive and unequivocal response. There is clearly a need to promote the use of bioassays by water pollution control authorities.

Some degree of standardisation in methods is desirable where the tests are to be used as part of statutory pollution control either nationally or internationally. The concept of standard methods of bioassay has attracted attention ever since the first test was put forward in 1969 (HMSO, 1969) and forms the basis for many international arguments. Standard methods for marine bioassays are especially attractive where a common sea is surrounded by several states such as the Mediterranean, North Sea or Persian Gulf. Foster (1984), reviewing United States environmental legislation, describes the operation of the Organisation for Economic Cooperation and Development (OECD) of which the UK is one of 24 member nations. In 1977 OECD identified the need for consistent data requirements and encouraged the harmonisation of testing guidelines. In 1981 it published a minimum Premarket Data Set which was a base set of ecotoxicological tests both lethal and sub-lethal.

International harmonisation of toxicity tests on salt water fish was proposed by Goldstein *et al.*, (1984) to meet the growing needs of joint international programmes. The authors listed the recommendations of various august bodies such as FAO (UN), OECD and United States EPA who had considered various species. The ideal fish should be convenient to maintain in the laboratory by way of size and requirements for oxygen and temperature. Availability in many countries and throughout the year were further plus points with stock animals sourced from fish farms or laboratories to ensure consistency. Having scanned the available literature, the data were computer-analysed to consider a range of 261 tests using 80 species of fish. Some 13 species conformed to some of the ideal requirements but, illustrating the difficulties of producing ideal tests, none fulfilled them all.

Most common species were:

Pleuronectes platessa (Linnaeus) – Plaice
Platichthys flesus (Linnaeus) – Flounder
Oncorhynchus kisutch (Walbaum) – Salmon (coho)
Fundulus heteroclitis (Linnaeus) – Mummichog
Cyprinodon variegatus (Lacepéde) – Sheepshead minnow

Tebo Jr. (1986) emphasises the need for common test species when listing his views of pressing needs for complex effluent monitoring as part of the licensing process in the United States. The author pleads for the use of a limited number of aquatic 'white rats' for use in developing a broad effluent data base.

Standardisation allows different bodies or Nations to 'speak the same language' in addition to providing a standard protocol for those needing to develop a new facility, to follow. However, no standard library of tests, no matter how comprehensive, would be of use without additional local tests based on particular relevant species. Any standardisation must be regularly checked by careful intercalibration to ensure reproducible and comparable data, only sound data will stand up in a Court of Law.

Considering the varied needs of statutory pollution control authorities, an ideal bioassay should be:

1. Reliable and reproducible
2. Economical of time and resources
3. Able to yield statistically robust data
4. Relevant, practicable and readily understood by the layman
5. Able to utilise test organisms available all the year round from reliable stock
6. Simple to emulate
7. Regularly intercalibrated

8. With a clearly defined endpoint
9. Sensitive to a wide range of pollutants.

Conclusions and future trends

Although there has been a general reluctance in the United Kingdom water industry to fully utilise bioassay techniques, a few authorities have found such tests to be a useful addition to their control and monitoring procedures. In addition, other statutory control bodies and research interests have continued to develop novel methods to address particular problems.

Looking towards the special needs of the water industry four areas of interest appear to be of special promise:

(1) Increased use of sub-lethal bioassays and ambient toxicity tests.

Increased interest could be usefully directed at developing tests that reflect toxic effects on reproductive capacity, embryogenesis and early development (Mount 1984). Sensitive laboratory and field bioassays should be developed to evaluate the toxicity of receiving water samples. Such tests could be used to verify the predictions of impact based on effluent toxicity tests and identify segments of water bodies exerting toxicity. Recently, several initiatives have been developed to use the responses of sensitive organisms to bioassay the quality of marine waters and associated discharges. Within the UK responses include lysosomal stability and scope-for-growth in mussels (Bayne, 1989), growth gonozooid production and stolon curving in hydroids (Stebbing & Brown, 1984) and embryonic development in mussels and oysters (Johnson, 1988). *In situ* monitoring of mussel scope-for-growth, combined with the use of sensitive laboratory bioassays to identify the cause of suspect pollution, appears to be a promising approach to future marine pollution monitoring.

(2) Increased use of sediment toxicity tests.

Pollutants may accumulate to toxic levels in sediments even though levels in the water column are non-toxic. The quantitative assessment of the benthos is both expensive and time consuming and, as such, benthic surveys are probably not the best means of detecting impact. Sediment toxicity tests may allow a better assessment of the extent of effluent impact on the benthos and help answer questions regarding the persistence of toxicants in complex mixtures.

(3) The development of ecotoxicological methods for assessing the impact of episodic pollution. Fish monitors, such as those developed at WRc, could be used to assess the impact of storm events and episodic pollution in rivers. Caged fish deployed in rivers may be used to detect short-term highly toxic discharges, long-term low levels of lethal toxicity and the possible bioaccumulation of pollutants by fish.

(4) Increased use of multi-species toxicity tests. A multi-species toxicity test examines the effects of toxicants at a level of biological organisation higher than a single-species. Parameters studied in such tests should primarily be those that cannot be studied in single-species tests, e.g. predation, competitive interactions, nutrient recycling. Single-species toxicity tests should continue to be used for regulatory purposes, but the use of multi-species tests to examine the influence of contaminants on ecological interactions (Cairns, 1985) should be expanded.

Toxicity tests can never fully replace chemical analyses or biological surveys, but used wisely can increase the cost efficiency of other monitoring tools. There is in existence a vast plethora of tests, perhaps it is time to concentrate on validating these rather than chasing new rainbows!

Acknowledgements

The authors wish to express their thanks to Mr. D. Hammerton, Director of the Clyde River Purification Board for permission to publish this paper.

86

References

Bayne, B. L., 1989. Measuring the biological effects of pollution: the Mussel Watch approach. Wat. Sci. Tech. 21: 1089–1100.

British Standard, 1983. B.S. 6068, Part 5, Section 5.1 British Standards Institution. London. 12 pp.

Cairns, J., Jr (Ed.), 1985. Multispecies toxicity testing. Pergamon Press, New York, 253 pp.

Cole, H. A. (Organiser), 1979. The assessment of sub-lethal effects of pollutants in the sea. A Royal Society discussion. The Royal Society, London. 235 pp.

Council of the European Communities, 1978. Directive on waste from the titanium dioxide industry. 78/176/EEC; OJ L 54, 20 February 1978. Brussels. 6 pp.

Cox, G. (Convener), 1974. Proceedings of a workshop on marine bioassays. Marine Technology Society, Washington D.C. 308 pp.

Curran, J. C., P. J. Holmes & J. E. Yersin, 1986. Moored shellfish cages for pollution monitoring; design and operational experience. Mar. Poll. Bull. 17: 464–465.

Evans, G. P., D. Johnson & C. Withell, 1986. Developments of the WRc MK III Fish Monitor: description of the system and its response to some commonly encountered pollutants. Technical Report TR 233, Water Research Centre, Medmenham, UK. 52 pp.

Foster, R. B., 1984. Environmental legislation. In G. M. Rand & S. R. Petrocelli (Eds.), Fundamentals of aquatic toxicology. Methods and applications. McGraw Hill International Book Company, Washington: 587–600.

Franklin, F. L., 1980. Assessing the toxicity of industrial wastes, with particular reference to variations in sensitivity of test animals. Fish. Res. Tech. Rep., MAFF Direct. Fish. Res., Lowestoft, 61, 10 pp.

Goldstein, E., R. Amavis, R. Cabridene, C. Gilliard & R. Schubert, 1984. Development of an international harmonisation scheme for salt water fish toxicity tests. In G. Persoone, E. Jaspers & C. Claus (Eds.), Ecotoxicological testing for the marine environment 1984. 1. State University of Ghent and Inst. Scient. Res., Bredene, Belgium: 689–732.

H. M. Government, 1974. Control of Pollution Act 1974, Chapter 40 Part 2. Her Majesty's Stationery Office, London: 43–78.

HMSO, 1969. Ministry of Housing and Local Government. Fish Toxicity Tests. Report of the Technical Committee. HMSO, London, 14 pp.

HMSO, 1983. Acute toxicity testing with aquatic organisms 1981. HMSO, London, 67 pp.

Haig, A. J. N., J. C. Curran, C. J. Redshaw & R. Kerr, 1989. Use of mixing zone to derive a toxicity test consent condition. J. Inst. Water Environ. Mgmt. (in press).

Hayward, P. A., 1984. Marine ecotoxicological testing in the framework of international conventions. In G. Persoone, E. Jaspers & C. Claus (Eds.), Ecotoxicological testing for the marine environment 1984. 1. State University of Ghent and Inst. Scient. Res., Bredene, Belgium: 15–37.

Johnson, D., 1988. Development of *Mytilus edulis* embryos: a bioassay for polluted waters. Mar. Ecol. Prog. Ser. 46: 135–138.

Lloyd, R., 1984. Marine ecotoxicological testing in Great Britain. In G. Persoone, E. Jaspers & C. Claus (Eds.), Ecotoxicological testing for the marine environment 1984. 1. State University of Ghent and Inst. Scient. Res., Bredene, Belgium: 39–55.

Mackay, D. W., A. J. N. Haig & R. Allcock, 1986. Licencing a major industrial discharge to coastal waters; the practical application of the EQO/EQS approach. Wat. Sci. Tech. 18: 287–295.

Morris, G. M. & F. T. Buckley, 1984. The role of the *Daphnia* bioassay in the assessment of the quality of effluent discharges. Wat. Pollut. Contr. 83: 539–546.

Moss, D., M. T. Furse, J. F. Wright & P. D. Armitage, 1987. The prediction of the macro-invertebrate fauna of unpolluted running-water sites in Great Britain using environmental data. Freshwat. Biol. 17: 41–52.

Mount, D. I., 1984. The role of biological assessment in effluent control. In Biological testing of effluents and receiving waters. Proceedings of an international workshop, Duluth, Minnesota, USA, September 1984. pp. 15–30.

Pearce, A. S., 1984. Biological testing of effluents and associated receiving waters – UK experience. In Biological testing of effluents and receiving waters. Proceedings of an international workshop, Duluth, Minnesota, USA, September 1984. 297–321.

Pearson, T. H., A. D. Ansell & L. Robb, 1986. The benthos of the deeper sediments of the Firth of Clyde, with particular reference to organic enrichment. Proc. Roy. Soc. Edin., 908: 329–350.

Stebbing, A. R. D. & B. E. Brown, 1984. Marine ecotoxicological tests with coelenterates. in G. Persoone, E. Jaspers & C. Claus, (Eds.), Ecotoxicological testing for the marine environment 1984. 1. State University of Ghent and Inst. Scient. Res., Bredene, Belgium: 307–335.

Tebo, Jr, L. B., 1986. Effluent monitoring: historical perspective. In H. C. Bergman, R. A. Kimerle & A. W. Maki (Eds.). Proceedings of a Pellston environmental workshop, Valley Ranch, Cody, Wyoming, August 1982. SETAC special publication series, Pergamon Press, New York: 13–31.

Tye, R. J., 1986. Mutagens in water sources: detection and risk assessment. J. Inst. Wat. Eng. Sci. 40: 541–548.

Water Authorities Association 1986. Mixing zones: guidelines for definition and monitoring. Water Authorities Association, London, 11 pp.

Hydrobiologia **188/189**: 87–91, 1989.
M. Munawar, G. Dixon, C. I. Mayfield, T. Reynoldson and M. H. Sadar (eds)
Environmental Bioassay Techniques and their Application.
© 1989 *Kluwer Academic Publishers. Printed in Belgium.*

The use of environmental assays for impact assessment

Donald C. Malins
Pacific Northwest Research Foundation, 720 Broadway, Seattle, Washington, 98122 U.S.A.

Key words: Sediment chemistry; bioassays; toxicity; risk assessment

Abstract

The assessment of impacts of chemically contaminated aquatic environments on animal systems has a number of shortcomings. These include problems with analyses for toxic chemicals and the relevance of bioassays for predicting risk to ecosystems. Research is urgently needed to find better ways to solve these problems, particularly with respect to chronic exposures.

The assessment of ecological impacts in chemically contaminated marine environments is vitally necessary for the evaluation of risk to ecosystems and the health of the human consumer of fish and shellfish. A major consideration in the assessment of risk is the availability of valid measurements of toxicity. Yet, it is recognized that toxicity, *per se*, cannot be divorced from an understanding of the types and concentrations of potentially toxic substances in sediments, water and the tissues of organisms. Neither can toxicity be adequately addressed in the absence of an understanding of bioavailability and metabolism (Buhler & Williams, 1988). The strengths and weaknesses of the approaches presently being used to assess the impact of toxic chemicals on organisms and ecosystems is the subject of this paper.

A major concern is 'how effective are procedures for diagnosing chemical contamination?' In my experience, we are often satisfied to simply apply a 'cook-book' approach in deciding upon the chemical analyses to be employed. In the United States, one such approach is to largely restrict our perspective to the 126 target chemicals on the U.S. Environmental Protection Agency's Priority Pollutant List. In doing this, we seem

oblivious to the fact that hundreds of other potentially toxic substances are often present in contaminated environments – compounds that through additive, antagonistic or synergistic interactions have the potential for markedly influencing toxicity. I will give a case in point: Everett Harbor, just north of Seattle in Washington State, took on a special significance recently. The U.S. Navy wanted to make it a home port for its vessels; however, many hurdles had to be overcome before the necessary permits could be issued. Among such hurdles was a requirement for the assessment of chemical contamination in the sediments from the inner harbor, coupled with an evaluation of sediment toxicity. The intention was to remove the sediments and deposit them in adjacent Port Gardner Bay which is a habitat for crabs, bottom fish and other economically important resource species. The chemicals selected for analysis were essentially the EPA's 126 pollutants. This selection was made, despite the fact that the area had been historically surrounded by pulp and paper industries that are known to produce a variety of chlorinated and other compounds, including many mutagens and carcinogens that are not on the EPA's list. Moreover, prior evidence indicated that the sediments in the area were substantially contaminated with chlorinated compounds, although little was known about the individual structures (Malins *et al.*, 1983). Ultimately, and late in the adjudication process, a consultant demonstrated that scores of chlorinated compounds were indeed present in the sediments from the inner harbor and that many were highly toxic (Draft Report, 1986). Mutagens and carcinogens were among the compounds identified. The evidence pertaining to the presence of the newly-identified toxic compounds was pivotal in the decision of a federal judge to rule that 'a permanent injunction must be issued' to enjoin the Navy from further activities in connection with the home port (Memorandum Decision, 1988).

My reason for citing this case is that it illustrates that care must be taken in assessing sediment contamination, as well as in taking into account historical uses of the environment in question. Clearly, the rote choice of target chemicals that, for one reason or another, catch our fancy can lead to misjudgments about environmental impacts.

We cannot hope to analyze for all the chemicals that are contained in sediments from polluted environments. Sometimes, in fact, compounds that are below normal detection thresholds can still be a problem for exposed organisms because they are extensively bioconcentrated. For example, sediment chemistry data from Eagle Harbor, Washington State, revealed that the pollution problem was caused by long-standing inputs of aromatic hydrocarbons and other compounds resulting from the use of creosote in the area (Malins *et al.*, 1985). The chemistry data showed that 'creosote hydrocarbons' predominated over other compounds. For example, chlorinated compounds were generally below the level of detection (Malins *et al.*, 1985), suggesting that chlorinated compounds are of little or no concern ecologically. However, studies of the brains of English sole from Eagle Harbor revealed complex profiles of chlorinated hydrocarbons (Malins, unpublished). Some of these compounds, such as PCBs and DDT derivatives, were present in con-

centrations of hundreds of parts per billion. In this regard, it is well known that chlorinated compounds tend to concentrate in neural tissues of exposed organisms. Great Lakes salmon, for example, have relatively high concentrations of DDT and Mirex in their brains (Hertzler, 1983). In addition, when contaminated salmon from the Great Lakes were fed to rats, these rodents also concentrated the chlorinated compounds in their brains (Hertzler, 1983). The brain concentrations correlated with alterations in grooming, rearing and other behaviors (Daly *et al.*, 1989; Hertzler, 1983). These and other studies support the conclusion that fish diets containing PCBs and other chlorinated compounds bring about deleterious alterations in animal systems (Fein *et al.*, 1984; Jacobson *et al.*, 1984; Reijnders *et al.*, 1986), although thresholds for such effects are poorly understood. For example, a dramatic reduction in the population of seals in the Wadden Sea has been attributed to the consumption of PCB-contaminated fish (Reijnders, 1986) and women who consumed PCB-contaminated fish from Lake Michigan had offspring with neuromuscular and other anomalies (Fien *et al.*, 1984).

Among the lessons to be learned from Eagle Harbor is that conclusions drawn from sediment data alone can lead to misconceptions of ecological and human risk. Clearly, the degree of uptake of potentially toxic chemicals – that is, the issue of bioavailability – is crucial in establishing risk, as the above illustrations suggest.

It is widely recognized that the nature of sediment contamination is pivotal in assessing ecological impact and nothing said so far was intended to detract from this well-known fact. However, sediment chemistry is often emphasized to the exclusion of the chemistry of the water when considering ecological impacts. This is an unfortunate omission because the water is a significant source of contamination for aquatic species. For example, fish readily concentrate toxic chemicals, such as hydrocarbons, from water (Krahn *et al.*, unpublished). When English sole were exposed to hydrocarbon-contaminated sea water in the absence of sediment, concentrations of aromatic compounds in the bile became substantially elevated over concentrations in the bile from control fish (Krahn *et al.*, unpublished). The surface microlayer, several hundred microns thick, is an additional source of contamination (Hardy *et al.*, 1987). Early life stages of a number of important vertebrate and invertebrate species spend part of their lives at the surface of the water where toxic chemicals concentrate in the microlayer.

Returning to the issue of the chemical analytical data, one has to admit that such data tells us very little, in specific terms, about toxicity – the particularly crucial issue in the assessment of risk. How have we attempted to overcome this problem?

Bioassays have been widely used as a means of assessing toxicity. Among the virtues of bioassays is that they, in effect, 'integrate' the impacts of all the chemicals with respect to the biological parameter being measured. However, despite their obvious importance in assessing toxicity, many of the bioassays routinely employed have little or no demonstrated relevance to impacts on ecosystems. This is a serious deficiency in that much of our concern in protecting marine environments rightfully focuses on injury at the level of the ecosystem.

Furthermore, most of the bioassays routinely employed are acute when much of the impact in contaminated marine environments is correctly perceived to be chronic. One might ask 'how relevant is the death of an amphipod or oyster larva to understanding the impact of contaminant stress on a complex ecosystem?' My answer is that the relevance has never been convincingly demonstrated through scientific studies.

Turning now to another approach for the assessment of toxicity, in studies of Puget Sound (Malins *et al.*, 1985; Malins *et al.*, 1987) we used pathologic and histopathologic techniques to provide insight into biological effects, such as liver cancer and 'precancerous' conditions of the liver. One of the conclusions drawn from this work was that histopathologic changes (and certainly pathologic changes) are almost exclusively associated with seriously impacted environments – that is, they occur mostly in fish from 'hot spots' and other areas having relatively high degrees of sediment contamination (Malins *et al.*, 1985; Malins *et al.*, 1987). Thus, it should be recognized that pathologic and histopathologic data obtained from field samples are often ineffective for predicting ecological damage, but rather reflect an 'eleventh hour' perspective. Besides, many target species migrate over large areas and sediment contamination may vary substantially, making it very difficult to interpret the data obtained, particularly in the absence of extensive sampling protocols.

There is a related issue of some importance – the use of cancer as an end-point in the assessment of biological effects. It seems that we have largely ignored the fact that many carcinogens require cell proliferation to produce cancer and that this process can be initiated by a wide variety of chemicals (Farber, 1980). That is, a host of chemicals, other than the carcinogen itself, often play obligatory roles in cancer formation. Thus, neglecting the role of 'the other chemicals' may be convenient, but it may cause us to severely miss the mark in evaluating ecological and human health risk. This is particularly so when dealing with environments that contain myriad individual, bioavailable chemicals.

An ability to predict the occurrence of significant biological change is a vital ingredient in the diagnosis of potential environmental impacts. In this regard, we have made little progress. There is clearly a pressing need to develop methods for revealing changes that precede events, such as rank tumor formation or other overt cellular alterations. Research is urgently required to solve this problem and I feel that one of our best hopes for success lies in the development of predictive tests focusing on alterations at the sub-cellular level. Recent developments in revealing modifications in DNA resulting from exposure of fish to aromatic hydrocarbons are a good example of what I mean (Varanasi *et al.*, 1989). Such methods should be readily applicable to the field using sessile organisms, such as molluscs, as indicator species. Clearly, however, a single 'probe' of a single system is insufficient. What is needed is a complementary array of probes that reflects changes in biochemical/physiological systems, thus providing a holistic understanding of altered health. At the organismal level, behavioral changes, to give one example, are promising as early warning signals (McDowell *et al.*, 1988). These types of events, as with those at the sub-cellular level, are likely to precede major alterations in the health of impacted organisms and are thus in the category of predictive indices.

Some attempts have been made to introduce greater meaning into the assessment of contamination and toxicity of marine environments. For example, 'sediment quality' has been assessed by interfacing chemistry and bioassay toxicity data with data on infaunal alterations (Chapman, 1986). The resulting indices are a step above earlier approaches employing single data sets. Yet, this 'Triad' approach still suffers from a number of shortcomings. The limitations mentioned with respect to the chemical analyses are still a problem; the relevance of the bioassays is still open to question; and from experience with Puget Sound (Malins *et al.*, 1987), infaunal alterations are almost always linked to severely impacted environments – that is, they are not a good predictive tool. Rather, they serve as indices of severe ecological damage. Unfortunately, the problems I am concerned about cannot be solved by simply attempting to establish various relationships between traditional ways of assessing environmental impacts – the imperfections in the techniques themselves persist and thus tend to preclude such a possibility.

In closing, one thought seems to overshadow all the others. In much of our attempt to understand and define marine pollution problems, we seem to have all but abandoned the scientific method! In the field, the approach most often taken is essentially methodological – the rote application of the same suites of tests, over and over again, regardless of the fact that many differences exist in both the chemical contamination and the nature of the organisms and ecosystems from area to area. Why are we not approaching these environmental problems from an experimental point of view, as we approach other problems, such as the cancer problem? We appear to have lost the ability to approach solutions through

concept formation and the validation of hypotheses! In many instances, we go into the field, collect data on the basis of some vague or ill-defined notion, and then try to sort out the findings after the fact. In doing this, we lose the opportunity to expand understanding and thus tend to repeat mistakes over and over again. We then proceed to the next problem little more informed than we were in the first place. I submit that solutions to aquatic pollution problems are by no means immune to the scientific method.

Our failures, as I see them, lie partly with the scientific community for standing too much on the sidelines, generally lacking incredulity and not demanding excellence. However, part of the blame, if that is the right word, also lies with the regulatory agencies who set the standards with which we all have to comply. The agencies should demand more relevant and meaningful assessments of marine environments – assessments that are truly credible for establishing ecological and human health risk. More specifically, we need increased effort directed toward understanding the underlying mechanisms and processes that govern the effects of pollution on marine ecosystems. We need this to provide a sounder basis for measuring and understanding toxicity and its implications for the health of organisms and the survival of ecosystems. Overall, we need to face up to our problems, recognize our deficiences and re-evaluate our whole approach to the assessment of risk. If this is not done with some urgency, progress will indeed be most difficult and protracted.

References

Buhler, D. R. & D. E. Williams, 1988. The role of bio-transformation in the toxicity of chemicals. Aquat. Toxicol. 11, 19–28.

Chapman, P. M., 1986. Sediment quality criteria from the sediment quality triad: An example, Envir. Toxicol. Chem., 5, 957–964.

Daly, H. B., D. R. Hertzler & D. M. Sargent. Ingestion of environmentally contaminated Lake Ontario salmon by laboratory rats increases the avoidance of unpredictable aversive non-reward and mild electric shock, Behavioral Neuroscience, 1989 (in press).

Draft report, Everett Harbor Action Program: Analysis of toxic problem areas, May 1988, Prepared for U.S. Environmental Protection Agency, Region X, Office of Puget Sound, Seattle, Washington (TC-3338-26), pp. 44–58.

Farber, E. & R. Cameron, 1980. The sequential analysis of cancer development. Adv. Cancer Res. 31: 125–226.

Fein, G. G., S. W. Jacobson, P. W. Schwartz & J. K. Fowler, 1984. Prenatal exposure to polychlorinated biphenyls: Effects on birth size and gestational age, Pediatr., 105: 213–220.

Hardy, J. T., E. A. Crecelius, E. Long, S. L. Kiesser, A. I. Stubin, J. M. Gurtisen & C. W. Apts, 1987. Contamination and toxicity of the sea-surface microlayer of Puget Sound, R. H. Gray, E. K. Chess, P. J. Mellinger, R. G. Riley and D. L. Springer (Eds), Health and Environmental Research on Complex Organic Mixtures, U.S. Department of Energy Symposium Series 62, pp. 643–655.

Hertzler, D. R., 1983. Behavioral evaluation of neurotoxic effects caused by the consumption of Lake Ontario salmon. Pap. pres. Eastern Psychological Assoc. Meeting, Philadelphia, PA.

Jacobson, J. L., S. W. Jacobson, P. W. Schwartz & J. K. Dowler, 1984. Prenatal exposure to an environmental toxin: A test of the multiple effects model., Dev. Psychol. 20: 523–532.

Malins, D. C., M. S. Krahn, M. S. Myers, M. L. Rhodes, D. W. Brown, C. A. Krone, B. B. Chain & S.-L. Chan, 1985. Toxic chemicals in sediments and biota from a creosote-polluted harbor; relationships with hepatic neoplasms and other hepatic lesions in English sole (*Paraphrys vetulus*), Carcinogenesis, 6: 1463–1469.

Malins, D. C., B. B. McCain, M. S. Myers, D. W. Brown & S.-L. Chan, 1983. Liver diseases in bottom fish from Everett Harbor, Washington, Coastal Ocean Assessment News, 2: 41–42.

Malins, D. C., B. B. McCain, M. S. Myers, D. W. Brown, M. M. Krahn, W. T. Roubal, M. H. Schiewe, J. T. Landahl & S.-L. Chan, 1987. Field and laboratory studies of the etiology of liver neoplasms in marine fish from Puget Sound, Envir. Health Persp. 71: 5–16.

McDowell Capuzzo, J., M. N. Moore & J. Widdows, 1988. Effects of toxic chemicals in the marine environment: Predications of impacts from laboratory studies, Aquatic Toxicol. 11: 303–311.

Memorandum Decision, No. C-88-380R, Barbara J. Rothstein, Chief United States District Judge, Seattle, Washington, August 11, 1988. pp. 1–87.

Reijnders, P. J. H., 1986. Reproductive failure in common seals feeding on fish from polluted coastal waters. Nature 324: 456–457.

Varanasi, U., W. L. Reichert & J. E. Stein, 1989. ^{32}P-post-labeling analysis of DNA adducts in liver of wild English sole (*Parophrys vetulus*) and winter flounder (*Pseudopleuronectes americanus*), Cancer Res. 49: 1171–1177.

Hydrobiologia **188/189**: 93–116, 1989.
M. Munawar, G. Dixon, C. I. Mayfield, T. Reynoldson and M. H. Sadar (eds)
Environmental Bioassay Techniques and their Application.
© *1989 Kluwer Academic Publishers. Printed in Belgium.*

Probing ecosystem health: a multi-disciplinary and multi-trophic assay strategy *

M. Munawar,[1] I. F. Munawar,[2] C. I. Mayfield[3] & L. H. McCarthy[1]
[1] *Fisheries & Oceans Canada, Great Lakes Laboratory for Fisheries and Aquatic Sciences, Canada Centre for Inland Waters, Burlington, Ontario, Canada L7R 4A6*; [2] *Plankton Canada, 685 Inverary Road, Burlington, Ontario, Canada L7L 2L8*; [3] *Biology Department, University of Waterloo, Waterloo, Ontario, Canada N2L 3G1*

Key words: ecosystem, health, multi-trophic, battery, structural, functional, bioassays

Abstract

The ecosystem health of stressed environments in the Great Lakes has been evaluated simultaneously by means of a battery of structural and functional tests based on current technology and involving various trophic levels. These tests attempt to assess ecosystem health at the organism level and simultaneously focus on water-borne and sediment-bound toxicities. The use of structural indicators has been successfully demonstrated. Similarly, functional tests were selectively chosen across various trophic levels and included size-fractionated primary productivity (filtered versus unfiltered assays), and *Colpidium*, *Daphnia*, *Hyalella*, and *Pontoporeia* assays. Some of the emerging techniques such as *in situ* plankton cages (I.P.C.), microcomputer-based chlorophyll fluorescence (Video Analysis System), and other assays are discussed. The multi-trophic and multi-disciplinary battery of tests followed in our laboratory adopts a field-to-laboratory approach. The availability of diverse bioassays have placed toxicologists and environmentalists in a position where they are now better equipped to probe the complexities of ecosystem health and its management.

Introduction

The ever-increasing development and industrialization in various parts of the world is a serious threat to the conservation of pristine ecosystems. The discharges originating from municipal and industrial sources continue to contaminate our environment. Around the North American Great Lakes, for example, 42 'Areas of Concern' which need immediate attention, have been identified and measures implemented for their decontamination (I.J.C., 1987). These environmental concerns have been reflected at the global level by the publication of the World Commission Report on Environment and Development (1987) with the aim of encouraging sustainable development. Several conferences have been conducted at international levels with this goal in mind.

It is apparent that environmental assessment and habitat evaluation are the first steps toward both the identification of this problem and the application of relevant solutions. Traditionally, environmental assessment has been mainly oriented towards bulk chemical characterization,

---

* Dedicated to the memory of my mother who was a great teacher, guide and an incredible source of inspiration.

94

which has resulted in the establishment of preliminary guidelines for regulatory purposes. Such parameters obviously lack the dynamic toxicity information needed to determine the bioavailability of contaminants to the biota residing in an ecosystem (Munawar & Munawar, 1987). The recognition of the need to base environmental management on the results of toxicological experimentation was an important step in the sound management of pollutants. As a result, several international endeavours such as the 'International Workshop on *in situ* Sediment Contaminants', Aberystwyth (1984), an international conference on 'Environmental Bioassay Techniques and Their Application', Lancaster (1988), and the 'European Conference on Ecotoxicology', Copenhagen (1988) have been convened with the focus on toxicology.

Attempts have also been made to develop rationale and approaches (Cairns & Pratt, 1989; Giesy *et al.*, 1988; Giesy & Graney, 1989; Calow, 1989) to measure and deal with environmental perturbation (Reish & Oshida, 1987; I.J.C., 1987; Ahlf & Munawar, 1988; Persoone & Van de Vel, 1987; Environment Canada, 1987; MISA, 1987). As a result of various initiatives, the concept of 'ecosystem health' has recently emerged in environmental toxicology in a clinical sense as opposed to approaches based on estimates of lethal dosages.

It is also increasingly apparent that an 'ecosystem health' approach must inevitably deal with whole biological communities in the ecosystem. In other words, a multi-trophic level and multi-bioassay approach has to be adopted to attain a holistic ecosystem-health assessment. The present paper represents one such attempt, and it endeavours to synthesize our efforts in the Great Lakes to deal with the ecosystem health of stressed environments by a battery of tests. Since the paper is techniques-oriented, the methods, results, and relevant discussion sections have been combined to provide a concise description of the protocol together with an example of its usefulness wherever necessary.

Results and discussion

Multi-trophic and Multi-Bioassay Approach

Toxicity testing today is not restricted to the acute lethality studies of fish since a variety of aquatic bioassays has been developed to evaluate acute toxicity, sublethal effects, bioconcentration, and nutrient/contaminant interactions. A complete assessment of contaminant effects in an aquatic ecosystem is basically achieved by using organisms from various trophic levels in comparative bioassays (Gächter, 1979; Maciorowski *et al.*, 1981; Bringmann & Kuhn, 1980).

In the North American Great Lakes area, our laboratory has been actively involved in the bioassessment of the 'Areas of Concern' and, over the years, has developed a suite of bioassays for environmental health assessment (Fig. 1; Table 1). The effects of contaminants on the biota have been measured under three main types of assessment (Munawar *et al.*, 1988.) which included structural, functional, and ultra-structural evaluation. Structural and functional responses of biota, especially in the lower trophic levels, will be emphasized in this paper.

Toronto Harbour – Ashbridges Bay and Lake Ontario have been used as a case study (Fig. 2) to demonstrate the applicability, use, and protocol of the battery of tests, although examples from other Great Lakes have also been given where necessary. The importance of Ashbridges Bay lies in the fact that it is the recipient of effluent from Toronto's main sewage treatment plant.

Structural response

The structural response of pelagic communities was determined by means of microscopic identification and enumeration. Water samples were collected from several stations (Fig. 2) around Toronto Harbour – Ashbridges Bay (Stations 12, 734, 909, 911, 419, 910 and 204) and an offshore transect in Lake Ontario (Stations A, B, C, D, E, & F). Samples were preserved in formalin and the abundance of bacteria, heterotrophic nannoflagellates (HNF), and autotrophic picoplankton

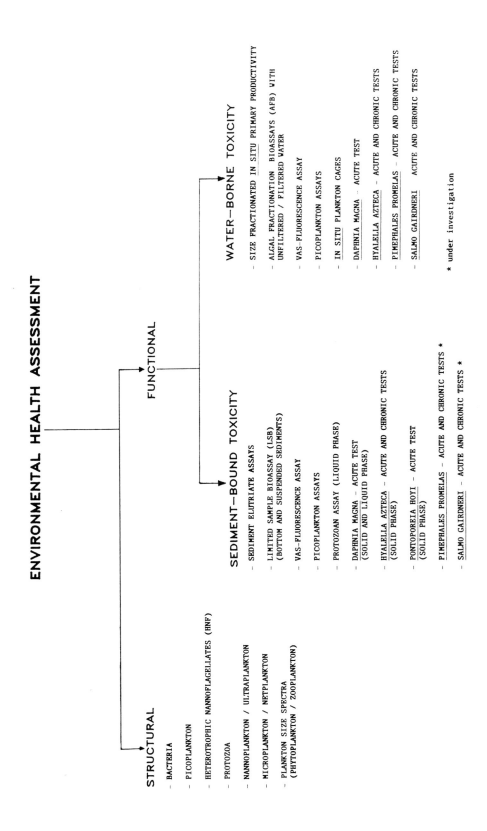

Fig. 1. Environmental health assessment strategy

Table 1. Choice and details of environmental health assessment tests

Evaluation procedure	Test organisms	Test parameter/response
Structural		
Bacteria	Natural bacteria	Epifluorescence microscopy/abundance
Picoplankton	Natural picoplankton	Epifluorescence microscopy/abundance
Heterotrophic nannoflagellates (HNF)	Natural HNF	Epifluorescence microscopy/abundance
Protozoa	Natural protozoa	Inverted microscopy/species abundance
Nannoplankton/ultraplankton	Natural phytoplankton	Inverted microscopy/species abundance
Microplankton/netplankton	Natural phytoplankton	Inverted microscopy/species abundance
Plankton size spectra	Natural phytoplankton	Microscopy/normalized size spectra
Functional (water-borne)		
Size-fractionated *in situ* primary productivity	Natural phytoplankton	C-14 assimilation enhancement-inhibition/acute sublethal
Algal fractionation bioassay – (AFB) filtered vs unfiltered	Natural phytoplankton	C-14 assimilation/enhancement-inhibition/acute sublethal
VAS – filtered/unfiltered	Natural phytoplankton	Fluorescence/fluorescence decay/acute sublethal
Picoplankton assay	Natural picoplankton	C-14 assimilation/enhancement-inhibition/acute sublethal
In situ Plankton Cage	Natural phytoplankton	C-14 assimilation/enhancement-inhibition of size assemblages/acute sublethal
Functional (sediment-bound)		
Sediment elutriate assay	Natural phytoplankton	C-14 assimilation/enhancement-inhibition of size assemblages/acute sublethal, EC50
Limited Sample Bioassay (LSB)	Mixed algal culture *Selenastrum capricornutum* Natural phytoplankton	C-14 assimilation/enhancement-inhibition/acute sublethal
VAS-fluorescence assay	Mixed algal culture *S. capricornutum* *Ankistrodesmus braunii* Natural phytoplankton	Fluorescence/fluorescence decay/acute sublethal
Protozoan assay	*Colpidium campylum*	Growth/growth inhibition/acute sublethal
Invertebrate assay	*Daphnia magna*	Mortality/acute lethal
	Hyalella azteca	Mortality/chronic lethal/growth
	Pontoporeia hoyi	Mortality/acute lethal
Fish assay	*Pimephales promelas*	Mortality/acute lethal/chronic lethal
	Salmo gairdneri	Mortality/acute lethal/chronic lethal

(APP) was determined using the DAPI staining technique (Weisse & Munawar, 1989) and epifluorescence microscopy. Samples for phytoplankton analyses were preserved in Lugol's solution. Identification and enumeration were carried out by the inverted microscope technique (Munawar *et al.*, 1974). Species biomass was size-fractionated using a computer program. Similarly, protozoa were identified and enumerated in Lugol-preserved samples by the inverted microscope technique.

The structural responses of natural biota in general and phytoplankton in particular have received little attention in toxicology (Rhee, 1988). They consist of changes in species composition (Munawar & Munawar, 1987) and size structure (Sprules & Munawar, 1986). These alterations of the community structure due to toxic stress might

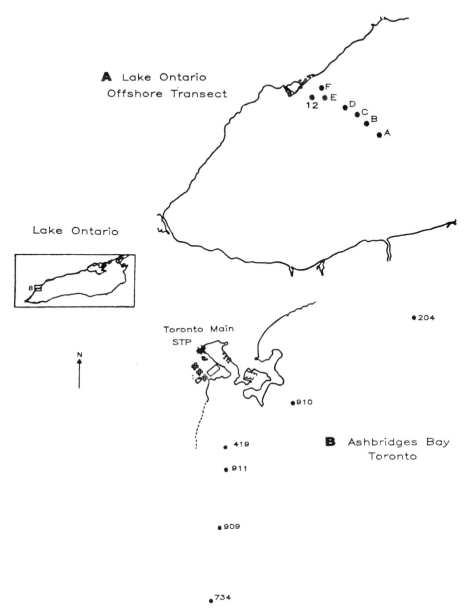

Fig. 2. Sampling locations in (A) Western Lake Ontario and (B) Ashbridges Bay, Toronto.

have serious implications for trophic interactions, nutrient cycling, the microbial loop, food web dynamics, fisheries food resources (Ross & Munawar, 1987), and fish yield of an ecosystem (Borgmann, 1987). However, such data for the North American Great Lakes phytoplankton are extremely limited due to the scarcity of long-term floristic records. Figure 1 indicates the various

components of the health assessment strategy including structural responses evaluation. Furthermore, Table 1 provides the choice and details of procedures which have been used in the Great Lakes and are currently available.

A recent example of the sensitivity of some components of the 'microbial loop' (Azam *et al.*, 1983; Munawar & Weisse, 1989) is shown in

98

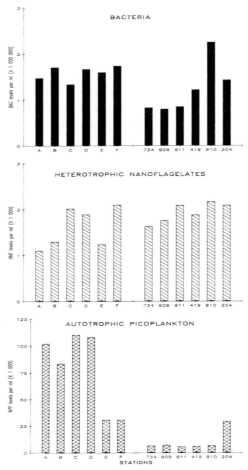

Fig. 3. Abundance of autotrophic picoplankton (APP), heterotrophic nannoflagellates (HNF), and bacteria (BACT) in Ashbridges Bay, Toronto, Ontario (modified from Munawar & Weisse, 1989).

Fig. 3. The pelagic microorganisms in Ashbridges Bay were compared to offshore stations across a transect in Lake Ontario (Fig. 2). The bacteria (Fig. 3) were relatively lower in number around the effluent discharge area (Stations 734, 909, 911, and 419 (the discharging zone)) compared to stations 910 and 204 and to the offshore transect (Stations A-F). Heterotrophic nannoflagellates (HNF), the potential major consumer of bacteria, showed no distinct trend. In contrast, the autotrophic picoplankton (APP) showed dramatic and well-pronounced distributional patterns between offshore stations and Ashbridges Bay. For example, APP concentrations were 14-fold lower in the Bay area than at the offshore stations. The

APP clearly avoided the effluent impact zone (Stations 734, 909, 911, 419, and 910) and appeared to recover slightly at Station 204 which was furthermost on the northern transect. The APP appear to be extremely sensitive to pollution in the Ashbridges Bay ecosystem (Munawar & Weisse, 1989). This is in agreement with the findings of Munawar & Munawar (1987) who compared the abundance of Niagara River picoplankton with that of Lake Ontario and concluded that their scarcity in the river was due to metal contamination.

Similarly, extensive studies have been undertaken in the Great Lakes to compare the phytoplankton composition of nearshore versus offshore areas, with the former area representing a perturbed or contaminated environment. These studies indicated significant differences between biomass, species, and size composition of the two areas (Munawar & Munawar, 1982; 1986; Munawar *et al.*, 1978). It is interesting to compare here the Great Lakes with the Experimental Lakes Area (ELA) in northwestern Ontario. It was found that the composition of small and rapidly-reproducing ELA phytoplankton species was the earliest indicator of stress (Schindler, 1987). Since ultraplankton/picoplankton dominate the phytoplankton of most of the Great Lakes (Munawar & Munawar, 1978; 1986), the impact of toxic substances can be serious by causing the elimination of minute organisms such as ultraplankton/picoplankton which play a major role in food chain dynamics. Indeed, it has been experimentally proven that ultraplankton and picoplankton are extremely sensitive to metal mixtures (Munawar *et al.*, 1987a; 1987b). Recently, the monitoring of phytoplankton and zooplankton has gained further momentum with the development of normalized biomass size spectra, and it is suggested that routine monitoring of aquatic communities by the normalized biomass spectrum could provide an early warning of nutrient or toxic stress (Sprules & Munawar, 1986; Sprules *et al.*, 1988).

Data concerning the abundance and composition of ciliated protozoa in the Great Lakes is scarce, although the effects of eutrophication and pollution on aquatic protists have been well documented and experimental results suggest protozoa to be useful indicators of water quality. In addition, it has been postulated that ciliates play a key role in the 'microbial loop' (Azam *et al.*, 1983; Sherr & Sherr, 1988). The ciliate data collected as part of the case study (Munawar & Gilron, 1989) indicated that the ciliate assemblage at Station 419 was quite distinct from those of the other stations. For example, peritrich ciliates (i.e. *Vorticella* sp., *Carchesium* spp.), known to be indicator organisms of waters containing sewage effluent (Cairns, 1978; Caron & Sieburth, 1981; Henebry & Ridgeway, 1979; Stossel, 1987), were found only at Station 419 where the effluent dis-

charge outlet is located. Species composition of the nearshore area was considerably different to that of the offshore transect.

Functional response

Size-Fractionated in situ Primary Productivity

Size-fractionated phytoplankton *in situ* primary productivity was estimated using the ^{14}C uptake technique (Vollenweider *et al.*, 1974). The phytoplankton samples were collected using an integrating sampler. A portion of the well-mixed sample was preserved in Lugol's solution for taxonomic identification and enumeration using the Utermohl inverted microscope technique (Munawar & Munawar, 1978). The rest of the test sample was inoculated in triplicate with 1 μCi of $NaH^{14}CO_3$ and incubated for four hours at constant light levels of 238 $\mu E \cdot m^{-2} \cdot sec^{-1}$ at 400–700 nm while being maintained at lake temperature. After the incubation period, the entire contents of each bottle were size-fractionated through a 20 μm Nitex screen and the retentate was back-washed onto a 0.45 μm Millipore membrane filter. This determined the microplankton/netplankton productivity ($> 20 \mu$m). The portion of the sample that passed through the 20 μm Nitex screen was filtered directly onto a 0.45 μm Millipore membrane filter which determined ultraplankton/picoplankton ($< 20 \mu$m) productivity. The membrane filters were then acidified with 10 ml of 0.1 N HCl and kept in a Phase Combining System (PCS) for liquid scintillation counting (Lind & Campbell, 1969). Statistical analysis included computation of means, standard errors, and t-test comparisons. Several stations were sampled this way in Ashbridges Bay.

The primary productivity of the size assemblages of phytoplankton has been used as an indicator of ecosystem health in our investigations (Munawar & Munawar, 1987; Munawar *et al.*, 1988). It is apparent (Fig. 4) that the effluent plume had significant inhibitory impact on the productivity of ultraplankton/picoplankton

100

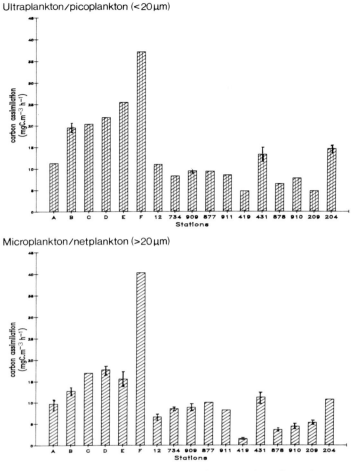

Fig. 4. Size-fractionated primary productivity in Ashbridges Bay, Toronto, Ontario and at an offshore transect of Lake Ontario during May, 1988.

($< 20 \, \mu$m) compared to the offshore control (Station A). The productivity rate was significantly ($P < 0.05$) lower at Station 419 (STP discharge pipe) and at northern Stations 878 and 910. Conversely, the furthermost Station (204) showed slightly higher productivity than the control. Similarly, the western Stations such as 911, 877, 909 and 734 also exhibited a significantly lower productivity rate than the control ($P < 0.05$).

The microplankton/netplankton productivity was also significantly inhibited at Station 419 as well as at the other northern stations ($P < 0.05$). However, no significant impact was observed at the southern stations.

Algal Fractionation Bioassays – AFB (Unfiltered and Filtered Assays)

A natural offshore Lake Ontario phytoplankton assemblage from Station 12 (control station) was concentrated by filtration onto a Millipore membrane filter and added to unfiltered and filtered water of six Ashbridges Bay Stations (734, 909, 419, 910, 204, and the control Station) to which ^{14}C was added. The samples were incubated for four hours (see above for details of size-fractionated primary productivity). All bioassays were conducted in triplicate. Phytoplankton was size-fractionated into ultraplankton/picoplankton ($< 20 \, \mu$m) and microplankton/netplankton ($> 20 \mu$m) after incubation to assess the impact of filtered and unfiltered water upon the offshore population.

Unfiltered bioassays were interpreted in the following manner:

Step 1: The primary productivity of indigenous phytoplankton from Station 12 was estimated.

Step 2: The primary productivity of concentrated phytoplankton from Station 12, when added to indigenous phytoplankton from the same station, was determined.

Step 3: The difference between the primary productivity of Steps 2 and 1 thus represented the total primary productivity of the concentrated phytoplankton which was used as a test assemblage.

The primary productivity rate of the concentrated phytoplankton was then divided by the overall primary productivity from Step 2, multiplied by 100 to generate the 'expected' percent primary productivity of the concentrated test assemblage.

Thus, when this concentrated natural phytoplankton was added to unfiltered water samples to be tested, the same productivity would be 'expected' as that observed when the test population was added to the control sample (i.e. unfiltered water from Station 12). The resulting rate was used as a zero baseline value (Figs. 5 and 6)

Any primary productivity rate which was higher or lower than the 'expected' rate was then interpreted as enhancement or inhibition respectively.

Filtered bioassays were interpreted directly by estimating the productivity of concentrated natural phytoplankton in the filtered water of each station and was compared with unfiltered bioassay results.

Figures 5 & 6 show the results of a primary productivity experiment with concentrated offshore phytoplankton added to the unfiltered water. Instead of an expected 42 percent enhancement, most of the tested stations showed considerable inhibition of ultraplankton/picoplankton (Fig. 5) and microplankton/netplankton (Fig. 6) productivity. On the other hand, the filtered-water bioassay experiment with the same population showed significantly higher productivity compared to the control ($P < 0.05$). This enhanced productivity, observed when offshore phytoplankton was exposed to filtered water, can be attributed to the removal by filtration of particulate matter which may be a carrier of both nutrients and contaminants.

Picoplankton Assay

Organisms which are less than 2 μm are designated as picoplankton (Johnson & Sieburth, 1982; Munawar *et al.*, 1987a, 1988). The Acroflux capsule (Gelman Sciences Inc., 1983) was used to fractionate picoplankton down to sizes as small as 1.2 μm (Munawar *et al.*, 1987a). With the Gelman 1.2 μm Acroflux capsule, nine litres of picoplankton sample were collected in approximately five minutes by gravity-filtration of fresh and natural phytoplankton samples. The 1.2 μm filtrate was further concentrated onto 0.45 μm Millipore filters at a vacuum pressure of 17 kPa (5 in. Hg). Each filtration head had one litre of stock picoplankton sample poured through it until approximately 150 ml of sample remained on the filtration head. Each filtration head containing the concentrated sample was then back-washed with 100 ml of filtered lake water into a large polycarbonate beaker until 2.5 litres of 2 $\times$ concentrated picoplankton stock population was prepared. This stock population was well mixed and used as test organisms in toxicity bioassays. Picoplankton has proven to be quite sensitive to contaminants. This has been demonstrated in numerous experiments involving samples from the Great Lakes in which picoplankton was exposed to metal mixtures and sediment elutriates (Munawar *et al.*, 1987a; Severn *et al.*, 1989).

102

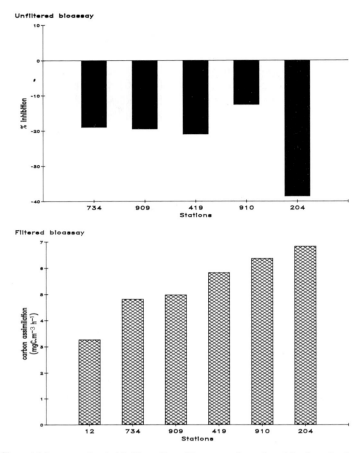

Fig. 5. Filtered versus unfiltered bioassays in Ashbridges Bay, Toronto, Ontario with ultraplankton/picoplankton ($<20 \mu$m) during May, 1988.

In situ Plankton Cage Experiment

In situ plankton cages (IPC) were placed at the sediment-water interface at a station in Hamilton Harbour near the Canada Centre for Inland Waters, Burlington, Ontario. The study was designed to determine the effects of the nutrients/contaminants upon caged offshore phytoplankton. An example of an experiment conducted during the summer of 1986 is presented here (Fig. 7).

The *in situ* plankton cage is made of plexiglass and has a main circular chamber with top and bottom plates affixed to it. Four stainless steel legs and a plexiglass base prevent the main chamber from sinking into the sediment. The chamber, equipped with a stainless steel nozzle to

facilitate filling and emptying, can hold 320 ml of test sample. Each endplate of the compartment has four circular 47 mm holes over which 47 mm 0.45 μm Millipore membranes are placed. These membrane filters allow nutrients and contaminants to enter the cage while preventing the 'escape' of the phytoplankton test assemblages. The control cage is fitted with non-leaching stainless steel discs instead of the membrane filters to prevent the movement of material in or out. After a pre-determined period of exposure (12–24 hours), the cages are removed from the test site and the contents analyzed both chemically and biologically. In our experiments, we have assessed the rate of primary productivity of caged phytoplankton compared to the control using the ^{14}C technique. The species composition

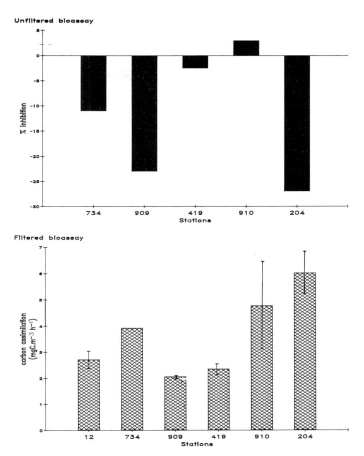

Fig. 6. Filtered versus unfiltered bioassays in Ashbridges Bay, Toronto, Ontario with microplankton/netplankton (<20 μm) during May, 1988.

of the caged phytoplankton was also analyzed before and after the exposure. However, other parameters such as chlorophyll *a* and ATP, etc. may also be determined. (For more details about the cages, see Munawar & Munawar (1987)).

The Hamilton Harbour experiment showed a significant enhancement of primary productivity (Fig. 7) for both ultraplankton (2–20 μm) and microplankton/netplankton (>20 μm) size assemblages. Although Hamilton Harbour is known to be an industrially contaminated ecosystem, the nutrient/contaminant interactions were apparently conducive to phytoplankton growth as shown by an overall enhancement of primary productivity at that particular time. However, seasonal response of phytoplankton has to be determined before any generalizations can be

made about the impact of the Harbour nutrients/contaminants on the offshore plankton. Similar experiments were conducted at the mouth of the Don River and Keating Channel in Toronto, Ontario, where a differential response of the phytoplankton size assemblages was observed due to nutrients/contaminants diffusing into the *in situ* plankton cage (Munawar & Munawar, 1987).

Sediment-Elutriate Bioassays with Phytoplankton

Sediment-elutriate bioassays have been conducted routinely in our laboratory with sediment collected from various Great Lakes sites and contaminated 'Areas of Concern' (I.J.C., 1987). Elu-

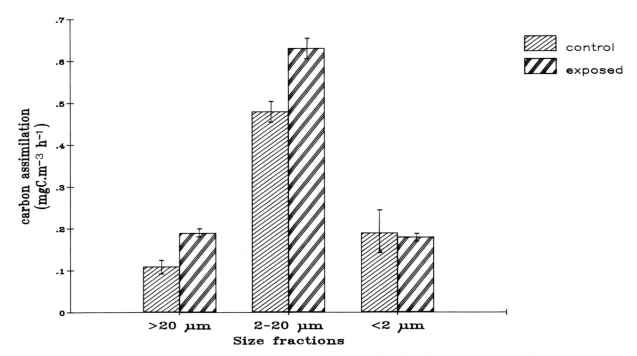

Fig. 7. *In situ* Plankton Cage experiment at a station in Hamilton Harbour, Ontario, 1986.

triates were prepared by mixing sediment (200 ml) with filtered offshore Lake Ontario water (800 ml). A more detailed account of elutriate preparation can be found in Daniels *et al.* (1989). Each bioassay included an offshore control (no elutriate added) with subsequent elutriate additions of 1, 5, 10, 20, and 40 percent dosages. Resultant primary productivity was once again measured by ^{14}C uptake, and size-fractionated after four hours of incubation. EC50s (Effective Concentration, i.e. the concentration at which 50 percent inhibition occurs) were calculated where possible (U.S.E.P.A., 1977). Several examples of elutriate bioassays in the Great Lakes have been published (Munawar *et al.*, 1983, 1985; Munawar & Munawar, 1987).

An example from a Station 419 experiment conducted in May, 1988 is presented here (Fig. 8). The sediment – elutriate was found to be toxic to both the size assemblages of an offshore Lake Ontario community with an EC50 of 15 percent.

Limited Sample Bioassays (LSB)

This technique was developed when only limited samples of sediment or suspended particulate matter were available for collection (Munawar *et al.*, 1989). Small aliquots of the sample did not allow for chemical analysis but were of great value as a toxicity screening technique. Sediment-to-water ratios were prepared with net concentrations equivalent to those normally used with conventionally-produced elutriates. Limited amounts of sample (usually no more than 50 g) were added in varying quantities to fixed amounts of water to achieve these concentrations. The water was obtained from the sample site and filtered through 0.45 μm filters before addition to the sample. The sediment or suspended particulates and filtered water were well mixed using a ferris-wheel type tumbler at a rate of five rpm for one hour. The mixture was then allowed to stand for six hours at 4 °C, decanted, and centrifuged at 10 000 rpm for 20 minutes. The supernatant was then filtered through 0.45 μm filters. One ml of a concentrated mixed laboratory-grown culture of algae was then

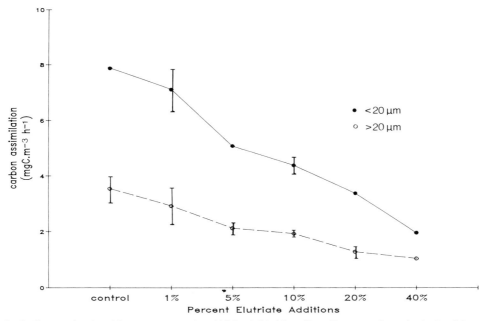

Fig. 8. Sediment-elutriate bioassays at station 419 in Ashbridges Bay, Toronto, Ontario during May, 1988.

added to the LSB elutriates. Enough elutriates were prepared to allow replication in order to make the experiment statistically sound. Test samples were then inoculated with $1 \mu Ci$ of $NaH^{14}CO_3$ and incubated for four hours at constant light levels of $238 \mu E \cdot m^{-2} \cdot sec^{-1}$ at 400–700 nm and constant temperature in an incubator. The entire contents of each bottle were then filtered through a $0.45 \mu m$ membrane filter for the determination of primary productivity rate.

The LSB example presented in this paper assessed the toxicity of particulate matter from the St. Clair River. The particulates were collected using a high-speed centrifuge and the experiments were conducted using a mixed culture of laboratory-grown phytoplankton. The mixture consisted of *Ankistrodesmus* sp., *Pandorina morum*, *Carteria olivieri*, *Synedra* sp., *Anacystis nidulans*, *Trachelomonas* sp., *Gloeocapsa* sp. and *Microcystis* sp. The experiments were then carried out as above. Figure 9 demonstrates the impact of St. Clair River particulate matter on the mixed algal culture. It is apparent that sediment : water ratios of 1 : 36 and 1 : 16 were toxic to laboratory-grown algae. Details of this technique are given in

Munawar *et al.* (1989) with several examples from the Great Lakes, Lake Diefenbaker (Saskatchewan), and the Northwest Territories of Canada. LSB permits the detection of toxicity of limited quantities of bottom or suspended sediments which would otherwise be left undetected. As indicated earlier, it is also a miniature version of the sediment elutriate bioassay. In addition, the LSB allows the bioassessment of suspended particulate matter and the role they play as carriers of nutrient and contaminants in various ecosystems, particularly rivers and harbours (Munawar *et al.*, 1988).

VAS – Fluorescence Assay

The algal culture *Ankistrodesmus braunii* was grown in 100 ml of CHU-10 medium in a 250 ml Erlenmeyer flask on a rotary shaker at 150 rpm and transferred every 11 days to fresh medium. A sample of these cultures was counted after 11 days of incubation and the cell density adjusted to 1×10^6 cells ml^{-1}. Subsamples of this adjusted culture were added to test-tubes and the heavy

108

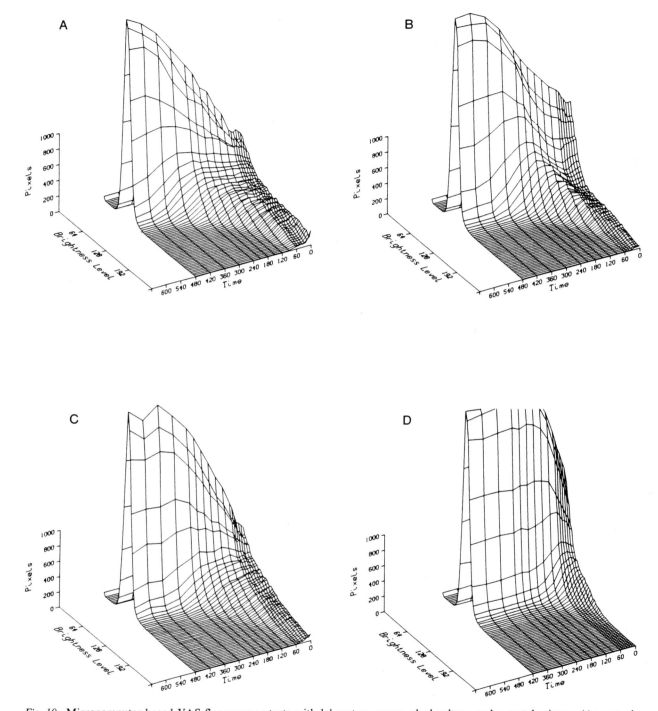

Fig. 10. Microcomputer-based VAS-fluorescence tests with laboratory grown algal culture and a metal mixture (A: control, B: 1 ppm, C: 10 ppm, D: 100 ppm).

the day of testing, young animals (less than 24 hours old) were removed from the breeding tank and used as test organisms.

Sediment-elutriate was prepared by adding sediment to filtered lake water in a 1 : 4 ratio and then mixed on a rotary shaker for one hour. The bottles were allowed to settle overnight; subsequently, the upper water layer was poured off into 250 ml centrifuge bottles and centrifuged for 30 minutes at 10 000 rpm. The supernatant was then filtered through a 0.45 μm Millipore filter. Two hundred ml aliquots of elutriate were placed in 250 ml zooplankton jars and allowed to stabilize to room temperature (approximately four hours). Ten *D. magna* of less than 24 hours old were placed into these randomly arranged jars and the tests were run under stable temperatures and received a photoperiod of 16 hours of light and 8 hours of dark. Oxygen was measured several times before animals were added and also during the test period. After 48 hours, the survival rate was then calculated (see results below).

Acute Solid Phase Test with Daphnia magna

Forty ml of 275 μm-filtered sediment was added to 250 ml zooplankton jars containing 160 ml of dechlorinated water. The jars were left to stand for 48 hours to allow settlement of sediment. Ten young (less than 24 hours old) *D. magna* were placed into each test vessel. The vessels were incubated for 48 hours, after which the number of survivors were counted.

Figure 11 shows the results of the *D. magna* assay for both solid and liquid phases of sediment collected during May/August 1988 from Stations 419, 911 and 910. An offshore Lake Ontario sediment was used as a control. It is obvious from Figure 11 that both the elutriate and solid phases were toxic to *Daphnia* in all experiments for Station 419 (May and August), and Stations 910 and 911 (August).

Chronic Toxicity Test with Hyalella azteca

Although Nebeker *et al.* (1984; 1986) suggested the use of *Hyalella* in chronic tests, no detailed protocols were given. Nebeker & Miller (1988) tested *Hyalella* for its tolerance to various levels of salinity and proposed its use in the bioassessment of estuarine sediments. Borgmann & Munawar (1989) provided the standard protocol for sediment testing with *Hyalella* for the first time.

Sediments were collected by Shipek sampler and stored temporarily in a refrigerator in polyethylene bags. An offshore Lake Ontario station was used as a control.

The procedure used to obtain young *Hyalella* is described in Borgmann *et al.* (1989). Amphipod cultures were maintained in straight-sided glass jars containing pieces of cotton gauze of the surgical bandage type (available in drug stores), an essential if large numbers of young are desired. No sediment or other substrate is required. Amphipods were fed Tetra-Min fish food flakes several times each week. The water was changed in the jars and the young were removed every week, even when no young were required. The age of the young obtained was then always zero to one week. Before being used in bioassays, the separated young amphipods were kept about two days in jars containing one litre of water mixed with 20 mg of Tetra-Min, each jar being covered with a 5 × 10 cm piece of gauze. This ensured that animals which normally die within a day or two due to rough handling during collection were not used in the bioassays. It also provided two days advance notice of the number of young animals available before bioassays were set up.

Sediment from each site was sifted through a 275 μm nylon screen and added to straight-sided Pyrex screwtop jars (15 cm diameter, approximately 2.5 litres capacity), giving a sediment layer of 1 to 1.5 cm depth. Sifting was carried out under water in the test jars. Dechlorinated tap water was then added to give a total volume of 1.5 litres and the jars were covered with a sheet of plexiglass. A tygon air line was passed through a hole in the plexiglass and an aquarium airstone attached to the end was suspended several cms above the sediment. Air was bubbled through the air line at a sufficiently slow rate so as to gently keep the water oxygenated but not to resuspend the sedi-

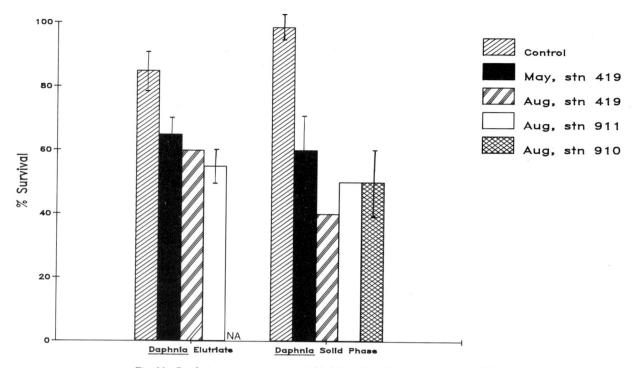

Fig. 11. Daphnia magna assays at Ashbridges Bay, Toronto, Ontario, 1988.

ment. The sediments were allowed to settle for several days and the water was given enough time to become oxygenated before the amphipods were added. This bioassay is restricted to oxygenated sediments since *Hyalella* cannot withstand extended periods of anoxia. Twenty young *Hyalella* were added to each jar. Tetra-Min (20 mg) was sifted through a 500 μm mesh screen and added as food twice each week. The jars were incubated at room temperature (20–22 °C) under fluorescent lights with a photoperiod of 16 hrs light and 8 hrs dark. Distilled water was added as needed to keep the water level constant. After four weeks, the contents of two jars with sediment from each location were sifted through a 275 μm nylon screen and the surviving amphipods were sorted, counted, and weighed on a microbalance (Borgmann & Munawar, 1989).

The results of solid phase tests are shown in Fig. 12 for Station 419 (May and August 1988) and Stations 910 and 911 (August 1988). Compared to the offshore control, *Hyalella azteca* survived well in Station 419 sediment whereas it failed to survive in the August sediment, thereby reflecting the toxic nature of the latter sediment. Similarly, Station 911 sediments collected in August were also toxic. On the other hand, Station 910 sediments showed no toxicity since *Hyalella*'s growth was better in this sediment than in the control sediment. These results clearly demonstrate the seasonality of sediment toxicity as well as the patchiness of toxicity distribution. Hence, caution should be exercised in the interpretation of results based on a single sample and must incorporate toxicity studies on a seasonal basis.

Acute Solid Phase Test with Pontoporeia hoyi

Pontoporeia hoyi are benthic amphipods (Crustacea) which live in the upper few inches of sediment. They are oligotrophic indicators, preferring well-oxidized conditions and temperatures of 10 °C or less. They have a long life cycle with a two-year maturity period. For bioassay experiments, the protocol developed by R. Dermott

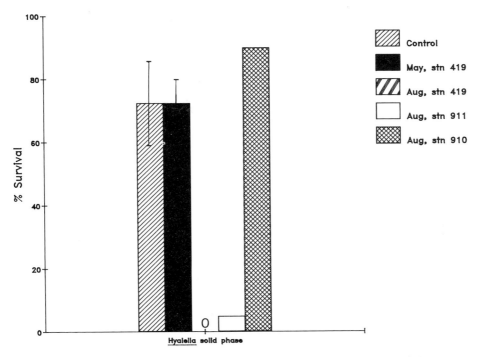

Fig. 12. *Hyalella azteca* assays at Ashbridges Bay, Toronto, Ontario, 1988.

(Fisheries & Oceans, personal communication) was adopted. Animals were collected in Lake Ontario at depths varying from 23 metres to 38 metres during April 1989. A 9-inch Ekman sampler was used to collect the bottom sediment which was gently sieved through a screen (500 μm) to collect the animals. The amphipods and mud were stored in a cooler at 8 °C.

Test sediments used were collected from Station 419 (Ashbridges Bay) in August 1988, and also from Station 41 (control) during the summer of 1987. The storage temperature was 8 °C. About 150 ml of each sediment were placed into distilled water-rinsed wide-mouthed 500 ml jars and left overnight at 8 °C to settle and cool to this temperature. The next morning, the *Pontoporeia* were removed from the dark cooler, gently sieved through a screen and washed into a tub for selection. The ideal animals for this experiment were healthy and medium-sized (4–6 mm). Pregnant females were not used since this condition could reduce their chances of survival in foreign sediment. During the selection process (using a 5 mm ID pipette), dechlorinated water

was dripped into the wide-mouthed jars at a rate which would not disturb the layer of sediment. When it was three-quarters full, ten animals were put into each jar and netting (which had been soaked in distilled water to eliminate previous contamination) was stretched across the top. These bottles were then placed in aquariums and dechlorinated water was added until a layer of water just covered the tops of the jars. Air was gently bubbled into the aquaria, allowing an exchange with the water in the jars. Care was taken to ensure that the animals did not stick to the netting, and daily observations were taken to determine activity and mortality. This was often difficult since one could not tell when animals were burrowing, etc. For this particular experiment, seven jars of each sediment were set up and left for seven days. After this period, the jars were removed from the aquaria, the water gently decanted, the remaining sediment sifted out, and the animals retained by the sieve were washed back into a beaker of water. Mortality was then estimated based on whether the animals moved or not. Since the animals tend to decompose in a

112

relatively short period, the number of animals left was also an indication of mortality. The results of the *Pontoporeia* assay for Station 419 August 1988 sediments showed that the test animals' survival was slightly lower than those in the control. Interestingly, these results based on acute testing are in contrast to the chronic mortality observed in *Hyalella* which totally perished after four weeks of experimentation.

Summary

The ecosystem health of stressed environments has been evaluated using structural and functional procedures. A battery of tests involving various trophic levels have been employed based on state-of-the-art technology. Our approach is in agreement with the current belief that the ecotoxicological assessment of stressed environments should be based on a series of tests to obtain a holistic evaluation for the development of effective Remedial Action Plans – RAPS (Reish & Oshida, 1987; Munawar & Ross, 1987; I.J.C., 1987, 1988).

The multi-disciplinary, multi-trophic assay approach adopted in our program appears to be balanced and systematic, and has resulted from several years of field experience. First of all, it attempts to tackle the environmental impact puzzle at the organism level with an emphasis on the use of natural populations. Secondly, it simultaneously focuses on both the structural and functional indicators of ecosystem health. Thirdly, it endeavours to employ a two-tier (water and sediment) strategy for assessing both water-borne and sediment-bound toxicity simultaneously, depending on the type of problem.

The experimentation provided some interesting insights and demonstrated the effectiveness of a multi-purpose bioassessment approach as shown in the case study of Ashbridges Bay. For instance, the structural analysis demonstrated the sensitivity of microorganisms such as the autotrophic picoplankton to nutrient/contaminant enrichment. The ciliated protozoa showed some characteristic indicator species like peritrichs at Ashbridges Bay. The results further demonstrated the usefulness of size-fractionated primary productivity as a screening tool for initial bioassessment, project planning, and program design (I.J.C., 1987; 1988; Munawar *et al.*, 1983, 1988).

The suite of tests dealing with sediment toxicity also provided interesting results in the Ashbridges Bay case study. The solid-phase sediment assays indicated that sediment was toxic to *Daphnia* and *Hyalella* during the summer but *Pontoporeia hoyi* survived well in the same sediment. Similarly, *Hyalella* thrived well during the spring compared to the lethality observed during the summer. These observations indicate a seasonal trend in the bioavailability of contaminants which may be due either to natural patchiness of sediment distribution or to the day-to-day fluctuations of municipal and industrial discharges. Consequently, more frequent toxicity testing is needed to obtain a better understanding of an ecosystem.

The multi-trophic battery of tests being developed in our laboratory adopts a field-to-laboratory approach with a preference for the use of natural populations as test organisms complemented by tests with cultured species. It is realized that facilities may not be available at a single laboratory to conduct all the tests proposed in this paper. Nevertheless, the main objective of this paper is to convey the concept of the multi-disciplinary, multi-trophic assay strategy for the bioassessment of both water and sediment toxicity. For instance, other agencies offer facilities for tests such as AMES, Microtox, Nematode, *Chironomus tentans*, *Hexagenia limbata*, etc. Similarly, some agencies conduct studies with parameters such as contaminant residues in biota, fish deformities, and paleolimnological analysis. All these techniques offer a tremendous choice from which a researcher may choose a battery of tests as needed, depending on the nature of contamination, type of problem, and project budget.

The field of environmental bioassessment is an area of rapidly developing and expanding research. Several techniques and strategies are emerging including enzymatic and genotoxic testing. A few examples of the developing procedures are microcomputer-based VAS-fluorescence, *Colpidium* (protozoan) bioassays, SOS chromotests, Dark mutant test, Microtest, Limited Sample Bioassay (LSB), Rotifer test (*Brachionus calyciflorus*), Oligochaete test, Microbial DNA assays, seed germination and plant growth (*Lolium multiflorum* and *Lepidium sativum*), *Lemna minor* and flow cytometry (Ahlf *et al.*, 1989; Dutka *et al.*, 1989; Sloterdijk *et al.*, 1989; Smith & Kwan, 1989; C. Blaise, Environment Canada, R. Dermott, Fisheries & Oceans, W. Gala and G.P. Giesy, Michigan State University – personal communications). Similarly, acute and chronic tests with fish such as *Pimephales promelas* and *Salmo gairdneri* are receiving fresh attention from researchers instead of the traditional lethality testing (A. Niimi, Fisheries & Oceans – personal communication) in the development of screening procedures. We hope to add these fish tests to the standardized battery in the near future.

Finally, the availability of diverse bioassay technologies have placed toxicologists and environmentalists in a position where they are better equiped now than ever before to confront the complexities of contaminant effects. Therefore a wise, cautiously planned and cost-effective application of rapid and sensitive technology seems to have great potential for the environmental assessment of stressed environments and continued toxicological monitoring of ecosystem health from a diagnostic, preventative, and remedial point of view.

Acknowledgements

We would like to thank the following colleagues and personnel who assisted in various aspects of the work towards the preparation of this manuscript: Uwe Borgmann, Shirland Daniels, Ron Dermott, Pauline Desroches, Guy Gilron, Art Niimi, Warren Norwood, Wendy Page and Thomas Weisse. We gratefully acknowledge the comments and suggestions of Dr. George Dixon (University of Waterloo) and Dr. Gary Leppard (National Water Research Institute) and the assistance of Harold Nicholson for technical editing.

References

Ahlf, W., W. Calman, J. Erhard & U. Förstner, 1989. Comparison of five bioassay techniques for assessing sediment-bound contaminants. In; M. Munawar, G. Dixon, C. I. Mayfield, T. Reynoldson & M. H. Sadar (Eds.), Environmental bioassay techniques and their application. Hydrobiologia (this volume).

Ahlf, W. & M. Munawar, 1988. Biological assessment of environmental impact of dredged material. In; W. Salomons & U. Förstner (Eds.), Chemistry and biology of solid wastes, dredged material and mine tailings. Springer-Verlag. pp. 127–142.

Azam, F., T. Fenchel, F. G. Field, J. S. Gray, L. A. Meyer-Reil & F. Thingstad, 1983. The ecological role of water-column microbes in the sea. Mar. Ecol. Prog. Ser. 10: 257–263.

Borgmann, U., 1987. Models on the slope of, and biomass flow up, the biomass size spectrum. Can. J. Fish. aquat. Sci. 44 (2): 136–140.

Borgmann, U. & M. Munawar, 1989. A new standardized sediment bioassay protocol using the amphipod *Hyalella azteca* (Saussure). In; M. Munawar, G. Dixon, C. I. Mayfield, T. Reynoldson & M. H. Sadar (Eds.), Environmental bioassay techniques and their application. Hydrobiologia. (This volume).

Borgmann, U., K. M. Ralph & W. P. Norwood, 1989. Toxicity test procedures for *Hyalella azteca* and chronic toxicity of cadmium and pentachlorophenol to *H. azteca*, *Gammarus fasciatus* and *Daphnia magna*. Arch. Envir. Contam. Toxicol. 18: 756–764.

Bringmann, G. & R. Kuhn, 1980. Comparison at the toxicity thresholds of water pollutants to bacteria, algae, and protozoa in the cell multiplication inhibition test. Wat. Res. 14: 231–241.

Cairns, J. R. Jr., 1978. Zooperiphyton (especially protozoa) as indicators of water quality. Trans. am. Micros. Soc. 97: 44–49.

Cairns, J. R. Jr. & J. R. Pratt, 1989. The scientific basis of bioassays. In; M. Munawar, G. Dixon, C. I. Mayfield, T. Reynoldson & M. H. Sadar (Eds.), Environmental bioassay techniques and their application. Hydrobiologia. (This volume).

Calow, P., 1989. The choice and implementation of environmental bioassays. In; M. Munawar, G. Dixon, C. I. Mayfield, T. Reynoldson and M. H. Sadar (Eds.), Environmental bioassay techniques and their application. Hydrobiologia. (This volume).

Caron, D. A. & J. M. Sieburth, 1981. Response of peritrichous ciliates in fouling communities to seawater-accommodated hydrocarbons. Trans. amer. Micros. Soc. 100: 183–203.

Daniels, S. A., M. Munawar & C. I. Mayfield, 1989. An improved elutriation technique for the bioassessment of sediment contaminants. In; M. Munawar, G. Dixon, C. I. Mayfield, T. Reynoldson and M. H. Sadar (Eds.), Environmental bioassay techniques and their application. Hydrobiologia. (This volume).

Dive, D., S. Robert, E. Angrand, C. Bel, H. Bonnemain, L. Brun, Y. Demarque, A. Le Du, R. El Bouhouti, M. N. Fourmaux, L. Guery, O. Hanssens & M. Murat, 1989. A bioassay using the measurement of the growth inhibition of a ciliate protozoan: *Colpidium campylum* Stokes. In; M. Munawar, G. Dixon, C. I. Mayfield, T. Reynoldson & H. Sadar (Eds.), Environmental bioassay techniques and their application. Hydrobiologia. (This volume).

Dutka, B. J., T. Tuominen, L. Churchland & K. K. Kwan, 1989. Fraser river sediments & waters evaluated by the battery of screening tests techniques. In; M. Munawar, G. Dixon, C. I. Mayfield, T. Reynoldson & M. H. Sadar (Eds.), Environmental bioassay techniques and their application. Hydrobiologia. (This volume).

Environment Canada, 1987. Recommendations on aquatic biological tests and procedures for environmental protection. Conservation & Protection. 102 pp.

Gächter, K., 1979. Effects of increased heavy metal loads on phytoplankton communities. Schweiz Z. Hydrol. 41: 228–246.

Gelman Sciences Inc., 1983. Filtration catalog and system design guide. Gelman Sciences Inc., Ann Arbor, Michigan, 98 pp.

Giesy, J. P. & R. L. Graney, 1989. Recent developments in and intercomparisons of acute and chronic bioassays and bioindicators. In; M. Munawar, G. Dixon, C. I. Mayfield, T. Reynoldson & M. H. Sadar (Eds.), Environmental bioassay techniques and their application. Hydrobiologia. (This volume).

Giesy, J. P., D. J. Versteeg & R. L. Graney, 1988. A review of selected clinical indicators of stress-induced changes in aquatic organisms. In; M. Evans (Ed.), Toxic contaminants and ecosystem health, a Great Lakes focus. pp. 169–200. John Wiley & Sons, New York.

Henebry, M. S. & B. T. Ridgeway, 1979. Epizoic ciliated protozoa of planktonic copepods and cladocerans and their possible use as indicators of organic water pollution. Trans. am. Micros. Soc. 98: 495–508.

International Joint Commission, 1987. Guidance on characterization of toxic substances problems in Areas of Concern in the Great Lakes basin. Report to the Great Lakes Water Quality Board, Windsor, Ontario. 179 pp.

International Joint Commission, 1988. Procedures for the assessment of contaminated sediment problems in the Great Lakes. Report from the Sediment Subcommittee and its Assessment Work Group to the Water Quality Board, Windsor, Ontario. 140 pp.

Johnson, P. W. & J. M. Sieburth, 1982. *In situ* morphology and occurrence of eucaryotic phototrophs of bacterial size in the picoplankton of estuarine and oceanic waters. J. Phycol. 18: 318–327.

Lind, O. T. & R. S. Campbell, 1969. Comments on the use of liquid scintillation for routine determination of ^{14}C activity in production studies. Limnol. Oceanogr. 14: 787–789.

Maciorowski, A. F., J. L. Simms, L. W. Little & F. O. Gerrard, 1981. Bioassays, procedures and results. J. Wat. Pollut. Control Fed. 53: 974–993.

Mayfield, C. I. & M. Munawar, 1988. Microcomputer-based measurement of algal fluorescence as a potential indicator of environmental contamination. Bull. Envir. Toxicol. 41: 261–266.

MISA, 1987. Municipal-Industrial Strategy for Abatement. The public review of the MISA white paper and the Ministry of the Environment's response to it. 55 pp.

Munawar, M. & G. Gilron, 1989. (Abstract). Ciliated protozoa as indicators of aquatic ecosystem health. Presentation at the International Association for Great Lakes Research (IAGLR) 1989 meeting, Madison, Wisconsin. May 1989.

Munawar, M. & I. F. Munawar, 1978. Phytoplankton of Lake Superior 1973. In; M. Munawar (ed). Limnology of Lake Superior. J. Great Lakes Res. 4 (3-4): 415–422.

Munawar, M. & I. F. Munawar, 1982. Phycological studies in Lakes Ontario, Erie, Huron and Superior. Can. J. Bot. 60 (9): 1837–1858.

Munawar, M. & I. F. Munawar, 1986. The seasonality of phytoplankton in the North American Great Lakes: A comparative synthesis. In; M. Munawar & J. F. Talling (Eds.). Seasonality of Freshwater Phytoplankton: A Global Perspective. Hydrobiologia. 138: 85–115.

Munawar, M. & I. F. Munawar, 1987. Phytoplankton bioassays for evaluating toxicity of *in situ* sediment contaminants. In; R. Thomas, R. Evans, A. Hamilton, M. Munawar, T. Reynoldson & H. Sadar (Eds.), Ecological effects of *in situ* sediment contaminants. Hydrobiologia 149: 87–105.

Munawar, M. & P. Ross, 1987. (Abstract). Toronto (Ont.) and Waukegan (Il.) harbours: Ecotoxicologial evaluation of sediment-bound toxicity in Great Lakes. Pap. pres. Amer. Soc. Limnol. Oceanogr. Madison, Wisconsin, 1987.

Munawar, M. & W. D. Taylor, 1985. Assessment of sediment-bound toxicity to protozoa. Great Lakes Lab. Fish. aquat. Sci., Burlington, Ontario, Canada. Report. 10 pp.

Munawar, M. & T. Weisse, 1989. Is the microbial loop an early warning indicator of anthropogenic stress? In; M. Munawar, G. Dixon, C. I. Mayfield, T. Reynoldson & M. H. Sadar (Eds.), Environmental bioassay techniques and their application. Hydrobiologia. (This volume).

Munawar, M., P. Stadelmann & I. F. Munawar, 1974. Phytoplankton biomass, species composition and primary production at a nearshore and a mid-lake station of Lake Ontario during IFYGL (IFYGL). Proc. 17th Conf. Great Lakes. Res. pp. 629–652.

Munawar, M., P. Wong & G. -Y. Rhee, 1988. The effects of contaminants on algae: an overview. In; N. W. Schmidtke (Ed.), Toxic contamination in large lakes. pp. 113–160. Lewis Publishers, Inc. Chelsea, Michigan.

Munawar, M., D. Gregor, S. A. Daniels & W. P. Norwood, 1989. A sensitive screening bioassay technique for the toxicological assessment of small quantities of contaminated bottom or suspended sediments. Hydrobiologia (76/177: 497–507.

Munawar, M., A. Mudroch, I. F. Munawar & R. L. Thomas, 1983. The impact of sediment-associated contaminants from the Niagara River mouth on various size assemblages of phytoplankton. J. Great Lakes Res. 9 (2): 303–313.

Munawar, M., I. F. Munawar, L. R. Culp & G. Dupuis, 1978. Relative importance of nannoplankton in Lake Superior phytoplankton biomass and community metabolism. In; M. Munawar (ed.). Limnology of Lake Superior. J. Great Lakes Res. 4 (3-4): 462–480.

Munawar, M., I. F. Munawar, W. Norwood & C. I. Mayfield, 1987a. Significance of autotrophic picoplankton in the Great Lakes and their use as early indicators of contaminant stress. In; M. Munawar (Ed.), Proc. Internat. Symp. on Phycology of Large Lakes of the World. Arch. Hydrobiol. Beih. Ergebn. Limnol. 25: 141–155.

Munawar, M., I. F. Munawar, P. Ross & C. Mayfield, 1987b. Differential sensitivity of natural phytoplankton size assemblages to metal mixture toxicity. In; M. Munawar (Ed.), Proc. Internat. Symp. on Phycology of Large Lakes of the World. Arch. Hydrobiol. Beih. Ergebn. Limnol. 25: 123–139.

Munawar, M., R. L. Thomas, W. Norwood & A. Mudroch, 1985. Toxicity of Detroit River sediment-bound contaminants to ultraplankton. J. Great Lakes Res. 11 (3): 264–274.

Nebeker, A. V., M. A. Cairns, J. H. Gakstatter, K. W. Malueg, G. S. Schuytema & D. K. Krawczyk, 1984. Biological methods for determining toxicity of contaminated freshwater sediments to invertebrates. Envir. Toxicol. Chem. 3: 617–630.

Nebeker, A. V. & C. E. Miller, 1988. Use of the amphipod crustacean *Hyalella azteca* in freshwater and estuarine sediment toxicity tests. Envir. Toxicol. Chem. 7: 1027–1033.

Nebeker, A. V., S. T. Onjukka, M. A. Cairns & D. K. Krawczyk, 1986. Survival of *Daphnia magna* and *Hyalella azteca* in cadmium-spiked water and sediment. Envir. Toxicol. Chem. 5: 933–938.

Persoone, G. & A. Van De Vel, 1987. Cost-analysis of 5 current aquatic ecotoxicological tests. Lab. Biol. Res. aquat. Pollut., State Univ. Ghent, Belgium. Report. 117 pp.

Reish, D. L. & P. S. Oshida, 1987. Manual for methods in aquatic environment research. FAO Fish. Techn. Pap. 247. 62 pp.

Rhee, G. -Y., 1988. Persistent toxic substances and phytoplankton in the Great Lakes. In; M. Evans (Ed.), Toxic Contaminants and ecosystem health, a Great Lakes focus. pp. 513–525. John Wiley & Sons, New York.

Ross, P. E. & M. Munawar, 1987. Zooplankton feeding rates at offshore stations in the North American Great Lakes. In; M. Munawar (Ed.), Proc. Internat. Symp. on Phycology of Large Lakes of the World. Arch. Hydrobiol. Beih. Ergebn. Limnol. 25: 157–164.

Schindler, D. W., 1987. Detecting ecosystem responses to anthropogenic stress. Can. J. Fish. aquat. Sci. 44: 6–25.

Severn, S. R. T., M. Munawar & C. I. Mayfield, 1989. Measurement of sediment toxicity of autotrophic and heterotrophic picoplankton by epifluorescence microscopy. Hydrobiologia. 176/177: 525–530.

Sherr, E. B. & B. F. Sherr, 1988. Role of microbes in pelagic food webs: a revised concept. Limnol. Oceanogr. 33 (5): 1225–1227.

Sloterdijk, H., L. Champoux, V. Jarry, Y. Couillard & P. Ross, 1989. Bioassay responses of microorganisms to sediment elutriates from the St. Lawrence River. In; M. Munawar, G. Dixon, C. I. Mayfield, T. Reynoldson & M. H. Sadar (Eds.), Environmental bioassay techniques and their application. Hydrobiologia. (This volume).

Smith, S. & M. K. H. Kwan, 1989. Use of aquatic macrophytes as a bioassay method to assess relative toxicity, uptake kinetics and accumulation forms of trace metals. In; M. Munawar, G. Dixon, C. I. Mayfield, T. Reynoldson & M. H. Sadar (Eds.), Environmental bioassay techniques and their application. Hydrobiologia. (This volume).

Sprules, G. & M. Munawar, 1986. Plankton size spectra in relation to ecosystem productivity, size, and perturbation. Can. J. Fish. aquat. Sci. 43: 1789–1794.

Sprules, G., M. Munawar & E. H. Jin, 1988. Plankton community structure and size spectra in the Georgian Bay and the North Channel ecosystems. In; M. Munawar (ed.), Limnology and Fisheries of Georgian Bay/North Channel ecosystems. Hydrobiologia 163: 135–140.

Stossel, F., 1987. Effect of the coefficients of discharge on ciliate populations of a running water contaminated by municipal wastewater. Arch. Hydrobiol. 108: 483–497.

Torrie, J., C. I. Mayfield & W. E. Inniss, 1987. A microcomputer-based image analysis system and applications to chlorophyll fluorescence studies. J. Microbiol. Methods. 6: 199–210.

U.S. Environmental Protection Agency, 1977. Technical Committee on Criteria for Dredged and Fill Material. Ecological evaluation of proposed discharge of dredged material into ocean waters. Environ. Effects Lab., U.S. Army Corp. of Engineers. Waterways Exp. Station, Vicksburg, Miss. 24 pp.

Vollenweider, R. A., M. Munawar & P. Stadelmann, 1974. A comparative review of phytoplankton and primary production in the Laurentian Great Lakes. J. Fish. Res. Bd. Can. 31 (5): 739–762.

Weisse, T. & M. Munawar, 1989. Evaluation of the microbial loop in the North American Great Lakes. *Can. Tech. Rep. Fish. Aquat. Sci.* 1709: i–vi. 1–30.

World Commission Report on Environment and Development. 1987. Our Common Future. Oxford University Press. 400 pp.

Hydrobiologia **188/189**: 117–121, 1989.
M. Munawar, G. Dixon, C. I. Mayfield, T. Reynoldson and M. H. Sadar (eds)
Environmental Bioassay Techniques and their Application.
© 1989 *Kluwer Academic Publishers. Printed in Belgium.*

Functional bioassays utilizing zooplankton: a comparison

Donald C. McNaught
Department of Ecology and Behavioral Biology, University of Minnesota, 318 Church St. SE,
Minneapolis, MN 55455 U.S.A. (Tel. (612)625-2765)

Key words: bioassay, zooplankton, crustaceans, functional, ingestion, reproduction, respiration, hormesis

Abstract

Functional zooplankton bioassays based on ingestion, reproduction and respiration are described, with methods for a new ingestion bioassay included. All bioassays are compared using three indices, including the variability of controls, the range of experimental responses, and a listing of contaminants causing inhibition/stimulation of response. The ingestion bioassay showed the greatest range of response, and was sensitive to pesticides, PCBs and heavy metals. It was also commonly characterized by a hormesis response. The reproduction bioassay showed the lowest variability, illustrated a reduced range of response, and was sensitive to nutrients and heavy metals. In one study, the respiration bioassay was sensitive only to PCBs.

Introduction

Functional or rate-based bioassays utilizing zooplankton may fall within five categories described by the equation for mass balance (1):

(1) Growth = Ingestion – Reproduction – Respiration – Defecation

Most heavily utilized zooplankton bioassays have used measures of reproduction or growth; Mirza (1968) was possibly the first to try an ingestion assay. I will describe one bioassay using a measure of ingestion, and then compare it to those in the other major categories (growth, reproduction, and respiration). Finally I wish to summarize briefly how useful this ingestion bioassay has been in studies of a natural ecosystem, the River Raisin (SE Michigan) tributary to Lake Erie.

Early zooplankton bioassays examined environmental factors through their impact on reproduction. Initially Nebeker *et al.* (1974) found that PCBs affected reproduction in *Daphnia*, measured readily by making egg counts. More recently the *Daphnia magna* bioassay (Gersich *et al.*, 1985) has been used to identify inhibition by contaminants. Possibly the most popular reproductive bioassay in current use employs *Ceriodaphnia* (Mount & Norberg, 1984). Detailed instructions on its application have been provided. This bioassay is useful over a wide temperature range (12 to 25 °C) and in a variety of trophic conditions (McNaught & Mount, 1986). In examining North American pollution problems, the *Ceriodaphnia* bioassay has been extensively used on three rivers and two of the Great Lakes (Norberg & Mount, 1986; Mount *et al.*, 1986a; Mount *et al.*, 1986b; Dolan, *et al.*, unpub.).

Zooplankton bioassays utilizing respiratory

rates are uncommon, but capable of detecting impacts of toxic contaminants as PCBs (Bridgham, 1988). Bridgham developed the zooplankton respiration bioassay, and provided instructions on its application.

Methods for ingestion bioassay

Classical methods for bioassays were followed; lake water was used as a resource control, while complex, mixed effluents added to lake water in varying amounts (10%, 25%, 50% by volume) constituted the experimental treatments. Grazing by common indigenous zooplankton was then measured in both control and experimental vessels. Seasonal data on the stimulation/inhibition of grazing can be constructed for dominant species occurring over an extended period. The bioassay requires the adaptation of grazers on control food resources for a long period (4 hrs) to minimize impact of handling on behavior (Bogdan & McNaught, 1975). Specifically the 8 steps involved in running a set of control and experimental vessels are:
1. Test animals (*Daphnia, Diaptomus, Cyclops*) acclimated to natural food levels of control water for 4 hours.
2. Dilutions of control water made with effluents from designated locations; effluent added at 10%, 25% and 50% by volume.
3. Particle density (3–40 μm ESD) determined for all control and experimental food containers, using an electronic particle counter.
4. Adult animals (10) transferred from acclimation container to experimental food container in carrier (acrylic tube with netting to allow passage of water and food).
5. Animals fed for 15 minutes.
6. Food density in control and experimental containers again determined.
7. Filtering rate calculated (within size-range of 3–40 μm ESD) using equation of Gauld (1951).
8. Inhibition/stimulation of grazing (relative to control) calculated.

One advantage of this bioassay is the speed of determination; controls and experimentals, with 3 replicates, can be completed in 1–2 days, enabling one collection of control water to be used. Both food and animals were maintained in a lighted, temperature-controlled cabinet. It should be noted however, that the initial results quoted here required two weeks to run, as water was shipped from a considerable distance.

Results

Three indices may be used to illustrate the usefulness of the ingestion bioassay, and to compare it to others using zooplankton. The variability of controls, measured by determining the coefficients of variation (C.V.) associated with each bioassay, tells something about precision and reproducibility. The range of experimental responses, relative to control fluxes, tells about the sensitivity of the flux to contaminants, and we further suggest that sensitivity of grazing to metals or organics would make this bioassay most useful. Finally, specific groups of contaminants to which the bioassay responded are listed, so that one may collect and disseminate such information.

The variability (C.V.) of controls for ingestion ranged from 3–38% of their mean (Table 1) and depended on the crustacean species employed and the ambient temperatures. For ingestion rates measured by visual counting, variability was low for *Daphnia* juveniles and adults (93–12%) at constant temperature (22 °C). Where electronic particle counting was used (McNaught *et al.*, 1988), variability between months, over a broad range of temperature 15–25 °C, was higher (37.5%). When the ingestion bioassay was used at constant temperature (15 °C), the C.V. was halved (15.3%).

For the ingestion bioassay, the range of experimental responses tells about sensitivity to specific compounds/ions. Rates of inhibition (by PCBs, lindane, toxaphene, Cu and Zn) ranged from − 25% to − 100%, or total inhibition of grazing (Table 2). Ranges of inhibition to unknown inhibitors in mixed toxic effluents ranged from − 96%

Table 1. Comparison of variability (C.V.) and range of response of *controls* for *filtration rates*.

Species		Mean filtering rate	C.V.	Range	Citation
Daphnia,	juv.	364 cpm	2.7	309–373	Gliwicz and Sienlowska, 1986
	adult	296 cpm	11.5	286–305	
Daphnia,	adult	0.57 ml/hr	–	0.33–1.02	McNaught, *et al.*, 1988
Diaptomus,	adult	0.40 ml/hr	37.5 *	0.14–1.02	McNaught, *et al.*, 1988
	adult	0.45 ml/hr	15.3 **	0.29–0.60	McNaught, *et al.*, 1988

At temperatures over range 15–25 °C*, versus 15 °C** constant.

to − 100%, illustrating the toxicity of these compounds (Table 2). Clearly this ingestion bioassay can measure impacts of chlorinated organics (PCBs), pesticides as lindane and toxaphene, and heavy metals (Cu, Zn).

Not surprisingly, we have frequently observed the phenomenon of hormesis, or stimulation of a process such as grazing by low concentrations of compounds, which are toxic at greater concentrations (Laughlin *et al.*, 1983). Stimulation of grazing by a calanoid copepod (*Diaptomus*), a cyclo-poid copepod (*Cyclops*) and two species of the cladoceran *Daphnia* ranged from 27 to 501% (Table 2). Specific compounds responsible for such a hormesis response are unknown, but as with other bioassays, it continues to be a real phenomenon. In running ingestion bioassays with pure PCB congeners, hormesis was also common (McNaught, 1984).

In comparison, the often used reproductive rate bioassay was characterized by low variability (Table 3); the coefficient of variation ranged from

Table 2. Zooplankton bioassays employing *filtration rates*; range of inhibition (%) and stimulation (%) of *experimental treatments* relative to controls; toxic substances identified in mixed effluents.

Bioassay	Species	Relative (% control)		Toxic substance	Range of concentration ($\mu g l^{-1}$)	Citation
		Inhibition	Stimulation			
Zooplankton Filtration Rate	*Daphnia magna*	− 87		Magnesium	75000	Mirza, 1968
		− 47	–	Nitrite	40000	
	Cyclops bicuspidatus	− 43		PCB	25	McNaught, *et al.*, 1984
	Eubosmina coregoni	− 96	–	Unknown		
	Chydorus sphaericus	− 99	–	Unknown		
	Diaptomus spp.	− 100	–	Unknown		
	Daphnia pulex	− 25	–	Lindane	50	Gliwicz and Sieniowska, 1986
	Diaptomus sicilis	− 65	–	Toxaphene	0.2–15	McNaught, *et al.*, 1988
	Limnocalanus marcurus	− 57	–	Toxaphene		
	Cyclops bicuspidatus	− 29	+ 501	Cu	0.2–21.9	McNaught, *et al.*, 1987
	Diaptomus spp.	− 100	+ 61	Zn	0.1–32.0	
	Daphnia retrocurva	− 71	+ 182			
	Daphnia schodleri	− 99	+ 27			
	Diaptomus spp.	− 90	–	Cu	1.2–21.9	McNaught, *et al.*, 1988
	Daphnia retrocurva	− 71	–	Zn	1.5–4.5	McNaught, *et al.*, 1988

Hydrobiologia **188/189**: 123–135, 1989.
M. Munawar, G. Dixon, C. I. Mayfield, T. Reynoldson and M. H. Sadar (eds)
Environmental Bioassay Techniques and their Application.
© *1989 Kluwer Academic Publishers. Printed in Belgium.*

A holistic approach to ecosystem health assessment using fish population characteristics

Kelly R. Munkittrick[1] & D. George Dixon[2]
[1] *Department of Zoology, College of Biological Science, University of Guelph, Guelph, Ontario, N1G 2W1, Canada; Present Address: E.V.S. Consultants, 195 Pemberton Ave., North Vancouver, B.C. V7P 2R4, Canada;* [2] *Dept. of Biology, University of Waterloo, Waterloo, Ontario, N2L 3G1 Canada*

Key words: ecosystem stress, fish populations, growth, fecundity, condition factor

Abstract

The status of a fish population is a reflection of the overall condition of the aquatic environment in which that population resides. As such, fish population characteristics can be used as indicators of environmental health. Simple and inexpensive methods to follow fish population responses to environmental degradation are lacking. This paper outlines a protocol whereby environmental impacts on fish populations are classified by five patterns based on characteristics such as mean age, fecundity and condition factor. The patterns summarize population changes and describe responses to exploitation, recruitment failure, the presence of multiple stressors, food limitation and niche shifts. Classification is best based on the selection, and appropriate sampling, of a comparable reference population. Population characteristics can be used to examine ecosystems exposed to stressors for evidence of long-term damage, and when used with biochemical indicators, can be a powerful tool for ecosystem health assessment. The five responses are illustrated using published data on a number of species challenged by increased predation pressure, acidification, eutrophication, mine waste and reservoir impoundment. Application of this scheme will aid in directing and focusing research efforts on crucial aspects impacted by changing conditions.

Introduction

The overall aim of most of the current research in environmental toxicology is to provide models or generalizations about contaminant action which can be used to decrease the costs of testing new chemicals, to predict health effects and to evaluate the impacts of effluents. The most common approach to wide-scale surveillance of aquatic ecosystems attempts to document the distribution and compartmentalization of contaminants, and to use this information to predict impacts on individual organisms, species or man. This approach is based largely on the assumption that toxicant impacts can be generalized, i.e. that

systems with similar characteristics will respond to similar toxicant exposures in similar ways.

This may not be a valid assumption. Variation has been seen in the responses of white sucker (*Catostomus commersoni*) populations to similar concentrations of elevated waterborne metals. McFarlane & Franzin (1978) examined a population of white sucker exposed to elevated copper (13 to 15 μg l^{-1}) and zinc (245 μg l^{-1}) associated with atmospheric metal deposition. The population exhibited an increased growth rate and fecundity and an earlier age to maturity. White sucker populations exposed to copper (12 to 15 μg l^{-1}) and zinc (209 to 253 μg l^{-1}) associated with mine waste discharge showed a decreased growth rate and fecundity and no change in age at maturity, relative to the reference site (Munkittrick & Dixon, 1988a, b).

It is still unclear why the response of the Manitouwadge white sucker population (Munkittrick & Dixon, 1988b) was so different from that described for Hamell Lake sucker (McFarlane & Franzin, 1978). White sucker are still abundant in our lakes, while collections at Hamell Lake in 1984 yielded only small numbers of older fish, and 1987 collection attempts were unsuccessful (Dutton *et al.*, 1988). Water metal levels are almost identical at the two sites and the deviations cannot be explained by differences in sediment metal levels or water hardness (Dutton, pers. comm.). Furthermore, liver metal burdens reported for the Hamell Lake site are only about 20% (McFarlane & Franzin, 1980; Dutton *et al.*, 1988) of those encountered in the Manitouwadge fish (Munkittrick and Dixon, 1988d).

Changes at Hamell Lake were related to compensatory responses to a decreased population size (McFarlane & Franzin, 1978; Trippel & Harvey, 1987a, b) while the changes at Manitouwadge were associated with the effect of increased sediment metals on food availability (German, 1971; Pugh & Maki, 1986; Munkittrick & Dixon, 1988d). The obvious differences exhibited by the two white sucker populations, in response to similar environmental stressors, complicates attempts to generalize ecosystem response to environmental perturbations. Variation has also been seen in the responses of species (Beggs & Gunn, 1986) and populations (Sun & Harvey, 1986) to acidification.

Inconsistencies such as these complicate attempts to develop comprehensive *a priori* sampling protocols and hypotheses for population based ecosystem health assessment. A sampling schedule designed to examine reproductive abnormalities associated with recruitment failure is not always the best design for measuring growth impairments. Until population responses, and the mechanisms governing their action, are better understood, contaminated lakes may require site by site evaluation. The apparent inability to provide general toxicant responses suggests that it is very important to identify factors associated with the responses of fish populations to stress, and to provide techniques for rapid assessment of population structure.

Holistic monitoring

Environmental health assessments attempt to relate changes evident in a population of free-living organisms to stressors acting on the ecosystem. A major drawback to this approach is however the virtual absence of data linking toxicant levels with meaningful whole organism, population or ecosystem responses. Attempts to assess ecosystem health can be accomplished through an overview of ecosystem activity from the 'top-down' (holistic approach) or through an examination of small components of the system (reductionist approach) (NRCC, 1985). The ability to rapidly detect significant, meaningful changes in populations must be in place before holistic monitoring can play a significant role in surveillance programs.

Holistic monitoring is retrogressive by nature and the system being monitored acts as the indicator of its own health. Several approaches are available (Cairns *et al.*, 1984; Ryder & Edwards, 1985; Schindler, 1987), but the time lag between contamination and the detection of alterations at the population or ecosystem level must be minimized. The ability to detect changes before damage is irreversible must be optimized, while avoiding the detection of changes which do not affect energy flow through the ecosystem. Monitoring of fish populations central to the aquatic food web is a compromise between the sensitivity and complexity of invertebrate responses and the prolonged time required for response of top-level predators (Munkittrick & Dixon, 1988c).

Population structure

Several generalized patterns describing changes in population structure have been identified. Population size is affected by a change in habitat and/or resource availability through the mechanisms of density-dependent regulation. Under normal situations, these changes would be associated with a change in the birth or death rate. The observer's ability to detect significant deviations from the norm is dependent upon the assumption that the population will exhibit certain general characteristics in an undisturbed environment.

Population characteristics such as fecundity, mean age and growth rate integrate the experience of the population over a relatively longer period of time into factors which can be easily measured. The degree to which population characteristics fluctuate in undisturbed environments is subject to speculation. Fish populations in northern single-species lakes, as well as the dominant species in more complex environments, are capable of maintaining age and frequency distributions which remain relatively stable (Elliott, 1985; Johnson, 1976; Mills, 1985; Donald & Alger, 1986b). Most changes in phenotypic characteristics can be detected by examining gross indicators of population status. Although it is possible to judge the degree of exploitation of a fishery from relatively simple measurements of mortality, growth, age-structure and maturity (Healey, 1975), attempts to do this accurately are not always successful. Attempts to describe such alterations mathematically do not always take into account the complexity of the factors regulating ecosystem activity or the interactions between various stressors (Schneider & Leach, 1977).

Growth, fecundity and maturity are phenotypic expressions induced by the environment; there is no evidence that environmentally-induced changes in growth, age to maturation, fecundity or longevity of walleye (*Stizostedion vitreum*) are heritable (Colby & Nepszy, 1981). Assuming that life history parameters are not treated independently, interrelationships can provide a rapid response to changing conditions of growth and survival (Jensen, 1985). Although changes in population structure may act as a sensitive indicator of changing environmental conditions, the timing, degree and nature of the feedback response to altered conditions will vary with the intensity, identity and the number of stressors, as well as the availability of energy. If the monitoring of fish population characteristics is to be of use in environmental health assessment, the nature of the population response must first be understood. This task will be easiest if we compare characteristics against a set of contrasting, generalized responses.

Descriptions of response patterns

Population responses to contaminants should be identical to any non-specific, density-independent stressor. Aquatic ecosystems can respond to changing conditions in only a limited number of ways, and the response will be the result of an altered performance or function as opposed to the direct action of a specific stressor. In the case of selective egg mortality, the population responds only to the decrease in recruitment and not selectively to the fact that the stressor happened to be abnormal temperature, decreased pH or high metal levels. The studies become increasingly complex when there are sublethal, or density-dependent impacts on survivors, and in these cases it is especially crucial to identify the impact point.

There are a limited number of ways in which stressors can affect fish population structure. Current data suggest that the response of adult populations to contaminants will be indistinguishable from the response to other stressors, such as overfishing or eutrophication (Ryder & Edwards, 1985). Direct impacts can be mediated through an increased adult mortality, decreased juvenile survival, or an artificial increase in food availability (eutrophication). In each case the population will respond to an increase in the relative availability of food and/or habitat. After the mortality of adults, the mean age of the population will decrease, resulting in a younger population showing better growth characteristics in response to the decreased population size (type I response). In the case of increased juvenile and/or egg mortality, the mean age of the population will increase, resulting in an older population. Since competition between adults would not change initially, growth characteristics of the adults could at first remain the same (type II response). Other responses correspond to impacts on juvenile performance (type III), the indirect effect of changes in predation pressure (type IV) or the indirect effect of food limitation (type V). Each alteration produces a characteristic change in the fish population, which has been described according to the response to fishing pressure (Colby, 1984) and contaminant impact (Munkittrick & Dixon, 1988c). The derivation of the patterns and their applicability to the use of white sucker populations for monitoring ecosystems stress has been fully described elsewhere (Munkittrick & Dixon, 1988c).

Application of the framework

The holistic population approach to ecosystem health assessment asks the question: 'How does this system differ from an unstressed system?' The description of changes must be made relative to a comparable reference population. The selection of an appropriate reference site, as well as a suitable sampling schedule, are essential for successful interpretation of changes (Munkittrick & Dixon, 1988c). It is crucial to select comparable reference data from sites with similar environmental conditions.

Information on many parameters can be incorporated, but the patterns can be separated based on the relative changes in mean age, condition factor and fecundity (Table 1). Although relevant

Table 1. Relative changes in population characteristics associated with the various response patterns.

Pattern	Description	Mean Age	Condition Factor	Fecundity
I	Exploitation	−	+	+
II	Recruitment	+	0	0
III	Multiple Stressors	+	−	−
IV	Limitation	0	−	−
V	Niche Shift	0	0	−

data should be available for any population of fish, the relative importance of characteristics will vary from species to species. Variation can be attributed to genetically-determined attributes of each species, as well as environmental restraint (Donald & Alger, 1986b). For example, size and age at maturity are not always related in lake trout (*Salvelinus namaycush*), and age at first spawning may vary with the age distribution of the population (Johnson, 1976; Donald & Alger, 1986a). The approach requires an understanding of the limitations of the species and system being investigated. Additional parameters, such as sex ratio may be defined in the future. In whitefish (*Coregonus* spp.), the sex ratio of the population may vary under stress (George, 1977), while the importance of changes in sex ratio for many temperate, freshwater fish have not been identified (Colby, 1984).

The description of the response pattern has the potential to offer the researcher important information early in the study, but conclusions should only be used to direct research and sampling efforts. It is important to realize that the assessment framework is meant to be descriptive, not predictive, and that the response patterns may overlap temporally. The Hamell Lake population of white sucker exposed to atmospheric metal deposition exhibited a type I response in the 1970s (McFarlane & Franzin, 1978) and a type II response in 1984 (Dutton *et al.*, 1988).

This does not mean that all populations will necessarily progress through the patterns sequentially from a type I to a type V response. The first response of a fish population to declining food availability would be a decline in fecundity, since the primary allocation of energy is to egg production for many species (Munkittrick & Dixon, 1988b). Theoretically, we can extend this example to show that patterns may in fact move sequentially from V to I. A further restriction of food availability would result in a further restriction of performance, and the condition factor of individuals could be expected to decline. Sampling of the population at this point in time would result in the classification of a type IV response (Table 1). The fall in fecundity and fish condition would result in a gradual increase in mean age, consistent with the drop in reproductive effort and recruitment. Sampling at this time would detect the characteristics of a type III response. A persistent decline in reproductive effort associated with poor food availability would eventually result in a fall in the population size to a new level. If this level was consistent with the altered (decreased) carrying capacity of the system, fecundity and condition factor would return to 'normal', and sampling would detect only an increased mean age (type II response). If the new population size was less than the carrying capacity, the relative increase in food availability would result in a type I response.

Since patterns can move in both directions it may not be possible to rank responses into degrees of deterioration. The impact site will vary with the identity and intensity of the stressor and sampling of the population will only provide a profile of population status at one specific time and place. Definition of the response is necessary for a complete understanding of the impact of stressors on the population, but until more information has been collected, it will not be possible to use holistic monitoring for predictive purposes. The following sections describe the derivation of the various patterns and outline some scenarios which may accompany them. The patterns were originally described by Colby (1984) for assessment of fishery impacts and we have retained much of this original terminology. An expansion of the framework together with a complete discussion of its applicability to case studies involving white sucker populations can be found in Munkittrick & Dixon (1988c).

Type I response (EXPLOITATION)

The best understood response pattern is the characteristic compensatory response of a previously unexploited fish population to adult removal (Colby, 1984). The removal of a significant number of adults results in a relative increase in the amount of food and habitat available for those surviving. This relative increase theoretically leads to an increased growth rate and fecundity, as well as an earlier age at maturation (McFarlane & Franzin, 1978; Trippel & Harvey, 1987b). Due to the shift in the age-structure of the population, the mean age of the population declines. This response pattern has been described for populations of whitefish (Healey, 1975; Jensen, 1981), walleye (Lysack, 1980 in Colby, 1984), rock bass (*Ambloplites rupestris*) and yellow perch (*Perca flavescens*) (Schneider & Leach, 1977) after exploitation.

A type I response should be found whenever a sudden decrease in the population size has occurred, and not just in response to man's harvest of a standing crop. Type I responses have been documented in response to increased mortality associated with predation on fish by harbour seals (*Phoca vitulina*) (Power & Gregoire, 1978), parasitization by *Ligula intestinalis* (Burrough & Kennedy, 1979) and *Petromyzon marinus* (Henderson, 1986) and the chronic effects of atmospheric metal deposition (McFarlane & Franzin, 1978) (Table 2).

A similar response would be expected if the amount of food or habitat available for utilization increased, a situation analogous to the removal of adults. Liming of acid lakes results in a restoration of habitat, and the increase in pelagic invertebrates is associated with an increase in the numbers of smaller fish, increased fish population size (Erikson & Tengelin, 1987) and an increase in the growth rate and condition of the fish (Rosseland, 1986; Trippel & Harvey, 1987a, b), all characteristic type I responses. Similar results have been reported after decreased acid deposition (Beggs & Gunn, 1986), experimental lake fertilization (Mills, 1985) and after removal of undesirable competing species (Johnson, 1975).

Type I responses have been documented for decreases in population size, or whenever the availability of food or habitat increases. An increase in food quality should result in the same response. Piscivorous feeding by lake trout is associated with an increased growth rate (Martin, 1951) and stocking of a lake with appropriate forage species resulted in an increased weight, growth rate, age to maturity, fecundity and egg

Table 2. Generalized type I response pattern (A) and summary of documented occurrences (B). Type I patterns represent a compensatory response to an increase in the availability of food or habitat.

Species	Stressor	Mean age	Age distribution	Growth rate	Condition factor	Age at maturation	Fecundity	Egg size	Population size	Reference
A. GENERAL PATTERN		−	shift to younger	+	+	−	+	−/+	−	Colby, 1984; Munkittrick & Dixon, 1988c
B. DOCUMENTED CASES										
Whitefish	Exploitation			+		−	+		−	Healey, 1975; Jensen, 1981
Whitefish	Eutrophication			+	+		0[1]		+	Mills, 1985
Lake trout	Predation pressure	−		+		−	+	−		Power & Gregoire, 1978.
Lake trout	Increased food supply			+		+	+	+		Martin, 1951, 1970
Walleye	Eutrophication	−		+		−				Lysack, 1980
Walleye	Removal of competitors			+	+					Johnston, 1975
White sucker	Atmospheric metal deposition	−	shift to younger	+	0	−	+	−	−	McFarlane & Franzin, 1978
White sucker	Acidification	−		+	+	+*	0*	+	−	Trippel & Harvey, 1987a, b
White sucker	Predation pressure	−	shift to younger	+	+				−	Henderson, 1986
Yellow perch	Acidification	−		+					+	Sun & Harvey, 1986
European perch	Liming of acid lakes	−	shift to younger	+	+		+		+	Rosseland, 1986; Erikson & Tengelin, 1987
Roach	Parasitization			+	+	−	+		+	Burrough & Kennedy, 1979

[1] Based on weight.

* Atypical responses, see text.

size, and an increased fertility relative to planktivorous lake trout (Martin, 1970).

Type II response (RECRUITMENT FAILURE)

A type II response is also characterized by an increased growth rate in response to a decreased population size. The response differs from a type I pattern in that there is an increase in the mean age of the population (Table 3), due to prolonged increases in egg mortality or recruitment failure (Colby, 1984). The response can be due to deterioration of spawning or nursery habitat, or to stressor-induced spawning failures, and is typical of a population approaching extinction.

The absence of younger year classes was evident in white sucker collected from Hamell Lake during the early 1980s, prior to unsuccessful collection attempts in 1987 (Dutton *et al.*, 1988). Similar changes have been reported in lake trout populations inhabiting water of pH < 5.5 (Beggs & Gunn, 1986), in a goldfish (*Carassius auratus*) population inhabiting a heavily polluted industrial basin (Munkittrick & Leatherland, 1984) and in a population of sauger (*Stizostedion canadense*) inhabiting a lake polluted by copper mining wastes (Black *et al.*, 1985). The pattern was also found in Lake Simcoe whitefish when the mean size and age of the population increased in association with successive recruitment failures (Evans, 1978). The whitefish reproductive failure was associated with a decline in water quality to levels which could not support natural reproduction (Colby, 1984).

Type III response (MULTIPLE STRESSORS)

In the absence of contaminants, a type III response is reflective of the persistence of marginal, adverse conditions for a prolonged period of time. Food supply problems are associated with a decline in growth rate and fecundity. The increased mean age can be related to a decline in reproduction and recruitment, the size-selective mortality of young fish, or to a prolonged decline in habitat or food supply.

The stocking of brook trout in small, high altitude lakes resulted in a decreased growth and condition factor and an increased age to maturation (Riemers, 1958, 1979). The absence of reproduction and failure of the fish to mature until an advanced age resulted in a gradual increase in mean age and typical type III characteristics (Table 4). Arctic charr (*Salvelinus alpinus*) living in a very deep, nutrient-poor, meteoritic crater also exhibited increased age at maturation and mean age, and decreased fecundity, growth rate and condition factor (Martin, 1955), all type III responses.

The invasion and establishment of a black crappie (*Pomoxis nigromaculatus*) population resulted in direct competition with existing walleye stocks in a small lake, and led to a decreased growth rate (Schiavone, 1981). Predation by the crappie on walleye larvae resulted in a decreased population size and an increase in the mean age (Schiavone, 1983).

Type III responses have also been associated with contamination events, and are suggestive of multiple stressors. Generally, factors associated with recruitment failure are responsible for increasing the mean age, while food availability problems prevent a characteristic compensatory response. The pattern was reported by Handford *et al.* (1977) for lake whitefish from Lesser Slave Lake. Siltation associated with oil exploration decreased both the survival of eggs and the feeding success of adult whitefish. These changes were combined with a prolonged harvest of fast-growing individuals by a gillnet fishery resulted in a characteristic type III response pattern. This pattern has also been reported for white sucker inhabiting a lake impacted by radionuclide mining waste. A decreased food abundance associated with the mine site resulted in a poorer condition of adults, while aluminum precipitation associated with the tailings adversely affected egg survival (Swanson, 1982).

Type IV response (LIMITATION)

This pattern is evident where a fish population has reached the carrying capacity of a system. The

Table 3. Generalized type II response pattern (A) and summary of documented occurrences (B). Type II responses are reflective of successive recruitment failures.

	Mean age	Age distribution	Growth rate	Condition factor	Age at maturation	Fecundity	Egg size	Population size	Reference
A. GENERAL PATTERN	+	shift to older	0/+	0	0	0	0	–	Colby, 1984; Munkittrick & Dixon, 1988c

B. DOCUMENT CASES

Species	*Stressor*	Mean age	Age distribution	Growth rate	Condition factor	Age at maturation	Fecundity	Egg size	Population size	Reference
Whitefish	Overfishing	+	shift to older						–	Evans, 1978
Lake trout	Acidification	+	shift to older						–	Beggs & Gunn, 1986
White sucker	Atmospheric metal deposition	+	shift to older						–	Dutton *et al.*, 1988
Goldfish	Pollution	+	shift to older						–	Munkittrick & Leatherland, 1984
Sauger	Copper mining waste	+	shift to older							Black *et al.*, 1985

Table 4. Generalized type III response pattern (A) and summary of documented occurrences (B). With respect to contaminants, a type III response usually is indicative of the presence of multiple stressors.

	Mean age	Age distribution	Growth rate	Condition factor	Age at maturation	Fecundity	Egg size	Population size	Reference
A. GENERAL PATTERN	+	shift to older	–	–	+	–	–/0	–	Colby, 1984; Munkittrick & Dixon, 1988c

B. DOCUMENTED CASES

Species	*Stressor*	Mean age	Age distribution	Growth rate	Condition factor	Age at maturation	Fecundity	Egg size	Population size	Reference
Whitefish	Oil exploration	+		–						Handford *et al.*, 1977
Brook trout	Poor food supply (overstocking)	+	shift to older	–	–	+		–		Reimers, 1958, 1967
Walleye	Competition with exotic species	+		–						Schiavone, 1981, 1982
White sucker	Radionuclide mining waste	+	shift to older	–	–		–			Swanson, 1982
Arctic char	Nutrient-poor lake	+		–	–	+	–			Martin, 1955

response is initiated by a decline in food and habitat availability, and the population does not show an increase in the mean age (Table 5). The response is often associated with an increased population size due to predator removal or over-stocking, or to a decline in habitat availability. A decline in food availability should result in decreased growth rate, condition factor and fecundity, and an increase in the age at maturity. The persistence of conditions will result in a gradual increase in mean age, characteristic of the type III response.

Reservoir impoundment has been associated with an increase in pike (*Esox lucius*) reproduction due to the increase in flooded vegetation. The subsequent increased population size resulted in decreased growth rate and condition of the year-class (Hassler, 1969; Nelson, 1974). It should be noted that there may be no detectable impact on the adult population, and that the increased production may be transient (Bodaly & Lesack, 1984).

Barnes *et al.* (1984) reported that whitefish accumulating below a hydroelectric dam exhibited decreased fecundity, condition factor and growth rate. The fish appeared to maintain their position in a suboptimal environment due to increased oxygen concentrations, with the changes in population characteristics being due, presumably, to increased population size and increased competition below the dam.

Similar changes have been seen when salmonids are stocked in high-altitude or alpine lakes with limited food availability. Lake trout in shallow, alpine ponds exhibited a very slow growth rate due to the absence of suitable food (Donald & Alger, 1986a). Identical changes have been reported for roach (*Rutilus rutilus*) (Burrough & Kennedy, 1979), brown trout (*Salmo trutta*), rainbow trout (*Salmo gairdneri*), Arctic charr and lake trout due to poor food availability (reviewed in Donald & Alger, 1986b).

Predator removal by overfishing can increase centrarchid population size, resulting in stunting of the fish (Anderson & Weithman, 1978; Colby, 1984). Stunted populations have been reported for yellow perch, Eurasian perch (*Perca fluviatilis*), Arctic charr, pumpkinseed sunfish (*Lepomis gibbosus*), bluegill sunfish (*Lepomis macrochirus*) and whitefish due to increased food competition (reviewed in Hanson & Leggett, 1986; Persson, 1987).

Type V response (NICHE SHIFT)

A type V response is characterized by a decline in fecundity of the fish without concomitant changes in condition or mean age (Table 6). This response is typically seen when a portion of the population is eliminated and a stressor prevents the population from regaining its former abundance. It can also be seen when there is a gradual change in food availability, or when the introduction of a competing species results in a niche shift.

A typical type V response has been described for situations where overfishing of walleye resulted in an initial decline in the number of adults. In several cases, pollution or competition from exotic species prevented the walleye from returning to their previous abundance (Schneider & Leach, 1977; Colby, 1984). The type V pattern was also reported for a white sucker population in response to a decline in food quality and quantity. Sucker collected at contaminated sites showed marked differences in stomach contents (Munkittrick & Dixon, 1988d) and growth (Munkittrick & Dixon, 1988b) characteristic of a niche shift after responding to changing food availability.

Eutrophication would be expected to produce a type I response due to the relative increase in nutrient availability, but Eurasian perch failed to show the expected response (Bregazzi & Kennedy, 1982). The failure was attributed to a niche expansion by a competing roach population, which prevented the perch from responding in the typical fashion. A limited decline occurred in the size of the perch population, with no change in descriptive parameters, characteristic of a type V response (Bregazzi & Kennedy, 1982).

Table 5. Generalized type IV response pattern (A) and summary of documented occurrences (B). A type IV response is generally associated with a decline in food availability.

	Mean age	Age distribution	Growth rate	Condition factor	Age at maturation	Fecundity	Egg size	Population size	Reference
A. GENERAL PATTERN	0	0	-	-	+	-	-	+	Colby, 1984; Munkittrick & Dixon, 1988c

B. DOCUMENTED CASES

Species	Stressor	Mean age	Age distribution	Growth rate	Condition factor	Age at maturation	Fecundity	Egg size	Population size	Reference
Whitefish	Hydro dam			-	-		-		+	Barnes et al., 1984
Lake trout	Stocking in alpine lakes			-		0			0	Donald & Alger, 1986
Pike	Lake impoundment			-	-				+	Bodaly & Lesack, 1984

Table 6. Generalized type V response pattern (A) and summary of documented occurrences (B). A type V response pattern represents a response to changes in food availability (niche shift).

	Mean age	Age distribution	Growth rate	Condition factor	Age at maturation	Fecundity	Egg size	Population size	Reference
A. GENERAL PATTERN	0	0	+	+	+	-	-	-/0	Colby, 1984; Munkittrick & Dixon, 1988c

B. DOCUMENTED CASES

Species	Stressor	Mean age	Age distribution	Growth rate	Condition factor	Age at maturation	Fecundity	Egg size	Population size	Reference
White sucker	Mixed metal mine waste	0		-	0		-	-		Munkittrick & Dixon, 1988a, b
Yellow perch	Eutrophication	-[1]		-[2]		+	-[3]			Breggazi & Kennedy, 1982

[1] Prevented from type I response by competition.
[2] Only in first three years.
[3] Absolute fecundity.

Summary

It is necessary to identify the population response so that impacts can be identified and sorted out. Response patterns can be identified with simple information collected from within many existing sampling programs. It is important to remember that the system is meant to be descriptive and not predictive. Mechanisms regulating population size will play a large role in the formulation of the response pattern, and will vary with both species and location. The loss of community diversity and stability evident in many damaged systems can lead to an inefficient operation of the mechanisms regulating population size, resulting in fluctuations in total population size (Adams & Olver, 1977). The effects of these fluctuations on general indicators of population status have not been well documented, but it is clear that changes become difficult to interpret when density-independent mechanisms are the dominant factors governing population size.

Perch populations are thought to be regulated predominantly by water temperatures and water level (Henderson, 1985). Perch populations have responded predictably to some stressors (Erikson & Tengelin, 1987; Persson, 1986, 1987), but have shown variation (Sun & Harvey, 1986) or failed to respond (Breggazi & Kennedy, 1982) in other situations. Salmonids respond predictably in most situations, but rainbow trout populations do not show stability in mountain lakes due to the large impact of density-independent factors such as temperature (Donald & Alger, 1986b). Fish population characteristics vary considerably from site to site and we need to understand the factors involved to provide indications of which data require collection, and what the resulting impacts have been on the population under examination.

If the population grows, reproduces and survives within the limits of a comparable reference population, we must conclude that there is no impact on the population. The identification or classification of a response pattern can lead researchers to ask the appropriate questions necessary to trace stressor impact. Although it may be impossible to separate stressor effects in complex situations, it may be possible to understand changes if the presence of stressors can be isolated or understood *a priori*. An assessment concluding with a type I response will suggest further study of factors associated with adult survival, while a type II response would suggest an investigation of reproductive performance and a type V response requires an investigation of food availability.

When comparisons are made against the extended pattern, not all examinations will fit a defined response pattern, although the contrasting setup can still direct research needs. Trippel & Harvey (1987a, b) describe a typical type I response of white sucker populations to decreased pH of lakes. However, the sucker failed to exhibit a decline in age to maturity and failed to show an increase in fecundity. The abnormal reproductive response to increased food availability obviously points to an area requiring further research.

Conclusions

Population characteristics can be used to examine populations experiencing stress for evidence of long-term damage, or evidence of responses to restoration (Ryder & Edwards, 1985). A comparison of population characteristics with historical data, with data from a reference population or with data collected from unstressed individuals within the population (Munkittrick & Dixon, 1988c) can offer insight into the chronic impacts of the alterations. Examination of population changes, in concert with sensitive, specific biochemical testing, can provide a powerful tool for demonstration of the impact of low-level stressors.

Response patterns can give clues to the selection of comparable reference sites and the development of suitable sampling schedules. Simultaneous analysis of biochemical and population responses can direct research efforts, identify sources of variability and provide information on the impact point of a stressor. Furthermore, the early identification of the mechanism of impact allows research resources to be concentrated on areas and tests which are crucial to understanding the response of the entire system.

Acknowledgements

The concepts put forward in this paper were developed and refined during research supported by a Natural Sciences and Engineering Research Council of Canada Operating Grant (A 8155) and two Ontario Ministry of the Environment Environmental Research Grants (193 RR and 331 RR) to D.G.D.

References

Adams, G. F. & C. H. Olver, 1977. Yield properties and structure of boreal percid communities in Ontario. J. Fish. Res. Bd. Can. 34: 1613–1625.

Anderson, R. O. & A. S. Weithman, 1978. The concept of balance for coolwater fish populations. Am. Fish. Soc. Spec. Publ. 11: 371–381.

Barnes, M. A., G. Power & R. G. H. Downer, 1984. Stress-related changes in lake whitefish (*Coregonus clupeaformis*) associated with a hydroelectric control structure. Can. J. Fish. Aquat. Sci. 41: 1528–1533.

Beggs, G. L. & J. M. Gunn, 1986. Response of lake trout (*Salvelinus namaycush*) and brook trout (*S. fontinalis*) to surface water acidification in Ontario. Wat. Air Soil Pollut. 30: 711–717.

Black, J. J., E. D. Evans, J. C. Harshbarger & R. F. Ziegel, 1985. Epizootic neoplasms in fishes from a lake polluted by copper mining wastes. J. Nat. Cancer Inst. 69: 915–926.

Bodaly, R. A. & L. F. W. Lesack, 1984. Response of a boreal Northern pike (*Esox lucius*) population to lake impoundment: Wupaw Bay, Southern Indian Lake, Manitoba. Can. J. Fish. Aquat. Sci. 41: 706–714.

Breggazi, P. R. & C. R. Kennedy, 1982. The response of perch, *Perca fluviatilis* L., population to eutrophication and associated changes in fish fauna in a small lake. J. Fish Biol. 20: 21–31.

Burrough, R. J. & C. R. Kennedy, 1979. The occurrence and natural alleviation of stunting in a population of roach, *Rutilus rutilus* (L.). J. Fish Biol. 15: 93–109.

Cairns, V. W., P. V. Hodson & J. O. Nriagu, 1984. Contaminant Effect on Fisheries, Advance. Envir. Sci. Technol. 16: 1–333.

Colby, P. J., 1984. Appraising the status of fisheries: rehabilitation techniques. In: V. W. Cairns, P. .V. Hodson and J. O. Nriagu (eds.), Contaminant Effects on Fisheries. Adv. Envir. Sci. Technol. 16: 233–257.

Colby, P. J. & S. J. Nepszy, 1981. Variation among stocks of walleye (*Stizostedion vitreum vitreum*): management implications. Can. J. Fish Aquat. Sci. 38: 1814–1831.

Donald, D. B. & D. J. Alger, 1986a. Stunted lake trout (*Salvelinus namaycush*) from the Rocky Mountains. Can. J. Fish. Aquat. Sci. 43: 608–612.

Donald, D. B. & D. J. Alger, 1986b. Dynamics of unexploited and lightly exploited populations of rainbow trout (*Salmo gairdneri*) from coastal, montane, and subalpine lakes in western Canada. Can. J. Fish. Aquat. Sci. 43: 1733–1741.

Dutton, M. D., H. S. Majewski & J. F. Klaverkamp, 1988. Biochemical stress indicators in fish from lakes near a metal smelter. Internat. Assoc. Great Lakes Res., 31st Conf., May 17, 1988, Hamilton, Ontario, p. A-14.

Elliott, J. M., 1985. Population regulation for different life-stages of migratory trout, *Salmo trutta* in a Lake District stream, 1966-83. J. Anim. Ecol. 54: 617–638.

Erikson, M. O. G. & B. Tengelin, 1987. Short-term effects of liming on perch, *Perca fluviatilis* populations in acidified lakes in South-west Sweden. Hydrobiologia 146: 187–191.

Evans, D. O., 1978. An overview of the ecology of the lake whitefish, *Coregonus clupeaformis* (Mitchill) in Lake Simcoe, Ontario with special reference to water quality and introduction of the rainbow smelt, *Osmerus mordax* (Mitchill). Ont. Min. Nat. Res. file rept., 132 pp.

George, C. J., 1977. The implication of neuroendocrine mechanisms in the regulation of population character. Fisheries 2(3): 14–19.

German, M. J., 1971. Effects of acid mine wastes on the chemistry and ecology of lakes in the Manitouwadge chain district of Thunder Bay, Ont. Water Res. Comm., Spec. Rept., Ont. Min. Env., Thunder Bay, Ontario, 19 pp.

Handford, P., G. Bell & T. Reimchen, 1977. A gillnet fishery considered as an experiment in artificial selection. J. Fish. Res. Bd. Can. 34: 954–961.

Hanson, J. M. & W. C. Leggett, 1986. Effect of competition between two freshwater fishes on prey consumption and abundance. Can. J. Fish. Aquat. Sci. 43: 1363–1372.

Hassler, T. J., 1969. Biology of the northern pike in Oahe reservoir, 1959 through 1965. U.S. Fish Wildl. Serv. Tech. Paper 29, 13 pp.

Healey, M. C., 1975. Dynamics of exploited whitefish populations and their management with special reference to the Northwest Territories. J. Fish. Res. Bd. Can. 32: 427–448.

Henderson, B. A., 1986. Effects of sea lamprey (*Petromyzon marinus*) parasitism on the abundance of white sucker (*Catostomus commersoni*) in South Bay, Lake Huron. J. Appl. Ecol. 23: 381–389.

Jensen, A. L., 1981. Population regulation in lake whitefish, *Coregonus clupeaformis* (Mitchell). J. Fish Biol. 19: 557–573.

Jensen, A. L., 1985. Relations among net reproductive rate and life history parameters for lake whitefish (*Coregonus clupeaformis*). Can. J. Fish. Aquat. Sci. 42: 164–168.

Johnson, F. H., 1975. Interspecific relationships of walleye, white sucker and associated species in a Northeastern Minnesota lake, with an evaluation of white sucker removal for increased walleye yield. Minn. Dept. Nat. Res., Div. Fish Wildl., Sect. Fisheries Investig. Rept. No. 338. 19 pp., Duluth, MN.

Johnson, L., 1976. Ecology of Arctic populations of lake trout, *Salvelinus namaycush*, lake whitefish, *Coregonus clupeaformis*, Arctic char, *Salvelinus alpinus*, and associated species in unexploited lakes of the Canadian Northwest Territories. J. Fish. Res. Bd. Can. 33: 2459–2488.

Lysack, W., 1980. Lake Winnipeg fish stock assessment program. Manitoba Dept. Nat. Res. rep. No. 80-30, 118 pp.

Martin, N. V., 1951. A study of the lake trout, *Salvelinus namaycush*, in two Algonquin Park, Ontario, lakes. Trans. Am. Fish. Soc. 81: 111–137.

Martin, N. V., 1955. Limnological and biological observations in the region of the Ungava or Chubb Crater, Province of Quebec. J. Fish. Res. Bd. Can. 12: 487–498.

Martin, N. V., 1970. Long-term effects of diet on the biology of the lake trout and the fishery in Lake Opeongo, Ontario. J. Fish. Res. Bd. Can. 27: 125–146.

McFarlane, G. A. & W. G. Franzin, 1978. Elevated heavy metals: a stress on a population of white suckers, *Catostomus commersoni*, in Hamell Lake, Saskatchewan. J. Fish. Res. Bd. Can. 35: 963–970.

McFarlane, G. A. & W. G. Franzin, 1980. An examination of Cd, Cu, and Hg concentrations in livers of Northern pike, *Esox lucius*, and white sucker, *Catostomus commersoni*, from five lakes near a base metal smelter at Flin Flon, Manitoba. Can. J. Fish. Aquat. Sci. 37: 1573–1578.

Mills, K. H., 1985. Responses of lake whitefish (*Coregonus clupeaformis*) to fertilization of Lake 226, the Experimental Lakes Area. Can. J. Fish. Aquat. Sci. 42: 129–138.

Munkittrick, K. R. & D. G. Dixon, 1988a. Evidence for a maternal yolk factor associated with increased tolerance and resistance of feral white sucker (*Catostomus commersoni*) to waterborne copper. Ecotox. Envir. Safety 15: 7–20.

Munkittrick, K. R. & D. G. Dixon, 1988b. Growth, fecundity and energy stores of white sucker (*Catostomus commersoni*) from lakes containing elevated levels of copper and zinc. Can. J. Fish. Aquat. Sci. 45: 1355–1365.

Munkittrick, K. R. & D. G. Dixon, 1988c. The use of white sucker (*Catostomus commersoni*) populations to assess the health of aquatic ecosystems exposed to low-level contaminant stress. Can. J. Fish. Aquat. Sci. in press.

Munkittrick, K. R. & D. G. Dixon, 1988d. In situ assessment of mixed copper and zinc impacts on white sucker populations in the Manitouwadge chain of lakes. Final Report, Ont. Min. Environ., Project 193RR, Rexdale, Ontario, 240 pp.

Munkittrick, K. R. & J. F. Leatherland, 1984. Abnormal pituitary-gonad function in feral populations of goldfish suffering epizootics of an ulcerative disease. J. Fish. Dis. 7: 433–447.

Nelson, W. R., 1974. Age, growth and maturity of thirteen species of fish from Lake Oahe during the early years of impoundment, 1963-1968. U.S. Fish Wildl. Serv. Tech. Paper 77, 29 pp.

NRCC, 1985. The role of biochemical indicators in the assessment of ecosystem health – their development and validation. Publ. No. NRCC 24371, Ottawa, 119 p.

Persson, L., 1986. Effects of reduced interspecific competition on resource utilization in perch (*Perca fluviatilis*). Ecology 67: 355–364.

Persson, L., 1987. The effect of resource availability and distribution on size class interactions in perch, *Perca fluviatilis*. Oikos 48: 48–160.

Power, G. & J. Gregoire, 1978. Predation by freshwater seals on the fish community of Lower Seal Lake, Quebec. J. Fish. Res. Bd. Can. 35: 844–850.

Pugh, D. M. & L. W. Maki, 1986. Benthic communities in the Manitouwadge Lakes chain, 1975. Tech. Support Section, Ont. Min. Env., NW Region, Thunder Bay, Ontario, 10 pp.

Reimers, N., 1958. Conditions of existence, growth and longevity of brook trout in a small, high-altitude lake of eastern Sierra Nevadas. Calif. Fish. Game 44: 319–333.

Riemers, N., 1979. A history of stunted brook trout population in an alpine lake: a lifespan of 24 years. Calif. Fish. Game 65: 196–215.

Rosseland, B. O., 1986. Ecological effects of acidification on tertiary consumers. Fish population responses. Wat. Air Soil Pollut. 30: 451–460.

Ryder, R. A. & C. J. Edwards, 1985. A conceptual approach for the application of biological indicators of ecosystem quality in the Great Lakes basin. Great Lakes Fish. Comm., Internat. Joint Comm., Windsor, Ontario, 169 pp.

Schiavone, A., 1981. Decline of the walleye population in Black Lake. N.Y. Fish Game J. 28: 68–72.

Schiavone, A., 1983. The Black Lake fish community: 1931 to 1979. N.Y. Fish Game J. 30: 78–90.

Schindler, D. W., 1987. Detecting ecosystem responses to anthropogenic stress. Can. J. Fish. Aquat. Sci. 44 (Suppl. 1): 6–25.

Schneider, J. C. & J. H. Leach, 1977. Walleye (*Stizostedion vitreum vitreum*) fluctuations in the Great Lakes and possible causes, 1800-1975. J. Fish. Res. Bd. Can. 34: 1878–1889.

Sun, J. & H. H. Harvey, 1986. Population dynamics of yellow perch (*Perca flavescens*) and pumpkinseed (*Lepomis gibbosus*) in two acid-stressed lakes. Wat. Air Soil Pollut. 30: 711–717.

Swanson, S. M., 1982. Levels and effects of radionuclides in aquatic fauna of the Beaverlodge area (Saskatchewan). Publ. C-806-5-E-82, Sask. Res. Council. 187 pp.

Trippel, E. A. & H. H. Harvey, 1987a. Abundance, growth and food supply of white suckers (*Catostomus commersoni*) in relation to lake morphometry and pH. Can. J. Zool. 65: 558–564.

Trippel, E. A. & H. H. Harvey, 1987b. Reproductive responses of five white sucker (*Catostomus commersoni*) populations in relation to lake acidity. Can. J. Fish. Aquat. Sci. 44: 1018–1023.

Hydrobiologia **188/189**: 137–142, 1989.
M. Munawar, G. Dixon, C. I. Mayfield, T. Reynoldson and M. H. Sadar (eds)
Environmental Bioassay Techniques and their Application.
© *1989 Kluwer Academic Publishers. Printed in Belgium.*

Environmental impact assessment: the growing importance of science in government decision making

Raymond M. Robinson
Federal Environmental Assessment Review Office, Government of Canada, Hull, Quebec K1A OH3
Canada

Key words: Brundtland report, environmental impact assessment

Abstract

The broad acceptance of the conclusions and recommendations contained in the Brundtland Report with its emphasis on environmentally sustainable development is a good indication that this concept is gaining world-wide recognition. Science and the ensuing technology must be credited for creating and sustaining our modern society and we must now apply our expertise to minimize the adverse impacts of industrial growth and preserve our environment.

Real progress towards the reconciliation of economic and environmental goals will require effective incorporation of environmental impact assessment (EIA) into the planning process and this paper describes the steps taken to improve the concept and practice of EIA in Canada. Public trust and participation in the process are key ingredients for its success as well as good science and meaningful dialogue between the scientists, the public and the decision makers. The availability of standard and reliable bioassessment techniques, which is the theme of this conference, should be helpful in strengthening the scientific basis for biophysical aspects of EIA.

Introduction

As a non-scientist, I was not certain that I could make a useful contribution to this international gathering of the bioassessment community but some members of the Organizing Committee assured me that my presence here would have some value. Besides, I am well aware of the good work done at Aberystwyth in 1984 during a similar conference on 'Ecological Effects of *In Situ* Sediment Contaminants' and I wanted to be a part of this present gathering. I am particularly sad, however, that I was not able to join you earlier in your proceedings because of a prior commitment to give the opening address at a conference on environmental assessment at the University of Aberdeen.

I am sure you have been told repeatedly by other non-scientists like me that scientific discoveries of the past century or so and ensuing technological innovations have revolutionized the way of life of the present generation of mankind. We also know that all these gains, substantial as they are, have had attendant costs.

Whereas the many benefits of scientific advances and the resultant technologies are widely known and recognized, the associated harmful effects, such as environmental deterioration, threats to public health, depletion of non-renewable resources, destruction and disappearance of plant and animal species and other valued ecosystem components, have often been understated, hidden, ignored or not anticipated. In general, modern industrial society has served man well in the sense that basic problems of scarcity of food and shelter, endemic disease and the need for crushing physical labour have been largely overcome in the developed world with the aid of expanded scientific knowledge and modern technology.

It is, however, fair to say that neither the scientists nor the supporters and users of scientific and technological advances have paid sufficient attention to the nature and causes of adverse impacts of industrial growth and the development of natural resources. The fact is that we still cannot claim full understanding, much less control, of life-supporting ecological systems; yet we continue to exploit them beyond the limits of natural renewal (Ehrlich *et al.*, 1979). Our past mistakes have brought home clearly the need to introduce ever more effective measures for protecting the environment through prudent planning and integration of economic growth and environmental considerations. Obviously, ecology and economy are intimately related: no economy can survive without a healthy resource base consisting of the basic natural components – soil, water, air and biota.

In Canada, as in many other parts of the globe, growing public and professional scientific concern for the environment is making itself known to decision makers at all levels with ever-increasing intensity. Time and again, according to the professional pollsters, Canadians have identified their desire for a healthy environment as one of their top priorities. Canada's participation in, and follow-up to, the report of the World Commission on Environment and Development (The Brundtland Report) is one of several indications of our collective resolve to introduce better means of protecting and managing our natural environment and to assist the global community in doing the same (CCREM, 1987).

The importance of the Bruntland Report cannot be overstated (World Commission on Environment & Development 1987). If world leaders pay attention to its warning and act on its advice, the report will indeed become a world-changing document. The Chinese have a delightful saying: 'If you do not change your direction, you will get where you are going'. The Bruntland Report spells out clearly where present policies are taking us – famine, widespread environmental disasters, man-induced climatic changes, the breakdown of ecological systems. However, unlike earlier predictions of doom, the report has a brighter side. While recognizing the existence of environmental limits and the fact that these are in some places now being reached or surpassed, the report argues that major changes in current development policies could yet avoid disaster. One Canadian journalist has characterized these recommended changes as being designed to push back the world's environmental limits by ensuring that future development is sensitive to the needs of the environment. The central concept in the report is environmentally sustainable development. In many instances, this means carefully planned development for the long term using the best available predictions of changes in the environment likely to occur as a result of that development. It also means seeking to introduce environmental objectives into basic policy right across the spectrum of government decision making.

Canada's initial response to the Brundtland Report is best reflected in the work of a task force made up of seven federal and provincial environment ministers, seven heads of large private corporations or industrial associations, and a number of individuals representing environmental and conservation organizations. The task force's report calls for a series of changes in government decision making that are designed to integrate environmental and economic concerns in all relevant sectors of government activity. The task force has remained in existence in order to comment on the reponse of governments to the report. The report is now under study but some governments in Canada have already begun to make some changes in response to it.

Clearly a major tool for achieving the goals outlined in the Brundtland Report is an effective environmental impact assessment (EIA) system. It is my belief that, in spite of differences in political and administrative structures, all governments face the same basic challenge in developing such a system, namely, that of striking a balance between public credibility, professional comprehensiveness, and political necessity. Although the appropriate balance will be different for every society, the ultimate success of any EIA procedure will depend largely upon the degree to which it (a) retains credibility with the general public, (b) is seen as reasonably comprehensive and rigorous by the scientific community, and (c) is viewed as a pragmatic and timely aid to decision making by political leaders.

The evolution of EIA in Canada

In the early part of the last decade, political leaders in Canada, like their counterparts in a number of other countries, were under some public pressure to develop a comprehensive approach to the protection of the natural environment. This came about as a result of the wave of 'environmentalism' which swept over much of the Western world in the 1960s, culminating in the Stockholm Conference in 1972. Between 1973 and 1980 the Canadian Federal Government and nine of our ten provinces established administrative procedures, policies or legislation specifically designed to require consideration of potential environmental consequences in the planning of resource exploitation and industrial development activities.

In 1973 the Canadian Cabinet instructed the Minister of the Environment to establish procedures whereby environmental consequences would be considered in the planning of projects, programs and activities of federal departments and agencies. The procedure that was put in place as a result of this policy directive is called the Environmental Assessment and Review Process (EARP) and it is administered by the Federal Environmental Assessment Review Office (FEARO) on behalf of the Environment Minister.

The purpose of the EARP is to ensure that the environmental consequences of federal governmental activities (including those private sector activities controlled by government decisions) are assessed early in the planning stage before any commitments or irrevocable decisions are made.

Depending upon the nature of the proposed activity which is being assessed, the EARP can involve up to three sequential review stages – screening, initial assessment, and public review. The third and most elaborate of these stages entails the appointment by the Environment Minister of an independent environmental assessment panel chaired and supported by FEARO to assess the proposal, hold public hearings, and provide a report with recommendations to the responsible ministers. Only those projects that engender significant public controversy or are likely to cause significant environmental effects are subjected to panel review. These projects constitute a very small percentage of the total number examined under the EARP.

The EARP policy was formally amended in 1977, primarily to deal with some financial aspects and the appointment of people from outside government to environmental assessment panels. At the administrative level, however, it has evolved significantly over the years. Considerable effort has been spent on streamlining procedures, providing advice and support to federal departments and agencies, increasing the degree of public consultation, and improving the scientific basis and relevance of information generated through the review process. At the same time, the scope of the process has been expanding. While initially the focus was on potential physical and biological effects, in recent years greater attention has been given to the socio-economic and cultural consequences of proposed developments.

In 1984 the Government of Canada issued an Order in Council containing Environmental Assessment and Review Process Guidelines. This latest stage in the evolution of the EARP provides clear direction to federal departments and agencies implementation procedures and also addresses a number of areas that had been considered deficient in the original policy. For example, the Order in Council:
- defines the scope of the EARP with respect to social and economic impacts;
- reduces the possibility of redundancy with other review procedures of government;
- directs initiating departments and agencies to

screen all of their projects and report the results to FEARO;

- directs initiating departments and agencies to consult more with the public during the screening and initial assessment stages of their review procedures; and
- directs the initiating departments and agencies to ensure that any post-assessment recommendations are carried out.

Many of the changes to the EARP over the years have been attempts to strike that critical balance needed to meet the legitimate expectations of the general public, scientists, and political leaders. The basic principles around which the EARP was designed have, however, remained intact.

FEARO, at the instruction of the Canadian federal cabinet, is engaging in a major effort to seek the views of interested and affected organizations and individuals across Canada on further improvements to the EARP. Several themes have emerged to date including calls for new environmental assessment legislation, greater public accessibility to departmental decision making, stronger powers for the Environment Minister, more comprehensive follow-up and monitoring procedures, and the effective application of environmental assessment concepts to policy decisions, not just to project proposals. Whatever the specific procedural mechanisms chosen to improve the EARP may be, it is clear that the reform of the EARP must be designed to enhance the ability of the Canadian government to respond effectively to the challenges of the Brundtland Report.

The challenging role of the scientific community in EIA

EIA as practised in Canada is fundamentally a process whereby social trade-offs can be assessed and discussed with a view to developing a consensus or, at least, well-informed decisions. As such, it is not primarily a scientific exercise. However, good scientific information and interpretation is crucial to the success of EIA. In general, developments in the scientific aspects of EIA have not kept pace with refinements in procedures. This has sometimes resulted in an inefficient use of time and resources in both the collection and review of information. In order to increase the scientific credibility of impact assessments in Canada, FEARO joined co-sponsors from the university community and private industry in supporting a major effort aimed at developing a consensus among those affected, especially within the scientific community, on an acceptable approach to EIA. The resulting report (Beanlands & Duinker, 1983) provided a basic scientific framework that has been adopted for many environmental assessments conducted in Canada. In general, the scientific standard of EIA studies has greatly improved over the last few years. However, difficulties still remain in the integration of the social and biophysical dimensions of most environmental problems.

In response to one of the recommendations of this report, the Canadian Environmental Assessment Research Council (CEARC) was established in 1984. The Council is charged with advising on ways to improve the practice of EIA in Canada and, with the assistance of FEARO, stimulates and guides research projects towards this end. The Council is composed of persons prominent in their various disciplines drawn from industry, the consultant community, universities, various levels of government, and conservation organizations. CEARC has funded or co-funded a number of studies on such topics as social impact assessment, cumulative environmental effects, mitigation and compensation in EIA, auditing of environmental impact statements, risk analysis, and the incorporation of health concerns in EIA. In one sense, CEARC was established to increase the credibility of EIA among scientists. Virtually no other research funding agency in Canada will support the interdisciplinary studies that are required to advance the state-of-the-art of EIA. CEARC has not only provided such support but has also attempted to convince other funding organizations of the complexity of the problems involved and the need to have our best scientific talents applied to their resolution. I believe that CEARC has been instrumental in establishing a more positive view of EIA among the scientific community and, thereby, in encouraging their contribution to the process. Nonetheless, much more remains to be done.

In a democratic society, the question of real and perceived fairness in decision-making processes in general and public review processes in particular is of fundamental importance. In more technologically developed societies, the decision makers are much influenced and guided by scientific evidence and technical information. The acceptability to the public of resulting decisions, however, depends heavily upon the level of public trust in the so-called 'scientific facts' and upon the perceived fairness with which the decision makers have assessed these facts. In this process, there is always room for subjective judgement but the scientist can, through rigorous analysis, contribute greatly to reducing the margin for error on the part of those exercising that judgement.

No one should argue that the pursuit of knowledge is not in itself a worthy goal, but a democratically governed and technologically-run society must be given the opportunity to understand the implications of particular directions in scientific activity, and of scientific and technical advances. What is involved is a mix of public educational efforts guided by knowledgeable scientists, as well as attitudinal changes within the scientific community itself, based on the recognition that a scientist is but one of many contributors to the building of the societies in which we live. Other non-scientific contributors can offer valuable guidance and insights to the scientist and should ideally be viewed and treated as partners. One result would be better public awareness of the importance and usefulness of scientific research activities and the resultant products, many of which are funded through the public purse.

One of several excellent background studies commissioned by the Science Council of Canada is R. W. Jackson's (1976) 'Human Goals and Science Policy.' The author touches upon one of the fundamental components of human nature: 'Human beings would hardly be human, nor would they develop their capacities, if they did not welcome an element of challenge in their lives'.

In my opinion, one of the greatest challenges facing the scientific community today is to find innovative ways to conduct meaningful multi-disciplinary research and to provide interpretation of its findings as an aid to a better understanding and management of the environment in which we live. Without intelligent, balanced interpretation, the vital biophysical component of environmental impact assessment may simply amount to massive data collection, adding more confusion to the process and possibly even masking the importance that such information ought to have.

Another factor worth mentioning is dealt with in a study done by the Science Council of Canada (1982) entitled, 'Regulating the Regulators: Science, Values and Decisions'. It deals with the need to redefine the relationship between science and law. Among other things, such redefinition is aimed at helping distinguish between truly scientific disagreements and controversies involving public values and science. The first involves the validity of scientific findings and the second is usually a dispute over the social, ethical, and political implications of such findings. For example, as the public becomes increasingly aware of and militant about its right to clean air and a healthy environment (e.g., smokers vs. non-smokers), the nature and strength of the science-law relationship will be tested with increasingly direct and practical application, often through the judicial process. The bottom line is that there has never been a time in human history when good science and good scientific management were more needed to help man govern himself.

Bioassay techniques and EIA

Your conference has been concerned with the application of bioassay techniques to the business of assessing and managing the environment. It would not be useful for me to try to add specific technical observations to what you have already been saying to one another. My purpose is to underline the importance of your work to society at large and to urge you to seek to make bioassessment ever more relevant to and useable by decision makers and planners alike. Most environmental controversies or disputes and environmental assessment exercises deal with the use or abuse of a water resource. This should hardly be surprising since water is essential not only to all forms of life but to most industrial activities as well. It follows that one cannot overemphasize the importance of monitoring and maintaining the health of our water resources. Various regulatory bodies in Canada, and I am sure in most other countries as well, are increasingly concerned with the huge costs associated with water pollution and are adopting appropriate remedial measures to minimize contaminant loadings from land and atmospheric sources. Since 1972, the International Joint Commission (IJC) of the United States and Canada has been emphasizing the need to investigate the role of contaminated sediments in the environmental degradation of several areas of concern in the Great Lakes Basin (IJC, 1984). Consequently, the umbrella groups working under the IJC, especially its Water Quality Board and Science Advisory Board, have been trying to untangle and understand the many complexities associated with this problem. One of the most significant steps toward that goal was the successful workshop held in Wales in 1984. The summary of conclusions and recommendations put forward by the participants in that workshop are worthy of serious consideration by scientists as well as those responsible for developing environmental protection strategies and implementing remedial programs (Thomas et al., 1987).

Based on my own experience as a regular recipient of scientific input for making decisions and/or advising the federal ministers of the Crown, I can hardly overemphasize the need for reliable, precise and accurate standard bioassessment methods. From my perspective, current bioassessment techniques must be improved to attain a greater degree of accuracy and reliability. We need to develop standard methods and criteria for choosing appropriate biological indicators and experimental species. As you are aware, a particular technique or methodology must have wide acceptability in order to be useful. It follows

that a simple and cost-effective methodology is preferable to one requiring complicated structures, sophisticated and expensive equipment, and highly specialized experts. I believe that the problems associated with the manipulation and interpretation of data collected by using current bioassessment techniques should also be critically evaluated. Ultimately, the quality of that interpretation determines the value of your activities to governments and to society at large.

In concluding I want to repeat that there has never been a time in human history when the contribution of science, especially those fields of science that relate to the life-sustaining processes of this planet, has been more important to the effective government of mankind. You are engaged in trying to improve that scientific contribution. For the sake of us all, I wish you every success.

References

Beanlands, G. E. & P. N. Duinker, 1983. An ecological framework for environmental impact assessment in Canada. Institute for Resource and Environmental Studies. Dalhousie University, Halifax, Nova Scotia, 132 pp.

Canadian Council of Resource and Environment Ministers (CCREM), 1987. Report of the National Task Force on Environment and Economy. Federal Environmental Assessment Review Office, Hull, Quebec, Canada. 18 pp.

Ehrlich, P. R., A. H. Ehrlich & J. P. Holdren, 1973. Human Ecology; Problems and Solutions. Freeman, San Francisco, California. 304 pp.

International Joint Commission (IJC) 1984. Second Biennial Report under the Great Lakes Water Quality Agreement of 1978 to the Governments of the United States and Canada and the States and Provinces of the Great Lakes Basin. IJC, Ottawa. 17 pp.

Jackson, R. W. 1976. Human Goals and Science Policy. Science Council of Canada, Ottawa, 134 pp.

Science Council of Canada, 1982. Regulating the Regulators: Science, Values and Decisions, Ottawa, 106 pp.

Thomas, R., R. Evans, A. Hamilton, M. Munawar, T. Reynoldson, and H. Sadar, (Eds), 1987. Ecological Effects of in situ Sediment Contaminants. Dr. W. Junk Publishers, Dordrecht, The Netherlands, 272 pp.

World Commission on Environment and Development, 1987. Our Common Future. Oxford University Press, U.K. 400 pp.

Hydrobiologia **188/189**: 143–147, 1989.
M. Munawar, G. Dixon, C. I. Mayfield, T. Reynoldson and M. H. Sadar (eds)
Environmental Bioassay Techniques and their Application.
© *1989 Kluwer Academic Publishers. Printed in Belgium.*

The role of microbial metal resistance and detoxification mechanisms in environmental bioassay research

J. T. Trevors
Department of Environmental Biology, University of Guelph, Guelph, Ontario, Canada N1G 2W1

Key words: bioassay, microbial, detoxification, metal-resistance mechanisms

Abstract

Numerous biological methods and bioassays have been developed for assessing toxicants in both environmental and laboratory samples. In many bioassays, microorganisms are used because of their rapid growth rates and ubiquitous distribution in aquatic and terrestrial environments. However, information is often lacking on the ecology, physiology and genetics of the test organism(s) or other organisms present in the test system. For example, Hg^{2+} can be volatilized *via* the mercuric reductase enzyme found in certain bacterial strains. Moreover, this detoxification/resistance mechanism may occur while the environmental sample is being used in a bioassay protocol. A fundamental knowledge of the mechanism(s) involved in microbial detoxification/resistance mechanisms is essential to understand how the bioassay organism(s) and toxicant(s) behave in environmental samples tested with bioassay protocols. This manuscript will review selected metal (arsenic, cadmium, mercury) detoxification/resistance mechanisms in bacteria with an emphasis on the manner in which the mechanisms may influence the bioassay results.

Introduction

The use of microorganisms in environmental toxicology research and monitoring is still in its developmental stages. Numerous microbial bioassays have been proposed in which a wide variety of organisms have been exposed to toxicants. In most of these bioassays, inhibition of growth, enzymatic reactions, metabolic heat evolution or changes in species numbers and diversity have served as the indicator of toxicity. For excellent reviews of bioassay monitoring using microorganisms, texts by Bitton & Dutka (1986) and Liu & Dutka (1984) are recommended. In addition, an excellent review of standard statistical procedures used to analyze data from biological assays has been published by Hubert (1980).

This research paper will deal with metal detoxification/resistance mechanisms in bacterial species that are often present in environmental samples being tested for the presence of toxicant(s) and their toxic effect(s). The desirable features of good microbial bioassays along with detailed descriptions have been previously published (Bitton & Dutka, 1986; Liu & Dutka, 1984). In addition, numerous manuscripts in this issue include bioassay design, sensitivity, and advantages/disadvantages of the bioassay system.

Results and discussion

Arsenic

Numerous environmental bioassays using bacteria, algae, fish and yeasts have been used to assess the effect(s) of toxicant(s) on these organisms.

However, in many environmental samples, such as sediment and water, the presence of high bacterial numbers (10^6 to 10^8) per g or ml sample is often overlooked. Microorganisms play a primary role in nutrient and mineral cycling in the biosphere, and at the same time have developed efficient mechanisms to detoxify or become resistant to a multitude of organic and inorganic pollutants. These detoxification/resistance mechanisms may be functioning while a bioassay is being conducted. This could have a significant effect on the bioassay.

Arsenic compounds found in aquatic and terrestrial environments and their toxic effects have been well documented. Williams and Silver (1984) showed that trivalent arsenicals (As^{3+}) are about 200-fold more toxic than pentavalent arsenicals (As^{5+}) and, therefore, the oxidation of arsenite to arsenate is a detoxification process. Osborne and Ehrlich (1976) reported that this reaction was catalyzed by an arsenite oxidoreductase enzyme.

Certain strains of *Staphylococcus* and *Escherichia coli* efflux arsenate, arsenite and antimony (III) *via* an energy-dependent mechanism (Mobley & Rosen, 1982; Silver & Keach, 1982). At least three plasmid-encoded proteins are required for the resistance mechanism in *E. coli*. The *ars*A protein (MW 63,219) is associated with the cytoplasmic membrane, and is required for arsenate and arsenite resistances (Mobley & Rosen, 1982; Silver & Keach, 1982; Silver *et al.*, 1986). The *ars*B protein (MW 45,598) is believed to be the membrane protein responsible for the ATP-dependent efflux of arsenite and arsenate ions (Mobley & Rosen, 1982). The *ars*C protein confers substrate specificity on the *ars*B and is also necessary for the efflux of arsenate ions (Silver *et al.*, 1986). Since arsenic compounds are analogs of phosphate, they enter microbial cells *via* the phosphate transport systems (*Pst, Pit*) where arsenicals can inhibit cellular kinases if not effluxed. This mechanism of arsenite and arsenate resistance is entirely different from the oxidation of arsenite to arsenate *via* an enzymatic transformation. If such a process was occurring in sediment samples used in long-term bioassays, a two-fold problem becomes apparent. First, the arsenite/arsenate ratio could be constantly changing, while the toxic effect is being reduced. In the efflux resistance system, no enzymatic mechanism is responsible for altering the form of the metal. In addition, the total arsenic concentration could remain relatively constant, except in the cells that are effluxing the metal to reduce the intracellular concentration before toxic effect(s) are produced.

Cadmium

Cadmium has been extensively studied in toxicity bioassays. It has no known biological function and is extremely toxic to living organisms at low concentrations (Babich & Stotzky, 1979; Gadd & Griffiths, 1978; Trevors *et al.*, 1985). Cadmium can exert a toxic effect by complexing to SH groups of proteins and amino acids and inhibiting cellular respiration (Trevors *et al.*, 1986; Foster, 1983). An example of Cd^{2+} complexation to cytoplasmic proteins from *E. coli* cells exposed to 1 mM Cd^{2+} is shown in Fig. 1. The lower scans at 255 and 279 nm show the elution patterns and size distribution (2000 to 300000 daltons) of the protein fractions. The top scans are the corresponding Cd^{2+} concentrations measured in the protein fractions. The Cd^{2+} concentrations are associated with components having M^r of approximately 50000, 120000 and 300000 daltons. Information of this type allows the location and concentration of the cellular metal ions to be studied.

Cadmium is not enzymatically transformed, but can be sequestered or excluded from microbial cells. A review by Trevors *et al.* (1986) deals extensively with Cd uptake, resistance and detoxification processes in bacteria, algae and fungi. From an environmental perspective, it is also useful to know that Cd can be precipitated by sulphate or phosphate. This has several implications in environmental bioassay research. First, the free Cd concentration is much lower than the total Cd concentration. Therefore, very little free Cd may actually be present to exert a toxic effect. Moreover, Cd-sulphide complexes are insoluble

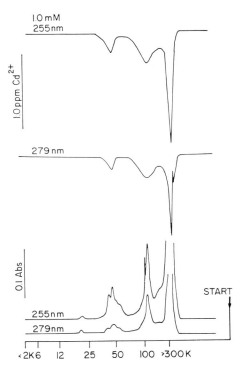

Fig. 1. Fast Protein Liquid Chromatography (FPLC) of cytoplasmic extracts from *E. coli* cells previously grown in the presence of 1 mM CdCl$_2$ (in LB broth) for 12 h at 37 °C. Cells were harvested at 7,000 × g for 15 min, and the pellets washed twice in 50 ml of 50 mM NH$_4$HCO$_3$ buffer. Finally, the pellets were resuspended in 4 ml of the same buffer and sonicated for 20 sec. Aliquots of 0.8 ml were centrifuged at 12000 × g for 5 min to remove cell debris. Two hundred μl samples of the supernatant were applied onto an FPLC Superose 12 column (Pharmacia) and eluted at a rate of 1 ml/min with 50 mM NH$_4$HCO$_3$ (lower scans). Eluent was monitored at 279 and 255 nm and then was passed directly into a flame atomic absorption spectrometer to continuously estimate Cd^{2+} concentrations (upper scans). A log (M^r) versus elution time calibration curve was used to estimate the M^r unknown protein components using the following standards: alkaline phosphatase (M^r 89 Kd), haemoglobin (M^r 67 Kd), cytochrome C (M^r 12.4 Kd) and streptomycin (Mmr 1.5 Kd).

and may precipitate out of solution. For example, a *Citrobacter sp.* was capable of detoxifying Cd^{2+} by forming insoluble Cd phosphate (Macaskie & Dean, 1984). Aiking *et al.* (1982, 1984) observed that a strain of *Klebsiella aerogenes* grown in the presence of Cd^{2+} used sulphide formation and Pi accumulation as a detoxification mechanism. This organism was later examined for the presence of extrachromosomal plasmid DNA in my laboratory. No plasmids were detected, indicating that the resistance mechanism was not encoded on extrachromosomal plasmids. It is possible that this type of mechanism is a general resistance response.

K. aerogenes also produces an extracellular polysaccharide capsule that acts as a metal trap

(Bitton & Freihofer, 1978) providing a significant degree of protection to the organism. However, in another bacterial strain, *Acaligenes sp.* the presence of a 34500 dalton membrane protein appeared to be associated with Cd^{2+} resistance. This protein was observed when Cd^{2+} levels exceeded 0.1 M; but was not detected in the presence of Cd^{2+} concentrations below this level. The exact mechanism of resistance is unknown. It was postulated that the new membrane protein may be involved with Cd^{2+} efflux or possibly prevent Cd^{2+} uptake (McEntee *et al.*, 1986).

Higham *et al.* (1984) observed the synthesis of low-molecular weight, cysteine-rich Cd^{2+} binding protein in *Pseudomonas putida*. It is possible that these proteins are closely related to metallothionein proteins, which bind metal ions and therefore reduce metal toxicity. Other research on Cd^{2+} uptake and resistance has demonstrated that *E. coli* accumulates Cd^{2+} *via* the Zn^{2+} active transport system (Laddaga & Silver, 1985). In addition, plasmid-encoded Cd^{2+} efflux has also been extensively studied (Novick & Roth, 1986; Dyke *et al.*, 1970; Tynecka *et al.*, 1981; Chopra, 1975). The presence of a specific R-plasmid can prevent a Cd^{2+}-resistant strain from becoming internally loaded with toxic Cd^{2+} ions (Tynecka *et al.*, 1981).

Mercury

Mercury resistance has been extensively investigated and reviewed by numerous researchers (Foster, 1983; Robinson & Tuovinen, 1984; Brown, 1985; Trevors *et al.*, 1985). Mercury-resistant bacteria have also been isolated from numerous environments including sediment samples (Trevors, 1986). Mercuric ions are extremely toxic since they bind to SH groups and inhibit enzyme activity. However, some bacterial strains have evolved very efficient mechanisms of volatilizing $Hg^{2+} \rightarrow Hg^{\circ}$. The genetics and physiology of this detoxification process have been extensively dealt with in the previously mentioned reviews, and will not be repeated here. There are also some mercury-resistant bacterial strains capable of detoxifying organomercurials *via* the plasmid-encoded organomercurial lyase enzyme. The Hg-C bond of methyl-, ethyl- and phenyl-mercury is cleaved forming Hg^{2+} and the corresponding methane, ethane, or benzene group is produced. The Hg^{2+} can then be further reduced to Hg^{0} *via* the mercuric reductase enzyme (Tezuka & Tonomura, 1978).

Another mercury detoxification/resistance process occurring in aquatic/sediment environments is methylation *via* bacteria (Trevors, 1986). Methyl mercury is considered more toxic than inorganic mercury. However, it is less toxic to the microorganisms responsible for the methylation process. Hg^{2+} resistance in a bacterial strain has also been attributed to an alteration of membrane permeability (Hidemitsu *et al.*, 1981). An *Enterobacter aerogenes* isolate produced two additional membrane proteins that apparently caused reduced Hg^{2+} uptake. This should be considered a resistance mechanism and not a detoxification process.

Throughout this manuscript, a true definition of resistant versus sensitive organisms has not been provided. This still presents a problem as standardized conditions are almost impossible to use in the multitude of research projects being conducted on metal-resistances. This problem is also complicated by the various forms of metals used, and the influence that complexing agents and environmental conditions have on the availability and form of the metals. For a review of this problem with examples of metal concentrations found in the literature, the reader is referred to a review by Trevors *et al.* (1985).

Summary

In the aquatic/sediment environment, bacteria are responsible for a wide range of metal detoxification/resistance mechanisms. In addition, they also act as surfaces or biofilms that can readily complex metal ions. Since most environmental bioassay research deals with aquatic and sediment samples, it is necessary to understand the mechanisms that microorganisms have evolved to protect selected strains from a lethal effect. It is noteworthy, that at extremely high levels of free metal ions, resistance/detoxification mechanism provides no protection and a lethal toxic effect can be produced. However, in many environmental samples sublethal effects are present, or only a percentage of the total microbial population is influenced. Under these conditions, resistance/detoxification mechanisms are probably very useful. It is also very probable that during many bioassays with higher organisms, numerous microbial processes are occurring simultaneously to bind or detoxify metal ions. This may in fact have an effect on the final outcome of the environmental bioassay.

Acknowledgements

This research was supported by a Natural Sciences and Engineering Research Council of Canada (NSERC) operating grant. Appreciation is expressed to B. McGavin for typing the manuscript. FPLC analysis was carried out at Birkbeck College, University of London, by Dr. P. Sadler.

References

Aiking H., A. Stijman, C. van Garderen, H. van Heekikhuizen and J. van 't Riet, 1984. Inorganic phosphate accumulation and cadmium dertoxification in *Klebsiella aerogenes* NCTC 418 growing in continuous culture. Appl. Envir. Microbiol. 47: 374–377.

Aiking H., K. Kok, H. van Heerikhuizen and J. van 't Riet, 1982. Adaptation to cadmium by *Klebsiella aerogenes* growing in continuous culture proceeds mainly *via* formation of cadmium sulphide. Appl. Envir. Microbiol. 44: 938–944.

Babich H. & G. Stotzky, 1979. Abiotic factors affecting the toxicity of lead to fungi. Appl. Envir. Microbiol. 38: 506–513.

Bitton, G. & V. Freihofer, 1978. Influence of extracellular polysaccharide on the toxicity of copper and cadmium towards *Klebsiella* sp. Microb. Ecol. 4: 119–125.

Bitton, G. & B. J. Dutka (eds.) 1986. Toxicity Teating Using Microorganisms. Vol. 1. CRC Press, Inc., Boca Raton, pp. 1–163.

Brown N. L., 1985. Bacterial resistance to *mercury-reductio ad absurdum*? Trends Biochem. Sci. 10: 400–403.

Chopra I., 1975. Mechanism of plasmid-mediated resistance of cadmium in *Staphylococcus aureus*. Antimicrob. Agents. Chemother. 7: 8–14.

Dyke K. G. H., M. T. Parker & M. H. Richmond, 1970. Penicillinase production and metal ion resistance in *Staphylococcus aureus* cultures isolated from hospital patients. J. Med. Microbiol. 3: 125–136.

Gadd G. M. & A. J. Griffiths, 1978. Microorganisms and heavy metal toxicity. Microb. Ecol. 4: 303–317.

Foster T. J. 1983. Plasmid-edetermined resistance to antimicrobial drugs and toxic metal ions in bacteria. Microbiol. Rev. 47: 361–409.

Higham D. P., P. J. Sadler & M. D. Schawen, 1984. Cadmium resistant *Pseudomonas putida* synthesizes novel cadmium binding proteins. Science. 225: 1043–1046.

Hubert, J. J. 1980, Bioassay. Kendall/Hunt Publishing Co., Toronto, pp. 1–164.

Hidemitsu S. P.-H., M. Nishimoto & N. Imura, 1981. Possible role of membrane proteins in mercury resistance of *Enterobacter aerogenes*. Arch. Mikrobiol. 130: 93–95.

Laddaga R. A. & S. Silver, 1985. Cadmium uptake in *Escherichia coli* K12. J. Bact. 162: 1100–1105.

Liu, D. & B. J. Dutka (eds.) 1984. Toxicity Screening Procedures Using Bacterial Systems. Marcel Dekker, Inc., New York, pp. 1–476.

Macaskie L. E. & A. C. R. Dean, 1984. Cadmium accumulation by a *Citrobacter* sp. J. Gen. Microbiol. 130: 53–62.

McEntee J. D., J. R. Woodrow, & A. V. Quirk, 1986. Investigation of cadmium resistance in an *Alcaligenes* sp. Appl. Envir. Microbiol. 51: 575–520.

Mobley H. L. T. & B. P. Rosen, 1982. Energetics of plasmid-mediated arsenate resistance in *Escherichia coli*. Proc. Natn. Acad. Sci. U.S.A. 79: 6119–6122.

Novick R. P. & C. Roth, 1968. Plasmid-linked resistance to inorganic salts in *Staphylococcus aureus*. J. Bact. 95: 1335–1342.

Osborne, F. H. & H. L. Ehrlich, 1976. Oxidation of arsenite by a soil isolate of *Alcaligenes*. J. Appl. Bacteriol. 41: 295–305.

Robinson J. B. & O. H. Tuovinen, 1984. Mechanisms of microbial resistance and detoxification of mercury and organomercury compounds: physiological, biochemical and genetic analyses. Microbiol. Rev. 48: 95–124.

Silver S. & D. Keach, 1982. Energy-dependent arsenate efflux: the mechanism of plasmid-mediated resistance. Proc. Natn. Acad. Sci. U.S.A. 79: 6114–6118.

Silver S., B. P. Rosen & T. K. Misra, 1986. DNA sequencing analysis of mercuric and arsenic resistance operons of plasmids from Gram-negative and Gram-positive bacteria. Fifth International Symposium on the Genetics of Industrial Microorganisms. Zagreb, Pliva. pp. 357–371.

Tezuka T. & K. Tonomura, 1978. Purification and properties of a second enzyme catalyzing the splitting of carbon-mercury linkages from mercury resistant *Pseudomonas* K-62. J. Bact. 135: 138–143.

Trevors J. T., K. M. Oddie & B. H. Belliveau, 1985. Metal resistance in bacteria. FEMS Microbiol. Revs. 32: 39–54.

Trevors J. T., 1986. Mercury methylation by bacteria. J. Basic Microbiol. 26: 499–504.

Trevors J. T., 1986. Mercury resistant bacteria isolated from sediment. Bull. Envir. Contam. Toxicol. 36: 405–411.

Trevors J. T., G. W. Stratton & G. M. Gadd, 1986. Cadmium transport, resistance and toxicity in bacteria, algae and fungi. Can. J. Microbiol. 32: 447–464.

Tynecka Z., Z. Gos & J. Jajac, 1981. Energy-dependent efflux of cadmium coded by a plasmid resistance determinant in *Staphylococcus aureus*. J. Bact. 147: 313–319.

Williams, J. W. & S. Silver, 1984. Bacterial resistance and detoxification of heavy metals. Enz. Microb. Technol. 6: 530–537.

Hydrobiologia **188/189**: 149–153, 1989.
M. Munawar, G. Dixon, C. I. Mayfield, T. Reynoldson and M. H. Sadar (eds)
Environmental Bioassay Techniques and their Application.
© *1989 Kluwer Academic Publishers. Printed in Belgium.*

Performances of three bacterial assays in toxicity assessment

C. Reteuna[1], P. Vasseur[1] & R. Cabridenc[2]
[1] *Centre des Sciences de l'Environnement, 1, rue des Récollets, 57000 Metz, France;* [2] *I.R.CH.A., Rue Lavoisier, 91710 Vert-Le-Petit, France*

Key words: bacterial assays, toxicity assessment, microtox test, oxygen consumption assay, glucose mineralization assay, bioavailability

Abstract

Three differing bacterial toxicity assays were compared: the 'Microtox' test, (*Photobacterium phosphoreum* luminescence inhibition assay), the 'oxygen consumption of activated sludge' assay (ISO 8192), and the 'Glucose U-^{14}C mineralization' assay (the rate of release of $^{14}CO_2$ by '*Escherichia coli*'). Metals, amines, halogenated alcans, chlorophenols, aromatic hydrocarbons, surfactants, and pesticides were screened for their toxic activity.

Results showed satisfactory repeatability of the three bacterial assays with variation coefficients between 5 and 32%. The 'Microtox' assay was the most sensitive test evaluated under our conditions. The lower sensitivity of the 'oxygen consumption' assay may have been due to high concentrations of substrates which modify toxicant bioavailability, and also to a high biomass/toxic substances ratio. The 'Glucose U-^{14}C mineralization' assay was selective, and low in sensitivity; but the specific species used in this test – *Escherichia coli* – may have been responsible for this selectivity.

The 'Microtox' test appears to be well adapted to the detection of aquatic environmental pollution, and to the toxicity screening of complex solid waste effluents and/or leachates. The 'oxygen consumption' assay can be advantageously used to measure the impact of sewage on activated sludge in biological treatment plants. The 'Glucose U-^{14}C mineralization' assay, which does not require high biomass, can be useful for *in situ* studies using field microorganisms.

Introduction

Biological toxicity assays are required for the detection of pollution in the aquatic environment and the assessment of toxicity of waste waters and chemical substances. Presently, toxicity tests are mainly conducted using crustacea, fish and algae. Bacterial toxicity assays are necessary since procaryotes play an important part in ecosystems. Bacterial tests are simple, fast and inexpensive, nevertheless they have not been widely used for toxicity assessment and their performances are relatively unknown.

In this study, three bacterial assays have been evaluated, the Microtox test, oxygen consumption of activated sludge, and glucose mineralization by *Escherichia coli*. Since these tests differ in their methodology (biological reagent, test medium and incubation time), it was of interest to compare their performances and try to relate these results to the specificity of each bioassay. Therefore the repeatability and sensitivity of these three tests have been studied using chemicals commonly found in the aquatic environment.

Materials and Methods

Microtox assay

The Microtox assay measures the inhibitory effects of toxicants on the luminescence emitted by a marine bacteria, *Photobacterium phosphoreum* (Bulich et al., 1981). Bioluminescence was measured in a saline medium (NaCl 2%), at a temperature of 15 °C using a thermostated photometer. The incubation time was 5, 15 or 30 min. according to the tested compound.

Oxygen consumption assay

The oxygen consumption assay was conducted according to the ISO Standard Method 8192, part A (1986). This assay evaluates the inhibitory effects of toxicants on the respiration of activated sludge in the presence of substrate. In this test, activated sludge was usually taken from a sewage treatment plant. In order to decrease the variability of the biological agent, a laboratory pilot was used to obtain the activated sludge following the French standard AFNOR T 73265 (1982).

The biological reagent was prepared as follows: activated sludge was removed from the pilot, centrifuged (3500 rpm × 30 min), washed and resuspended in an isotonic solution (NaCl 5 g/l – $MgSO_4$, 7 H_2O 0.12 g/l). The final concentration of activated sludge in the test medium was approximately 50 mg/l of suspended solids, allowing oxygen measurements for 3 h with a Clark electrode.

Glucose mineralization assay

A new method was developed for the glucose mineralization assay. The principle of the test involves the evaluation of the inhibitory effects of toxicants on the rate of labelled CO_2 released by *E. coli* in the presence of glucose U-^{14}C.

The biological reagent was prepared as follows: 3 successive transfers of *E. coli* (ATCC 11303) were performed on Plate Count Agar (0479011, OSI) for 24 h at 37 °C. Then *E. coli* was suspended in a sterile NaCl solution (7 g/l) to obtain approximately 50×10^5 bact./ml. The bacterial suspension was shaken for 1 h before use to ensure homogeneity.

Glucose mineralization was determined from the kinetics of released $^{14}CO_2$, established in toxic and control media simultaneously. Twenty-four μg/l glucose U-^{14}C (CB1-D glucose, CEA, 300 mci/mmole) were added to the saline bacterial suspension in air-tight flasks. The flasks were equipped with a central well receiving a tube with the CO_2 absorbant, 200 μl phenylethylamine (7342, Merck). The mineralization reaction was performed for 12 min. at 20 ± 2 °C in a shaking incubator, and it was stopped by the injection of H_2SO_4 0.1 N (9912, Merck) into the test medium. The flasks were shaken for a further 4 h to permit complete absorption of CO_2, then the radioactivity of phenylethylamine was measured using a scintillation counter. The same calculations as in the oxygen consumption assay were used for the determination of the IC_{50}.

Potentiometric measurements of free copper

The speciation of copper was studied in the case of the oxygen consumption assay. Each component of the nutrient medium was determined for its influence in the complexation of copper. To tested solutions were added 40 mg/l $CuSO_4$, and $NaNO_3$ 0.1 M for ionic strength stabilization; and residual free copper was measured using a specific electrode (9429, ORION).

Tested substances

Eleven chemical substances were used for the toxicity screening: $CuSO_4$ (2791, Merck); Cd Cl_2, H_2O (2011, Merck); $ZnCl_2$ (8816, Merck); $K_2Cr_2O_7$ (26781297, Prolabo); 3,5-dichlorophenol (821768, Merck); Benzene (21803291, Prolabo); Dibromo 1,2-ethane (34418, Labosi);

diethylamine (803010, Merck); dodecylbenzene sulfonic acid (803911, Montanoir); atrazine (35702, Riedel de Haën); malathion (38736, Riedel de Haën).

Substances displaying low solubility in water, such as atrazine and malathion, were dissolved with a co-solvent (DMSO, up to 2%), that had been shown to have no inhibitory effects on bacteria.

Results and discussion

The repeatability of the three test procedures was studied in depth with 3,5-dichlorophenol and copper sulphate. It was expressed by the variation of the IC_{50} values (Table I). The coefficients of variation obtained for dichlorophenol and copper were respectively 5% and 28% with the Microtox test, 15% and 18% with the oxygen consumption assay, and 5% and 32% with the glucose mineralization assay. On the basis of these values the repeatability of the three tests can be considered as satisfactory. These results are not surprising for the Microtox and glucose-mineralization assays, since they used pure bacterial strains grown under standard conditions. The good repeatability of the Microtox test was also confirmed in many studies (Bulich et al., 1981; Bazin et al., 1987; Curtis et al., 1982; Ferard et al., 1983; Qureshi et al., 1984; Reteuna et al., 1986; Vasseur et al., 1984a, 1984b, 1986). On the other hand our results for the oxygen consumption assay were better than those observed in the literature (Brown et al., 1981; King & Painter, 1986). The good repeatability was probably due to the relative stability of activated sludge taken from the laboratory pilot.

Secondly, we studied the sensitivity of the three bacterial tests expressed by the IC_{50} of different compounds (Fig. 1). Copper and zinc were toxic in the three assays with an IC_{50} of less than 10 mg/l. Cadmium and chromium were toxic to luminescent bacteria and to activated sludge, but these two compounds displayed no toxicity to E. coli ($IC_{50} > 100$ mg/l). In the case of the organic substances, in particular dichlorophenol,

diethylamine and dodecylbenzene sulfonic acid, the results were different according to the methodology used. Lastly, no toxicity to bacteria was observed with benzene, dibromo 1,2-ethane, atrazine and malathion; their IC_{50} were close to or greater than 100 mg/l.

The lack of toxicity of the two pesticides can be easily explained: atrazine is a herbicide which acts specifically on photosynthesis; and malathion is a cholinesterase-inhibiting insecticide (Reteuna et al., 1987).

The ranking of the IC_{50} values on a logarithmic scale, as in Fig. 1, clearly shows that the Microtox assay was the most sensitive testing procedure followed by the oxygen consumption assay. The glucose mineralization assay was the least sensitive.

The different sensitivities of these three tests can be explained by the characteristics of each bioassay. With regard to the oxygen consumption assay, two factors can account for its lower sensitivity compared to the Microtox test. First, a high inoculum of activated sludge was used and this biomass is known to adsorb toxicants, resulting in a decrease of their effective concentration in the test medium. Secondly, the medium itself, which contained a number of complexing substances, can modify the speciation of tested compounds, and consequently their toxicity.

Therefore an experiment was conducted with copper, in order to evaluate the importance of metal complexation by the nutrient medium and by each component of this substrate.

The results in Table 2 are expressed as the percentage of residual free copper in relation to added copper. The residual free copper represented only 0.25% added metal in the presence of the nutrient medium. So, nearly all the added copper was complexed by the substrate. The organic fraction (with 0.4% residual free copper) interfered greatly in these complexation mechanisms, especially the peptone with 0.6% and the meat extract with 1.7%. The mineral fraction adsorbed 11.1% residual free copper, with the complexation due mostly to the phosphates with 10.1%.

The bioavailability of metal ions is directly correlated to their toxicity to bacteria (Gillepsie &

152

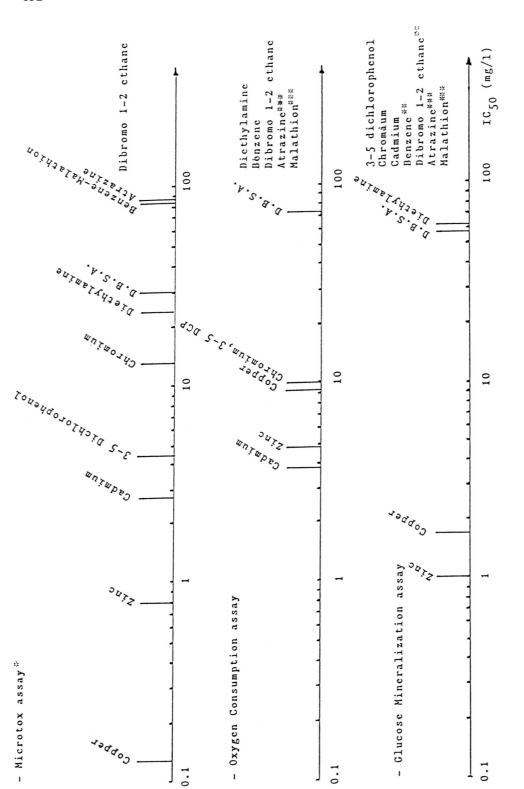

Fig. 1. Sensitivity of Microtox, Oxygen consumption of activated sludge and Glucose Mineralization by *E. coli* assays.

* The incubation time was : 30 min. for mineral substances
                                  15 min. for dodecylbenzene sulfonic acid
                                  5 min. for the other organic substances

**NT : Not toxic at 1 g/l
***NT : Not toxic at 100 mg/l

Vaccaro, 1978). Therefore, the use of a substrate which does not complex toxicants, would certainly improve the sensitivity of the oxygen consumption assay. The lack of sensitivity of the glucose mineralization assay is more difficult to explain; while *E. coli* was specifically sensitive to zinc and copper, it was unaffected by the other tested compounds. These results indicate that *E. coli* is not a good organism for use with the glucose mineralization assay.

Some specific applications can now be considered according to the performances of each bioassay. The Microtox test which is fast (it takes less than one hour) and is easy to perform, is valuable in toxicity screening of effluents and leachates. As it is also sensitive, it may be used in controlled studies to detect pollution in the aquatic environment. However, the microtox test does not seem to be well suited for predicting the possible toxic effects of waste waters on the activated sludge of sewage treatment plants. The oxygen consumption assay which uses activated sludges would certainly be a better alarm system. The use of the glucose mineralization assay should only be considered, if it can be shown that it has a better sensitivity with a biological agent other than *E. coli*. As its methodology does not require high biomass, it could be suited for *in situ* studies using field microorganisms.

Conclusion

It can be concluded that there is no ideal assay. Thus the choice of a bacterial assay in toxicity assessment, will depend on the objectives of the research project, and equally on the performances and the limits of each bioassay.

References

AFNOR T 73265, 1982. Agents de surface, détergents, agents de surface anioniques, détermination de la bio-dégradabilité. 12 pp.

Bazin, C., P. Chambon, M. Bonnefille & G. Larbaigt, 1987. Comparaison des sensibilités du test de luminescence bactérienne (*Photobacterium phosphoreum*) et du test Daphnie (*Daphnia magna*) pour 14 substances à risque toxique élevé. Sciences de l'Eau 6: 403–413.

Brown, M. J., H. R., Hitz & L. Schafer, 1981. The assessment of the possible inhibitory effect of dye-stuffs on aerobic waste-water bacteria, experience with a screening test. Chemosphere 10: 245–261.

Bulich, A. A., M. W. Greene & D. L. Isenberg, 1981. Reliability of the bacterial luminescence assay for determination of toxicity of pure compounds and complex effluents. Aquatic Toxicology and Hazard assessment. ASTM STP 737: 338–347.

Curtis, C., A. Lima, S. J. Lozano & G. D. Veith, 1982. Evaluation of a bacterial bioluminescence bioassay as a method for predicting acute toxicity of organic chemicals to fish. Aquatic Toxicology and Hazard Assessment. ASTM STP 766: 170–178.

Ferard, J. F., P. Vasseur, L. Danoux & G. Larbaigt, 1983. Application d'un test d'inhibition de luminescence bactérienne à l'étude toxicologique d'effluents complexes et de substances chimiques. Revue Française des Sciences de l'Eau 2: 221–237.

Gillepsie, P. A. & R. G. Vaccaro, 1978. A bacterial bioassay for measuring copper chelation capacity of sea water. Limnol. Oceanogr. 23: 543–548.

ISO, 1986. Standard Method 8192, part A, Water quality: test for inhibition of oxygen consumption by activated sludge. 10 pp.

King, E. F. & H. A. Painter, 1986. Inhibition of respiration of activated sludge: variability and reproducibility of results. Toxic. Assess. INT. Q. 1: 27–39.

Qureshi, A. A., R. N. Coleman & J. H. Paran, 1984. Evaluation and refinement of the Microtox test for use. In: D. Liu and B. J. Dutka (Eds.), Toxicity Screening Procedure using Bacterial Systems, Marcel Dekker, New-York. pp. 1–22.

Reteuna, C., P. Vasseur, R. Cabridenc & H. Lepailleur, 1986. Comparison of respiration and luminescent tests in bacterial toxicity assessment. Toxic. Assess. INT. Q. 2: 159–168.

Reteuna, C., P. Vasseur, R. Cabridenc, H. Lepailleur & D. Briand, 1987. Le test Microtox pour l'évaluation de la toxicité des pesticides. Annales A.N.P.P. 5: 209–220.

Vasseur, P., F. Bois, J. F. Ferard & C. Rast, 1986. Influence of physico-chemical parameters on the Microtox test response. Toxic Assess. INT. Q. 1: 283–300.

Vasseur, P., J. F. Ferard, C. Rast & G. Larbaigt, 1984a. Luminescent marine bacteria in ecotoxicity screening tests of complex effluents. In: D. Liu and B. J. Dutka (Eds.), Toxicity Screening Procedures using Bacterial Systems, Marcel Dekker, New-York. pp. 23–26.

Vasseur, P., J. F. Ferard, J. Vial & G. Larbaigt, 1984b. Comparaison des tests Microtox et Daphnie pour l'évaluation de la toxicité aiguë d'effluents industriels. Envir. Pollut. 34: 225–235.

Hydrobiologia **188/189**: 155–162, 1989.
M. Munawar, G. Dixon, C. I. Mayfield, T. Reynoldson and M. H. Sadar (eds)
Environmental Bioassay Techniques and their Application.
© *1989 Kluwer Academic Publishers. Printed in Belgium.*

Luminescent bacteria toxicity assay in the study of mercury speciation

J. M. Ribo[1]*, J. E. Yang[2] & P. M. Huang[2]
*Toxicology Research Centre[1] and Dept. of Soil Science[2] University of Saskatchewan, Saskatoon Saskatchewan, S7N OWO Canada; *Present address: National Hydrology Research Institute, Environment Canada, Saskatoon, SK. S7N 3H5*

Key words: Mercury speciation, pH effect, effect of ligands, toxicity, luminescent bacteria bioassay.

Abstract

The toxicities of solutions of 10 mercury compounds to luminescent bacteria were measured using the Microtox Toxicity Bioassay. The aim of this study was to assess the influence that the counter-ions have on the aquatic toxicity of mercury salts. The toxicities of these mercury compounds were very similar, except for mercurous tannate and mercuric salicylate. This can be attributed to differences in the ionization and speciation patterns of these compounds relative to the other compounds tested. In general, the toxicity of the solutions at pH 5 was not significantly different from the toxicity of these solutions at pH 6, but a clear reduction in toxicity was observed when the pH of the solution was adjusted to pH 9. Significant differences were found between the toxicity of Hg(I) and Hg(II) salts of the same anion at pH 9. When cysteine was added to a mercuric nitrate solution (at pH 6), a reduction in the toxicity was observed. This can be explained in terms of the strong binding of mercury to cysteine, thus reducing the concentration of mercury species available to cause an observable toxic effect to the bioluminescent bacteria.

Introduction

Mercury and mercury derivatives are toxic to living organisms. Because of the peculiar behaviour of mercury in aqueous solution, the chemical nature of the mercury containing compounds present in the environment will be influenced by the physicochemical properties of the aqueous medium and the presence of other substances. As a result, the prediction of the toxic effects of mercury to aquatic biota, and higher organisms requires a thorough knowledge of the speciation patterns of mercury and the influence that other chemicals will have on this speciation.

Bioaccumulation of mercury in different organs of rainbow trout is affected by the concentration of chloride in the media (Walczak *et al.*, 1985).

Mercury speciation is expected to be influenced by the presence in the solution of organic and inorganic compounds of very different chemical properties (Huang, 1988). As a consequence, bioavailability and bioaccumulation of mercury in aquatic biota will be affected also by organic and inorganic solutes.

In aqueous solutions, mercury may co-exist in three oxidation states (0, +1 and +2) including in each one of them, a large number of different species associated with water and hydroxyl ions (OH$^-$). Unfortunately, the mechanisms of speciation of mercury in the freshwater environment are still obscure (Ramamoorthy *et al.*, 1983).

The form of mercury found most commonly in fish is methyl mercury. Potential methylating agents are abundant in the aquatic ecosystem

(Moore & Ramamoorthy, 1984). However, the concentration of available mercury for methylation will be dependant on its speciation and will be also affected by adsorption-desorption phenomena, complexation and precipitation induced by colloidal materials, humic acids, etc.

The aquatic toxicity of mercury will be, therefore, ultimately dependent on the concentration of the different mercury-containing species in solution and this will be influenced by other chemical entities present and also by the physico-chemical properties of the solution (such as temperature and pH).

The luminescent bacteria bioassay (Bulich et al., 1981) provides a very adequate tool to study experimentally the influence of other chemical species on the aquatic toxicity of mercury. This toxicity bioassay, known as the Microtox test*, uses the luminescent marine bacteria *Photobacterium phosphoreum* as the test organism, to determine the total toxic effect of an aqueous sample. The response measured is the reduction in the light emitted by the bacteria, after the organisms have been exposed to a toxic sample for a determined period of time. The reduction in light emission is directly related to the toxicity, which is expressed as the 'Effective Concentration' of the sample that causes a 50% reduction in light emission (EC50).

Previous work demonstrated that, for a wide range of organic chemicals, there is a good correlation between the toxicities determined using the luminescent bacteria bioassay and toxic effects measured using other toxicity bioassays (Ribo & Kaiser, 1983). The Microtox toxicity bioassay has been used extensively for the quick and reliable assessment of aquatic toxicity of organic and inorganic chemicals and complex mixtures of potential contaminants (Bulich *et al.*, 1981; Kaiser & Ribo, 1988).

Light emitted by the luminescent bacteria is independent of the pH of the solution (in the absence of toxicant) when the pH is kept within pH 4.5 and pH 9.5 (Krebs, 1983). In other words,

---

* Microtox is a Trademark of Microbics Corp, Carlsbad, CA.

if the pH of an aqueous solution of a toxic substance happens to be lower than 4.5 or higher than 9.5, the observed reduction in light output may be due to the effect of pH, and not to the toxicity of the sample. On the other hand, the pH of a test sample may induce variations in the relative concentration of one or more species present in the solution, ultimately affecting its toxicity.

The type of response of the test organisms to mercury can be qualified as slow (increasing toxic effect with exposure time) and is similar to the response of the bacteria to other inorganic contaminants such as cadmium, zinc, copper and nickel. However, the response to mercury is unique in that there is a very narrow range of concentrations within which a transition from no-effect (or a slight stimulation) to total inhibition occurs. The reported range of concentrations at which mercury causes total light loss within five to thirty minutes is 0.05 to 0.1 mg of Hg per liter (Beckman instruments, 1981).

The toxicity of mercury to the luminescent bacteria has been reported by several authors (Table 1), in the course of the evaluation of the Microtox test and comparison of its performance related to other microbial bioassays.

The variation observed in the EC50 values in Table 1 is not surprising. These discrepancies are expected when comparing data from different laboratories, obtained using different bacteria batches, and following perhaps different operational procedures. In the case of mercury, small variations in sample concentration, and in the performance of the bacteria, from batch to batch, may be responsible for such a range of experimental values for the same toxicant.

In all the reported cases (Table 1) the toxicity of mercury was determined using mercuric chloride ($HgCl_2$) as the source of mercury, with no pH adjustment of the solutions, prior to the test.

In the case of mercury, pH changes will result in changes in the concentration of one or more species in solution, hence, the final toxicity can be different, and dependant on the pH of the sample.

Moreover, when a sample contains more than one potentially toxic chemical, synergism and/or antagonism may result in an observed toxicity

Table 1. Toxicity of mercury (HgCl$_2$) according to several authors.

EC50 values (mg Hg/L)							
Exposure time	A	B	C	D	E	F	G
5 min	0.065	0.07	0.064	0.06	0.04	0.032	0.056
15 min	–	0.05	0.046	0.02	–	0.024	0.036
30 min	–	0.022		0.01	–	–	–

A: Bulich *et al.* (1981).
B: Beckman Instruments (1981).
C: Dutka *et al.* (1983).
D: Greene *et al.* (1985).
E: McFeters *et al.* (1983).
F: De Zwart & Sloof (1983).
G: this work.

which is different from the simple combination of the toxicity of each individual species. Also, the mechanism(s) of toxic action may be very different for the various species present, thus complicating the interpretation of the experimental results.

The objective of the present study was to assess the influence of the counter ion on the toxicity of mercury salts, under different pH conditions, using the luminescent bacteria bioassay.

Materials and methods

Mercury salts were reagent grade and were used without further purification. Hg(II) chloride and Hg(I) chloride were obtained from J.T. Baker Chemical Co.; Hg(II) nitrate from BDH Canada Ltd.; and Hg(II) acetate, Hg(I) acetate, Hg(II) oxalate, Hg(I) oxalate, Hg(II) salicylate, Hg(II) lactate and Hg(I) tannate from K & K Labs Division, ICN Pharmaceuticals, Inc. Solutions were prepared using double distilled water. The concentration of mercury ranged from 0.12 to 0.34 mg/L. Since in some cases the solubility of the mercury derivative was very low, the actual concentration of total mercury present in the solutions was determined spectroscopically in a UV Mercury Analyzer (Model Hg^{-3}) utilizing the cold vapour technique at 253.7 mm wavelength

(U.S. Environmental Protection Agency, 1974). The pH of the solutions prepared, prior to any pH adjustment, was very close to pH 5.

Luminescent bacteria used in the bioassay was obtained, in the standard freeze-dried form, from Microbics Corp., Carlsbad, California, exclusive supplier of the Microtox Toxicity Analyzer. Microtox Reconstitution solution (double distilled water), Microtox Diluent (2% NaCl solution) and Microtox Adjusting solution (22% NaCl solution) were also obtained from this supplier.

The toxicity of each sample was determined following the experimental procedure described elsewhere (Ribo & Kaiser, 1987), after 5 and 15 minutes of exposure.

Each one of the solutions was tested first without any pH adjustment, and then after adjustment to pH 6 and pH 9 by dropwise addition of dilute 0.1 M NaOH.

The luminescent bacteria used in the Microtox test, *Photobacterium phosphoreum* are marine organisms. As part of the standard operational procedure of the Microtox bioassay, 10% (v/v) of a Microtox Osmotic Adjusting Solution of sodium chloride was added to the sample, to reach a 2% saline concentration in the test solution, which provides the necessary osmotic protection to the bacteria.

Results

All compounds tested were highly toxic to the luminescent bacteria. Although stability constants of these mercury compounds differ substantially (Table 2), their toxicities were very similar in the pH range of 5 to 6 (Table 3). At pH 5, the observed EC50 values for these compounds do not differ significantly, at both exposure times, except for the toxicity of Hg(I) tannate and Hg(II) salicylate. At pH 6 only the toxicity of Hg(I) tannate is significantly different from the toxicities of the other compounds, both at 5 and 15 min. of exposure. Except for Hg(II) salicylate, Hg(I) tannate and Hg(II) lactate, whose stability constants are not available in the literature, the theoretical calculations by GEOCHEM computer program (Sposito & Mattigod, 1979; Parker et al., 1987) show that all Hg compounds studied are totally converted to the Hg-chloride complex in the test solutions, in the concentration of Hg and pH ranges studied, due to the presence of 2% NaCl, which is required for the toxicity bioassay.

The toxicities of the compounds tested, at the same pH and for each exposure time, have a coefficient of variation of about 22%, which is in agreement with the variation found in Microtox round-robins and interlaboratory comparison studies (Qureshi et al., 1987; Moynihan et al., 1988). At pHs of 5 and 6, the observed toxicity does not seem to be related to the state of oxidation of the mercury.

When the pH of the solutions was adjusted to 9 prior to the bioassay, the toxicities of all compounds were reduced, in comparison to the ECS50 values observed at lower pHs, with the exception of mercuric salicylate. Moreover, at pH 9 and at both exposure times, the average EC50 values of mercuric oxalate, mercuric acetate, mercuric chloride and mercuric nitrate, are significantly different from the average of the EC50 values for mercurous oxalate, mercurous

Table 2. Stability constants (log values) of selected mercury compounds at zero ionic strength at 25 °C.

Compound		Stability constant[+]	Reference
Hg(II)	chloride		Smith & Martell (1976)
	$HgCl^+$	7.2	
$HgCl_2$		14.0	
$HgCl_3^-$		14.9	
$HgCl_4^{2-}$		15.6	
HgClOH		10.5	
Hg(I)	chloride		Smith & Martell (1976)
	Hg_2Cl_2	17.9	
Hg(II)	nitrate		Smith & Martell (1976)
	1:1 Hg(II)-nitrate	0.9	
Hg(II)	acetate		Martell & Smith (1982)
	1:1 Hg(II)-acetate	4.2	
	1:2 Hg(II)-acetate	8.4	
Hg(I)	acetate		Martell & Smith (1982)
	1:1 Hg(I)-acetate	3.6	
	1:2 Hg(I)-acetate	6.6	
Hg(II)	oxalate		Perrin (1979)
	1:1 Hg(II)-oxalate	10.1	
Hg(I)	oxalate		Martell & Smith (1977)
	1:2 Hg(I)-oxalate	8.0	
	Hg(I)-oxalate-OH	13.2	

[+] The stability constants have been converted to zero ionic strength from the original data found in the literature.

Table 3. Microtox toxicity of mercury compounds.

pH non-adjusted (pH ≈ 5)

| Hg compound | pH | EC50 (mg Hg/L) | | EC50 (μmol Hg/L) | |
		5-min	15-min	5-min	15-min
Hg(I) tannate	4.2	0.230	0.075	1.15	0.37
Hg(I) acetate	5.2	0.062	0.032	0.31	0.16
Hg(II) lactate	5.2	0.069	0.034	0.34	0.17
Hg(II) oxalate	5.0	0.062	0.030	0.31	0.15
Hg(II) acetate	5.1	0.060	0.037	0.30	0.18
Hg(II) chloride	5.3	0.056	0.036	0.28	0.18
Hg(II) nitrate	5.4	0.052	0.032	0.26	0.16
Hg(I) oxalate	5.1	0.050	0.025	0.25	0.12
Hg(I) chloride	5.1	0.050	0.033	0.25	0.17
Hg(II) salicylate	4.8	0.012	0.010	0.06	0.05

(pH adjusted at pH ≈ 6)

| Hg compound | pH | EC50 (mg Hg/L) | | EC50 (μmol Hg/L) | |
		5-min	15-min	5-min	15-min
Hg(I) tannate	6.0	0.650	0.360	3.25	1.80
Hg(I) acetate	5.9	0.070	0.048	0.35	0.24
Hg(II) lactate	5.9	0.058	0.039	0.29	0.20
Hg(II) oxalate	6.0	0.053	0.025	0.27	0.13
Hg(II) acetate	6.2	0.077	0.017	0.39	0.09
Hg(II) chloride	6.0	0.059	0.027	0.29	0.13
Hg(II) nitrate	6.0	0.054	0.038	0.27	0.19
Hg(I) oxalate	6.0	0.039	0.025	0.20	0.14
Hg(I) chloride	6.0	0.048	0.039	0.24	0.19
Hg(II) salicylate	6.0	0.052	0.038	0.26	0.19

(pH adjusted at pH ≈ 9)

| Hg compound | pH | EC50 (mg Hg/L) | | EC50 (μmol Hg/L) | |
		5-min	15-min	5-min	15-min
Hg(I) tannate	9.0	0.300	0.250	1.50	1.25
Hg(I) acetate	9.2	0.580	0.470	2.90	2.35
Hg(II) lactate	9.0	0.460	0.310	2.30	1.55
Hg(II) oxalate	9.2	0.180	0.090	0.90	0.45
Hg(II) acetate	9.3	0.190	0.110	0.95	0.55
Hg(II) chloride	9.1	0.110	0.070	0.55	0.35
Hg(II) nitrate	9.1	0.140	0.090	0.70	0.45
Hg(I) oxalate	9.3	0.550	0.380	2.75	1.91
Hg(I) chloride	9.1	0.490	0.300	2.45	1.50
Hg(II) salicylate	9.1	0.014	0.011	0.07	0.06

chloride and mercurous acetate, suggesting that, at high pH, there is some difference in toxicity due to the state of oxidation of mercury.

Student-t test applied to the mean of the EC50 values obtained confirmed that, at pHs of 5 and 6, the average of the 5 min-EC50 values are significantly different from the 15 min-EC50 values, at the 95% confidence level. The mean of 5 min-

EC50 values at pH 5 [excluding Hg(I) tannate and Hg(II) salicylate] was not significantly different from the mean of 5 min-EC50 values at pH 6, and, also, no statistically significant difference was observed between the mean of the 15 min-EC50 values at pHs of 5 and 6, at the 95% confidence level.

Three organic compounds, citric acid, glycine and cysteine, were added to the solution of mercuric nitrate, to evaluate their influence on the observed toxicity (Table 4). The concentration of mercury was 2.0 mg Hg/L and all solutions were tested at pH 6. Under these conditions, neither glycine nor citric acid had any significant effect on the toxicity of Hg(II) nitrate, in the presence of 2% NaCl, due to the formation of Hg-Cl complexes as the predominant Hg species.

After the addition of cysteine, which has a higher stability constant when complexed with mercury (Simpson, 1961), the toxicity of the solution appeared to decrease, and this effect was greater when the molar ratio of cysteine/mercuric nitrate was increased.

Discussion

All the compounds tested are, in aqueous solution, dissociated to some degree. Their aquatic toxicity can be interpreted, in a first approximation, as a result of the integration of the mercury ion (I or II), and the toxicity of the undissociated salt, and assuming that the counter ion has no toxic effect.

Table 4. Influence of organic ligands on the toxicity of mercury.

Compounds	EC50 (μmol Hg/L)	
	5-min	15-min
Hg(II) (NO$_3$)$_2$	0.260	0.160
Hg(II) (NO$_3$)$_2$ + cysteine 1 : 1	0.500	0.290
Hg(II) (NO$_3$)$_2$ + cysteine 1 : 1000	1.220	0.800
Hg(II) (NO$_3$)$_2$ + citric acid 1 : 1	0.130	0.070
Hg(II) (NO$_3$)$_2$ + citric acid 1 : 1000	0.260	0.160
Hg(II) (NO$_3$)$_2$ + glycine 1 : 1	0.190	0.140
Hg(II) (NO$_3$)$_2$ + glycine 1 : 1000	0.230	0.180

Mercury in aqueous solution can be present in a large number of different species; their chemical structure and relative concentration will depend on the nature and concentration of other chemical species in the same solution. In the test solutions used to determine the toxicity of mercury, 342.2 mmol/L of chloride was added to provide osmotic protection to the test organisms (see Materials and Methods). This represents an excess of chloride concentration of 2.0 to 5.7×10^5 times the concentration of mercury in the test solutions, and will obviously affect the speciation of mercury.

Without taking into account other speciation patterns nor association with water, OH$^-$ or H$^+$, the dissociation of any of mercury salts in aqueous solution may be represented by:

$$Hg_n\text{-}A_m = nHg^{+m} + mA^{-n}$$

If the degree of dissociation is represented by α, then the concentration of each of these species can be represented, respectively, by

$$(1\text{-}\alpha)[Hg_nA_m], \quad \alpha n[Hg^{+m}], \text{ and } \alpha m[A^{-n}]$$

The toxicity will be due to the presence of any chemical species containing mercury, which are:

$$[Hg^{+m}] \text{ and } [Hg_nA_m]$$

and the observed toxic effect (assuming that no synergism or antagonism occurs) can be expressed as:

$$T(\text{observed}) = \alpha n Tox(Hg^{+m}) + (1\text{-}\alpha)Tox(Hg_n\text{-}$$

α being the fraction of total concentration of mercury present in the form of $[Hg^{+m}]$ ion, and, therefore, $(1\text{-}\alpha)$ the fraction of total mercury present as undissociated form $[Hg_nA_m]$.

In an aqueous solution, Hg species would be more properly represented as a combination of terms, each term being the contribution of each one of the possible species containing mercury (such as Hg^{+2}, $[Hg(OH)]^+$, $Hg(OH)_2$, and $[HgA(H_2O)]^{n+}$), and each one of these species having its own toxicity.

The presence of the high concentration of chloride in the test solutions would direct the speciation of mercury towards predominant species, Hg-Cl complexes, common to all solutions tested. Wang *et al.* (1988) reported that chloride has a prominent effect on mercury speciation. If the anion itself does not contribute significantly to the observed toxicity, the toxicities of all solutions tested should be similar within the expected coefficient of variation of the test.

The results obtained, presented in Table 3, suggest that, at pH 5 and pH 6, one or more common species are responsible for the toxicity, and the observed toxic effect is very similar for most of the compounds under the test conditions. A clear exception is the case of mercurous tannate, and mercuric salicylate. The toxicity of Hg(I) tannate is, at all pH tested, the lowest of all compounds. The toxicity of Hg(II) salicylate at pH 5 and pH 9 appears to be similar; its toxicity is lower at pH 6. In terms of chemical structure, the phenolic structure of the salicylate anion can account for its different behaviour from the other salts. Phenolic compounds are known to be toxic to micro-organisms (Kuwahara *et al.*, 1970). In the case of tannate, the structure of the anion is, again, very different. The Hg(I)-tannate complex apparently has a higher stability constant which may account for the lower toxicity.

The toxicity of the compounds tested appears to be independent of the oxidation state of the mercury ion, at pHs of 5 and 6. For alkaline solutions, it appears that, with the exception of Hg(II) lactate, the toxicity of Hg(II) solutions is higher than the toxicity of Hg(I)-containing solutions. There is no precise explanation, at this time, for this behaviour. Nevertheless, the differences in the toxicity between Hg(II) and Hg(I) at alkaline conditions may be attributed to the difference in the extent of hydrolysis. Mercurous hydroxide cannot be formed (Baes & Mesmer, 1976). On the other hand, $Hg(OH)_2$ is the predominant species of Hg(II) in the pH range of 4 to 9 (Singh & Huang, 1988). The toxicities at pH 9 were lower than the toxicities at pHs of 5 or pH 6. This suggests that the change of 3 orders of magnitude in the activity of hydrogen ion, from pH 6 to pH 9

affects significantly the speciation of mercury, even in the presence of excess of chloride, resulting in the increase in the relative concentration of a less toxic species.

When cysteine was added to the test solution of mercuric nitrate, a reduction in toxicity was observed (Table 4), and this effect was enhanced when the concentration of cysteine was increased. This effect is corroborated by other studies of toxicity of mercury using amphipods as test organisms (Merkowsky & Hammer, 1988). The decrease in toxicity of mercury can be attributed to the high stability constant of the complex Hg-cysteine (Simpson, 1961). As a result of the complexation, there is a reduction in the relative concentration of the more toxic mercury-containing species in the test solution, thus resulting in a reduction of the observed toxicity.

Conclusions

The results do not indicate significant differences in toxicity between the compounds tested, except for Hg(I) tannate and Hg(II) salicylate. These exceptions appear to be attributable to the very different chemical structure of the anion.

The contribution to the observed toxicity of the counter ion appears to be negligible compared to the toxicity of the mercury-containing species.

From the similarity in the observed toxicities, it can be inferred that the speciation of mercury follows similar patterns in most of the solutions tested, possibly due to the large concentration of chloride present. If the speciation is different, the combined toxic effect of the different species is similar for all compounds tested, at these pH values.

Except for tannate and salicylate the observed toxicity was not significantly different for the compounds tested at the same pH. This suggests a similarity in the speciation patterns of these substances. Moreover, the speciation is not influenced by the chemical nature of the counter ion, but strongly influenced by the high concentration of chloride and the pH of the aqueous medium.

L. I. H. E.
THE MARKLAND LIBRARY
STAND PARK RD., LIVERPOOL, L16 9JD

162

At alkaline pH, there is differentiation between the observed toxicity of Hg(I) and Hg(II). With the exception of Hg(II) lactate, the solutions of Hg(II) are more toxic than the Hg(I) solutions.

The results of the toxicity determinations described in this work suggest that the elucidation of the aquatic toxicity of mercury will necessitate a more comprehensive study of the speciation patterns of mercury in aqueous solutions, focusing specifically on the influence of organic ligands and inorganic species.

Acknowledgement

This study was supported by Natural Sciences and Engineering Research Council of Canada, Strategic Grant G1994 - Huang. SIP Publication No. R602.

References

Baes, C. F. Jr. & R. E. Mesmer, 1976. The hydrolysis of cations. J. Wiley and Sons, New York. 489 pp.

Beckman Instruments, 1981. Advantages of using several test times. Microtox Application Notes, Number M102.

Bulich, A. A., M. W. Greene & D. L. Isenberg, 1981. Reliability of the bacterial luminescence assay for determination of the toxicity of pure compounds and complex effluents. In D. R. Branson and K. L. Dickson (eds.), Aquatic Toxicology and Hazard Assessment: 4th conference. ASTM STP 737, American Society for Testing and Materials, pp. 338–347.

De Zwart, D. & W. Sloof, 1983. The microtox as an alternative assay in the acute toxicity assessment of water pollutants. Aquatic Toxicology 4: 129–138.

Dutka, B. J., N. Nyholm & J. Petersen, 1983. Comparison of several microbiological toxicity screening tests. Water Res. 17: 1363–1368.

Greene, J. C., W. E. Miller, M. K. Debacon, M. A. Long & C. L. Bartels, 1985. A comparison of three microbial assay procedures for measuring toxicity of chemical residues. Arch. Environ. Contam. Toxicol. 14: 659–667.

Huang, P. M., 1988. Limnology: Dispersion of toxic substances. In Systems and Control Encyclopedia, pp. 2772–2782. M. G. Singh (ed.). Pergamon Press, Oxford.

Kaiser, K. L. E. & J. M. Ribo, 1988. Photobacterium phosphoreum Toxicity Bioassay. II. Toxicity Data Compilation. Toxicity Assessment 3: 195–237.

Krebs, F., 1983. Toxicity Test Using Freeze-dried Luminescent Bacteria. Gewässerschutz, Wasser, Abwasser 63: 173–230.

Kuwahara, M., N. Shindo & K. Munakata, 1970. The photochemical reaction of pentachlorophenol (part 4). J. Agric. Chem. Soc. Japan 44: 169–174.

Martell, A. E. & R. M. Smith, 1977. Critical Stability Constants. Vol. 3. Other Organic Ligands. Plenum Press, New York. 495 pp.

Martell, A. E. & R. M. Smith, 1982. Critical Stability Constants. Vol. 5. First Supplement. Plenum Press, New York. 604 pp.

McFeters, G. A., P. J. Bond, S. B. Olson & Y. T. Tchan, 1983. A comparison of microbial bioassays for the detection of aquatic toxicants. Water Res. 17: 1757–1762.

Merkowsky, A. & U. T. Hammer, 1988. (Abstract) Influence of Organic Compounds on the Toxicity of Mercury to Amphipods. Presented at the 'Workshop on mercury speciation as influenced by ligands and sediment particulates and the impact on freshwater toxicology.' University of Saskatchewan, Saskatoon, Canada.

Moore, J. W. & S. Ramamoorthy, 1984. Heavy metals in natural waters. Springer Verlag, New York. 268 pp.

Moynihan, K. J., J. B. MacLeod, Y. V. Hardy, A. R. Teal, M. L. Korchinski, D. C. Roberts, I. B. Zaborski & J. R. Creasy, 1988. (Abstract) Toxicity Evaluations of Drilling Sump Fluids: Microtox versus Fish Toxicity Tests. Proc. 1988 Intl. Conf. on Drilling Wastes, Calgary, Alberta, Canada.

Parker, D. R., L. W. Zelazny & T. B. Kindraide, 1987. Improvements to the program GEOCHEM. Soil Sci. Soc. Am. J. 51: 488–491.

Perrin, D. D., 1979. Stability Constants of Metal-ion Complexes. Part B. Organic Ligands (IUPAC). Pergamon Press, Toronto. 1263 pp.

Qureshi, A. A., A. K. Sharma & J. H. Paran, 1987. (Abstract) Microtox quality control collaborative study: a unique and enlightening experience. 3rd. Intl. Symp. on Toxicity Testing Using Microbial Systems. Valencia, Spain.

Ramamoorthy, S., T. C. Cheng & D. J. Kushner, 1983. Mercury speciation in water. Can. J. Fish. Aq. Sci. 40: 85–89.

Ribo J. M. & K. L. E. Kaiser, 1983. Effects of selected chemicals to photoluminescent bacteria and their correlations with acute and sublethal effects on other organisms. Chemosphere 12: 1421–1442.

Ribo J. M. & K. L. E. Kaiser, 1987. Photobacterium phosphoreum Toxicity Bioassay. I. Test Procedures and Applications. Toxicity Assessment 2: 305–323.

Simpson, R. B., 1961. Association constants of methylmercury with sulfhydryl and other bases. J. Am. Chem. Soc. 83: 4711–4717.

Singh, J. & P. M. Huang, 1988. (Abstract) Effect of glycine on mercury speciation under varying pH, temperature and ionic strength conditions. 23rd Canadian Symp. on Water Pollution Research, Burlington, Ontario, Canada. Abstract 43 pp.

Smith, R. M. & A. E. Martell, 1976. Critical Stability Constants. Vol 4. Inorganic complexes. Plenum Press, New York. 257 pp.

Sposito, G. & S. V. Mattigod, 1979. GEOCHEM: A computer program for the calculation of chemical equilibria in soil solution and other natural water systems. Kearney Foundation of Soil Science, University of California, Riverside. 106 pp.

U.S. Environmental Protection Agency, 1974. Methods for Chemical Analysis of Water and Wastewater. Office Technol. Transfer. EPA-62576-74-003. Washington, D.C. 298 pp.

Walczak, B. Z., U. T. Hammer & P. M. Huang, 1985. Ecophysiology and mercury accumulation of rainbow trout (Salmo gairdneri) when exposed to mercury in various concentrations of chloride. Can. J. Fish. Aq. Sci. 43: 710–714.

Wang, J. S., P. M. Huang, U. T. Hammer & W. K. Liaw, 1988. Influence of chloride/mercury molar ratio and pH on the adsorption of mercury by poorly crystalline oxides of Al, Fe, Mn, and Si. Verh. Internat. Verein. Limnol. 23: 1594–1600.

Hydrobiologia **188/189**: 163–174, 1989.
M. Munawar, G. Dixon, C. I. Mayfield, T. Reynoldson and M. H. Sadar (eds)
Environmental Bioassay Techniques and their Application.
© 1989 *Kluwer Academic Publishers. Printed in Belgium.*

Is the 'microbial loop' an early warning indicator of anthropogenic stress?

M. Munawar[1] & T. Weisse[2]
[1] *Fisheries & Oceans Canada, Great Lakes Laboratory for Fisheries and Aquatic Science, Canada Centre for Inland Waters, Burlington, Ontario, Canada L7R 4A6*; [2] *Limnological Institute, University of Konstanz, P.O. Box 5560, D-7750 Konstanz, West Germany*

Key words: microbial loop, indicator, stress, contaminants, nutrients

Abstract

Various components of the 'microbial loop' such as bacteria, heterotrophic nanoflagellates and autotrophic picoplankton were analyzed, for the first time across the Great Lakes, during a cruise in the summer of 1988. In addition, the size fractionated primary productivity using carbon-14 techniques was also determined. The statistical analysis indicated that bacteria, autotrophic picoplankton and ultraplankton/picoplankton productivity were significantly higher in Lakes Ontario and Erie than Lakes Huron and Michigan. The autotrophic picoplankton and ultraplankton/picoplankton productivity was higher in Lake Erie compared to Lake Ontario.

The autotrophic picoplankton showed sensitivity to nutrients and contaminants in various types of environments. A dramatic decrease of autotrophic picoplankton in eutrophic-contaminated areas, such as Ashbridges Bay, Hamilton Harbour and western Lake Erie was observed. Conversely, in Saginaw Bay, another eutrophic environment, the autotrophic picoplankton were significantly higher than in Lake Huron. The sensitivity of autotrophic picoplankton to nutrients/contaminants might have implications to trophic interactions. Our results suggest that structural and functional characteristics of the 'microbial loop' may be operating differently in stressed versus unstressed ecosystems. The possibility of using autotrophic picoplankton as an early warning indicator of environmental perturbation is proposed.

Introduction

The significance of the 'microbial loop' in both marine and freshwater aquatic ecosystems is well known (Azam *et al.*, 1983; Pomeroy & Wiebe, 1988; Sherr & Sherr, 1988; Stockner, 1988). In spite of this, very little is known about the 'microbial loop' in the North American Great Lakes, where only certain aspects have been investigated. The autotrophic picoplankton (APP, 0.2–2.0 μm) and its productivity have received the most attention (Munawar & Fahnenstiel, 1982; Munawar &

Munawar, 1986; Munawar *et al.*, 1987; Leppard *et al.*, 1987; Caron *et al.*, 1985; Fahnenstiel *et al.*, 1986; Pick & Caron, 1987). The North American Great Lakes range in their trophic status from oligotrophy to eutrophy and some of them are also contaminated (Vollenweider *et al.*, 1974; I.J.C., 1987). They provide an excellent opportunity for a survey of the structure of the 'microbial loop' in environments of various trophic states and degrees of contamination. The present paper reports such a study for the first time conducted in the summer of 1988 to focus on the relative sensitivity of pelagic microorganisms to nutrient enrichment and contamination. The details of the dynamics of the 'microbial loop' will be published elsewhere (Weisse & Munawar, unpublished).

Methods

Biological

Water samples across the Great Lakes (Figs. 1 & 2) were collected by an integrating water sampler from the euphotic zone during the months of August to October, 1988 (Munawar & Munawar, 1986) and preserved in formalin to determine the microbial community structure, namely bacteria (BACT), heterotrophic nanoflagellates (HNF) and autotrophic picoplankton (APP) by epifluorescence microscopy according to the Dapi staining technique (Porter & Feig, 1980; Weisse,

1988, 1989a). Approximately 2 to 5 ml of samples were filtered on 0.2 μm Nuclepore membrane filters and enumerated under 1250 × magnification with a Nikon microscope. Size-fractionated primary productivity was estimated by the carbon-14 technique according to Vollenweider *et al.* (1974) and Munawar *et al.* (1987) for the following size assemblages:

Ultraplankton-picoplankton .... <20 μm
Microplankton-netplankton .... >20 μm

Statistical

Only offshore stations were used to compare the status of the four lakes. For this purpose the following stations were chosen:
Lake Michigan
1–8, 11–16
Lake Huron
1, 5, 6, 8, 9, 12, 14, 15, 17, 27, 29, 30, 31, 32, 35,

Fig. 1. Sampling locations, August 1988.

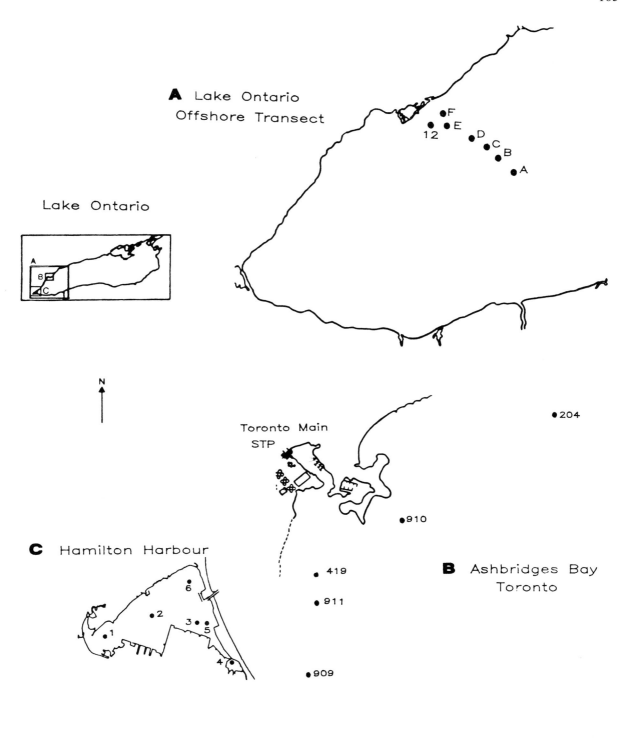

Fig. 2. Sampling locations of Ashbridges Bay and Hamilton Harbour.

41, 42, 43, 45, 48, 55, 61, 65, 69, 71, 77, 79, 82, 89.
Lake Erie
9, 23, 25, 55, 84
9, 23, 55, 84 (Eastern)
25, 29, 357, 358 (Western)
Lake Ontario
A–F, 1, 2, 3

A one-way ANOVA (analysis of variance) (Box *et al.*, 1978) was done to determine if there were any differences among the lakes. Contrasts were used to determine the significance of the differences of interest.

Lake Huron stations were identified as follows in order to test for differences among locations within the lake.

North Channel	stations 69, 71, 77, 79, 82, 89
Georgian Bay	stations 6, 9, 15, 17, 27, 29, 31, 35, 42, 43, 45
Lake Huron	stations 1, 5, 8, 12, 14, 32, 41, 55, 61, 65
Saginaw Bay	stations 95, 98, 100, 101, 101A

A one-way ANOVA analysis was made to determine if there were any differences among the different locations within the lake. Contrasts were used to test for differences between locations.

Lake Ontario was analyzed using the same methods as for Lake Huron. The locations within the lake were identified as follows:

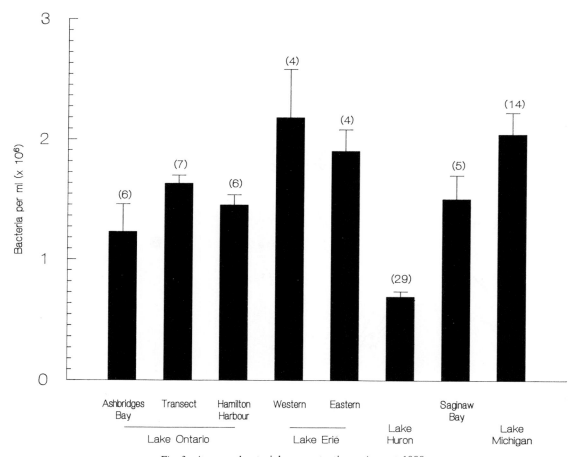

Fig. 3. Average bacterial concentrations, August 1988.

Ashbridges Bay stations 734, 909, 419, 910, 911, 204
Lake Ontario Transect stations A, B, C, D, E, F, 12
Hamilton Harbour stations 1, 2, 3, 4, 5, 6

Results

A comparative overview of the abundance of various pelagic organisms, namely BACT (Fig. 3), HNF (Fig. 4) and APP (Fig. 5), is presented. Similarly the ultraplankton-picoplankton (UPP; Fig. 6) and microplankton-netplankton (MNP; Fig. 6) productivity was also compared. The results of various types of statistical comparisons are described below:

Comparison between lakes:
BACT, APP and UPP (Table 1) were significantly ($\alpha \leqslant 0.05$) higher in the lower lakes (Lakes Ontario and Erie) than in the upper lakes (Lakes Huron and Michigan). Lake Michigan had significantly higher BACT and HNF ($\alpha \leqslant 0.05$) than Lake Huron.

Lake Erie had significantly higher ($\alpha \leqslant 0.05$) APP and UPP than Lake Ontario. (Table 1).

Comparison between different sections of Lake Huron:
All five tests, namely BACT, HNF, APP, UPP and MNP, gave similar results when comparing the different locations (Table 2). There were no significant differences between the North Chan-

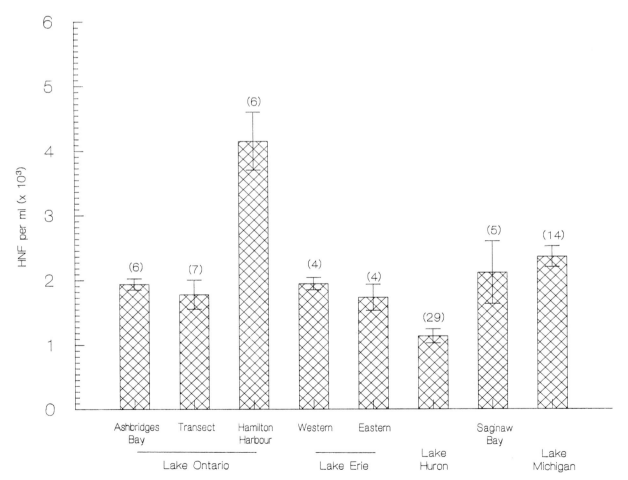

Fig. 4. Average heterotrophic nanoflagellate concentration, August 1988.

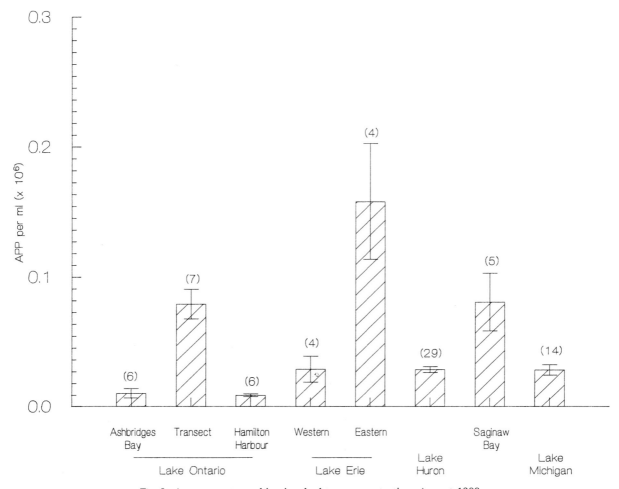

Fig. 5. Average autotrophic picoplankton concentration, August 1988.

nel, Georgian Bay and main Lake Huron locations; however Saginaw Bay showed significantly ($\alpha \leqslant 0.05$) higher values than the other
three locations combined for all the test variables.

Comparison between Lake Ontario transect, Ash-
bridges Bay & Hamilton Harbour:
Ashbridges Bay possessed significantly lower
APP and UPP when compared to the offshore
Lake Ontario transect (Fig. 2). Hamilton
Harbour was found to have significantly higher
HNF, UPP and MNP than the offshore transect.
On the other hand the number of APP in
Hamilton Harbour was significantly lower than

the offshore Lake Ontario transect (Table 3). Of
the five tests, HNF, UPP and MNP values were
significantly lower ($\alpha = 0.05$) in Ashbridges Bay
than they were in Hamilton Harbour (Table 3).

Discussion

The distributional patterns of various pelagic organisms reveal some characteristic features of the
Great Lakes. For example the bacterial abundance (Fig. 3) was lowest in the oligotrophic
(Vollenweider *et al.*, 1974; Munawar & Munawar, 1982) Lake Huron (including Georgian Bay

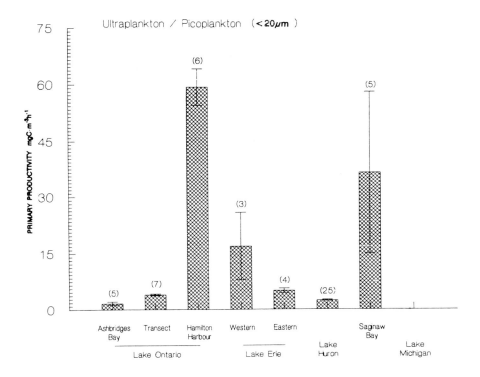

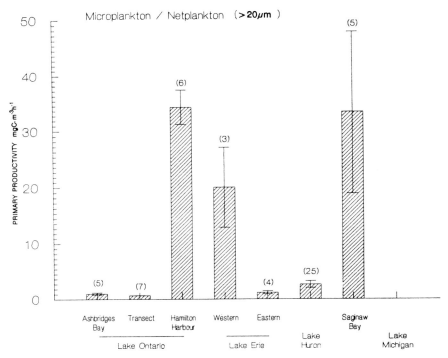

Fig. 6. Average size-fractionated productivity of ultraplankton-picoplankton ($<20\,\mu$m) and microplankton-netplankton ($>20\,\mu$m), August 1988.

Table 1. ANOVA results for comparing all five lakes using the offshore stations.

	Lower lakes–Upper lakes	L. Michigan^–L. Huron	L. Erie L. Ontario
Bacteria	*** +	***	NS
Heterotrophic Nanoflagellates	NS	***	NS
Autotrophic Picoplankton	***	NS	**
Primary productivity Ultraplankton/Picoplankton ($>20 \mu m$)	***	–	***
Primary productivity Microplankton/Net plankton ($>20 \mu m$)	NS	–	NS

+   Without brackets the given difference is positive and with brackets the difference is negative.
^   No primary productivity data for Lake Michigan was available.
**   Significant at the 1% level.
***   Significant at the 0.1% level.
NS Not significant for $\alpha = 5\%$.

and the North Channel), whereas the eutrophic Lake Erie and its three basins had similar levels to those observed in Lake Michigan. The bacterial levels recorded in the highly eutrophic Saginaw Bay were not different from those observed in Lake Ontario. Thus no typical trends for bacterial abundance in relation to the contamination were observed.

The HNF revealed similar concentrations across the Great Lakes although there was a ten-

Table 2. ANOVA results for different sections of Lake Huron.

	Lake Huron–North Channel	Lake Huron–Georgian Bay	North Channel–Georgian Bay	Saginaw Bay–Lake Huron, North Channel, Georgian Bay
Bacteria	NS +	NS	NS	**
Heterotrophic Nanoflagellates	NS	NS	NS	***
Autotrophic Picoplankton	NS	NS	NS	***
Primary productivity Ultraplankton/ Picoplankton ($<20 \mu m$)	NS	NS	NS	**
Primary productivity Microplankton/ Net plankton ($>20 \mu m$)	NS	NS	NS	***

+   Without brackets the given difference is positive and with brackets the difference is negative.
**   Significant at the 1% level.
***   Significant at the 0.1% level.
NS Not significant for $\alpha - 5\%$.

Table 3. ANOVA results for comparing different sections of Lake Ontario.

	Ashbridge Bay– Hamilton Harbour	Ashbridge Bay– Offshore Transect[†]	Hamilton Harbour– Offshore Transect
Bacteria	NS [+]	NS	NS
Heterotrophic Nanoflagellates	(***)	NS	***
Autotrophic Picoplankton	NS	(***)	(***)
Primary productivity Ultraplankton/ Picoplankton ($<20\ \mu m$)	(***)	(***)	***
Primary productivity Microplankton/ Netplankton ($<20\ \mu m$)	(***)	NS	***

[+]   Without brackets the given difference is positive and with brackets the difference is negative.
[†]   Lake Ontario offshore transect.
** Significant at the 1% level.
*** Significant at the 0.1% level.
NS Not significant for $\alpha = 5\%$.

dency of elevated HNF abundance with increasing trophic status. We observed lowest HNF numbers in oligotrophic Georgian Bay, North Channel and Lake Huron. It must be considered that bacteria and HNF do not vary independently from each other. HNF are known as the major bacterial consumers in the ocean (Azam *et al.*, 1983), and typical predator-prey relationships have also been demonstrated in lakes (Bloem & Bar-Gilissen, 1989; Weisse, 1989). HNF in turn are subject to heavy grazing pressure by ciliates and other microzooplankton (Bloem & Bar-Gilissen, 1989; Finlay *et al.*, 1988; Weisse & Muller, 1989). Therefore, the observed inconsistency of bacteria and HNF numbers in relation to contamination might be explained by changing grazing pressure within the microbial loop. It is also important to note that HNF abundance gives only a crude indication of their ecological significance since the average cell size might vary considerably (Weisse, 1989b). A more detailed analysis of the HNF community structure is in preparation (Weisse & Munawar, unpubl.). Yet, the overwhelming dominance of HNF in the eutrophic-contaminated Hamilton Harbour is apparent (Fig. 4), but cannot be explained solely based on predator-prey relationship at the present time.

The APP exhibited the most interesting trend and revealed a diversity of response to nutrients and contaminants (Fig. 5). For example their abundance was lowest in the eutrophic-contaminated ecosystems such as Hamilton Harbour, Ashbridges Bay and the western Lake Erie, compared to Lake Ontario offshore transect and Lake Erie stations (central and eastern basins). In fact, it is interesting to note that when a hierarchial clustering analysis was made using APP from all the Great Lakes, the group with lowest APP measurements was made up of stations from some of the most eutrophic and contaminated areas such as Ashbridges Bay, Hamilton Harbour, western Lake Erie, and Green Bay etc.

On the other hand the APP levels were found to be higher in Saginaw Bay (Fig. 5) than Lake Huron. Although Saginaw Bay is known to be an extremely eutrophic and contaminated ecosystem, APP seem to thrive well in it. This may be attributable to high concentrations of nutrients and prevailing nutrient-contaminant interactions which might be conductive to the development of these microorganisms.

Our results show an interesting pattern of sensitivity of APP to anthropogenic stress. It showed extremely low concentrations in Ashbridges Bay compared to the offshore transect. The Ashbridges Bay is a eutrophic environment which receives municipal effluents from sewage treatment plants and other industrial pollutants (MISA, 1987; Munawar, 1989). The APP were very low in Hamilton Harbour which is an extremely eutrophic ecosystem and a recipient of large amounts of industrial waste products originally from the steel industry (Harlow & Hodson, 1988) and sewage effluents. The western end of Lake Erie is also known to be eutrophic, (Vollenweider et al., 1974; Munawar & Munawar, 1986; Weiler, 1981) receiving municipal and industrial loadings from the Maumee and Detroit Rivers. Here, as well, the numbers of APP were relatively low. On the other hand, Saginaw Bay is extremely eutrophic and contaminated with both municipal and industrial effluents (I.J.C., 1977) but had a larger number of APP compared to Lake Huron.

Some interesting similarities could be noticed when APP's sensitivity is related with size fractionated productivity. For example APP was significantly lower in Ashbridges Bay compared to the offshore transect. UPP ($< 20 \mu m$) productivity was significantly lower in the Bay than the offshore transect (Table 3). However in Hamilton Harbour the APP were low and yet the UPP was extremely high (Fig. 6). Another scenario is that of Saginaw Bay in which both APP and UPP were higher than in Lake Huron. The MNP ($> 20 \mu m$) productivity showed no significant differences between Ashbridges Bay and the transect whereas MNP was significantly higher in Hamilton Harbour than that observed for the Lake Ontario offshore transect (Table 3). Similarly, as bacteria and HNF, the population dynamics of APP is largely controlled by grazing within the microbial loop (Weisse, 1988). However, the differences in microzooplankton biomass in the various parts of the Great Lakes

(Weisse & Munawar, unpubl. data) appear to be too small to explain the extreme variations in APP numbers along the transects.

The sensitivity of the autotrophic picoplankton to various types of anthropogenic stress, as observed in this study, is not new although various components of the 'microbial loop' were evaluated in the Great Lakes for the first time. Earlier work was confined mainly to APP and its productivity and ecology. For example, their distribution and ecology has been discussed by Munawar et al. (1978); Munawar & Fahnenstiel (1982); and Munawar & Munawar (1986). From a toxicological point of view, autotrophic picoplankton have been used as test organisms in bioassays in our laboratory and elsewhere with heavy metals and sediment elutriates. For instance, a mixture of heavy metals was found to be extremely toxic to picoplankton productivity (Munawar & Munawar, 1982; Munawar et al., 1987). Assays with sediment elutriates indicated both enhancement (Munawar & Munawar, 1987) and reduction of picoplankton density (Severn et al., 1989) as determined by epifluorescence microscopy. In the present study the dramatic avoidance of contaminated environments by APP confirms earlier reports of their sensitivity. The anomalous increase of the APP concentration in Saginaw Bay could be attributable to the availability of excess nutrients which may have complexing effects resulting in the detoxification of contaminants. Since these data are preliminary, however, caution should be exercised in their interpretation until more detailed data from Saginaw Bay is available.

The monitoring of pelagic microorganisms which comprises the 'microbial loop', particularly the APP, could be an easy, sensitive, inexpensive and rapid bioassessment tool for evaluating structural changes in an ecosystem. This has a great potential for inclusion in the emerging multi-trophic battery of tests strategy proposed by Munawar et al. (1989). Since these microorganisms are effective early warning indicators of environmental stress it will be logical to direct future research towards them for the evaluation of the 'microbial loop' in stressed versus unstressed environments. Such an approach might enhance our understanding of the trophic interactions. This ultimately may facilitate the elucidation of factors governing the food web dynamics of various ecosystems and their possible conservation by appropriate remedial action plans.

Acknowledgements

We would like to thank Iftekhar F. Munawar, Lynda McCarthy, Wendy Page, Lisa Keeler and Janis Nawrocki for their assistance in various aspects of the work. We are grateful to Dr. Gary G. Leppard for his constructive criticism of the manuscript. This study was partially supported by Volkswagen Foundation (Grant 1/63 699 to T.W.).

References

Azam, F., T. Fenchel, J. G. Field, J. S. Gray, L. A. Meyer-Reil & F. Thingstad, 1983. The ecological role of water-column microbes in the sea. Mar. Ecol. Prog. Ser. 10: 257–263.

Bloem, J. & M. -J. Bar-Gilissen, 1989. Bacterial activity and protozoan grazing potential in a stratified lake. Limnol. Oceanogr. 34: 297–309.

Box, G. E. P., W. G. Hunter & J. S. Hunter, 1978. *Statistics for Experimenters*. Toronto: John Wiley & Sons.

Caron, D. A., F. R. Pick & D. R. S. Lean, 1985. Chroococcoid cyanobacteria in Lake Ontario: vertical and seasonal distributions during 1982. J. Phycol. 21: 171–175.

Fahnenstiel, G. L., L. Sicko-Goad, D. Scavia & E. F. Stoermer, 1986. Importance of picoplankton in Lake Superior. Can. J. Fish. Aquat. Sci. 43: 235–240.

Finlay, B. J., K. J. Clarke, A. J. Cowling, R. M. Hindle, A. Rogerson & U. -G. Berninger, 1988. On the abundance and distribution of protozoa and their food in a productive freshwater pond. Europ. J. Protistol. 23: 205–217.

Harlow, H. E. & P. V. Hodson, 1988. Chemical contamination of Hamilton Harbour: A review. Can Tech. Rep. Fish. Aquat. Sci. 16031–91.

International Joint Commission, 1977. The waters of Lake Huron and Lake Superior. Vol. II (part A): Lake Huron, Georgian Bay and the North Channel. Report to the International Joint Commission by the Upper Lakes Reference Group. Windsor, Ontario. 292 pp.

International Joint Commission, 1987. Guidance on characterization of toxic substances problems in Areas of Concern in the Great Lakes basin. Report to the Great Lakes Water Quality Board. Windsor, Ontario. 179 pp.

Leppard, G. G., D. Urciuoli & F. R. Pick, 1987. Characterization of cyanobacterial picoplankton in Lake Ontario by transmission electron microscopy. Can. J. Fish. Aquat. Sci. 44: 2173–2177.

MISA, 1987. Municipal-Industrial Strategy for Abatement. The public review of the MISA white paper and the Ministry of the Environment's response to it. 55 pp.

Munawar, M., 1989. Ecosystem health evaluation of Ashbridges Bay environment using a battery of tests. Fisheries and Oceans Report to MISA, Ministry of the Environment. 43 pp.

Munawar, M. & G. L. Fahnenstiel, 1982. The abundance and significance of ultraplankton and micro-algae at an off-shore station in central Lake Superior. Can. Tech. Rep. Fish. Aquat. Sci., 1153.(I–IV): 1–13.

Munawar, M. & I. F. Munawar, 1982. Phycological studies in Lakes Ontario, Erie, Huron and Superior. Can. J. Bot. 60 (9): 1837–1858.

Munawar, M. & I. F. Munawar, 1986. The seasonality of phytoplankton in the North American Great Lakes: A comparative synthesis. In; M. Munawar & J. F. Talling (Eds.), Seasonality of Freshwater Phytoplankton: A Global Perspective. Hydrobiologia. 138: 85–115.

Munawar, M. & I. F. Munawar, 1987. Phytoplankton bioassays for evaluating toxicity of *in situ* sediment contaminants. Hydrobiologia. 149: 87–105.

Munawar, M, I. F. Munawar, L. R. Culp & G. Dupuis, 1978. Relative importance of nanoplankton in Lake Superior phytoplankton biomass and community metabolism. In: M. Munawar (Ed.) Limnology of Lake Superior. J. Great Lakes Res. 4 (3–4): 462–480.

Munawar, M., I. F. Munawar, C. I. Mayfield & L. H. McCarthy, 1989. Probing ecosystem health: a multi-disciplinary and multi-trophic assay strategy. In; M. Munawar, G. Dixon, C. Mayfield, T. Reynoldson & M. H. Sadar (eds.), Environmental Bioassay Techniques and Their Application. Hydrobiologia. (This volume).

Munawar, M., I. F. Munawar, W. P. Norwood & C. I. Mayfield, 1987. Significance of autotrophic picoplankton in the Great Lakes and their use as early indicators of contaminant stress. Arch. Hydrobiol. Beih. Ergebn. Limnol. 25: 141–155.

Pick, F. R. & D. A. Caron, 1987. Picoplankton and nanoplankton biomass in Lake Ontario: relative contribution of phototrophic and heterotrophic communities. Can. J. Fish. Aquat. Sci. 44: 2164–2174.

Pomeroy, L. R. & W. J. Wiebe, 1988. Energetics of microbial food webs. Hydrobiologia. 159: 7–18.

Porter, K. G. & Y. S. Feig, 1980. The use of Dapi for identifying and counting aquatic microflora. Limnol. Oceanogr. 25: 943–948.

Severn, S. R. T., M. Munawar & C. I. Mayfield, 1989. Measurements of sediment toxicity of autotrophic and heterotrophic picoplankton by epifluorescence microscopy. Hydrobiologia. 176/177: 525–530.

Sherr, E. B. & B. F. Sherr, 1988. Role of microbes in pelagic food webs: a revised concept. Limnol. Oceanogr. 33 (5): 1225–1227.

Stockner, J. G., 1988. Phototrophic picoplankton: an overview from marine and freshwater ecosystems. Limnol. Oceanogr. 33 (4): 765–775.

Vollenweider, R. A., M. Munawar & P. Stadelmann, 1974. A comparative review of phytoplankton and primary production in the Lauretian Great Lakes. J. Fish. Res. Bd. Canada. 31 (5): 739–762.

Weiler, R. R., 1981. Chemistry of the North American Great Lakes. Verh. Internat. Verein Limnol. 21: 1681–1694.

Weisse, T., 1988. Dynamics of autotrophic picoplankton in Lake Constance. J. Plankton Res. 10: 1179–1188.

Weisse, T., 1989a. The microbial loop in the Red Sea: Dynamics of pelagic bacteria and heterotrophic nanoflagellates. Mar. Ecol. Prog. Ser. 55: 241–250.

Weisse, T., 1989b. Trophic interactions among heterotrophic microplankton, nanoplankton, and bacteria in Lake Constance. Hydrobiologia. (In press).

Weisse, T. & H. Muller, 1989. Significance of heterotrophic nanoflagellates and ciliates in large lakes: Evidence from Lake Constance. In: M. M. Tilzer & C. Serruya (eds.), Ecological structure and function in large lakes, Sci. Tech. Publ. (in press).

Hydrobiologia **188/189**: 175–179, 1989.
M. Munawar, G. Dixon, C. I. Mayfield, T. Reynoldson and M. H. Sadar (eds)
Environmental Bioassay Techniques and their Application.
© 1989 *Kluwer Academic Publishers. Printed in Belgium.*

On the accuracy and interpretation of growth curves of planktonic algae

Gerda Bolier & Marcel Donze
Delft University of Technology, Department of Civil Engineering, P.O. Box 5048, 2600 GA Delft, The Netherlands

Key words: algae, growth curves, *Scenedesmus*, accuracy, reproducibility

Abstract

About 100 growth curves of *Scenedesmus quadricauda* in batch cultures on different growth media are analysed. In many cases after a retardation phase, a second exponential growth phase appears. The accuracy of measurements of the different growth phases as effected by the growth medium is discussed.

Introduction

In several countries tests are developed to determine the algal growth potential, the AGP, of a water. Basically the experiment determines the maximum concentration of algae that can grow in a water sample under standardized conditions. These tests are used to judge:
- the degree of eutrophication of surface water (Thomas, 1953; Skulberg, 1964)
- the eutrophication potential of the effluent of sewage treatment plants (Forsberg, 1972)
- the possible effects of environmental measures on the degree of eutrophication of water systems (Van der Does & Klapwijk, 1987).

Very little is known about accuracy and reproducibility of the measurements of growth curves in batch cultures both within and between laboratories.

In addition to the growth yield, often a complete growth curve of the population is measured. This curve can give more information than AGP-value alone. Accuracy and reproducibility are not known.

Monod (1949) distinguished a succession of six phases in the growth of a bacteria culture, charac-terized by variations in the growth rate as illus-trated in Fig. 1. This interpretation of these phases is summarized in Table 1. For example, a lag-phase may indicate the presence of growth inhibiting factors. Lag phase and acceleration

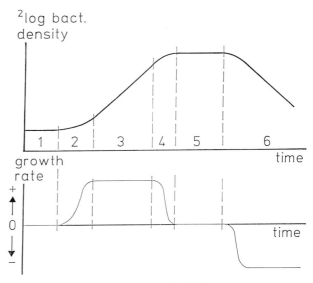

Fig. 1. Phases of growth (Monod, 1949).
Upper curve: log bacterial density
Lower curve: variations in growth rate
Vertical dotted lines mark the limits of phases. Figures refer to phases as defined in table 1.

Table 1. Description and interpretation of the different phases of the growth curves, characterized by Monod (1949).

nr	phase	growth	interpretation
1	lag	zero	a) presence of toxic substances b) physiological adaptation of the inoculum to changing conditions
2	acceleration	increasing	trivial
3	exponential	constant	population growth changes the environment of the cells a) the organism is insensitive b) physiological adaptation is faster
4	retardation	decreasing	effects of changing conditions appear
5	stationary	zero	one or more nutrients (or light) are exhausted down to the threshold level of the cells
6	decline	negative	The duration of stationary phase and the rate of decline are strongly dependent on the kind of organisms. Both have hardly been studied

phase are often absent in experiments. The retardation phase can be very short or it can show more complicated dynamics (Pardee, Jacob & Monod, 1959). Most experiments are terminated before decline occurs.

The different phases can be measured by duration and slope of the growth curve. Also the parameters of a population dynamic model can be fitted to the measurements (Kooijman, 1983).

Materials and methods

During the investigation to standardize the AGP-test for Dutch surface water regularly test series were carried out in 7- or 8-fold, to collect informa-

tion on the reproducibility of these tests. The growth media used were:
– artificial medium Z8 (Skulberg, 1964, modified)
– surface water
– diluted effluent from sewage treatment plants
– mixtures of 90% surface water and 10% effluent.

Scenedesmus quadricauda, isolated from the eutrophic Lake Tjeukemeer in 1974 (Steenbergen, 1975), was used as the test alga. Cultures were kept at 20 °C ± 1 °C on a rotary shaker at 100 rpm and continuously illuminated by 24 W/m² fluorescent light (TL 33). The algal concentration (optical density at 750 nm/cm (OD_{750})) was measured daily. The different phases of growth were determined graphically from the growth curves.

Results

Some typical growth curves on different media are shown in Fig. 2. It appears that growth kinetics

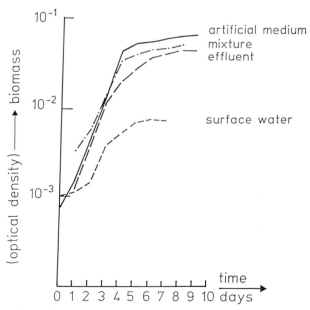

Fig. 2. Illustrations of growth curves of *S. quadricauda* on different media. Biomass expressed as optical density at 750 nm/cm is plotted against time.

It appears that the shape of a growth curve strongly depends on medium composition.

on artificial medium is the simplest. All phases except exponential and stationary phase are absent or very short, while in surface water all five phases are prominent.

The results of graphical analysis of about 100 growth curves are given in Tables 2–4. In these data, the lag phase could not be studied, since it rarely appeared. Exponential growth rates and their duration are given in Table 2. Daily sampling appears to be insufficiently frequent to allow a reasonably accurate estimate of the duration of the different phases of the growth curve. This probably is a reason why lag phases escape attention. The accuracy of growth rate determination is highest in artificial medium, lowest in surface water and intermediate for the other cases. Part of this effect may be explained by the longer duration of the exponential phase in artificial medium although this did not account for the variation in the series grown on surface water.

Table 2. Duration (days) and slope (μ-max) of the exponential growth phase of a culture of *S. quadricauda* cultivated on different media.

medium	month	n	D	μ-max	% s.d.
Z8	5	3	2.00	9	
surface water:					
A	12	8	2	1.14	41
A	5	7	1	0.35	29
B	2	7	1	0.63	20
B	6	7	1–2	0.36	21
effluent of sewage treatment plants:					
C	5	7	1	1.46	27
C	6	7	2	1.66	10
D	1	7	1	3.19	11
D	4	7	(1)	1.57	9
mixtures of surface water (90%) and effluent (10%):					
AC	5	7	1	1.90	12
BC	6	7	2	1.27	13

Surface water:
A: Put van Broeckhoven; B: Akkerdijkse plassen.
Sewage treatment plants:
C: at Delft University of Technology; D: at Nieuwveen.

$$\mu\text{-max:} \quad \frac{\log_2 (x_2/x_1)}{t_2 - t_1}$$

in which x_1: beginning of the exp. phase at time t_1
$\quad\quad\quad x_2$: end of the exp. phase at time t_2

Table 3. Duration (days) and slope (μ_2) of the second exponential growth phase of a culture of *S. quadricauda* cultivated on different media.

medium	month	n	D	μ_2	% s.d.
Z8	5	3	0.10	20	
surface water:					
A	12	3	3	0.34	48
B	2	4	2.3	0.12	64
B	6	3	3.8	0.10	12
effluent of sewage treatment plants:					
C	5	7	3.4	0.11	27
C	6	5	3	0.11	25
D	1	7	2.7	0.15	52
D	4	6	2.7	0.12	16
mixtures of surface water (90%) and effluent (10%):					
AC	5	4	2.3	0.19	46
BC	6	6	2.3	0.09	40
BD	4	4	2.3	0.15	25

Surface water:
A: Put van Broeckhoven; B: Akkerdijkse Plassen
Sewage treatment plants:
C: at Delft, University of Technology; D: at Nieuwveen.

In the last column standard deviation of the sample is given; to obtain these of the main, the value has to be divided by $\sqrt{n-1}$.

In many experiments, after a retardation phase, a second exponential growth phase with a lower growth rate occurs. This new phase may persist for about 3 days. In Fig. 2 this phenomenon appears to some degree in all curves. Its characteristics are given in Table 3. These effects also occurred in most cultures grown on the diluted effluent of sewage treatment plants. Cultures grown on surface water showed it only occasionally. Total increase in optical density during this phase ranges from 20% to 50%, so it significantly affects the interpretation of bioassays. The accuracy of these bioassays would be considerably increased if proper account could be taken of the secondary exponential growth as reported here.

Data on the stationary phase are tabulated in Table 4. Maximal biomass, measured as OD_{750}, appears to be the most accurately determined property of growth curves. This may be due to the fact that the time-axis does not play a role. The accuracy of rate measurements is limited by the

Table 4. Maximum biomass (measured as optical density) of a culture of *S. quadricauda* grown on different media.

medium	month	n	day	max. biomass	% s.d.
Z8		5	9	0.63	5
surface water:					
A	12	8	6–7	0.07	7
A	4	7	6–7	0.08	6
A	5	7	5–6	0.08	13
B	2	7	7	0.17	5
B	4	7	6–7	0.06	7
B	6	7	–	0.14	4
effluent of sewage treatment plants:					
C	5	7	8	0.50	5
C	6	7	9	0.44	5
D	1	7	7	0.33	2
D	4	7	8	0.33	2
mixtures of surface water (90%) and effluent (10%):					
AC	5	7	9	0.64	6
BC	6	7	8	0.51	5
AD	4	7	9	0.32	6
BD	4	7	7–8	0.30	6

Surface water:
A: Put van Broeckhoven; B: Akkerdijkse Plassen
Sewage treatment plants:
C: at Delft, University of Technology; D: at Nieuwveen.

low resolution in time of daily measurements as compared to the speed of changes in growth rate. Also, probably for the same reason, the accuracy of the measurement of the stationary state did not depend on the source of test water.

Discussion

The nature of the second exponential growth phase requires more investigation. Some explanations for its interpretation come to mind. It may be related to the phenomenon of diauxic growth (Monod, 1942). In our opinion the explanation is that cell number remains constant during this phase while average cell volume increases. This increase would be recorded by our technique of measurement. Optical density is based on light scattering and increases with particle size and the amount of internal structures. More detailed research on biochemical changes during the stationary phase may help to resolve longstanding debates concerning the measument of biomass in the field with different techniques. Experiments using flowcytometry are in progress.

In general it is to be expected that the definition of 'biomass' in the stationary phase depends on the kind of measurement taken. Especially during the retardation phase and following phases, different properties of the cells begin to vary in different ways; e.g. different properties of a culture are no longer closely coupled by constant conversion factors. For example we regularly observed in other experiments that optical density at 750 nm still increased while chlorophyll content was decreasing. Knowledge about the duration of this stationary phase and about the shape of the decline phase is virtually nonexistent. Observations indicate that these vary between different species and depend on the particular limiting factor. In an ecological context such knowledge would be highly desirable. In models of plankton dynamics the unspecified 'loss rate' of phytoplankton often is much higher than can be accounted for by grazing and sedimentation. Study of the natural death processes might well help to explain the situation.

Acknowledgements

The results discussed in this paper are based on work carried out under contract with the Foundation of Applied Wastewater Research (STORA). Skillful technical assistance by Mrs. Anke Brouwer is greatly appreciated.

References

Does, J. van der & S. P. Klapwijk, 1987. Effects of phosphorus removal on the maximal algal growth in bioassay experiments with water from four Dutch lakes. Int. Revue ges. Hydrobiol. 72: 27–39.

Forsberg, C., 1972. Algal assay procedure. J. Wat. Pollut. Cont. Fed. 44: 1623–1628.

Kooijman, S. A. L. M., A. O. Hansveit & H. Oldersma, 1983. Parametric analyses of population growth in bioassays. Wat. Res. 17: 527–538.

Monod, J., 1942. Diauxie et respiration au cours de la croissance des cultures de E. coli. Ann. d'Inst. Pasteur. 68: 548–549.

Monod, J., 1949. The growth of bacterial cultures. Ann. Rev. Microbiol. 3: 371–394.

Pardee, A. B., F. Jacob & J. Monod, 1959. The genetic control and cytoplasmic expression of 'inducibility' in the synthesis of β-galactosidase by *E. coli*. J. Mol. Biol. 1: 165–178.

Skulberg, O. M., 1964. Algal problems related to the eutrophication of European water supplies, and a bioassay method to assess fertilizing influences of pollution on inland waters. in: Algae and Man, D. F. Jackson, New York, Plenum Press: 262–299.

Steenbergen, C. L. M., 1975. Light-dependent morphogenesis of unicellular stages in synchronized cultures of *Scenedesmus quadricauda* (Turp.) Bréb. (Chlorophyceae). Acta Bot. Neerl. 24: 391–396.

Thomas, E. A., 1953. Zur Bekämpfung des See-Eutrophierung: Empirische und experimentelle Untersuchungen zur Kenntnis der Minimumstoffen in 46 Seen der Schweiz und angrenzender Gebiete. Monatsbull. Schweiz. Ver. Gas-Wasserfachm. 33: 25–32; 71–79.

Hydrobiologia **188/189**: 181–188, 1989.
M. Munawar, G. Dixon, C. I. Mayfield, T. Reynoldson and M. H. Sadar (eds)
Environmental Bioassay Techniques and their Application.
© *1989 Kluwer Academic Publishers. Printed in Belgium.*

A bioassay using the measurement of the growth inhibition of a ciliate protozoan: *Colpidium campylum* Stokes

D. Dive, S. Robert, E. Angrand, C. Bel, H. Bonnemain, L. Brun, Y. Demarque, A. Le Du, R. El Bouhouti, M.N. Fourmaux, L. Guery, O. Hanssens & M. Murat
INSERM U.146, Domaine du CERTIA, 369 rue Jules Guesde, 59651 Villeneuve d'Ascq Cedex, France.
(Request for reprint: DIVE Daniel, INSERM U.42, Domaine du CERTIA, 369 rue Jules Guesde, B.P. 39, 59651 Villeneuve d'Ascq Cedex, France)

Key words: bioassay, protozoan, standardization, intercalibration

Abstract

A bioassay method using the ciliate protozoan *Colpidium campylum* is presented in a standardized form. The influence of the initial cell concentration on the potassium dichromate EC50 values was determined. Two intercalibration experiments between two laboratories were performed on ten toxicants in two different conditions. The potassium dichromate EC50 determinations performed by eight different people are also presented. All results are discussed in terms of feasibility and reproducibility of the method, fields of application, and limitations.

Introduction

While there have been many attempts to develop a standard method using protozoa to measure toxicity, no test is presently accepted at the international level. Many problems must be solved to develop a protozoan test, such as selection and conservation of the strains and development of the test as pointed out by some authors (Persoone & Dive, 1978; Dive, 1982; Dive & Persoone, 1984; Parker, 1983). The examination of the literature shows that two types of protozoan tests can be distinguished:
– The methods using protozoa in axenic culture (peptone based media), which are well standardized, but not adapted to test products in conditions representative of aquatic toxicology. Such a method was proposed by Greenberg, Connors & Jenkins (1980).

– The methods using protozoa in mineral medium. Slabbert *et al.* (1983) have proposed a method based on the respiration of *Tetrahymena*, which is very sensitive and rapid. Amoebae (Bogaert *et al.*, 1982), flagellates (Bringmann & Kühn, 1980), ciliates (Berk *et al.*, 1978; Burbank & Spoon, 1967; Gray & Ventilla, 1973; Parker, 1979; Persoone & Uyttersproot, 1975) were used for tests based on the inhibition of growth in bacterial cultures. Since 1974, we have developed a method using *Colpidium campylum* for the measurement of toxicity. This method was applied to the study of pure mineral and organic toxicants (Dive & Leclerc, 1977; Dive *et al.*, 1980). Furthermore, we have tried to adapt the test in a protocol suitable for standardization, on the basis of material used, the examination of each step of the method, and intercalibration on several chemicals.

Material and methods

Organism and its cultivation

Maintenance of the strain
Colpidium campylum Stokes (Ciliophora, Protozoa) is axenically cultivated in PPYS medium (Plesner *et al.*, 1964) enriched with bovine serum albumin (Sigma A4503) according to Dive & Rasmussen (1978). The cultures are incubated at 28°C in the dark and subcultured each week.

Preparation of ciliates for the bioassay
To prepare the ciliates for the bioassay, it is necessary to acclimate them to monoxenic cultivation. They can be grown successfully with lyophilized *Escherichia coli*, strain ATCC 11303 (Sigma EC11303), strain ATCC 9637 (Sigma EC9637) or strain K12 (Sigma EC1).

The MM medium used for the test ($CaCl_2$, $2H_2O$: 107 mg; NaCl: 14.5 mg; $NaNO_3$: 4.5 mg; $MgSO_4$, $7H_2O$: 75.7 mg; Na_2SO_4: 39.5 mg; $NaHCO_3$: 135 mg; distilled, de-ionized or Milli-Q grade water added to 1000 ml; pH 8.15 + 0.02) is filtered on a 0.45 μm membrane filter and stored at 4°C.

For the preparation of the inoculum, two 125 ml sterile borosilicate Erlenmeyer or sterile cell culture bottles (Nunclon, 50 ml, 25 cm^2) containing 10 ml of MM medium and 0.4 ml of *E. coli* suspension (2.5 mg/ml in MM medium) are inoculated with two drops of axenic culture of *C. campylum*.

After 48 h of incubation at 28°C, the cells are counted and the definitive inoculum is prepared in 500 ml sterile borosilicate Erlenmeyer or cell culture bottles (Corning, 270 ml, 75 cm^2) with 50 ml of MM medium, 2 ml of *E. coli* (2.5 mg/ml in MM medium) and inoculated with 1000 ciliates/ml. After 48 h of growth at 28°C, the inoculum can be used for the bioassay.

Bioassay method

The test is based on the growth inhibition of *C. campylum* as a function of the initial concentration of toxicant added in the medium. The number of cells produced during 24 h in the presence of toxicant is compared to the value obtained in a control culture.

Counting method
The cells are counted electronically with a Coulter counter fitted with a 200 μm aperture probe, after fixation by glutaraldehyde and dilution with a 1% NaCl electrolyte solution filtered through a 0.45 μm membrane filter.

Bioassay protocol
The inoculum is counted (a 3 ml sample is fixed with 1 ml of 5% glutaraldehyde solution and diluted with 10 ml of 1% NaCl solution). A dilution containing 3333 cells/ml is prepared.

The test is performed routinely in 30 ml crystal polystyrene screw-capped vials. It can also be performed directly in vials commonly used for electronic counting (Coulter accuvettes). If products tested can adsorb on plastic or alter it, borosilicate glass or teflon vials can be used.

The vials are prepared for the test as follows:

– Toxicant solution in MM
  1.25 final concentration in the vial   4   ml
– *E. coli*, suspension 2.5 mg/ml
  in MM medium   0.25 ml
– *C. campylum* dilution, 3333 cells/ml   0.75 ml

The toxicant dilution is first placed in the vials, followed by the bacterial suspension. The ciliates must be added last. The moment of addition of protozoa is considered as time 0 of the test. The final volume is 5 ml and initial cell concentration is 500 cells/ml. For each test on a substance, one flask per concentration tested and three control flasks (without toxicant) are used.

The true value of the inoculum (500 cells/ml in theory) is verified by distribution of 3×0.75 ml aliquots of the 3333 cells/ml dilution in three separate vials, addition of 0.25 ml of MM, and fixation with 1 ml of 2.5% glutaraldehyde. The samples are then counted and densities calculated. The mean of the three values is considered as the initial concentration (N_0) for the experiment.

At the same time, a standard reference toxicant test with potassium dichromate is performed to verify the sensitivity of the biological material.

The vials are incubated in the dark at 28°C for 24 h.

At the end of the incubation, each flask is fixed with 1 ml of 5% glutaraldehyde solution. The population is counted (N) and the number of cells produced (CP) is given by the relation $CP = N - N_0$.

Calculation of the EC50

The cells produced in each concentration of toxicant are estimated in % of the control (mean of the three flasks). The percentages are plotted on a log-probit paper and the EC50 is determined. The EC50 of potassium dichromate is obtained in the same way. The EC50 can also be calculated using a computer program.

Table 1. Analysis of variation of potassium dichromate EC50's related to technical errors and fluctuations of the biological material. The coefficients of variations of five values of the EC50 are given in percentages of the mean value. Condition A: Five different concentrated solutions of product were prepared and tested with the same ciliate culture, without replicates. Condition B: From the same concentrated solution, five parallel dilution series were prepared and tested with the same ciliate culture. Condition C: The same concentrated solution is tested in five replicates with the same ciliate culture. Condition D: The same concentrated solution is tested with five different ciliate cultures. The experimental results are arranged in chronological order.

Conditions of test	Successive experiments				
	1	2	3	4	5
A	6.68	3.67	1.69		
B	16.26	14.18	8.46	6.1	1.24
C	4.89	3.15	7.22		
D	11.78	9.83	11.52		

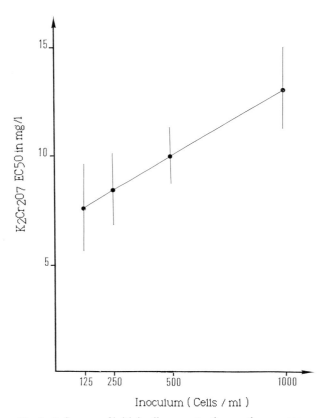

Fig. 1. Influence of initial cell concentration on the response of *Colpidium campylum* to potassium dichromate.

Intercalibration exercise organization

For the intercalibration exercise, chemicals of the same lot No. were used. All products were analytical grade. The products were divided between two labs. Each laboratory used its own water: Milli-Q grade water in lab no. 1 and de-ionized water in lab no. 2. Test preparation was performed in each lab but all counts were undertaken in lab no. 1, as it was equipped with an electronic counter.

The products tested were $K_2Cr_2O_7$, $CdSO_4$, $NaBO_2$, $HgCl_2$, $ZnCl_2$, Acrylnitril, Anilin, Cetrimide (cetyl-trimethylammonium bromide), Tetrapropylbenzene sulfonate (T.P.B.S.) and Phenol.

Four of five products were tested simultaneously and, for each experiment, a potassium dichromate control was included.

Results

'Performance appraisal' of the test

The different steps of the method were evaluated with potassium dichromate. The points systematically studied were the influence of the initial concentration of ciliates, the errors occurring during the weighing of the toxicant and during the dilution of concentrated solutions, repetitions in the same experiment or variations with different precultures.

The influence of the initial cell concentration was studied between 125 and 1000 cells/ml with the same inoculum culture and the same dilutions of potassium dichromate. The results of twelve experiments (Fig. 1) show that the EC50 increases with initial cell concentration. The smal-

Table 2. EC50's measured in the two intercalibration experiments performed with *Colpidium campylum*. Results in mg/l. The first experiment was performed with an inoculum of 1000 cells/ml. The second experiment was performed with a theoretical inoculum of 500 cell/ml and with a verification of the initial cell number.

Product	Intercalibration 1			Intercalibration 2		
	x	s	n	x	s	n
$K_2Cr_2O_7$						
Lab 1	12.7	2.9	35	13.54	3.4	25
Lab 2	13.39	3.31	38	14.94	4.48	17
$CdSO_4$						
Lab 1	0.075	0.013	15	0.105	0.058	14
Lab 2	0.079	0.015	15	0.171	0.058	7
$NaBO_2$						
Lab 1	199.3	21.9	14	140.6	36.9	12
Lab 2	193.7	28.6	12	190.2	51.3	10
$HgCl_2$						
Lab 1	0.197	0.084	12	0.537	0.262	14
Lab 2	0.276	0.059	12	0.233	0.061	9
$ZnCl_2$						
Lab 1	7.1	3.87	11	2.03	0.82	9
Lab 2	15.7	8.8	11	3.24	1.53	9
Acrylnitril						
Lab 1	75.5	17.0	15	32.0	9.1	16
Lab 2	261	91	14	60.6	16.3	9
Anilin						
Lab 1	358	60	11			
Lab 2	329	86	14			
Cetrimide						
Lab 1	1.69	0.22	15	0.96	0.29	13
Lab 2	2.03	0.26	14	1.58	0.44	9
T.P.B.S.						
Lab 1	62–79		9	80.2	19.2	16
Lab 2	62–79		11	117	17.2	9
Phenol						
Lab 1	242	24.7	14	116.5	26.8	16
Lab 2	218	21.7	14	148	20.3	9

lest dispersion is obtained with 500 cells/ml and this value was selected for the standard protocol.

The variability due to technical operations were evaluated by various experiments (reproducibility for different weighings, different dilutions, repetitions in the same experiment, and different pre-cultures). The coefficients of variation for five replicates were calculated. The results of three to

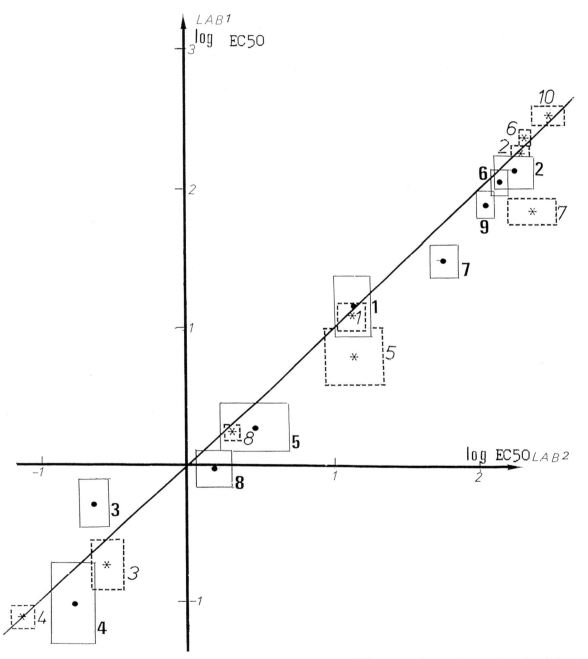

Fig. 2. Results of two intercalibration experiments. First experiment: * mean EC50; --- standard deviation. Second experiment: ● mean EC50; —— standard deviation. 1. $K_2Cr_2O_7$; 2. $NaBO_2$; 3. $HgCl_2$; 4. $CdSO_4$; 5. $ZnCl_2$; 6. Phenol; 7. Acrylnitril; 8. Cetrimide; 9. Tetrapropylbenzenesulfonate; 10. Anilin.

Table 3 Correlation between the results of the two laboratories on the basis of log of mean values of EC50 observed during the two intercalibration experiments.

Experiment no.	Correlation coefficient	Slope	Constant term	EC50 ratios Lab 2/Lab 1
1				
Without acrylnitril	0.9937	1.0012	− 0.10	1.25
With acrylnitril	0.9886	0.9610	− 0.10	1.27
2				
Without mercury	0.9968	1.017	− 0.194	1.56
With mercury	0.9907	0.971	− 0.113	1.30

five successive experiments (Table 1) show that the classic 'technical variability', such as the weighing of the product and the preparation of the dilutions, can be reduced easily to a very low level (2 to 3%) but can be much more important than commonly admitted. On the other hand, the variation occurring from one inoculum to another is about 10% and cannot be substantially reduced (Table 1). It can be considered to be the optimal reproducibility of the test when technical errors are reduced to a minimum. The variation between the repetitions of the same experiment is 3 to 7% (Table 1). Under these conditions a coefficient of variation of 15% for the EC50 of a given product can be considered very satisfactory.

Intercalibration between two laboratories

Two intercalibration experiments were conducted between two laboratories and 9 chemicals were investigated during each experiment. The first was based on a theoretical inoculum of 1000 cells/ml. No measurement of the initial cell concentration was done and the N_0 value used for the calculation of cells produced was 1000. The second experiment, performed after the 'performance appraisal' of the test, was based on a 500 cells/ml inoculum with a determination of the mean value of N_0.

The results are shown in Table 2 and the correlation between the EC50's measured in the two laboratories is shown in Fig. 2. Means and standard deviation are plotted in log. The re-

gression line corresponding to identical results between the two labs is represented. It corresponds to an equation of slope = 1, with a constant term = 0 and with a correlation coefficient = 1. We calculated also the regression equations obtained between the two labs during the two intercalibration experiments (Table 3).

During the first intercalibration, the results were in agreement between the two labs except for acrylnitril. Concerning this last product, verifications showed that the higher EC50 value in lab 2 can be attributed to a poor dissolution of the product during the preparation of the concentrated solution. During the second experiment, significant differences were found between the two labs for mercuric chloride and again acrylnitril. The results of acrylnitril cannot be related to a poor dissolution of the product in the concentrated solution because special attention was devoted to this step. No reasonable hypothesis can explain these differences, except for the quality of water used in the two labs.

Table 4 Comparison of the EC50 of potassium dichromate measured by 7 different trainees in lab 1 and 1 trainee (§) in lab 2. vc% is the coefficient of variation expressed in % of the mean.

	A	B	C	D	E	F	G	H§
x	14.4	12.2	10.0	11.6	11.8	13.5	14.0	14.9
s	2.1	2.1	1.3	2.0	3.4	3.4	4.9	4.7
vc%	15	17	13	17	28	25	35	31
n	9	20	12	48	53	24	14	17

The regression equations obtained with the results of the two laboratories during the two experiments are shown in Table 3 and indicate:
- an excellent correlation between the results obtained in the two labs
- a slope very close to 1, indicating that the relative toxicity of the products is the same in the two labs; and
- a systematic difference between the results of lab 1 and lab 2. The EC50 values obtained in lab 2 were 1.25- to 1.5-fold those measured in lab 1.

Intercalibration of potassium dichromate among operators

Eight different trainees performed the test with potassium dichromate between 1984 and 1988 (Table 4). Four of them (A to D) worked without verification of the initial concentration of ciliates at T_0, and four (E to H) with verification. The results were recorded when trainees had about one week of experience with the method and obtained a satisfactory reproducibility.

Discussion

During our study, we have tried to develop a bioassay using a ciliated protozoan and to test how technical aspects can increase the dispersion of the results. We have also compared the results obtained by two laboratories on ten products.

Our results show that some technical aspects, such as weighing and dilutions, are more important sources of variation than is generally supposed. The fluctuations due to the biological material were estimated and, taking into account the intra-experiment error, a coefficient of variation of 15% in EC50 appears to be optimal. During the intercalibration tests, the coefficients of variation observed were larger (20 to 30%) and a significant technical error persists. The experiments performed on potassium dichromate with eight operators gave similar results.

A good bioassay should be developed which can be performed in a large number of laboratories by many different people. Our study shows that this test can be performed in different laboratories with high coherence in the response integrated over nine products (with a slope very close to 1). The comparison of the EC50's, based on several products, allows the determination of a response ratio between the laboratories. The exercise performed on potassium dichromate by several people shows that consistent results can be obtained by different operators. In this respect, the bioassay presented here fulfills two important prerequisites for its application.

Generally speaking, our approach to the development and to the assessment of the method allowed us to determine more accurately all the technical problems that can occur in the protocol. A larger intercalibration experiment will be necessary to verify the results presented here and to complete the information on this bioassay.

This method can be applied to the study of pure compounds, mixtures of toxicants and interactions (Vasseur *et al.*, 1988; Dive *et al.*, 1983, 1988), complexation of metals (Dive *et al.*, 1982), industrial waters and leachate products (unpublished results). The limitation of the method is the possible occurrence of insoluble particles in the sample, which can interfere with electronic counting. Chemicals can be tested, if they are water soluble at a concentration equal to at least the EC50. For industrial waters and leachates, filtration is necessary and only soluble toxicity is estimated.

Many studies have pointed out the importance of multi-species bioassays in estimating hazards due to chemicals. Surprisingly, only very few classes of organisms are represented by standardized bioassays currently applied in aquatic toxicology (mainly fishes, crustaceans, algae and luminescent bacteria). Our method provides a complementary test representative of unicellular animals. The duration of this acute test is short (24 h) and its methodology has been devised to be performed easily, with commercially available material. Its use as a standardized technique is not limited to pure toxicants but can be applied in a versatile fashion.

188

Acknowledgements

Comments & suggestions of three anonymous regimes were quite useful in the revision of the manuscript.

References

Berk S. G., A. L. Mills, D. L. Hendricks & R. R. Colwell, 1978. Effect of mercury containing bacteria on mercury tolerance and growth rates of ciliates. Microb. Ecol. 4: 319–330.

Bogaert T., G. Persoone & D. G. Dive, 1982. Research on the development of a short term bioassay for water and soil, with amoebae, starting from cysts. In: H. Leclerc & D. Dive (eds), Acute aquatic ecotoxicological tests: Methodology, standardization and significance. Bull. INSERM, 106: 565 pp.

Bringmann G. & R. Kuhn, 1980. Comparison of toxicity thresholds of water pollutants to bacteria, algae and protozoa in the cell-multiplication inhibition test. Wat. Res. 14: 231–242.

Burbank W. D. & D. H. Spoon, 1967. The use of sessile ciliates collected in plastic Petri dishes for rapid assessment of water pollution. J. Protozool. 14: 739–744.

Dive D. G., 1982. Perspectives d'utilisation des protozoaires en écotoxicologie aquatique. In 'Acute aquatic ecotoxicological test: In: H. Leclerc & D. Dive (eds), Acute aquatic ecotoxicological tests: Methodology, standardization and significance. Bull. INSERM, 106: 273–294.

Dive D. G. & H. Leclerc, 1977. Utilisation du protozoaire cilié Colpidium campylum pour la mesure de la toxicité et de la bioaccumulation des micropolluants: Analyse critique et applications. Envir. Pollut. 14: 169–185.

Dive D. G. & G. Persoone, 1984. Protozoa as test organisms in marine ecotoxicology: Luxury or necessity? In: G. Persoone, E. Jaspers & C. Claus, (eds), Ecotoxicological testing for the marine environment, pp. 281–306. State Univ. Ghent, Belgium.

Dive D. G., G. Persoone & H. Leclerc, 1980. Pesticide toxicity on the ciliate protozoan Colpidium campylum:
Possible consequences of the effect of pesticides in the aquatic environment. Ecotoxicol. Envir. Safety 4: 129–133.

Dive D. G., N. Pommery, M. Lalande & F. Erb, 1982. Cadmium complexation by humic substances: Chemical and ecotoxicological study with ciliate protozoan Colpidium campylum. Can. Tech. Rep. Fish. Aquat. Sci. 1163: 9–21.

Dive D. G. & L. Rasmussen, 1978. Growth studies on Colpidium campylum under axenic conditions. J. Protozool. 25 (3): 42A.

Dive D. G., P. Vasseur, C. Bel & C. Danoux, 1983. Cadmium-zinc and cadmium-selenium interactions. A comparative study with Colpidium campylum and Photobacterium phosphoreum. J. Protozool. 30 (3): 65A–66A.

Dive D. G., P. Vasseur, O. Hanssens & P. J. Gravil, 1988. Studies on interactions between components of electroplating industrial wastes. Can. Tech. Rep. Fish. Aquat. Sci. 1607: 23–34.

Gray J. S. & R. F. Ventilla, 1973. Growth rates of sediment-living marine protozoans as a toxicity indicator for heavy metals. Ambio 2: 118–120.

Greenberg A. E., J. J. Connors & D. Jenkins (eds), 1980. Standard Methods for the examination of water and wastewaters. Publ. Office, Amer. Publ. Hlth Assoc., 15th Edit.

Parker J. G., 1979. Toxic effects of heavy metals upon cultures of Uronema marinum (Ciliophora, Uronematidae). Ecotoxicol. Envir. Safety 7: 172–178.

Persoone G. & D. Dive, 1978. Toxicity tests on ciliates: A short review. Ecotoxicol. Envir. Safety 2: 105–114.

Persoone G. & G. Uyttersprot, 1975. The influence of inorganic and organic pollutants on the rate of reproduction of a marine hypotrichous ciliate: Euplotes vannus. Rev. Intern. Oceanogr. Méd. 37–38: 125–152.

Plesner P., L. Rasmussen & E. Zeuthen, 1964. Techniques used in the study of synchronous Tetrahymena. In: E. Zeuthen (ed.), Synchrony in cell division and growth, pp. 543–563. John Wiley & Sons, New York.

Slabbert J. L., R. Smith & W. S. G. Morgan, 1983. Applications of a Tetrahymena pyriformis bioassay for the rapid detection of toxic substances in water. Water S.A. 3: 81–87.

Vasseur P., D. G. Dive, Z. Sokar & H. Bonnemain, 1988. Interactions between copper and some carbamates used in phytosanitary treatments. Chemosphere 17: 767–782.

Hydrobiologia **188/189**: 189–199, 1989.
M. Munawar, G. Dixon, C. I. Mayfield, T. Reynoldson and M. H. Sadar (eds)
Environmental Bioassay Techniques and their Application.
© 1989 Kluwer Academic Publishers. Printed in Belgium.

The application of algal growth potential tests (*AGP*) to the canals and lakes of western Netherlands

S.P. Klapwijk[1], G. Bolier[2] & J. van der Does[1]
[1] *Waterboard of Rijnland, Technical Service, P.O. Box 156, 2300 AD Leiden, The Netherlands*; [2] *Delft University of Technology, Department of Civil Engineering, P.O. Box 5048, 2600 GA Delft, The Netherlands*

Key words: bioassays, limiting nutrient factor, nitrogen, phosphorus, eutrophication

Abstract

Four hundred and forty bioassays with *Scenedesmus quadricauda* (Turp.) Bréb. as a test organism have been carried out with samples from canals and lakes in the western part of the Netherlands. The results are used to assess the algal growth potential (AGP) and to determine the limiting nutrient(s) for maximum biomass production. Special attention has been paid to the effects of deep-freezing and autoclaving as pretreatment of water samples on pH and nutrient concentrations.

The AGP ranged from very low in the relatively isolated polder lakes to very high in canals and lakes, which form part of the basin system of Rijnland. The lowest yields are observed in nitrogen and phosphorus co-limited waters, while the highest are found in waters limited by nitrogen alone. AGP proved to be primarily determined by the amount of nitrogen, especially nitrate, in the samples and only secondarily by the amount of phosphorus.

The observed ranges indicating phosphorus limitation, > 50 for inorganic and > 30 for total N/P ratios, lie considerably higher than reported so far. It is concluded that, once the relations between AGP and nutrients are established, AGP tests do not have to be carried out routinely, but still can be very useful in special studies, e.g. in lake restoration projects.

Introduction

Algal growth potential tests (AGP-tests) have been applied in limnology since the 1960's (Skulberg, 1964; Bringmann & Kühn, 1965; E.P.A., 1971). Now they are used over the whole world (Nordforsk, 1973; Gargas & Pedersen, 1974; Chiaudani & Vighi, 1974, 1975, 1976; Miller *et al.*, 1974, 1978; E.P.A., 1975; S.I.L., 1978; Marvan *et al.*, 1979; Youngman, 1980; Ryding, 1980; Raschke & Schulz, 1987).

In the Netherlands, AGP-tests have been introduced in water quality studies in 1981 for testing surface waters and effluents of sewage treatment plants (Bolier *et al.*, 1981; Bolier & van Breemen, 1982). They are applied in special water quality studies, e.g. to investigate the effects of phosphorus removal at sewage treatment plants (Klapwijk, 1981; Hoogheemraadschap van Rijnland, 1984; van der Does & Klapwijk, 1985, 1987). They are also used to measure the growth potential as a general water quality parameter, and to

190

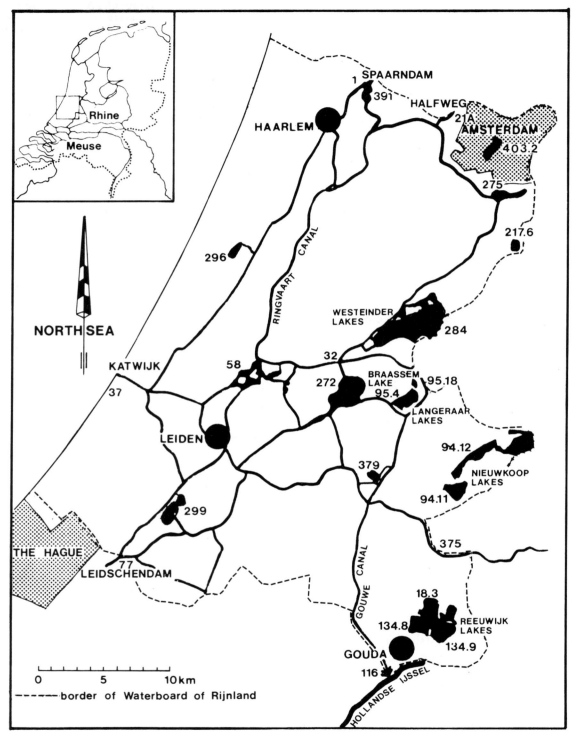

Fig. 1. The area of the Rijnland Waterboard and the position of the sampled locations (location numbers conform to the Waterboard scheme).

determine the limiting nutrient factors (Veenstra, 1981; de Haan *et al.*, 1982; de Vries & Klapwijk, 1987). Recently, Bolier and co-workers standardized an AGP method for Dutch conditions (Anonymous, 1986).

Results of 440 AGP-tests which are presented here, were carried out on 24 sampling locations in the west central part of the Netherlands over a four-year period (1983–1986). They have been used in this study with the following aims:
1. To check the effects of the pretreatment applied to the water samples on the bioassay results.
2. To evaluate advantages of AGP-tests compared to chemical methods.
3. To correlate AGP results with nutrient and chlorophyll-*a* concentrations and to evaluate the predictability of limiting factors, based on AGP-tests and on nutrient concentrations and ratios.

Materials and methods

Water samples were collected in the Rijnland Waterboard area at a 0.5 m depth in 24 sampling locations with different trophic degrees (Fig. 1). Sample collection was generally conducted four times per year in March, June, September and December from 1983–1986. At two sampling locations in the Nieuwkoop Lakes (94.11 and 94.12) and two in the Langeraar Lakes (95.04 and 95.18) samples were collected six or seven times per year in 1985 and 1986. At two locations in the Reeuwijk Lakes (134.08 and 134.09) sampling took place every four weeks from 1983–1986. General characteristics of the sampling locations are given in Table 1. The first fifteen sites represent canals and lakes in the basin system of Rijnland, an interconnected system of canals, lakes and ditches with a constant water level of 0.6 m below mean sea level. The last nine locations are situated in polder lakes at about 2 m

Table 1. General characteristics of the sampled locations in the Rijnland Waterboard area.

Location nr.	Name:	Type:	Mean depth: (m)	Surface (km^2)
1	Noorder Buiten Spaarne	canal	3.9	
21A	Canal at Halfweg pumping station	canal	3.0	
32	Canal Haarlemmermeerpolder	canal	3.0	
37	Canal at Katwijk pumping station	canal	6.0	
58	Kaag Lakes	lake	3.0	3.2
77	Rijn-Schie canal	canal	3.6	
116	Canal at Gouda pumping station	canal	3.4	
272	Lake Braassem	lake	3.5	4.6
275	Lake Nieuwe Meer	lake	5.0	1.4
284	Lake Westeinder	lake	2.9	8.9
296	Lake Oosterduin	deep sand pit	12.0	0.1
299	Lake Vlietlanden	deep sand pit	34.5	0.6
375	Oude Rijn canal	canal	2.9	
379	Lake Zeegerplas	deep sand pit	30.5	0.4
391	Lake Mooie Nel	lake	3.4	0.1
18.03	Lake Broekvelden/Vettenbroek	deep sand pit	27.0	0.9
94.11	Nieuwkoop Lakes (south)	peat lake	2.7	1.0
94.12	Nieuwkoop Lakes (north)	peat lake	2.5	1.5
95.04	Langeraar Lakes (north)	peat lake	2.0	0.8
95.18	Langeraar Lakes (Geerplas)	peat lake	2.5	0.2
134.08	Reeuwijk Lakes (Elfhoeven)	peat lake	2.4	1.1
134.09	Reeuwijk Lakes (Nieuwenbroek)	peat lake	1.7	1.1
217.06	Lake Amstelveense Poel	peat lake	1.5	0.6
403.02	Lake Sloterplas	deep sand pit	25.0	0.8

below mean sea level. They are separated from the basin system but depend on it for their water-supply, except for sampling location 18.03, which depends only on precipitation.

Nitrate + nitrite nitrogen (NO_2 + NO_3-N), ammonium nitrogen (NH_4-N), Kjeldahl-nitrogen (Kj-N), orthophosphate (PO_4-P), total phosphorus (TP) and chlorophyll-a (Chl-a) analyses were carried out according to standard methods of the Nederlands Normalisatie Instituut (1981–1987). Total (TN), particulate (PN) and dissolved inorganic nitrogen (DIN) as well as particulate phosphorus (PP) was calculated as NO_2 + NO_3-N plus Kj-N, Kj-N minus NH_4-N, NH_4-N plus NO_2 + NO_3-N and TP minus PO_4-P respectively.

Algal growth potential tests were carried out to determine the availability of nutrients for algal growth, and to identify the growth-limiting nutrients (van der Does & Klapwijk, 1987). All

water samples except those from the sampling locations in the Nieuwkoop, Langeraar and Reeuwijk Lakes, were preserved by deep-freezing until the bioassays were conducted. The samples were thawed in a waterbath at 45 °C and poured into 150 ml Erlenmeyer flasks filled up to 75 ml during 1983 and 1984. In 1985 and 1986 a larger volume of 100 ml was put into 500 ml Erlenmeyer flasks to improve the CO_2-supply (Anonymous, 1986). After autoclaving the flasks to kill the natural phytoplankton population and to liberate nutrients (Miller *et al.*, 1978), some flasks were enriched with 10 mg N l^{-1} (as KNO_3) or 0.5 mg P l^{-1} (as K_2HPO_4/KH_2PO_4) or 10 mg N l^{-1} + 0.5 mg P l^{-1}. After autoclaving, the water was analyzed for pH, Kj-N, NH_4-N, NO_3 + NO_2-N, PO_4-P and TP in order to measure possible side-effects of the pretreatment. The experiments were run in duplicate (1983, 1984) or triplicate (1985, 1986). Two or three

Table 2. Averaged values ($\pm$ s.d.) over a four-year period (1983–1986) of pH, nutrients and chlorophyll-*a* concentrations at different sampling locations (n = number of observations).

Location nr.	pH	DIN (mg l^{-1})	TN (mg l^{-1})	PO_4-P (mg l^{-1})	TP (mg l^{-1})	Chl-a (mg m^{-3})	n
1	7.9 ± 0.2	6.9 ± 2.4	9.7 ± 2.6	2.08 ± 0.26	2.25 ± 0.27	97 ± 85	8
21A	7.8 ± 0.2	4.6 ± 2.0	8.4 ± 5.5	0.51 ± 0.12	0.83 ± 0.14	40 ± 22	8
32	8.0 ± 0.2	3.2 ± 1.8	5.3 ± 1.9	0.54 ± 0.13	0.73 ± 0.15	40 ± 31	15
37	7.8 ± 0.2	5.7 ± 2.2	7.7 ± 2.3	1.08 ± 0.37	1.23 ± 0.40	30 ± 24	15
58	8.2 ± 0.3	2.9 ± 1.8	5.0 ± 1.8	0.66 ± 0.55	0.77 ± 0.15	34 ± 32	16
77	7.7 ± 0.2	5.1 ± 1.2	7.2 ± 2.0	0.54 ± 0.27	0.80 ± 0.18	30 ± 33	8
116	7.7 ± 0.2	6.0 ± 1.9	7.9 ± 2.2	0.60 ± 0.19	1.15 ± 0.74	27 ± 22	16
272	8.3 ± 0.4	2.5 ± 1.4	4.3 ± 1.5	0.43 ± 0.11	0.52 ± 0.08	40 ± 46	16
275	8.0 ± 0.3	2.8 ± 0.8	4.4 ± 1.0	0.42 ± 0.06	0.48 ± 0.08	22 ± 24	16
284	8.5 ± 0.3	0.9 ± 0.8	2.8 ± 0.8	0.18 ± 0.05	0.31 ± 0.06	52 ± 36	16
296	8.4 ± 0.3	0.6 ± 0.4	2.2 ± 0.3	0.69 ± 0.21	0.79 ± 0.18	26 ± 28	16
299	8.0 ± 0.2	5.4 ± 1.5	7.2 ± 1.7	0.22 ± 0.10	0.28 ± 0.11	9 ± 8	16
375	7.6 ± 0.2	4.1 ± 1.6	6.2 ± 1.9	0.69 ± 0.25	0.92 ± 0.21	39 ± 30	16
379	8.1 ± 0.5	5.1 ± 1.3	6.8 ± 1.4	0.91 ± 0.24	0.98 ± 0.23	30 ± 37	16
391	8.1 ± 0.2	5.8 ± 2.2	8.1 ± 2.2	1.81 ± 0.31	1.92 ± 0.30	58 ± 54	16
18.03	8.1 ± 0.2	1.1 ± 0.6	2.0 ± 0.6	0.01 ± 0.01	0.04 ± 0.01	3 ± 2	16
94.11	8.5 ± 0.6	0.2 ± 0.2	3.4 ± 0.6	0.03 ± 0.03	0.26 ± 0.10	223 ± 123	21
94.12	8.3 ± 0.5	0.2 ± 0.1	2.6 ± 0.3	0.01 ± 0.01	0.08 ± 0.02	102 ± 34	21
95.04	8.7 ± 0.4	0.3 ± 0.3	3.6 ± 0.7	0.12 ± 0.11	0.37 ± 0.11	177 ± 76	21
95.18	8.7 ± 0.4	0.2 ± 0.1	3.6 ± 1.0	0.23 ± 0.13	0.47 ± 0.12	199 ± 60	12
134.08	8.1 ± 0.4	0.2 ± 0.1	2.1 ± 0.5	0.02 ± 0.02	0.14 ± 0.07	75 ± 45	51
134.09	8.2 ± 0.4	0.3 ± 0.3	3.0 ± 0.4	0.01 ± 0.01	0.09 ± 0.03	98 ± 37	51
217.06	8.7 ± 0.5	0.5 ± 0.8	4.3 ± 1.1	0.09 ± 0.10	0.29 ± 0.07	220 ± 115	16
403.02	8.4 ± 0.4	1.1 ± 0.9	2.2 ± 0.9	1.23 ± 0.41	1.31 ± 0.36	27 ± 29	16

Erlenmeyer flasks without addition served as controls.

All Erlenmeyer flasks were inoculated with approximately 10 000 cells per ml of *Scenedesmus quadricauda* (Turp.) Bréb., which had been nitrogen and phosphorus starved for about 12 days. They were derived from a batch culture acquired from the Limnological Institute, Nieuwersluis. The culture vessels, closed with cotton plugs, were placed on a shaking table (120–150 rpm) at a temperature of 20 ± 1 °C and at a constant light intensity of $70–100$ mE m^{-2}s^{-1}. The optical density (OD) at 680, corrected with the OD at 750 nm, was measured on a Philips/Pye Unicam PU 8800/2 Spectrophotometer as an indicator of the biomass. Maximal biomass was calculated from three measurements during the stationary growth phase. When the growth was two times less than 5% per day, the resulting and the preceding data were used to calculate maximal biomass after rejection of outliers (E.P.A., 1971). A scheme of the assay technique is given in de Vries and Klapwijk (1987).

Results

The results of the pH measurements and chemical analyses in the original water samples are summarized in Table 2. Great variation in nutrient concentrations was observed between the different sampling locations. Average DIN and TN concentrations, for instance, ranged from 0.2 and 2.1 mg N l^{-1} at sampling location 134.08 to 5.8 and 8.1 mg N l^{-1} at sampling location 391 respectively. Orthophosphate and TP ranged from approximately 0.01 and 0.04 mg P l^{-1} (location 18.03) to 2.08 and 2.25 mg P l^{-1} (location 1). This observed variation was dependent on the degree of isolation of the various lakes from the basin system of Rijnland and on the local degree of pollution, especially from effluent discharges of sewage treatment plants.

Chlorophyll-*a* was highest in the shallow peat lakes (locations 94.11, 94.12, 95.04, 95.18, 134.08, 134.09, 217.06), relatively isolated from Rijnland's basin system and which have rather long residence times (usually > 1 year).

A comparison of the analyses of the original water samples with the results after deep-freezing and autoclaving showed that at most sampling stations the pH, and the TN, PO$_4$-P and TP concentrations had decreased. This comparison was made with all samples using paired Student's t-test (Table 3). Also, Kj-N, NH$_4$-N and PN decreased significantly, while PP increased. Nitrite + nitrate-N and DIN did not change significantly, nor did the inorganic N/P and the total N/P ratio, although the mean inorganic-N/P ratio has halved. Obviously, the pretreatment caused a

Table 3. Results of t-tests (means $\pm$ s.d.) between pH and nutrient concentration before and after deep-freezing and autoclaving.

		Before pretreatment	After pretreatment	Difference	n	r	t	p
pH		8.23 ± 0.48	8.72 ± 0.42	0.49	313	0.15	14.75	<0.001
Kj-N	(mg l^{-1})	2.94 ± 1.70	2.74 ± 1.47	-0.20	416	0.95	-7.64	<0.001
NH$_4$-N	(mg l^{-1})	0.68 ± 1.20	0.62 ± 1.02	-0.06	418	0.95	-3.14	0.002
NO$_2$ + NO$_3$ – N	(mg l^{-1})	1.47 ± 1.73	1.49 ± 1.74	0.02	418	0.98	1.42	0.157 ns
DIN	(mg l^{-1})	2.14 ± 2.51	2.10 ± 2.36	-0.04	418	0.98	-1.57	0.117 ns
PN	(mg l^{-1})	2.26 ± 1.10	2.12 ± 0.89	-0.14	416	0.88	-5.30	<0.001
TN	(mg l^{-1})	4.41 ± 2.67	4.23 ± 2.45	-0.18	416	0.98	-6.08	<0.001
PO$_4$-P	(mg l^{-1})	0.42 ± 0.53	0.18 ± 0.20	-0.24	417	0.83	-13.13	<0.001
PP	(mg l^{-1})	0.15 ± 0.19	0.30 ± 0.32	0.15	417	0.16	8.75	<0.001
TP	(mg l^{-1})	0.57 ± 0.56	0.48 ± 0.47	-0.09	417	0.91	-8.13	<0.001
Inorg-N/P		38.3 ± 267.2	19.7 ± 40.5	-18.60	417	0.10	-1.42	0.155 ns
Total-N/P		16.2 ± 16.3	17.5 ± 19.8	1.30	417	0.68	1.78	0.075 ns

n = number of observations; r = product-moment correlation coefficient; t = Student's t-value; p = two-tailed probability; ns = not significant.

194

pH increase due to the loss of CO_2, and a loss of about 4% total nitrogen and 16% total phosphorus.

It is not possible to show in detail all the results of 440 bioassays, but Fig. 2 shows two typical examples of frequently occurring results. In both examples, the addition of N resulted in a higher yield than in the control, while the addition of P did not. In the example given in Fig. 2A the addition of N *plus* P resulted in an even higher yield than the addition of N alone. In the Fig. 2B example, the combined addition (N + P) resulted in about the same yield as the N-addition alone. In both bioassays the AGP (= yield without addition) was primarily N-limited, but the yield with the N-addition example in Fig. 2B was not limited by P but by another factor, while the yield

with the N-addition example in Fig. 2A was presumably limited by P.

Table 4 shows the mean values (± s.d.) of the bioassay results from the different sampling locations ranked in increasing order of yield. The mean yields (controls) ranged from 0.0015 to 0.046 (O.D. 680–750 nm); they were lowest in the polder lakes. From the nitrogen and phosphorus enrichment bioassays it can be seen that in most lakes nitrogen clearly stimulated the algal yield. Only in location 18.03 was the yield increased by the addition of phosphorus alone. In several other lakes, e.g. locations 299, 94.11, 94.12, 95.04, 134.08, 134.09 and 217.06, the yield was stimulated by the addition of nitrogen, but even more so by the enrichment with N + P. This indicated that the algal growth in these lakes was limited primar-

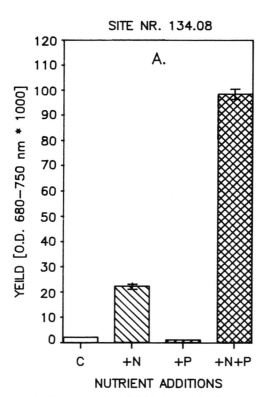

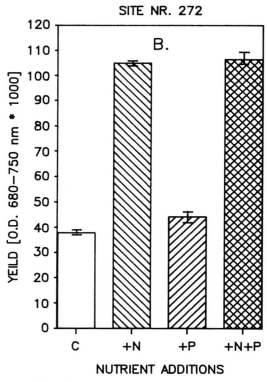

Fig. 2. Two examples of yields (± s.d.) in bioassay experiments with nitrogen and phosphorus enrichments carried out with water from the Reeuwijk lakes (Fig. 2A; site nr. 134.08; date 860624) and lake Braassem (Fig. 2B; site nr. 272, date 840306).

C = control (no enrichment);
+ N = with nitrogen addition;
+ P = with phosphorus addition;
+ N + P = with nitrogen *and* phosphorus addition.

Table 4. Mean values (± s.d.) as optical density (680–750 nm) × 1000 of bioassay results from the different sampling locations ranked from low to high yield.

| Location nr. | Control | (n) | With additions of | | | | | | | Generally limiting nutrient (s) |
			Nitrogen	(n)	Phosphorus	(n)	N + P	(n)	
134.08	1.5 + 1.7	(51)	19 + 12	(50)	1.5 + 1.9	(50)	65 + 21	(47)	N, p
94.11	2.2 + 1.6	(21)	32 + 12	(8)	1.7 + 1.1	(8)	54 + 11	(8)	N, p
94.12	2.3 + 2.2	(21)	16 + 5.9	(8)	2.2 + 0.8	(8)	48 + 13	(8)	N, p
134.09	2.5 + 2.0	(51)	8.9 + 5.6	(50)	2.2 + 1.9	(50)	62 + 25	(48)	N, p
95.18	2.6 + 1.5	(12)							
95.04	3.1 + 2.6	(21)	50 + 18	(8)	3.8 + 2.6	(8)	70 + 8.5	(8)	N, p
18.03	4.3 + 3.1	(16)	5.2 + 3.9	(8)	8.8 + 3.8	(8)	36 + 14	(8)	P, n
217.06	4.3 + 2.8	(16)	46 + 13	(8)	2.5 + 1.6	(8)	56 + 16	(8)	N, p
296	5.3 + 3.0	(16)	88 + 29	(8)	6.1 + 2.8	(8)	86 + 11	(8)	N
284	6.9 + 5.3	(16)	46 + 23	(8)	7.5 + 5.5	(8)	46 + 24	(8)	N
403.02	7.0 + 6.4	(16)	68 + 18	(8)	10 + 8.1	(8)	68 + 16	(8)	N
272	19 + 11	(16)	85 + 20	(8)	23 + 12	(8)	96 + 26	(8)	N
275	20 + 6.5	(16)	69 + 15	(8)	19 + 4.3	(8)	65 + 22	(8)	N
58	21 + 13	(16)	101 + 23	(8)	26 + 15	(8)	111 + 29	(8)	N
299	25 + 13	(16)	53 + 22	(8)	33 + 7.0	(8)	83 + 29	(8)	N, p or P, n
32	27 + 18	(15)	113 + 12	(7)	34 + 20	(7)	119 + 19	(7)	N
77	32 + 10	(8)							
375	32 + 15	(16)	138 + 14	(8)	33 + 14	(8)	134 + 18	(8)	N
21A	33 + 17	(8)							
379	39 + 13	(16)	98 + 25	(8)	37 + 7.9	(8)	104 + 21	(8)	N
379	40 + 15	(16)	106 + 23	(8)	41 + 13	(8)	100 + 17	(8)	N
37	41 + 17	(16)	125 + 16	(8)	43 + 18	(8)	130 + 24	(8)	N
1	46 + 15	(8)							
116	46 + 15	(16)	147 + 33	(8)	47 + 13	(8)	155 + 35	(8)	N

(n) = number of observations;
N, p = primarily nitrogen, secondarily phosphorus limited;
P, n = primarily phosphorus, secondarily nitrogen limited;
N = nitrogen limited.

ily by nitrogen, but also by phosphorus. At the other sampling locations, including all the canals, enrichment with N + P did not generally produce a significantly higher yield than enrichment with N alone. Therefore, algal growth was limited only by nitrogen at these sampling stations, where phosphorus was obviously present in excess. By the ranking of the bioassay results in increasing order of yield it is clear that those with the lowest yields are mostly phosphorus, or nitrogen and phosphorus limited, while the highest algal yields are generally limited exclusively by nitrogen.

In order to determine whether, and to what degree bioassay results are related to nutrient and chlorophyll-*a* concentrations and nutrient ratios

in the original water samples, correlation coefficients were calculated (Table 5). Very high correlations were found between bioassays without enrichments (controls) and DIN and TN (0.91 and 0.85 respectively; $n = 440$). Therefore, it can be concluded that the AGP in these waters is predominantly determined by the amount of nitrogen in the samples. Phosphorus was only a secondary limiting factor as can be concluded from the even higher correlation between the yields in the bioassays with P-enrichment and DIN and TN (0.98 and 0.92 respectively; $n = 243$). This is illustrated in Fig. 3, in which the yields without enrichments and with P-addition are plotted against DIN. Fig. 3A shows a positive

196

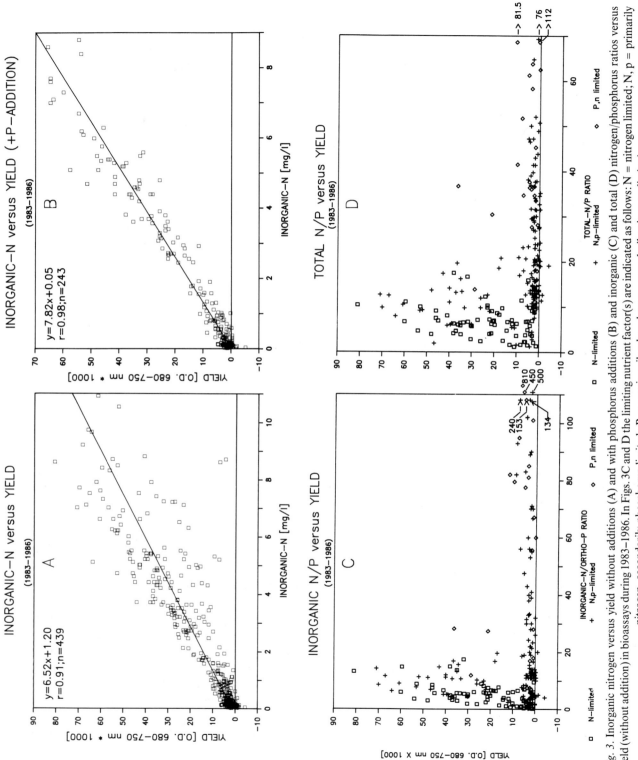

Fig. 3. Inorganic nitrogen versus yield without additions (A) and with phosphorus additions (B) and inorganic (C) and total (D) nitrogen/phosphorus ratios versus yield (without addition) in bioassays during 1983–1986. In Figs. 3C and D the limiting nutrient factor(s) are indicated as follows: N = nitrogen limited; N, p = primarily nitrogen, secondarily phosphorus limited; P, n = primarily phosphorus, secondarily nitrogen limited.

Table 5. Product-moment correlation coefficients between AGP results and chlorophyll-*a* with nutrient concentrations and ratios in the original water samples.

	DIN N	TN	PO$_4$-P	TP	inorg-N/P	total N/P	(*n*)
Control (no addition)	0.91*	0.85*	0.63*	0.65*	− 0.07 ns	− 0.31	440
Nitrogen addition	0.73*	0.70*	0.69*	0.71*	− 0.23*	− 0.60*	243
Phosphorus addition	0.98*	0.92*	0.63*	0.63*	− 0.12 ns	− 0.33*	243
N + P addition	0.64*	0.65*	0.45*	0.52*	− 0.05 ns	− 0.40*	238
Chlorophyll-*a*	− 0.38*	− 0.11*	− 0.29*	− 0.20*	− 0.04 ns	0.01 ns	434

* = p < 0.001; ns = not significant; (*n*) = number of paired observations.

correlation between yield and DIN in the sampled waters, with a relatively high variability between 3 and 10 mg N l^{-1}. Addition of P (Fig. 3B) diminishes the variability, probably by removing the P-limitation.

Applying multiple regression analysis (King, 1969) to 'explain' the AGP due to the NO$_2$ + NO$_3$-N, NH$_4$-N and PO$_4$-P concentrations gave a highly significant multiple correlation coefficient of $R = 0.914$ ($n = 439$; $P < 0.001$) with the following equation:

$$Y = 5.43 X_1 + 6.50 X_2 + 3.06 X_3 + 0.72$$

where:

Y = AGP (as O.D. 680−750 nm * 1000),
X_1 = concentration of NH$_4$ (in mg N l^{-1}),
X_2 = concentration of NO$_2$ + NO$_3$ (in mg N l^{-1}),
X_3 = concentration of PO$_4$-P (in mg N l^{-1}).

The standard partial regression coefficients (not shown here) indicated that NO$_2$ + NO$_3$-N was responsible for most of the total variance in the AGP, followed by NH$_4$-N and PO$_4$-P. With this

formula, the AGP can be precisely calculated on the basis of nutrients.

It should also be noted (Table 5) that the chlorophyll-*a* concentrations in the field samples are inversely related to those of DIN, TN, PO$_4$-P and TP. Obviously, the actual algal biomass in the sampled canals and lakes is generally more dependent upon other factors, such as turbidity, residence time and zooplankton grazing, than upon nutrient concentrations.

The negative correlations with N/P ratios, especially the negative correlation between the N-enriched yields with total N/P ratio, indicate that lower growth potentials are found at higher N/P ratios, which mostly implicates low P-concentrations. The correlation coefficients with the nutrient ratios are not very high due to the fact that the relationships are not linear. In Figs. 3C and 3D, where the yields (without enrichments) are plotted against inorganic and total N/P ratios, critical ranges for nitrogen and/or phosphorus limitation can be derived (Table 6).

Discussion and conclusions

The measurements before and after deep-freezing and autoclaving show clearly that the chemical composition of the water samples is altered by the pretreatment. This is consistent with the findings of Filip & Middlebrooks (1975), Eloranta & Laitinen (1981), de Vries & Ouboter (1985) and Anonymous (1986). Obviously, the speciation of phosphorus, which can be very important in considering its availability to algae, is altered by pretreatment due amongst other things to pH

Table 6. Critical ranges of N/P ratios indicating different nutrient limitations according to this study.

Limitation:	Inorg-N/P ratio	Total N/P ratio
Nitrogen	0–20	0–20
Primarily nitrogen, sec. phosphorus	10–70	10–50
Primarily phosphorus, sec. nitrogen	> 50	> 30

changes. Whether or not this seriously affects the bioassay results is not known. The decrease due especially to the PO_4-P concentration is presumably reversible at least partially, when the pH is again decreased during the bioassays. Van der Does (pers. comm.) found with samples from location nr. 391, a reversible reaction at pH levels <8.5 in the original samples and irreversible reactions at pH levels >8.5. It is obvious that the pretreatment by autoclaving does not solubilize all nutrients, as suggested by Miller et al. (1974, 1978).

The results have shown that AGP is a good parameter for the potential algal growth in surface waters. It is not directly related to field measurements of Chl-a. On the contrary, AGP yields are inversely related to actual Chl-a measurements in the canals and lakes ($r = -0.33$; $n = 437$). Therefore, both parameters should be considered as supplementary.

With the nitrate enrichments, an additional amount of 27.9 mg K l^{-1} was added to the test waters, whereas only 1.23 mg K l^{-1} was added with the phosphorus additions. The question can therefore be raised if the growth stimulation observed by the nitrate additions are really due to the N-addition, or instead to the addition of potassium. Because of the high concentration of potassium in the surface waters of the Rijnland Waterboard area (maximum about 30 mg K l^{-1}), and the relative low K concentrations needed for optimal growth (1.1–2.3 mg K l^{-1}) of algal species (Leentvaar, 1980), it is very unlikely that growth stimulation caused by K-addition occurs in these waters. Moreover, in two bioassay experiments with water from the Reeuwijk Lakes in June, 1983 we found no difference in the maximal algal growth after addition of 10 mg N as KNO_3 or NH_4Cl (unpublished data van der Does & Klapwijk).

The ranges found in this study for N/P ratios indicating a primary phosphorus limitation, >50 and >30 for inorganic and total N/P ratios respectively, lie considerably higher than the ranges for P-limitation reported to date in the literature (cf. Forsberg et al. 1978; Claesson & Forsberg, 1980; de Vries & Klapwijk, 1987). This means that in the Rijnland surface waters, phosphorus limitation can only be achieved at relatively high N/P ratios and low P-concentrations.

Correlation and regression analyses with AGP and nutrients, measured in the original water samples, prove that the algal growth potential measured in bioassays can be calculated with high precision based on nutrient concentrations. Therefore, once the relationship between AGP and nutrient concentrations is established, as in this case, AGP tests do not have to be carried out on a routine basis to monitor the algal growth potential of surface waters. However, they can be very useful in special studies dealing with the effects of phosphorus reduction programs and in lake restoration studies (Klapwijk, 1981; van der Does & Klapwijk, 1985, 1987).

Acknowledgements

The authors thank Prof. Dr. M. Donze and Prof. Dr. W.H.O. Ernst for valuable suggestions with respect to the manuscript, Mr. P. Nieuwpoort for compilation of the data and graphical presentations and Miss. A. Honnef for correcting and Miss C. van Dijk for typing the English text.

References

Anonymous, 1986. Ontwikkeling van een algengroeipotentietoets voor oppervlaktewater en afvalwater. Rapport Stichting Toegepast Onderzoek Reiniging Afvalwater, Rijswijk, 77 pp.

Bolier, G., A. N. van Breemen & G. Visser, 1981. Eutrophication tests: a possibility to estimate the influence of sewage discharge on the biological quality of the receiving water. H_2O 14: 88– 92 (in Dutch with an English summary).

Bolier, G. & L. W. C. A. van Breemen (eds.), 1982. Standaardisatie van de algengroeipotentietoets voor Nederland: verkenning en afbakening. Laboratorium voor Gezondheidstechniek, TH-Delft. Rapport nr. 82–03, 49 pp.

Bringmann, P. G. & R. Kühn, 1965. Nitrat oder Phosphat als Begrenzungsfaktor des Algenwachstums. Gesundheitsing. 7: 210–214.

Chiaudani, G. & M. Vighi, 1974. The N : P ratio and tests with Selenastrum to predict eutrophication in lakes. Wat. Res. 8: 1063–1069.

Chiaudani, G. & M. Vighi, 1975. Dynamics of nutrient limitation in six small lakes. Verh. int. Ver. Limnol. 19: 1319–1324.

Chiaudani, G. & M. Vighi, 1976. Comparison of different techniques for detecting limiting or surplus nitrogen in batch cultures of *Selenastrum capricornutum*. Wat. Res. 10: 725–729.

Claesson, Å. & Å. Forsberg, 1980. Algal assay studies of polluted lakes. Arch. Hydrobiol. 89: 208–224.

Does, J. van der & S. P. Klapwijk, 1985. Phosphorus removal and effects on water quality in Rijnland. H₂O 18: 381–387 (in Dutch with an English summary).

Does, J. van der & S. P. Klapwijk, 1987. Effects of phosphorus removal on the maximal algal growth in bioassay experiments with water from four Dutch lakes. Int. Revue ges. Hydrobiol. 72: 27–39.

Eloranta, V. & O. Laitinen, 1981. Evaluation of sample preparation for algal assays on waters receiving cellulose effluents. Verh. int. Ver. Limnol. 21: 770–775.

Environmental Protection Agency (E.P.A.), 1971. Algal assay procedure bottle test, National Eutrophication Research Program. Corvallis, Oregon, 82 pp.

Environmental Protection Agency (E.P.A.), 1975. Proceedings: Biostimulation and nutrient assessment. Workshop, Oct. 16–17, 1973. Corvallis, Oregon. EPA-660/3-75-034.

Filip, D. S. & E. J. Middlebrooks, 1975. Evaluation of sample preparation techniques for algal bioassays. Wat. Res. 9: 581–585.

Forsberg, C., S. -O. Ryding, Å. Claesson & Å. Forsberg, 1978. Water chemical analyses and/or algal assay? – Sewage effluent and polluted lake water studies. Mitt. int. Ver. Limnol. 21: 352–363.

Gargas, E. & J. S. Pedersen, 1974. Algal assay procedure batch technique. Contribution from the Water Quality Institute. Danish Academy of Technical Science, 1 (2nd ed.), 48 pp.

Haan, H. de, J. B. W. Wanders & J. R. Moed, 1982. Multiple addition bioassay of Tjeukemeer water. Hydrobiologia 88: 233–244.

Hoogheemraadschap van Rijnland, 1984. Rapport betreffende het onderzoek naar de effecten van fosfaatverwijdering op de a.w.z.i.'s Gouda, Bodegraven en Nieuwveen. Rapport technische dienst van Rijnland, Leiden, Ch. 5, 37 pp.

King, L. L., 1969. Statistical analysis in geography. 2nd ed. Prentice Hall. Inc., Englewood Cliffs, N.J., pp. 135–148.

Klapwijk, S. P., 1981. Limnological research on the effects of phosphate removal in Rijnland. H₂O 14: 472–483 (in Dutch with an English summary).

Leentvaar, P., 1980. Eutrophication, nature management and the role of potassium. Hydrobiol. Bull. 14: 22–29.

Marvan, P., S. Přibil & Lhotský, 1979. Algal assays and monitoring eutrophication. E. Schweizerbart'sche Verlagsbuchhandlung, Stuttgart, 261 pp.

Miller, W. E., T. E. Maloney & J. C. Greene, 1974. Algal productivity in 49 lake waters as determined by algal assays. Wat. Res. 8: 667–679.

Miller, W. E., J. C. Greene & T. Shiroyama, 1978. The *Selenastrum capricornutum* Printz algal assay bottle test. Environmental Protection Agency (EPA). Corvallis, Oregon, 126 pp.

Nederlands Normalisatie Instituut, 1981–1987. Test methods for water. NEN 6474, 6481, 6520, 6646, 6663 (in Dutch).

Nordforsk, 1973. Algal assays in water pollution research. Proceedings from a Nordic symposium, Oct. 1972, Oslo. Nordforsk secretariat of Environmental Sciences Publ.: 2, 128 pp.

Raschke, R. L. & D. A. Schulz, 1987. The use of algal growth potential test for data assessment. J. Wat. Pollut. Cont. Fed. 59: 222–227.

Ryding, S. -O. (ed), 1980. Monitoring of inland waters. Report from the working group for eutrophication research. Nordforsk publication 1980: 2, Helsingfors, Finland, 207 pp.

S. I. L., 1978. Symposium: Experimental use of algal cultures in limnology. Mitt. int. Ver. Limnol. 21: 1–607.

Skulberg, O. M., 1964. Algal problems to the eutrophication of European water supplies, and a bioassay method to assess fertilizing influences of pollution on inland water. In: D. F. Jackson (ed.): Algae and Man, Plenum Press, New York, pp. 262–269.

Veenstra, S., 1981. De algengroei-potentie-toets. Een nieuw instrument ter beoordeling van het eutrofiërend karakter van een water. L. H. Wageningen, Vakgroep Waterzuivering, sektie Hydrobiologie. Doktoraal verslagen serie nr. 81–1, 61 pp. (in Dutch).

Vries, P. J. R. de & P. S. H. Ouboter, 1985. Watersample treatments and their effects on bioassays using *Stigeoclonium helveticum* Vischer. Aquat. Bot. 22: 177–185.

Vries, P. J. R. de & S. P. Klapwijk, 1987. Bioassays using *Stigeoclonium tenue* Kütz. and *Scenedesmus quadricauda* (Turp.) Bréb. as testorganisms; a comparative study. Hydrobiologia 153: 149–157.

Youngman, R. E., 1980. The prediction of algal crops by bioassay. Technical Report TR 148, Water Research Center, Environmental Protection, Medmenham, UK, 55 pp.

Hydrobiologia **188/189**: 201–209, 1989.
M. Munawar, G. Dixon, C. I. Mayfield, T. Reynoldson and M. H. Sadar (eds)
Environmental Bioassay Techniques and their Application.
© 1989 *Kluwer Academic Publishers. Printed in Belgium.*

A study of phosphate limitation in Lake Maarsseveen: phosphate uptake kinetics versus bioassays

E. Van Donk,[1,*] L.R. Mur[2] & J. Ringelberg[3]
[1] *Provincial Waterboard of Utrecht, Postbox 80300, 3508 TH Utrecht, The Netherlands (*author for correspondence)*; [2] *Laboratory of Microbiology, University of Amsterdam, Nieuwe Achtergracht 127, 1018 WS Amsterdam, The Netherlands*; [3] *Department of Aquatic Ecology, University of Amsterdam, Kruislaan 320, 1098 SM Amsterdam, The Netherlands*

Key words: bioassays, phytoplankton, physiological indicators, phosphate uptake experiments

Abstract

In order to assess possible phosphate limitation for the phytoplankton community of Lake Maarsseveen, two techniques (phosphate uptake experiments and bioassays) were employed simultaneously in February–March 1982. In that period the ambient phosphate concentration of the lake water was less than 0.03 μM P and the diatom *Asterionella formosa* constituted more than 90% of the phytoplankton population. The phosphate uptake experiments showed relatively high uptake capacities and low cell phosphorus contents for the natural phytoplankton community. This suggested phosphate limitation throughout the test period. The growth stimulation of the phytoplankton after enrichment with phosphate, however, only revealed phosphate limitation from the beginning of March and bioassays may therefore be regarded as a less sensitive method.

Introduction

There are three common approaches to the study of nutrient limitation in phytoplankton populations: they are ambient nutrient concentrations of the water, physiological indicators, and the use of bioassays.

The first approach is simply based on measurements of the nutrient content of the water, inferring limitation of those nutrients that are in short supply. However, this method by itself does not yield reliable results if fluxes of nutrients are unknown and if no information is available on the nutrient requirements of the species involved.

The second approach, the use of physiological indicators (e.g. cellular nutrient contents, short-term nutrient uptake kinetics), may give a better insight, especially when physiological characteristics of nutrient-limited reference cultures are available for comparison. In the study of nutrient uptake kinetics, two aspects may be considered. One may examine the steady state specific nutrient uptake rate (q), i.e. nutrient uptake rate proportional to growth rate (μ) (at steady state the amount of nutrient taken up is equal to the amount used to make new cells). This steady state specific nutrient uptake rate (q) can be calculated according to the equation (Droop, 1973)

$$q = \mu \cdot Q, \tag{1}$$

where Q is the amount of internal nutrient per unit population. The other possibility for studying nutrient uptake is to determine the nutrient uptake curve, by measuring the short-term (initial) nutrient uptake rates (V) at different nutrient concentrations (S) (equation 2):

$$V = V_{max} \cdot \frac{S}{K_{s \cdot u} + S},\qquad(2)$$

where V and V_{max} are the velocity and maximum velocity (uptake capacity) of nutrient uptake; S is the initial nutrient concentration; $K_{s \cdot u}$ is the half-saturation constant for uptake.

Nutrient limitation generally induces the potential for a high uptake capacity (V_{max}) of the limiting nutrient compared with the maximum specific nutrient uptake rate (q_{max}) (e.g. Gotham & Rhee, 1981; Riegman & Mur, 1984a). Therefore short term nutrient uptake experiments may be useful for determining the growth rate limiting nutrient of phytoplankton communities. A lowered cellular content of the limiting nutrient is also a general response to nutrient limitation (e.g. Droop, 1974; Rhee, 1978).

The third approach, the use of bioassays, is based on measuring the *growth* of the phytoplankton after enrichment with nutrients. This technique has been widely applied to assess the nutrient limitation of phytoplankton growth in natural waters. Two major variants to the method can be distinguished: (1) nutrients are added to filtered lake water in which laboratory cultured species are inoculated (e.g. Paasche, 1978; Reynolds & Butterwick, 1979; De Vries, 1983 and 1985) and (2) nutrients are added to lake water, containing the natural phytoplankton community (e.g. Schelske *et al.*, 1974; Frey & Small, 1980; Van der Does & Klapwijk, 1987; Van Donk *et al.*, 1988; Munawar *et al.*, 1988). Variant 2 has been chosen to study the growth limitation of the phytoplankton in Lake Maarsseveen.

The objective of the study is to compare the results obtained simultaneously with physiological indicators and natural community bioassay experiments, applied to the phytoplankton community of Lake Maarsseveen.

Methods

Site description

Lake Maarsseveen is situated in the centre of The Netherlands, near the city of Utrecht. It is a man-made lake formed around 1960 by excavation of sand in a peat-bog area. The oligo-mesotrophic, trough-shaped lake (70 ha; 30 m max. depth) is replenished essentially by precipitation and ground water and drained by an outlet. A more comprehensive description of the lake is given by Van Donk (1987). Every year during winter and spring, diatoms are predominant. In the spring of 1982, the diatom *Asterionella formosa* was predominant over a period of more than three months. This alga was able to reach a high abundance (3800 cells/ml), whereas the other species (*Fragilaria crotonensis*, *Stephanodiscus astraea* and *S. hantzschii*) appeared only in small numbers. In previous years (1980–1981) *A. formosa* was heavily infected by a chytrid fungus and the other diatoms were able to bloom. However, in 1982 this fungus was temporarily inhibited in its activity due to low water temperatures (Van Donk & Ringelberg, 1983). It was expected that in 1982 the diatoms were growing under phosphate limitation because the phosphate concentration in the lake was very low ($< 0.03\ \mu M$ P) and nitrate ($> 30\ \mu M$ N) and silicate ($> 50\ \mu M$ Si) were relatively high.

Short-term P-uptake with the natural phytoplankton community

From February 1, 1982, until March 31, 1982, once a week, around nine 'o clock in the morning, composite water samples were collected of the upper 10 metres of the open-water zone of the lake, taken with a 3-litre van Dorn sampler and concentrated through a 55 μm-mesh plankton net. The sample was transported to the laboratory within one hour and filtered over 150 μm mesh plankton net to remove zooplankton. The concentrated sample consisted of more than 90% of *A. formosa*.

The cell phosphorus content was measured according to Menzel & Corwin (1965). To determine the relation between the short term phosphate uptake rate and the phosphate concentration of the water, the sample was divided among six 1-litre Pyrex Erlenmeyer flasks containing 500 ml of sterile medium (Guillard, 1975) with varying concentrations of K_2HPO_4 (0.24–10 μM P). The concentrated sample was divided among the uptake flasks with an initial cell concentration of 1.5×10^5 cells ml^{-1}. Incident light intensity during the experiment was optimal (25 Watt.m^{-2}) and the incubation temperature was 10 °C ± 1 °C. Since the cell density was low the incident light intensity was about equal to the intensity experienced by the algae in the flasks. In each flask, orthophosphate was measured immediately after cell addition and subsequently every 15 or 30 minutes onward during the period that the orthophosphate concentration declined as a linear function of time (2–4 hours). Orthophosphate was determined by the Murphy and Riley method (1962), with all the samples filtered through 0.45 μm filters presoaked in distilled water. Spectrophotometric readings were made using 4 cm cuvettes, allowing determinations of as little as 0.03 μM P. The flasks were stirred constantly. For each flask the initial phosphate uptake rate, V, was calculated from the initial disappearance of the phosphate by least squares linear regression analysis of the observed phosphate concentrations. Thus,

$$V = \frac{-\Delta S}{\Delta t} \cdot \frac{1}{n}, \qquad (3)$$

where $-\Delta S$ is the decrease in phosphate concentration during the time interval Δt and n is the number of *A. formosa* cells per ml. Calculated values of V were fitted by an iterative, non-linear regression (Hanson *et al.*, 1967) to the Michaelis-Menten equation: (see equation 2).

Short-term P-uptake with reference laboratory cultures of A. formosa

A. formosa was isolated from Lake Maarsseveen using the pipette technique (Guillard, 1973). The cells were cultured in freshwater medium 'WC' (Guillard, 1975) at 10 °C and with 25 Watt.m^{-2} (12 h light – 12 h dark) illumination, provided by cool-white fluorescent tubes. Exponentially growing *A. formosa* cells then transferred to flasks with fresh 'WC' medium without phosphate. KNO_3 was added to supplement potassium, which was low due to the omitted K_2HPO_4. The cells were allowed to grow until they were depleted of phosphorus, as indicated by the culture reaching a stationary phase. To determine the uptake capacity (V_{max}) at various degrees of phosphate limitation, the *A. formosa* cells were preloaded with different amounts of phosphorus.

One day before each uptake experiment, cells from the stationary phase where inoculated in sterile medium with a particular amount of phosphate. By the time the cells were used for the uptake experiments, all added phosphorus had been taken up by the cells. No free phosphorus could be measured in the medium. Before incubation the cell phosphorus content was measured according to Menzel and Corwin (1965). By varying the added amount of phosphate it was possible to perform uptake experiments with cells having different internal phosphorus contents. The uptake experiments were done according to the procedure described for the natural phytoplankton community.

In order to compare the maximum value of the steady state specific nutrient uptake rate (q) (equation 1) with the uptake capacity (V_{max}), q_{max} ($= \mu_{max} \cdot Q_{max}$) was calculated using the internal storage model of Droop (1973) (equations 1, 4 and 5). The Droop model describes the relationship between growth rate and internal phosphorus content, during steady state conditions. It is probable that the Droop model can also be used for non steady state conditions:

$$\mu = \mu'_{max} \cdot \frac{Q - Q_o}{Q}, \qquad (4)$$

where Q_o is the minimum cell quota and μ'_{max} refers to 'infinite' internal phosphorus content. μ'_{max} is higher than the true μ_{max} value (Droop, 1973).

$$\mu_{max} = \mu'_{max}\left(1 - \frac{Q_o}{Q_{max}}\right). \qquad (5)$$

For μ_{max}, Q_o and Q_{max} (the maximum cell quota) we used the values measured by Van Donk (1983) and Van Donk and Kilham (1989) at $10\,°C$ ($\mu_{max} = 0.58\,d^{-1}$, $Q_o = 6 \times 10^{-9}\,\mu M$ P cell^{-1} and $Q_{max} = 124 \times 10^{-9}\,\mu M$ P cell^{-1}). To get an upper limit for q (q_{max}), μ'_{max} instead of μ was used in equation 1.

Bioassays

Mixed water samples of the upper seven metres of Lake Maarsseveen, taken with a 3-litre van Dorn sampler, were brought back to the laboratory within one hour after sampling. The samples were filtered through $150\,\mu m$ gauze to remove the crustacean zooplankton. Once a week from February 17 until March 31 the natural phytoplankton cells were incubated in three 1-litre Pyrex Erlenmeyer flasks and placed in the laboratory under optimal light conditions (25 Watt. m^{-2}, 12 h light – 12 h dark) at $5\,°C$ (ambient lake temperature) and $10\,°C$ (standard temperature). The flasks were manually shaken twice a day. The nutrient combinations tested were 'All', 'All-P' and 'LW'. 'All' indicates the lake water (LW) enriched with all nutrients listed in Table 1 and 'All-P' all nutrients added except phosphorus. The growth of the different algal species in the flasks was followed for seven days by counting the number of cells each day. The growth rates were calculated by a linear least squares regression of log transformed data.

Results

Table 2 gives the percentages of the phytoplankton community consisting of *A. formosa*

Table 1. Range and concentration of nutrients added to the bioassay enclosures.

Nutrients	Compound	Concentration
P	K_2HPO_4	3.20 (μM)
N	$NaNO_3$	71.00
Si	$Na_3SiO_3 \cdot 9H_2O$	35.70
B	H_3BO_3	0.10
Vitamin mix		
Biotin		0.05 ($\mu g\,l^{-1}$)
B_{12}	Cyanocobalamin	5.0
B_1	Thiamine $\cdot$ HCl	100.0
Trace metals		
Cu	$CuSO_4 \cdot 5H_2O$	0.039 (μM)
Zn	$ZnSO_4 \cdot 7H_2O$	0.077
Co	$CoCl_2 \cdot 6H_2O$	0.042
Mn	$MnCl_2 \cdot 4H_2O$	0.914
Mo	$Na_2MoO_4 \cdot 2H_2O$	0.026
Fe	$FeCl_3 \cdot 6H_2O$	16.00
EDTA	$Na_2EDTA \cdot 2H_2O$	13.00

Table 2. Percentage of the phytoplankton community consisted of *A. formosa* and the ambient concentration of PO_4-P, NO_3-N and SiO_2-Si in early spring of 1982.

Date	*A. formosa* (%)	PO_4-P (μM)	NO_3-N (μM)	SiO_2-Si (μM)
30–01–82	60	0.05	35	63
28–02–82	90	<0.03	31	58
31–03–82	99	<0.03	32	57
30–04–82	20	<0.03	31	52

during the spring of 1982. Also included in the table are the ambient concentrations of silicon, phosphate and nitrate. The details about the abundance and succession of the diatom species and nutrient concentrations in Lake Maarsseveen have been published elsewhere (Van Donk & Ringelberg, 1983; Van Donk *et al.*, 1988).

Figure 1 illustrates the relationship between the cell phosphorus content (Q) and the kinetic parameter (V_{max}) of the natural phytoplankton community as well as the laboratory cultures of *A. formosa*. Also depicted in the graph is the relationship between Q and q_{max}, i.e. the maximum specific uptake rate of *A. formosa*, calculated according to equations 1, 4 and 5.

From the experiments with the laboratory cul-

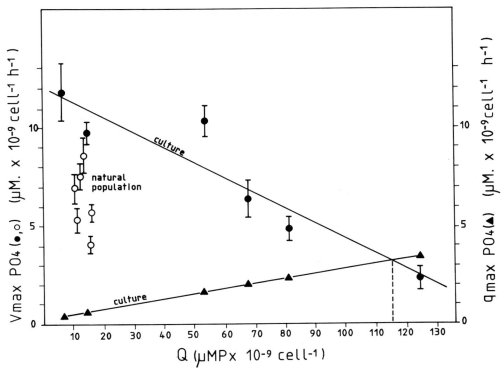

Fig. 1. Maximum specific uptake rate (q_{max} PO$_4$, ▲) and uptake capacity for phosphate (V_{max} PO$_4$, ○, ●) as function of the cell phosphorus content (Q) of the cultured *A. formosa* population (closed symbols) and of the natural phytoplankton community (open symbols). The 95% confidence intervals are given for the uptake capacity.

tures of *A. formosa*, a linear declining slope ($r^2 = 0.87$) between V_{max} and Q was observed. High V_{max} values ($> 9 \times 10^{-9} \mu$M P cell^{-1} h^{-1}), were measured for cells with internal P-contents ranging from 6 to $53 \times 10^{-9} \mu$M P cell^{-1}. A low V_{max} of $2.2 \times 10^{-9} \mu$M P cell^{-1} h^{-1} was found for cells with a Q-value of $124 \times 10^{-9} \mu$M P cell^{-1} (Q_{max} at 10 °C). For Q-values higher than $115 \times 10^{-9} \mu$M P cell^{-1}, q_{max} exceeded V_{max}; these *A. formosa* cells were not P-limited. Cells with Q-values lower than $115 \times 10^{-9} \mu$M P cell^{-1} were growing under P-limitation.

Table 3 presents the P-uptake kinetics V_{max} and $K_{s \cdot u}$ obtained from experiments with the

Table 3. Phosphate uptake data of the natural *Asterionella formosa* population. 95% confidence intervals are given in parentheses. Also given are the percentage of cells infected by *Zygorhizidium planktonicum* (% I) at the start of the experiment.

Date	Q (μM P $\times 10^{-9}$ cell^{-1})	V_{max} (μM P $\times 10^{-9}$ cell^{-1} h^{-1})	$K_{s \cdot u}$ (μM)	% I
01–02–82	23.0	–	–	–
05–02–82	19.0	–	–	–
17–02–82	12.5	7.54 (6.99–8.10)	1.48 (1.03–1.92)	15
24–02–82	–	7.12 (6.57–7.66)	0.73 (0.47–1.00)	16
03–03–82	12.6	8.52 (7.47–9.56)	1.28 (0.76–1.70)	8
10–03–82	10.6	6.90 (6.07–7.73)	0.64 (0.41–0.88)	14
17–03–82	15.4	5.71 (5.38–6.05)	0.65 (0.54–0.76)	32
24–03–82	10.9	5.40 (4.96–5.83)	0.47 (0.35–0.60)	71
31–03–82	14.8	4.07 (3.74–4.39)	0.57 (0.39–0.75)	88

natural phytoplankton community; also compiled in the table are the cell phosphorus contents and the percentage of cells infected by the fungus *Zygorhizidium planktonicum*.

Before March 10, V_{max} was high ($>7 \times 10^{-9}\,\mu M$ P cell^{-1} h^{-1}; about the same magnitude of P-uptake was observed for *A. formosa* cultured under P-limitation) and the cell phosphorus content was low, decreasing from $23 \times 10^{-9}\,\mu M$ P cell^{-1} (February 1) to $10.6 \times 10^{-9}\,\mu M$ P cell^{-1} (March 10). After March 10, the number of cells, infected by *Z. planktonicum*, increased up to a percentage of 88 (March 31) and V_{max} decreased to 4.1 (March 31). V_{max} of the natural phytoplankton community was consistently higher than q_{max} (Fig. 1) – an indication suggesting that the phytoplankton was growing under phosphate limitation from the start of the experiments. The decrease in V_{max} after March 10, probably caused by fungal parasitism, lowered the P-uptake rate of the phosphate limited cells.

Table 4 presents the growth rates of the three

dominant diatom species, i.e. *A. formosa*, *S. hantzschii* and *S. astraea* in the bioassays, performed at 5 °C and 10 °C. The growth rates at 10 °C were consistently higher than those observed at 5 °C and the rates in 'All' higher than those recorded in 'All-P' and 'LW'. According to the growth responses *A. formosa* appeared to be not significantly phosphorus limited before March 10, while both *Stephanodiscus* species showed signs of phosphorus limitation. It was in the middle of March the three diatom species revealed the largest growth rate discrepancies between 'All' and 'All-P' treatments. At the end of March the growth rate of *A. formosa* in 'All' was very low as more cells became infected by the fungus *Z. planktonicum* (Table 3).

Discussion

P-uptake capacity (V_{max})

In many studies it has been found that phytoplankton growth, limited by essential growth ele-

Table 4. The growth rates (μ) of three dominant species from the lake at 5 °C and 10 °C, as response to nutrient addition. LW = pure lake water; All = addition of all nutrients (see Table 1); All-P = addition of all nutrients except phosphorus. The 95% confidence intervals are in parentheses.

Date	Species	LW $\mu(d^{-1})$		All-P $\mu(d^{-1})$		All $\mu(d^{-1})$	
		5 °C	10 °C	5 °C	10 °C	5 °C	10 °C
17–2	*A. formosa*	0.23($\pm$0.06)	0.45($\pm$0.08)	0.22($\pm$0.05)	0.42($\pm$0.07)	0.31($\pm$0.05)	0.51($\pm$0.08)
24–2		0.23($\pm$0.05)	0.43($\pm$0.05)	0.26($\pm$0.04)	0.44($\pm$0.04)	0.32($\pm$0.07)	0.48($\pm$0.09)
3–3		0.20($\pm$0.03)	0.42($\pm$0.05)	0.21($\pm$0.02)	0.40($\pm$0.06)	0.26($\pm$0.03)	0.46($\pm$0.07)
10–3		0.08($\pm$0.04)	0.20($\pm$0.03)	0.12($\pm$0.03)	0.23($\pm$0.02)	0.24($\pm$0.06)	0.44($\pm$0.06)
17–3		–	0.12($\pm$0.04)	0.03($\pm$0.02)	0.08($\pm$0.03)	0.15($\pm$0.04)	0.35($\pm$0.08)
24–3		–		0.02($\pm$0.02)	0.04($\pm$0.02)	0.06($\pm$0.02)	0.16($\pm$0.03)
31–3		–		–		0.07($\pm$0.02)	0.08($\pm$0.03)
17–2	*S. hantzschii*	0.20($\pm$0.02)	0.36($\pm$0.07)	0.16($\pm$0.04)	0.38($\pm$0.04)	0.40($\pm$0.07)	0.60($\pm$0.09)
24–2		0.19($\pm$0.04)	0.29($\pm$0.08)	0.15($\pm$0.05)	0.26($\pm$0.05)	0.30($\pm$0.08)	0.52($\pm$0.07)
3–3		0.17($\pm$0.04)	0.28($\pm$0.04)	0.18($\pm$0.05)	0.27($\pm$0.05)	0.35($\pm$0.04)	0.57($\pm$0.05)
10–3		0.02($\pm$0.02)	0.07($\pm$0.02)	0.03($\pm$0.02)	0.08($\pm$0.02)	0.27($\pm$0.03)	0.48($\pm$0.06)
17–3		–	–	–	–	–	–
17–2	*S. astraea*	0.14($\pm$0.03)	0.30($\pm$0.04)	0.16($\pm$0.06)	0.32($\pm$0.06)	0.30($\pm$0.04)	0.52($\pm$0.07)
24–2		0.15($\pm$0.05)	0.32($\pm$0.05)	0.18($\pm$0.04)	0.31($\pm$0.05)	0.29($\pm$0.03)	0.49($\pm$0.05)
3–3		0.12($\pm$0.04)	0.28($\pm$0.06)	0.14($\pm$0.02)	0.25($\pm$0.04)	0.22($\pm$0.03)	0.44($\pm$0.06)
10–3		0.07($\pm$0.02)	0.14($\pm$0.03)	0.06($\pm$0.01)	0.09($\pm$0.02)	0.15($\pm$0.02)	0.36($\pm$0.04)
17–3		–	–	–	–	0.12($\pm$0.03)	0.33($\pm$0.02)

ments such as P, can be characterized by a high uptake capacity (V_{max}) for the limiting nutrient compared to the maximum specific uptake rate (q_{max}) (e.g., Gotham & Rhee, 1981; Zevenboom, 1980; Riegman & Mur, 1984a).

The short-term phosphate uptake has been examined in detail in steady state cultures of *Scenedesmus* sp. (Rhee, 1973; 1974), *Anabaena flos-aquae* and *Microcystis* sp. (Gotham & Rhee, 1981) and *Oscillatoria agardhii* (Riegman & Mur, 1984b). Phosphate uptake in these organisms being a function of both internal and external phosphate concentrations, can be described by an equation resembling non-competitive enzyme inhibition:

$$V = \frac{V_{max}}{\left(1 + \dfrac{K_s \cdot u}{s}\right)\left(1 + \dfrac{i}{K_i}\right)}, \qquad (6)$$

where i is the internal total inorganic poly-phosphate concentration and K_i is a constant which expresses the degree of inhibition by the internal phosphorus concentration. Rhee (1973) found that V_{max} depends on K_i. A similar negative effect of cellular polyphosphate on phosphate uptake was reported for batch culture studies of *Chlorella* (Jeanjean, 1969; Aitchison & Butt, 1973).

However, in some species this type of feedback control for P-uptake has apparently not been observed. Burmaster & Chrisholm (1979) found that the V_{max} of *Monochrysis lutheri* remains constant with increasing Q. Healey & Hendzel (1975) and Nyholm (1977) found a decrease in V_{max} only when Q approaches Q_{max}.

Our study with *A. formosa* demonstrated a negative linear relationship between V_{max} and Q:

$$V_{max} = 12.1 - 0.08Q. \qquad (7)$$

Using the *A. formosa* reference curve as a guide to define conditions where P is not limiting, the high phosphate uptake capacities of the natural phytoplankton community of Lake Maarsseveen compared to that of the reference curve indicated phosphate limitation throughout the entire study period.

Cell phosphorus content

Rhee (1974) and Harrison *et al.* (1976) found for some phytoplankton species, that under nitrogen or silicon limited conditions, the cell phosphorus contents remained high and generally constant regardless of growth rate. Tilman and Kilham (1976) demonstrated this for *A. formosa* growing under silicon limitation.

The cell phosphorus content of the *A. formosa* population in the lake was 6–12 times lower than the maximum value measured in the laboratory. The values found for natural populations were comparable with the minimum cell phosphorus content measured in the laboratory at 5 °C (9.5×10^{-9} μM P cell^{-1}) (Van Donk, 1983; Van Donk & Kilham, 1989). These low values strongly pointed to a phosphate limitation.

Bioassays

Despite the evidence showing P-limitation in the lake water, as identified by the low cell phosphorus contents and high phosphate uptake capacities over the whole experimental period, the results of the bioassay experiments, somehow, were less supportive. Although the growth of *A. formosa* was consistently higher in 'All' than in 'All-P' and 'LW', a significantly higher growth rate was observed with addition of phosphorus only after March 3. Bioassays, as indicated in our tests, have been considered less sensitive compared to short-term nutrient uptake experiments because the difference between the growth response in the enriched and the control samples was too small to be statistically significant. Furthermore, because the growth responses observed at 5 °C and 10 °C were almost the same, temperature has not been considered to be the factor responsible for the low sensitivity.

Conclusions

According to the phosphate uptake capacity and the cell phosphorus content, we may draw the

208

conclusion that *A. formosa* was growing under phosphate limitation from the beginning of the experimental period onwards (half February). However, according to the bioassays, *A. formosa* became limited by phosphate only from the beginning of March. At March 10, *A. formosa* was strongly limited by phosphate, while only 15% of the cells were infected by *Z. planktonicum*. Thus it seems likely that phosphate limitation was the primary cause of the end of the spring bloom in 1982 and the fungus infection might have only accelerated the collapse of *A. formosa*.

Acknowledgements

This investigation was financially supported by the Foundation of Fundamental Biological Research (BION), which is subsidized by the Netherlands Organization for the Advancement of Pure Research (ZWO).

We thank W. van Doesburg for his study of the short-term phosphate uptake. L. Hakkert for his help in preparation of the figures and L. Matulessya for typing the manuscript. Dr. H. de Haan, Prof. Dr. N. Daan and three anonymous referees made very useful suggestions and comments on drafts which significantly improved the quality of the paper.

References

Aitchison, P. A. & V. S. Butt, 1973. The relation between the synthesis of inorganic polyphosphate and phosphate uptake by *Chlorella vulgaris*. J. Exp. Bot. 24: 497–510.

Burmaster, D. E. & S. W. Chrisholm, 1979. A comparison of two methods for measuring phosphate uptake by *Monochrysis lutheri* Droop growing in continuous culture. J. Exp. Mar. Biol. Ecol. 39: 187–202.

De Vries, P. J. R., 1983. Bioassays with *Stigeoclonium* Kütz (Chlorophyceae) to identify nitrogen and phosphorus limitations. Aquatic Botany, 17: 95–105.

De Vries, P. J. R., 1985. Effect of phosphorus and nitrogen enrichment on the yield of some strains of *Stigeoclonium* Kütz (Chlorophyceae). Freshwat. Biol. 15: 95–103.

Droop, M. R., 1973. Some thoughts on nutrient limitation in algae. J. Phycol. 9: 264–272.

Droop, M. R., 1974. The nutrient status of algal cells in continuous culture. J. Mar. Biol. Assoc. U.K. 54: 825–855.

Frey, B. E. & L. F. Small, 1980. Effects of micronutrients and major nutrients on natural phytoplankton population. Journal of Plankton Research, 2: 1–22.

Gotham, I. J. & G. Y. Rhee, 1981. Comparative kinetic studies of phosphate-limited growth and phosphate uptake in phytoplankton in continuous culture. J. Phycol. 17: 257–265.

Guillard, R. R. L., 1973. methods for microflagellates and nannoplankton. In; Stein, J. R. (Ed.), Handbook of Phycological Methods; Culture methods and growth measurements, Cambridge New York. pp. 69–85.

Guillard, R. R. L., 1975. Culture of phytoplankton for feeding marine invertebrates. In; Smith, W. L. & M. H. Chanley (Eds.), Culture of marine invertebrate animals. Plenum, New York. pp. 29–60.

Hansen, K. R., R. Ling & E. Havir, 1967. A computer program for fitting data to the Michaelis-Menten equation. Biochem. Biophys. Res. Commun. 29: 194–197.

Harrison, P. J., H. L. Conway & R. C. Dugdale, 1976. Marine diatoms grown in chemostats under silicate and ammonium limitation. I. Cellular chemical composition and steady-state growth kinetics of *Skeletonema costatum*. Marine Biology 35: 177–186.

Healy, F. P. & L. L. Hendzel, 1975. Effects of phosphorus deficiency in two algae growing in chemostats. J. Phycol. 11: 303–309.

Jeanjean, R., 1969. Influence de carence en phosphore sur les vitesses d'absorption du phosphate par les Chorelles. Bulletin de la Societe Française de Physiologie Vegetale: 159–171.

Menzel, D. W. & N. Corwin, 1965. The measurement of total phosphorus in seawater based on the liberation of organically bound fraction by persulfate oxidation. Limnol. Oceanogr. 10: 280–283.

Munawar, M., P. T. S. Wong & G-Y. Rhee, 1988. The effects of contaminants on algae: An overview. In; N. W. Schmidtke (ed.), Toxic Contamination in Large Lakes. Lewis Publishers Inc. Chelsa, Michigan. 113–160 pp.

Murphy, J. & J. P. Riley, 1962. A modified single solution method for the determination of phosphate in natural waters. Anal. Chim. Acta 26: 31–36.

Nyholm, N., 1977. Kinetics of phosphate limited algal growth. Biotechnol. Bioengineering 19: 467–492.

Paasche, E., 1978. Growth experiments with marine plankton algae: the role of 'waterquality' in species succession. Mitt. Int. Ver. Limnol. 21: 521–527.

Reynolds, C. S. & C. Butterwick, 1979. Algal bioassays of unfertilized and artificially fertilized lake water, maintained in Lund Tubes. Archiv. für Hydrobiologie Supplement, 56: 166–183.

Rhee, G. Y., 1973. A continuous culture study of phosphate uptake, growth rate and polyphosphate in *Scenedesmus* sp.. J. Phycol. 9: 495–506.

Rhee, G. Y., 1974. Phosphate uptake under nitrate limitation by *Scenedesmus* sp. and its ecological implications. J. Phycol. 10: 470–475.

Rhee, G. Y., 1978. Effect of N: P atomic ratios and nitrate limitation on algal growth, cell composition and nutrient uptake. Limnol. Oceanogr. 23: 10–25.

Riegman, R. & L. R. Mur, 1984a. Phosphate uptake by P-limited *Oscillatoria agardhii*. FEMS. Microbiol. Letters. 21: 335–339.

Riegman, R. & L. R. Mur, 1984b. Regulation of phosphate uptake kinetics in *Oscillatoria agardhii*. Arch. Microbiol. 139: 28–32.

Schelske, C. L., E. D. Rotham, E. F. Stoermer & M. A. Santiago, 1974. Responses to phosphorus limited Lake Michigan phytoplankton to factorial enrichments with nitrogen and phosphorus. Limnol. Oceanogr. 19: 409–419.

Tilman, D. & S. S. Kilham, 1976. Phosphate and silicate growth and uptake kinetics of the diatoms *Asterionella formosa* and *Cyclotella meneghiniana* in batch and semi-continuous culture. J. Phycol. 12: 375–383.

Van der Does, J. & S. Klapwijk, 1987. Effects of phosphorus removal on the maximal growth in bioassay experiments with water from four Dutch lakes. Int. Revue ges. Hydrobiol. 72: 27–39.

Van Donk, E., 1983. Factors influencing phytoplankton growth and succession in Lake Maarsseveen (I). Ph. D. Thesis. Univ. Amsterdam. 148 pp.

Van Donk, E., 1987. The water quality of the two Maarsseveen Lakes in relation to their hydrodynamics. Hydrobiol. Bull., 21: 17–24.

Van Donk, E. & J. Ringelberg, 1983. The effects of fungal parasitism on the succession of diatoms in Lake Maarsseveen (The Netherlands). Freshwat. Biol. 13: 241–251.

Van Donk, E. & S. S. Kilham, 1989. Temperature effects on silicon- and phosphorus-limited growth and competitive interactions among three diatoms. J. Phycol. (in press.).

Van Donk, E., A. Veen & J. Ringelberg, 1988. Natural community bioassays to determine the abiotic factors that control phytoplankton growth and succession. Freshwat. Biol. 20: 199–210.

Zevenboom, W., 1980. Growth and nutrient uptake kinetics of *Oscillatoria agardhii*, Ph.D. thesis, Univ. Amsterdam. 178 p.

Hydrobiologia **188/189**: 211–228, 1989.
M. Munawar, G. Dixon, C. I. Mayfield, T. Reynoldson and M. H. Sadar (eds)
Environmental Bioassay Techniques and their Application.
© 1989 *Kluwer Academic Publishers. Printed in Belgium.*

Evidence from algal bioassays of seasonal nutrient limitations in two English lakes

Francisco A. R. Barbosa
Freshwater Biological Association, The Ferry House, Far Sawrey, Ambleside, Cumbria LA22 OLP,
England; present address: Federal University of Minas Gerais, P.O. Box 2486, 30.161 Belo
Horizonte-MG, Brazil

Key words: algal bioassays, nutrient limitations, English lakes, *Asterionella formosa, Rhodomonas lacustris*

Abstract

Comparative laboratory bioassays using *Asterionella formosa* and *Rhodomonas lacustris* as test organisms were performed from March to November 1987 on filtered water samples from two English lakes, to assess their potential fertility and to identify possible limiting nutrients. The relative growth responses ($\log_2$ increments) per week, were measured after additions of P, Fe, Si, N, and K singly and in combinations in comparison with unenriched (control) samples. Phosphate appeared to be the major limiting element for both species throughout the year, except during the spring diatom maxima when silicon usually becomes limiting. On most occasions chelated iron increased the growth increments, particularly in combination with phosphate. In general, the bioassay results showed correspondence with the nutrient concentrations in the test waters, which showed low ($< 1 \mu g\, l^{-1}$) levels of soluble reactive phosphate during all or most of the year and depleted silicon levels in late spring. Comparison between relative (incremental) ratio and absolute (cell concentration) response was made.

Introduction

Chemical analysis has been largely used to evaluate the quality of a water-body and to assess its trophic status. However, as pointed out by Marvan (1979), algal assays constitute the most direct means of such assessment giving quantitative information about the growth potential of the water and allowing predictions of its potential fertility.

Seasonal depletion of nutrients – particularly C, Si, N and P – induced by algal growth are well established in some lakes of the English Lake District and some have been suggested to limit natural population increase.

Examples are given by Lund (1950), Mackereth (1953), Lund *et al.* (1963), Talling (1976, 1985).

Sutcliffe *et al.* (1982), Reynolds (1986). Experimental work, including systematic bioassays (Lund, Jaworski & Butterwick, 1975; Reynolds & Butterwick, 1979; Box, 1983), has produced evidence that these elements plus complexed iron may be seasonally limiting to algal growth.

In the present study, comparative laboratory bioassays were tested as a method of evaluating the increasing potential fertility of the waters of two contrasting English lakes – the deep, base-poor Windermere South Basin and the shallow, base-rich Malham Tarn. The assays used the most common diatom in Windermere, *Asterionella formosa* and the dominant phytoplanker in Malham Tarn, *Rhodomonas lacustris*, and were carried out in a seasonal sequence that involved known nutrient depletions, with the aim of identifying one or more limiting nutrient(s) for these algae.

Comparison was made with growth responses of natural populations of both species and with indications of potential limitation from seasonal nutrient depletion. Tests of the use of absolute (cell number) and relative (incremental) measures of population increase, and of removal of the added nutrients during exposures with and without cells, are used to assess the physiological interpretation and ecological applicability of the bioassay results.

Description of the Lakes

Windermere South Basin (area 6.6 km^2) is a terminal basin in the associated drainage system, thus receiving drainage that includes sewage inputs and upland run-off from a largely mountainous and base-poor catchment area (Talling & Heaney, 1988). The lake is monomictic, with the stratification period extending from mid-April until late October and temperature differences between top and bottom as high as 11 °C (Fig. 1). A strong silicon depletion develops during and after the spring diatom maximum, with relatively high values during the winter. Soluble reactive phosphate is usually present in low concentrations ($<10\ \mu g\,l^{-1}$) although during the winter higher values have been recorded. Nitrate-nitrogen is usually present in fairly high concentrations (ca. $500\ \mu g\,l^{-1}$) except in late summer and early autumn. The concentration of ammonium-nitrogen is usually low ($<10\ \mu g\,l^{-1}$) although higher values ($30-50\ \mu g\,l^{-1}$) have been recorded during summer and especially after the autumnal overturn.

Seasonal succession of the phytoplankton assemblage includes a spring (May) diatom maximum; colonial chrysophytes and green algae, which form a late spring to early-summer pulse; diatoms, dinoflagellates and desmids, which are conspicuous during the summer; and, during the autumn, mainly diatoms, dinoflagellates, chryptomonads and blue-green algae (Reynolds, 1984a, b).

In contrast, Malham Tarn is a shallow, largely unstratified and calcareous lake in the Pennine uplands (Lund, 1961). The concentration of silicon is depleted each spring, increases from mid-summer and reaches its maximum in winter (Fig. 2). Nitrate-nitrogen levels are high during the winter but show a strong depletion during summer. Ammonium-nitrogen is typically below $40\ \mu g\,l^{-1}$ except in autumn when there is a net release, probably from decaying vegetation. Soluble reactive phosphate is characteristically low ($<1\ \mu g\,l^{-1}$) throughout the year.

The phytoplankton community in the Tarn is never very abundant, partly as a consequence of the Tarn morphometry, which promotes the growth of benthic macrophytes, especially *Chara* spp. and benthic algae (Lund, 1961). Nevertheless, moderate densities of phytoplankton were occasionally recorded, including large diatoms such as *Asterionella formosa* or *Diatoma elongatum*, and a predominance of *Rhodomonas lacustris* v. *nanoplanctica*, *Cryptomonas* spp. and in summer, *Gloetrichia echinulata* (Talling, pers. comm.).

Methods

Details of the bioassay technique and sample pretreatments are described in Lund, Jaworski & Bucka (1971) and Lund, Jaworski & Butterwick (1975).

For the work on Windermere integrated lake water samples (0–7 m) were collected using a 7 m weighted polyethylene tube; for those on Malham Tarn samples of surface water were taken near the lake outflow.

The samples were in part preserved with Lugol's iodine solution for later algal enumeration and in part brought back to the laboratory for chemical analysis. Filtration was by suction through Whatman glass-fibre GF/C filters. Analyses for soluble reactive phosphate, total phosphorus, nitrate- and ammonium-nitrogen, and silicon were carried out on the day of collection, by spectrophotometric determinations, whose principles are described in Mackereth, Heron &

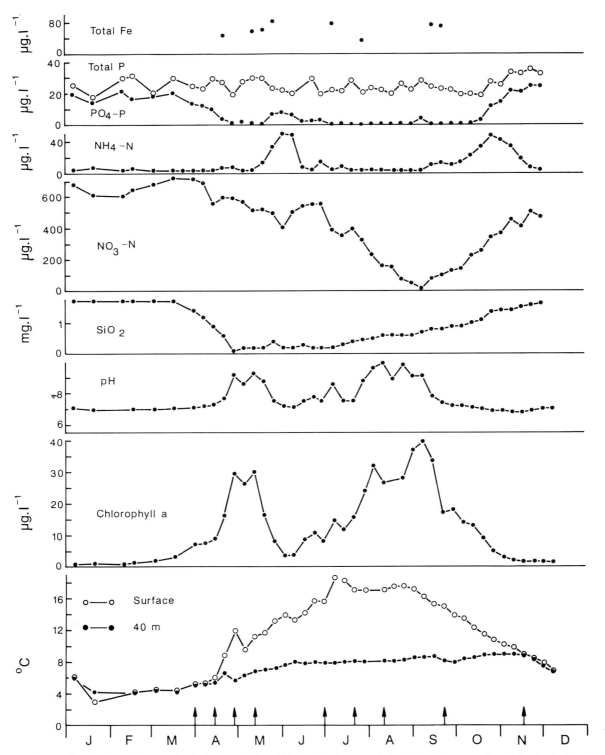

Fig. 1. Seasonal variations of some physical, chemical and biological characteristics in Windermere, South Basin during 1987 (data by courtesy of the Freshwater Biological Association). The arrows indicate the sampling dates and the bioassay experiments.

214

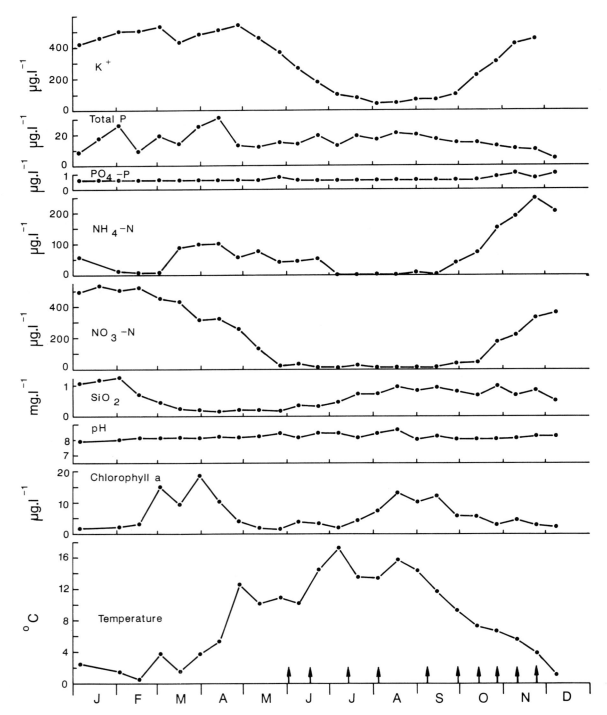

Fig. 2. Seasonal variations of some physical, chemical and biological characteristics in Malham Tarn, outflow during 1987 (data by courtesy of the Freshwater Biological Association). The arrows indicate the sampling dates and the bioassay experiments.

Talling (1978). The limits of detection were $3 \mu g \, l^{-1}$ for total phosphorus and soluble reactive phosphate; $11 \mu g \, l^{-1}$ for nitrate-nitrogen; $4 \mu g \, l^{-1}$ for ammonium-nitrogen and $9 \mu g \, l^{-1}$ for silicon. Total iron concentrations were determined by the wet-oxidations procedure and measured spectrophotometrically in 5.93 M hydrochloric acid (Davison & Rigg, 1976), and potassium was determined by atomic absorption spectroscopy.

Algal enumeration was performed using sedimentation in Lugol's iodine and the inverted microscope technique (Lund, Kipling & Le Cren, 1958) for *Asterionella formosa* and the Lund chamber technique (Lund, 1959) for *Rhodomonas lacustris*.

For bioassays, water samples were collected in 10 l polyethylene containers and filtered with weak suction through pre-acidified and rinsed Whatman GF/C filters into silica flasks, where the chemical additions were made. The experiments were carried out with treatments tested in duplicate in 150 ml silica flasks covered by silica caps.

For the *A. formosa* assays, enrichments of the filtered lake water were made with sodium hydrogen phosphate (50 μg PO$_4$-P l^{-1}), the ferric complex of the disodium salt of ethylenediaminetetracetic acid – hereafter referred to as FeEDTA made by adding 1 mg FeCl$_3$ 6 H$_2$O to 2 mg disodium EDTA (100 μg Fe l^{-1}), sodium silicate (1.5 mg Si l^{-1}), ammonium chloride (200 μg N H$_4$-N l^{-1}), alone and in combinations. Following the sodium silicate additions, 1 N HCl was added to produce a final pH as close as possible to that of the filtered lake water.

For the *R. lacustris* experiments the chemicals and concentrations were the same, with the exception of potassium chloride (400 μg K l^{-1}) in substitution for sodium silicate.

The cells of *A. formosa* were those of the non-axenic clone L313 of the Freshwater Biological Association's algal culture collection, and those of *R. lacustris* were from a non-axenic clone L164 isolated by Mr G. Jaworski.

The stock cultures were maintained in a diatom culture medium (DM) developed by Beaks, Canter & Jaworski (in press, 1988).

Exponentially growing populations were used for the inocula, after being washed 3 times with filtered lake water, by centrifugation. Sizes of inocula ranged between 12 and 194 cells ml^{-1} for *Asterionella formosa* and between 304 and 1850 cells ml^{-1} for the smaller *Rhodomonas lacustris*.

The chemical additions and controls were done in duplicate and the experimental flasks were placed in a constant temperature cabinet (19 °C $\pm$ 1 °C) and continuous illumination from below (147 μmol m^{-2} s^{-1} PAR) from March to August 1987 and (223 μmol m^{-2} s^{-1} PAR) from September to November 1987, by four 40 and 58 watt daylight fluorescent lamps, respectively. These irradiances were measured at the bottom of the flasks by a spherical quantum sensor (Biospherical Instr., San Diego).

All flasks were shaken daily by hand. The incubation time was 7 days, after which cell density was estimated and the growth increment expressed as doublings (log$_2$ units) per week. The statistical significance of the growth rates in the controls, and after the additions, was obtained by a F test of the geometric means of the cell numbers at the beginning and the end of each experiment.

The utilization of the natural populations (indigenous cells) of the two species followed the same procedure but using admixtures of unfiltered lake water with its filtrate where necessary to reduce the initial population density.

Results

Changes of algal population and medium during bioassays.

Algal numbers

Although the experiments were designed to obtain relative growth measurements (equivalent to cell-divisions) after an incubation time of one week, measures of the absolute responses in terms of cell yield were made for both species, to test if the results were comparable. Figure 3 shows growth-time curves for *Asterionella formosa* with additions

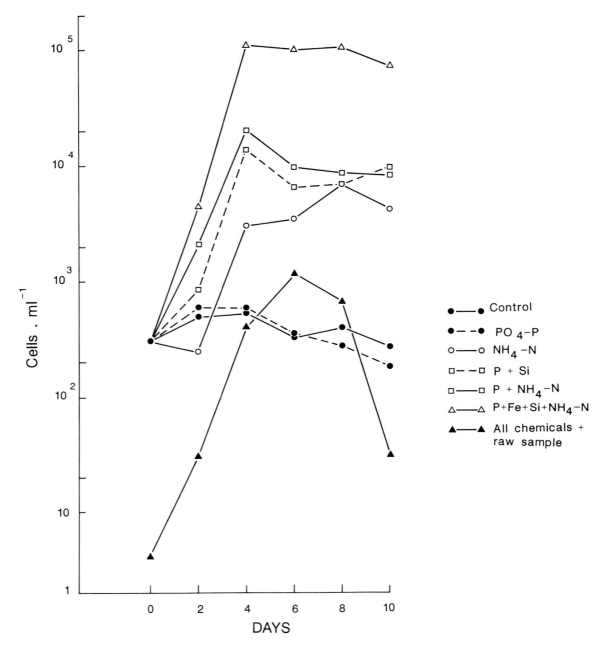

Fig. 3. Growth of *Asterionella formosa* (cells ml⁻¹) during 10 days of incubation under controlled conditions. The growth of the indigenous cells is also shown.

of phosphate, nitrogen, phosphate plus nitrogen and the combination of all these nutrients to the cultured and the indigenous cells, in relation to the control. With the exception of the nitrogen addition to the cultured cells, the maximum yield

was reached after 4 days of incubation. Indigenous cells reached maximum yield after 6 days of incubation, after which cell density declined rapidly.

Similar tests with *Rhodomonas lacustris* (cultured and indigenous cells) are shown in Fig. 4. The maximum yield in the control was reached after 2 days of incubation, and followed by a strong decline. Additions of phosphate alone or in combination with nitrogen, chelated iron and potassium resulted in significant increases of the growth rate and the yield was maximal after 6 days of incubation, declining afterwards. The indigenous cells reached the maximum yield after 4 days of incubation, also declining later.

The experiments illustrate the differences possible between maximal and 7 days yields as cell concentrations, which varied at ratios of 1:1 to 3:1 in the bioassays with *Asterionella formosa* and 2:1 to 40:1 in the *Rhodomonas lacustris* ones, and that will be influenced by inocula density. A comparison between these two types of measurement of the growth increments is hindered by the much bigger variability recorded for the absolute cell

yields which usually showed their maximal yields after 4 days of incubation, for both species. However, better comparisons would be possible with shorter incubations and possibly a 5-day incubation period would be the best for the usual inocula utilized in the present study.

Nutrient removal

Concentration of nutrients were measured after the incubations and compared with the initial concentrations plus additions. Phosphate analysis of media with and without cells showed that most (90%) of the phosphate can be adsorbed to the glassware, although not necessarily unavailable for cell consumption. Complexation with various ions (e.g. Fe^{3+}) is another important source of rendering phosphorus unavailable for

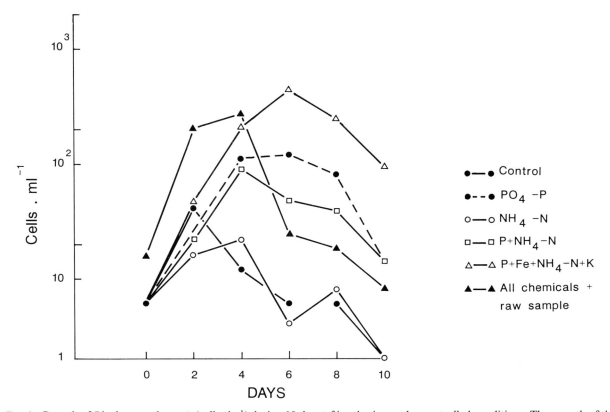

Fig. 4. Growth of *Rhodomonas lacustris* (cell ml^{-1}) during 10 days of incubation under controlled conditions. The growth of the indigenous cells is also shown.

the analytical reaction after filtration as is its utilization by bacteria.

Figure 5 shows the percentage removal of phosphate, silicon, nitrogen and chelated iron in the bioassays with *Asterionella formosa*. With few exceptions, more than 90% of the added phosphate was removed in the experiments, including the controls. The consumption of silicon varied widely (0–100%); especially when added in combination with phosphate, chelated iron and nitrogen, its removal was usually above 60%.

In the controls and most of the single additions, less than 50% of the added nitrogen was removed; however, added in combination with phosphate, chelated iron and silica, its consumption varied between 50% and 97%, apart from three occasions. In the unfiltered test water its removal was usually over 90%, with two exceptions.

In only one experiment was chelated iron slightly removed in the controls (7%). The addition of phosphate increased its consumption up to 30% in two experiments and after the addition of silicon, 75% was removed in one experiment. Higher removal by the indigenous cells were recorded in most experiments (>50%) with the unfiltered test water.

A correlation between the loss of the added nutrients and the maximal cell density showed much scatter, although it could be noted that even small crops can remove high amounts of silicon and nitrogen and high crops were obtained with low removal of nitrogen but not silicon. Furthermore, synergistic effects among the added nutrients were present and possibly explain the low correlations recorded.

The percent removal of phosphate, chelated iron, nitrogen and potassium in the bioassays with *Rhodomonas lacustris* is shown in Fig. 6. As for *A. formosa*, more than 90% of the added phosphate was removed in most experiments, including the controls, Chelated iron was lost in quantities varying from 10 to 90%. Nitrogen consumption varied between 10 and 98%, and particularly when added in combination with phosphate, chelated iron and potassium its removal was over 75% in all experiments, a pattern also

recorded with the indigenous cells. Potassium removal was usually low although it would reach 75% or more of the low summer concentrations.

Bioassay responses, Windermere South Basin

Growth of *Asterionella formosa* (cultivated and indigenous cells) in unenriched and enriched water samples is shown in Fig. 7. Only increments greater than 2 divisions per week (dotted line) were considered significant allowing for the existence of internal content of nutrients of the cells although other authors (e.g. Lund *et al.*, 1975) had suggested higher values (ca. 5 div. week^{-1}) mainly considering the phosphorus content of the cells (see Mackereth, 1953). However, this will not be applicable to other nutrients (e.g. silicon).

Growth increase of *A. formosa* in the control samples was generally high in mid-spring (3.1 to 8.2 div. week^{-1}), low during summer (2.1 to 3.7 div. week^{-1}) and poor (1.8 div. week^{-1}) during autumn. Phosphate additions were inhibitory in spring but produced significant increases ($P < 0.05$) over controls in summer and early autumn, having no effect in mid-November. Chelated iron additions produced significant increase in early spring, summer and late autumn but were inhibitory on all other occasions except late September.

Silicon produced significant increases in late spring and mid-summer, inhibited in early spring and early autumn and had no effect on the other occasions. Ammonium-nitrogen, hereafter referred to as nitrogen, produced significant increases in early and late spring and late autumn, was inhibitory in mid-spring and mid-summer and had no effect in early summer and early autumn.

Phosphorus plus chelated iron produced significant increases in all experiments except in mid-spring when it was inhibitory. Similar effects were recorded for the additions of phosphorus plus silicon and phosphorus plus nitrogen, with the exception of early summer when the last combination had no effect. Each of these three elements

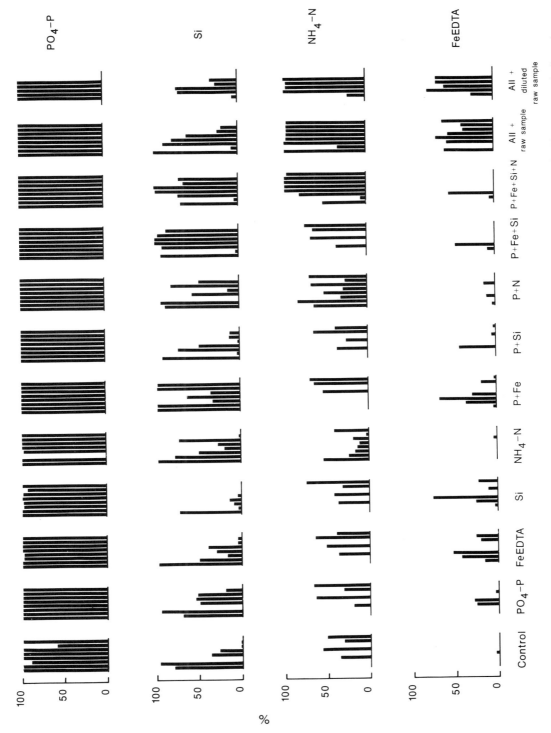

Fig. 5. Percentage removal of added P, Si, N, and Fe in the series of experiments with *Asterionella formosa* under the 12 treatments indicated.

220

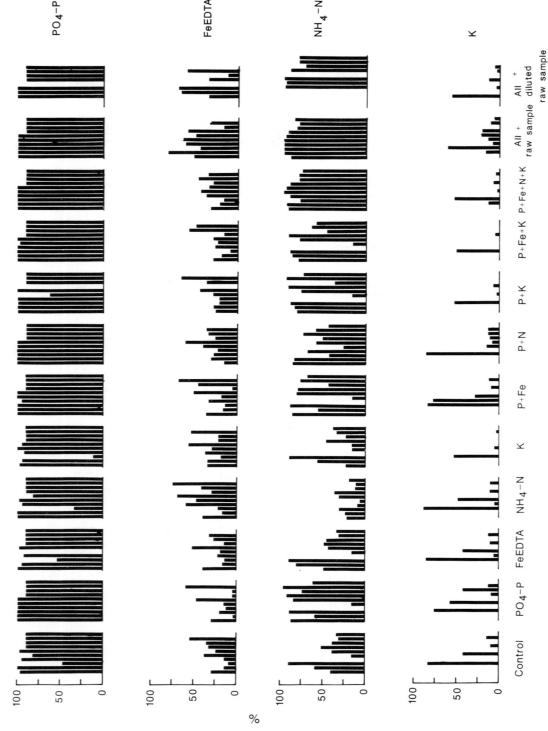

Fig. 6. Percentage removal of added P, Fe, N, and K in the series of bioassay experiments with *Rhodomonas lacustris* under the 12 treatments indicated.

221

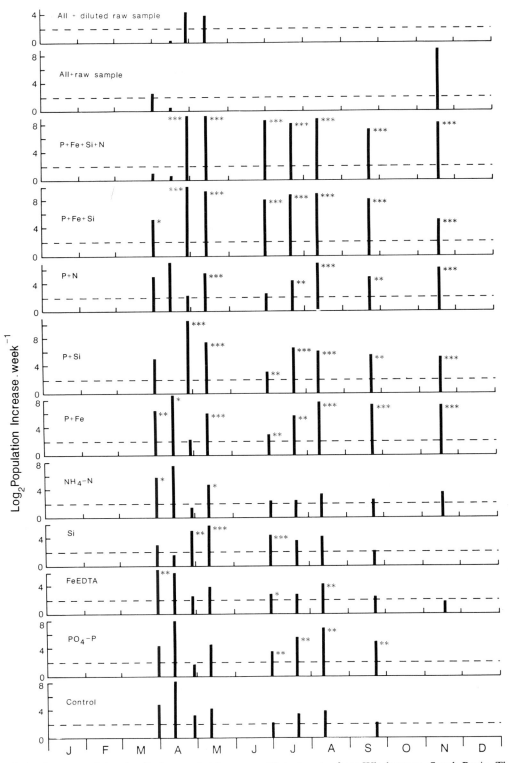

Fig. 7. Population increase of *Asterionella formosa* in bioassays with test water from Windermere, South Basin. The growth of the indigenous cells in raw (unfiltered) samples is also shown. Significance differences from controls: * P < 0.10; ** P < 0.05; *** P < 0.01.

combined with nitrogen, produced significant increases in all experiments, except in early and mid-spring, when inhibition resulted.

The combined addition of all these nutrients to the indigenous cells had an inhibitory effect in all experiments except one in late autumn when 9.1 div. week^{-1} was recorded. Similar effects were obtained after the inoculum density was reduced by dilution of the unfiltered test water, except in mid-spring when a positive response was recorded (4.5 div. week^{-1}).

Figure 8 summarizes relative effects of the nutrient additions on the growth of *Asterionella formosa*. Growth was increased by phosphorus additions in five experiments, by chelated iron in four, by silicon in three and by nitrogen in three. Phosphate plus chelated iron increased growth in seven experiments; phosphate plus silicon in six; phosphate plus nitrogen in five; phosphate plus chelated iron plus silicon in seven, and phosphate plus chelated iron plus silicon plus nitrogen in six experiments.

Predictions of nitrogen and/or phosphorus limitations in lake waters can be done by estimates of the relative availability of soluble inorganic nitrogen and total soluble phosphorus through the N:P ratio; if this ratio is greater than 13:1 the test water is susceptible to phosphorus limitation; if it is between 9:1 and 12:1, nitrogen and phosphorus are likely to limit and if this ratio is less than 8:1, the test water is susceptible to nitrogen limitation (Weiss, 1976). Estimates of the N:P concentration ratios as total dissolved inorganic nitrogen and soluble reactive phosphate in filtered water from Windermere South Basin, at the times of the bioassays suggest that this lake was phosphorus limited (N:P ratios ranged between 20.7 and 64).

The bioassay results agreed with this prediction and other limiting factors were suggested. The first experiment (early spring) suggested chelated iron firstly to limit growth; once providing this limitation, phosphorus became limiting and added in combination with chelated iron produced the best response (6.5 div. week^{-1}). At this time also silicon became limiting although added alone did not produce any effect.

The second bioassay (mid-spring) did not suggest any limitation, judging by the high growth recorded in the control. Some inhibition was recorded after the additions of silicon and nitrogen alone or in combination with phosphorus and chelated iron. The high levels of these nutrients in the test water support this hypothesis. During the spring diatom maxima silicon was suggested as the primary limiting nutrient (third and fourth bioassays) which agreed with its low levels in the test water; after its addition phosphorus became limiting. During the summer and early autumn, phosphorus was suggested as the major limiting nutrient and after its addition, chelated iron became limiting.

The last bioassay (late autumn) suggested a nitrogen limitation, judging by the positive growth recorded after its addition and the lack of response after the additions of phosphate and silicon separately. Chelated iron addition also showed a positive increase of the growth rate. However, considering that this experiment was conducted just after the autumnal overturn, when all nutrients except ammonium-nitrogen were in high concentrations in the test water, it is unlikely a nitrogen limitation. Further interpretation is difficult as no measurable growth rate was recorded in the control at this time.

Bioassay responses, Malham Tarn

Growth of *Rhodomonas lacustris* (cultured and indigenous cells) in unenriched and enriched water samples, from summer and autumn only, is shown in Fig. 9. As for the assays with *Asterionella formosa*, only increments greater than two cell divisions were considered 'positive' responses.

In none of the control experiments was positive response recorded possibly due to the considerable size of the inocula and the general state of the cells. However, in one experiment (mid-autumn) 0.9 div. week^{-1} was recorded, demonstrating that the cells were alive.

Phosphate additions produced significant increases in early summer and mid-autumn but not

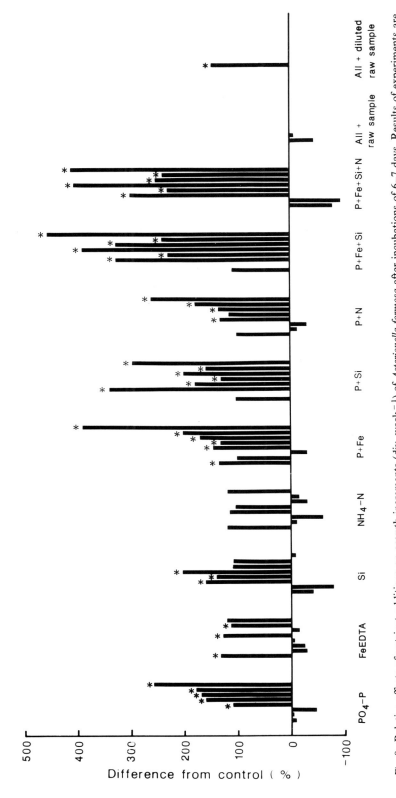

Fig. 8. Relative effects of nutrient additions on growth increments (div. week^{-1}) of *Asterionella formosa* after incubations of 6–7 days. Results of experiments are ordered sequentially from left to right in each set of histograms. * Results significantly different from controls (P < 0.05).
The units of the ordinate are just the percent difference in relation to the base line (0) taked as from the control. The asterisks show the significance level for these differences. The F-test was the only possible test which compared the two replicates for each test in relation to the controls, through the geometric mean.

224

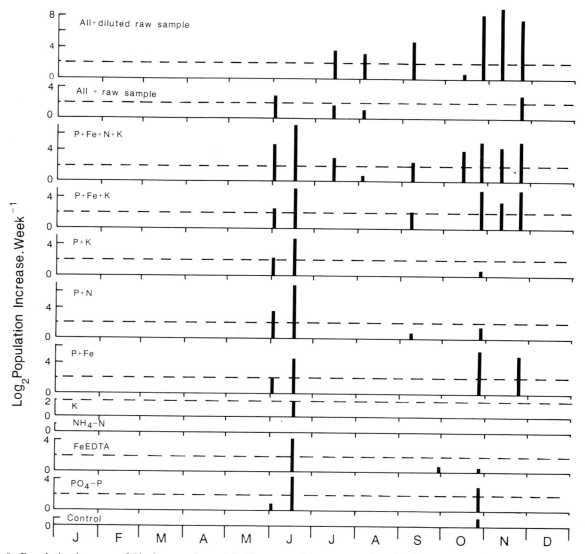

Fig. 9. Population increase of *Rhodomonas lacustris* in bioassays with test water from Malham Tarn, outflow. The growth of the indigenous cells in raw (unfiltered) samples is also shown.

at other times. The response pattern for the additions of chelated iron was similar, except for an inhibitory effect in mid-autumn. Nitrogen additions had no effect on the growth, apart from an inhibitory effect recorded in mid-autumn. Potassium additions produced significant increase only in early summer, with inhibition in mid-autumn, and no response at other times.

Additions of phosphorus plus nitrogen produced significant increases in early summer, early and

mid-autumn, and no response at other times. The pattern was similar to that recorded for the additions of phosphorus plus potassium which, however, inhibited in mid-autumn.

In combination, phosphorus, chelated iron and potassium produced significant increases in early summer and autumn, but not at other times. However, the addition of nitrogen to this combination resulted in positive effects in all experiments, except in late summer and early autumn, when no effect was recorded. Indigenous cells at the original (undiluted) concentrations had their growth increased by all these nutrients in combination during summer and late autumn; after dilutions (10–20 times) positive response was recorded in all experiments, except two in early summer and mid-autumn.

The estimated soluble inorganic nitrogen to soluble reactive phosphate-phosphorus concentration ratios in this test water also suggested phosphorus limitation (N : P ratio ranged between 33.3 and 786) and the bioassay results were consistent, although measurable growth increments in the controls were not recorded. Low concentrations of phosphate prevailed in the Tarn throughout the year. Nitrogen might also become limiting during the summer, judging by the low concentrations of nitrate- and ammonium-nitrogen at this time. However there is no supporting evidence from the bioassays.

Discussion

Seasonal incidence of nutrient limitations

The present bioassays on Windermere, although not covering all the year, showed high yields in the unenriched samples during late winter, and a reduction during the summer. An enrichment with phosphate was to be demonstrated necessary to increase the growth rate at this time. In all the bioassays with test waters from both lakes, the addition of chelated iron had a positive effect, particularly in combination with phosphate. On two occasions silicon was the major limiting nutrient followed by phosphate and chelated iron, in the bioassays with *Asterionella formosa*. From a series of similar bioassays conducted during 1971–74 on Blelham Tarn, a small lake of the Windermere catchment, Lund *et al.* (1975) concluded that phosphorus, silicate and chelated iron were important limiting nutrients for *A. formosa*, which showed maximum yields in winter samples and low ones during summer. Reynolds & Butterwick (1979), in comparative bioassays between unenriched and enriched waters from

large enclosures in the same lake, demonstrated phosphorus deficiencies in the unfertilized water throughout and occasionally chelated iron and silica deficiencies. They concluded that phosphorus occasionally limited growth in the fertilized water but that it was not necessarily a controlling factor.

Responses to additions of chelated iron are known from earlier bioassay work. In some English lakes Lund *et al.* (1975) found that FeEDTA was needed for the growth of *A. formosa* once phosphorus and silicon had been added during the summer, and Box (1983) showed that *Microcystis aeruginosa* had its yields enhanced throughout the year by additions of EDTA or FeEDTA. Elsewhere, Clasen & Bernhardt (1974) demonstrated that for *Oscillatoria rubescen* the growth-stimulating effect of iron, given sufficient phosphate, was particularly significant; they suggested that the iron requirement for this species was high as compared to that for phosphorus. Wurtsbaugh & Horne (1983) concluded that iron limited algal growth and particularly nitrogen fixation in Clear Lake, California, and that daily additions of iron maintained the algal population of their experiments at high levels for several days. All these results agree with the present findings where chelated iron was demonstrated to be a controlling factor, particularly in combination with phosphate. However, Storch & Durham (1986) suggested that it would be erroneous to assume that iron uptake by algae is always enhanced in the presence of EDTA, as their study for Lake Erie with *Anabaena flos-aquae* indicated that this species can incorporate iron from an unchelated system, and that EDTA can reduce the rate of iron uptake. Further comparisons cannot be made with the present study as unchelated iron was not tested. FeEDTA alone increased the growth of *Asterionella formosa* in three out of nine occasions, and that of *Rhodomonas lacustris* in two out of ten occasions.

Lund *et al.* (1975) correlated the levels of phosphorus and chelated iron in Blelham Tarn with dry or wet weather conditions during their study period, and generalized from their bioassay experiments as follows, if the weather was wet phosphorus was the major limiting factor, and if it was dry complexed iron was the primary limiting factor. Similar correlations can be observed from the present results for both lakes, although for Malham Tarn such comparisons can only be made for July (driest month in 1987) when chelated iron additions resulted in a positive increase of growth. As observed by Lund *et al.* (1975) for Blelham Tarn, silicon was the major limiting factor at the time of the diatom maximum, followed by phosphorus and chelated iron.

Potassium is an important element in the general physiology of algae, and there is some recent evidence from cultures of *Rhodomonas lacustris* that it can be a limiting factor for growth in the range of 25–50 μg l^{-1} (Jaworski, pers. comm.). In the present experiments, only one test conducted in mid-June with this element added separately (400 μg l^{-1}) produced a positive increase in the growth of *R. lacustris*. The increase was greater when potassium was added in combination with phosphate. It is doubtful that there was a potassium limitation at that time since such positive response was absent in later summer bioassay involving lower concentrations of potassium ($< 100 \mu$g l^{-1}).

Judging from the seasonal distribution of nutrients recorded for Windermere South Basin and neglecting recycling, silicon was expected to limit growth from mid-spring until mid-summer; nitrogen in early autumn and soluble reactive phosphate from mid-spring until the autumnal overturn. In Malham Tarn, silicon was expected to limit in early and mid-spring, nitrogen during the summer and phosphorus throughout the year. Except for the nitrogen additions in the test water from Malham Tarn, the bioassay results were broadly consistent; additions of phosphate, silicon and chelated iron resulted in significant increases of the algal growth.

The present results for Windermere South Basin conformed with the seasonal trend suggested for Blelham Tarn by Lund *et al.* (1975) in which the size of the vernal maximum of *Asterionella formosa* is controlled by the amount of silicon in the lake in February when the population begins to increase. Phosphate-phosphorus de-creases rapidly in the beginning of the vernal increase, and the bioassays suggested that phosphorus is probably the major limiting factor from the vernal maximum until the autumnal overturn, followed by silicon and chelated iron. The latter is likely to be a major limiting nutrient in summer. The combined addition of all three nutrients produced significant increases in all the experiments.

After the autumnal overturn and the consequent replenishment of nutrients, phosphorus additions made singly did not produce positive response although in combination with chelated iron, silicon and nitrogen they resulted in significant increases. In the control, however, no growth was recorded.

Bioassay interpretation and limitations

A basic function of growth bioassays is to determine how the algae in a given test water would respond to increases in certain nutrients, thereby indicating which nutrients might be considered most likely to limit growth (Fitzgerald, 1972). A bioassay result can be founded on the response of the specific growth rate (μ) or of the maximal population density or yield achieved.

The present bioassays do not represent solely one or the other. From the growth curves shown in Figs. 3 and 4 one can use the maximal population recorded as an index, or alternatively calculate the specific growth rate from the slope of each curve. Actually, as most curves also include the stationary phase, what had been measured was probably determined more by the maximal population density (yield) than the specific growth rates (μ). The maximal population densities recorded are likely to be affected by non-algal depletion of nutrients *in vitro*, loss of the original recycling rates and growth-inhibition throughout the experimental time. All these influences could potentially introduce artifacts.

The present results were obtained by the use of relative growth measurements based on the ratio of cell numbers at the beginning and end of the incubation period, unlike the more commonly used absolute values in which growth is expressed as a concentration yield (usually given in terms of dry or wet weight, optical density or packed cell volume) after a period of time. This last approach has the major limitation as an indicator of growth in that the course of growth within the period for which yields is estimated remains unknown, and during this period the growth rate may decline and net growth absent or even 'negative'.

The present bioassay results were potentially affected by the development during incubations of some artifacts of wide application. The adsorptive removal of some nutrients, particularly phosphorus, and the modification of a recycling supply could constitute limitations to algal growth and thus to the interpretation of the results. Further, the primary action of the additions of nutrients on the growth increments can be confounded with other experimental limitations such as competition for the nutrients by the presence of filter-passing contaminants.

The so-called 'inhibition' of growth by some added nutrient may not represent an ecologically relevant suppression of growth but rather a result of an interaction with other external factors such as light conditions and temperature changes.

In doing the bioassays, the major technical difficulties found were in controlling the size of the inocula and in the preparation procedures (filtration and centrifugation), which might have affected the viability of the cells, particularly of *Rhodomonas lacustris* due to its small size and fragility. For this species, temperature differences between the stocks of cultured cells and the test water during late autumn, might have affected growth adversely, although temperature has been suggested to exert little direct control over the densities of natural populations (Lund, 1962; Taylor & Wetzel, 1984).

Particular limitations of data from Malham Tarn were mainly the common occurrence of 'negative growth', mainly in the control experiments of which none showed net growth rates equal to or greater than 2 divisions per week, rendering the interpretation of positive effects of nutrient additions less reliable.

Acknowledgements

I am grateful to Mr George Jaworski and Ms Christine Butterwick for help with the bioassay technique, to Mr Eric Rigg for the iron, ammonium- and nitrate- nitrogen analyses, to Mr T. R. Carrick for the potassium determinations, to Mrs M. A. Hurley for statistical advice, to Mrs J. Waterhouse for typing the manuscript, and to Dr J. F. Talling FRS for scientific suggestions and comments and corrections on the manuscript. The work was supported by the Brazilian Research Council-CNPq. (Proc. 20.1065/85-ZO) and the Freshwater Biological Association.

References

Beakes, G., H. M. Canter & G. H. M. Jaworski, 1988. Zoospore ultrastructure of *Zygorhizidium affluens* Canter and *Z. planktonicum* Canter, two chytrids parasitizing the diatom *Asterionella formosa* Hassal. Can. J. Bot., 66: 1054–1067.

Box, J. D., 1983. Temporal variation in algal bioassays of water from two productive lakes. Arch. Hydrobiol. Suppl. 67, 1: 81–103.

Clasen, J. & H. Bernhardt, 1974. The use of algal assays for determining the effect of iron and phosphorus compounds on the growth of various algae species. Wat. Res. 8: 31–44.

Davison, W. & E. Rigg, 1976. Performance characteristics for the spectro-photometric determination of total iron in fresh water using hydrochloric acid. Analyst 101: 634–638.

Fitzgerald, G. P., 1972. Bioassay analysis of nutrient availability. In: H. E. Allen & J. R. Kramer (eds.). Nutrients in Natural Waters, Wiley-Interscience Publ., N.Y. John Wiley & Sons, Inc. 147–169.

Lund, J. W. G., 1950. Studies on *Asterionella formosa* Hass. II. Nutrient depletion and the spring maximum. J. Ecol. 38: 1–35.

Lund, J. W. G., 1959. A simple counting chamber for nannoplankton. Limnol. Oceanogr. 4: 57–65.

Lund, J. W. G., 1961. The algae of the Malham Tarn District. Field Studies 1 (3): 85–119.

Lund, J. W. G., 1962. A rarely recorded but very common British alga, *Rhodomonas minuta* Skuja. Br. Phycol. Bull., 2 (3): 133–139.

Lund, J. W. G., F. J. H. Mackereth & G. H. Mortimer, 1963. Changes in depth and time of certain chemical and physical conditions and of the standing crop of *Asterionella formosa* Hass. in the north basin of Windermere in 1947. Phil. Trans. Roy. Soc. Lond. (B) 246: 255–290.

Lund, J. W. G., G. H. M. Jaworski & H. Bucka, 1971. A

228

technique for bioassay of freshwater, with special reference to algal ecology. Acta Hydrobiol. 13: 235–249.

Lund, J. W. G., C. Kipling & E. D. Le Cren, 1958. The inverted microscope method of estimating algal numbers and the statistical basis of estimations by counting. Hydrobiologia 11: 143–170.

Lund, J. W. G., G. H. M. Jaworski & C. Butterwick, 1975. Algal bioassay of water from Blelham Tarn, English Lake District and the growth of planktonic diatoms. Arch. Hydrobiol./Suppl. 49: 49–69.

Mackereth, F. J. H., 1953. Phosphorus utilization by *Asterionella formosa* Hass. J. Exp. Bot. 4: 296–313.

Mackereth, F. J. H., J. Heron & J. F. Talling, 1978. Water analyses: some revised methods for limnologists. Sci. Publ. Freshwat. Biol. Ass. 36: 124 pp.

Marvan, P., 1979. Algal assays – an introduction into the problem. In: P. Marvan, S. Pribil, O. Lhotsky (eds), Algal assays and monitoring Eutrophication, pp. 17–22, Třeboň, Czechoslovakia, E. Schweizerbart'sche Verlagsbuchlandlund, Stuttgart.

Reynolds, C. S., 1984a. The ecology of freshwater phytoplankton. Cambridge University Press, Cambridge. 384 pp.

Reynolds, C. S., 1984b. Phytoplankton periodicity: the interactions of form, function and environmental variability. Freshwat. Biol. 14: 111–142.

Reynolds, C. S., 1986. Diatoms and the geochemical cycling of silicon. In: B. S. C. Leadbeater, J. R. Riding (eds.), Biomineralization in lower plants and animals, pp. 269–289. Oxford University Press, Oxford.

Reynolds, C. S. & C. Butterwick, 1979. Algal bioassay of unfertilized and artificially fertilized lake water, maintained in Lund tubes. Arch. Hydrobiol. Suppl. 56: 166–183.

maintained in Lund tubes. Arch. Hydrobiol. Suppl. 56: 166–183.

Storch, T. A. & V. L. Durham, 1986. Iron-mediated changes in the growth of Lake Erie phytoplankton and axenical algal cultures. J. Phycol. 22: 109–117.

Sutcliffe, D. W., T. R. Carrick, J. Heron, E. Rigg, J. F. Talling, C. Woof & J. W. G. Lund, 1982. Long-term and seasonal changes in the chemical composition of precipitation and surface waters of lakes and tarns in the English Lake District. Freshwat. Biol. 12: 451–506.

Talling, J. F., 1976. The depletion of carbon dioxide from lake water by phytoplankton. J. Ecol. 64: 79–121.

Talling, J. F., 1985. Inorganic carbon reserves of natural waters and ecophysiological consequences of their photosynthetic depletion: micro algae. In: W. J. Lucas, J. Lucas & J. Berry (eds.), Bicarbonate utilization and transport in plants. Amer. Soc. Plant. Physiol.: 403–420.

Talling, J. F. & S. I. Heaney, 1988. Long-term changes in some English (Cumbrian) lakes subjected to increased nutrient imputs. In: F. E. Round (ed.), Contributions to algal biology and environments in honour of J. W. G. Lund. Biopress, Bristol (in press).

Taylor, W. D. & R. G. Wetzel, 1984. Population dynamics of *Rhodomonas minuta* var. *nannoplanctica* Skuja (Cryptophyceae) in a hardwater lake. Verh. Int. Ver. Limnol. 22: 536–541.

Weiss, C. M., 1976. Evaluation of the algal assay procedure. U.S. Envir. Protect. Agency Corvallis, Oregon, 58 pp.

Wurtsbaugh, W. A. & A. J. Horne, 1983. Iron in eutrophic Clear Lake, California: its importance for algal nitrogen fixation and growth. Can. J. Fish. Aquat. Sci. 40: 1419–1429.

Hydrobiologia **188/189**: 229–235, 1989.
M. Munawar, G. Dixon, C. I. Mayfield, T. Reynoldson and M. H. Sadar (eds)
Environmental Bioassay Techniques and their Application.
© *1989 Kluwer Academic Publishers. Printed in Belgium.*

Examination of the effect of wastewater on the productivity of Lake Zürich water using indigenous phytoplankton batch culture bioassays

Christa Lehmacher & Ferdinand Schanz
Hydrobiologisch-limnologische Station, Seestr. 187, CH-8802, Kilchberg, Switzerland

Key words: bioassays, batch cultures, indigenous phytoplankton, wastewater, net microbial production

Abstract

Phytoplankton batch cultures were used to study the effect of biologically and chemically treated wastewater on algal growth in water from Lake Zürich. The question of whether the influence of sewage on the biomass production corresponds solely to the nutrient content of the sewage was also considered. The relationship between net microbial production (y, in 0.1 N $KMnO_4$ consumption) and total phosphorus concentration (x, in μg P/l) was found to be characterized by the equation $y = 2.12 \ln x - 4.12$. The fact that net microbial production is strongly dependent on total phosphorus concentration emphasizes the significance of the latter for the trophic state of Lake Zürich. We recommend the introduction of additional purification of all sewage entering the lake.

Introduction

If wastewater enters a lake with water poor in nutrients, the nutrients and trace elements introduced will lead to an increase in phytoplankton and periphyton biomass. However chemical analysis of the nutrient input gives too little information for an interpretation with regard to ecological effects (Bombówna & Bucka, 1972). Thomas & Munawar (1985), Gaur & Kumar (1986) and Hecky & Kilham (1988) demonstrated that bioassays are needed to allow us to assess the biological availability of the nutrients introduced.

Minitest experiments carried out by Claesson & Forsberg (1978) involving the addition of 2.5% unfiltered sewage to pure and mixed algal cultures resulted in a biomass increase of about 300% (expressed in terms of optical density) in three days. In experiments with unfiltered wastewater, Forsberg & Claesson (1981) obtained lower algal production values than expected considering the phosphorus concentrations present in the culture water, perhaps as a result of inhibition of algal growth by unknown substances. Couture *et al.* (1985) have shown the usefulness of the algal assay procedure in determining the toxic activity of urban wastewater: metal concentration present in the wastewater produced toxicity in the algal culture. In laboratory tests, Källqvist (1975) found that 5% mechanically treated sewage will result in the production of 11% to 20% more biomass in lake water than 5% biologically treated sewage. Experiments with water from mesotrophic Lake Lucerne (Bossard & Ambühl, 1974) have shown that both wastewater after secondary treatment and wastewater after tertiary treatment have a fertilizing effect. Spiking lake water with inorganic phosphate and nitrate had the same fertilizing effect as the addition of secondary or tertiary treated wastewater containing the same amounts of phosphorus and nitrogen. It thus seems that the stimulation of algal growth by wastewater can be explained by its P and N concentration alone. Thomas (1953) has shown this to be the case for numerous lakes in Switzerland and nearby regions. Similar results have been obtained by Miller & Maloney (1971) in cultures with *Selenastrum capricornutum* as the test organism. Wastewater which had undergone tertiary treatment was found by them to cause an increase in algal growth in nutrient-poor culture water only when spiked with phosphorus, whereas sewage treated mechanically and biologically (secondary treatment) was found to lead to an increased algal yield in the same culture water without spiking. Schanz & Pleisch (1981) compared algal production in batch cultures containing wastewater from the tertiary and quaternary stages of a sewage treatment plant. They found a significantly smaller algal production in cultures containing wastewater from the quaternary stage than in those containing wastewater from the tertiary stage. This was due to the lower total phosphorus concentrations caused by quaternary purification.

The aim of our experiments was to ascertain the effect on microbial growth of adding wastewater which had undergone tertiary treatment to Lake Zürich water. Bioassays were carried out on batch cultures containing phytoplankton from the surface water of Lake Zürich with the intention of determining the proportion of biologically available nutrients. Based on the results of these experiments, it should be possible to make assumptions concerning the ecological effects of treated wastewater phosphorus on the lake, thus allowing us to draw conclusions with respect to future water protection measures.

Materials and methods

A number of 300 ml Erlenmeyer flasks were cleaned as described by Schanz (1974), plugged with cotton wool and subjected to dry sterilization for 2 hours at 180 °C. Wastewater from the teritary stage of the municipal sewage works in Thalwil, Canton Zürich, Switzerland (phosphorus elimination by co-precipitation with $FeSO_4$) and unfiltered surface water from Lake Zürich (0.1 m depth) were mixed in varying proportions (Table 1), and 100 ml of the resulting mixed water were added to each flask. Five parallel cultures were set up for each water mixture. The 0% loading of sewage can also be considered as the control.

The inoculum consisted of phytoplankton taken from 0.1 m depth in Lake Zürich using a 50 μm-mesh plankton net. In order to eliminate large particles and macroinvertebrates, the phytoplankton were subsequently filtered in the laboratory through a 250 μm-mesh net (Dunstan & Menzel, 1971). Each of the Erlenmeyer flasks was inoculated with 1 ml of the homogeneous phytoplankton suspension (1 ml had a consumption of 0.6 ml of 0.1 N $KMnO_4$ solution, a mean of three determinations). The flasks were kept at a temperature of 20 °C and light intensity of 51 μE m^{-2} s^{-1} ($\sim$2600 lux, 9 Philips TL20/33W fluorescent tubes) in a culture chamber (KKB 600 L, Heraeus GmbH, D-6450 Hanau, FRG) 24 hours per day for a period of 21 days. After that period the microbial production was at its stationary growth phase as shown in detail for several similar experiments with different nutrient concentrations (Lehmacher, in preparation). At the beginning and end of the bioassays, we determined the biomass directly in the culture vessel by measuring the consumption of 0.1 N $KMnO_4$ (Deutsche Einheitsverfahren, 1981, method H4). The difference between the two values gives the net microbial production. It is well known that the $KMnO_4$ method does not allow all the organic material of a water sample to oxidize completely (Schanz, 1974). However the method is easy to handle and several replicates can be done simultaneously. Therefore, numerous experiments were made to check the usefulness of the method for microbial biomass determination (Schanz & Juon, 1983). Compared to other possibilities for biomass measurements (chlorophyll, protein, dry weight, total phosphorus), the $KMnO_4$ consumption gave similar or better results (Betschart, 1979; Lehmacher, in preparation). Before commencing the growth tests, the water samples were analysed for orthophosphate and total phosphorus (Eidgenössische Richtlinien, 1982, method

37), ammonium nitrogen (Eidgenössische Richt-linien, 1982, method 30), nitrate nitrogen (UV method: American Public Health Association, 1975) and nitrite nitrogen (Eidgenössische Richt-linien, 1982, method 36; Wagner, 1969).

Results

Table 1 lists the nutrient content of Lake Zürich culture water with increasing sewage loading. The addition of 10% wastewater to pure Lake Zürich water caused a mean increase in the total phosphorus content of 85% (all tests) and a mean increase in the inorganic nitrogen content (the sum of ammonium nitrogen, nitrate nitrogen and nitrite nitrogen) of about 82% (tests 3 and 4). Figure 1 shows that the addition of wastewater,

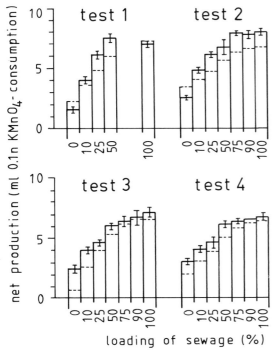

Fig. 1. Net microbial production in Lake Zürich water containing various proportions of biologically and chemically treated wastewater (four tests: see Table 1 for chemical characteristics of culture media). Inoculum: phytoplankton from Lake Zürich. Experiments begun on October 5, 1986 (test 1); April 13, 1987 (test 2); May 11, 1987 (test 3); July 4, 1987 (test 4). Duration of experiments: 21 days. Column: means (n = 3 and 5) and standard deviations; dashed lines: calculated algal biomass (see Fig. 2).

Table 1. Nutrient concentrations (in $\mu g\,l^{-1}$) in culture water from Lake Zürich with increasing amounts of tertiary treated sewage used for tests 1 to 4 (Fig. 1). P = orthophosphate phosphorus; T-P: total phosphorus; NH_4^+ = ammonium nitrogen; NO_3^- = nitrate nitrogen; NO_2^- = nitrite nitrogen. All tests carried out using sewage from the sewage treatment plant in Thalwil, Canton Zürich, Switzerland.

Proportion of sewage	Test 1					Test 2		Test 3					Test 4				
	P	T-P	NH_4^+	NO_3^-	NO_2^-	P	T-P	P	T-P	NH_4^+	NO_3^-	NO_2^-	P	T-P	NH_4^+	NO_3^-	NO_2^-
0%	3	29	50	409	13	33	50	2	15	57	634	9	2	27	218	696	12
10%	–	56	–	–	–	–	68	18	37	568	661	55	10	45	704	922	44
25%	–	97	–	–	–	–	96	50	71	783	930	94	26	64	1758	1132	87
50%	78	169	4315	481	113	–	141	106	130	1116	1459	219	52	112	3027	1498	151
75%	–	–	–	–	–	–	187	164	195	3198	1751	355	80	165	15000	1868	214
90%	–	–	–	–	–	–	215	200	235	5411	1829	428	97	196	15000	2743	264
100%	143	300	20000	540	227	100	233	226	233	5090	2023	491	108	202	15000	2300	292

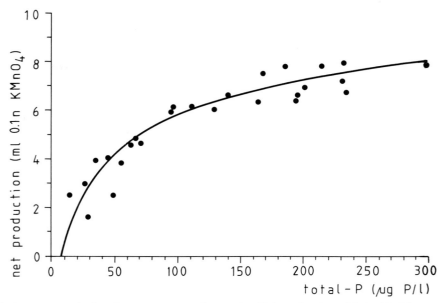

Fig. 2. Relationship between total phosphorus content and net microbial production of Lake Zürich water with increasing amounts of tertiary treated sewage. Data from tests presented in Fig. 1. Dots: mean values. Solid line: regression equation $y = 2.12 \ln x - 4.12$ (x = total P in μg P l^{-1}, y = consumption of 0.1 N KMnO$_4$ in ml, n = 26, r = 0.88, p < 0.001).

which had undergone tertiary treatment, to cultures of Lake Zürich water caused an increase in the net microbial production in all cases. The addition of 10% wastewater resulted in a biomass increase of 157% (s.d. = 23%) compared to the controls containing lake water alone (Fig. 1: test 1). In a culture containing 50% wastewater, a mean increase in the net microbial production (all four experiments averaged) of more than 100% compared with the controls was obtained. Figure 2 illustrates the relationship between net microbial production and total phosphorus content of the culture water. At low total phosphorus concentrations ($< 50 \,\mu$g P l^{-1}), the addition of small amounts of phosphorus caused a considerable increase in the microbial standing crop; phosphorus can therefore be considered to be the limiting nutrient in this range. At higher total phosphorus concentrations ($> 50 \,\mu$g P l^{-1}), the addition of the same amount of phosphorus resulted in only a slight biomass increase. With the aid of the Student t-test (Sokal & Rohlf, 1969) we checked to see if the increase in the net microbial production at various sewage loadings was significant at the 95% significance level. Our results (Table 2) show that there is no significant

Table 2. Statistical comparisons of net microbial production values in cultures of Lake Zürich water to which increasing amounts of wastewater have been added. NP$_x$ = net production in culture containing x% wastewater; t = critical t-value at 95% significance level; $\hat{t}$ = computed t-value (Sokal & Rohlf, 1969).

Date of bioassay	DF	t	NP$_0$—NP$_{10}$ $\hat{t}$	NP$_{10}$—NP$_{25}$ $\hat{t}$	NP$_{25}$—NP$_{50}$ $\hat{t}$	NP$_{50}$—NP$_{75}$ $\hat{t}$	NP$_{75}$—NP$_{90}$ $\hat{t}$	NP$_{90}$—NP$_{100}$ $\hat{t}$
05.10.1986	8	2.306	12.23	10.31	5.55	–	–	–
13.04.1987	8	2.306	13.34	6.40	1.74	3.66	– 0.29	1.13
03.06.1987	4	2.776	4.04	2.77	4.65	0.91	0.74	0.81
25.07.1987	8	2.306	5.07	1.91	5.89	1.34	2.30	2.28

biomass increase at sewage loadings exceeding 50% (with the sole exception of the bioassay on April 13, 1987, where the net production at a sewage loading of 75% was found to be significantly higher than that at a loading of 50%).

Discussion

Bioassays with water from a Danish river-lake system and sewage effluent from various treatment plants were carried out by Gargas (1978). His results based on batch cultures with water from Lake Brass showed that a loading of 10% biologically and chemically treated wastewater can cause a mean increase in the biomass of about 390% (dry weight). In our tests, the addition of 10% wastewater after tertiary treatment increased the mean algal growth by 84% (s.d. = 48%; minimum increase = 35%; maximum increase = 145%). In the above-mentioned tests by Gargas (1978), phosphorus and nitrogen concentrations in the culture water after spiking with 10% sewage varied between 0.052 and 1.456 mg P l^{-1} and between 1.175 and 3.022 mg N l^{-1}. The nutrient concentrations in our culture water at a sewage loading of 10% were considerably lower: the total phosphorus concentration lay between 0.037 and 0.068 mg P l^{-1} and the mean inorganic nitrogen concentration was 1.5 mg N l^{-1} (Table 1). The addition of 10% sewage to water from Lake Brass caused a 500% increase in the total phosphorus concentration. Assuming all phosphorus to be utilized in biomass production, a biomass increase of 500% was to be expected. This was however not the case, and therefore we assume that phosphorus was not always limiting. Either growth was controlled by elements other than phosphorus, or algicides were present in the water used (Forsberg & Claesson, 1981). Biotests in culture waters with high nutrient concentrations carried out under the same conditions as the experiments mentioned above resulted in much higher net microbial production values (up to 20 ml 0.1 N $KMnO_4$ consumption; Lehmacher, in preparation), showing that the light intensity as well as the CO_2 concen-

tration were sufficient enough. Experiments in which the culture water was spiked with inorganic phosphorus and nitrogen showed that phosphorus was the first limiting nutrient in filtered Lake Zürich water. This result is supported by the fact that both biomass and total phosphorus concentration increased by the same order of magnitude after spiking Lake Zürich water with 10% tertiary treated wastewater.

The results can therefore be interpreted on the basis of total phosphorus alone. Statistical investigations have shown that total phosphorus is highly correlated with inorganic nitrogen (Table 1, data from tests 3 and 4: $r = 0.96$, $n = 11$, $p < 0.001$). The relationship between net microbial production (y, in ml 0.1 N $KMnO_4$ consumption) and total phosphorus (x, in μg P l^{-1}) presented in Fig. 2 was found to be characterized by the expression $y = 2.12 \ln x - 4.12$ (least squares regression: $r = 0.88$, $n = 26$, $p < 0.001$). Based on this relationship, a microbial yield was predicted for the nutrient concentrations determined in the culture water. In tests 1 and 2, the calculated yield for pure Lake Zürich water (0% sewage loading) was greater than that found in the bioassays, with deviations from the expected values of 50% (test 1) and 36% (test 2). Net production values in tests 1 and 2 lay outwith the 95% confidence intervals of the curve illustrated in Fig. 2 (Sachs, 1974). Apparently the phosphorus present could not be completely utilized for the production of biomass. We may assume in the case of tests 1 and 2 that factors other than phosphorus were growth limiting factor and/or, as it is well known, total phosphorus values are not reliable for evaluating the availability of phosphorus. Deviations from the values expected on the basis of the logarithmic curve were not found in culture water containing 10% sewage. It is therefore unlikely that reasons other than those mentioned above are responsible for the deviations.

On closer examination, Fig. 2 shows that a linear relationship between total phosphorus and net microbial production can be assumed only between 0 and 50 μg P l^{-1}, confirming the results obtained by Schanz and Pleisch (1981). In this concentration range, the N/P atomic ratio (where $N = NH_4^+ + NO_3^- + NO_2^-$; P = total phosphorus; Chiaudini & Vighi, 1974) calculated are always above 20 which indicates that P is the chemical limiting factor (see the Redfield ratio, 106 C : 16 N : 1 P; Redfield, 1958). At higher concentrations, the proportion of non-assimilable phosphorus increases. This is surprising in view of the fact that phosphorus was not added in pure solution, but in sewage along with considerable amounts of other nutrients, trace elements and growth-promoting organic substances (e.g. vitamins), which, according to the laws of Mitscherlich (1909), also influence algal growth to varying degrees. Further research into this phenomenon is needed and experiments where available forms of phosphorus are taken into account must be conducted with wastewater.

Conclusions

Our tests show that the addition of even slight amounts of tertiary-treated sewage to water from Lake Zürich causes a considerable increase in net microbial production. The discharge of wastewater into the surface water of the lake, especially in summer, can be expected to result in easily detectable wastewater concentrations up to several hundred metres from the outfall point (P. Klöti, pers. comm.). The resulting increase in algal biomass in the lake is directly related to the phosphorus content of the wastewater. It is to be expected that any reduction in the phosphorus input will cause a reduction in algal growth and in the occurrence of related symptoms of eutrophication (e.g. an increase in transparency and a decrease in hypolimnetic oxygen consumption). We assume that the planned introduction of the quaternary purification stage with a high phosphorus removal in the sewage plants will accelerate the oligotrophication of the Lake Zürich.

Acknowledgements

Mr. H. P. Mächler helped with sampling and advised us on chemical determination methods. Mr. D. M. Livingstone corrected the English text. We thank Len L. Hendzel, P. Stadelmann and S. Daniels for their comments.

References

American Public Health Association, 1975. Standard methods for the examination of water and wastewater. Washington, 1193 pp.

Betschart, B., 1979. Die Verwendung von Biotesten zur Untersuchung des eutrophierenden Einflusses der Zürcher Abwässer auf die Limmat. Ph.D. Thesis, University of Zürich, Switzerland, 84 pp.

Bómbowna, M. & H. Bucka, 1972. Bioassay and chemical composition of some Carpathian rivers. Verh. int. Ver. Limnol. 18: 735–741.

Bossard, P. & H. Ambühl, 1974. Der Einfluss von gereinigtem Abwasser auf das Phytoplankton in Seen. Schweiz. Z. Hydrol. 36: 187–200.

Chiaudini, G. & M. Vighi, 1974. The N : P ratio and tests with *Selenastrum* to predict eutrophication in lakes. Wat. Res. 8: 1063–1069.

Claesson, A. & A. Forsberg, 1978. Algal assay procedure with one or five species. Minitest. Mitt. int. Ver. Limnol. 21: 21–30.

Couture, P., S. A. Visser, R. Van Coillie & C. Blaise, 1985. Algal bioassays: their significance in monitoring water quality with respect to nutrients and toxicants. Schweiz. Z. Hydrol. 47: 127–158.

Deutsche Einheitsverfahren zur Wasser-, Abwasser- und Schlammuntersuchung, 1981. Lieferungen 1–9. Verlag Chemie, Weinheim/Bergstr., BRD, 257 pp.

Dunstan, W. & D. Menzel, 1971. Continuous cultures of natural populations of phytoplankton in dilute, treated sewage effluent. Limnol. Oceanogr. 16: 623–632.

Eidgenössische Richtlinien für die Untersuchung von Abwasser und Oberflächenwasser, 1982. Eidgenössisches Departement des Innern, Bern, 68 pp.

Forsberg, A. & A. Claesson, 1981. Algal assays with wastewater to determine the availability of phosphorus for algal growth. Verh. int. Ver. Limnol. 21: 763–769.

Gargas, E., 1978. The effect of sewage (mechanically, biologically and chemically treated) on algal growth. Mitt. int. Ver. Limnol. 21: 110–124.

Gaur, J. P. & H. P. Kumar, 1986. Effects of oil refinery effluents on *Selenastrum capricornutum* Printz. Int. Rev. Ges. Hydrobiol. 71: 271–281.

Hecky, R. E. & P. Kilham, 1988. Nutrient limitation of phytoplankton in freshwater and marine environments: a review of recent evidence on the effects of enrichment. Limnol. Oceanogr. 33: 796–822.

Källqvist, T., 1975. Algal growth potential of six Norwegian waters receiving primary, secondary and tertiary sewage effluents. Verh. int. Ver. Limnol. 19: 2070–2081.

Miller, W. E. & T. E. Maloney, 1971. Effects of secondary and tertiary wastewater effluents on algal growth in a lake-river system. J. Wat. Pollut. Cont. Fed. 42: 2361–2365.

Mitscherlich, E. A., 1909. Das Gesetz vom Minimum und das Gesetz des abnehmenden Bodenertrags. Landw. Jahrb. 38: 537–552.

Redfield, A. C., 1958. The biological control of chemical factors in the environment. Amer. Sci. 46: 205–221.

Sachs, L., 1974. Angewandte Statistik. Springer Verlag, Berlin, BRD, 548 pp.

Schanz, F., 1974. Wachstumsansprüche der Cladophoracee Rhizoclonium hieroglyphicum Kütz. in Reinkultur. Ph.D. Thesis, University of Zürich, Switzerland, 140 pp.

Schanz, F. & P. Pleisch, 1981. Bioteste zum Vergleich von drei neuen, parallel betriebenen Verfahren zur weitestgehenden Phosphor-Elimination (4. Reinigungsstufe). Gas Wass. Abwass. 61: 6–11.

Schanz, F. & H. Juon, 1983. Two different methods of evaluating nutrient limitations of periphyton bioassays using water from the river Rhine and eight of its tributaries. Hydrobiologia 102: 187–195.

Sokal, R. R. & F. J. Rohlf, 1969. Biometry. W. H. Freeman & Comp., San Francisco, 776 pp.

Thomas, E. A., 1953. Empirische und experimentelle Untersuchungen zur Kenntnis der Minimumstoffe in 46 Seen der Schweiz und angrenzender Gebiete. Monatsbull. Schweiz. Ver. Gas Wasserfachmänner 9: 1–15.

Thomas, R. L. & M. Munawar, 1985. The delivery and bioavailability of particulate bound phosphorus in Canadian rivers tributary to the Great Lakes. In J. N. Lester & P. W. W. Kirk (eds.), Proc. Internat. Conf.: Management Strategies for phosphorus in the Environment. Lisbon, Portugal, 462–469 pp.

Wagner, R., 1969. Neue Aspekte in der Stickstoff-Analytik. Vom Wass. 36: 263–328.

Hydrobiologia **188/189**: 237–246, 1989.
M. Munawar, G. Dixon, C. I. Mayfield, T. Reynoldson and M. H. Sadar (eds)
Environmental Bioassay Techniques and their Application.
© 1989 *Kluwer Academic Publishers. Printed in Belgium.*

Early warning assays: an overview of toxicity testing with phytoplankton in the North American Great Lakes

M. Munawar [1], I. F. Munawar [2] & G. G. Leppard [3]
[1] *Department of Fisheries and Oceans, Great Lakes Laboratory for Fisheries and Aquatic Sciences, Ecotoxicology Division, 867 Lakeshore Rd., P.O. Box 5050, Canada Centre for Inland Waters, Burlington, Ontario, Canada L7R 4A6*; [2] *Plankton Canada, 685 Inverary Rd., Burlington, Ontario, Canada L7L 2L8*; [3] *Environment Canada, National Water Research Institute, Lakes Research Branch, Canada Centre for Inland Waters, Burlington, Ontario, Canada L7R 4A6*

Key words: algae, overview, bioassessment, sediment, bioassays, contaminants

Abstract

The use of phytoplankton as test organisms in bioassays has recently gained momentum due to their simplicity, availability, sensitivity, rapidity of analysis, and cost-effectiveness. Increasing emphasis is currently being given to field and *in situ* experiments using indigenous populations, particularly ultra-plankton/picoplankton (2–20 μm) which play a key role in the 'microbial loop' and food chain dynamics. Impact evaluation can be determined at the structural, ultra-structural, and functional level. An array of techniques is available for toxicity testing including the use of either algal cultures or natural assemblages in laboratory or *in situ* experiments, the selection of which depends on the objectives, precision required, and project budget of the particular study. An overview is presented of the various procedures using algae in toxicity testing with a focus on the Great Lakes and an emphasis on field techniques. The effective use and application of such sensitive technology has tremendous potential for early warning detection of ecosystem perturbations in concert with a multi-trophic battery of tests.

Introduction

There is a growing global concern over the increasing contamination of our aquatic environments from municipal, industrial, urban, and agricultural sources. This problem has been further complicated by atmospheric pollutants crossing international boundaries and spreading as far as the Arctic regions. However, it is encouraging to see that concern for our environment and a desire to seek solutions has been voiced on an international level (World Commission Report on Environment and Development 1987) and on a national level in Canada (Canadian Environmental Protection Act 1988). The traditional pollution control programs thus far have been confined mostly to the reduction of nutrient enrichment, and very little has been accomplished on the decontamination of toxic substances such as organics and metals in the aquatic environments.

It is now well established that the impact of contaminants cannot be effectively evaluated solely from chemical analysis, because this approach does not provide the vital data concerning their bioavailability. It is also accepted that an ecosystem approach must be adopted to achieve

238

a more holistic perspective of both the environment and its inhabitants. Such an assessment of the impact of contaminants in an ecosystem is best carried out using aquatic organisms from various trophic levels in comparative bioassays (Gächter, 1979; Maciorowski *et al.*, 1981; Bringmann & Kuhn, 1980). In the Great Lakes, a multi-trophic battery of tests has been standardized for the bioassessment of Great Lakes 'Areas of Concern' (Munawar *et al.*, 1989c; I.J.C., 1987) which includes algal toxicity testing. Although the use of algae in bioassays is not new (Allen & Nelson, 1910; Schreiber, 1927), it is only recently that these organisms have attracted the attention of toxicologists and regulatory agencies (Rai *et al.*, 1981; Munawar & Munawar, 1987). This increased awareness of algae is due to their vital role in the 'microbial loop', food chain dynamics, and trophic interactions (Sherr & Sherr, 1988; Ross & Munawar, 1987; Munawar *et al.*, 1989c). Furthermore, the suitability of algae as test organisms is gaining support due to their structural simplicity, ubiquitous abundance in nature, and the ease of obtaining commercially available algal cultures for laboratory testing (Munawar *et al.*,

1988b). Also, algal toxicity tests are rapid, inexpensive, and sensitive, and can be used effectively to assess those toxic substances which are found in concentrations too low for effective detection by higher trophic level organisms (Munawar & Munawar, 1987; Wong & Couture, 1986). Phytoplankton in their natural environments, unlike other organisms, are affected directly by both nutrients and contaminants.

This paper resulted from an invitation by the Conference Committee to review the state-of-the-art use of algae in toxicity testing, with an emphasis on field/*in situ* investigations. Here we provide a general overview of techniques currently being used for the assessment of environmental hazards caused by both metals and organic contaminants. A more detailed overview is available elsewhere (Munawar *et al.*, 1988b).

The phytoplankton health assessment strategy in this paper has been conveniently divided into the following categories (Fig. 1):
1. Structural indicators (community structure)
2. Ultra-structural indicators (cytological)
3. Functional indicators (physiological/biochemical)

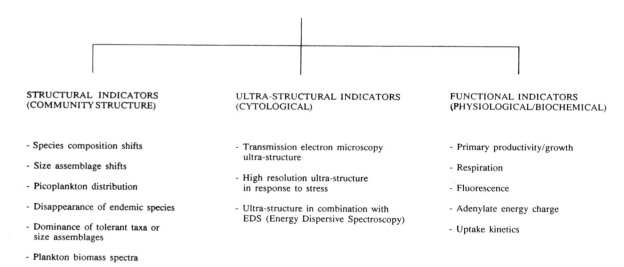

Fig. 1. Strategy for phytoplankton health assessment using indicators such as community structure, cytology, and physiological and biochemical parameters.

In addition, we also describe new techniques having considerable potential for future environmental hazard assessment. Examples from the North American Great Lakes have been focussed on to demonstrate the usefulness of some of the techniques discussed in this review.

Structural Indicators

Some of the common structural indicators are given in Fig. 1. It is apparent that the structural aspect is heavily dependent on standard, microscopical, analytical techniques that utilize all size components of phytoplankton such as picoplankton (< 2 μm), ultraplankton (2-20 μm) and microplankton/netplankton (> 20 μm). It has been demonstrated that picoplankton and ultraplankton are sensitive to contaminants such as metal mixtures (Munawar et al., 1987c). However, the time-consuming, meticulous nature of taxonomic identification, the scarcity of trained phycologists (Munawar et al., 1987a; Munawar & Munawar, 1980) and a lack of standardized, data-processing procedures are all limiting factors in generating long-term and consistent floristic records to evaluate changes in the community structure. Such long-term species data sets are generally limited in the Great Lakes (Vollenweider et al., 1974; Munawar & Munawar, 1986; Stoermer, 1978) and as a result, the structural response of phytoplankton to contaminants has received very little attention (Rhee, 1988). However, it is encouraging that such studies of structural response are being initiated in the Great Lakes 'Areas of Concern' (I.J.C., 1987) such as Toronto Harbour (Munawar, 1989). Furthermore, the size assemblage shifts (Munawar et al., 1987a; 1988a) and biomass spectral changes in contaminated (nearshore) versus offshore areas (Sprules & Munawar, 1986; Sprules et al., 1988) are emerging. Traditional microscopy is also now combining with microcomputer-assisted devices to measure size structure. Also, knowledge concerning the 'microbial loop' and trophic interactions has rapidly increased since the classical papers by Azam et al. (1983), Porter et al. (1985),

Stockner (1988), and Sherr & Sherr (1988). In the North American Great Lakes, Munawar et al. (1987c) and Munawar & Weisse (1989) have demonstrated the sensitivity of autotrophic picoplankton to nutrients and contaminants.

Ultra-structural Indicators

Ultra-structural indicators are worthy of discussion because they represent a field of considerable potential for bioassay technology. It has been recognized since the early years of electron microscopy that the structure of cell components such as organelles, microtubule arrays, membrane systems, fibrils, and various granules can be interpreted in terms of the compartmentalization of cell function (Brinkley & Porter, 1977). In some cases, at a resolution approaching 0.001 μm, the structural-functional relationships can be confirmed down to the level of identifiable macromolecules such as enzymes (phosphatases) (Blum et al., 1965; Dodge, 1973), individual structural components (cellulose microfibrils) (Preston, 1974), and storage materials (starch granules) (Dodge, 1973; Gibbs, 1971). Thus, an ultrastructural response to an environmental insult can yield more than just a correlation. Because it is a sensitive measure of a physiological change, such a response can be used inferentially as a guide for selecting measures of both physiological and anabolic responses which might otherwise not be considered. Also, it has potential for permitting development of a structural index of health of unicellular and pauci-cellular organisms. In the case of picoplankton, such a research thrust has already begun (Leppard et al., 1987).

Let us examine the basis for the general statements above and the potential opportunities that they present through technology transfer from cell biology. Several cell compartments of relevance to algal productivity have been 'dissected' in such a way that an organelle dysfunction can be diagnosed by an examination of ultra-structure. In addition to the large specialized literature on this subject, several decades of general reference works have been produced which adequately introduce the subject to the non-specialist who has an interest in transferring the technology into the aquatic sciences. Among these are early general works on structure-function relationships in cell membranes (Rothfield, 1971), chloroplasts (Gibbs, 1971), and mitochondria (Lehninger, 1965). The abundant literature of general findings on these major structures (see Journal of Cell Biology, Journal of Cell Science and Protoplasma) is applicable to algal cell membranes, chloroplasts, and mitochondria; some of these findings were based directly on algal experiments. With respect to the normal ontogeny and/or experimental perturbation of algal intracellular compartments there have been revealing studies on: (A) intracellular membranes in relation to the spatial orientation of organelles (Bouck, 1965); (B) assembly and disassembly of the microtubular cytoskeleton (Brown & Bouck, 1973); (C) extracellular secretion by the Golgi apparatus (Brown et al., 1973); (D) flagellar structure in relation to flagellar motion (Gibbons, 1977; Bouck, 1971); (E) the regulation of gas vacuole activity (Walsby, 1972); (F) structure-chemistry-function associations in cell wall growth (Preston, 1974); and (G) analyses of many other compartments and sub-compartments of relevance to algal productivity in nature (Dodge, 1973). As Dodge (1974) has shown, one can classify, at least on a crude scale, many unicellular and small algae by an examination of ultra-structure at high resolution.

The literature above from cell biology provides limited information to the ultra-structural biologist in diagnosing physiological alterations caused by environmental insults. Despite the limitations, the microscopical technology (a combination of optical microscopy, scanning electron microscopy, transmission electron microscopy, cytochemistry, and energy-dispersive spectroscopy) can be used to assist the aquatic biologist in selecting assays to delineate a biochemical, physiological, and ecotoxicological response by algae to contaminants. This technology transfer has been little exploited in the past, despite the fact that the limitations are not severe.

At this point, two notes of caution are necessary. Firstly, the specialized literature created by the cell biologists is based mainly on whichever algal species was most amenable, for a given kind of study, to the analytical techniques of the moment. Consequently, an incomplete effort was made to provide a coverage of algal types based on their importance to natural ecosystems. Technology transfer may require some adaptation before it can be applied to a species that was of little cytological interest in the past. Secondly, many of the most helpful descriptive works were done several decades ago. Thus, a descriptive work upon a species whose cytological literature is incomplete might require a manual expertise which is great and a time frame which is long.

Functional Indicators

Most toxicity testing falls into the functional indicators category since the assays must be rapid, sensitive, and cost-effective and serve as early warning indicators of ecosystem health. A large variety of physiological and anabolic tests are available to choose from depending on the type of problem, precision required, and budget. Figure 2 summarizes bioassays that are commonly done in laboratory and field/in situ situations. The experiments are conducted using algal cultures, natural assemblages, and in situ procedures such as cages, enclosures, etc.

Impact of metals:

Assays conducted with cultures grown in the laboratory are useful in providing basic information on physiological limits for individual species (Braek et al., 1976; Wong et al., 1978), but the data are extremely difficult to extrapolate to a natural situation due to inherent environmental interactions in each ecosystem. Consequently, our laboratory has adopted a field-to-laboratory approach and an effort has been made to use fresh, natural phytoplankton assemblages as test organisms wherever possible. Several laboratory techniques are available in evaluating the impact of toxicants to phytoplankton. These include

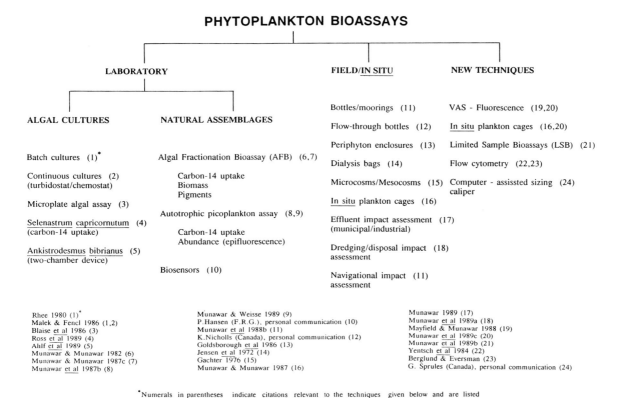

Fig. 2. Phytoplankton bioassays commonly employed in laboratory and field/*in situ* situations.

in vitro batch and continuous cultures, with the latter using turbidostats or chemostats. Details of these techniques are provided by Malek & Fencl (1966), Rhee (1982), Wong *et al.*, (1983) and Munawar *et al.* (1988b).

Very little information is available on the interactive effects of metal mixtures on algae, although there is abundant literature available concerning the toxicity of individual metals (Gächter, 1976; Rai *et al.*, 1981; Lustigman, 1986). Synergistic and antagonistic effects have been reported which shed considerable light on the complexity of metal mixture toxicity (Gächter, 1976; Bartlett *et al.*, 1974; Hutchinson & Stokes, 1975; Wong *et al.*, 1978; Munawar & Munawar, 1982).

The most common assay used with natural assemblages in the Great Lakes was either ^{14}C size-fractionated primary productivity monitoring, or Algal Fractionation Bioassays (AFB). The impact of various toxic materials, such as sedi-

ment-bound or water-borne contaminants and synthetic metal mixtures, on phytoplankton size assemblages was assessed (Munawar & Munawar, 1982; 1987; Munawar *et al.*, 1983; 1987c). In total phytoplankton community assays, the impact of contaminants on various components of the test community was either missed or masked (Munawar, 1982; Lane & Goldman, 1984). In contrast, the AFB simultaneously focussed on the effects of toxic substances on various size fractions by isolating a natural assemblage of diverse species with a wide variety of sizes, physiological requirements, and environmental tolerances. Several papers cited above demonstrate the observed differential response of various size assemblages to contaminant stress and indicate the usefulness, convenience and ease of performing this rapid and sensitive procedure. We have successfully used size-fractionated, primary productivity as an overall indicator of ecosystem health and as a biomonitoring tool for initially understanding an ecosystem (Munawar *et al.*, 1989c). A large number of assays with sediment elutriates and filtered versus unfiltered water was done using offshore phytoplankton as a test population (Munawar *et al.*, 1989c).

As a result of an extensive AFB program, techniques concerning differential filtration, isolation, and concentration were developed. These techniques were used in conjunction with epifluorescence microscopy. Johnson & Sieburth (1982) fractionated organisms to sizes less than 2 μm (picoplankton) while we isolated and concentrated organisms with a Gelman 1.2 μm Acroflux capsule (Munawar *et al.*, 1987b). These 1.2 μm organisms were used in ^{14}C picoplankton bioassays with metal mixtures and the resulting primary productivity was found to be extremely inhibited by the addition of these toxicants. Recent work by Munawar & Munawar (1987) and Munawar & Weisse (1989) indicated that autotrophic picoplankton were scarce or lacking in contaminated environments of the Great Lakes such as the Niagara River, Toronto Harbour, and Hamilton Harbour. The autotrophic picoplankton, on the contrary, were abundant in oligotrophic and mesotrophic systems such as Lakes Superior and Ontario (Munawar *et al.*, 1987a).

Impact of Organic Contaminants:

Very little information is available concerning the impact of organic contaminants on phytoplankton although these toxicants pose a serious threat to fisheries. Field data is lacking and even laboratory work is limited (Rhee, 1982; Munawar *et al.*, 1988b). There is a paucity of historical records of phytoplankton abundance and it is difficult to isolate the simultaneous impact of other factors such as nutrient enrichment, food chain changes, and the introduction of exotics which may alter the predator-prey relationship (Rhee, 1988). These problems are dealt with in a detailed review of Great Lakes research by Munawar *et al.* (1988b). Both enhancement and inhibition of primary productivity due to organic contaminants

was observed (Rhee, 1982; Wright, 1978; Lal & Saxena, 1982) although the mechanism of these impacts is not clearly understood. Like the bioavailability of metals, the susceptibility of algal species to organo-chlorine compounds results in the community structural changes of relatively resistant species. The impact of synergistic, antagonistic, or additive effects is not well-known for the organic contaminants (Rhee, 1988) and much needs to be done in this area of research.

The results of the impact of organic contaminants should receive the immediate attention of researchers due to the obvious human health risks. For example, in the Great Lakes, data has been generated over a number of years for thousands of organic contaminants and a disproportionate number of 'non-detectable' values were observed (D. Sergeant, Fisheries & Oceans Canada – Personal Communication). The cost and time to perform the analyses were considerable for minimal gain. Furthermore, other factors such as detection limits need resolution for various organics. In essence, the lower the detection limits, the greater the time and cost of analysis. The 'non-detectable' data of the field samples speaks for itself and suggests that the issue of chemical analysis, particularly organic substances, needs a review and change of approach. For example, an alternate but scientifically and economically sound approach could be to have the field samples pre-screened by means of rapid, sensitive, and cost-effective bioassays to identify samples with toxic effects which then could be subjected to intensive organic analysis. This could be followed by testing with chronic assays. Then, the pre-defined, compound-specific screening of organics could be changed since many acutely or chronically toxic compounds might not be detected and may be treated as interference (D. Sergeant & M. Munawar, Fisheries & Oceans Canada, unpublished data).

Field/*In situ* and New Techniques

The use of natural assemblages is favoured over laboratory-grown cultures because extrapolation of laboratory data to natural conditions is often difficult and misleading. Toxicity tests using natural assemblages compared to laboratory-grown cultures have yielded results showing enhancement of primary productivity in natural phytoplankton, while the same test indicated inhibition in mixed cultures (Munawar & Munawar, 1987).

Our laboratory differentiated between 'field' and 'in situ' experiments. The former tests utilize natural, as opposed to cultured phytoplankton but the experiments were conducted in incubators aboard ships. The latter, or in situ experiments, were conducted in the original test site, without the benefit of incubators or other simulating devices. In the past, field/in situ work in the Great Lakes (Fig. 2) consisted mainly of incubations in polycarbonate bottles under constant light regimes at ambient lake temperatures (Vollenweider et al., 1974). For some assays, the bottles were exposed in the lake on moorings to assess the impact of contaminants on the phytoplankton assemblages at various depths in the water column (Munawar et al., 1988b). Flow-through bottles with Nitex nets have been used in Ashbridges Bay, Ontario, by the Ministry of the Environment to assess the impact of effluents on phytoplankton biomass and species composition (K. Nicholls & L. Heintsch, personal communication). Other methods have included various types of enclosures, bags, microcosms/mesocosms, and cages (Goldsborough et al., 1986; Jensen et al., 1972; Gächter, 1976; Munawar & Munawar, 1987; Munawar et al., 1989c).

The field/in situ assessment of contaminants in the Great Lakes utilizing algae as test organisms included the following effluent impact assessment, dredging/disposal impact assessment, and navigational impact assessment. The effluent impact assessment was conducted in Ashbridges Bay (Munawar, 1989) using size-fractionated primary productivity and filtered versus unfiltered assays with offshore phytoplankton. The dredging and disposal assessment dealt with monitoring productivity for pre-dredging and post-dredging activities (Munawar et al., 1989a). Navigational impacts were evaluated using a similar approach

with bioassays being conducted before and after ships' passage in Toronto Harbour (Munawar et al., 1989a).

The environmental bioassay technology is advancing rapidly as a multi-trophic and multi-disciplinary approach and bioassessment is no longer confined to fish lethality testing. Algal assays, ignored in the past, are again receiving considerable attention from both academic and applied researchers. The assays are proving to be sensitive, rapid, and cost-effective. Moreover, the resolution of food chain dynamics along with the microbial loop concept is another valid reason for the study of algae as an effective and early warning indicator of contamination and ecosystem health. Several new techniques (Fig. 2) are developing such as Video Analysis Systems, In situ Plankton Cages, Limited Sample Bioassays, Epifluorescence Microscopy, and Flow Cytometry (Munawar et al., 1989b; 1989c; Munawar & Munawar, 1987; Weisse, 1989a; 1989b; Yentsch et al., 1984; Berglund & Eversman, 1988). These new techniques, together with a wide variety of existing procedures and computer-assisted methodologies, provide an excellent array of tests to assist in environmental protection and conservation of endangered aquatic environments.

Acknowledgements

We would like to thank the organizing committee of the 'International Conference on Environmental Bioassay Techniques and their Application' for inviting this review. We also thank L.H. McCarthy, P. Desroches, and J. Milne for their hard work and assistance in the preparation of the manuscript. We are grateful to Drs. G.-Y. Rhee, C.I. Mayfield and T. Edsall for their comments and review and Mr. H.F. Nicholson for technical editing.

References

Ahlf, W., W. Calmano, J. Erhard & U. Forstner, 1989. Comparison of five bioassay techniques for assessing sediment-bound contaminants. In: M. Munawar, G. Dixon, C. I.

Mayfield, T. Reynoldson & M. H. Sadar (Eds.), Environmental Bioassay Techniques and their Application. Hydrobiologia. 188/189: 285–289.

Allen, E. J. & E. W. Nelson, 1910. On the artificial culture of marine plankton organisms. J. Mar. Biol. Assoc. 8: 421–474.

Azam, F., T. Fenchel, F. G. Field, J. S. Gray, L. A. Meyer-Reil & F. Thingstad, 1983. The ecological role of water-column microbes in the sea. Mar. Ecol. Prog. Ser. 10: 257–263.

Bartlett, L., F. W. Robe & W. H. Funk, 1974. Effect of copper, zinc and cadmium on *Selenastrum capricornutum*. Wat. Res. 8: 179–185.

Berglund, D. L. & S. Eversman, 1988. Flow cytometric measurement of pollutant stresses on algal cells. Cytometry. 9: 150–155.

Blaise, C., R. Legault, N. Bermingham, R. Van Coillie & P. Vasseur, 1986. A simple microplate algal assay technique for aquatic toxicity assessment. Toxicity assessment: An International Quarterly. 1: 261–281.

Blum, J. J., J. R. Sommer & V. Kahn, 1965. Some biochemical, cytological, and morphogenetic comparisons between *Astasia longa* and a bleached *Euglena gracilis*. J. Protozool. 12 (2): 202–209.

Bouck, G. B., 1965. Fine structure and organelle associations in brown algae. J. Cell Biol. 26: 523–537.

Bouck, G. B., 1971. The structure, origin, isolation, and composition of the tubular mastigonemes of the *Ochromonas* flagellum. J. Cell Biol. 50: 362–384.

Braek, G. S., A. Jensen & A. Mohus, 1976. Heavy metal tolerance of marine phytoplankton. III. Combined effects of copper and zinc ions on cultures of four common species. J. Exp. Mar. Biol. Ecol. 25: 37–50.

Bringmann, G. & R. Kuhn. 1980. Comparison at the toxicity thresholds of water pollutants to bacteria, algae, and protozoa in the cell multiplication inhibition test. Wat. Res. 14: 231–241.

Brinkley, B. R. & K. R. Porter, (Eds.) 1977. International Cell Biology 1976-1977. Rockefeller Univ. Press. 694 pp.

Brown, D. L. & G. B. Bouck, 1973. Microtubule biogenesis and cell shape in *Ochromonas*. II. The role of nucleating sites in shape development. J. Cell Biol. 56: 360–378.

Brown, R. M. Jr., W. Herth, W. W. Franke & D. Romanovicz, 1973. The role of the Golgi apparatus in the biosynthesis and secretion of a cellulosic glycoprotein in *Pleurochrysis*: A model system for the synthesis of structural polysaccharides. In: F. Loewus (Ed.), Biogenesis of Plant Cell Wall Polysaccharides. Academic Press, New York. pp. 207–257.

Canadian Environmental Protection Act (CEPA). Enforcement and Compliance Policy. Environment Canada. May, 1988. 58 pp.

Dodge, J. D., 1973. The Fine Structure of Algal Cells. Academic Press, London. 261 pp.

Dodge, J. D., 1974. Fine structure and phylogeny in the algae. Sci. Prog. Oxf., 61: 257–274.

Gächter, R., 1976. Untersuchungen über die Beerinflussung der Planktischer durch anorganische Metallsilze in eutropher Alpanchescee und der mesotrophen Horwerbrucht. Schweiz. Z. Hydrol. 38: 97–119.

Gächter, R., 1979. Effects of increased heavy metal loads on phytoplankton communities. Schweiz. Z. Hydrol. 41: 228–246.

Gibbons, I. R., 1977. Structure and function of flagellar microtubules. In: B. R. Brinkley & K. R. Porter (Eds.) International Cell Biology 1976-1977. Rockefeller Univ. Press. pp. 348–357.

Gibbs, M., 1971. (Ed.), Structure and Function of Chloroplasts. Springer-Verlag, Berlin. 286 pp.

Goldsborough, L. G., G. G. C. Robinson & S. E. Gurney, 1986. An enclosure/substratum system for *in situ* ecological studies of periphyton. Arch. Hydrobiol. 106: 373–393.

Hutchinson, T. C. & P. M. Stokes, 1975. Heavy metal toxicity and algal bioassays. In: Water Quality Parameters. Amer. Soc. Test. Materials, ASTM STP Philadelphia, pp. 320–343.

International Joint Commission, 1987. Guidance on characterization of toxic substances problems in Areas of Concern in the Great Lakes basin. Report to the Great Lakes Water Quality Board, Windsor, Ontario. 179 pp.

Jensen, A., B. Rystad & L. Skoglund, 1972. The use of dialysis culture in phytoplankton studies. J. Exp. Mar. Biol. Ecol. 8: 241–248.

Johnson, P. W. & J. M. Sieburth, 1982. *In situ* morphology and occurrence of eucaryotic phototrophs of bacterial size in the picoplankton of estuarine and oceanic waters. J. Phycol. 18: 318–327.

Lal, R. & D. M. Saxena, 1982. Accumulation, metabolism and effects of organochlorine insecticides on microorganisms. Microbiol. Rev. 46: 95–127.

Lane, J. L. & C. R. Goldman, 1984. Size fractionation of natural phytoplankton communities in nutrient bioassay studies. Hydrobiol. 118: 219–223.

Lehninger, A. L., 1965. The Mitochondrion – Molecular Basis of Structure and Function. W. A. Benjamin, Inc., New York. 263 pp.

Leppard, G. G., D. Urciuoli & F. R. Pick, 1987. Characterization of cyanobacterial picoplankton in Lake Ontario by transmission electron microscopy. Can. J. Fish. aquat. Sci. 44: 2173–2177.

Lustigman, B. K., 1986. Enhancement of pigment concentrations in *Dunaliella tertiolecta* as a result of copper toxicity. Bull. Envir. Contam. Toxicol. 37: 710–713.

Maciorowski, A. F., J. L. Simms, L. W. Little & F. O. Gerrard, 1981. Bioassays, procedures and results. J. Wat. Pollut. Control Fed. 53: 974–993.

Malek, I. and Z. Fencl, 1966. Theoretical and Methodological Basis of Continuous Culture of Microorganisms. Acad. Press. New York. 655 pp.

Mayfield, C. I. & M. Munawar, 1988. Microcomputer-based measurement of algal fluorescence as a potential indicator of environmental contamination. Bull. Envir. Contam. Toxicol. 41: 261–266.

Munawar, M., 1982. Toxicity studies on natural phytoplankton assemblages by means of fractionation bioassays. Can. Tech. Rep. Fish. aquat. Sci. 1152. pp. i–vi, 1–17.

Munawar, M., 1989. Ecosystem health evaluation of Ashbridges Bay environment using a battery of tests (MISA). Unpublished report. Great Lakes Laboratory for Fisheries and Aquatic Sciences, Burlington, Ontario.

Munawar, M. & I. F. Munawar, 1980. The importance of using standard techniques in the surveillance of phytoplankton indicator species for the establishment of long range trends in the Great Lakes: a preliminary example, Lake Erie. Proc. First Biological Surveillance Symposium, 22nd Conference on Great Lakes Research, Rochester, New York. May, 1979. Can. Tech. Rep. Fish. aquat. Sci. 976: 59–89.

Munawar, M. & I. F. Munawar, 1982. Phycological studies in Lakes Ontario, Erie, Huron and Superior. Can. J. Bot. 60 (9): 1837–1858.

Munawar, M. & I. F. Munawar, 1986. The seasonality of phytoplankton in the North American Great Lakes: A comparative synthesis. In; M. Munawar & J. F. Talling (Eds.), Seasonality of Freshwater Phytoplankton: A Global Perspective. Hydrobiol. 138: 85–115.

Munawar, M. & I. F. Munawar, 1987. Phytoplankton bioassays for evaluating toxicity of in situ sediment contaminants. In; R. Thomas, R. Evans, A. Hamilton, M. Munawar, T. Reynoldson & M. H. Sadar (Eds.), Ecological effects of in situ sediment contaminants. Hydrobiol. 149: 87–105.

Munawar, M. & T. Weisse, 1989. Is the microbial loop an early warning indicator of anthropogenic stress. In: M. Munawar, G. Dixon, C. Mayfield, T. Reynoldson & M. H. Sadar (Eds.), Environmental Bioassay Techniques and their Application. Hydrobiol. 188/189: 163–174.

Munawar, M., I. F. Munawar & L. H. McCarthy, 1987a. Phytoplankton ecology of large eutrophic and oligotrophic lakes of North America: Lakes Ontario and Superior. In: M. Munawar (Ed.), Proc. Internat. Symp. on Phycology of Large Lakes of the World. Arch. Hydrobiol. Beih. Ergebn. Limnol. 25: 51–96.

Munawar, M., I. F. Munawar & L. H. McCarthy, 1988a. Seasonal succession of phytoplankton size assemblages and its ecological implications in the North American Great Lakes. Verh. int. Ver. Limnol. 23: 659–671.

Munawar, M., W. P. Norwood & L. H. McCarthy, 1989a. In situ bioassessment of dredging and disposal activities in a stressed ecosystem: Toronto Harbour, Ontario, Canada. In: M. Munawar, G. Dixon, C. Mayfield, T. Reynoldson & M. H. Sadar (Eds.), Environmental Bioassay Techniques and their Application. Hydrobiol. 188/189: 601–618.

Munawar, M., P. T. S. Wong & G. -Y. Rhee, 1988b. The effects of contaminants on algae: an overview. In: N. W. Schmidtke (Ed.), Toxic Contamination in Large Lakes. Lewis Publishers Inc. Chelsea, Michigan. pp. 113–160.

Munawar, M., D. Gregor, S. A. Daniels & W. P. Norwood, 1989b. A sensitive screening bioassay technique for the toxicological assessment of small quantities of contaminated bottom or suspended sediments. Hydrobiol. (In press).

Munawar, M., A. Mudroch, I. F. Munawar & R. L. Thomas, 1983. The impact of sediment-associated contaminants from the Niagara River mouth on various size assemblages of phytoplankton. J. Great Lakes Res. 9 (2): 303–313.

Munawar, M., I. F. Munawar, C. I. Mayfield & L. H. McCarthy, 1989c. Probing ecosystem health: a multidisciplinary and multi-trophic assay strategy. In: M. Munawar, G. Dixon, C. I. Mayfield, T. Reynoldson & M. H. Sadar (Eds.), Environmental Bioassay Techniques and their Application. Hydrobiol. 188/189: 93–116.

Munawar, M., I. F. Munawar, W. Norwood & C. I. Mayfield, 1987b. Significance of autotrophic picoplankton in the Great Lakes and their use as early indicators of contaminant stress. In: M. Munawar (Ed.), Proc. Internat. Symp. on Phycology of Large Lakes of the World. Arch. Hydrobiol. Beih. Ergebn. Limnol. 25: 141–155.

Munawar, M., I. F. Munawar, P. Ross & C. Mayfield, 1987c. Differential sensitivity of natural phytoplankton size assemblages to metal mixture toxicity. In; M. Munawar (Ed.), Proc. Internat. Symp. on Phycology of Large Lakes of the World. Arch. Hydrobiol. Beih. Ergebn. Limnol. 25: 123–139.

Porter, K. G., E. B. Sherr, B. F. Sherr, M. Pace & R. W. Sanders, 1985. Protozoa in planktonic food webs. J. Protozool. 32: 409–415.

Preston, R. D., 1974. The Physical Biology of Plant Cell Walls. Chapman and Hall, London. 491 pp.

Rai, L. C., J. P. Gaur & H. D. Kumar, 1981. Phycology and heavy metal pollution. Biol. Rev. 56: 99–151.

Rhee, G. -Y., 1980. Continuous culture in phytoplankton ecology. In: M. R. Droop & H. W. Jannasch (Eds.), Advances in Aquatic Microbiology. Vol. 2. Academic Press. New York. pp. 151–203.

Rhee, G. -Y., 1982. Overview of phytoplankton contaminant problems. J. Great Lakes Res. 8 (2): 326–327.

Rhee, G. -Y., 1988. Persistent toxic substances and phytoplankton in the Great Lakes. In: M. Evans (Ed.), Toxic Contaminants and Ecosystem Health: A Great Lakes Focus. John Wiley & Sons, New York. pp. 513–525.

Ross, P., V. Jarry & H. Sloterdijk, 1988. A rapid bioassay using the green alga *Selenastrum capricornutum* to screen for toxicity in St. Lawrence River sediment elutriates. In: J. Cairns & J. R. Pratt (Eds.), Functional testing of aquatic biota for estimating hazards of chemicals. STP 988. American Society for Testing and Materials. Philadelphia. pp. 68–73.

Ross, P. E. & M. Munawar, 1987. Zooplankton feeding rates at offshore stations in the North American Great Lakes. In; M. Munawar (Ed.), Proc. Internat. Symp. on Phycology of Large Lakes of the World. Arch. Hydrobiol. Beih. Ergebn. Limnol. 25: 157–164.

Rothfield, L. I., 1971. (Ed.), Structure and Function of Biological Membranes. Academic Press, New York. 486 pp.

246

Schreiber, R., 1927. Die Reinkultur von marinem Phytoplankton und deren Bedeutung für die Erforschung der Produktionsfähigkeit des Meerwassers. Wiss. Meeresunters. Abt. Helgoland. 16: 1–34.

Sherr, E. B. & B. F. Sherr, 1988. Role of microbes in pelagic food webs: a revised concept. Limnol. Oceanogr. 33 (5): 1225–1227.

Sprules, G. & M. Munawar, 1986. Plankton size spectra in relation to ecosystem productivity, size, and perturbation. Can. J. Fish. aquat. Sci. 43: 1789–1794.

Sprules, G., M. Munawar & E. H. Jin, 1988. Plankton community structure and size spectra in the Georgian Bay and the North Channel ecosystems. In: M. Munawar (Ed.), Limnology and Fisheries of Georgian Bay/North Channel Ecosystems. Hydrobiol. 163: 135–140.

Stockner, J. G., 1988. Phototrophic picoplankton: an overview from marine and freshwater ecosystems. Limnol. Oceanogr. 33 (4): 765–775.

Stoermer, E. F., 1978. Phytoplankton as indicators of water quality in the Laurentian Great Lakes. Trans. am. Microsc. Soc. 97: 2–16.

Vollenweider, R. A., M. Munawar & P. Stadelmann, 1974. A comparative review of phytoplankton and primary production in the Laurentian Great Lakes. J. Fish. Res. Bd. Can. 31 (5) 739–762.

Walsby, A. E., 1972. Structure and function of gas vacuoles. Bacteriol. Rev. 36: 1–32.

Weisse, T., 1989a. Dynamics of autotrophic picoplankton in Lake Constance. J. Plank. Res. 10: 1179–1188.

Weisse, T., 1989b. Trophic interactions among heterotrophic microplankton, nannoplankton, and bacteria in Lake Constance (FRG). Hydrobiol. (in press).

Wong, P. T. S., Y. K. Chau & P. L. Luxon, 1978. Toxicity of a mixture of metals on freshwater algae. J. Fish. Res. Bd. Can. 35: 479–481.

Wong, P. T. S., Y. K. Chau & D. Patel, 1983. The use of algal batch and continuous culture techniques in metal toxicity study. In: J. O. Nriagu (Ed.) Aquatic Toxicology. John Wiley & Sons Inc. pp. 449–466.

Wong, P. T. S. & P. Couture, 1986. Toxicity screening using phytoplankton. In: B. J. Dutka & G. Bitton (Eds.). Toxicity Testing Using Microorganisms. Vol. II. CRC Press, Boca Raton, Florida. pp. 79–100.

World Commission Report on Environment and Development, 1987. Our Common Future. Oxford University Press. 400 pp.

Wright, S. J. L., 1978. Interactions of pesticides with microalgae. In: I. R. Hill & S. J. L. Wright (Eds), Pesticide Microbiology. Academic Press. London, New York. pp. 535–602.

Yentsch, C. M., L. Cucci & D. A. Phinney, 1984. Flow cytometry and cell sorting: problems and promises for biological ocean science research. In; O. Holm-Hansen, L. Bolis & R. Gilles (Eds.), Lecture Notes on Coastal and Estuarine Studies. Proceedings organized within the 5th conference of the European Society for Comparative Physiology and Biochemistry – Taormina, Sicily, Italy, Sept. 1983.

Hydrobiologia **188/189**: 247–257, 1989.
M. Munawar, G. Dixon, C. I. Mayfield, T. Reynoldson and M. H. Sadar (eds)
Environmental Bioassay Techniques and their Application.
© *1989 Kluwer Academic Publishers. Printed in Belgium.*

Continuous culture algal bioassays for organic pollutants in aquatic ecosystems

G-Yull Rhee
Wadsworth Center for Laboratories and Research, New York State Department of Health and School of Public Health, State University of New York at Albany, Albany, N.Y. 12201–0509

Abstract

Short-term responses of phytoplankton to organic pollutants are highly transitory. Time-course studies of non-steady state cells in continuous culture showed varying growth or photosynthetic responses such as enchancement, inhibition, adaptation (or development of resistance) or rebound, depending on the direction of changes in the intracellular toxicant concentration and the duration of exposure. However, steady-state cells in a two-stage chemostat system exhibited an increased tolerance to toxicants and subtle physiological effects such as photosynthetic enhancement which was accompanied by a considerable leakage of photosynthesates. It is important to understand such steady-state responses for the prediction and assessment of ecological impact by organic pollution on phytoplankton, since the time scale of changes in the toxicant/biomass ratio in most natural waters is long enough to approximate an equilibrium state.

Introduction

Bioassays have been widely used to assess the impact of toxicant pollution on the aquatic ecosystem (Wong *et al.*, 1987; Rhee, 1988). The principal concern of the pollution is its biological effect. Therefore, the qualitative or quantitative information provided by chemical analysis alone is insufficient, although it is essential for establishing a causal relationship for the biological impact. Planktonic algae are well suited for aquatic bioassays, since (a) they are the ultimate food sources for all aquatic organisms, (b) their growth rate is fast relative to other aquatic organisms, and (c) their growth, especially that of unicellular forms, is easily amenable to kinetic analyses.

Aquatic pollution affects the ecosystem at population as well as community levels. Ideally, bioassays should be simple and easy to carry out and at the same time yield information for both levels, and the information should also be predictive of potential impacts on other food chain organisms or the ecosystem as a whole. However, there is no single method which meets these requirements. Thus, one has to choose or devise a method tailored for the particular objectives. It is relatively simple and straightforward to determine the effects at the population level. This can be accomplished directly by testing a species of concern with known or unknown toxicants or indirectly by using a standard assay organism which has been well studied for its physiological attributes and extrapolate the results to other species. At the community level pollutants may be examined for their effects on community metabolism (McNaught, 1982) or differential impairment of the competitive ability of various species or size classes (Fisher *et al.*, 1974).

Whether it is a unialgal or community assay, the type of system used is highly important, because the ecological pertinence of the results depends on the system as much as experimental designs. This point has been frequently overlooked in the past. Algal assay systems, large or small, can be divided into two general types: closed and open systems. Most commonly used are batch cultures because of their simplicity. They are closed systems with no material input or output. The population size increases continuously until growth ultimately ceases by the exhaustion of a limiting factor while the concentrations of assay substrates change in the opposite direction. Any products formed during growth also accumulate in the system. Therefore, closed systems are inherently transient and highly dynamic, with cell physiology and assay substrate concentrations changing continuously.

For organic pollutants such as chlorinated hydrocarbons, which are strongly hydrophobic and bioaccumulate by simple adsorption and/or absorption (sorption), the bioconcentration is determined by factors such as particle size, organic content, and surface area. Therefore, the cellular concentration of those pollutants in a closed system varies with the inoculum size and population growth, and toxic effects may vary accordingly even at the same toxicant concentration (Harding & Phillips, 1978; Rhee & Kane, unpublished data). In such a system, biological responses observed over a period of time represent an integration of transitory reactions. Even a single point measurement of physiological responses is no more than a snapshot of transient reactions.

Conditions equivalent to those in batch culture are rarely found in the open system of natural waters. In the aquatic ecosystem, there is always continuous material input and output. The time scale for changes in pollutant concentrations is also long relative to the generation time of phytoplankton. Thus, populations may be considered to be in states approximating an equilibrium with respect to toxicant concentrations. Therefore, to predict the ecological impact of aquatic contamination, it is essential to understand long-term steady-state responses. It is even more imperative since organisms can develop resistance to toxic contaminants over time (Rhee, 1988; see Fig. 3 of Kayser, 1976).

Continuous culture systems

Continuous cultures have been widely used in microbial ecology (Tempest, 1970; Veldkamp & Jannasch, 1972; Jannasch & Mateles, 1974; Meers, 1974; Veldkamp, 1976a, b, 1977; Rhee, 1980; Rhee et al., 1982). They have also been employed for toxic algal bioassays albeit infrequently (Fisher et al., 1974; Kayser, 1976; Lederman & Rhee, 1982; Wong et al., 1983; Rhee et al., 1988). Continuous cultures may be largely divided into two types in terms of the flow control. One type uses a measuring pump to maintain a constant flow of fresh medium from a medium reservoir into the culture vessel, simultaneously removing an equal volume of the culture. This is called a chemostat. In the other type, called turbidostat, the flow is most often controlled by a photocell sensing the culture density; whenever the density increases above a preset level, the culture receives fresh medium and the same amount of the culture is removed, thus maintaining a constant culture density. The main virtues of continuous culture systems are their ability to control microbial growth and maintain a defined culture and physiological condition for, in theory, an indefinite period of time. The systems are also easily amenable to mathematical analysis and results are highly reproducible.

There are also variants of chemostats and turbidostats such as chemostats in series, or multistage chemostats (see Pirt, 1975). Multistage chemostats are highly valuable for toxic bioassays because of their ability to maintain steady-state growth under growth inhibition. In a simple chemostat the steady state tends to be highly unstable under growth inhibition and only a limited range of toxicity can be studied (see below). There are also in situ chemostats designed to investigate natural communities in the natural environment (deNoyelles et al., 1980). However, continuous cultures are not capable of recreating the enormously complex conditions that exist in nature regardless of their sophistication (see Rhee, 1980). In fact, they were not designed for it. However, they may be used to investigate one or a few environmental factors in isolation with the aim of reconstructing more complex systems.

(A) *Theory of the chemostat*

The objective of the theory is to establish a quantitative relationship among growth rates (or physiological conditions), biomass production, and substrate concentrations under different conditions (Herbert *et al.*, 1956; Herbert, 1958; Tempest, 1969; Pirt, 1975). In a chemostat fresh growth medium with growth-limiting substrate is introduced continuously into a culture vessel at a constant rate and the culture medium, including organisms, is removed concomitantly at the same rate (Fig. 1). Thus, when fresh medium is introduced into a culture vessel with a volume V (ml) at a rate F (ml/hr) the change in population dX over an infinitely small time interval (dt) can be expressed as

$$dX/dt = \mu \cdot X - D \cdot X = X(\mu - D), \qquad (1)$$

where X is biomass, μ is the specific growth rate, and D is dilution rate, F/V. Equation 1 may be rewritten as

$$\mu = D - (\ln dX)/dt . \qquad (2)$$

At steady state, $dX/dt = 0$ and therefore from Equation 1

$$\mu = D . \qquad (3)$$

Thus, in a chemostat the growth rate or the nutritional status of cells can be determined by the dilution rate selected by the operator. This is a major advantage of the system.

The balance of the growth-limiting nutrient may be expressed as

$$dS/dt = D \cdot S_r - D \cdot S - \mu \cdot X/Y, \qquad (4)$$

where S_r and S are the concentration of the limiting nutrient in the reservoir and the culture

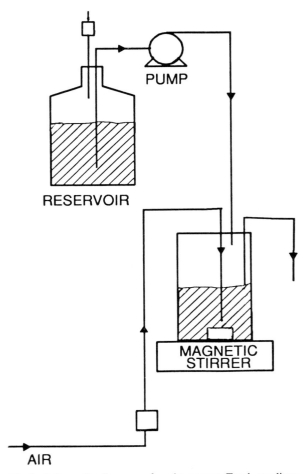

Fig. 1. Schematic diagram of a chemostat. Fresh medium with a growth-limiting substrate is continuously introduced into the culture vessel at a constant rate and the culture volume is kept constant by simultaneous removal of culture.

vessel, respectively, and Y is the growth yield, dX/dS. In the steady state $dS/dt = 0$ and therefore,

$$D(S_r - \tilde{s}) - \mu\tilde{x}/Y = 0 , \qquad (5)$$

where $\tilde{s}$ and $\tilde{x}$ are the steady-state concentration of the limiting nutrient and biomass, respectively.

The relationship between the growth rate and the limiting nutrient concentration is expressed by the model given by Monod (1942)

$$\mu = \mu_m \cdot S/(K_s + S) , \qquad (6)$$

250

Where μ_m is the maximum growth rate and K_s is the half saturation constant or S when $\mu = \mu_m/2$.

Substituting μ with D in Equation 6 for steady state,

$$\tilde{s} = K_s \cdot D/(\mu_m - D) . \qquad (7)$$

Substituting μ and $\tilde{s}$ in Equation 5 with D and Equation 7, respectively,

$$\tilde{x} = Y(S_r - \tilde{s}) =$$
$$= Y\{S_r - K_s \cdot D/(\mu_m - D)\} . \qquad (8)$$

Equation 8 shows that, in the chemostat, the steady-state biomass $\tilde{x}$ is a dependent function of the dilution rate D.

The steady-state cellular concentration of the assay toxicant $\tilde{q}_a$ may be expressed as

$$\tilde{q}_a = (A_r - \tilde{a})/\tilde{x} , \qquad (9)$$

where A_r and $\tilde{a}$ are the reservoir concentration of the toxicant and its steady-state concentration in the culture vessel, respectively. Equation 9 assumes the absence of adsorption to surfaces during the medium delivery. In chemostats, if assay substrates reduce the growth rate below the dilution rate, the culture would be washed out. [If the dilution rate is reduced to prevent washout, this would introduce a new variable by increasing the degree of nutrient limitation.] Therefore, only a limited range of toxicity may be examined in chemostats.

The chemostat theory requires that (1) one nutrient must be limiting in the culture medium, (2) the culture must be homogeneously mixed without surface growth or clumping, and (3) a constant proportion of cells must be viable.

(B) Theory of the turbidostat

The turbidostat may be considered as a chemostat with a photocell to continuously monitor the turbidity of the culture and add fresh medium when the biomass increases above a predetermined level. The turbidity control may also be accom-plished by sensing metabolic products linked to growth such as O_2, CO_2, organic acids, or changes in pH. The culture volume is kept constant by either gravity overflow or some withdrawal devices. Thus, the dilution rate is allowed to adjust itself (Fig. 2). The system eventually reaches a steady state. Since the system maintains a constant population density, it is useful for investigating those toxic contaminants which require the use of a constant population size.

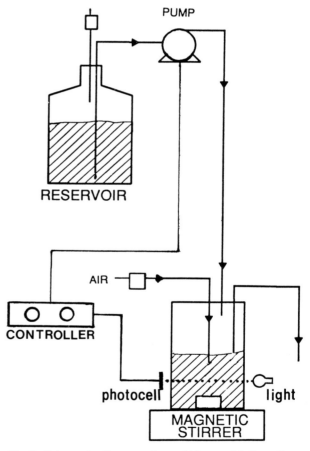

Fig. 2. Schematic diagram of a turbidostat. Medium flow from the reservoir into the culture vessel is controlled by the signal of a photocell sensing the turbidity of the culture. When the culture opacity exceeds a preset value, the reservoir medium flows into the culture and at the same time the same volume of culture is removed by a constant level device or gravity overflow.

A major difference between the chemostat and the turbidostat is that while dilution rate is an independent variable and biomass is a dependent one in a chemostat, in a turbidostat biomass is an independent variable and dilution rate is a dependent one. However, the population kinetics for a turbidostat is the same as that for a chemostat (Equations 1 to 9). In the chemostat, there is an upper limit of the dilution rate corresponding to μ_m, and culture washout will take place at or above this value (Equation 1). Therefore, although it can attain a steady state at values below μ_m controlled by a single growth-limiting nutrient (Equation 6), it cannot obtain a steady state near or at the maximum growth rate. In the turbidostat, the culture density controls the dilution rate and thus, the culture can attain a steady state even at the maximum growth rate in nutrient sufficiency. Thus, the chemostat and turbidostat are complementary. In the bioassay of toxic contaminants, the turbidostat may be utilized to investigate toxicity in healthy nutrient sufficient cells whereas the chemostat may be used to find the effects on cells stressed by various degrees of nutrient limitation. A comparative study of the two culture systems may be used to elicit the sensitivity to toxic contaminants as affected by the nutritional status of cells (Lederman & Rhee, 1981a).

(C) Theory of the two-stage chemostat

There are many different derivations of the chemostat such as those with biomass feedback for a maximum biomass production and multi-stage systems connecting chemostats in series (see Pirt, 1975). Among other applications, multi-stage chemostats are used to obtain stable steady-state growth under growth inhibition when it is difficult to attain a steady state in the simple chemostat or turbidostat. Thus, this system is highly valuable for toxic bioassays, although its application has so far been limited (Jones *et al.*, 1973; Rhee *et al.*, 1988).

In two chemostats joined in series with the medium fed to both the first and second stages

(Fig. 3), the dilution rate of the second stage, D_2, is given by

$$D_2 = D_{02} + D_{12}, \qquad (10)$$

where D_{02} is the dilution rate of the second stage chemostat by the input from the second stage reservoir and D_{12} is the dilution rate by the overflow from the first stage. The biomass balance in the second stage is

$$dX_2/dt = \mu_2 \cdot X_2 + D_{12} \cdot X_1 - D_2 \cdot X_2, \quad (11)$$

where X_1 and X_2 are biomass in the stage 1 and 2 respectively and μ_2 is the growth rate in the second stage. In the steady state when $dX_2/dt = 0$,

$$\mu_2 \cdot \tilde{x}_2 + D_{12} \cdot \tilde{x}_1 - D_2 \cdot \tilde{x}_2 = 0, \qquad (12)$$

thus,

$$\mu_2 = D_2 - D_{12} \cdot \tilde{x}_1/\tilde{x}_2 \qquad (13)$$

and

$$\tilde{x}_2 = D_{12} \cdot \tilde{x}_1/(D_2 - \mu_2). \qquad (14)$$

Unlike in the single-stage chemostat, the growth rate in the second stage of two-stage chemostats, μ_2, is a dependent variable and $\mu_2 < D_2$ (Equation 13). Equation 14 also shows that as long as there is a finite value of $\tilde{x}_1$, $\tilde{x}_2$ will have a positive value regardless of how large D_2 is. Thus, there cannot be a washout. This relationship enables the second stage to attain a steady state under nutrient limitation as well as nutrient sufficiency in contrast to a single-stage chemostat.

Since the second stage cannot have a washout as long as $\tilde{x}_1$ is finite, steady-state growth can also be obtained in the second stage at various degrees of growth inhibition by toxic chemicals. This can be readily accomplished by growing the culture in a toxicant-free medium in the first stage and introducing a medium containing an assay toxicant to the second stage. In a single-stage chemostat, on the other hand, if inhibitory effects reduce the

252

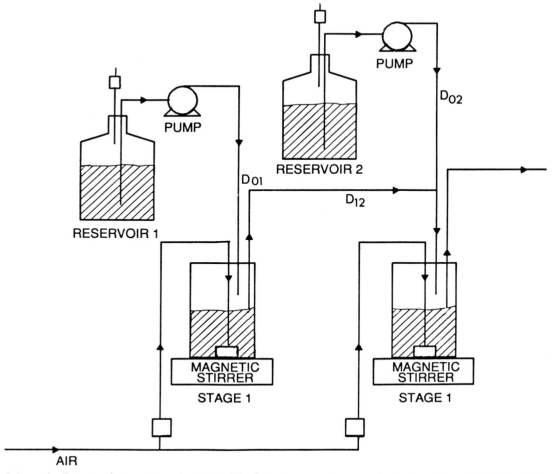

Fig. 3. Schematic diagram of a two-stage chemostat. The first stage operates as a single-stage chemostat in Fig. 1. However, its overflow goes to the second stage culture vessel, which also receives fresh medium from its own reservoir at a constant rate. The culture volume of the second stage is kept constant by simultaneous removal of the culture at the same rate as the input rate.

growth rate below the dilution rate, washout will take place and as mentioned above, even when μ is higher than D the steady state tends to be unstable.

As Equation 13 shows, μ_2 can be regulated by controlling D_2 while holding D_{12} constant and vice versa, or controlling both simultaneously. The balance of limiting nutrients for the second stage is

$$dS_2/dt = D_{12} \cdot S_1 + D_{02} \cdot S_{02} - \\ - D_2 \cdot S_2 - \mu_2 \cdot X_2/Y. \qquad (15)$$

Substituting μ_2 with Equation 13 and putting $\tilde{x}_1 = Y(S_{01} - \tilde{s}_1)$ for the steady state, we obtain

$$\tilde{x}_2 = Y\{(D_{12}/D_2)S_{01} + \\ + (D_{02}/D_2)S_{02} - \tilde{s}_2\}. \qquad (16)$$

If the assay toxicant enters from the second-stage reservoir, the balance of toxicants in the absence of any sorption to surfaces can be expressed as

$$dA_2/dt = D_{02} \cdot A_{02} - D_2 \cdot A_2 - \\ - \mu_2 \cdot \tilde{x}_2 \cdot Q_a, \qquad (17)$$

---

where A_{02} and A_2 are the toxicant concentration in the second stage reservoir and the second stage vessel, respectively, and Q_a is the cellular toxicant concentration. In the steady state,

$$D_{02} \cdot A_{02} - D_2 \cdot \tilde{a}_2 - \mu_2 \cdot \tilde{x}_2 \cdot Q_a = 0 , \quad (18)$$

where $\tilde{a}_2$ is the steady-state toxicant concentration in the culture vessel. Thus,

$$\tilde{a}_2 = (D_{02} \cdot A_{02} - \mu_2 \cdot \tilde{x}_2 \cdot \tilde{q}_a)/D_2 , \quad (19)$$

$$\tilde{a}_2 = (D_{02} \cdot A_{02} - D_2 \cdot \tilde{a}_2)/\mu_2 \cdot \tilde{x}_2 . \quad (20)$$

Effects of a hexachlorobiphenyl in the chemostat

Some unexpected long-term responses were revealed in chemostat bioassays of PCBs. Growth enhancement, inhibition, adaptation (or development of resistance), and a rebound phenomenon were observed as cellular PCB concentrations changed. *Fragilaria crotonensis, Ankistrodesmus falcatus*, and *Microcystis* sp., all isolates from the Great Lakes, were investigated in a P-limited chemostat using 2, 4, 5, 2′, 4′, 5′-hexachlorobiphenyl (HCBP) as a toxicant (Lederman & Rhee, 1982a). The culture was first grown to a steady state in a HCBP-free medium and when a steady state was reached, the HCBP-free reservoir medium was replaced with one containing HCBP. The ^{14}C-labeled HCBP was solubilized to saturation without a carrier solvent. Although Teflon tubing was used to the maximum extent possible for the medium inflow to minimize loss due to adsorption, surface adsorption reduced the concentration of HCBP considerably before reaching the culture vessel. Therefore, HCBP concentrations in the liquid and algal phases in the culture vessel had to be directly determined. The effects were examined at three different states of P limitation represented by dilution rates 0.22, 0.39, and 0.60 d^{-1} corresponding to the relative growth rates (μ/μ_m) of 0.28, 0.50, and 0.77.

Cellular HCBP in *Fragilaria crotonensis* increased exponentially until it reached maximum (Fig. 4). Most HCBP was partitioned to cells due to its high partition coefficient (in the order of 10^6, Lederman & Rhee, 1982b). The intracellular level was a function of both the dilution rate and the culture density. Therefore, even with the same reservoir HCBP concentration, different intracellular concentrations resulted. (The varying cell HCBP contents at different dilution rates made it impossible to determine the sensitivity of cells as affected by different degrees of nutrient limitation.]

Daily growth rates of *F. crotonensis* were calculated by Equation 2. The general pattern of changes in the growth rate relative to the rate in the HCBP-free state (μ_{HCBP}/D) was very similar at all dilution rates. There was an initial enhancement on day 1 or day 2. It was followed by inhibition with further intracellular accumulation. It was most interesting to note, however, that as the intracellular concentration began to stabilize at maximum values, there was a trend for recovery. This strongly suggested the importance of understanding long-term effects. The cultures were unable to attain a stable steady state before cells started to grow on the wall and experiments had to be terminated. Therefore, we failed to observe true steady-state responses.

At the dilution rate 0.6 d^{-1}, the reservoir with HCBP medium was replaced with HCBP-free medium when cells attained a high steady intracellular HCBP concentration. As the intracellular concentration decreased, μ_{HCBP}/D showed a recovery and then a rebound to values much higher than the HCBP-free steady state ratio. This indicated that the transient responses could vary even at the same intracellular concentration, depending on the direction of changes.

Both *Ankistrodesmus falcatus* and *Microcystis* sp. did not exhibit any enhancement, inhibition, or rebound effects in the course of intracellular HCBP accumulation at comparable concentrations, exhibiting higher resistance than *F. crotonensis*. Similar wide interspecies differences in sensitivities have also been found with pentachlorophenol (Gotham & Rhee, 1982).

254

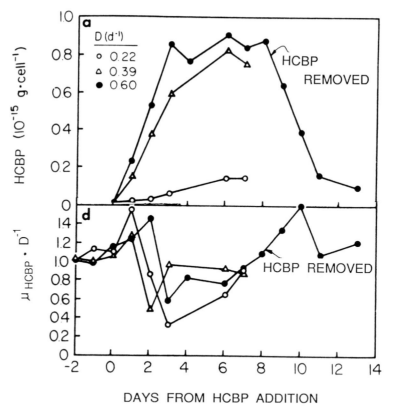

Fig. 4. Time course of changes in cellular HCBP concentrations and growth rate in P-limited *Fragilaria crotonensis* at three different dilution rates. μ_{HCBP} was calculated by Equation 2 and was normalized by the dilution rate. (Adapted from Lederman & Rhee, 1982a).

Effects of hexachlorobiphenyl in the nutrient-sufficient turbidostat

To investigate whether nutrient limitation might increase sensitivity to toxicants, nutrient-sufficient turbidostats were used in parallel with the P-limited chemostats. As with chemostat studies, the culture was allowed to reach a steady state in HCBP-free medium before the reservoir medium was replaced with a HCBP-containing medium. Since a constant culture density was maintained by feedback regulation of the dilution of the culture, the growth rate was monitored simply as a change in overflow volume (Fig. 5).

The general response pattern was strikingly similar to that of the chemostats. In one experiment there was an initial increase in the growth rate as HCBP accumulated. After this initial in-

crease, the relative growth rate (μ_{HCBP}/μ_m) varied by and large as a mirror image of cellular HCBP concentrations until the HCBP reservoir medium was replaced with HCBP-free medium. Upon replacement of the medium, cell HCBP decreased while the growth rate started to recover. As cellular HCBP decreased further, the growth rate not only recovered but rebounded to exceed the maximum rate. As cellular HCBP decreased to values near 0, it appeared to return to the μ_m. We were unable to determine the effects of nutrient limitation on the organisms' sensitivity to the congener, because intracellular HCBP concentrations were different between turbidostat and chemostat cultures.

As with chemostat cultures, the turbidostat culture exhibited a high instability before the experiments were terminated due to wall growth. Wall

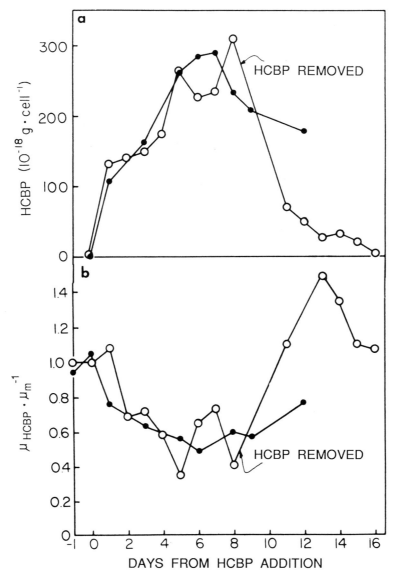

Fig. 5. Time course of changes in cellular HCBP concentrations and growth rate for two nutrient-sufficient *Fragilaria crotonensis* cultures. For one culture (o) the HCBP medium in the reservoir was replaced with HCBP-free medium on the day 8. (From Lederman & Rhee, 1982a).

growth in the turbidostat results in inaccurate photocell readings, which in turn make overflow unrelated to true growth.

Toxic bioassay in two-stage chemostat and steady-state effects

A two-stage chemostat was used to examine the kinetics of growth inhibition by PCBs and true steady-state responses (Rhee *et al.*, 1988). Jones *et al.* (1973) elegantly demonstrated the application of two-stage chemostats in their study of the growth kinetics of phenol-oxidizing bacteria. Phenol was a growth-limiting substrate at low concentrations but a growth inhibitor at high concentrations. Both growth limitation and inhibition were fit by the Haldane's noncompetitive enzyme inhibition kinetics. Thus, substrate affinity and toxicity could be defined quantitatively in terms of a species specific half-saturation constant, K_s, and inhibition constant K_i, respectively.

The green alga *Selenastrum capricornutum* was chosen for this study mainly because of the ease of their counting by a particle counter (Rhee *et al.*, 1988). They were studied in a P-limited two-stage chemostat (Fig. 3) using ^{14}C-labeled 2, 5, 2′, 5′-tetrachlorobiphenyl (2, 5, 2′, 5′-TCBP) as a growth inhibitor. The first stage was only P-limited and the second stage received the P-limited medium saturated with the TCBP (Rhee *et al.*, 1988). In parallel control experiments, the second stage received TCBP-free medium. At cellular TCBP concentrations ranging from $12–17 \times 10^{-8}$ ng cell^{-1}, the growth rate was not affected. However, both the photosynthetic capacity and photosynthetic efficiency measured by the ^{14}C uptake technique were higher in the TCBP cultures [64.9×10^{-9} µmol C·cell^{-1}·h^{-1} and 0.91×10^{-9} µmol C·cell^{-1}·h^{-1}·(µEin·m^2·s^{-1})$^{-1}$ vs. 32.9×10^{-9} µmol C·cell^{-1}·h^{-1} and 0.49×10^{-9} µmol C·cell^{-1}·h^{-1}·(µEin·m^2·s^{-1})$^{-1}$].

Carbon balance for cell C may be expressed as

$$dQ_c/dt = V_c - \mu \cdot Q_c , \qquad (21)$$

where Q_c is the cell carbon concentration and V_c is the short-term gross carbon fixation rate. At steady state when $dQ_c/dt = 0$,

$$V_c = \mu \cdot \tilde{q}_c , \qquad (22)$$

where $\tilde{q}_c$ is the steady-state cell C concentration. $\mu \cdot \tilde{q}_c$ represents the *net* C fixation rate. Since the light intensity in the chemostat was saturating for photosynthesis, the calculated net photosynthetic rate should be the same as the photosynthetic capacity which may be considered as the short-term or gross photosynthetic rate if there is no significant loss due to respiration or leaking. For the control culture, the calculated net rate was little different from the rate determined by ^{14}C uptake, but the net rate for the steady-state TCBP cells (41.4×10^{-9} µmol C·cell^{-1}·h^{-1}) was about 36% less than the photosynthetic capacity.

There was little difference in the respiration rate between the control and TCBP cultures. Therefore, about 36% of the fixed carbon appeared to have been excreted by TCBP cells (Rhee *et al.*, 1988). This is a good illustration of the advantages of chemostat cultures.

We failed to observe growth inhibition in this study because it was not possible to raise sufficiently the TCBP concentration in the culture vessel. The difficulty was due to (a) the low aqueous solubility of the congener and (b) its absorption loss to surfaces before reaching the culture vessel from the reservoir.

To increase the toxicant concentration in the culture vessel, the toxicant delivery method was improved. Instead of dissolving directly in the reservoir medium, a measured quantity of 2, 4, 2′, 4′-tetrachlorobiphenyl (2, 4, 2′, 4′-TCBP) with the ^{14}C-labeled tracer was dissolved in hexane and fine glass beads were coated with the TCBP by soaking them in the hexane solution and evaporating the hexane. [No more ^{14}C-labeled 2, 5, 2′, 5′-tetrachlorobiphenyl was available at this time.] A glass cartilage filled with these glass beads was fitted to the medium inflow tubing nearest to the culture vessel so that the medium would pass through it with a minimum loss to surface sorption. The total concentration of the TCBP in the culture vessel stabilized at a constant level after about 3 days. Using this method, its cellular concentration in *S. capricornutum* could be raised to between 6.1 and 9.6×10^{-5} ng·cell^{-1}, with the total concentration in the vessel ranging from 45–68 µg·l^{-1}. Even at these high concentrations, no growth inhibition was noted. However, the photosynthetic rate (at 130 µEin·m^2·s^{-1}) was slightly inhibited, unlike the rate obtained with steady-state cells in the 2, 5, 2′, 5′-tetrachlorobiphenyl medium. These results show a high resistance of the alga to the congener.

Conclusions

The acute toxicity of environmental contaminants is relatively easy to detect by a simple bioassay. It is the long-term sublethal toxicity which is difficult to assess. The long-term steady state effects are of primary ecological concern, since the time scale of changes in the ratio of the toxicant to the biomass in most aquatic environments is sufficiently long to consider them to be approximately in equilibrium.

Closed system bioassays are in general satisfactory for the cursory examination of lethal and acute toxicity, but are of little value in assessing long-term chronic impact. The problem stems largely from the highly transient nature of the system and the bioconcentration behavior of most major organic pollutants, which is strictly governed by physical sorption. The open system bioassay of continuous cultures is far better suited for the evaluation of environmental impact. The dynamics of continuous culture systems tends to move toward an equilibrium state. Since there is also continuous input of fresh medium, the assay can be performed at low environmental concentrations while maintaining a constant biomass. Theoretically, steady state can be maintained indefinitely unless a deviation of ideal growth such as mutation, wall growth or cell clumping occurs. Admittedly, continuous culture bioassays are time-consuming and require skilled operators. However, this disadvantage may in part be offset with highly reproducible results and information which is ecologically vital, but otherwise unobtainable with other approaches. Continuous culture algal bioassays of PCBs, for example, have shown a variety of sublethal effects such as enhancement of growth, photosynthesis, and P uptake as well as their inhibition, growth rebound, the development of resistance, and the leakage of photosynthesates.

References

DeNoyelles, F., Jr., R. Knoechel, D. Reinke, D. Treanor & C. Altenhofer, 1980. Continuous culturing of natural phytoplankton communities in the experimental lake area: effects of enclosure, *in situ* incubation, light, phosphorus, and cadmium. Can. J. Fish. Aquat. Sci. 37: 424–433.

Fisher, N. S., E. J. Carpenter, C. C. Remsen & C. F. Wurster, 1974. Effects of PCB on interspecies competition in natural and gnotobiotic phytoplankton communities in continuous and batch culture. Microb. Ecol. 1: 39–50.

Gotham, I. J. & G-Y. Rhee, 1982. Effects of a hexachlorobiphenyl and a pentachlorophenol on growth and photosynthesis of phytoplankton. J. Great Lakes Res. 8: 328–335.

Harding, L. W. Jr. & J. H. Phillips Jr., 1978. Polychlorinated biphenyls (PCB) effects on marine phytoplankton photosynthesis and cell division. Mar. Biol. 49: 93–101.

Herbert, D., 1958. Some principles of continuous culture. *In* G. Tunevall (Ed.), Recent Progress in Microbiology, 7th International Congress of Microbiology Symposium. Almquvist and Wiksell, Stockholm: 381–396.

Herbert, D., R. Elsworth & R. C. Telling, 1956. Continuous culture of bacteria; a theoretical and experimental study. J. gen. Microbiol. 14: 601–622.

Jannasch, H. W. & R. I. Mateles, 1974. Experimental bacterial ecology studied in continuous culture. *In* A. H. Rose & D. W. Tempest (Eds.), Advances in Microbial Physiology, Vol. 11. Academic Press, London/N.Y.: 165–212.

Jones, G. L., F. Jansen & A. J. McKay, 1973. Substrate inhibition of the growth of bacterium NCIB 8250 by phenol. J. gen. Microbiol. 74: 139–148.

Kayser, H., 1976. Waste-water assay with continuous algal cultures: the effect of mercuric acetate on the growth of some marine dinoflagellates. Mar. Biol. 36: 61–72.

Lederman, T. C. & G-Y. Rhee, 1982a. Influence of a hexachlorobiphenyl in Great Lakes phytoplankton in continuous culture. Can. J. Fish. Aquat. Sci. 39: 388–394.

Lederman, T. C. & G-Y. Rhee, 1982b. Bioconcentration of a hexachlorobiphenyl in Great Lakes planktonic algae. Can. J. Aquat. Sci. 39: 380–387.

McNaught, D. C., D. Griesmer, M. Buzzard & M. Kennedy, 1980. Inhibition of Productivity by PCB's and Natural Products in Saginaw Bay, Lake Huron. U.S. EPA, Environ. Res. Lab., Duluth.

Meers, J. L., 1974. Growth of bacteria in mixed cultures. *In* A. Laskin and H. Lechevalier (Eds.), Microbial Ecology. CRC Press: 136–181.

Monod, J., 1942. Recherches sur la Croissance des Cultures Bactériennes. 2nd ed. Hermann, Paris.

Pirt, S. J., 1975. Principles of Microbe and Cell Cultivation. John Wiley & Sons, N.Y.

Rhee, G-Y. (1980). Continuous culture in phytoplankton ecology. *In* M. R. Droop and H. W. Jannasch, (Eds.), Advances in Aquatic Microbiology, vol. 2. Academic Press, London/N.Y.: 151–203.

Rhee, G-Y., 1988. Persistent toxic substances and phytoplankton in the Great Lakes. *In* M. S. Evans (Ed.), Toxic Contaminants and Ecosystem Health: A Great Lakes Focus, John Wiley & Sons, N.Y.: 513–525.

Rhee, G-Y., I. J. Gotham & S. W. Chisholm, 1982. Use of cyclostat cultures to study phytoplankton ecology. *In* P. C. Calcott (Ed.), Continuous Culture of Cells, Vol. 2. CRC Press, Boca Raton, Fl. 159–186.

Rhee, G-Y., L. Shane & A. DeNucci, 1988. Steady-state effects of 2, 5, 2′, 5′-tetrachlorobiphenyl on growth, photo-

258

synthesis, and P uptake in *Selenastrum capricornutum*. Appl. Envir. Microbiol. 54: 1394–1398.

Tempest, D. W., 1969. The continuous cultivation of microorganisms. I. Theory of chemostat. *In* J. R. Norris & D. W. Ribbons (Eds.), Methods in Microbiology, Vol. 2. Academic Press, London/N.Y.: 259–276.

Tempest, D. W., 1970. The place of continuous culture in microbiological research. *In* A. H. Rose & J. F. Wilkinson (Eds.), Advances in Microbial Physiology, Vol. 4. Academic Press, London/N.Y.: 223–250.

Veldkamp, H., 1976a. Mixed culture study with the chemostats. *In* A. C. R. Dean, D. C. Elwood, C. G. T. Evans & J. Melling (Eds.), Continuous Culture 6: Application and New Fields, Ellis Horwood, Chichester: 315–328.

Veldkamp, H., 1976b. Continuous Culture in Microbial Physiology and Ecology. Meadowfield Press, Ltd., Durham, England.

Veldkamp, H., 1977. Ecological studies with chemostats. *In* M. Alexander (Ed.), Advances in Microbial Ecology, Vol. 1. Plenum Press, N.Y.: 59–94.

Veldkamp, H. & H. W. Jannasch, 1972. Mixed culture studies with the chemostat. J. Appl. Chem. Biotech. 22: 105–123.

Wong, P. T. S., Y. K. Chau & D. Patel, 1983. The use of algal batch and continuous culture techniques in metal toxicity study. *In* J. O. Nriagu (Ed.), Aquatic Toxicology. John Wiley & Sons, N.Y.: 449–466.

Hydrobiologia **188/189**: 259–268, 1989.
M. Munawar, G. Dixon, C. I. Mayfield, T. Reynoldson and M. H. Sadar (eds)
Environmental Bioassay Techniques and their Application.
© 1989 *Kluwer Academic Publishers. Printed in Belgium.*

Round Robin testing with the *Selenastrum capricornutum* microplate toxicity assay

C. Thellen[1], C. Blaise[2], Y. Roy[3] & C. Hickey[4]

[1]*Environnement Québec, 2700 rue Einstein, Ste-Foy, Québec, Canada, G1P 3W8;* [2]*Environnement Canada, Centre Saint-Laurent, 1001 Pierre Dupuy, Longueuil, Québec, Canada, J4K 1A1;* [3]*Eco-Recherches (Canada) Inc., 121 Blvd Hymus, Pointe Claire, Québec, Canada, H9R 1E6;* [4]*Department of Scientific and Industrial Research, P.O. Box 11–115, Hamilton, New Zealand*

Key words: *Selenastrum capricornutum*, microplate bioassay, round robin, CdCl$_2$, phenol

Abstract

Three Quebec-based ecotoxicological laboratories participated in an intercalibration exercise to assess the performance of a recently-published cost-efficient algal microplate toxicity assay. Three test series were carried out with six operators (2 from each laboratory) and two reference toxicants (Cd^{2+} as CdCl$_2$ and phenol). Variables included algal cultivation technique (series 1), presence or absence of Na$_2$EDTA in the growth medium (series 2), and passive or active gas exchange during incubation (series 3). Control growth variability conferred an overall test precision of 8.7% (coefficient of variation obtained for 204 microplate tests). Cadmium (96 h EC50 = 56 μg·l^{-1}) and phenol (96 h EC50 = 69.7 mg·l^{-1}) toxicity test reproducibility was reflected by coefficients of variation of 24.3% and 34.9%, respectively. Algal cultivation technique, whether standardized or 'in house', had no effect on toxicity results. Na$_2$EDTA, as part of the growth medium, significantly ameliorated algal growth and toxicity. While active gas exchange during microplate incubation significantly improved growth, toxicity results were unaffected. Phenol volatility was found to have a marked influence on algal growth. This effect can be offset, however, by providing appropriate modifications to better seal individual wells and to improve experimental design.

Introduction

The use of simple, practical and cost-efficient microbioassays, whose popularity began in the 1970's (Maciorowski *et al.*, 1980), has significantly expanded during the present decade (Bitton & Dutka, 1986). Indeed, their attractive features offer many advantages for a variety of ecotoxicological investigations (Blaise *et al.*, 1988). In particular, tests with 96-well microtiter plates, extensively employed in the field of clinical microbiology for many years, are now being applied increasingly for environmentally-related studies (Blaise *et al.*, 1986; Blanck *et al.*, 1984;

Dive, 1981; Hassett *et al.*, 1981; Lukavsky, 1983; Maul & Block, 1983; Xu *et al.*, 1987). More specifically, microplate techniques have been developed successfully to assess the toxicity of aqueous solutions to micro-algae (Blaise *et al.*, 1986; Blanck *et al.*, 1984; Lukavsky, 1983), whose trophic importance in aquatic ecosystems is critical (Christensen & Scherfig, 1979).

In this paper, we present the results of an intercalibration exercise conducted on the *Selenastrum capricornutum* algal microplate toxicity assay developed by Blaise *et al.*, 1986. The basic objectives were to 1) verify test reliability and reproducibility, 2) verify the influence of specific test

variables and 3) optimize the experimental protocol following data assessment.

Materials and methods

Experimental plan

Three Quebec-based laboratories, engaged in ecotoxicological activities, participated in the round
robin exercise. Each laboratory provided two
operators for this study which involved three test
series and two reference toxicants (Cd^{2+} as
$CdCl_2$ and phenol). For each of the latter, microplate tests were conducted in triplicate by each
operator. Toxicant stock solutions were prepared
by the Environment Quebec Laboratory and distributed to the other two laboratories.

For each of the test series, the standardized
protocol was that described by Blaise et al., 1986
and Blaise, 1986. For the purposes of this work,
the algal cultivation technique, from which test
cell inocula were drawn, was also standardized
and consisted of an Environment Quebec-design-
ed system ensuring a constant supply of exponen-
tially-growing cells (details concerning system set-
up and operation can be requested from the first
author). Each laboratory used its own test culture
of Selenastrum capricornutum (American Type
Culture Collection No. 22662, 12301 Parklawn
Drive, Rockville, Maryland 20852, U.S.A.).

In each test series, a variable was introduced to
assess its influence on algal growth and toxicity
results. The 'in house' algal cultivation technique
followed by each laboratory was the variable
chosen for the first test series: this rudimentary
cultivation technique essentially involves weekly
transfers of cells growing in one culture flask to
another and does not necessarily ensure that ex-
ponentially-growing cells will be harvested for
inoculating purposes. Test series 2 variable was
Algal Assay Procedure (AAP) growth medium
(Chiaudani & Vighi, 1978) without its Na_2EDTA
(Na ethylenedinitilotetra-acetate) supplement.
Finally, test series 3 variable sought to promote
active gas exchange in test microplates by repetitive multichannel pipetting of well contents (3 to

4 quick volume uptakes and releases) on days 1,
2 and 3 of the 96 h test exposure period. In the
standardized test protocol, microplates are incubated without shaking.

Experimental disposition of microplates

The experimental set-up of microplates is shown
in Fig. 1. A total of 72 wells were employed to
accommodate 8 serially-diluted toxicant concentrations, each with 9 replicates, while 24 wells
were reserved for controls. After 96 h, three 100 μl
aliquots from each of three wells were pooled to
eventually generate one cell count with the help of
an electronic particle counter (Blaise, 1986).
Hence, three and eight cell counts were obtained
for each toxicant concentration and control
growth, respectively, which were then averaged.

Data analysis

Per cent growth inhibition in relation to controls
were calculated and associated EC50's were
determined by interpolated regression analysis
with the help of a computer program. Student t,
one-way ANOVA and Scheffé tests were performed to indicate significant differences linked to
specific variables (Sokal & Rohlf, 1981).

Results and discussion

Reliability of the microplate technique

The overall performance of the microtest undertaken with the standardized experimental protocol described is presented in Tables 1, 2 and 3. An
estimate of test precision is reflected in calculated
coefficients of variation (C.V.'s), derived from intra-microplate algal control growths (Table 1). All
six laboratory operators were able to display adequate precision limits which corroborate earlier
results (reported C.V. range of 1.2% to 18.7% in
96 h cell count-derived control growths for 23
tests: Blaise et al., 1986). Highest variation came

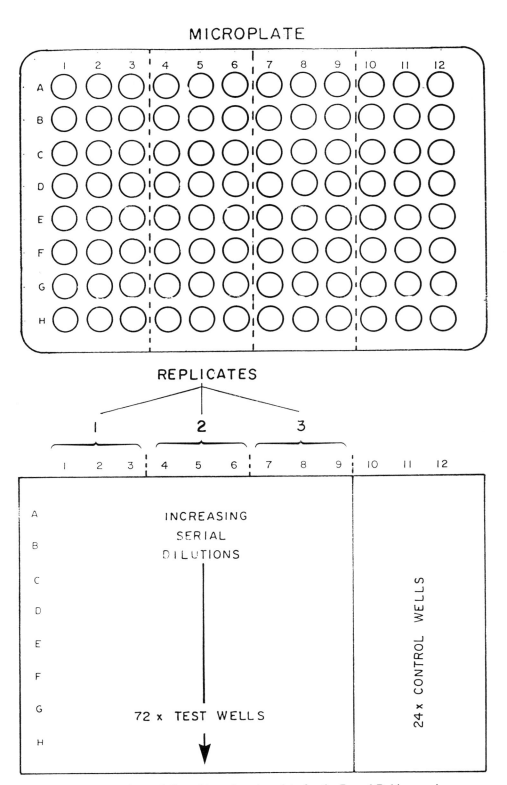

Fig. 1. Experimental disposition of a microplate for the Round Robin exercise.

Table 1. Intra-microplate algal control growth variability[a] with the standardized protocol.

		Lab 1		Lab 2		Lab 3	
		Operator		Operator		Operator	
		1	2	1	2	1	2
Phenol	$\overline{X}$	6.5	7.5	11.8	10.4	16.2	12.2
	S.D.	2.0	3.1	5.5	6.2	5.8	4.1
	n	18	18	15	15	18	18
Cd	$\overline{X}$	7.9	8.5	6.4	8.3	5.8	4.8
	S.D.	3.8	2.4	2.6	2.9	2.6	2.6
	n	18	18	15	15	18	18
Phenol + Cd	$\overline{X}$		7.6		9.2		9.4
	S.D.		3.0		5.0		5.8
	n		72		60		72
Total	$\overline{X}$		8.7				
	S.D.		4.8				
	n		204				

[a] Coefficient of variation = $\dfrac{(\text{standard deviation}) \times 100}{\text{mean}}$

from the phenol microplate tests. This is apparent for both operators of laboratories 2 and 3 (Table 1). This is likely due to the volatile properties of phenol which can affect growth in control wells under the experimental design adhered to in this exercise. The influence of phenol volatility on algal growth was confirmed experimentally, as will be seen later on in this section.

Variation for 96 h algal control growth **between** microplates with the standardized protocol is shown in Table 2. The overall C.V. is around 20%, irregardless of the toxicant assayed in the microplates. While not significant, 96 h control biomass was lower in the phenol microplates.

Additionally, one-way ANOVA and Scheffé tests (not shown) indicated significant intragroup variations linked to specific operators. Both of these observations, again, are likely attributable to phenol volatility effects. Barring the influence of phenol *per se*, inter-microplate control biomass variation resulting after 96 h can be tied to 1) inoculum precision, to 2) AAP growth medium preparation variations and to 3) the physiological state of algal cells at the time of inoculum which can, in turn, influence the time of lag phase. Further exploration of the actual factor(s) responsible for this variation was clearly outside the scope of this study. It is of interest to note, however, that an early eight-laboratory evaluation

Table 2. Inter-microplate algal control growth and variability with the standardized protocol.

Toxicant	n	$\overline{X}$ (10^6 cells $\cdot$ ml^{-1})	C.V.[a]
Phenol[b]	48	1.73	18.8
Cd	48	2.0	21.1

[a] Coefficient of variation = $\dfrac{(\text{standard deviation}) \times 100}{\text{mean}}$

[b] ANOVA and Scheffé tests indicate significant intra-group variations linked to specific operators.

of the AAP eutrophication bottle test, from which flask and microplate toxicity tests eventually evolved, reported a coefficient of variation of 43.3% for cell numbers produced after 21 days of incubation starting from an initial inoculum of 1000 cells·ml^{-1} (USEPA, 1971).

Phenol and cadmium EC50 results generated with the standardized protocol are displayed in Table 3. An average EC50 of 69.7 mg·l^{-1} was obtained for phenol with a reproducibility reflected by a C.V. of 34.9%. As mentioned earlier, volatile effects owing to phenol are likely responsible for the greater variability observed for phenol in comparison to cadmium, whose overall C.V. of 24.3% indicates better repetitive test precision. Once again, ANOVA and Scheffé tests conducted on the phenol data showed significant variation linked to specific laboratories and operators.

Influence of test variables

Growth of algae and toxicity results in relation to different algal cultivation techniques are presented in Table 4. While it can be seen that resulting growth appears to be more abundant when standardized cultivation is employed instead of 'in house' cultivation, the difference between the two techniques is not significant. Furthermore, cultivation technique had no influence on toxicity results for both cadmium and phenol. It appears, therefore, that no further consideration need be given to this variable for improved standardization of the microassay protocol.

The presence or absence of Na$_2$EDTA in the algal growth medium has a marked influence on assay control growth and toxicity results (Table 5). The dual role which this chelating compound plays in 1) promoting algal growth by

Table 3. Toxicant EC50 results and variation with the standardized protocol.

Toxicant		n	$\overline{X}$ (EC50)[a]	C.V.[b]
Phenol[c]	Lab 1	18	84.5	25.4
	Lab 2	15	54.5	14.3
	Lab 3	17	67.6	40.8
	Total	50	69.7	34.9
Cd	Lab 1	18	58.7	11.7
	Lab 2	14	63.2	19.3
	Lab 3	18	47.8	32.1
	Total	50	56.0	24.3

[a] In mg·l^{-1} for phenol and in μg·l^{-1} for Cd.

[b] Coefficient of variation = $\dfrac{(\text{standard deviation}) \times 100}{\text{mean}}$

[c] ANOVA and Scheffé tests indicate significant intra-group variations linked to specific operators.

Table 4. Influence of algal cultivation technique on assay control growth and EC50 results.

Algal cultivation technique	Control growth (10^6 cells·ml^{-1})			EC50 – Cd (μg·l^{-1})			EC50 – phenol (mg·l^{-1})		
	$\overline{X}$	S.D.	n	$\overline{X}$	S.D.	n	$\overline{X}$	S.D.	n
In house	1.74	0.53	36	53.6	11.2	15	68.7	28.8	14
Standardized	2.01	0.53	36	56.4	9.9	15	68.2	16.3	15
t test (0.05)	ns			ns			ns		

ns = non significant difference.

Table 5. Influence of growth medium composition on assay control growth and EC50 results.

Growth medium composition	Control growth (10^6 cells $\cdot$ ml^{-1})			EC50 – Cd (μg $\cdot$ l^{-1})			EC50 – phenol (mg $\cdot$ l^{-1})		
	$\overline{X}$	S.D.	n	$\overline{X}$	S.D.	n	$\overline{X}$	S.D.	n
AAP	1.88	0.47	33	55.0	15.1	18	77.2	21.6	15
AAP – EDTA	0.85	0.61	30	23.2	2.2	6	55.2	13.3	11
t test (0.001)	s						s[a]		

s = significant difference.

[a] Significant difference at 0.01 level.

allowing better assimilation of essential macro- and micro-nutrients and in 2) ameliorating toxicity by complexing metals (Miller *et al.*, 1978) is well borne out here. Indeed, algal growth is significantly enhanced in the presence of Na$_2$EDTA and cadmium toxicity is significantly reduced. Surprisingly perhaps, phenol toxicity was also decreased significantly in the presence of Na$_2$EDTA. Healthier cells because of improved assimilation of EDTA-mediated nutrients (Miller *et al.*, 1978) coupled with the presence of higher cell densities being exposed to the chemical (Albertano *et al.*, 1978) may have contributed in making the algae more resistant to phenol aggression.

The relative importance of gas exchange within microplate wells during incubation can be seen in Table 6. Significantly more growth is attained when active gas exchange is applied during incubation. CO$_2$-limitation, which may prevail near the latter stages of the 96 h incubation period under a passive regime as growth and pH increase (Miller *et al.*, 1978), may explain growth differences. Toxicity results, however, are unaffected

by the gas exchange variable which does not appear to be an essential factor to control in future refinements of the microassay.

Volatile effects of phenol

A series of additional microplate experiments were performed in an attempt to demonstrate the suspected volatile contaminating effects of phenol which the round robin exercise results suggested.

In a first experiment, 1000 mg $\cdot$ l^{-1} of phenol were placed in wells of columns 1, 5 and 9 with algae inoculated in all other columns under identical round robin experimental conditions (Fig. 2). Ten microplates were covered with their lids as before ('non-sealed wells'). Ten others were snugly overlaid with the bottom of a non-lidded microplate, whose U-shaped well bottoms afforded tightly-capped conditions to each individual well ('sealed wells'). It is clear from 96 h algal growth measurements that microplate wells sealed in the fashion described were unaffected by phenol. On the other hand, microplate wells

Table 6. Influence of gas exchange during incubation on assay control growth and EC50 results.

Gas exchange	Control growth (10^6 cells $\cdot$ ml^{-1})			EC50 – Cd (μg $\cdot$ l^{-1})			EC50 – phenol (mg $\cdot$ l^{-1})		
	$\overline{X}$	S.D.	n	$\overline{X}$	S.D.	n	$\overline{X}$	S.D.	n
Passive	2.12	0.64	30	64.6	9.0	14	68.1	23.5	15
Active	2.54	0.78	30	67.3	9.8	14	65.8	29.0	15
t test (0.05)	s			ns			ns		

s = significant difference.

ns = non significant difference.

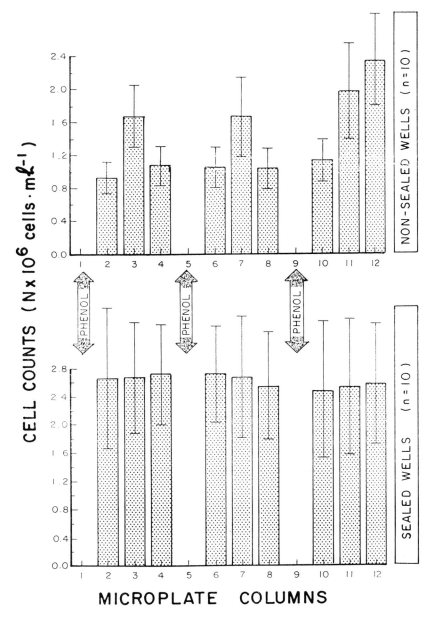

Fig. 2. Phenol volatile effects: sealed *versus* non-sealed wells.

merely covered with a standard microplate lid are subject to considerable growth-inhibiting effects owing to phenol volatility. This experiment confirms the marked role which phenol may have played insofar as result variability is concerned during the intercalibration exercise. It also points out the need for improved microplate covering in order to protect individual wells from the undesirable effects of volatile chemicals.

In a second experiment, different microplate dispositions with phenol were investigated to assess their influence on algal growth and toxicity (Fig. 3). Again, this one-time experiment reveals that results are affected by test configuration. For

266

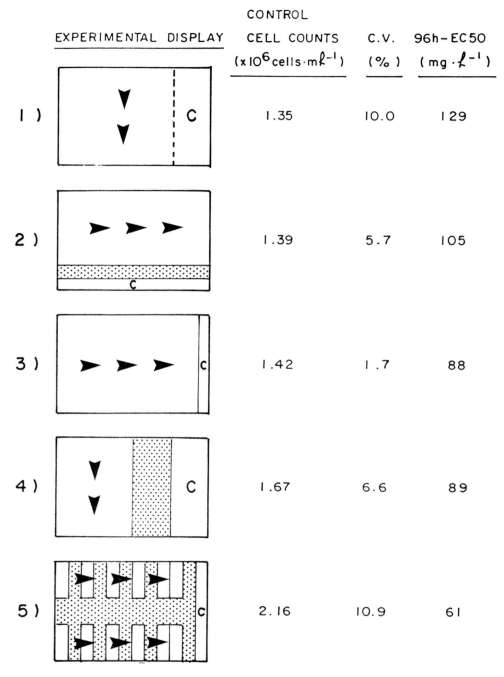

	EXPERIMENTAL DISPLAY	CONTROL CELL COUNTS (x10^6 cells·ml^{-1})	C.V. (%)	96h—EC50 (mg·l^{-1})
1)		1.35	10.0	129
2)		1.39	5.7	105
3)		1.42	1.7	88
4)		1.67	6.6	89
5)		2.16	10.9	61

: H_2O

: DECREASING SERIAL CONCENTRATIONS OF 1000 mg·l^{-1} PHENOL

C : CONTROL

Fig. 3. Algal growth and toxicity results based on different microplate experimental dispositions with phenol.

example, arrangement 1 produced a biomass corresponding to 1.35×10^6 cells·ml^{-1} and an EC50 of 129 mg·l^{-1} phenol, while arrangement 5, less prone to phenol volatility, produced a biomass corresponding to 2.16×10^6 cells·ml^{-1} and an EC50 of 61 mg·l^{-1} phenol.

A third experiment showed that microplate assays initiated with lower phenol concentrations could minimize the volatile effects of this chemical (Fig. 4). In the 1000 mg·l^{-1} phenol microplate, it is clear that the algal control growth values generated from wells adjacent to the first four highest concentrations of the toxicant (i.e. 1000, 500, 250, and 125 mg·l^{-1} of phenol) have been affected. No such effect was observed in the 250 mg·l^{-1} phenol microplate. In this comparison, the higher initial phenol concentration tends to increase control variability and to decrease the reported toxicity result. The information derived from this experiment is valuable and actually confers an additional advantage on the microassay in that it can show the presence of volatile toxicants in liquid samples. Indeed, using the experimental disposition of microplates outlined in Fig. 4, one could assume the presence of volatile toxicants in an unknown sample by observing skewness in control cell counts, such as those witnessed in the 1000 mg·l^{-1} phenol microplate. A more precise EC50 could then be reported by repeating the assay at a lower initial toxicant concentration to lessen or eliminate volatile effects.

Summary and conclusions

A three-laboratory intercalibration exercise conducted with the *Selenastrum capricornutum* microplate technique for acute aquatic toxicity assessment sought to 1) verify test reliability and reproducibility, 2) verify the influence of specific test variables and 3) further optimize the experimental

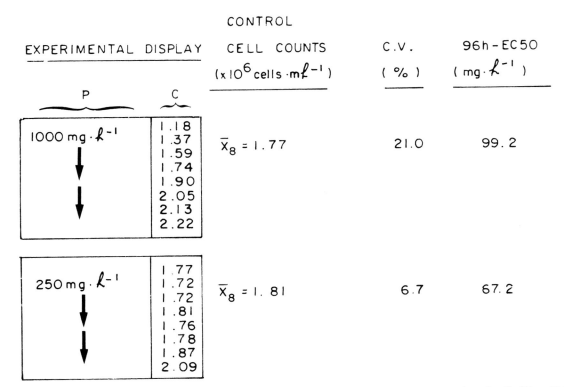

Fig. 4. Influence of initial phenol concentration on 96 h control algal growth in the microplate assay: phenol wells (P) and highest phenol test concentration; control wells (C) and 96 h mean cell counts x 10^6·ml^{-1} for each series of three wells.

protocol based on the overall data generated. This study provided interesting information which will form a basis for future refinement of this simple microassay. General findings and conclusions are outlined below:

1. Intra-microplate 96 h control biomass variation (C.V. = 8.7%, n = 204) reflected adequate methodological precision of the microassay.
2. Inter-microplate comparisons indicated that close to 2×10^6 cells $\cdot$ ml^{-1} are generated in control wells after 96 h.
3. Average EC50's generated for cadmium and phenol were 56 μg $\cdot$ l^{-1} (C.V. = 24.3%) and 69.7 mg $\cdot$ l^{-1} (C.V. = 34.9%), respectively.
4. Algal cultivation technique (standardized versus 'in house') had no significant influence on cadmium and phenol toxicity results.
5. Growth medium composition with and without Na$_2$EDTA had a significant influence on both algal growth and bioassay results.
6. Active gas exchange significantly promoted algal growth, but did not affect toxicity results.
7. Phenol volatility was found to have a marked influence on algal growth in a microplate. Better sealed wells, optimization of experimental design, and/or undertaking the assay with lower test toxicant concentrations can reduce/eliminate volatile effects.
8. The results of this study provide an incentive to plan a future larger scale inter-laboratory exercise aimed at further standardization of the algal microplate toxicity assay.

Acknowledgements

The authors thank their respective managements for supporting this experimental initiative. Appreciation is also extended to all those who participated in the round robin in a technical capacity. We are also indebted to Angèle Cameron for typing of the manuscript.

References

Albertano, P., G. Pinto & R. Taddei, 1978. Evaluation of the toxic effects of heavy metals to unicellular algae: the influence of inoculum concentration on the evaluation of toxicity. Delpinoa 20: 76–85.

Bitton, G. & B. Dutka (eds.), 1986. Toxicity testing using microorganisms, volumes I and II. CRC Press, Inc., Boca Raton, Florida, U.S.A., 163 pp. (Volume I) and 202 pp. (Volume II).

Blaise, C., 1986. Micromethod for acute aquatic toxicity assessment using the green alga Selenastrum capricornutum. Tox. Assess. 1: 377–385.

Blaise, C., R. Legault, N. Bermingham, R. van Coillie & P. Vasseur, 1986. A simple microplate algal assay technique for aquatic toxicity assessment. Tox. Assess. 1: 261–281.

Blaise, C., G. Sergy, P. Wells, N. Bermingham & R. van Coillie, 1988. Biological testing – development, application, and trends in Canadian environmental protection laboratories. Tox. Assess. 3: 385–406.

Blanck, H., G. Wallin & S. A. Wängberg, 1984. Species-dependent variation in algal sensitivity to chemical compounds. Ecotox. Environ. Safe. 8: 339–351.

Chiaudani, G. & M. Vighi, 1978. The use of Selenastrum capricornutum batch cultures in toxicity studies. Mitt. int. Ver. Limnol. 21: 316–329.

Christensen, E. R. & J. Scherfig, 1979. Effects of manganese, copper and lead on Selenastrum capricornutum, Chlorella stigmatophora. Wat. Res. 13: 79–82.

Dive, D., 1981. Nutrition et croissance de Colpidium campylum: contribution expérimentale et possibilités d'application en écotoxicologie. Thèse de doctorat présentée à l'Université des Sciences et Techniques de Lille (France), No. 527, 285 pp.

Hassett, J. M., J. C. Jennett & J. E. Smith, 1981. Microplate technique for determining accumulation of metals by algae. Appl. Envir. Microbiol. 41: 1097–1106.

Lukavsky, J., 1983. Evaluation of algal growth potential by cultivation on solid media. Wat. Res. 17: 549–558.

Maciorowski, A. G., J. L. Sims, L. W. Little & E. D. Gerrard, 1980. Bioassays – procedures and results. J. Wat. Pollut. Cont. Fed. 53: 974–993.

Maul, A. & J. C. Block, 1983. Microplate fecal coliform method to monitor stream water pollution. Appl. Envir. Microbiol. 46: 1032–1037.

Miller, W. E., J. C. Greene & T. Shiroyama, 1978. The Selenastrum capricornutum Printz algal assay bottle test: Experimental design, application, and data interpretation protocol. US Environmental Protection Agency Report No. EPA-600/9-78-018, Corvallis, OR, 126 pp.

Sokal, R. R. & F. J. Rohlf, 1981. Biometry. W. H. Freeman & Co., San Francisco, 859 pp.

USEPA (United States Environmental Protection Agency), 1971. The interlaboratory precision test: an eight-laboratory evaluation of the provisional algal assay procedure bottle test. National Eutrophication Research Program, Corvallis, OR, EPA Water Quality Office Project 16010DQT, 70 pp.

Xu, H., B. J. Dutka & K. K. Kwan, 1987. Genotoxicity studies on sediments using a modified SOS Chromotest. Tox. Assess. 2: 79–87.

Hydrobiologia **188/189**: 269–276, 1989.
M. Munawar, G. Dixon, C. I. Mayfield, T. Reynoldson and M. H. Sadar (eds)
Environmental Bioassay Techniques and their Application.
© *1989 Kluwer Academic Publishers. Printed in Belgium.*

Phytoplankton recovery responses at the population and community levels in a hazard and risk assessment study

P. Couture,[1] C. Thellen[2] & P.-A. Thompson[1]
[1]*INRS-EAU, C.P. 7500, Ste-Foy, Qc, Canada, G1V 4C7;* [2]*Ministère de l'Environnement du Québec, Direction des Laboratoires, 2700 Einstein, Ste-Foy, Qc, Canada G1P 3W8*

Key words: effluent, toxicity, recovery, algae, population, community

Abstract

Both structural and functional relationships were investigated in experiments using *S. capricornutum* populations and an indigenous microbial community. Our aims were to diagnose cellular stress and to predict recovery during exposures to a chlor-alkali effluent.

Laboratory experiments demonstrated that the effluent was toxic at concentrations greater or equal to 4%, v/v. It appears that during the exposure period, the functional parameters, particularly the intracellular adenylates ratios were reliable in predicting algal population recovery.

On the other hand, the river gradient experiments failed to demonstrate a toxic effect on community structure over the time scale studied. Functional parameters revealed a significant effect on photosynthetic activity while adenylate energy charge was an insensitive indicator.

Finally, our results tend to demonstrate that functional responses, particularly intracellular adenylates ratios (ATP/cell; ATP/AMP) are appropriate to predict recovery responses to a toxicant at the population and community levels. This would prove useful in enhancing the ecological significance of toxicity tests in hazard assessment.

Introduction

Biological effects of pollutants may result in structural changes in the microplanktonic communities of receiving systems. This implies that some species are sensitive to these pollutants while others become tolerant. A series of laboratory methodologies have been developped (Cairns, 1983) to assess the hazards of local pollutions. In particular, screening toxicity tests using planktonic organisms are recognized to be sensitive to detect potential effects at the population level (Wong & Couture, 1986). They appear, however, to be of low ecological significance (Cairns & Pratt, 1987; Hellawell, 1988).

In short, living organisms tend to develop compensatory mechanisms to react against stresses. Therefore primary effects may overestimate resulting effects. This theory is not widely applied in the design of methodologies supporting hazard and risk assessments.

Since compensatory reactions may take place during a short term shock exposure, our aim is to verify potential recovery responses using a cultured algal population (laboratory experiments) and an indigenous microbial community (field experiments). More specifically, we attempted to find indicators for predicting recovery during an exposure to an industrial effluent.

Material and Methods

Laboratory experiments

The analytical procedures are detailed in Thompson *et al.* (1987). Briefly, a *Selenastrum capricornutum* clone (ATCC 22662) was aseptically cultured in 6 l carboys containing AAP medium without EDTA (Chiaudani & Vighi, 1978). Algae were maintained at laboratory conditions (T°: 22 ± 2 °C; pH: 7.8 ± 0.1 using a pH controller; continuous light: $100 \mu E \sec^{-1} m^{-2}$ and agitation by filtered air) for at least 2 weeks prior to using. These conditions ensured asynchronously growing algae.

For toxicity tests, effluent concentrations (0; 1; 4; 10 and 25%, v/v) were prepared according to Joubert (1980). For each concentration, an inoculum of 10^4 cells ml^{-1} was introduced in each 6 l carboys for the initiation of the tests. The experimental conditions were identical to the above. The exposure period was 96 h during which cell counts, chl-*a*, nucleotide adenylates and ^{14}C assimilation were measured at different intervals. Recovery experiments were carried out after 24 h and 96 h exposure periods. For this, exposed algae were collected, concentrated by filtration and re-inoculated (10^4 cells ml^{-1}) in fresh culture medium for another 96 h. These 24 h and 96 h recovery experiments were performed under the same incubation conditions, except for automatic pH control. The same parameters were measured.

Field experiments

The site was located on a river system receiving the industrial effluent (St-Louis River, Québec, Canada). The river gradient study comprised 4 cross-sections (1 upstream and 3 downstream) spaced at increasing distances in the 1.2 km following the effluent discharge and corresponding to the mouth of the river. Three stations were located at each cross-section. At each station, 4 l samples were taken at the surface and immediately prepared to measure the parameters identified in the laboratory study. In addition taxonomic identification was carried out. Analytical procedures are described in Couture *et al.* (1987).

Results and discussion

Laboratory experiments

Cell counts showed a significant reduction of *S. capricornutum* growth caused by the effluent (Fig. 1). At t_{24} and t_{96} of the exposure experiment, cell counts remained significantly different than the controls ($\alpha = 0.05$) from 4 to 25%. The 24 h recovery experiments showed that cell counts remained lower at $\geq 4\%$ of the effluent. This was not the case at the 96 h recovery experiment where cell counts were similar to the controls except for the 25% concentration. The last was not tested due to a lack of viable cells at the end of the exposure period.

At all effluent concentrations the light-saturated rate of photosynthesis per unit chlorophyll (P/B ratio: μg carbon per μg chl-*a* per hour) were significantly lower than the control at t_{24} of the exposure experiment (Table 1) while adenylate energy charges (EC_A: $([ATP] + \frac{1}{2}[ADP])/([ATP] + [ADP] + [AMP])$) remained lower at concentrations $\geq 4\%$. At t_{24}, for concentrations $\geq 4\%$, low cell counts were accompanied by low P/B and EC_A ratios. P/B and EC_A are considered qualitative indicators of the physiological state of the cells: P/B reflects the photosynthetic rate per unit chl-*a* and EC_A is considered to be a measure of the energy momentarily stocked in cells (Atkinson, 1968).

During the exposure experiments, EC_A varied from 0.04 to 0.70 at t_{24} and stabilized at > 0.60 at t_{96}. The last represents the stationary phase of the population (Riemann & Wium-Andersen, 1982; Falkowski, 1977). From cell counts, P/B ratios and EC_A values, it appeared that the 25% effluent concentration was lethal to the *S. capricornutum* population. In fact, no recovery was observed at this concentration. On the other hand, at lower effluent concentrations (1 to 10%), P/B ratios and EC_A were restored between t_{24}

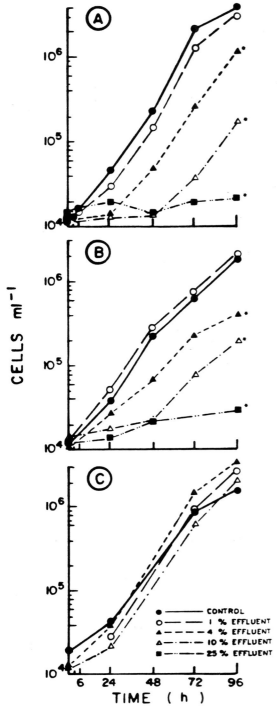

Fig. 1. Growth variation of *S. capricornutum* populations exposed to the chlor-alkali effluent during the 96 h exposure (A), the 24 h recovery (B) and the 96 h recovery (C) experiments. (*) indicates that population responses are significantly lower than the control.

and t_{96}. This suggested a certain acclimation of the population. This recovery response may also be linked to chlorine evaporation during the exposure period.

The EC_A appeared to be a useful stress indicator. In fact, at t_{24} EC_A values were low (< 0.5) for populations having P/B ratios lower than 0.80 mgC mg chl-a^{-1} h^{-1}. Côté (1983) determined P/B ratios never exceeding 0.1 in Saguenay river phytoplankton assemblages submitted *in vitro* to more than 10 μg Cu l^{-1}. Couture *et al.* (1987) also observed P/B ratios < 0.4 in enclosure experiments with the St-Louis River phytoplankton community exposed for 20 h to a 25% concentration of the same effluent. Low EC_A values (Table 1) were transitory and this indicates that the 0.5 EC_A level suggested by Atkinson (1977) as the viability threshold is not applicable at the population level. The experiments showed that low EC_A values must be interpreted as representative of a mixed population of cells having different EC_A ratios. In this perspective, the resulting acclimation may be due to a more

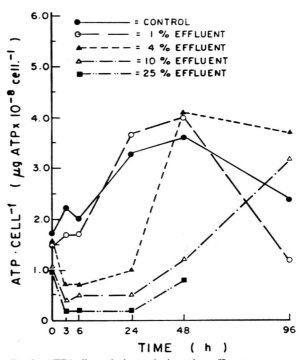

Fig. 2. ATP/cell variations during the effluent exposure experiments.

Table 1. Variations in photosynthetic efficiency (P/B) and adenylate energy charge (EC_A) of *S. capricornutum* populations during the 96 h exposure period and the recovery experiments (adapted from Thompson *et al.*, 1987).

			Exposure		24 h recovery	96 h recovery
			t_{24}	t_{96}	t_{96}	t_{96}
PB (μg C $\cdot$ μg chl $- a^{-1} \cdot h^{-1}$)						
Control	Mean (n)		1.31 (4)	1.67 (4)	1.19 (4)	1.95 (6)
	LL		0.94	0.77	0.65	1.56
	UL		1.68	2.57	1.74	2.27
Effluent ($\%$, v $\cdot$ v$^{-1)}$		1	0.96*	1.74	2.64	2.45
		2	1.07*	1.51	–	2.88
		4	0.73*	1.50	–	3.17
		10	$\leq$0.001*	2.67	–	2.83
		25	$\leq$0.001*	nt	0.90*	nt
CE$_A$						
Control	Mean (n)		0.45 (4)	0.69 (6)	0.73 (6)	0.69 (6)
	LL		0.27	0.58	0.60	0.59
	UL		0.63	0.82	0.86	0.79
Effluent ($\%$, v $\cdot$ v^{-1})		1	0.36	0.63*	0.63*	0.72
		2	0.70	0.80	–	0.63*
		4	0.04*	0.74	–	0.63*
		10	0.04*	0.72	–	0.63*
		25	0.12*	nt	0.27*	nt

LL: lower 95$\%$ confidence limit
UL: upper 95$\%$ confidence limit
nt: not tested due to lack of algae
*: significantly lower than the control ($\alpha \leq 0.05$)

tolerant portion of the exposed population. Therefore, at the population level, EC_A appears to be an unreliable stress indicator.

Recovery at the population level could be predicted from ATP/cell ratios. Two patterns were observed (Fig. 2). The first, where ATP/cell remained above 1, prevailed at the 1$\%$ effluent concentration. When this value was observed during the first 24 h of exposure, a physiological response was taking place that allowed population growth to be restored after the cells were transferred to fresh medium (Fig. 1).

The second pattern was observed when growth was not restored during the 24 h recovery experiment and was indicative of more acute sublethal effects during exposure to the toxicant. For example, during the first 24 h of exposure to 4 and 10$\%$ effluent, ATP/cell remained lower than 1

and as a consequence in the 24 h recovery experiment cell counts revealed no population growth. On the other hand, between 24 and 96 h of the exposure experiment, ATP/cell ratios increased above 1 and consequently, population growth was restored during the 96 h recovery experiment.

Our results suggest that functional variables are useful to predict recovery following exposure to a pollutant. For example, high P/B ratios and high EC_A values were observed at the end of the exposure period, even though cell counts were still low. Moreover, the ATP/cell ratio is seen as a useful indicator of the recovery of an algal population.

Field experiments

Taxonomic composition reflected the eutrophic status of the river. Bacillariophyceae were dominant, *Cyclotella* sp. being most important. Community structure was not significantly modified by the effluent discharge. The diversity index (H') varied between 1.09 bitt cell^{-1} at cross-section A, 0.92 at B, 0.94 at C and 1.03 at D.

According to Harris (1984), the short effluent exposure period observed in the St-Louis River (between 0.5 and 6.6 h depending on river flow: Moulins, 1983) cannot impair community structure; such changes may appear after exposure periods greater or equal to the time required for physiological acclimation, probably between 12 and 200 h. The time scale necessary to induce physiological effects is contrastingly shorter. For example, cellular responses such as ATP content, total adenylate content and EC_A may occur in less than 10 min (Atkinson, 1977). Our results are in agreement with these suggested time scales. In fact, while chl-*a* contents remained constant ($\approx 3\ \mu g\,l^{-1}$), P/B ratios decreased from 2.6 at cross-section A to 0.77, 0.83 and 0.69 $\mu gC\,l^{-1}$ at cross-sections B, C and D respectively.

Variations in total adenylates (A_T) between cross-section A and B (1.38–1.62 to 0.84–1.36 nM) are thought to be mainly due to dilution by the effluent rather than to its toxicity. In fact chl-*a*/total adenylate ratio remained relatively constant at all cross-sections (0.0047 ± 0.0009). The structural biomass indicators were therefore mainly modified by dilution due to the effluent discharge.

Analysis of variance and Scheffe's tests showed that functional relationships at the upstream cross-section were significantly different from the downstream cross-sections: 70–90% decrease of the P/B ratio. From this parameter, no recovery response seemed initiated in the time scale studied (Fig. 3). Longer experiments would be necessary to observe any recovery response in this community.

Although EC_A values remained low along the river gradient (Fig. 3), no significant difference was revealed by the ANOVA. The EC_A values were slightly lower than those observed by Falkowski (1977) for a stationary growing population. Low EC_A's may result from an increase in cellular AMP (Fitzwater *et al.*, 1983; Heath,

1984). This situation reflects an increase in energy demand on the adenylate pool (Chapman & Atkinson, 1977). At downstream cross-sections the ATP/AMP ratios revealed such an energy demand (Fig. 3); values remained significantly lower than those of the upstream cross-section (ANOVA and Scheffe's tests). It must be pointed out that when the $ATP \rightarrow AMP + P_i - P_i$ reaction is favored over the $ATP \rightarrow ADP + P_i$ reaction, the direct cleavage of the pyrophosphate is accompanied by a greater release of energy. For example, this type of reaction is favored for the enzymatic activation of fatty acids in the form of esters CoA (Lehninger, 1977). We hypothesize that the algal cells are rapidly taking the large quantity of energy required to respond to stress conditions by using an analogous reaction ($ATP \rightarrow AMP + P_i - P_i$).

During the time scale studied (<6.6 h), and considering the photosynthetic activity (P/B), the phytoplanktonic community did not acclimatize to the toxic effluent despite the fact that variations in EC_A values observed between cross-sections were not significant. It is therefore possible that the energy loss due to the decreased photosynthetic activity and as a consequence of the reduced photophosphorylation yield, was compensated by an increase in the activity of the oxydative phosphorylation pathway.

Conclusions & research perspectives

The hazard assessment approach using an algal batch assay showed that the effluent was lethal at 25% concentration. At lower concentrations, although sublethal effects were observed, structural, mainly cell counts, and functional variables such as P/B ratios and cellular energy state parameters (EC_A; ATP/cell) revealed that recovery was taking place during the exposure period.

The use of these structure-function relationships seems promising to relate the information obtained from the laboratory approach with the one resulting from the field study performed on the indigenous microbial community of the

274

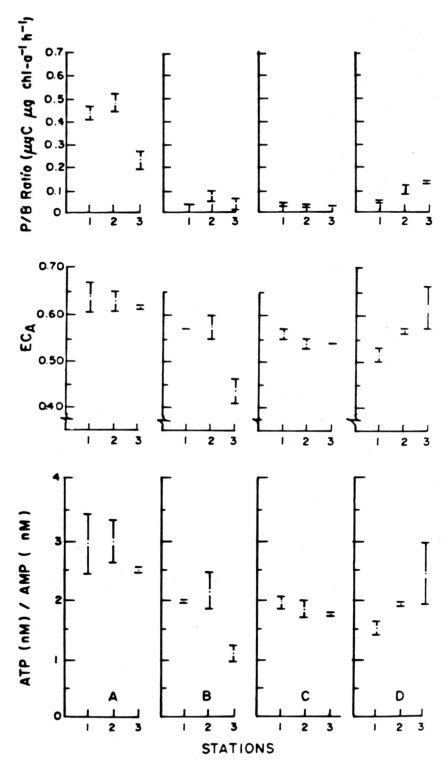

Fig. 3. Variations of P/B ratios, EC_A and ATP/AMP ratios at upstream (A) and downstream (B, C and D) cross-sections for each station (1,2,3).

St-Louis river. In summary, 100% inhibition of photosynthesis was observed during experiments with *S. capricornutum* exposed to 10 and 25% effluent for 24 h. Although the exposure period in the river gradient was relatively short (<6.6 h) and the effluent dilution certainly greater than 25%, toxic effects were observed. In fact, marked reductions of the P/B ratios occurred at the downstream cross-sections (B, C and D) even if the phytoplankton biomass and the energy state of the cells (EC_A) were not significantly affected. It seems that the adenylates metabolism expressed in term of EC_A (Karl, 1980) was not affected despite the fact that photosynthetic processes were inhibited. This is in agreement with the laboratory recovery responses observed at concentrations lower than 25%.

Our results demonstrate that functional variables are useful in the detection of the potential effects of toxic substances. It seems that lasting EC_A reductions are more indicative of lethal effects whereas the ATP/AMP ratio and more particularly the P/B ratio are sensitive to sublethal toxicant concentrations. Furthermore, these parameters should be considered for the prediction of biological acclimation/recovery responses. To better understand its physiological significance, the ATP/AMP ratio should be further investigated; it could constitute a sensitive screening tool for the detection of acclimatory processes.

The environmental significance of biological parameters used in monitoring networks of industrial discharges should be improved by more research conducted on acclimation/recovery processes. In this perspective, the following hypotheses will be investigated:

H1: An increase in energy requirements and/or an inhibition of photosynthesis may produce adenylates ratio (EC_A; ATP/ADP; ATP/AMP) disequilibria which may induce an increase in respiratory rates to restore energy equilibrium;

H2: The presence of toxic substances may promote the utilization of recently synthesized carbohydrates in respiratory pathways rather than protein synthesis;

H3: The respiratory capacity; carbohydrate stocks and the formation of intracellular inclusions may ameliorate a species competitiveness among the community;

H4: The presence of toxic substances may initiate cellular protection mechanisms such as the synthesis of metallothionein-type complexing molecules.

Acknowledgements

This study was supported by Environment Canada and the Ministry of Supplies and Services (UP-1ST83-00313), and by the Natural Sciences and Engineering Research Council of Canada (grant no A8123).

References

Atkinson, D. E., 1968. The energy charge of the adenylate pool as a regulatory parameter: interaction with feedback modifiers. Biochem. J. 7: 4030–4034.

Atkinson, D. E., 1977. Cellular energy metabolism and its regulation., Academic Press, New York, NY, 293 pp.

Cairns, J. Jr., 1983. The case of simultaneous toxicity testing at different levels of biological organization. In W. E. Bishop, R. D. Cardwell & B. Heidolph (eds.), Aquatic Toxicology and Hazard Assessment: Sixth Symposium, STP 802. American Society for Testing and Materials, Philadelphia, Pa., 111–127.

Cairns, J. Jr & J. R. Pratt, 1987. Ecotoxicological effect indices: a rapidly evolving system. Wat. Sci. Tech. 19: 1–12.

Chapman, A. G. & D. E. Atkinson, 1977. Adenine nucleotide concentrations and turnover rates. Their correlation with biological activity in bacteria and yeast. Adv. Microbiol. Physiol. 15: 253–306.

Chiaudani, G. & M. Vighi, 1978. The use of *Selenastrum capricornutum* batch cultures in toxicity studies. Mitt. Int. Ver. Limnol. 21: 316–329.

Côté, R., 1983. Aspects toxiques du cuivre sur la biomasse et la productivité du phytoplancton de la rivière Saguenay, Québec. Hydrobiol. 98: 85–95.

Couture, P., C. Thellen, P. A. Thompson & J. C. Auclair, 1987. Structure and function of phytoplankton and microbial communities in relation to industrial wastewater discharge: an ecotoxicological approach in a lotic system. Can. J. Fish. Aquat. Sci. 44: 167–175.

Falkowski, P. G., 1977. The adenylate energy charge in marine phytoplankton: the effect of temperature on

276

physiological state of *Skeletonema costatum* (Grev.) Cleveland. J. Exp. Mar. Biol. Ecol. 27: 37–45.

Fitzwater, S. E., G. A. Knauer & J. H. Martin, 1983. The effects of Cu on the adenylate energy charge of open ocean phytoplankton. J. Plankton Res. 5: 935–938.

Harris, G. P., 1984. Phytoplankton productivity and growth measurements: past, present and future. J. Plankton Res. 6: 219–237.

Heath, A G., 1984. Changes in tissue adenylates and water contents of bluegill, *Lepomis macrochirus*, exposed to copper. J. Fish Biol. 24: 299–309.

Hellawell, J. M., 1988. Toxic substances in rivers and streams. Envir. Pollut. 50: 61–85.

Joubert, G., 1980. A bioassay application for quantitative toxicity measurements using the green algae *Selenastrum capricornutum*. Wat. Res. 14: 1759–1763.

Karl, D. M., 1980. Cellular nucleotide measurements and applications in microbial ecology. Microb. Rev. 44: 739–796.

Lehninger, A. L., 1977. Biochemistry, the molecular basis of cell structure and function, 2nd ed., Worth Publishers Inc., New York, NY, 1088 pp.

Moulins, L. J., 1983. Impact des effluents d'une usine de chlore-soude caustique sur l'environnement bio-physique du lac et de la rivière St-Louis., Mémoire de maitrise, Ecole Polytechnique, Université de Montréal, 189 pp.

Riemann, B. & S. Wium-Andersen, 1982. Predictive value of adenylate energy charge for metabolic and growth states of planktonic communities in lakes. Oikos 39: 256–260.

Thompson, P. A., P. Couture, C. Thellen & J. C. Auclair, 1987. Structure-function relationships for monitoring cellular stress and recovery responses with *Selenastrum capricornutum*. Aquat. Toxicol. 10: 291–305.

Wong, P. T. S. & P. Couture, 1986. Toxicity screening using phytoplankton. In B. J. Dutka & G. Bitton (eds.), Toxicity testing using microorganisms, Vol. 2, CRC Press Inc, Boca Raton, Fl.: 79–100.

Hydrobiologia **188/189**: 277–283, 1989.
M. Munawar, G. Dixon, C. I. Mayfield, T. Reynoldson and M. H. Sadar (eds)
Environmental Bioassay Techniques and their Application.
© 1989 *Kluwer Academic Publishers. Printed in Belgium.*

Functional response of *Fucus vesiculosus* communities to tributyltin measured in an *in situ* continuous flow-through system

C. Lindblad, U. Kautsky, C. André, N. Kautsky & M. Tedengren
Department of Zoology and Askö Laboratory, University of Stockholm, S-106 91 Stockholm, Sweden

Key words: macroalgae, disturbance, primary production, community metabolism, hazard assessment

Abstract

The effects of antifouling paint leachate containing tributyltin on community metabolism and nutrient dynamics were measured *in situ* on natural communities dominated by *Fucus vesiculosus*. The measurements were made in two areas with different salinities and at various TBT concentrations up to about $5 \mu g \, l^{-1}$. A portable continuous flow-through system was used in which the communities were incubated for a week. Continual measurements of oxygen, temperature, light and flow rate of water were made. A Perturbation Index (PI) and an Absolute Disturbance Index (ADI) were used to describe the changes due to treatment relative to the control, and to obtain a total picture of disturbance using all measured parameters. Photosynthesis was particularly strongly affected and changes were obvious in oxygen production and nutrient uptake at TBT levels as low as $0.6 \mu g \, l^{-1}$.

Introduction

The study of natural or induced changes in functional responses of ecosystems or communities under naturally fluctuating environmental conditions is an important component of bioassay studies. Well controlled laboratory tests offer many advantages (e.g. constant light, temperature, salinity, the use of cultured organisms), but meaningful extrapolation of such results to real complex field situations is almost impossible (Cairns, 1983). Field observations yield important information about local changes in the ecosystem, but for an understanding of the mechanisms, long time-series and reference sites are needed. Well designed enclosures to study the functional processes at the ecosystem level (e.g. primary production, community respiration, nutrient cycling) offer a convenient bridge between laboratory tests and field observations.

In the Baltic sea the salinity increases from about 0.5‰ in the northern Bothnian Bay to around 9‰ in the southern Baltic Proper (Kullenberg, 1981). The northern limit of the *Fucus* community is in the Bothnian Bay where salinity falls below 2.8‰ (Kautsky, 1988).

On shallow hard substrata *Fucus vesiculosus* is the dominant biotic element. With some 30 species of macrofauna and epiflora it is the richest algal community in the Baltic Sea and an important foraging and nursery area for fish. *Fucus* is sensitive to changes in the environment, for example increased sedimentation (Rönnberg *et al.*, 1985) and decreased light penetration in the water (Kautsky *et al.*, 1986). Paper mill wastes have resulted in large areas devoid of *Fucus* around the effluent outlets (Lindvall, 1984). *Fucus* can also be more sensitive to disturbance in areas with low salinity as shown for other Baltic species (Tedengren *et al.*, 1988).

Tributyltin (TBT) is commonly used as a biocide in antifouling paint. Its concentrations in coastal areas range from < 30 to $100 \, \text{ng} \, l^{-1}$ (Laughlin *et al.*, 1988). Laboratory studies of acute TBT toxicity have been carried out on several groups of organisms (Valkirs *et al.*, 1987; Strømgren *et al.*, 1987; Hall *et al.*, 1988). There have been *in situ* investigations on benthic suspension feeders (André *et al.*, 1989) but no studies have been conducted on functional responses of macroalgal communities under field conditions.

We report here results of a study designed to address the functional response of *Fucus* communities, at the ecosystem level, to low levels of TBT in two areas with different salinities.

Material and methods

The experiments were conducted during July 1986 at the Askö Laboratory in the northern Baltic Proper (mean water temperature 20 °C, salinity 6.3‰ during the experiments) and at the Tjärnö Marine Biological Laboratory on the Swedish west coast (mean water temperature 18 °C and salinity 26.5‰ during the experiments).

We employed the *in situ* flow-through system of Kautsky (1984) and Lindblad *et al.* (1986, 1988) to study the functional response of *Fucus* communities. The system consisted of a submersible pump which filled a tank which in turn provided a constant head of water to each of ten transparent acrylic jars (volume 20 l). Introduction of water ($1 \, l \, \text{min}^{-1}$) to each jar through a 3 mm diam. nozzle created a current which ensured good mixing. The water was later discharged from the jars via solenoid valves either directly to waste or to a measurement chamber. The measurement chamber contained a stirrer, a polarographic oxygen electrode, a pelton flow meter, and a silicon temperature sensor. Sensor outputs were recorded on a microcomputer using a 12 bit analog-digital converter. The means of 100 readings of oxygen, temperature, flow rate and light intensity were stored in the computer each minute. Two jars without organisms served as controls to mea-

sure changes in the characteristics of the water alone. Since the same electrode was used for all measurements, the effects of such changes as long term drift of membrane properties and temperature changes were minimized because all jars were affected equally.

The difference between the inflow and outflow concentrations of dissolved substances (compensated for residence time and flow rate) was used to calculate net fluxes of oxygen and nutrients.

Daily respiration (R) was calculated from the mean of readings in darkness multiplied by 24 hours. Gross primary production (GP) was defined as the area under the production curve integrated over a diurnal period. These values were divided by the dry weight (DW) determined at the end of the experiment, to obtain weight-specific respiration or production. Single values that differed by at least an order of magnitude from the mean because of electrical interference or occasional air bubbles in the system were automatically discarded. The coefficient of variation for measurements was less than 2% ($n > 200$). Water samples for determination of NH_4-N and PO_4-P were taken from the waste discharge every six hours during the two days before treatment and during the first and third days after treatment. The analyses were carried out in the laboratory using standard colorimetric methods (Carlberg, 1972).

To minimize disturbance, stones with intact *Fucus* plants were placed into jars. The entire epifaunal (e.g. *Idotea baltica*, *Gammarus* spp., *Hydrobia* spp., *Theodoxus fluviatilis*, *Mytilus edulis*), and – algal (e.g. *Dictyosiphon foeniculaceus*, *Elachista fucicola*, *Ectocarpus siliculosus*, *Ceramium tenuicorne*) communities were included. The biomass of epiphytic algae was between 10 and 20%, and of animals about 10% of the total biomass.

After the communities were placed in the jars, the parameters were measured over 36 hours to obtain undisturbed values before treatment. The treatment started in five of the jars when the incoming water was led through PVC pipes of two different lengths (23 cm and 70 cm) painted on the inside with the antifouling paint 'Interracing'

(International Paint Company) containing Tributyltin (TBT) as the active agent. Assuming leaching rates of 22.5 μg TBT cm^{-2} day^{-1} (Laughlin et al., 1984), we calculated TBT concentrations to be 1.6 μg l^{-1} and 5 μg l^{-1}, in the two treatments. After three days, water samples were taken, immediately frozen and sent for later TBT-analyses to Harbor Branch Oceanographic Institution USA.

Three jars where left untreated and served as controls during the seven-day experiment.

Indices

Ecosystem functional changes were measured by different indices based on basal processes such as gross primary production (GP), respiration (R), biomass (B), oxygen content (O), phosphorus (P) or nitrogen (N) excretion. The GP/R ratio is a measure of the amount of self-maintenance of a community and Giddings and Eddlemon (1978) used this as an indicator of stress in the ecosystem; a change in the ratio indicates disturbance.

Another index, the Perturbation Index (PI) gives a measure of the disturbance (Lindblad et al., 1986, 1988). Each measurement value after disturbance is divided by the mean of values before perturbation and then divided by the corresponding measurement of the controls. Thus PI also normalizes differences in biomass and physiological status in the tested communities. The formula for PI is:

$$\mathrm{PI(X)} = \frac{\sum \dfrac{T_j}{C_j}}{n_a \sum \dfrac{T_i}{C_i n_b}},$$

T = treated jars before (i) and after (j) treatment,
C = control jars time before (i), after (j) treatment,
n = number of measurements before (b) and after (a) treatment,
X = oxygen or any other parameter used.

To get a clear picture of the total effect of the treatments the PI values for individual parameters were plotted in a multidimensional space, where the origin represents an undisturbed community and each PI one dimension. The absolute distance from the origin divided by the number of parameters used (n) summarizes the total effect of all measured parameters (PI(X)), for example PI (R), PI (GP/R). We call this the Absolute Disturbance Index (ADI) (Lindblad et al., 1988) which has the following formula:

$$\mathrm{ADI} = \sqrt{\frac{\sum (\mathrm{PI(X}_i) - 1)^2}{n}}.$$

Results

The analyses of the TBT in water samples are shown in Table 1. The calculated concentrations did not correspond exactly with the analyzed TBT concentrations. Evidently, TBT was taken up by the biota and particles. The leaching rates were probably slower than the theoretical values, particularly at low salinities.

Net community production (NP) before treatment was in the range of 10–30 mg O$_2$ g$^{-1} \cdot$ day^{-1} in the two study areas. NP decreased strongly after treatment in both areas; at the low salinity it decreased to below zero (Fig. 1a, 2a).

The diurnal range in respiration rate (R) was 10–30 mg O$_2$ g$^{-1} \cdot$ day^{-1} before treatment at both salinities. Respiration of all communities increased slightly after treatment and no differences

Table 1. Results from TBT analysis at different salinities and theoretical concentrations

	TBT conc. (μg l^{-1})		
	Tube length		Background conc. in the water
	23 cm	70 cm	
Salinity 6.3‰	n.a.	2.8	0.1
Salinity 26.5‰	0.6	4.7	0.3
Theoretical conc.	1.6	5.0	–

n.a. = not analyzed.

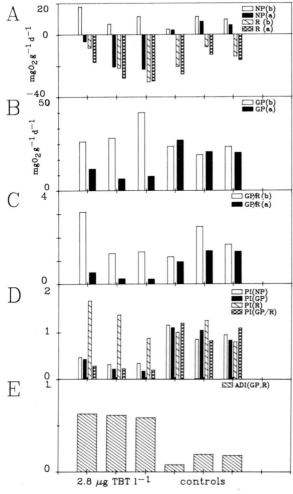

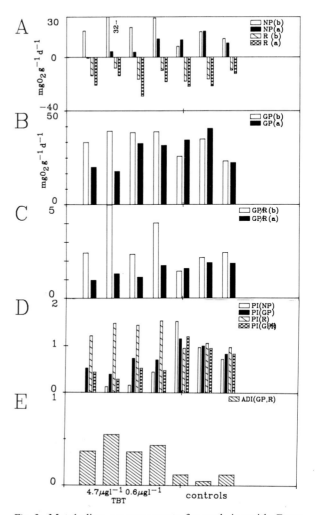

Fig. 1. Metabolism measurements for each jar with *Fucus* communities in low salinity (6‰). Three replicates treated with 2.8 μg TBT l⁻¹ and three controls. A. Net production (NP) and respiration (R) before (b) and after (a) treatment. B. Gross production (GP) before (b) and after (a) treatment. C. GP/R ratios before (b) and after (a) treatment. D. Perturbation Index for PI (NP), PI (GP), PI (R), PI (GP/R). E. Absolute Disturbance Index (ADI) calculated from PI (GP) and PI (R).

Fig. 2. Metabolism measurements for each jar with *Fucus* communities in high salinity (26.5‰). Two replicates treated with 4.7 μg TBT l⁻¹ and two replicates treated with 0.6 μg TBT l⁻¹ and three controls. A–E are explained in Fig. 1.

were visible between the different salinities (Fig. 1a, 2a).

Gross production (GP) was 20–40 mg O_2 g⁻¹·day⁻¹ at both sites before treatment. After treatment the strongest effects were found at the low salinity where GP fell to a mean value of 8.6 mg O_2 g⁻¹·day⁻¹. At the high salinity GP was reduced to about 18 mg O_2 g⁻¹·day⁻¹ at

4.7 μg l⁻¹ TBT while the low concentration of TBT (0.6 μg l⁻¹) gave a mean value of 32 mg O_2 g⁻¹ (Fig. 1b, 2b). The decrease in GP shows that the decrease in NP is not only an effect of increased respiration.

The GP/R ratios ranged between 1.3 and 10. The ratio was generally higher at the high salinity (mean 2.8) than at the low (mean 1.9) before treatment. After treatment the GP/R mean value decreased to 1.3 at the high salinity and 0.3 at the low salinity (Fig. 1c, 2c). When calculating the

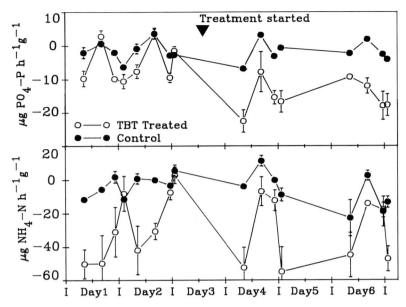

Fig. 3. PO4-P and NH4-N excretion or uptake. Means calculated on samples taken from three treated *Fucus* communities and three controls in low salinity (6‰).

Perturbation Index of NP, GP, R and GP/R there are clear differences between treated communities and the controls. For the controls, PI was around 1 which means that these parameters did not change during the experiment. The PI (R) increased after TBT treatment at both sites while PI (NP), PI (GP) and PI (GP/R) decreased. The largest changes were visible at the low salinity. The higher concentration of TBT ($4.8\,\mu g\,l^{-1}$) used in the high salinity tests showed stronger effects than the low dose ($0.6\,\mu g\,l^{-1}$). Communities at the low salinity, however, were still most affected (Fig. 1d, 2d).

The Absolute Disturbance Index calculated on PI (GP) and PI (R) gave a disturbance mean value of 0.61 for communities in low salinity treated with $2.8\,\mu g\,$ TBT l^{-1}. At the high salinity, $4.7\,\mu g\,$ TBT l^{-1} gave an ADI mean value of 0.46 and the low concentration of $0.6\,\mu g\,$ TBT l^{-1} gave a mean value of 0.40. The controls in both study areas gave mean values of 0.12 (Fig. 1e, 2e).

Although the nutrient data were variable and sampling rather spaced out in time, there was a tendency towards decreased nutrient uptake and increased release in all TBT treatments.

Ammonium (NH_4-N) and phosphorus (PO_4-P) showed diurnal fluctuations with uptake during the day and release at night. Control communities at the low salinity had a PO_4-P uptake in the afternoon (15.00 h), of up to $4\,\mu g\,g^{-1}$ DW h^{-1} and release values of $5-10\,\mu g\,g^{-1}$ DW h^{-1} during the night. TBT-treated communities had a release of $10-20\,\mu g\,g^{-1}\,h^{-1}$ during the whole diurnal cycle. The NH_4-N uptake ranged from $0-10\,\mu g\,g^{-1}$ before treatment. After treatment no uptake was measured while the release was between 10 and $50\,\mu g\,g^{-1}\,h^{-1}$ (Fig. 3).

Discussion

In general our results show that the *Fucus*-based ecosystem is very sensitive to TBT and the flow-through method used was adequate for measuring the functional response to these disturbances. Bioassay studies at the community and ecosystem levels are more relevant for predicting the impact of disturbance on the environment if the communities used in the study include several species with interactions at different trophic levels. Also important for adequate results in all bioassay studies is the choice of ecologically relevant communities (Cairns & Niederlehner, 1987; Giesy & Odum, 1980; Underwood & Peterson, 1988). We have used the *Fucus vesiculosus* community, the dominating biomass on littoral hard bottoms along the Swedish east coast.

It is difficult to establish experimental protocols for ecosystem monitoring purposes. The use of flow-through systems such as ours largely reduces errors due to oxygen saturation during experiments. Temperature, light and nutrients follow the natural variations. Another advantage is that we are able to use communities taken from the field directly for measurements. This avoids disturbance arising from storage of organisms or possible artifacts due to the use of laboratory cultivated species. Our small enclosure system allows adequate replication and makes it possible to obtain statistically valid results on the effects of disturbance.

When measuring several community parameters at different experimental sites with different treatments, problems arise due to differences in basal metabolism between individual samples, and changes in light or temperature making direct comparisons difficult. The use of indices such as the Perturbation Index (PI) and the Absolute Disturbance Index (ADI) make such comparisons possible because they eliminate non-treatment effects. Even metabolic differences due to size are normalized which makes long term comparison possible without destructive weighing after each measurement. Tributyltin was shown to change all measured parameters of the *Fucus* community at both low and high salinities. Respiration and production were strongly affected, in addition to decreases in nutrient uptake. A difference in sensitivity at the two salinities was observed. The same reduction in functional response was evident for the 2.8 μg TBT l^{-1} treatment in the low salinity Baltic Sea as for the 4.7 μg TBT l^{-1} treatment in the high salinity North Sea. This difference in sensitivity is probably due to the fact that organisms living in low salinity close to their distribution limit are less resistant to toxicants (Tedengren *et al.*, 1988).

At the high salinity site, there was only a small difference between the effects at the low and high TBT doses. This probably means that even a dose of 0.6 μg l^{-1} is significant and effects would be visible at even lower concentrations.

The apical parts of all the treated *Fucus vesiculosus* plants changed colour to red-brown compared to the green-brown control algae. This has also been observed by Schonbeck & Norton (1980). This is probably due to the photo oxidation of the polyphenols and to destroyed cell membranes. Laboratory investigations with microalgae exposed to tin compounds have shown that primary production was inhibited and growth was reduced at 0.1 μg TBT and TBTO l^{-1} (Beaumont & Newman, 1986; Wong *et al.*, 1982). Experiments with the Gastropod *Nucella lapillus* exposed to TBT concentrations of 0.01–0.02 μg l^{-1} for 4 months showed a high degree of imposex, the introduction of male sex in the female (Bryan *et al.*, 1986). It is evident that TBT is active at very low concentrations for consumers as well as for primary producers. Our investigations show such drastic changes in functional response of the *Fucus* communities that TBT additions would probably result in a decline of the *Fucus* community, followed by a change in the ecosystem to favor opportunistic species.

Acknowlegements

We are thankful to Dr. Klaus Koop and Dr. Arno Rosemarin for giving valuable comments on the manuscript. We also thank three anonymous referees for their comments and suggestions.

References

André, C., N. Kautsky, U. Kautsky, C. Lindblad & M. Tedengren, 1989. *In situ* measurements of the functional response of benthic suspension feeders exposed to Cadmium and anti-fouling paint. Kieler Meeresforschung (in press).

Beaumont, A. R. & P. B. Newman, 1986. Low levels of tributyltin reduced growth of marine micro-algae. Mar. Pollut. Bull. Vol. 17, 10: 457–461.

Bryan, G. W., P. E. Gibbs, L. G. Hummerstone & G. R. Burt, 1986. The decline of the gastropod *Nucella lapillus* around south-west England; evidence for the effect of tributyltin from antifouling paints. J. Mar. Biol. Ass. U.K. 66: 611–640.

Cairns, J. Jr., 1983. Are single species toxicity tests alone adequate for estimating environmental hazards? Hydrobiologia 100: 47–57.

Cairns, J. Jr. & B. R. Niederlehner, 1987. Problems associated with selecting the most sensitive species for toxicity testing. Hydrobiologia 153: 87–94.

Carlberg, S., 1972. New Baltic Manual, International Council for the Exploration of the Sea Cooperative Research Report. Series A. Vol. 29: 1–145.

Giddings, J. & G. K. Eddlemon, 1978. Photosynthesis/respiration ratios in aquatic microcosms under arsenic stress. Wat. Air Soil Pollution 9: 207–212.

Giesy, J. P. Jr. & E. P. Odum, 1980. Microcosmology: introductory comments. In: J. P. Giesy Jr. (ed.), Microcosm in Ecological Research. Dept. of Engineering (DOE) Symposium Series: 52 conf-781101. Department of Energy. National Technical Information Service, Springfield, VA. pp. 1–13.

Hall, L. W. Jr., S. J. Bushong, W. S. Hall & W. E. Jonson, 1988. Acute and chronic effects of tributyltin on a Chesapeake Bay copepod. Envir. Toxicol. Chem. 7: 41–46.

Kautsky, H., 1988. Factors structuring phytobentic communities in the Baltic Sea. Ph. D. dissertation, University of Stockholm, pp. 1–29.

Kautsky, N., 1984. A battery operated, continuous-flow enclosure for metabolism studies in benthic communities. Mar. Biol. 81: 47–52.

Kautsky, N., H. Kautsky, U. Kautsky & M. Waern, 1986. Decreased depth penetration of *Fucus vesiculosus* (L) since the 1940's indicates eutrophication of the Baltic Sea. Mar. Ecol. Prog. Ser. 28: 1–8.

Kullenberg, G., 1981. Physical Oceanography. In: A. Voipio (ed.), The Baltic Sea. Elsevier Oceanography Ser. 30, Amsterdam, pp. 135–181.

Laughlin, R. B., R. Gustavson & P. Pendoley, 1988. Chronic embryo-larval toxicity of tributyltin (TBT) to the hard shell clam *Mercenaria mercenaria*. Mar. Ecol. Prog. Ser. 48: 29–36.

Laughlin, R., K. Nordlund & O. Linden, 1984. Long-term effects of tributyltin compounds on the Baltic amphipod, *Gammarus oceanicus*. Mar. Envir. Res. 12: 243–271.

Lindblad, C., N. Kautsky & U. Kautsky, 1986. An *in situ* method for bioassay studies on functional response of littoral communities to pollutants. Ophelia Suppl. 4: 159–165.

Lindblad, C., U. Kautsky & N. Kautsky, 1988. An *in situ* system for evaluating effects of toxicants to the metabolism of littoral communities. In J. Cairns Jr. & J. R. Pratt (eds.), Functional Testing of Aquatic Biota for Estimating Hazards of Chemicals, ASTM STP 988. American Society for Testing and Materials, Philadelphia, pp. 97–105.

Lindvall, B., 1984. The condition of a *Fucus vesiculosus* community in a polluted archipelago area on the east coast of Sweden. Ophelia Suppl. 3: 147–150.

Rönnberg, O., J. Lehto & I. Haahtela, 1985. Recent changes in the occurrence of *Fucus vesiculosus* in the Archipelago Sea, SW Finland. Ann. Bot. Fenn. 22: 231–244.

Schonbeck, W. M. & T. A. Norton, 1980. The effects on intertidal fucoid algae of exposure to air under various conditions. Botanica Marina 23: 141–147.

Strømgren, T. & T. Bongard, 1987. The effects of tributyltin oxide on growth of *Mytilus edulis*. Mar. Pollut. Bull. 18: 30–31.

Tedengren, M., M. Arnér & N. Kautsky, 1988. Ecophysiology and stress response of marine and brackish water *Gammarus* species (Crustacea, Amphipoda) to changes in salinity and exposure to cadmium and diesel oil. Mar. Ecol. Prog. Ser. 47: 107–116.

Underwood, A. J. & C. H. Peterson, 1988. Towards an ecological framework for investigating pollution. Mar. Ecol. Prog. Ser. 46: 227–234.

Valkirs, A. O., B. M. Davidson & P. F. Seligman, 1987. Sublethal growth effects and mortality to marine bivalves from long-term exposure to tributyltin. Chemosphere 16(1): 201–220.

Wong, P. T. S., Y. K. Chau, O. Kramar & G. A. Bengert, 1982. Structure-toxicity relationship of tin compounds on algae. Can. J. Fish. Aquat. Sci. 39: 483–488.

Hydrobiologia **188/189**: 285–289, 1989.
M. Munawar, G. Dixon, C. I. Mayfield, T. Reynoldson and M. H. Sadar (eds)
Environmental Bioassay Techniques and their Application.
© *1989 Kluwer Academic Publishers. Printed in Belgium.*

Comparison of five bioassay techniques for assessing sediment-bound contaminants

Wolfgang Ahlf, Wolfgang Calmano, Judith Erhard & Ulrich Förstner
Technische Universität Hamburg-Harburg, D-2100 Hamburg 90, F.R.G.

Key words: Acute bioassays, sediment, contaminants

Abstract

Biological response could not be predicted based on chemical concentration of the sediment contaminants. Bioassays integrate the response of test organisms to contaminants and nutrients. Comparative results of five acute bioassays indicated that Neubauer phytoassay was the most sensitive. The microbial biomass and algal growth tests indicated a response to the availability of contaminants and nutrients. These results suggest the usefulness of a diversity of bioassays in toxicity testing of sediment contamination.

Introduction

Many pollutants are preferentially associated with sediments in aquatic systems. Although chemical analyses provide valuable supplementary information they cannot replace direct bioassay measurements in establishing sediment quality criteria for contaminated sediments.

Environmental hazard assessments require rapid, inexpensive screening tests to characterize the extent of contamination. Preference should be given to short-term tests which are simple and reproducible. The major advantage of this approach is that the biological response to a complex mixture of compounds integrates the effects of environmental variables such as solubility, pH, antagonism and synergism, all of which affect toxicity to organisms (Dutka & Bitton, 1986; Munawar & Munawar, 1987).

The comparability of bioassays is somewhat limited by the different sensitivities of test organisms to individual contaminants (Williams *et al.*, 1986). The biological response depends on the metabolic activity of the organisms and is closely tied to the availability of contaminants from sediments.

Algal assays have proven to be very sensitive indicators of contaminant stress (Miller *et al.*, 1985). Experimental difficulties arise from the need to provide an interacting system of sediments and algae, while retaining the ability to analyze each system separately. An approach to overcome these problems was made by Ahlf (1985), who used an apparatus in which a membrane separated the algae and sediments. The objective of our study was to compare this algal assay with four sediment bioassays using bacteria, algae and higher plants as indicators.

Materials and methods

Sediments

Three different sediments from the Hamburg area were used for the investigations. The sediments were freeze-dried, sifted through a 2-mm sieve, and adjusted with H_2O to 60% of the maximum water holding capacity. Heavy metals were analyzed using flame and carbon furnace atomic absorption spectrometry (ZAAS, Hitachi model 180–70 (Table 1). Sediment samples were Soxhlet extracted for 16 h with 1 : 1 (v/v) acetone-hexane, preconcentrated and selected organochlorine compound concentrations were determined by GC (Lohse, 1988).

Biological methods

There are numerous methods available for assessing the environmental impact of sediment-associated contaminants (Ahlf & Munawar, 1988). We used the following assays.

a) Elutriate bioassays were initially designed for monitoring of dredging projects (Keeley & Engler, 1974). For each sediment sample an elutriate was prepared by mixing one volume of

Table 1. Measured chemical characteristics of sediments used.

	Sediment I	Sediment II	Sediment III	
As	11	45	5	(mg/kg)
Cd	6.1	10.4	0.7	
Cu	172	191	27	
Hg	1.5	11	<0.1	
Pb	457	172	61	
Ni	30	83	6	
Zn	1208	1178	127	
Heptachlor-				
Expoxid	37	109	<2	(µg/kg)
DDT	68	10	<2	
DDT	27	86	<2	
PCB's	738000	170000	<5	

sediment with four volumes of test medium described by Miller *et al.* (1978). The mixture was agitated 30 minutes by air, settled for one hour and then passed through a 0.45 µm membrane. The short-term impacts of sediment elutriates were assayed with the alga *Ankistrodesmus bibraianus*, using techniques developed by Miller *et al.* (1978).

b) While this elutriate test simulates the immediate impact of resuspended sediments on the water column, the recirculating systems (Hoke & Prater, 1980) determine to which extent the contaminants are available within a certain period of time and how they affect the test organisms. Similarly, the exchange between the water/organism and sediment compartments can be studied in a two-chambered device in which a 0.45 µm pore diameter membrane separated the algae and sediment (Ahlf, 1985). This method allowed an interaction of contaminants and algae, but prevented mixing of the algae and sediments. The inoculum contained 5×10^5 cells $\times$ ml^{-1} of *A. bibraianus*. The incubation period was 72 h.

c) The reactions of the microbial biomass were described with DNA content. The method applied has been described by Benndorf *et al.* (1977) for the measurement of DNA.

d) A second series of experiments were conducted in order to evaluate the bioactivity of microorganisms. For each measurement 0.5 g wet sediment was used. The method for determining dehydrogenase activity was described by Liu & Strachan (1981).

e) The Neubauer phytoassay has been used to evaluate seed germination and initial plant growth of *Lolium multiflorum* and *Lepidium sativum* (Thomas & Cline, 1985). The sediments were progressively diluted by mixing of contaminated material with washed silica sand or an uncontaminated humic sediment (20% org. C).

All experiments were conducted in triplicate.

Results and discussion

Algal toxicity has been monitored using both sediment elutriates and sediment suspensions.

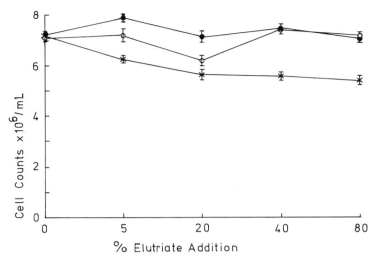

Fig. 1. Effect of standard elutriate addition on the growth of *Ankistrodesmus bibraianus*; bars represent the s.e.m. ● = Sediment 1, ○ = Sediment 2, x = Sediment 3.

Increasing additions of the standard elutriate to test medium resulted in a slight inhibition of biomass production only with the Sediment 3 elutriate (Fig. 1). A more differentiated interpretation was possible with the data of the two-chamber device (Fig. 2). Once again Sediment 3 affected negatively the biomass production. The addition of Sediment 1 showed a slight increase of algae growth, whereas Sediment 2 stimulated

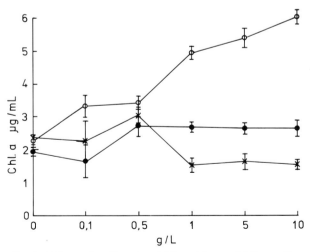

Fig. 2. Effect of sediment addition (dry weight) on the growth of *A. bibraianus*; bars represent the s.e.m. ● = Sediment 1, ○ = Sediment 2, x = Sediment 3.

growth of algae with increasing sediment concentration.

Mobilization of nutrients resulted in a significant enhancement of biomass compared to elutriates. The test system allowed a continous leaching during the exposure time, which seemed to be more sensitive in characterization of biological effects. The large number of chemicals associated with suspended sediments implies that water extraction alone is insufficient to detect toxic chemicals in aquatic systems (Ongley *et al.*, 1988). The use of sediments themselves in bioassays with algae is recommended over elutriates prepared from sediments.

Direct effects of sediments on microbial biomass were conducted by estimation of DNA content after an exposure time of 6 days. Contaminants existed on sediments but did not produce a toxic response in bacteria comparable to chemical analysis (Fig. 3). The ranking of sediments by DNA content were in good agreement with the results from the two-chamber device. Additional studies on the dehydrogenase activity in sediments were performed by mixing subsamples with the artificial electron acceptor INT. However, the extracts of sediment 1 and 2 were coloured and disturbed the background control data. The estimation of dehydrogenase activities are considered

288

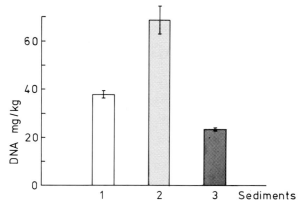

Fig. 3. DNA content in sediments after 6 days exposure time; bars represent the s.e.m.

to be an inappropriate tool for toxicity measurements in our sediment subsamples.

The Neubauer phytoassays with higher plants showed that Sediment 1 and 2 caused significant inhibition of growth. The effects of contaminated sediment were modified by the additions of humic soil or silica sand. In a sediment with high organic matter content differences to the uncontaminated control were lower than in sediments with silica sand additions. The test organisms were able to detect the presence of toxic contaminants even in dilute mixtures (Ahlf, 1988). However, data for a single biological end point (fresh weight after 10 days growth) were sufficient to demonstrate toxic effects (Fig. 4).

These results suggest that a soil contact phytoassay could be the most sensitive bioassay among the five techniques studied. However, Marschner *et al.* (1986) demonstrated that roots decreased pH in the rhizosphere to a pH-value of 4–5. We believe that this particular mechanism for the mobilization of mineral nutrients is also important for the mobilization of heavy metals from Sediments 1 and 2. Thus, the toxic response of higher plants in our experiments does not reflect the entire range of all other possible contaminants. This interpretation is supported by data from former studies with algae, where the effect of decreasing pH was simulated in the two-chamber device under controlled conditions. A sediment, which stimulates algal growth at pH 7–8, caused high toxic response at pH 5-6 within the algae/water/suspended sediment system (Ahlf, 1985).

Acknowledgment

P. M. Chapman (E.V.S. Consultants) and an anonymous referee reviewed the manuscript and provided many helpful comments, and Dr. M. Munawar (Fisheries & Oceans Canada) provided editorial advice.

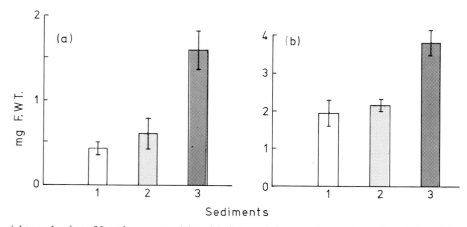

Fig. 4. Fresh weight production of *Lepidum sativum* (a) and *Lolium multiflorum* (b) grown in sediments for 10 days; bars represent the s.e.m.

References

Ahlf, W., 1985. Behaviour of sediment-bound heavy metals in a bioassay with algae: Bioaccumulation and toxicity. Vom Wasser 65: 183–188.

Ahlf, W., 1988. Correlations between chemical and biological evaluation procedures for the determination of heavy metal availability from soils. In: K. Wolf, W. J. van den Brink & F. J. Colon (Eds.), Contaminated Soil '88 pp. 67–69. Kluwer Academic Publishers Dordrecht.

Ahlf, W. & M. Munawar, 1988. Biological assessment of environmental impact of dredged material. In: W. Salomons & U. Förstner (Eds.), Chemistry and Biology of Solid Waste, pp. 127–142. Springer Verlag, Heidelberg.

Benndorf, A., J. Benndorf, W. Horn & W. Stelzer, 1977. Biochemische Charakteristika des Sestons. Acta Hydrochim. Hydrobiol. 5: 33–42.

Dutka, B. J. & G. Bitton, 1986 (Eds.). Toxicity Testing Using Microorganisms. Vol. 2 CRC Press, Boca Raton, Fla. pp. 186.

Hoke, R. A. & B. L. Prater, 1980. Relationship of percent mortality of four species of aquatic biota from 96-hour sediment bioassays of five Lake Michigan harbors and elutriate chemistry of the sediments. Bull. envir. Contam. Toxicol. 25: 394–399.

Keeley, J. W. & R. M. Engler, 1974. Discussion of regulatory criteria for ocean disposal of dredged materials; Elutriate test rationale and implementation guidelines. U.S. Army Corps of Engineers, DMRP, Vicksburg, Miss. Technical Report D-74-14. pp. 39.

Liu, D. & G. W. Strachan, 1981. A field method for determining the chemical and biological activity of sediments. Wat. Res. 15: 353–359.

Lohse, J., 1988. Ocean incineration of toxic wastes: A footprint in North Sea sediments. Mar. Pollut. Bull. 19 (8): 366–371.

Marshner, H., V. Römheld, W. J. Horst & P. Martin, 1986. Root-induced changes in the rhizosphere: Importance for the mineral nutrition of plants. Z. Pflanzenernaehr. Bodenk. 149: 441–456.

Miller, W. E., J. C. Greene & T. Shiroyama, 1978. The Selenastrum capricornutum Printz. algal assay bottle test: Experimental design application, and data interpretation protocol. EPA 600/9-78-018, U.S. Environmental Protection Agency, Corvallis OR.

Miller, W. E., S. A. Peterson, J. C. Greene & C. A. Callahan, 1985. Comparative toxicology of laboratory organisms for assessing hazardous waste sites. J. Envir. Qual. 14: 569–574.

Munawar, M. & I. F. Munawar, 1987. Phytoplankton bioassays for evaluating toxicity of in situ sediment contaminants. In: R. L. Thomas, R. Evans, A. Hamilton, M. Munawar, T. Reynoldsen & H. Sadar (Eds.) Ecological effects of in situ sediment contaminants. Hydrobiologia. 149: 87–105.

Ongley, E. D., D. A. Birkholz, J. H. Carey & M. R. Samoiloff, 1988. Is water a relevant sampling medium for toxic chemicals? An alternative environmental sensing strategy. J. Envir. Qual. 17: 391–401.

Thomas, J. M. & J. F. Cline, 1985. Modification of the Neubauer technique to assess toxicity of hazardous chemicals in soils. Envir. Toxicol. Chem. 4: 201–207.

Williams, L. G., P. M. Chapman & T. C. Ginn, 1986. A comparative evaluation of marine sediment toxicity using bacterial luminescence, oyster embryo and amphipod sediment bioassays. Mar. Envir. Res. 19: 225–249.

Hydrobiologia **188/189**: 291–300, 1989.
M. Munawar, G. Dixon, C. I. Mayfield, T. Reynoldson and M. H. Sadar (eds)
Environmental Bioassay Techniques and their Application.
© 1989 *Kluwer Academic Publishers. Printed in Belgium.*

Assessing toxicity of Lake Diefenbaker (Saskatchewan, Canada) sediments using algal and nematode bioassays

D. J. Gregor[1] & M. Munawar[2]
[1]*Water Quality Branch, Environment Canada, 1901 Victoria Avenue, Regina, Saskatchewan, Canada, S4P 3R4;* [2]*Great Lakes Laboratory for Fisheries and Aquatic Sciences, Fisheries and Oceans Canada, Canada Centre for Inland Waters, P.O. Box 5050, Burlington, Ontario, Canada, L7R 4A6*

Key words: bioassays, sediment toxicity, Lake Diefenbaker

Abstract

Lake Diefenbaker, on the South Saskatchewan River, Saskatchewan, Canada, receives, on average, 90% of its inflow from snowmelt and rainfall in the Rocky Mountains. The inflowing rivers also receive irrigation return flows and municipal and industrial effluents which may result in the contamination of lake sediments. The sediments were assessed by nematode and algal bioassays.

The toxicity of five chemical fractions of the sediment was determined using the nematode *Panagrellus redivivus* as the test organisms. The results suggest that the sediment chemical fractions frequently inhibit growth and maturation, while lethality was observed at 4 of 12 sites.

Samples from 3 of these sites were further evaluated using conventional elutriate Algal Fractionation Bioassays (AFB) with both natural Lake Diefenbaker phytoplankton and a mixed laboratory grown algal culture. The natural phytoplankton showed inhibition at sediment: water ratios of 10:1; whereas the algal cultures showed both enhancement and inhibition. Evidently, the sediments are frequently toxic to the species tested except for the algal culture. The AFB assesses the mitigative and synergistic effects of contaminants and nutrients and being a conventional elutriate, is more realistic and potentially more acceptable than the chemical fractionation/nematode bioassay technique which essentially considers potential trace organic contaminant effects.

Introduction

Sediment bioassays have tended to focus on higher trophic organisms promoted by the U.S. Fish and Wildlife Service and the U.S. Environmental Protection Agency (I.J.C., 1986). Consequently, the intricate balance between various trophic levels, such as bacteria, protozoa, algae and zooplankton has not been studied satisfactorily (Munawar *et al.*, 1984). Many of these higher trophic level tests are time-consuming and usually not as sensitive as tests with some of the lower trophic species with high metabolic rates

and shorter half-life cycles. In addition, phytoplankton bioassays have been found to be useful for evaluating the suitability of dredged sediments for in-lake disposal where known toxic substances, including metals and industrial organic compounds have accumulated (Munawar & Thomas, 1989). In many lakes where dredging and disposal may not be a concern, toxicity testing of bottom sediments is useful to determine whether or not toxic substances are entering the lake and may be harmful to the aquatic ecosystem. Detection and biological screening of the presence of toxic substances in this manner is

292

made more essential in view of the high cost and difficulties of sampling and chemically analysing a wide range of pesticides and toxic industrial substances.

Lake Diefenbaker is a large and important multi-purpose reservoir in the heart of Canada's semi-arid grainbelt. Created in the mid 1960's, the lake is entirely situated within the province of Saskatchewan. However, more than 80% of its drainage basin area is located in the province of Alberta (Fig. 1). Two major river systems, the South Saskatchewan and the Red Deer, provide more than 90% of the total annual water supply to the lake. These rivers, which have their head waters in the Rocky Mountains, receive municipal and industrial discharges as well as agricultural runoff and irrigation return flows as they traverse several hundred kilometres across the province of Alberta. Although monthly water quality monitor-

ing programs have existed on both of these rivers for more than a decade, the sampling has generally not been appropriate to identify many dilute and potentially toxic substances, which may nevertheless contribute significant loads to the downstream lake. Further, the sampling has tended not to address the critical transportation times, such as the high flow prairie snow melt season, when sediment-bound contaminants are likely to be transported. Previous studies in the basin have been directed at the presence of nutrients in the South Saskatchewan River upstream of the lake (Cross *et al.*, 1984) and nutrient and metal pathways in the Bow and Oldman Rivers, two major tributaries to the South Saskatchewan system, during 1981 and 1982 (Ongley & Blachford, 1984).

In 1984, the Water Quality Branch, Environment Canada and the Saskatchewan Department

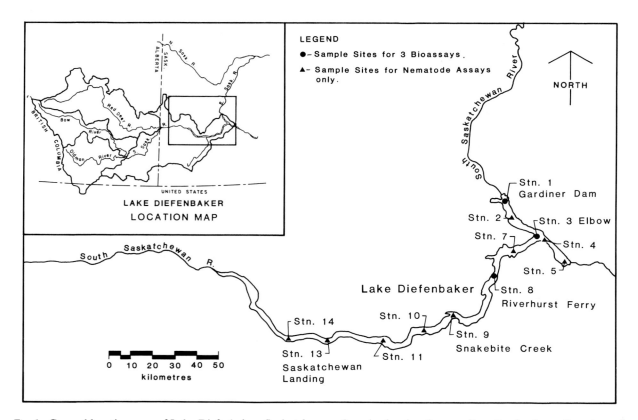

Fig. 1. General location map of Lake Diefenbaker, Saskatchewan, Canada showing the sampling sites for the sediments used in the comparative bioassays.

of Environment undertook an intensive study of Lake Diefenbaker and the South Saskatchewan and Red Deer Rivers. This study was designed to provide additional information regarding nutrient loadings to the lake, the trophic status of the lake and, essentially for the first time, to investigate the status of the system with respect to toxic substances. This assessment utilized biological screening techniques in association with chemical analyses of suspended and bottom sediments.

A variety of bioassay techniques are available for lake sediments. The principles of these have been reviewed by Munawar *et al.* (1984); Munawar & Munawar (1987); Samoiloff & Bogeart (1984); Blaise *et al.* (1985) and Cairns (1985). In this study a nematode (*Panagrellus redivivus*) test has been used for sediment samples collected the length of the lake. This technique has been applied widely in western Canada, but the general approach is one that emphasizes potential toxicity of trace organic contaminants rather than natural in-lake effects which integrate mitigating influences such as diversified populations, nutrients and substance bio-availability. Consequently, three of the sites were selected to perform Algal Fractionation Bioassays (AFB) with sediment elutriates using natural phytoplankton assemblages and a mixed culture of laboratory grown algae (Munawar & Munawar, 1982; Munawar *et al.*, 1983). This paper considers the comparison of the results from these published techniques and their applicability as a screening tool for Lake Diefenbaker.

Methods

Nematode Bioassays

The bottom sediments of Lake Diefenbaker were sampled at a total of 12 locations throughout the length of the reservoir (Fig. 1) in the summer of 1984. All sediment samples were collected using a Ponar dredge from the deepest point of each cross-section as determined by echo-sounding. The sediments were placed on aluminum foil which had been pre-rinsed with pesticide grade acetone and hexane, described and then sub-sampled to provide a number of samples representative of the top 2 to 3 cm of sediment. Two sub-samples were placed in pre-cleaned Teflon containers and stored frozen.

The samples were chemically fractionated following the procedure described by Birkholz *et al.* (1983) and Ongley *et al.* (1986). The BASE/ NEUTRAL (B/N) Fraction was first extracted into dichloromethane (DCM) and then sulphuric acid was added to the residual aqueous phase from the B/N extraction to give a pH < 2.0. The Acid Fraction was extracted with DCM. The B/N extract was further fractionated with a silica-gel column eluted with 80 ml of hexane for Fraction 1, 20% DCM in hexane for Fraction 2, 60% DCM in hexane for Fraction 3, and finally Fraction 4 with 50 ml of DCM followed by 50 ml of methanol (CH_3OH). Each of the five fractions was concentrated to 5 ml. A 2 ml aliquot of each fraction was retained for chemical analysis. The remaining 3 ml aliquot was exchanged into dimethyl sulfoxide (DMSO) and transferred to amber glass vials.

The nematode bioassays, using *Panagrellus redivivus*, were conducted by Bioquest International Inc. using each of the chemical fractions exchanged into DMSO following the method described by Samoiloff *et al.* (1980, 1983). The test is performed with replicate sets of 10 animals at the earliest free-swimming stage in the test medium. After a 96-hour exposure period, the number and size distribution (growth) of these test populations relative to control populations grown in 1% DMSO added to the test medium, are determined. During this time period, the nematode undergoes three moults to attain the adult stage.

Algal Bioassays

Algal Fractionation Bioassays (AFB)
The AFB follows the procedure described in Munawar and Munawar (1982, 1987). Sediments for these tests were collected at stations 1, 3 and 8 (Gardiner Dam, Elbow and Riverhurst Ferry,

respectively) in September 1986 using a Ponar dredge. These samples, consisting of at least 1 kg of sediment were placed in clean stainless steel containers, frozen and air freighted to the laboratory in Burlington, Ontario. Dilution water was also collected at this time for the preparation of the elutriates. This water was pre-filtered in 0.45 μm filters. The sediment was mixed in 1 l glass bottles using a volume ratio for sediment to water of 1:4. This mixture was agitated using a ferris-wheel type tumbler at a rate of 4-5 revolutions per minute for 1 hour (Daniels et al., 1989). The sediments were then allowed to settle overnight at 4 °C, decanted and then spun down in a refrigerated centrifuge. The water was then filtered through pre-washed 0.45 μm filters and collected and stored in pre-cleaned glass containers.

The phytoplankton used for the bioassays consisted of natural assemblages collected from station 8 (Fig. 1). The raw water samples were collected in cleaned and triple-rinsed 2 l polyethelene containers in the morning, stored in coolers and air-freighted to the laboratory in Burlington, Ontario, for use in bioassays the following morning. Subsamples of the plankton assemblages were collected and preserved in Lugol's Solution for species identification and size analysis (Munawar et al., 1974).

Treatments included 1, 5, 10, 20, and 40% additions of elutriate to phytoplankton samples with 4 replicates for each treatment as well as controls and dark controls. The test bottles were spiked with 1 μCi of ^{14}C-labelled NaHCO$_3$, mixed and incubated at 12 °C (the temperature of the lake at the time of sediment sampling) for 4 hours. The entire contents of each bottle were separated by filtering through a 20 μm Nitex screen and subsequently a 0.45 μm Millipore membrane filter at low vacuum. The filters were then acidified with 10 mL of 0.1 N HCl and placed in a scintillation vial and treated with 10 ml of PCS II (a toluene-based scintillation fluor) and counted in an automatic liquid scintillation counter. This provided production estimates for ultraplankton (<20 μm) and microplankton plus netplankton (>20 μm). The resulting count data were corrected and used to compute carbon assimilation rates (mgC m^{-3} hr^{-1}) (Munawar & Munawar, 1982).

Limited Sediment Bioassay (LSB)

This procedure, designed for use where only small quantities of sediment are available, as discussed by Munawar et al. (1989) used a split of the same sediment sample collected for the nematode bioassays. Samples for stations 1, 3 and 8 were stored frozen at -35 °C until the bioassays were performed. A 200 ml LSB elutriate was prepared to provide enough medium to produce four 50 ml replicates. Filtered lake water was used for the elutriate preparation in a manner similar to that used for the AFB except that the elutriates were prepared with sediment to water ratios as shown in Table 1.

A concentrated mixed algal culture (Munawar et al., 1989) containing approximately 3×10^6 cells in 1 ml, was added to each LSB elutriate so as not to change significantly the sediment: water ratios. Each bottle was then treated with 1 μCi of ^{14}C as NaHCO$_3$ and incubated for 4 hours at 15 °C. Thereafter the samples were treated in a manner identical to that used for the AFBs except that the algal size fractionation was not performed.

Table 1. Sediment and water ratios used for the preparation of elutriates in the Limited Sample Bioassay (LSB) compared with the Algal Fractionation Bioassay (AFB)

Limited Sample Bioassay		Algal Fractionation Bioassay	
Sediment/ water (g ml^{-1})	Sediment/ water ratio	Treatment (% elutriate)	Sediment/ water ratio
0:200	0:1	0	0:1
2.6:200	1:76	5	1:80
5.6:200	1:36	10	1:40
12.5:200	1:16	20	1:20
		40	1:10

Results

Nematode bioassays

Three distinct types of effects were detected by the nematode bioassay:

a) *Survival* – the number of animals that survived the 96-hour period of exposure to the test material relative to that of controls containing 1% DMSO in the growth medium;

b) *Growth* – the proportion of the test population that grew through a minimum of two moults compared to the control; and,

c) *Maturation* – the proportion of the test population that completed growth to the adult stage compared to that of the control population.

The *survival* measurement indicates the extent of effects of the tested material on the biochemical and physiological processes essential for nematode life. Decreased *survival* is the most severe toxic effect and samples producing lethal effects are considered the most toxic. The inhibition of *growth* indicates the extent of effects on non-essential biological processes, due to exposure to the samples. *Growth* may also be enhanced as a result of the presence of nutrients. *Maturation* reflects specific inhibition of the normal genetic functions suggesting a potential long term risk to individual members of a population but not to the population as a whole (Samoiloff *et al.*, 1983).

The combined effects on *survival, growth* and *maturation* are combined as a summary weighted value termed fitness (Ongley *et al.*, 1986). This provides a means of comparing the total toxicity of samples for the purpose of ranking a set of samples (Table 2). The sample with the lowest *fitness* is considered the most toxic. A Chi-square analysis of the stage distribution of each test population compared to the control population was performed. In those cases where significant ($p < 0.005$) differences between test and control populations were detected, the type of observed effect is designated in Table 2 by a letter as follows, L = lethality, I = inhibition of growth, and i = inhibition of maturation.

In general, the Acid Fraction and Fraction 1 of the chemically fractionated sediment extracts can be expected to contain the chlorinated compounds such as pesticides and polychlorinated biphenyls (PCB) while Fraction 2 can be expected to contain polycyclic aromatic hydrocarbons (PAH). Fractions 3 and 4 will contain oxygenated, more polar compounds, not normally identified in target compound identification.

The *fitness* results for each chemical fraction

Table 2. Summary of Fitness results determined by the *Panagrellus redivivus* bioassay for Lake Diefenbaker sediment samples showing only the most severe effect that was significantly different from the control population (i.e. *Lethality > Inhibition of growth > inhibition of maturity*). (Significant enhancement is indicated by S)

Station Number (name)	Distance (km)	Fitness Acid	Fitness F1	Fitness F2	Fitness F3	Fitness F4
14 (Cabri)	32	93i	86i	90i	94	88i
13 (Sask. Landing)	52	82i	53L	86I	90	95
12 (Swift Current Cr.)	65	101	66I	89i	100	76I
11	75	94i	79i	87I	88I	86i
10 (Herbert Ferry)	95	86	99	96I	86I	95i
9 (Snakebite Cr.	110	79I	72L	96I	12L	81I
8 (Riverhurst Ferry)	142	69L	101	100	89I	95I
7 (Hitchcock Bay)	157	82I	68L	84I	97I	92I
3 (Elbow)	177	91i	65I	89	97	87i
5 (Douglas Park)	160	87	72i	99	91I	66I
4 (Elbow)	175	98	88	100	102	99
2	190	77I	78I	88i	93i	86I
1 (Gardiner Dam)	200	91i	102	95	101	101S

296

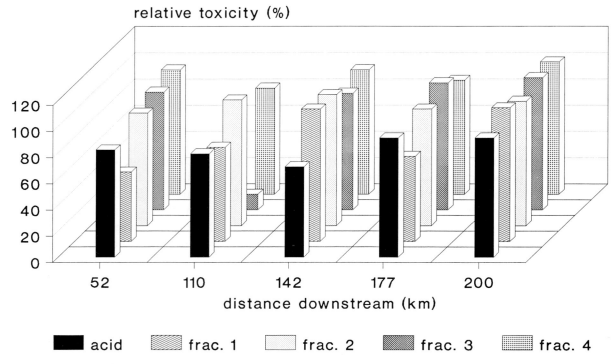

relative toxicity (%)

Fig. 2. Comparison of results of *Panegrellus redivivus* bioassays for selected stations of Lake Diefenbaker summarized as Fitness and expressed as relative toxicity (100 equals no toxicity) for the five chemical fractions tested.

used in the nematode tests are plotted for selected sites in Fig. 2. The stations are plotted as distance downstream to the Gardiner Dam. There is no clear downstream trend, and there is considerable variability among the sites although station 1 at the Gardiner Dam appears to be one of the stations with least toxicity. Four of the 12 stations (Table 2) showed lethality; specifically, stations 13 and 7 in Fraction 1, station 9 in Fractions 1 and 3, and station 8 in the Acid Fraction. Inhibition of *growth* was evident in all chemical fractions but was most frequently observed (8 stations) in Fraction 1. Only station 1 showed significant enhancement over the control populations. Inhibition of *maturation* was widely distributed among the fractions but again was frequently observed in the Acid Fraction, Fraction 1, and Fraction 4 (8, 9 and 9 stations respectively). Note that only the most serious effect has been shown in Table 2 (i.e. *Lethality > Inhibition of growth > inhibition of maturity*).

Algal bioassays

The results for the AFBs and the LSBs are presented in Table 3. All tests are reported as a percentage relative to the controls with the 'dark uptake' subtracted. A negative sign indicates inhibition while a positive sign indicates enhancement relative to the untreated control. Only in cases where the change is significantly different from the controls at $> 95\%$ has the change been quantified, otherwise only the direction of response is shown. These directional responses have been retained for the purpose of qualitative comparisons among the three tests.

For the AFBs, the ultraplankton showed significant inhibition at sediment to water ratios of $1:10$ at stations 1, 3 and 8 and at $1:20$ at station 1 (Gardiner Dam). No effects, including enhancement, were noted for these plankton at lesser additions. The micro/netplankton similarly showed inhibition at all three sites for the $1:10$

Table 3. Results for Algal Fractionation Bioassay (AFB), and Limited Sample Bioassay (LSB) for Riverhurst Ferry, Elbow and Gardiner Dam stations of Lake Diefenbaker, Saskatchewan (all 'changes' shown numerically are significant at 95% or better whereas '+' or '−' signs show trends relative to the control which are not significant at this level)

Limited Sample Bioassay (culture population)			Algal Fractionation Bioassay (L. Diefenbaker population)		
Station	Sed: H_2O	Change (%)	% Elutriate (Sed: H_2O	Ultra-plankton Change (%)	Micro/Net Plankton Change (%)
Riverhurst	1:76	+ 17	5 (1:80)	+	−
Ferry	1:36	+ 21	10 (1:40)	−	+
	1:16	+ 27	20 (1:20)	−	− 24
			40 (1:10)	− 47	− 42
Elbow	1:76	− 11	5 (1:80)	−	+
	1:36	−	10 (1:40)	−	−
	1:16	− 10	20 (1:20)	−	−
			40 (1:10)	− 34	− 47
Gardiner	1:76	+ 4	5 (1:80)	−	−
Dam	1:36	+ 11	10 (1:40)	−	−
	1:16	+ 27	20 (1:20)	− 24	−
			40 (1:10)	− 53	− 48

ratios. In contrast to the ultraplankton, station 8 (Riverhurst) showed inhibition at a 1:20 addition rather than station 1 (Gardiner Dam). Again, enhancement was not observed.

The results of the LSBs using the mixed algal culture were notably different from the AFBs. Station 8 (Riverhurst) showed increasing enhancement as the sediment to water ratio decreased (i.e., as a greater proportion of sediment is present). In contrast the sample from station 3 (Elbow) showed inhibition at all ratios although that observed for a 1:36 sediment to water ratio was not statistically significant. The Gardiner Dam site showed increasing enhancement as the quantity of sediment elutriate added increased.

Discussion

The nematode *Panagrellus redivivus*, is known to be a hardy organism tolerant of its environment (Samoiloff *et al.*, 1980). Nevertheless, the results of the nematode bioassays, although highly variable among stations and the various fractions,

clearly show negative effects on this organism. Indeed, four stations showed lethality:
- station 13 (Saskatchewan Landing) with Fraction 1;
- station 9 (Snakebite Creek) with Fractions 1 and 3;
- station 8 (Riverhurst Ferry) with Acid Fraction; and,
- station 7 (Hitchcock Bay) with Fraction 1.

Inhibition of *growth* was most frequently observed in Fraction 1 whereas the less severe effect of inhibition of *maturation* was widespread but most common in the Acid Fraction and Fractions 1 and 4. Significant enhancement of *growth* was noted only at station 1 (Gardiner Dam).

The sediment extraction/fractionation procedure, a chemical technique using organic solvents, effectively separates many trace organic substances from the sediments for analyses (Birkholz *et al.*, 1983) but this does not characterize the biological availability of these substances in a natural aquatic environment. Additionally, this technique does not consider the possible effects on the test medium of nutrients, metals and inorganic trace contaminants present in the sediment as these are not efficiently extracted in this procedure. Consequently, bioassays using water based elutriates and natural and cultured phytoplankton populations were conducted for comparison purposes even though it is recognized that the differences between the methods precludes anything but general, qualitative comparisons. The elutriate based tests should indicate the synergistic effects of contaminants plus mitigating influences of available nutrients and since they utilize either real lake phytoplankton populations or a representative culture, may be more acceptable to lake managers. The results of the three methods are summarized in Fig. 3.

Station 8 (Riverhurst), the only station of the three stations compared to show lethality in the nematode bioassay, was highly toxic to the Lake Diefenbaker phytoplankton but enhanced the growth of the cultured population of the LSB. Perhaps a larger sediment to water ratio in the LSB would have eventually produced inhibition, as there was no significant effect in both size fractions of the AFB until a sediment to water ratio of 1:10 was used. However, further treatments were precluded by the LSB sample size. Station 3 (Elbow) showed inhibition of growth for

all three tests. Interestingly, the LSB for station 3 showed evidence of inhibition at all treatment ratios, while the AFB showed no significant effect for either ultraplankton or micro/netplankton fraction until the sediment to water ratio of 1:10 was reached. Station 1 (Gardiner Dam), although found to be the only station with significant enhancement in the nematode bioassay and with enhancement at all treatments in the LSB, was highly inhibitory to both AFB algal size fractions at the 1:10 ratio.

Munawar & Munawar (1982) and Munawar et al. (1987) observed that the smaller size fraction, the ultraplankton, tended to be more susceptible to heavy metal mixtures. Similar effects were expected for the mixtures of metals and organic contaminants in the Lake Diefenbaker sediments; however, the effects on the two algal size fractions were more or less comparable for the three stations. All stations showed similar inhibition for both fractions at 1:10 sediment to water additions, whereas two stations showed inhibition at 1:20 but in different size fractions.

Using the AFB method for Great Lakes sediment, Munawar & Munawar (1987) have shown that the inhibition observed at the 40% elutriate treatment is not much greater than that observed at the 20% addition. Munawar & Thomas (1989) have shown that Great Lakes harbour sediments from Toronto and Toledo had maximum inhibition of the carbon assimilation rates at the 20% elutriate treatments. Evidently, the Lake Diefenbaker sediment elutriates have a similar inhibitory effect on Lake Diefenbaker phytoplankton; however, the effect seems to be mitigated at the lower elutriate additions. This could be due to the presence of high concentrations of available nutrients in the Lake Diefenbaker sediments as well as differences in the nature and availability of the toxic substances.

In conclusion, although the methods are only qualitatively comparable, the nematode bioassay and the Algal Fractionation Bioassay using Lake Diefenbaker phytoplankton show that the Lake Diefenbaker sediments contain quantities of toxic substances sufficient to cause significant inhibi-

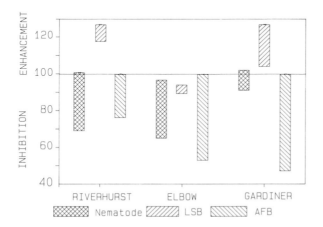

Fig. 3. Comparison of the range of relative toxicity (100 equals no toxicity) for Riverhurst (142 km), Elbow (177 km) and Gardiner Dam (200 km) for the *Panegrellus redivivus* bioassay (nematode), algal fractionation bioassay (AFB) and the limited sample bioassay using algal cultures (LSB).

tion to these bioassay organisms. Because on-lake contaminant sources are non-existent, the major source for these toxic substances must be the two tributaries, the South Saskatchewan River and the Red Deer River. To better understand the nature of the toxic substances, chemical analyses of the suspended sediments of these inflowing rivers is required. The absence of significant effects at sediment elutriate additions less than 20% in the AFB method suggests mitigating effects of biologically available nutrients in these sediments.

The Limited Sample Bioassay, using a cultured algal population, showed enhancement at two stations. The possible reason for this contrasting response between the LSB and the AFB may be attributable to the expected differential response experienced between natural assemblages and laboratory grown cultures exposed to the same source of contaminants by Munawar & Munawar (1987).

In general, the nematode bioassay and the AFB results are comparable with respect to the bioassay response. It is noted that the elutriate preparation for the AFB is simpler, covers a broader range of toxicants and is arguably more comparable to the natural environment than the chemical extract used for the nematode bioassays. Also, the AFB elutriate better integrates the synergistic effects of nutrients and contaminants and utilizes natural organisms from the ecosystem under investigation as test species. Therefore the AFB results are concluded to be the most realistic of the techniques compared and are therefore likely to be more acceptable to environmental managers.

Based on this experience, we conclude that the chemical fractionation technique applied in the nematode bioassay is useful for subsequent chemical analyses of trace organic substances and for ranking of samples (Ongley et al., 1988). As indicated earlier the elutriate-based AFB is broader in scope and more representative of actual in-lake effects. Chemical analyses of the water-based elutriate is more appropriate than the chemical extraction of the sediment as carried out for the nematode bioassay, since the objective is to do a holistic assessment of the potential environmental hazard of the sediment-bound contaminants rather than just trace organic substances. Hence the most appropriate procedure for screening sediments would be to prepare a water-based elutriate as done for Algal Fractionation Bioassays (AFB). This elutriate could then be split for comparative bioassays with phytoplankton, nematode, and for complete chemical analyses including nutrients, metals and trace metals plus extraction and fractionation following the method of Birkholz et al. (1983) for organic chemical characterization. Additional controls need to be incorporated in the tests to determine the relative importance of the various contaminant groups (i.e. organic versus inorganic) as well as the role of nutrient availability in mitigating contaminant effects on the test organisms.

Acknowledgements

We would like to thank W. P. Norwood, L. H. McCarthy (Fisheries and Oceans Canada) and S. Daniels (National Water Research Institute) for their assistance. The constructive reviews of W. Ahlf and L. Burnett were very beneficial and we are grateful to them for their comments and suggestions.

References

Birkholz, D. A., G. Nelson, S. Daignault & L. Mulligan, 1983. Measurements of organics in petroleum refinery waste waters. Environmental Protection Services Report, Environment Canada, Edmonton, 83 p.

Blaise, C., N. Bermingham & R. Van Coillie, 1985. The integrated ecotoxicological approach to assessment of ecotoxicity. Wat. Qual. Bull. 10: 3–10, 60.

Cairns, J., 1985. Bioassays as they relate to water quality. Wat. Qual. Bull. 10: 17–20, 58, 59.

Cross, P. M., H. R. Hamilton & S. E. D. Charlton, 1984. The limnological characteristics of the Bow, Oldman and South Saskatchewan Rivers: 1. nutrient and water chemistry. Water Quality Control Branch, Alberta Environment, unpublished report.

Daniels, S. A., M. Munawar & C. I. Mayfield, 1989. An improved elutriation technique for testing the bioavailability of contaminants in sediments. In: M. Munawar, G.

Dixon, S. I. Mayfield, T. Reynoldson and M. H. Sadar, Environmental Bioassay Techniques and Their Application. Hydrobiologia. (This volume).

International Joint Commission (I.J.C.), 1986. Evaluation of sediment bioassessment techniques. Report of the Dredging Subcommittee to the Great Lakes Water Quality Board, Windsor, Ontario, 123 pp.

Munawar, M., P. Stadelman & I. F. Munawar, 1974. Phytoplankton biomass, species composition and primary production at a nearshore and midlake station of Lake Ontario during IFYGL. Proc. 17th Conf. Great Lakes Res., Internat. Assoc. Great Lakes Res. pp. 629–652.

Munawar, M. & I. F. Munawar, 1982. Phycological studies in Lakes Ontario, Erie, Huron and Superior. Can. J. Botany 60: 1837–1858.

Munawar, M., A. Mudroch, I. F. Munawar & R. L. Thomas, 1983. The impact of sediment-associated contaminants from the Niagara River mouth on various size assemblages of phytoplankton. J. Great Lakes Res. 9: 303–313.

Munawar, M., R. L. Thomas, H. Shear, P. McKee & A. Mudroch, 1984. An overview of sediment-associated contaminants and their bioassessment. Can. Tech. Rep. Fish. Aquat. Sci. 1253 (i-vi): 1–136.

Munawar, M. & I. F. Munawar, 1987. Phytoplankton bioassays for evaluating toxicity of in situ sediment contaminants. In: R. L. Thomas, R. Evans, A. Hamilton, M. Munawar, T. Reynoldson & H. Sadar (Eds.), Proc. Internat. Workshop on Ecological Effects of In Situ Sediment Contaminants. Hydrobiologia. 149: 87–105.

Munawar, M., I. F. Munawar, P. E. Ross & C. Mayfield, 1987. Differential sensitivity of natural phytoplankton size assemblages to metal mixture toxicity. In M. Munawar (Ed.), Proc. Internat. Symp. on Phycology of Large Lakes of the World. Arch. Hydrobiol., Beih. Ergebn. Limnol., 25: 123–139.

Munawar, M., D. Gregor, S. A. Daniels & W. P. Norwood, 1989. A sensitive screening bioassay technique for the toxicological assessment of small quantities of contaminated bottom or suspended sediments. In: P. G. Sly & B. T. Hart (eds.), Sediment/Interactions. Hydrobiologia. (In press).

Munawar, M. & R. L. Thomas, (1989). Sediment toxicity testing in two areas of concern of the Laurentian Great Lakes: Toronto (Ontario) and Toledo (Ohio) harbours. In: P. G. Sly & B. T. Hart (eds.), Sediment/Interactions. Hydrobiologia. (In press).

Ongley, E. D. & D. Blachford, 1984. Tributary effects on main stream suspended sediment chemistry. Wat. Poll. Res. J. Can. 19: 37–46.

Ongley, E. D., D. A. Birkholz, J. Carey & M. R. Samoiloff, 1988. Is water a relevant sampling medium for toxic chemicals?: an alternative environmental sensing strategy. J. Envir. Qual. 17: 391–401.

Samoiloff, M. R., S. Schulz, Y. Jordan, K. Denich & E. Arnott, 1980. A rapid simple long-term toxicity assay for aquatic contaminants using the nematode Panagrellus redivivus. Can. J. Fish. Aquat. Sci. 37: 1167–1174.

Samoiloff, M. R., J. Bell, D. A. Birkholz, G. R. B. Webster, E. G. Arnott, R. Pulak & A. Madrid, 1983. Combined bioassay-chemical fractionation scheme for the determination and ranking of toxic chemicals in sediments. Envir. Sci. Tech. 17: 329–334.

Samoiloff, M. R. & T. Bogaert, 1984. The use of nematodes in marine ecotoxicology. In: G. Persoone, E. Jaspers & C. Claus (Eds.), Ecotoxicological Testing for the Marine Environment, State University, Ghent and Inst. Mar. Scient. Res., Bredene, Belgium, vol. 1, 407–425.

Hydrobiologia **188/189**: 301–315, 1989.
M. Munawar, G. Dixon, C. I. Mayfield, T. Reynoldson and M. H. Sadar (eds)
Environmental Bioassay Techniques and their Application.
© *1989 Kluwer Academic Publishers. Printed in Belgium.*

Fraser river sediments and waters evaluated by the battery of screening tests technique

B. J. Dutka,[1] T. Tuominen,[2] L. Churchland[2] & K. K. Kwan[1]
[1] *Rivers Research Branch, National Water Research Institute, Canada Centre for Inland Waters, Burlington, Ontario, L7R 4A6;* [2] *Water Quality Branch, Environment Canada, 502-1001 West Pender Street, Vancouver, B.C. V6E 2M9*

Key words: river, sediments, microbiological, biochemical

Abstract

The suitability of a variety of microbiological, biochemical and toxicant screening tests to become part of a battery of test procedures to identify degraded or degrading water bodies are evaluated in this report. Data were collected from 40 sampling sites within the Fraser River Basin in British Columbia. These data re-emphasize that individual toxicant, biochemical or microbiological screening tests do not provide a sufficient data base upon which realistic management decisions can be made. This study also confirms that the fecal sterol tests do not seem amenable to a 'battery of tests' approach and that the *Daphnia magna* test continues to be the most sensitive procedure for indicating the presence of contaminants with toxicant activity.

Introduction

In previous publications and reports, Dutka *et al.* (1986, 1986a, 1987, 1987a) described the results of studies to evaluate the suitability of various microbiological, biochemical and bioassay tests to become part of a 'battery of test procedures' which could be used to designate, nationally and internationally, water bodies or sediments that are degraded or are being degraded due to toxic chemical discharges, excessive nutrient inputs or microbiological contamination. This 'battery of tests' could also be used to monitor the effectiveness of remedial actions or the effect of specific discharges on ambient riverine or lacustrine ecology.

In this paper, we examine waters and sediments and conditions very different from those previously used to evaluate the tests which might be included in the 'battery of tests'. The sampled waters in this study are those of the Fraser River in the Canadian province of British Columbia. The Fraser River drains an area of 230 000 sq km and has a length of approximately 1400 km from its headwaters in the Rocky Mountains to the Strait of Georgia (Fig. 1 and 1a). The Fraser River estuary receives municipal effluent and storm water originating from the largest population centre in the province (Vancouver) and points upstream. The river is also subject to a multiplicity of industrial discharges. The Fraser is used for commercial shipping, recreational boating and for transporting log booms, which are stored along much of its shoreline. The Fraser also supports a large commercial salmon fishery and is known worldwide for its sport fishing. The river is a migration route for juvenile and adult salmon and a rearing area for various salmon and

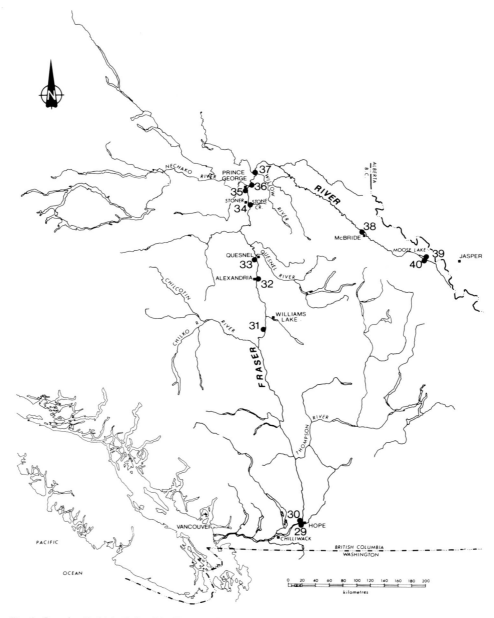

Fig. 1. Interior British Columbia Fraser River Sampling Sites using Battery of Tests Approach.

trout. The estuary is one of the world's most productive fish, wildlife and agricultural areas. The wetlands support an annual catch of eight million adult salmon and support over one million migratory birds using the Pacific Flyway. Farmland in the Fraser flood plain provides most of western Canada's fresh vegetable and berry crops. The waters of the Fraser River are not used for public water supplies but do influence swimming areas in the outer estuary (Kwiatkowski, 1986). Data from 40 water and sediment samples collected from the Fraser River and its estuary are presented and the results discussed.

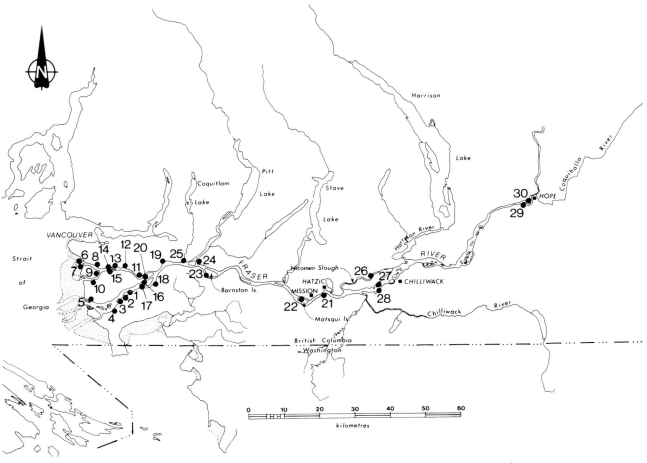

Fig. 1a. Lower Fraser River Sampling Sites using Battery of Tests Approach.

Methods

Sampling site

During June 1987 a total of 40 samples, water and sediment, were collected from sites where fine sediment deposits were expected (Fig. 1, 1a). Twenty of the samples were from sloughs and arms of the lower Fraser River in or near Vancouver. Some of the estuarine samples were affected by salt water intrusion. Sample site latitudes and longitudes and sediment descriptions are shown in Table 1.

Sample collection

Sediments were collected with an Ekman dredge or shovel. Frequently, it was necessary to use the dredge many times before sufficient surface (1–3 cm layer) sediment was collected. At each site, the surface layers were pooled, well mixed, dispensed into aliquots for each testing procedure and refrigerated. To obtain sediment extract for toxicant screening tests, sediments were extracted with Milli Q water (four cartridge system – one Super C carbon cartridge, two Ion-Ex[tm] cartridges, one Organet-Q[r] cartridge and a Milli-Stak[tm] filter, with a glass-distilled water feed) by mixing sediment and Milli Q water in a 1 : 1 ratio and shaking vigorously for two minutes,

Table 1. Fraser river basin, British Columbia, Canada. Sampling site locations and sediment description.

Station name and number	Latitude	Longitude	Sediment description and Shepard classification
1. Cove just upstream of Tilbury Dock	49°08'54" N	123°00'36" W	sand 6.24%, silt 78.05%, clay 15.71% SILT
2. Tilbury Slough west side of Hopcott Rd. (off culvert)	49°08'18" N	123°01'30" W	sand 78.73%, silt 16.81%, clay 4.46% SAND
3. Deas Slough	49°07'14" N	123°03'30" W	sand 12.75%, silt 62.66%, clay 24.59% CLAYEY SILT
4. Ladner Harbour	49°05'31" N	123°05'24" W	sand 5.29%, silt 68.19%, clay 26.52% CLAYEY SILT
5. Steveston, Cannery Channel	49°07'09" N	123°10'00" W	sand 19.67%, silt 61.05%, clay 19.28% SANDY SILT
6. North Arm Jetty, north shore of jetty	49°13'26" N	123°12'54" W	sand 30.37%, silt 55.92%, clay 13.71%, organic material present SANDY SILT
7. McDonald Slough	49°12'45" N	123°11'11" W	sand 20.48%, silt 62.03%, clay 17.49% SANDY SILT
8. Just upstream of Richmond Island, near mouth of storm sewer	49°12'20" N	123°08'51" W	sand 27.89%, silt 59.28%, clay 12.84% SANDY SILT
9. Middle Arm, west shore downstream of marinas	49°11'21" N	123°08'19" W	sand 12.04%, silt 61.30%, clay 26.65%
10. Middle Arm downstream of Dinsmore Bridge, south shore	49°10'32" N	123°09'22" W	sand 17.69%, silt 63.92%, clay 18.39% CLAYEY SILT
11. North Arm, north shore at Burnaby Bend	49°10'52" N	122°58'33" W	sand 51.85%, silt 38.23%, clay 9.91% SILTY SAND
12. North Arm, north shore at dock of MacMillan Bloedel, White Pine Division	49°12'11" N	123°02'03" W	sand 14.47%, silt 66.71%, clay 18.82% CLAYEY SILT
13. Mitchell Island north shore across channel from Aero Trading	49°12'18" N	123°05'25" W	sand 23.55%, silt 61.63%, clay 14.82%, organic material present SANDY SILT
14. Mitchell Island, north shore ≃400 m upstream of transmission line crossing	49°12'16" N	123°06'33" W	sand 28.93%, silt 58.81%, clay 12.27% SANDY SILT
15. Mitchell Island, south shore at Western Canada Steel	49°11'59" N	123°06'03" W	sand 37.21%, silt 53.08%, clay 9.71%, organic material present SANDY SILT
16. Annacis Channel, north shore downstream end near Shelter Island Marina	49°10'04" N	122°58'08" W	sand 30.28%, silt 49.15%, clay 20.57% SANDY SILT CLAY
17. Southwest shore of Annacis Island, near Purfleet 23Pt.	49°09'34" N	122°58'53" W	sand 28.90%, silt 54.22%, clay 16.88% SANDY SILT
18. Gundersen Slough	49°10'20" N	122°55'05" W	sand 16.2%, silt 56.15%, clay 27.65% CLAYEY SILT
19. Mouth of Brunette River	49°13'11" N	122°53'26" W	sand 15.27%, silt 66.09%, clay 18.64% CLAYEY SILT
20. Tree Island Slough	49°11'02" N	122°57'44" W	sand 21.23%, silt 58.25%, clay 20.51%, organic material present SANDY SILT CLAY
21. Across Fraser River from Haztic	49°08'29" N	122°15'46" W	sand 70.44%, silt 27.40%, clay 2.16% SILTY SAND
22. Matsqui Island, northwest shore	49°07'44" N	122°21'05" W	sand 42.76%, silt 53.77%, clay 3.47% SANDY SILT

Table 1. (continued).

Station name and number	Latitude	Longitude	Sediment description and Shepard classification
23. Barnston Island, Parsons Channel at bend in Island	49°11′15″ N	122°43′00″ W	sand 71.81%, silt 25.80%, clay 2.38%, organic material present SILTY SAND
24. North Shore of Fraser just upstream of mouth of Pitt River	49°13′31″ N	122°45′30″ W	sand 31.71%, silt 57.86%, clay 10.43%, organic material present SANDY SILT
25. Mouth of Coquitlam River	49°13′32″ N	122°48′13″ W	sand 19.56%, silt 60.33%, clay 20.11% CLAYEY SILT
26. Nicomen Slough downstream of Highway 7 Bridge at Deroche	49°11′08″ N	122°04′03″ W	sand 32.24%, silt 62.32%, clay 5.45%, organic material present SANDY SILT
27. South end of Yaalstrick Indian Reserve	49°09′45″ N	122°02′53″ W	sand 63.50%, silt 33.49%, clay 3.01%, organic material present SILTY SAND
28. South shore of river, on downcurrent base of Chilliwack Mountain	49°08′47″ N	122°03′10″ W	sand 71.70%, silt 25.27%, clay 3.03%, organic material present SILTY SAND
29. Hope, backwater behind Bristol Island	49°22′24″ N	121°28′34″ W	sand 30.51%, silt 43.83%, clay 24.66%, organic material present SANDY SILTY CLAY
30. Hope, backwater behind Croft Island	49°22′39″ N	121°27′28″ W	sand 73.21%, silt 23.48%, clay 3.31%, organic material present SILTY SAND
31. Upstream of Chilcotin Highway Bridge, west bank	51°59′12″ N	122°16′30″ W	sand 95.03%, silt and clay 4.97% SAND
32. East shore of Fraser River at site of old Alexandria Ferry	52°39′00″ N	122°29′22″ W	sand 66.60%, silt 32.38%, clay 1.02%, organic material present SILTY SAND
33. Quesnel, east side of river, 2 km downstream of Westply lumber mill	52°54′41″ N	122°28′40″ W	sand 33.17%, silt 51.38%, clay 15.45% SANDY SILT
34. Stoner, just downstream of mouth of Stone Creek	53°38′10.5″ N	122°40′07″ W	gravel 0.12%, sand 65.40%, silt 32.94%, clay 1.55%, organic material present SILTY SAND
35. Prince George, west bank of river, 1.4 km downstream of Hwy 97 Bridge	53°52′53″ N	122°45′40″ W	sand 54.21%, silt 40.32%, clay 5.47% SILTY SAND
36. Prince George, east bank of Fraser, downstream of mouth of Bittner Creek	53°55′48″ N	122°39′59″ W	sand 58.00%, silt 39.43%, clay 2.57%, organic material present SILTY SAND
37. North shore of Fraser, downstream of mouth of Willow River	54°04′56″ N	122°32′15″ W	sand 59.64%, silt 35.66%, clay 4.70% SILTY SAND
38. McBride, west bank of river just downstream of Hwy 16 bridge	53°18′12″ N	120°08′27″ W	sand 76.48%, silt 21.48%, clay 2.04%, organic material present SAND
39. Moose Lake, north of Fraser River entrance to lake	52°56′11″ N	118°50′38″ W	sand 61.74%, silt 34.19%, clay 4.07%, organic material present SILTY SAND
40. Moose Lake, south of Fraser River entrance to lake	52°55′56″ N	118°51′06″ W	sand 69.15%, silt 28.72%, clay 2.14%, organic material present SILTY SAND

then centrifuging at 10 000 rpm in a refrigerated centrifuge for 20 minutes. The supernatant was used in toxicity screening tests.

A second portion of the sediment was sieved for size distribution, following the procedure outlined by Duncan & LaHaie (1979). The sample was sieved at $1/2$ or $1/4$ PHI scale intervals (Krumbein & Pettijohn, 1938). The size distribution was determined with SIZDIST, a programme used in conjunction with the IBM PC computer (Sandilands & Duncan, 1980).

Surface water samples (1 L) were collected at each site for fecal coliform, fecal streptococci and coliphage tests. These tests were usually processed within eight hours of collection. Also at each site another 1 L sample of water was collected and preserved at 4 °C for toxicant screening tests. Toxicant screening test samples were tested after being concentrated 10 × by flash evaporation at 45 °C.

A one-litre surface water sample was also collected at each site, for coprostanol and cholesterol analyses. The sample was preserved with 1 ml concentrated H_2SO_4 and refrigerated at 4 °C.

Microorganism tests

Fecal coliform MF, fecal streptococci MF and coliphage tests were performed on all water samples as described by Dutka *et al.* (1986). Enterococci population estimates were also performed on water samples using the 48 hour, 35 °C incubation Azide Dextrose Broth and the five-tube MPN technique with positive tubes being confirmed on Bile Esculin Agar (35 °C for 24 hours).

Fecal coliform populations in sediments were estimated using A-1 broth and the five tube MPN technique with 24 hours incubation at 44.5 °C. Sediment *Clostridium perfringens* populations were estimated by using the MPN technique described by Bonde (1963) and Dutka *et al.* (1986a).

Biochemical and toxicity screening tests

Coprostanol and cholesterol analyses were performed on water samples and the Microtox test was performed on water and sediment extracts as described by Dutka *et al.* (1986). SOS genotoxicity tests on water and sediment extracts were performed as described by Xu *et al.* (1987) without S-9 addition. ATP-TOX system, a new toxicity screening test based on toxicant inhibition of bacterial growth and luciferase activity, was applied to water and sediment extracts (Xu & Dutka, 1987). *Spirillum volutans*, a large aquatic bacterium with a rotating fascicle of flagella at each pole was also used to test water and sediment extract samples for toxicity following procedures described by Dutka & Kwan (1982). An algal-ATP toxicant screening test was also performed on water and sediment extracts. This test is based on the inhibition of ATP production in cultures of the green alga *Selenastrum capricornutum* (Blaise *et al.*, 1984). The ATP content of the stressed *Selenastrum* was measured by the procedure described in Luminescence Review (1983). The results are reported as a percentage of Relative Light Units (RLU) output by the tested sample compared to the non-stressed control which is 100%. A 48-hour *Daphnia magna* test using ten organisms per sample and sample dilution was also carried out to assess toxicant activity (APHA, 1985) on natural water samples and sediment extracts.

Results

Latitudes, longitudes and brief sampling site descriptions are presented in Table 1. Also shown in this Table is the composition of each sediment sample based on particle size distribution by sieve analyses (Salisbury, 1987) and sediment sample classifications (Shepard, 1954). The majority of lower Fraser River sediments (sites 1–25) with few exceptions (sites #2, #11, #21 and #23), were composed mainly of silt while the sediments of the upper reaches of the Fraser were composed predominantly of sand with organic material.

The format used to award points for specific data values, in order to rank the sampled waters and sediments from those of most concern to least, is presented in Table 2. The point allocation scheme is biased and not scientifically defensible, but it reflects the author's evolving experience with data accumulated from the application of a variety of toxicant screening tests to waters and sediments throughout Canada, as well as the distribution patterns of health related indicator bacteria in Canadian waters, sediments and effluent discharges.

The present point allocation scheme has evolved over a three year period and is an ongoing viable process which may change with increased data accumulations.

Samples with the most points are deemed to contain the greatest potential hazard to man and organisms found in the aquatic ecosystem. High toxicant levels may have reduced microbial levels/activity in some sediment samples. Cause and effect relationships however, were not investigated.

Table 3 is a complex table which presents all of the microbiological, biochemical and toxicological data obtained from the water samples. Examination of the microbiological data in this table reveals that ten sampling sites had fecal coliform densities greater than 200/100 ml, four sites had fecal streptococci densities greater than 100/100 ml and 12 sites had enterococci counts greater than 100/100 ml. Only three sites had these elevated indicator levels in all three tests, #2 (Tilbury Slough), #19 (mouth of Brunette River) and #32 (near old Alexandria Ferry). The highest fecal coliform counts ($> 3000/100$ ml) were found in samples 12, 13, 14 and 15, around Mitchell Island and the North Arm of the Fraser River.

Only nine samples were found to contain coliphage, with the highest count being 15 plaque forming units per 100 ml at sampling sites #17 (southwest shore of Annacis Island) and #24 (upstream of Pitt River mouth). The majority of the sites positive for coliphage (eight out of nine) were located downstream of site #27 (south end of Yaalstrick Indian Reserve).

Coprostanol and cholesterol levels were generally negative with only three sites positive, for each test. Coprostanol positive sites were #5 (Steveston), #17 (southwest shore of Annacis Island) and #19 (mouth of Brunette River) and cholesterol positive sites were #5, #19 and #39 (Moose Lake, north).

The toxicological screening tests using $10 \times$ concentrated water samples generally indicated that toxicant levels at most sites were below the sensitivity level of the tests applied. Both the *Spirillum volutans* and Algal-ATP screening tests were negative in all samples tested. The ATP-TOX System indicated that only 13 samples were completely negative for toxicant activity; however, the samples which produced a positive effect were only slightly above background and none were found to produce an EC_{50} effect (50% inhibition). The most toxic sample by this test was sample #13 (Mitchell Island, north shore) with 35% inhibition.

The SOS Chromotest which was performed without S-9 addition indicated that only three samples, #21 (Fraser River at Hatzic), #28 (Chilliwack Mountain) and #34 (downstream of Stone Creek) produced a genotoxic effect. Only six water samples were positive in the Microtox test [#6 (North Arm jetty), #8 (upstream of Richmond Island), (#15 Mitchell Island, south shore), #26 (Nicomen Slough), #29 (Bristol Island, Hope) and #35 (Prince George, west bank)] with sample #35 showing the highest degree of toxicant activity in this test.

The *Daphnia magna* test using natural water samples proved to be the most sensitive screening test for toxicant activity.

Only 11 samples were found to be completely free of toxicant effects as measured by the death of *Daphnia magna* organisms over a 48 hr period. Within these negative samples, there were seven that had EC_{10} values, but were considered to be borderline values and for evaluation purposes were reported as negative. The samples with the highest concentration of toxicants as measured by *Daphnia magna* reactions were #12 (North Arm by MacMillan Bloedel, White Pine Division) and #26 (Nicomen Slough).

Table 2. Point awarding scheme for sample ranking, based on suspected contained hazards.

Fecal coliform fecal streptococci enterococci per 10 g sediment/100 ml water	Coliphage per 100 ml water	Clostridium perfringens per 10 g sediment	Points	Coprostanol ppb	SOS chromotest genotoxicity induction factor per ml (10× water sample, 1:1 milli Q water sediment extract)	Algal-ATP % relative light units per ml (10× water sample, 1:1 milli Q water sediment extract)	Points	Cholesterol ppb	Points
1- 100	5- 24	1- 25	1	<1.0	1.0 -1.29	100 -50	1	<2.0	1
101- 500	25- 100	26- 100	2	1.0-3.0	1.30-1.50	49 -20	3	2.1-4.0	2
501- 2500	101- 250	101- 500	3	3.1-5.0	1.51-2.0	19 - 1.0	5	4.1-6.0	3
2501- 16000	251- 1000	501- 2500	4	5.1-7.0	2.1 -3.0	0.9 - 0.1	7	6.1-8.0	4
16001-160000	1001-5000	2501-10000	7	7.1+	3.1 +	0.09+	10	8.1+	5
160000 +	5001 +	10000 +	10						

ATP-TOX system % inhibition per ml (10× water sample, 1:1 milli Q water sediment extract)	Microtox EC_{50} % per ml (10× water sample, 1:1 milli Q water sediment extract)	Points	Daphnia magna EC % per ml (1:1 milli Q water sediment extract)	Points	EC % per ml water sample	Points	Spirillum volutans (10× water sample, 1:1 milli Q water sediment extract)	Points
1-30	40.0+	1	EC_{20} at 100%	1	EC_{20} at 100%	1	negative	0
31-60	40.0-25.0	3	EC_{40} at 100%	2	EC_{40} at 100%	2	positive	5
61-90	24.0-10.0	5	EC_{50} at 100%	4	EC_{50} at 100%	5		
91-99	9.0- 1.0	7	at 75%	5	at 75%	7		
100	0.9	10	at 50%	6	at 50%	8		
			at 25%	8	at 25%	10		
			at 10%	10				

Table 3. Results of Fraser River Basin water analyses by battery of tests approach.

Sample number	Fecal coliform MF-mFC /100 ml	Fecal streptococci MF-KF /100 ml	Entero-cocci MPN /100 ml	Coliphage /100 ml	Fecal sterols		Microtox EC$_{50}$ % ml	Algal-ATP RLU[1] %	Spirillum volutans[2] 120 min test	SOS chromotest[2] induction factor	ATP-TOX % inhibition[2]	Daphnia[3] magna EC$_{50}$ % sample	Points	Rank
					Coprostanol ppb	Cholesterol ppb								
1	150	42	79	<5	<0.05	0.3	NEG	S[2]	NEG	0.77	NEG	EC$_{20}$	6	12
2	400	490	220	<5	<0.05	0.6	NEG	S	NEG	0.54	19	NEG	8	10
3	60	18	11	<5	<0.05	0.6	NEG	S	NEG	1.00	16	NEG	6	12
4	198	57	49	<5	<0.05	0.3	NEG	S	NEG	0.92	NEG	NEG	5	13
5	43	30	49	<5	0.5	2.2	NEG	S	NEG	0.92	NEG	NEG	6	12
6	100	45	110	<5	<0.05	0.4	49.8	S	NEG	0.69	17	NEG	7	11
7	200	39	110	<5	<0.05	0.6	NEG	S	NEG	0.85	14	NEG	7	11
8	100	34	170	<5	<0.05	0.4	18.1	S	NEG	0.92	14	NEG	11	7
9	70	59	49	10	<0.05	0.1	NEG	S	NEG	0.92	NEG	EC$_{30}$	6	12
10	210	30	130	<5	<0.05	0.2	NEG	S	NEG	0.92	8.7	EC$_{40}$	9	9
11	100	30	49	<5	<0.05	0.2	NEG	S	NEG	0.92	NEG	NEG	4	10
12	4600	37	79	<5	<0.05	0.2	NEG	S	NEG	0.92	NEG	50%	15	3
13	6400	43	49	<5	<0.05	0.1	NEG	S	NEG	0.92	35	88%	15	3
14	3100	25	33	<5	<0.05	0.1	NEG	S	NEG	1.15	NEG	NEG	8	10
15	4100	47	31	5	<0.05	0.2	19.8	S	NEG	0.92	NEG	NEG	13	5
16	120	220	170	10	<0.05	0.2	NEG	S	NEG	0.92	18	EC$_{40}$	11	7
17	130	76	310	15	0.3	0.9	NEG	S	NEG	0.93	1.0	88%	14	4
18	110	89	140	<5	<0.05	0.6	NEG	S	NEG	0.86	28	100%	11	7
19	910	227	170	10	0.2	1.5	NEG	S	NEG	0.86	16	100%	15	3
20	20	17	49	10	<0.05	0.3	NEG	S	NEG	1.14	18	NEG	8	10
21	90	30	49	5	<0.05	0.5	NEG	S	NEG	1.43	29	EC$_{40}$	11	7
22	40	49	49	<5	<0.05	0.6	NEG	S	NEG	1.14	NEG	EC$_{40}$	7	11
23	21	27	49	<5	<0.05	0.6	NEG	S	NEG	1.14	3.0	100%	11	7
24	130	42	49	15	<0.05	0.1	NEG	S	NEG	1.14	NEG	EC$_{20}$	8	10
25	100	16	17	<5	<0.05	0.2	NEG	S	NEG	1.14	NEG	EC$_{30}$	7	11
26	60	28	17	<5	<0.05	0.5	24.9	S	NEG	1.00	17	50%	19	1
27	120	22	79	5	<0.05	0.1	NEG	S	NEG	1.14	2.4	82%	13	4
28	300	32	70	<5	<0.05	0.1	NEG	S	NEG	1.43	7.1	78%	14	4
29	10	29	12	<5	<0.05	0.1	46.5	S	NEG	1.14	17	84%	12	6
30	110	49	130	10	<0.05	0.4	NEG	S	NEG	0.86	18	100%	12	6
31	150	17	49	<5	<0.05	0.2	NEG	S	NEG	1.29	14	100%	12	6
32	1500	117	280	<5	<0.05	0.1	S	S	NEG	1.00	NEG	82%	14	4
33	310	40	110	<5	<0.05	0.1	NEG	S	NEG	1.14	24	82%	13	5
34	45	10	79	<5	<0.05	0.1	NEG	S	NEG	1.43	NEG	80%	12	6
35	12	4	23	<5	<0.05	0.1	5.7	S	NEG	0.14	1.2	100%	18	2
36	15	25	43	10	<0.05	0.2	NEG	S	NEG	1.14	1.7	92%	12	6
37	25	4	13	<5	<0.05	0.3	30.8	S	NEG	1.14	1.9	92%	14	4
38	4	1	18	<5	<0.05	0.5	NEG	S	NEG	1.08	21	84%	11	7
39	2	<1	12	<5	<0.05	1.8	NEG	S	NEG	1.00	19	84%	10	8
40	<1	<1	<2	<5	<0.05	0.8	NEG	S	NEG	1.00	21	84%	8	10

RLU[1] = Relative light units.
S[2] = Growth stimulatory.
Daphnia[3] = Performed on unconcentrated water sample.

Microbiological examination of Fraser River sediments (Table 4) produced an interesting set of observations. Only four sites were found to have very low concentrations (< 100/10 gm) of fecal coliform, #2 (Tilbury Slough), #20 (Tree Island Slough), #29 (Bristol Island, Hope) and #40 (Moose Lake, south), while another three had moderate numbers (< 500/10 gm), #3 (Deas Slough), #31 (by Chilcotin Highway bridge) and #39 (Moose Lake, north). All the other sediment samples contained more the 2300 fecal coliform per 10 gm wet weight of sediment with stations #11 (North Arm by Burnaby Bend), #18 (Gundersen Slough), #26 (south end of Yaalstrick Indian Reserve), #33 (Quesnel, east side of Fraser River), #34 (downstream of Stone Creek) and #35 (Prince George, west bank) having 160 000 or more per 10 gm sediment. The Cl. perfringens data indicate that the whole length of the Fraser River has been subjected to fecal pollution at one time or another with sampling sites #26 (Nicomen Slough), #31 (by Chilcotin Highway bridge), #36 (Prince George, east bank), #37 (north shore Fraser downstream of Willow River), #38 (McBride) and #40 (Moose Lake, south) being impacted the least by fecal pollution.

Microtox and Spirillum volutans toxicity tests were found to be negative for toxicant activity with the 1 : 1 sediment water extracts. The majority of the ATP-TOX System results were negative. However, a gradation of effects was noted from background noise levels to levels indicating the presence of low grade toxicants (site 9, Middle Arm; site 17, southwest shore of Annacis Island; site 32, near old Alexandria Ferry). The SOS Chromotest indicated that sample #13 (Mitchell Island, north shore) produced a genotoxic effect and sample #12 (North Arm by MacMillan Bloedel, White Pine Division) produced a border-line genotoxic effect.

The Algal-ATP test indicated that the majority of the sediment extracts ($^{35}/_{40}$) contained a chemical/nutrient balance which was stimulating to the growth of algae. Only one sample #31 (by Chilcotin Highway bridge) contained sufficient contaminants to produce a toxic effect in the Algal ATP toxicant screening test.

Similar to tests on the water samples, the Daphnia magna tests on sediment extract indicated that this procedure was the most sensitive indicator of toxicant activity with 50% of the sediment samples showing some toxicant activity, with samples #33 (Quesnel, east side of Fraser) and #30 (Hope behind Croft Island) having the highest toxicant levels.

Discussion

Microbiological studies of Fraser River water samples were generally indicative that the further upstream the samples were collected the better the microbiological water quality with few exceptions; #32 (Alexandria Ferry), #33 (Quesnel) and #28 (Chilliwack Mountain). Sites #32 and #33 are believed to be responding to bacteria originating in Quesnel sewerage discharges and lumber mill wastes. Similarly site #28 may also be impacted by bacteria originating from Chilliwack sewerage discharges. Cattle herds were noted at, and upstream of site #32 and their droppings may also play a role in the elevated bacterial population found here.

At sites #12, #13, #14 and #15 (North Arm of Fraser and Mitchell Island), extremely high fecal coliform counts were found in conjunction with very low fecal streptococci/enterococci counts; a finding suggesting that these organisms do not primarily originate from human fecal material. Unfortunately, isolates were not collected for identification procedures. It is surmised that if this extra step had been carried out, combined with the knowledge of the heavy concentration of forestry-related industries in this area, the majority of these fecal coliform would have been shown to be Klebsiella species and other non-E. coli which respond to organic pollution. In fecal coliform enumeration procedures, Klebsiella respond similarly to E. coli, although health hazard implications are different.

The fecal streptococci counts (MF using KF agar) were generally very low with only three sites having fecal streptococci counts greater than 100/100 ml. Enterococci counts which usually

Table 4. Results of Fraser River Basin sediment analyses by battery of tests approach.

Sample number	Fecal coliform Al broth 10g/100 ml MPN	Clostridium perfringens 10g/100 ml MPN	Microtox EC$_{50}$/ml water extract	Algal-ATP % RLU/ml water extract	Spirillum volutans 120 min test water extract	SOS chromotest induction factor/ml water extract	ATP-TOX % inhibition /ml water extract	Daphnia magna EC$_{50}$ water extract	Points	Rank
1	7900	170	S	S^1	NEG	1.00	NEG	75%	13	9
2	9	1,600	NEG	S	NEG	1.00	NEG	EC$_{20}$	7	15
3	280	240	NEG	S	NEG	0.92	NEG	NEG	5	16
4	7,000	430	NEG	S	NEG	1.00	NEG	NEG	8	14
5	7,900	>16,000	NEG	S	NEG	1.08	NEG	NEG	15	7
6	3,300	2,800	NEG	S	NEG	0.83	NEG	NEG	12	10
7	3,500	1,600	NEG	S	NEG	0.67	NEG	NEG	8	14
8	11,000	16,000	NEG	S	NEG	0.83	8.3	EC$_{30}$	15	7
9	7,900	1,100	NEG	S	NEG	0.92	28	NEG	9	13
10	4,900	430	NEG	S	NEG	1.00	NEG	NEG	8	14
11	160,000	95	NEG	S	NEG	0.92	NEG	100%	13	9
12	2,300	840	NEG	NEG	NEG	1.34	0.8	NEG	11	11
13	92,000	430	NEG	NEG	NEG	1.47	19	NEG	14	8
14	9,500	2,100	NEG	S	NEG	1.15	21	NEG	10	12
15	4,900	700	NEG	S	NEG	0.89	2.5	NEG	9	13
16	7,000	1,200	NEG	S	NEG	1.00	13	NEG	10	12
17	7,000	16,000	NEG	S	NEG	1.00	23	NEG	16	6
18	>160,000	3,500	S	S	NEG	0.86	NEG	NEG	17	5
19	3,300	1,200	S	S	NEG	0.78	2.7	NEG	12	10
20	95	430	S	S	NEG	1.00	NEG	NEG	5	16
21	92,000	240	S	S	NEG	1.00	NEG	Ec$_{30}$	12	10
22	35,000	280	S	S	NEG	0.78	NEG	Ec$_{20}$	11	11
23	4,900	70	S	S	NEG	1.00	NEG	Ec$_{30}$	8	14
24	13,000	140	NEG	S	NEG	0.86	1.0	NEG	8	14
25	3,500	16,000	NEG	S	NEG	0.93	7.1	31%	21	3
26	7,160,000	40	NEG	S	NEG	1.00	NEG	Ec$_{40}$	15	7
27	4900	540	NEG	S	NEG	0.93	NEG	60%	13	9
28	2300	2,200	S	S	NEG	1.00	12	40%	13	9
29	46	1,100	S	S	NEG	1.00	NEG	45%	12	10
30	28,000	700	NEG	S	NEG	1.07	NEG	6.4%	22	2
31	220	11	NEG	25.9	NEG	1.17	9	18%	16	6
32	35,000	950	NEG	NEG	NEG	1.17	25	30%	19	4
33	160,000	3,500	NEG	NEG	NEG	0.92	NEG	24%	24	1
34	>160,000	700	NEG	S	NEG	1.17	NEG	Ec$_{40}$	17	5
35	>160,000	1,400	S	S	NEG	0.92	NEG	Ec$_{40}$	16	6
36	35,000	17	S	S	NEG	1.17	NEG	NEG	9	13
37	22,000	39	S	S	NEG	1.17	NEG	NEG	9	13
38	92,000	15	S	S	NEG	1.08	NEG	Ec$_{20}$	10	12
39	490	35,000	NEG	S	NEG	1.17	10	NEG	14	8
40	5	49	NEG	S	NEG	1.00	NEG	Ec$_{30}$	5	16

S^1 = Growth stimulatory.

paralleled fecal streptococci counts, were higher with 12 water samples having counts greater than 100/100 ml. Again as no isolates were collected for identification procedures, it is not certain which of the *Streptococci* species were actually enumerated by the two procedures and also whether the differences in counts are real or merely reflect normal variations in microbiological population estimates.

Fecal sterol test results, as noted in earlier reports (Dutka *et al.*, 1986, 1986a) were low and not associated with predicted sources of fecal material. Sites positive for coprostanol, #5 (Steveston), #17 (southwest shore of Annacis Island) and #19 (mouth of Brunette River), also showed high fecal coliform and *Clostridium perfringens* in their sediments. Sites 5 and 19 are noted as mooring sites for fishing boats and site 17 is directly downstream of the Annacis sewage treatment plant outfall. Thus at these sites coprostanol and bacteriological concentrations suggest the presence of fecal contamination. Not withstanding the above, it is suspected that preservation procedures were inadequate and the coprostanol was biodegraded, thus perhaps accounting for the low number of positive findings.

Of all the toxicant screening tests used on the water samples, the *Daphnia magna* 48 hr test proved to be the most responsive to contaminants in the Fraser. The *Spirillum volutans* and Algal-ATP tests responded the least with all results indicating that no toxicants were present. The Algal-ATP test results indicated a stimulatory effect on algal growth by the Fraser River water samples. In 29 of the samples, the *Daphnia* test indicated the presence of toxicant activity. In contrast only seven samples were positive for toxicant effects by the Microtox test, of which four samples, #26 (Nicomen slough), #29 (Bristol Island, Hope), #35 (Prince George, west bank) and #37 (north shore Fraser, downstream of Willow River), were also positive by the *Daphnia* test.

Only three water samples, #21 (Fraser River at Hatzic), #28 (Chilliwack Mountain) and #34 (downstream of Stone Creek) could be considered positive for genotoxic activity as measured by the SOS Chromotest. These samples were also positive by the *Daphnia* test but were negative by the Microtox test. In 27 water samples some degree of toxicant activity was indicated by the ATP-TOX system and there also appears to be a relationship between ATP-TOX System values greater than 20% inhibition and the finding of a toxic response with the *Daphnia* test. However, none of the water samples contained sufficient toxic contaminants to produce an EC_{50} value with the ATP-TOX system.

Based on the point scheme proposed in Table 2, the nine water sample sites of the greatest potential concern are:

1. Sample #26; Nicomen Slough, ranking due mainly to toxicity load,
2. Sample #35; Prince George, west bank, ranking due mainly to toxicity load,
3. Sample #12; North Arm, near MacMillan Bloedel, White Pine Division, ranking due to bacteria and *Daphnia* test,
   Sample #13; Mitchell Island across channel from Aero Trading, ranking due to bacteria and toxicity loads,
   Sample #19; Brunette River mouth, ranking due to bacteria and toxicity load,
4. Sample #17; southwest shore of Annacis Island, ranking due to bacteria and toxicity load,
   Sample #28; Chilliwack Mountain, ranking due to toxicity load,
   Sample #32; Alexandria Ferry site, ranking due to bacteria and toxicity load.
   Sample #37; Willow River, ranking due to toxicity load,

When water column and sediment microbiological data are examined, some interesting patterns are observed. For instance, based on *Clostridium perfringens* and fecal coliform counts, site #40 (Moose Lake, south) would be assumed to be a site which is rarely if ever impacted by human fecal pollution and geographically this is borne out, as the lake is in a pristine area near the headwaters of the Fraser. Similarly, it is also believed that site #26 (Nicomen Slough) as well as site #11 (North Arm by Burnaby Bend), #23 (Barnston Island), #31 (by Chilcotin Highway bridge), #36 (Prince George, east bank), #37 (north shore Fraser, downstream of Willow River) and #38 (McBride) are also minimally impacted by human fecal pollution and that the high fecal coliform sediment counts are not related to *E. coli* levels but rather to *Klebsiella* and *Enterobacter* species. However, since no isolates were collected, this cannot be proven, although with the level of *Clostridium perfringens* found here and the very low densities of indicator organisms in the water column, the evidence supports a nonfecal source for these fecal coliform.

There are other sites, e.g. #5 (Steveston), #25 (mouth of Coquitlam River), #35 (Prince George, west bank) and #39 (Moose Lake, north) where water column data do not reflect the degree of fecal pollution indicated by the sediment data. In making these assumptions with respect to fecal pollution sites versus organic pollution sites based on *Clostridium* data, one always must be cognisant that *C. perfringens* spores can survive for years and represent not only fresh pollution but also past pollution patterns which may have changed.

The water sediment extracts used to test for toxicant and genotoxic activity were found, generally, to contain very low levels of compounds promoting these activities. The results seen in Table 4 make it obvious that (a) there are little or no chemicals in the Fraser River samples with toxic or genotoxic activity or (b) that the water extraction procedure used is not able to extract the organic and/or heavy metals contaminants from the sediments.

The Algal-ATP procedure was positive at only one site, #31 (by Chilcotin Highway bridge) while the sediment water extract from 35 other sites showed stimulatory effects, an effect also noted in the 10× water samples. Presence of genotoxic activity at site #13 (Mitchell Island, north shore) and suspected genotoxic activity at site #12 (North Arm by MacMillan Bloedel Mill) were not confirmed or supported by water column results which is not surprising because of the volume and rate of water movement and probable intermittent nature of contaminant inputs.

The *Daphnia magna* test on the sediment extract was the only toxicant screening test which frequently indicated the presence of toxicant activity in the water extracts. These results were very similar to the water column results. Interestingly, the only sediment extract positive by the Algal-ATP test (#31) was also positive in the *Daphnia magna* test while the two sites #12 and #13 which indicated the presence of genotoxic activity were both negative when tested by the *Daphnia magna* test.

Using the point scheme shown in Table 2, the 11 sediments of the greatest concern based on their point score are:

1. Sample #33; Quesnel area, ranking due to bacteria and toxicity load,
2. Sample #30; Hope behind Croft Island, ranking due to bacteria and toxicity load,
3. Sample #25; Coquitlam River mouth, ranking due to bacteria and toxicity load,
4. Sample #32; Alexandria ferry, ranking due to bacteria and toxicity load,
5. Sample #18; Gundersen Slough, ranking due to bacterial load,
   Sample #34; Stoner area, downstream of Stone Creek, ranking due to bacteria and toxicity load,
6. Sample #17; Southwest shore of Annacis Island, ranking due to bacterial load,
   Sample #31; Upstream Chilcotin Highway bridge, ranking due to toxicity load,
   Sample #35; Prince George, west bank, ranking due to bacteria and toxicity load,
7. Sample #5; Steveston Cannery Channel, ranking due to bacterial load,
   Sample #26; Nicomen Slough, ranking due to bacteria and toxicity load.

Surprisingly, the majority of the sediments of potential concern based on their point score totals are upstream of Coquitlam River. These concerns are primarily microbiological in sediments collected in the lower Fraser and both microbiological and toxicological in sediments collected upstream of Coquitlam River.

Examination of the top nine water column and nine sediment sites revealed that there were only three sites common to each list. These are listed here below:

Water column rank	Sediment extract rank	Site
2	6	Site 35, Prince George, West bank
4	4	Site 32, Alexandria Ferry
4	6	Site 17, Southwest shore Annacis Island

The results of this study are very illustrative of and supportive of the need for a battery of tests, the composition of which should be very carefully selected to reflect local conditions. Of the toxicant screening tests evaluated, the *Daphnia magna* test was the most sensitive procedure for indicating the presence of contaminants with toxicant activity.

Three major shortcomings of this study became obvious as the data were being analyzed. One was the need for fecal coliform isolate identification to clarify the sources of the large bacterial populations found in some water column and sediment samples. The second need was for the testing of solvent and acid extracted sediments for toxicological activity. Testing the water extracted sediments provided information on the toxic effects of contaminants which were likely to be biologically available in the aqueous environment; whereas testing of solvent and acid extracts would have provided information on potential toxic effects of more firmly bound contaminants. The third need was for the inclusion of an *in situ* organism test e.g., chironomids or a laboratory test for chronic effects, e.g., Ceriodaphnia, to assess accumulative effects. Thus, it is believed that these missing features would have produced a much clearer picture of the potential hazardous sites within the Fraser River.

Use of the 'battery of tests' approach re-emphasizes that individual toxicant, biochemical and microbiological screening tests do not provide a sufficient data base for realistic management deci-

sions to be made. This study also further confirms that the fecal sterol tests are not amenable to a 'battery of tests' approach and that their cost benefit ratio is very high.

Further refinement of the present 'battery of tests' will continue with emphasis on the inclusion of more vigorous extraction procedures and evaluating their effect on acute and chronic test results. The eventual goal will be to select a maximum of two microbiological tests and three toxicant screening tests as a core group. The ranking scheme will be reviewed after each study to ensure the points allocated to various response levels continue to reflect country wide conditions.

Acknowledgements

We should like to thank A. Jurkovic for technical and ekmaning support in the field, mobile laboratory and home laboratory. Toxicant screening test support from K. Jones and R. McInnis are gratefully acknowledged. We are also indebted to Keith Salisbury for sediment particle size distributions.

References

APHA., 1985. Standard Methods for the Examination of Water and Wastewater. 16th Ed. American Public Health Association, Washington, D.C. 1268 pp.

Blaise, C., R. Legault, N. Bermingham, R. Van Coillie & P. Vasseur, 1984. Microtest mesurant l'inhibition de la croissance des algues C150 par le dosage de l'ATP. Sc. et Tech. de l'Eau, 17: 245–250.

Bonde, G. J., 1963. Bacterial Indicators of Water Pollution. A study of Quantitative Estimation. Tuknisk Furlag, Copenhagen. 422 pp.

Duncan, G. A. & G. G. LaHaie, 1979. Size Analysis Used in the Sedimentology Laboratory. Hydraulics Division Manual, NWRI, CCIW, Burlington, Ontario, Canada, September. 7 pp.

Dutka, B. J. & K. Kwan, 1982. Application of Four Bacterial Screening Procedures to Assess Changes in the Toxicity of Chemicals in Mixtures. Envir. Pollut. Series A, 29: 125–134.

Dutka, B. J., K. Walsh, K. K. Kwan, A. El-Shaarawi, D. L. Liu & K. Thompson, 1986. Priority Site Selection of Degraded Areas Based on Microbial and Toxicant Screeting Tests. Wat. Poll. Res. J. Canada, 21: 267–282.

Dutka, B. J., K. Jones, H. Xu, K. K. Kwan & R. McInnis, 1986(a). Phase II. Priority Site Selection for Degraded Areas Based on Microbial and Toxicant Screening Tests. NWRI Contribution No. 86-174, NWRI, CCIW, Environment Canada, Burlington, Ontario, Canada. 31 pp.

Dutka, B. J., K. Jones, K. K. Kwan, H. Bailey & R. McInnis, 1987. Use of Microbial and Toxicant Screening Tests for Priority Site Selection of Degraded Areas in Water Bodies. Wat. Poll. Res. J. Canada, 22: 326–339.

Dutka, B. J., K. K. Kwan, K. Jones & R. McInnis, 1987a. Application of the Battery of Screening Tests Approach to Sediments in the Port Hope Area of Lake Ontario, Canada. NWRI Contribution #87–134, NWRI, CCIW, Environment Canada, Burlington, Ontario, Canada, 19 pp.

Krumbein, W. C. & F. J. Pettijohn, 1938. Manual of Sedimentary Petrography. Appleton-Century-Crofts. New York.

Kwiatkowski, R. E., 1986. Water Quality in Selected Canadian River Basins – St. Croix, St. Lawrence, Niagara, Souris and the Fraser Estaury. IWD Scientific Series No. 150, IWD, Water Quality Branch, Environment Canada, Ottawa. 67 pp.

Luminescens Review, 1983. Bulletin No. 240, Turner Designs. Mountain View, Cal., U.S.A. 7 pp.

Salisbury, K., 1987. Particle size data report, Fraser River, B.C. NWRI Research on Applications Branch Technical Report # RAB-87-07b. Canada Centre for Inland Waters, Burlington, Ontario, Canada, 48 pp.

Sandilands, R. G. & G. A. Duncan, 1980. SIZDIST – A Computer Programme for Size Analysis. Hydraulics Division Technical Note No. 80–08. CCIW, Burlington, Ontario, Canada. 6 pp.

Shepard, F. P., 1954. Nomenclature Based on Sand-silt Ratios. J. Sed. Petrology 24: 151–158.

Xu, H., B. J. Dutka & K. K. Kwan, 1987. Genotoxicity Studies on Sediments Using a Modified SOS Chromotest. Toxicity Assessment 2: 79–88.

Xu, H. & B. J. Dutka, 1987. ATP-TOX System – A New rapid Sensitive Bacterial Toxicity Screening System Based on the Determination of ATP. Toxicity Assessment 2: 149–166.

Hydrobiologia **188/189**: 317–335, 1989.
M. Munawar, G. Dixon, C. I. Mayfield, T. Reynoldson and M. H. Sadar (eds)
Environmental Bioassay Techniques and their Application.
© 1989 *Kluwer Academic Publishers. Printed in Belgium.*

Bioassay responses of micro-organisms to sediment elutriates from the St. Lawrence River (Lake St. Louis)

Harm Sloterdijk [1], L. Champoux [1,2], V. Jarry [2], Y. Couillard [2,3] & P. Ross [2,4]
[1] *Centre Saint-Laurent, Environment Canada, 105 McGill, Montreal, Quebec. H2Y 2E7*; [2] *Biological Sciences Department, University of Montreal, Quebec, H3C 3J7*; [3] *INRS-Eau, Complexe scientifique, 2700 rue Einstein, C.P. 7500, Ste-Foy, Quebec. G1V 4C7*; [4] *State Natural History Division, Illinois Department of Energy and Natural Resources, Champaign, Illinois 61820.*

Key words: microorganisms, sediments, bioassays

Abstract

A sediment study, involving both chemical and biological analyses, was carried out in the St. Lawrence River near Montreal (Lake St. Louis). About 60 stations were sampled during 1984-85, and the sediments were analyzed for support variables, heavy metals, and organochlorinated compounds. Subsamples were elutriated using a 1 to 4 sediment/water ratio. The resulting elutriates were analyzed for several chemical variables, while toxicity was measured using the Microtox test, algal ^{14}C assimilation, and lethality/developmental inhibition in cladocerans, rotifers, and nematodes. The results showed a great variety of responses and sensitivity, and correlations between the tests were non-significant. In terms of toxic responses, the algal and Microtox tests were the most sensitive. Toxic responses could not be explained in simple terms of contaminant concentrations. Therefore, the chemistry of elutriates is not predictive of the toxic potential of contaminated sediments. Biotests can give an insight into the hazard assessment of sediments, but no single test will be sufficient; the use of a battery of standardized biotests, representing different levels of organization/food chain, including representative natural species, is highly recommended.

Introduction

Chemical analyses of sediments can give useful information on the nature and extent of the contamination of a particular aquatic system (Sly *et al.*, 1981). However, they are insufficient for the evaluation of environmental risks and the biological significance associated with the presence of these contaminants in sediments (Levin & Kimball, 1984; Marquenie, 1985). Therefore, more and more attention is being given to the toxic potential of contaminated sediments by carrying out bioassays, either directly on the sediments or on sediment extracts (Munawar *et al.*, 1984; Samoiloff *et al.*, 1983).

Contamination of sediments in the St. Lawrence River has been studied on a few occasions since the early 1970s (Sérodes, 1978; Sloterdijk, 1985; Champoux & Sloterdijk, 1988). These results have not yet been published in detail in the primary literature; most of them, however, have been included in published reviews (Allan, 1988; Couillard, 1983). Particular attention has been paid to the various 'lake' systems within the river (St. François, St. Louis & St. Pierre; Fig. 1). These, however, are not really true lakes but widenings of the river. Fluvial, and at the most semi-lacustrine, conditions prevail, with frequent high energy periods due to spring run-off and storm-induced wave action. This, combined with

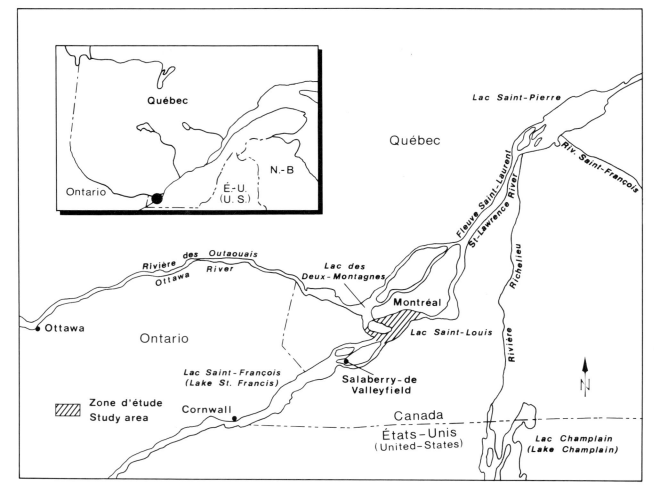

Fig. 1. St. Lawrence River system.

shallow water depths, allow only temporary deposition of finer particles, and resuspension is frequent (Allan, 1986).

In 1984, a joint study between Environment Canada and the University of Montreal was initiated on the contamination of sediments in Lake St. Louis. In addition, the potential liberation of contaminants during resuspension, and its toxic effects, were also evaluated, using the elutriation technique developed by the USEPA/ Corps of Engineers (1977).

This paper presents a brief overview of the bioassay results which were carried out on the sediment elutriates. A comparison between the tests results and water quality guidelines is made to see whether non-compliance with guidelines implies toxicity of the sample and vice versa. Results of the sediment study have been published in three master theses (Champoux, 1986; Couillard, 1987; Jarry, 1986). Part of the results have been published in the secondary scientific literature (Champoux *et al.*, 1987; Couillard *et al.*, 1987; Jarry *et al.*, 1985), while the complete study has been published in two technical reports (Champoux & Sloterdijk, 1988; Champoux *et al.*, 1989).

Material and methods

Study area

Lake St. Louis (Fig. 1) is formed by the confluence of the St. Lawrence River and the Ottawa River, the former having a flow rate of 10 to 12 times larger. The Ottawa River flow does, however, increase significantly during a short period at spring runoff. The two water types are quite distinct, the former being highly mineralized (conductivity: 300 to 350 μSiemens $\cdot$ cm^{-1}) and clear green-blue in colour, while the latter is brownish and much less mineralized (70 to 100 μS $\cdot$ cm^{-1}). The two waters practically do not mix within the lake system, the Ottawa River becoming somewhat mixed with St. Lawrence River water only along the north shore.

The lake is considered to be one of the most contaminated water bodies within the St. Lawrence River system. It receives contaminated waters from Lake Ontario and the international section of the river (border between New York State and the Province of Ontario), which are highly industrialized. Locally, it is the site of sewage discharges and industrial effluents from the Montreal West Island and the south shore community of Beauharnois. The latter is notorious for its Hg discharges from a chlor-alkali plant.

Sampling and elutriation

Sediment samples were taken from 52 stations in Lake St. Louis (Fig. 2) using an Ekman dredge

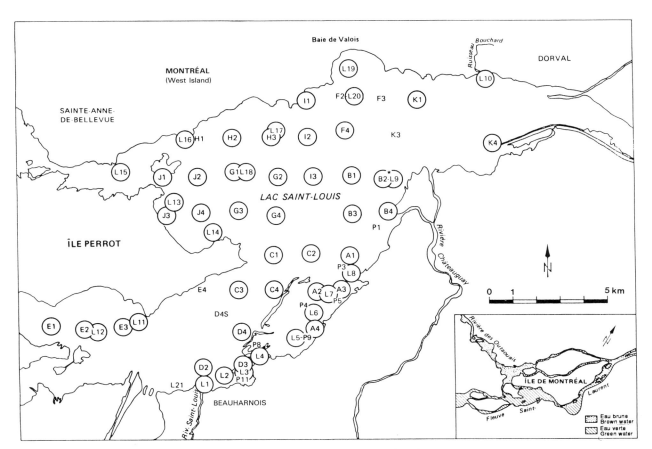

Fig. 2. Lake St. Louis and station location.

(30 × 30 cm), at a rate of two stations per week during the 1984 and 1985 field seasons. After homogenizing, subsamples were taken for chemical analyses, while the remainder was conserved at 4 °C in darkness until elutriation.

After a standardized storage time of one week, the sediments were elutriated following the method outlined by the U.S. E.P.A./Corps of Engineers (1977). St. Lawrence River water was taken at 0.1 m depth near the outlet of Lake St. Louis every week (Station B2), at the same time as the sediment samples were taken. This water was used for the elutriation in a ratio of one volume of sediments to four volumes of water. After vigorous mixing for 30 minutes, the slurry was allowed to settle for one hour. The supernatant was then centrifuged by means of a continuous-flow solid-liquid separator (DeLaval, Gyro Tester) at 12 000 RPM and a flowrate of about $0.5 \, l \cdot min^{-1}$. The effluent was filtered on cellulose acetate-nitrate filter (Millipore^{TM}) of 0.45 μm pore size, combined with a glass-fibre prefilter, using a positive pressure continuous-flow system (Sartorius, model SM 16505).

The final elutriate was kept at 4 °C in the dark until analysis. The bioassays were carried out not later than 48 hrs after elutriation. As for the chemical analyses, standard preservation techniques were applied, with allowable delays as set out in the Environment Canada (1979) manual. Heavy metals, nutrients, and major ions were analyzed following standard Environment Canada (1979) methods. Five types of bioassays were carried out using bacteria, algae, cladocerans, nematodes, and rotifers (Table 1).

Microtox test

The bacterial test consisted of the Microtox system (Microtox Toxicity Analyser Model 2055) which uses a bioluminescent bacteria, *Photobacterium phosphoreum* Cohn, according to the standard method described by Bulich and Isenberg (1981). Test temperature was 15 °C and endpoint reading, inhibition of bioluminescence, was taken after 5 and 15 minutes. In 1984 and 1985, Microtox tests were carried out at the National Water Research Institute in Burlington by Kwan & Dutka (1985). Their elutriates differed from ours in that 50 g of sediments were mixed with 50 ml of Milli-Q demineralized water for five minutes and subsequently centrifuged at 3500-4000 rpm. Their results were submittted to us as EC_{50}'s (reading

Table 1. Bioassays carried out on Lake St. Louis elutriates.

Test	Toxic response	Test time	Test temp.	References
BACTERIAL: *Photobacterium phosphoreum*	Inhibition of bioluminescence	5 min. 15 min.	15 °C	Bulich and Isenberg (1981)
ALGAL: *Selenastrum capricornutum*	Inhibition of photosynthetic ^{14}C assimilation	3 hrs	26 °C	Jarry (1986) Ross *et al.* (1988)
CLADOCERAN: *Daphnia pulex* *Simocephalus vetulus*	a) Mortality b) Inhibition of reproduction	a) 48 hrs b) 21 days	15 °C	Champoux (1986)
ROTIFER: *Brachionus calyciflorus*	Mortality	24 hrs	20 °C	Couillard (1987)
NEMATODE: *Panagrellus redivivus*	Mortality Inhibition of molting Phenotoxicity	96 hrs	22 °C	Samoiloff (1980)

after 15 minutes). Our laboratory carried out the Microtox tests in 1985 only, where samples consisted of serial dilutions (100, 75, 66.6, 50, 25, 12.5, and 0%) of our standard elutriate with the Microtox diluant, to determine a dose-response curve, and were carried out in triplicate (readings after five minutes).

Algal bioassay

The algal test consisted of measuring the inhibition of ^{14}C uptake rates. A laboratory-cultured species, *Selenastrum capricornutum* Printz, which is widely used in North America as a test organism (Munawar *et al.*, 1988; Wong & Couture, 1986) has been used in the algal bioassays. Although there is increasing emphasis on using natural phytoplankton populations (Munawar *et al.*, 1988; Munawar & Munawar, 1987), we were limited to the standard species *S. capricornutum* because of its availability within our laboratory, and logistic and technical restraints for field collection and subsequent culturing.

The principle of the test is based on the method by Steeman-Nielsen (1952), as described by Vollenweider *et al.* (1974) and Fitzwater *et al.* (1982). To obtain a dose-response curve, the algae were exposed to six elutriate concentrations (three replicates per concentration): 0, 12.5, 25, 50, 75 and 100 percent (diluant: filtered (0.45 μm) St. Lawrence River water, used for the elutriate). Exposure time of cells (log-phase) was three hrs at 26 °C, after 5 μCi of ^{14}C was added to each replicate. After incubation, the algae were filtered and rinsed with 0.1 N HCl to expel any excess ^{14}C. The filters were dissolved in methyl cellusolve™ and suspended in a scintillation cocktail (Econofluor™). Radioactivity was measured as disintegrations per minute on a Searle Mark III scintillation counter. More details of the tests are found in Jarry (1986) and Ross *et al.* (1988).

Cladoceran bioassay

The cladoceran tests were carried out using two indigenous species from the St. Lawrence River

(Lake St. Louis): *Daphnia pulex* Leydig and *Simocephalus vetulus* Muller. Specimens were reared at 15 °C with a 16L : 8D photoperiod and fed green algae three times a week (1×10^5 cells·ml^{-1}). Two types of bioassays were run: 48 h acute toxicity using a flow-through system described by Smith and Hargreaves (1983); three weeks chronic toxicity (reproductive success) using semi-static conditions, where at the beginning of each week individuals were transferred into a fresh test solution. During the acute test individuals were not fed, while for the chronic tests they were fed three times a week with green algae. Nine dilutions of elutriate with St. Lawrence water were used: 0, 10, 25, 40, 50, 60, 75, 90 and 100 percent were carried out in quadruplicate with 20 individuals per replicate for the acute tests and 5 for the chronic tests. Endpoint for the acute test was immobility after tactile stimulus, and the percentage of adults bearing eggs, the number of young produced and the percentage of young surviving for the chronic tests. More details of the test can be found in Champoux (1986).

Nematode bioassay

The nematode test, using *Panagrellus redivivus* as test species, was carried out by a private consultant laboratory (Bioquest International), and is based on exposing 2nd instar organisms to a 10 percent dilution of the elutriate with the nematode growth medium for a period of 96 hrs at 22 °C, during which the animals go through three moltings. Toxic responses are measured as mortality and inhibition of molting. The test has been described in detail by Samoiloff *et al.* (1980).

Rotifer bioassay

The acute toxicity test using rotifers was carried out with *Brachionus calyciflorus* Pallas, which were obtained from laboratory cultures maintained at Concordia University in Montreal. Cultures in log-phase of growth were used for the bioassays, which were done with undiluted elu-

Table 2. Mean contaminant concentrations in elutriates (52 stations: 34 in 1984 and 18 in 1985), compared to water quality guidelines[1] and St. Lawrence River Water (mean values for Lake St. Louis[2]); all concentrations expressed as $\mu g \cdot l^{-1}$.

Contaminant	St. Lawrence Water	Guidelines	Elutriates		Number of stations over guidelines		
			1984	1985	1984	1985	Total
As	0.7	50	2.2	1.0	0	0	0
Se	0.3	10	0.4	0.3	0	0	0
Hg	0.02	0.10	0.11	0.11	4	2	6
Co	<1	–	2	1	–	–	–
Cr	2	40	–	3	–	0	0
Cu	2	2	7	6	32	16	48
Ni	2	25	7	4	0	0	0
Pb	<1	5	6	3	13	3	16
Zn	4	50	35	284	7	18	25

[1] CCREM (1987).
[2] Environment Canada (1985).

triates. The sediment elutriates used for these tests were different from those used for the other bioassays in that synthetic water (demineralized water to which are added certain amounts of nutrients and growth factors) was used. Test temperature was held at 20 °C (room temperature) and exposure time was 24 hrs, at which mortality was determined by immobility after tactile stimulation. Details of the test can be found in Couillard (1987) and Couillard *et al.* (1987).

Results and discussion

Concentration average of contaminants detected in the elutriates for all stations are presented in Table 2, which also lists average concentrations in St. Lawrence River water used for the elutriation (Environment Canada, 1985), and water quality guidelines for the protection of aquatic life (Canadian Council of Resource and Environment Ministers, 1987). Generally, there is significant desorption of most of the contaminants, except PCB/OC (not presented in the table; concentrations in the elutriate were below the detection limit of 0.003 $\mu g \cdot l^{-1}$), and Se and Cr. Mercury, copper, lead, and especially zinc, showed significant desorption. In a study on the mobility of heavy metals, De Groot *et al.* (1971) showed that Hg

can be easily mobilized, and to a lesser degree, Cu, Pb and Zn (cited in Golterman, 1975).

The elutriate concentrations of these four metals (Hg, Cu, Pb and Zn) were also found to be above the water quality guidelines recommended by Canadian Council of Resource and Environment Ministers (1987) (Table 2 and Fig. 3). Copper was most frequently above the $2 \mu g \cdot l^{-1}$ guideline. The increased liberation of Zn in 1985 as compared to 1984 may be due to the change in elutriation technique from mechanical stirring (1984) to compressed air bubbling (1985), causing changes in redox potentials. Comparison trials for

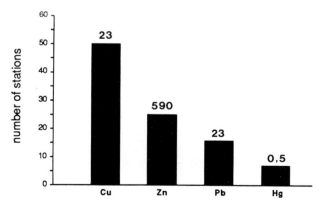

Fig. 3. Frequency histogram of stations with Cu, Hg, Pb and Zn levels above water quality guidelines (CCREM, 1987). Maximum observed concentrations ($\mu g\,l^{-1}$) indicated above bars.

the two techniques on sediments from the same station showed that Zn desorption doubles when air mixing is used, as compared to mechanical mixing, while Hg, Cu, and Pb desorbed to the same extent under both conditions.

A contamination index was calculated based on the concentrations of these four metals divided by the guideline value, using the following formula:

$$C.I. = \sum_{i=1}^{4} \frac{[Cont.]_i}{[G.L.]_i} ,$$

where C.I. = contamination index
    Cont. = contaminant (Cu, Pb, Zn, Hg)
    C.L. = guideline.

It must be understood, however, that all contaminants, whether above or below guidelines, may contribute to the overall toxicity. It is, however, simplistic to assume that the effects are additive, since it is well known that synergistic and antagonistic effects can be quite important (Munawar *et al.*, 1988; Wong *et al.*, 1978). Since organic contaminants desorb very little, it is felt that toxicity of elutriates is mainly due to metals (see also Munawar & Munawar, 1987). Accepting the use of water quality guidelines, the contribution to overall toxicity should be largely due to these four metals, their concentrations being above the guidelines.

The underlying assumption of the use of water quality guidelines in the calculation of a contamination index (C.I.) is that the sum of the ratios should give us an indication of the toxic potential of the elutriate sample. The extension of this assumption is that a station with a higher C.I. should also show increased toxicity. Later on, we shall come back to the contamination indices, when we compare them to toxic responses obtained from the bioassays.

The sensitivity of the various bioassays is presented in Table 3, which indicates the percentage of tests (stations) with a toxic response. The sublethal tests (algae, microtox, and cladoceran reproduction) were the most sensitive, while acute lethality tests (rotifers and cladoceran-acute) were the least. The nematode test is in between, which

Table 3. Percentage of stations showing toxic response obtained for the various bioassays.

	Algae	Microtox		Cladocerans		Nema-todes	Rotifers
		EC_{50}	Inh.	Mort.	Repr.		
1984	67	96	–	11	40	24	6
1985	–	65	53	–	–	15	–

Inh. = inhibition test (see text).
Mort. = mortality.
Repr. = reproductive success.

could be expected, since the responses are a mixture of survival and inhibition of development. There were no stations where the elutriate was toxic to all bioassay organisms.

Microtox test

Results of the Microtox tests (Tables 4 and 5) from Kwan & Dutka (1985) are presented as EC_{50}'s (elutriate concentration with the Microtox diluant producing 50 percent reduction in light emission). Our data (percent reduction in light emission at 75 percent elutriate concentration) for

Table 4. Microtox bioluminescence inhibition results expressed as EC_{50} (elutriate concentrations) for the 1984 survey.

Station	EC_{50}	Station	EC_{50}
A1	8	E2	3
A2	5	E3	7
A3	12	F3	1
A4	4	G1	11
B1	8	G2	5
B2	ST	G3	10
B3	2	G4	5
B4	47	H2	5
C1	27	H3	5
C3	32	I1	21
C4	20	I3	23
D2	16	J1	9
D3	3	J2	12
D4	8	J4	40
E1	3	K1	10

ST = stimulation.

Table 5. Microtox test results for the 1985 survey: EC_{50} (Kwan & Dutka, 1985) compared to % inhibition at 75% elutriate concentration (our laboratory).

Station	EC_{50}	% inh. at 75%
L1	38	6
L2	ST	−10
L3	29	−2
L4	29	7
L5	19	9
L6	18	−17
L7	16	18
L8	18	41
L9	ST	13
L10	ST	34
L11	ST	−2
L12	23	18
L13	ST	−7
L14	19	24
L15	39	−
L16	49	30
L17	ST	7
L18	21	−5
L19	ST	−14
L20	13	−38

ST = stimulation.

neg. sign = stimulation.

1985 are included to make a comparison with Kwan & Dutka's reading.

The 1984 data indicate the Microtox test to be extremely sensitive: 97 percent of the tests showed toxicity. The 1985 EC_{50} data show less toxic responses (65 percent of total tests), and is quite comparable to our 1985 results (toxic responses in 53 percent of the tests). Correlation between the two sets of responses was found to be non-significant. This indicates that desorption of toxic elements is greatly influenced by elutriate conditions. A 50 : 50 mix of water : sediments with de-mineralized water seems to result in higher contaminant levels than the standard EPA method (1 : 4 of sediment : lake-water mix) which was used by our laboratory. Unfortunately, Kwan and Dutka did not obtain any chemical data on their elutriates to make a direct comparison. The differences in response time allowed for end-point reading (Kwan and Dutka: 15 minutes; our laboratory: 5 minutes) may also be responsible for the

lack of correlation. Our inhibition data for 1985 showed significant correlations with some elements in the elutriate, namely As, Cr, Cu, and Pb ($p < 0.05$). No correlation was found between the EC_{50} data and contaminant concentrations in the elutriate.

Algal bioassay

In 1984, of 18 stations tested with the algal bioassay, 11 showed toxicity (Table 6). Of these stations, two showed straight inhibition, while nine showed stimulation and inhibition. From these results, three typical responses are shown in Fig. 4: a) inhibition, b) inhibition − stimulation, and c) no dose-response relationship. Since the algal test gave the most interesting results, data treatment has been here somewhat more extensive than for the other tests. Therefore, a summary of its conclusions will be presented.

Stimulation of photosynthesis is thought to take place when the presence of nutrients desorbed from the sediments cause an increase in the

Table 6. Algal bioassay results presented as % inhibition observed at 100% elutriate and as slopes of the dose-response curve with Kendall's Tau to indicate significance (* $p < 0.05$, ** $p < 0.01$).

Station	% inhibition	Slope	Tau
A1	100	0.86	0.90**
A2	71	0.67	0.72**
A3	37	0.39	0.62**
B2	−	0.94	0.69**
B3	99	0.76	0.67**
C3	3	0.04	−0.02
C4	−52	0.01	−0.08
D3	34	0.35	0.34*
D4	−11	0.37	0.14
E2	47	0.24	0.19
E3	62	0.20	0.13
G3	18	0.14	−0.11
G4	22	0.07	−0.02
H3	65	0.83	0.63**
I1	52	0.53	0.65**
I2	57	1.22	0.71**
I3	45	0.66	0.65**
J2	66	0.59	0.51**

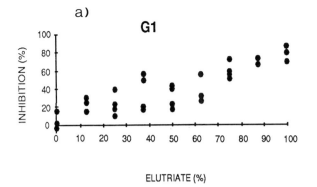

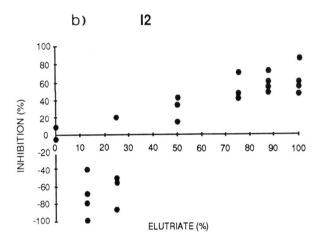

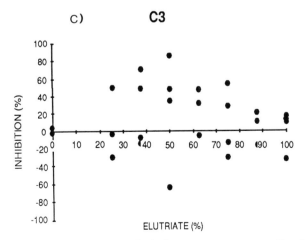

Fig. 4. Representative algal dose-response curves for 3 stations showing a) inhibition, b) stimulation-inhibition, c) no dose-response relationship.

rate of photosynthesis. (Munawar *et al.*, 1986). Nutrient levels in our elutriates increased significantly compared to original St. Lawrence water used for the control. Stimulation is often concentration (% elutriate)-dependent, in which case we obtain a significant dose-response curve with a negative slope. The presence of contaminants, however, causes a decrease in stimulation with higher elutriate concentrations. This is illustrated in Fig. 4b for a station where at low concentrations contaminants are not high enough to inhibit the stimulatory effect of nutrients on photosynthesis. At 50 percent and higher, the inhibitory effect is so great that no stimulus is observed anymore, only significant inhibition with respect to the control. Figure 4a shows a dose-response curve where, at 100 percent elutriate concentration, there is almost complete inhibition of photosynthesis.

To explain these toxicity curves with contaminant concentrations, correlations (Kendall's tau; Legendre & Legendre, 1984) between the dose-response curves (slope and inhibition at 100% elutriate) and contaminant concentrations in the elutriate were carried out. Only 5 of the 26 chemical variables analyzed in the elutriate showed significant correlations (Table 7). Positive correlations indicate toxicity and negative correlations indicate stimulation. As, Ni, and Pb in the elutriates may cause significant inhibition of photosynthesis, as probably do other metals. Only variations in these three could explain the differences in toxic response. Interestingly, Mg and NO_2 -NO_3 were found to be stimulatory, since an

Table 7. Correlations (Kendall's Tau) between selected chemical variables and dose-response parameters (algal bioassay).

Chemical	Slope	Inhibition at 100% elutriate
As	0.23	0.27*
Ni	0.27	0.31**
Pb	0.27*	0.35*
Mg	−0.37*	−0.47*
NO_2-NO_3	−0.39*	−0.29*

* $p < 0.05$.
** $p < 0.01$.

increase in their concentration was concomitant with a decrease in toxicity or an increase in stimulation.

Phosphorus concentrations in the elutriate were not related to algal responses in the same manner, correlation being not significant. This is because total P was analyzed, including forms of phosphorus, which are not all bioavailable; generally, only orthophosphates are directly usable by algae.

It is not surprising that NO_2-NO_3 stimulates algal growth (or decreases toxicity), since it is an essential nutrient, but for Mg it is not so clear. It is known that Mg is necessary for photosynthesis, since it is the central element of the chlorophyll a molecule. Generally, concentrations in water are much higher than what is needed, so that it is rarely a limiting factor (Wetzel, 1979). It may be that the presence of Mg, as related to water hardness, makes toxic metals less bioavailable.

Multiple regressions were also carried out using the same five variables that showed significant simple correlations with the biological variables (Table 8). Again Mg is a significant element as a stimulant for photosynthesis. As before, arsenic also plays an important role as a toxic element inhibiting photosynthesis, while Ni and Pb are not significant in this case. NO_2-NO_3 has a stimulatory effect, but at higher elutriate concentrations (up to 100 percent) it cannot 'outcompete' the contaminants anymore. Therefore, at 100 percent elutriate it does not influence toxic response, but it

does influence the general trend of the slope, since at lower concentrations it has a significant stimulatory effect, pulling down the curve and increasing the slope in this particular case. In general, however, increase in NO_2-NO_3 causes a decrease (negative increase) in slope.

These observations were confirmed by a stepwise regression model, which used the same five chemical variables. Only three chemical variables were found to be significant for the slope variable, and two for the 100 percent elutriate inhibition variable. This resulted in the following pre-model, which seems to explain in the best manner the effects of the elutriates on photosynthesis in *S. capricornutum*:

$$\text{Slope} = -1.8 \times 10^{-4} + 0.283\,[\text{As}] - 0.526\,[\text{Mg}] - 0.322\,[NO_2\text{-}NO_3]$$

$$\text{Inhibition at } 100\% \text{ elutriate} = 1.9 \times 10^{-4} + 0.435\,[\text{As}] - 0.736\,[\text{Mg}].$$

Cladoceran bioassay

Results of the acute cladoceran test which were statistically significant, are presented in Table 9. Only two tests out of 19 (10%) provoked a significant decrease in survival (stations C3 and K1) while four stations (20%) showed a significant increase in survival. Correlation analyses (Kendall's Tau) with chemical variables carried out by

Table 8. Multiple regression of algal photosynthesis inhibition with 5 chemicals as the dependent variables.

Chemical	Independent variable					
	Slope			Inhibition at 100% elutriate		
	Equation coefficient	Standard deviation	Probability	Equation coefficient	Standard deviation	Probability
As	0.365	0.205	0.090	0.438	0.188	0.030
Ni	-0.417	0.334	0.226	-0.363	0.306	0.249
Pb	0.420	0.239	0.166	0.448	0.269	0.111
Mg	-0.486	0.214	0.033	-0.611	0.195	0.005
NO_2-NO_3	-0.434	0.219	0.068	-0.180	0.201	0.382
	$R^2 = 42\%$, $p = 0.014$			(b) $R^2 = 56$, $p = 0.003$		

Table 9. Cladoceran acute bioassay: survival percentages at 100% elutriate as estimated by the dose-response regression equation when significant (Kendall's Tau; *p < 0.05, **p < 0.01) for acute toxicity tests. Only stations with significant response are presented.

Station	Species	% survival	Tau
B3	Daphnia	102	0.32*
D4	Simocephalus	106	0.23*
G1	Simocephalus	105	0.40*
G3	Simocephalus	98	− 0.47*
H3	Simocephalus	102	0.53**
K1	Daphnia	84	− 0.54**

Champoux (1986) showed that Zn favours survival for *D. pulex* while chlorides inhibit it. Since metal concentrations in elutriates are below published LC_{50} for *D. magna* Strauss (Biesinger & Christensen, 1972), it is postulated that some of the essential metals may even have a stimulatory effect. This is confirmed by the observation that an increase in metal concentration correlated with an increase in survival percentage, even if they are above water quality guidelines.

Results of the chronic cladoceran test (reproductive success) are shown in Table 10. Of the seven chronic tests conducted, three (43%) showed significant toxicity (stations E3, I2 and J2), while three others showed a significant stimulation (A2, D3 and G4). Chronic toxicity tests seem essential to correctly evaluate toxic hazard of contaminated sediments. Number of young produced seemed to be the most sensitive measure of elu-

Table 10. Cladoceran reproductive success at 100% elutriate, estimated from the dose-response regression (Kendall's Tau significance: *p < 0.05, **p < 0.01, ***p < 0.001).
D = *Daphnia*   S = *Simocephalus*

Station	Species	Adult survival %	Adults with eggs %	Young survival %	Reproductive impairment
A2	D	133*	91	104	+ 70
D3	D	105	97*	98	+ 47
E3	S	0***	–	–	− 68
G4	D	101*	91**	100	+ 270
H2	S	90	60	110	− 30
I2	S	93	76	91**	− 20
J2	S	61**	36*	73*	− 74

triate effects on reproduction, since it provoked six responses out of the seven tests. Reproductive impairment represents the difference in number of young produced in 100 percent elutriate and the control (0% elutriate). Champoux (1986) showed that Hg concentrations are negatively correlated with young survival. However, in general, no chemical explanation of the variations in reproductive success can be found.

Nematode Bioassay

Tests results are presented in Table 11 as fitness indices (control value: 100) which is a composite

Table 11. Test results of the nematode bioassay (*P. redivivus*).

1984		1985	
Station	Fitness	Station	Fitness
A1	103	L1	103 S
A2	102	L2	97
B1	113	L3	105 S
B2	87 L	L4	90 I
B3	105 P	L5	98
C1	115	L6	99
C2	124 S	L7	62 L
C3	103 S	L8	99 S
C4	104 S	L9	117 S
D2	101 P	L10	99
D3	89 P	L12	98 S
D4	83 I	L13	102 S
E1	102	L14	100
E2	99	L15	101 S
E3	109	L16	100
F3	100	L17	90
G1	100	L18	100
G2	96	L19	94 I
G3	99	L20	97
G4	95		
I1	99		
I3	115		
J3	102		
K1	104		
K4	104 P		

L: lethality.
P: phenotoxicity.
S: stimulation.
I: inhibition.

of survival (mortality), growth (molting), and maturation (final molt to the adult stage). The inhibition of maturation is called phenotoxicity, since it measures the expression of the genetic activity involved in the final molting. The most significant toxic effect, if present, is indicated next to the fitness index. Significance of effects with respect to controls (0% elutriate) was determined statistically by chi-square calculations at the 95 percent level (Samoiloff et al., 1980).

For the 1984 survey, six of the 25 stations analyzed showed significant toxicity as inhibition of molting (D4), phenotoxicity (B3, D2, D3, and K4), and lethality (B2). Except D3, stations exhibiting phenotoxicity nevertheless show a fitness value of 100$^+$, since stimulation in growth before maturation occurred (increased rate of molting). Such stimulation, including maturation, was observed at all C stations, which are located in the same area of sandy substrate (low contaminant concentration). No phenotoxicity was observed for the 1985 survey, where two stations showed inhibition (L4, L19), one lethality (L7), and seven stimulations (L1, L3, L8, L9, L12, L13, L15).

Correlation analyses for the 1984 data indicate that Ba, Cu, Pb, and Zn levels had a toxic effect. However, the 1985 data did not show any of these effects; on the contrary, Zn was even found to be stimulatory.

Rotifer bioassay

Test results of the rotifer bioassay (B. calyciflorus) are presented in Table 12. The test was one of the least sensitive bioassays used, as only one station (I3) showed significant toxicity while several elutriates caused stimulation of rotifer survival. Station I3 is an interesting case. In terms of the chemical composition of the elutriate, this station stands out for the high conductivity (600 $\mu S \cdot cm^{-1}$, compared to 200–300 $\mu S \cdot cm^{-1}$ at the other stations), K (16 mg $\cdot l^{-1}$; around 5 mg $\cdot l^{-1}$ for the others) and Cl (33 mg $\cdot l^{-1}$; around 20 mg $\cdot l^{-1}$ at the other stations). Statistical analyses (correlation, regression, multivariate) of the complete results of the test indicated that

Table 12. Test results of the rotifer bioassay using B. calyciflorus.

Stations	% mortality	Stations	% mortality
B1	0	D4	−13
B3	−6	G2	−4
B4	−7	G3	3
C1	−6	G4	2
C2	−3	I1	−1
C3	−21	I3	56
C4	−36	K1	−5
D2	−6	K4	−5

K was toxic to rotifers, which was confirmed by laboratory exposure tests (Couillard et al., 1987). It was also found that stations with higher concentrations of Zn and NO_2-NO_3 showed more toxicity (or lower stimulation), while increased magnesium concentrations stimulated rotifer growth. The toxic effects of NO_2-NO_3 on rotifers has also been observed by Schlüter (1980), who found a 90 percent reduction in reproductive success of B. rubens when exposed to 10 ppm nitrates (as $NaNO_2$).

Synopsis of results and ranking of stations

Table 13 presents a synopsis of the bioassays as presence or absence of toxicity (including stimulation) for a selected number of stations from the 1984 survey (those where all or most of the various bioassays were carried out). Also included is a ranking of these stations according to the contamination index which was calculated based on the concentrations of four contaminants in the elutriates as indicated previously.

Most of the stations (elutriates) were above the guideline for Cu (2 $\mu g \cdot l^{-1}$), while those with the highest C.I. (stations E3, E2 and B2) are significantly contaminated by Pb (all 3 stations), Zn (E3), and Hg (B2). There is no geographical pattern to this ranking of stations and for the purpose of this paper no attempt is being made to explain ranking with respect to station location. We will, however, compare this ranking to the

Table 13. Contamination index (C.I.) and toxicity ranking of 1984 elutriates for a number of selected stations.

Station	C.I.	Tox.	No tox.	Sti	No test	Mi	Al	Cl	Ro	Ne
E3	9.3 (Pb, Zn)	1	2	0	2	T	N	–	–	N
E2	9.1 (Pb)	2	1	0	2	T	T	–	–	N
B2	9.0 (Hg, Pb)	2	1	1	1	S	T	N	–	T
D4	6.6	2	1	2	0	T	N	S	S	T
A1	5.4	2	1	0	2	T	T	–	–	N
D3	5.3	3	1	0	1	T	T	N	–	T
G3	4.5	1	4	0	0	T	N	N	N	N
I3	4.5	3	2	0	0	T	T	N	T	N
B3	3.8	2	1	2	0	S	T	S	N	T
G4	3.8	1	3	1	0	T	N	S	N	N
C3	3.8 (Hg)	1	2	2	0	T	N	N	S	S
A2	3.3	2	1	0	2	T	T	–	–	N
I1	2.3	2	3	0	0	T	T	N	N	N
C4	2.7	1	1	3	0	T	N	S	S	S
B1	1.1	1	3	0	1	T	–	N	N	N

Tox = toxic
No tox. = non-toxic
Sti = stimulation
No test: number of type of tests not done
T = Toxic
N = non-significant
S = stimulation

Mi = microtox
Al = algal test
Cl = cladoceran test (acute)
Ro = rotifer test
Ne = nematode test

number of toxic responses from the various bioassays for each station.

It is quite evident that there is no relationship between the contamination index (C.I.) and the number of toxic responses obtained. Furthermore, if we look at the individual bioassays and the presence or absence of toxicity, no relationship with the C.I. is evident. Statistical analyses showed that only the rotifer and the cladoceran tests were correlated.

Champoux *et al.* (1989) calculated contamination indices for the elutriates using all contaminant concentrations and their respective water quality guidelines, as well as a toxicity ranking value for each station based on bioassay responses. For several contaminants in sediments a similar C.I. was calculated (Champoux & Sloterdijk, 1988) using Ontario Ministry of the Environment guidelines for the disposal of dredged material in open water (Levings, 1983). The resulting values have been mapped out geographically (Fig. 5), using a computer software package SYMAP (Dougenik & Sheenan, 1977) which uses a point distribution coefficient by linear interpolation between two neighbouring stations (Clark & Evans, 1954).

The distribution of sediment C.I.'s (Fig. 5a) shows 3 to 4 zones with high values, which corroborates observations made on most individual contaminants, and corresponds in general to deposition zones of finer particles. The elutriate C.I. distribution (Fig. 5b) shows a similar pattern, meaning that sediments with relatively high contaminant concentrations will result in elutriates having also relatively high levels of these contaminants. Therefore, sediment contamination is somewhat predictive of elutriate contaminant concentrations. Based on these results, one might predict that resuspension of sediments from these deposition zones (e.g. storm-induced wave action) present a relatively high potential for the desorption of contaminants (especially metals) into the water column.

The geographical distribution of the toxicity rank values of all stations (Fig. 5c) does not correspond to the C.I.'s pattern of the elutriate. Highly

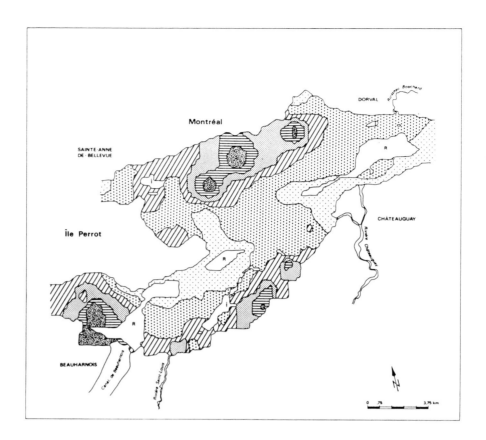

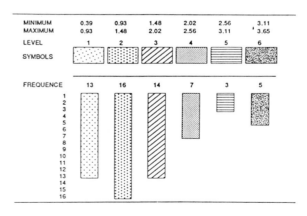

Map a. Sediments CI.

Fig. 5. Distribution maps of contamination indices (CI) and toxicity ranking (TR). Stations are distributed in six classes from lowest (light shading) to highest (dark shading) contamination or toxicity. The histogram gives the number of stations in each class with the limiting values. On the map, white zones are islands (I) or rocky areas (R). Map a. Sediments CI. Map b. Elutriates CI. Map c. Bioassays TR.

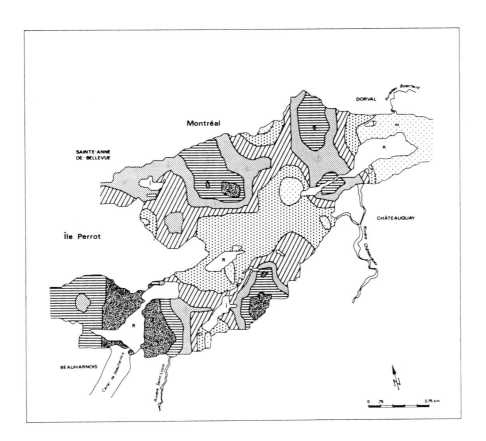

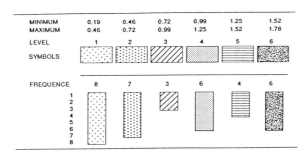

Map b. Elutriates CI.

Fig. 5 (continued).

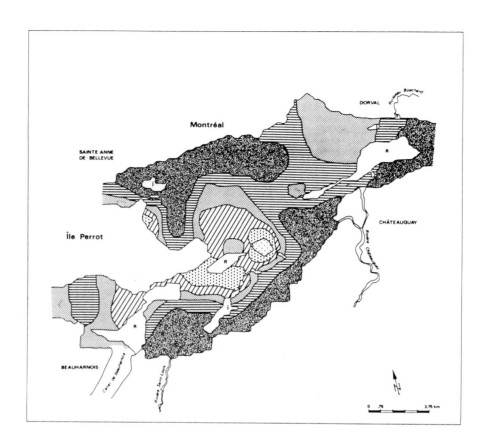

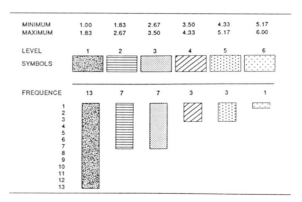

Map c. Bioassays TR.

Fig. 5 (continued). See p. 330.

contaminated sediments and their corresponding elutriates do not necessarily result in higher toxicity as measured by our bioassays. This indicates that perhaps during elutriation, and perhaps similarly during resuspension under natural conditions, many other substances than contaminants are liberated into the water column, such as nutrients (NO_2-NO_3, various forms of P) and major ions; these are responsible for stimulatory effects, and decreasing toxicity through complexing and chelating, and making certain metals less bioavailable. All these interactions are complex and not yet well understood.

Summary

Elutriation of Lake St. Louis sediments caused the desorption of heavy metals. Organic contaminants were not desorbed in any significant way, while there was some slight desorption of nutrients. Desorption of four metals, Cu, Pb, Zn, and Hg, resulted in concentrations higher than water quality guidelines for the protection of aquatic life.

Although sediment and elutriate chemistry showed fairly well-defined geographical patterns (depositional areas), this was not so well reflected by the toxicity data. Based on Microtox data, however, it seems that the areas corresponding to A, D, and E were the most toxic. These are the depositional areas under the influence of the Beauharnois industrial sector. The 'E' region receives sediments from the Ottawa River basin, but backflow of the St. Lawrence River 'proper' extends the Beauharnois influence. The least toxic region is represented by the C stations. Interestingly, this is where the rotifer and nematode tests showed significant enhancement of growth.

Cladoceran results were too sporadic to confirm or deny these observations, but the nematode test also indicated the D region as toxic. The algal tests showed both inhibition and stimulation for the C and D region, but the A and E region showed significant inhibition.

Stations within the Ottawa River influence (north shore) showed a variety of responses. Algal inhibition was not severe, but the H region was fairly toxic to Microtox; cladoceran reproduction was inhibited at H, I, and J stations.

Sediment chemistry was found to be somewhat predictive of elutriate chemistry. Elutriate chemistry, however, was not predictive of toxic responses, therefore it does not seem to be a useful measure of biohazards associated with contaminated sediments. Sediment contaminants and elutriation contaminants as a function of the appropriate guidelines do not have any predictive value of biohazard assessment. Elutriate chemistry is too complex, where stimulatory and inhibitory forces interact through bioavailability, antagonism, synergism, etc. Relationships between toxicity tests were rare or non-existent.

Toxicity hazard associated with contaminated sediments should be evaluated by using appropriate biotests, the 'battery approach' is recommended, where various trophic and organizational levels are represented (see Ahlf et al., 1989; Munawar et al., 1989). It should be realized that the use of 'surrogate species' only gives an indication of a potential hazard at best. Within a study, these tests should be well standardized, with a rigid protocol, assuring a sustained and steady effort with respect to the number of analyses and different types of tests being carried out.

Acknowledgements

This research was supported by a grant from 'Toxfund' of Environment Canada (H. Sloterdijk) and a strategic grant from the Natural Sciences and Engineering Research Council of Canada (NSERC–G1571) (P. Ross et al.). The personnel of the Captain Bernier Laboratories (Longueuil, Que.) and the Canada Centre for Inland Waters have provided support through their operational budget.

We would like to specifically thank F. Blanchette, J.F. Doyon, B. Laplante, D. Léger and J. Parent who, as summer students, have provided much of the technical support in the field and in the laboratory. We also would like to thank Dr.

M. Munawar for his support and patience regarding this publication.

Special thanks go to C. Bois for her assistance in word processing the text; without her assistance and her patience this article would not have been possible. The assistance of Carto-Media (Daniel Cloutier) in the preparation of the figures is gratefully acknowledged.

References

Ahlf, W., W. Calmano, J. Erhard & U. Forstener, 1989. Comparison of seven bioassay techniques for assessing sediment-bound contaminants. In; M. Munawar, G. Dixon, C. Mayfield, T. Reynoldson and M. H. Sadar (eds.), Environmental Bioassay Techniques and their Application. Hydrobiologia. This volume.

Allan, R. J., 1986. The limnological units of the Lower Great Lakes – St. Lawrence River corridor and their role in the source and aquatic fate of toxic contaminants. Wat. Pollut. Res. J. Can. 21(2): 168–186.

Allan, R. J., 1988. Toxic chemical pollution of the St. Lawrence River (Canada) and its upper estuary. Wat. Sci. Tech., 20 (6/7): 77–88.

Biesinger, K. E. & G. M. Christensen, 1972. Effects of various metals on survival, growth, reproduction, and metabolism of *Daphnia magna*. J. Fish Res. Bd. Can. 29: 1691–1700.

Bulich, A. A. & D. L. Isenberg, 1981. Use of the luminescent bacterial system for the rapid assessment of aquatic toxicity. Trans. Am. Inst. Soc., 20: 29–33.

Canadian Council of Resource and Environment Ministers, 1987. Canadian Water Quality Guidelines. Water Quality Branch, Inland Waters Directorate, Environment Canada. Ottawa. 504 pp.

Champoux, L., 1986. Réponses aigües et chroniques de deux cladocères indigènes aux contaminants des sédiments du lac Saint-Louis (fleuve Saint-Laurent). Mémoire de maîtrise, Université de Montréal. 101 pp. et annexes.

Champoux, L. E., P. E. Ross, V. Jarry, H. Sloterdijk, A. Murdroch & Y. Couillard, 1987. Libération par élutriation des contaminants des sédiments du lac Saint-Louis (fleuve Saint-Laurent). Rev. Int. Sci. Eau, 2(4): 95–107.

Champoux, L. & H. Sloterdijk, 1988. Etude de la qualité des sédiments du lac Saint-Louis, 1984-1985. Rapport technique No. 1: Géochimie et contamination. Environnement Canada, Direction des eaux intérieures, Région du Québec. 177 pp.

Champoux, L., H. Sloterdijk, Y. Couillard, V. Jarry & P. Ross, 1989. Etude de la qualité des sédiments du lac Saint-Louis (fleuve Saint-Laurent) 1984-1985. Rapport technique no. 2: Contamination et toxicité des élutriats. Direction des eaux intérieures, Environnement Canada, région du Québec. 106 pp.

Clark, P. J. & F. C. Evans, 1954. Distance to neighbour as a measure of spatial relationships in populations. Ecology, 35: 445–453.

Couillard, D., 1983. BPC et pesticides organochlorés dans le système Saint-Laurent. Can. Wat. Res. J. 8: 32–63.

Couillard, Y., 1987. Variabilité psysico-chimique, et évaluation toxicologique à l'aide de *Brachionus calyciflorus* (Rotifera), d'élutriats de sédiments du lac Saint-Louis (fleuve Saint-Laurent, Québec). Mémoire de maîtrise, Université de Montréal, 143 pp.

Couillard, Y., B. Pinel-Alloul, P. Ross, H. Sloterdijk, L. Champoux & V. Jarry, 1987. Evaluation toxicologique, par élutriation des sédiments du lac Saint-Louis (fleuve Saint-Laurent, Québec) à l'aide du rotifère *Brachionus calyciflorus*. Revue internationale des sciences de l'eau, 3 (3-4): 84–94.

De Groot, A. J., J. J. M. De Goeij & C. Zegers, 1971. Contents and behaviour of mercury as compared with other heavy metals in sediments from the rivers Rhine and Ems. Geol. Mijnb. 50 (3): 393–398.

Dougenik, C. A. & D. E. Sheenan, 1977. Symap user's reference manual. Laboratory for computer graphics and spatial analysis. Graduate School of Design, Harvard University.

Environment Canada, 1979. Analytical Methods Manual. Water Quality Branch, Inland Waters Directorate, Environment Canada, Ottawa, Canada.

Environment Canada, 1985. National Water Quality Data Bank (NAQUADAT), 1981-1984. Water Quality Branch, Inland Waters Directorate, Ottawa.

Fitzwater, S. E., A. K. Knawer & J. H. Martin, 1982. Metal contamination and its effects on primary production measurements. Limnol. Oceanog., 27: 544–551.

Germain, A. & M. Janson, 1984. Qualité des eaux du fleuve Saint-Laurent de Cornwall à Québec (1977-1981). Rapport technique, Environnement Canada, Direction générale des eaux intérieures, Région du Québec. 232 pp.

Golterman, H. L., 1975. Physiological Limnology. Elsevier Scientific Publishing Company, New York. 429 pp.

Jarry, V., 1986. Répartition spatiale et effet phytotoxique des contaminants dans les sédiments du lac Saint-Louis (fleuve Saint-Laurent). Mémoire de maîtrise, Université de Montréal. 90 pp. et annexes.

Jarry, V., P. E. Ross, L. Champoux, H. Sloterdijk, Y. Couillard, A. Mudroch & F. Lavoie, 1985. Répartition spatiale des contaminants dans les sédiments du lac Saint-Louis. Wat. Pollut. Res. J. Can. 20 (2): 75–99.

Kwan, K. K. & B. K. Dutka, 1985. Microbiological studies of Lake St. Louis sediments, 1984. Microbiology Laboratories Section, Analytical Methods Division, National Water Research Institute, Canada Centre for Inland Waters, Burlington, Ontario.

Legendre, L. & P. Legendre, 1984. Ecologie numérique. Masson (Paris) et les Presses de l'Université du Québec. Tome I: 260 pp. et Tome II: 335 pp.

Levin, S. A. & K. D. Kimball (eds.), 1984. New perspectives in ecotoxicology. Envir. Mgt. 8 (5) 375–442.

Levings, C. D., 1983. Les conséquences écologiques du dragage et de l'élimination des résidus de dragage dans les eaux canadiennes. Conseil national de recherches du Canada, Comité associé sur les critères scientifiques concernant l'état de l'environnement No. 181, publ. 18131. 150 pp.

Marquenie, J. M., 1985. Bioavailability of micropollutants, Envir. Technol. Letters 6: 351–358.

Munawar, M., R. L. Thomas, H. Shear, P. McKee & A. Mudroch, 1984. An overview of sediment-associated contaminants and their bioassessment. Can. Tech. Rep. Fish. Aquat. Sci., 1253. 136 pp.

Munawar, M., R. L. Thomas, W. P. Norwood & S. A. Daniels, 1986. Sediment toxicity and production-biomass relationships of size-fractionated phytoplankton during on-site simulated dredging experiments in a contaminated pond. In: P. C. Sly (ed.), Sediment and Water Interaction, Springer-Verlag. pp. 407–426.

Munawar, M. & I. F. Munawar, 1987. Phytoplankton bioassays for evaluating toxicity of *in situ* sediment contaminants. Hydrobiologia, 149: 87–105.

Munawar, M., Munawar, I. F., Mayfield, C. & L. H. McCarthy, 1989. Probing ecosystem health: a multi-disciplinary and multitrophic assay stategy. In; M. Munawar, G. Dixon, C. Mayfield, T. Reynoldson & M. H. Sadar (Eds.), Environmental Bioassay Techniques and their Application. Hydrobiologia. This volume.

Munawar, M., P. T. S. Wong & G.-Y. Rhee, 1988. The effects of contaminants on algae: an overview. In: W. Schmidtke (ed.) Toxic Contamination in Large Lakes. Vol. I, Chronic Effects of Toxic Contaminants in Large Lakes. Lewis Publishers, Chelsea, Mich. pp. 113–160.

Ross, P., V. Jarry & H. Sloterdijk, 1988. A rapid bioassay using the green algal *Selenastrum capricornutum* to screen for toxicity in St. Lawrence River sediment elutriates. In: J. Cairns, Jr. & J. R. Pratt (Eds.) Functional testing of aquatic chemicals, ASTM STP 1988. American Society for Testing and Materials, Philadelphia. pp. 68–73.

Samoiloff, M. R., S. Schulz, Y. Jordan, K. Denich & E. Arnott, 1980. A rapid simple long-term toxicity assay for aquatic contaminants using the nematode *Panagrellus redivivus*. Can. J. Fish. aquat. Sci. 37: 1167–1174.

Samoiloff, M. R., J. Bell, A. Birkholz, G. R. B. Webster, E. G. Arnott, R. Pulak & A. Madrid, 1983. Combined bioassay – chemical fractionation scheme for the determination and ranking of toxic chemicals in sediments. Envir. Sci. Technol. 17: 329–334.

Schlüter, M., 1980. Mass culture experiments with *Brachionus rubens*. Hydrobiologia, 73: 45–50.

Sérodes, J.-B., 1978. Qualité des sédiments de fond du fleuve Saint-Laurent entre Cornwall et Montmagny. Rapport technique No. 15. Comité d'étude sur le fleuve Saint-Laurent, Environnement Canada et Service de la protection de l'environnement du Québec, l'éditeur officiel du Québec. 497 pp.

Sloterdijk, H., 1985. Substances toxiques dans les sédiments du lac Saint-François (fleuve Saint-Laurent, Québec). Rapport Technique, Environnement Canada, Direction générale des eaux intérieures, Région du Québec. 77 pp.

Sly, P. G., H. L. Golterman & R. L. Thomas, 1981. Importance des sédiments dans l'étude de l'environnement aquatique. Bull. Qual. des Eaux, 6 (2): 29–33 & 53–54.

Smith, R. L. & B. R. Hargreaves, 1983. A simple toxicity apparatus for continuous flow with small volumes. Bull. Envir. Contam. Toxic. 30(4): 406–413.

Steeman-Nielsen, E., 1952. The use of radioactive carbon (^{14}C) for measuring organic production in the sea. J. Cons. Int. Explor. Mer, 18: 117–140.

United States Environmental Protection Agency/Corps of Engineers, Technical Committee on Criteria for Dredged and Fill Material, 1977. Ecological Evaluation of Proposed Discharge of Dredged Material into Ocean Waters: Implementation Manual for Section 103 of Public Law 92-532 (Marine Protection, Research and Sanctuaries Act of 1972) (Second Printing, April 1978) Environm. Effects Lab., U.S. Army Engin. Waterways Exper. Stat. Vicksburg, Mississippi.

Vollenweider, R. A., M. Munawar & P. Stadelman, 1974. A comparative review of phytoplankton and primary production in the Laurentian Great Lakes. J. Fish. Res. Bd. Can., 31: 739–762.

Wetzel, R. G., 1975. Limnology. Saunders, Philadelphia. 743 pp.

Wong, P. T. S. & P. Couture, 1986. Toxicity screening using phytoplankton. In: B. J. Dutka & G. Bitton (eds.). Toxicity Testing Using Microorganisms. Vol. II. CRC Press, Boca Raton, Florida. pp. 79–100.

Wong, P. T. S., Y. K. Chau & P. L. Luxon, 1978. Toxicity of a mixture of metals on freshwater algae. J. Fish. Res. Bd. Can., 35: 479–481.

Hydrobiologia **188/189**: 337–343, 1989.
M. Munawar, G. Dixon, C. I. Mayfield, T. Reynoldson and M. H. Sadar (eds)
Environmental Bioassay Techniques and their Application.
© 1989 *Kluwer Academic Publishers. Printed in Belgium.*

Metal contamination in sediments and biota of the Bay of Quinte, Lake Ontario, Canada

Adele Crowder [1], William T. Dushenko [1], Jean Greig [1] & John S. Poland [2]
[1] *Dept. of Biology and* [2] *Department of Chemistry, Queen's University, Kingston, Ontario, Canada, K7L 3N6*

Key words: shorelines, biomonitors, macrophytes, snails, neutron activation, atomic absorption spectrophotometry

Abstract

The Bay of Quinte receives drainage from several large river systems, including the Moira River which carried sediment from mines into the Bay from the 1880s to the 1960s. We are investigating possible metal contamination of submerged weed beds and marsh biota which may contribute to the low diversity and biomass of macrophyte beds and *Typha* marshes in the Bay. In 1987, sediment, macrophytes and snails were sampled in wetlands close to the Moira River and at Hay Bay (part of the Bay of Quinte presumably unaffected by mine effluents) located 20 km from the Moira. Some element concentrations in sediment and biota were determined by neutron activation analysis (NAA) including Al, As, Br, Ca, Co, Cl, Cr, Cs, Fe, Hf, K, La, Na, Mg, Sb, Sc, Rb, Ta, Th, Ti, U, V and Zn. Other elements were analysed by acid dissolution and atomic absorption spectrophotometry (AAS) including Ag, As, Cu, Hg, Ni, Pb, and Zn. Levels of As in sediments and plants were higher close to the Moira River, whereas Cu and Ni showed the opposite pattern in sediments. The usefulness of species as bioassays differed: *Stagnicola elodes* Say accumulated significantly higher levels of Cu (35 vs 18 ppm) and V (1.1 vs 0.5 ppm) than *Planorbella trivolvis* Say collected from the same sites. The macrophyte, *Myriophyllum spicatum* L. acted as an accumulator of Pb (up to 9.6 ppm), whereas Pb in *Vallisneria americana* Michx. at the same sites was undetectable.

Introduction

The Bay of Quinte, located on the north shore of Lake Ontario, Canada (Fig. 1), is a complex water system receiving drainage from several large tributaries. A number of urban, industrial and commercial developments are found within its drainage basin; Minns *et al.* (1986) describe it in further detail. The Bay reached a hypereutrophic state during the 1960's resulting in a loss of submergent macrophyte cover and area. Despite a

reduction of point-source loadings of phosphorus into the Bay (Minns *et al.*, 1986), the recovery of macrophytic vegetation has been poor in some areas, particularly in the shallow upper Bay. Standing crop and species diversity of wetland plants and fauna in the early 1980's were considerably lower than in similar aquatic systems in the area and estimates suggest that submerged macrophytes in the Bay do not occur in much of their potential habitat. In 1979, the area occupied by emergent vegetation was approximately

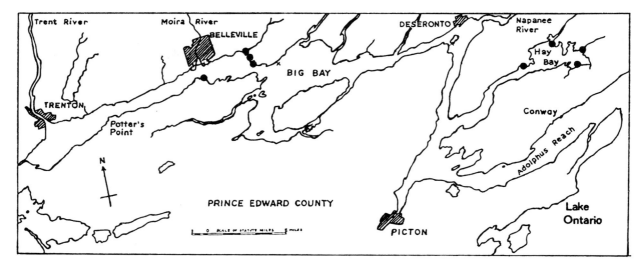

Fig. 1. Map of the Bay of Quinte showing major tributaries and eight sampling areas (adapted from Crowder & Bristow, 1986).

2800 ha, while submergents occupied an area of approximately 600 ha (Crowder & Bristow, 1986).

The Bay has elevated levels of Ag, As, Co, Cu, Hg, Ni, and Pb, in deep water sediments near Belleville (Sly, 1986; Mudroch & Capobianco, 1979; 1980) originating from old mining sites in the Moira River drainage basin which were active from the 1880's to 1960's (Paehlke *et al.*, 1982). Other contaminants such as PCB's and PCP's from industrial and urban discharges in the Trent and Moira Rivers have also been found in deep water sediments in the Bay (Fox & Joshi, 1984; Frank *et al.*, 1980). The possibility exists that the observed decline in wetland and shoreline habitat may be due in part to the long-term effects of contaminants deposited in upland and nearshore areas.

Trace elements were measured in selected species of submerged macrophytes and molluscs, and in sediments, to indicate possible contamination of shoreline marshes. It was hypothesized that trace element levels in sediments and plants would be higher in sites at Belleville, close to potential contaminant sources, than in Hay Bay which was thought to be relatively isolated. Aquatic macrophytes are capable of bioaccumulating a number of trace elements (Mayers *et al.*, 1977; Agami & Waisel, 1986) and therefore

serve as potential biomonitors. Hydrophytes in the Moira River valley north of the Bay of Quinte have been found to contain high levels of As, Co, Cu, and Ni (Mudroch & Capobianco, 1979). Aquatic plants with elevated levels of contaminants in the Bay may affect other trophic levels.

Molluscs have frequently been used in studies of environmental contamination of aquatic systems (Williamson, 1978; Newman & MacIntosh, 1982; Fantin *et al.*, 1985; Ewell *et al.*, 1986; Nebeker *et al.*, 1986; Catsiki & Arnoux, 1987). In these marshes, snails were the most abundant molluscs and were therefore used as representative consumers. They are important in the diet of higher consumers, particularly waterfowl (McCullough, 1981; Barton, 1986; Hoppe *et al.*, 1986) and fish (Sheldon, 1987). The potential exists therefore for transfer of contaminants from water and sediments to higher wetland consumers through snails.

Methods

Sampling sites

Sediment and submerged macrophytes were collected during June of 1987 just offshore of marshes in the Bay. Ten sites in total were sampled from four wetland areas near Belleville (close to contaminant sources) and four in Hay Bay, a separate arm of the middle Bay located approximately 20 km southeast of the Moira River (Fig. 1).

Sediment and plant samples
A composite sample of six sediment cores, 10 cm in depth, was collected at each site at a water depth of 0.5 m. Sediment samples were air-dried and sieved to 2 mm to remove large particles. The two most abundant submergent macrophyte species in the nearshore sites, *Myriophyllum spicatum* L. (Eurasian milfoil) and *Vallisneria americana* Michx. (water celery) were selected as biomonitors. A composite sample of ten plants of each species was collected at each wetland site (where available) adjacent to the sediment core samples. Plant shoots were oven-dried at 70 °C for 24 h and ground to pass through a 1 mm screen.

Snail samples
Planorbella trivolvis and *Stagnicola elodes* were collected in September, 1987, from two marshes in Hay Bay. Samples of approximately 35 individuals, collected in close proximity, were taken by hand from pools and channels within the cattail mats. The snails were kept in deionized water for 24 h for digestive clearance before freezing; tissue was later removed from the shell and dried at 90 °C. Samples were analyzed unground due to their small size.

Analytical methods
Two available analytical techniques were employed for measuring total element levels in sediment, plant and animal samples, (i) neutron activation analysis (NAA) using the Slowpoke 2 reactor at the Royal Military College at Kingston, Ontario, (Ryan *et al.*, 1978, Campbell & Bewick, 1978) and (ii) atomic absorption spectrophotometry (AAS) using techniques outlined in 'Analytical Methods' (Ontario Ministry of the Environment, 1983). Methods were developed for both techniques and validated for a selected number of elements using sediment, plant and animal tissue

standards. The following certified reference materials were analysed by AAS and NAA to validate analytical methods: BCSS (NRCC marine sediment), Citrus Leaves (NBS SRM 1572) and TORT-1 (NRCC invertebrate tissue).

Neutron activation analysis
Two aliquots of each dried sediment, plant or animal sample were weighed out in 7 ml plastic vials and heat-sealed. Samples were irradiated in the Slowpoke-2 reactor located at the Royal Military College of Canada in Kingston, Ontario, using an automated handling system. Optimal irradiation, decay and counting times were developed to obtain determinations of the widest range of elements possible. One set of samples was subjected to 1 to 3 minute irradiations at a neutron flux of 5×10^{11} n cm^{-2} sec^{-1} followed by 1 to 10 m decay times to detect the short-lived nuclides including Al, Ca, Cl, Cu, K, Mg, Mn, Na, Ti, and V. The second set of samples were irradiated for 2 h and left to decay for a) 100 h to detect As, Br, K, La, Na, Sb, and Sc, and b) 250 h to detect Ba, Co, Cr, Cs, Fe, Hf, Rb, Ta, Th, U and Zn.

Atomic absorption spectrophotometry
Two preparatory methods were employed for sediment and plant tissue samples. a) One to two grams of dried sample were weighed into teflon beakers and subjected to sequential acid dissolution using HNO_3, $HClO_4$ and HF, followed by re-acidification using HCl (20%). b) A second set of samples for AAS was prepared by digestion with HNO_3 and H_2O_2 in teflon bombs at 130 °C for 12 h. Standards were prepared in 20% HCl and the elements Ag, Cd, Co, Cu, Ni, Pb, and Zn were determined in samples prepared by each method using flame AAS. Results from each method were compared. Determinations of Hg were performed using a cold vapour technique and As in samples was determined using a hydride generation system. (Ontario Ministry of the Environment, 1983).

Statistical analysis
Elemental levels were compared in sediment and plants using two-sample testing (Statgraphics,

Statistical Graphics System, 1986). Significance was indicated by P values less than 5 percent.

Results and discussion

Distribution of element concentrations in sediment
Sediment samples collected in marsh sites at Belleville (Table 1) had significantly higher levels of As ($P = 0.002$), Co ($P = 0.003$) and Na ($P = 0.01$) than those from Hay Bay; levels of these elements were at least twice as great. These data were consistent with the proposed hypothesis that wetland sediments in sites closest to contaminant sources would yield higher levels of certain trace elements. Ag, Cd and Hg were not detectable in sediment samples from either area of the Bay. Unanticipated findings of elevated Na in sediments at the Belleville sites may be partly attributable to introduction of road salts from runoff.

Sediment at Hay Bay contained significantly higher levels of Cr ($P = 0.03$), Cu ($P = 0.001$), Fe ($P = 0.03$), and Ni ($P = 0.03$) than sites at Belleville (Table 1) which is contrary to the distribution hypothesis. These values were comparable to those in Lake Ontario (Mudroch *et al.*, 1988, Table 1). A number of possible explanations may exist for this opposite trend. Higher levels in Hay

Bay may be due to the introduction of contaminants from local or non-point sources which might be associated with the more intense agricultural land use in this region (authors' pers. obs.). Differences in sediment properties such as particle size, organic content and mineralogy may also have an affect on the total levels of elements occurring in the two different regions.

Relationships occurred between the distribution of different elements. Nickel and Cu were positively correlated in sediment ($r = 0.88$ $P = 0.008$, as were levels of As and Pb ($r = 0.99$ $P = 0.001$); As and Cu, however, were marginally negatively correlated ($r = -0.60$ $P = 0.06$). These patterns suggest that Ni and Cu, for example, have the same source or that their sediment chemistry is very similar in the Bay; Cu and Ni have been shown to associate predominantly with sediment organic matter in Quinte (authors' unpublished data). Relationships of these elements to sediment organic content and particle size are being examined further to account for these patterns.

Sediment toxicity
A majority of elements determined in sediment samples from the Bay, including As, Co, Cr, Cu, Mn, Ni, Pb, were within the soil ranges usually considered toxic to plant life (Bohn *et al.*, 1979;

Table 1. Selected mean total element values determined for nearshore sediments of wetland sites in the upper (Belleville) and middle (Hay Bay) Bay of Quinte. Ranges in Lake Ontario sediments and toxic soils are included where possible.

Region	As µg/g	Co µg/g	Cr µg/g	Cu µg/g	Fe %	Hg µg/g	Na %	Mn µg/g	Ni µg/g	Pb µg/g
Belleville ($n = 5$)	4.5	6.5	29.2	10.3	1.77	<0.2	1.70	373	12.8	4.3
Hay Bay ($n = 5$)	2.1	2.5	46.6	25.1	2.53	<0.2	0.55	291	24.7	2.1
Lake Ontario:[a]										
lower	0.2	–	8.0	26	2.41	0.14	–	–	29	2.4
upper	17.0	–	133	109	9.62	3.95	–	–	99	9.6
Toxic levels[b]	5	8	20	20	–	0.05	–	850	40	10
Range: lower	1	1	5	2	–	0.02	–	100	10	2
upper	50	40	1000	100	–	0.20	–	4000	1000	200

[a] Values from Mudroch *et al.* (1988).
[b] Values from Bohn *et al.* (1979).

Table 1) but most were close to the lower end of the range. Compared with sediment levels in the Moira River (Mudrock & Capobianco, 1980), values reported here were relatively low for some elements. In the long term, however, these elements may have contributed to declining macrophytic vegetation. Campbell *et al.* (1988) suggest that some macrophytes have been found to tolerate high levels of trace elements, but their communities generally decline in biomass and species diversity. A similar pattern of decline may have occurred in the Bay of Quinte.

Element concentrations in biota

Plant shoot tissue levels of As were significantly greater at the Belleville sites (Table 2) for both *Myriophyllum spicatum* ($P = 0.003$) and *Vallisneria americana* ($P = 0.02$). This trend is consistent with differences in sediment As values between the two areas (Table 1) and suggests both species are reliable monitors of such contamination on a regional basis. *Myriophyllum* also accumulated higher levels of Mn at the Belleville sites than in Hay Bay (Table 2) ($P = 0.05$), despite the lack of significant differences in sediment levels ($P = 0.40$) between the two areas (Table 1). No differences were found for tissue levels of other elements between regions for either species; Cd, Co and Hg were not detectable.

Differences in accumulation between macrophyte species were also found for some elements (Table 2). *Myriophyllum spicatum* was found to be a bioaccumulator of Pb whereas levels in *Vallisneria americana* were not detectable. *Vallisneria americana*, conversely, accumulated five times more Na than *M. spicatum* implying that physiological differences in the uptake of some elements exist between these species.

Significant differences in accumulation were also found between snail species from Hay Bay marshes. *Stagnicola elodes* had significantly higher concentrations of Al, Cu, V, K, and Na while *Planorbella trivolvis* had significantly higher concentrations of Mg and Ca (Table 3). Since they occurred in similar habitats, these differences probably reflected physiological differences between the two species. *Stagnicola elodes* accumulated up to 62 μg/g of Cu, almost four times as much as in *P. trivolvis*. It may be more useful, therefore, to collect *S. elodes* as a monitoring organism since elevated levels of Al, V, and Cu are all considered toxic and Cu is an element of particular interest in the Bay of Quinte.

Comparison of analytical methods

Both analytical methods were found to give reliable results for the reference materials except for As by AAS in TORT-1 for which low values

Table 2. Selected mean total element values determined for shoots of *Myriophyllum spicatum* and *Vallisneria americana* from the upper (Belleville) and middle (Hay Bay) Bay of Quinte. Some toxic values are given for terrestrial plants.

Region/Species	Mean tissue levels							
	As μg/g	Cr μg/g	Cu μg/g	Hg μg/g	Mn μg/g	Na %	Ni μg/g	Pb μg/g
Belleville ($n = 4$)								
M. spicatum	4.6	33	4.0	<0.2	703	0.66	11.9	6.0
V. americana	6.8	28	5.3	<0.2	656	2.43	13.3	<1.0
Hay Bay ($n = 4$)								
M. spicatum	1.2	23	4.2	<0.2	420	0.60	10.3	6.5
V. americana	1.9	29	5.0	<0.2	522	2.32	9.5	<1.0
Toxic range [a]								
lower	25	–	4	–	15	–	1	0.1
upper	50	–	15	–	100	–	–	10

[a] Values from Bohn *et al.* (1979).

Table 3. Comparison of selected element concentrations per unit dry weight of tissue for *Planorbella trivolvis* ($n = 10$) and *Stagnicola elodes* ($n = 10$) from marshes in Hay Bay using t-testing.

	Average concentration in tissue								
	Ca %	Mg %	Al μg/g	Cu μg/g	V μg/g	K %	Na %	Cl %	Mn μg/g
P. trivolvis	8.87	0.67	230.3	18.31	0.47	0.40	0.31	0.28	131.8
S. elodes	4.40	0.53	1204.3	34.93	1.06	0.62	0.38	0.31	148.1
p* =	0.000	0.020	0.000	0.001	0.001	0.005	0.009	0.246	0.635

* *p*-values less than 0.05 are significant.

were obtained. Different sample digestion methods had little or no effect on final determinations by AAS. The short sample preparation time, small sample size required, use of automated systems and ready application to a large number of trace elements made NAA a suitable analytical technique for this study. High levels of Al and Na in sediments, however, often precluded the determination of some elements (ie. Cu, Ni, and Zn) due to high compton background. Levels of some elements in a few samples were below the detection level of the method. Mercury and Pb were not readily determined by NAA which necessitated the use of AAS in combination with NAA to include these elements.

Conclusions

Sediment contaminant patterns in the Bay of Quinte were more complex than originally anticipated by the hypothesis and did not all exhibit a simple downstream effect. Further investigations of distributional patterns of elements in sediments and plants are required, particularly in Hay Bay. Comparison with sediment characteristics may further explain the distribution of these elements in the two areas.

Both *Vallisneria americana* and *Myriophyllum spicatum* were reliable indicators of total As levels in the sediments. Care must be exercised, however, in the selection of biomonitors given the differences in accumulation observed between species. Both NAA and AAS were reliable analytical methods, but were best used in combination for detecting a wide spectrum of elements in sediments and biota.

Acknowledgements

The authors would like to thank Dr. P. Beeley for co-ordinating the NAA analysis, Dr. G. L. Mackie of Guelph for assistance with snail identifications and Ms. Catherine Vardy for field assistance. This research is funded by grants from the Ontario Ministry of the Environment and the World Wildlife Fund.

References

Agami, M. & Y. Waisel, 1986. The ecophysiology of roots of submerged vascular plants. Physiol. Veg. 5: 607–624.

Barton, D. R., 1986. Nearshore benthic invertebrates of the Ontario waters of Lake Ontario. J. Great Lakes Res. 12: 270–280.

Bohn H. L., B. L. McNeal & G. A. O'Connor, 1979. Soil chemistry. J. Wiley & Sons, N.Y., 329 pp.

Campbell, J. A. & M. W. M. Bewick, 1978. Neutron activation analysis – review of the method and its present and potential use in agriculture and soil science. Spec. Publ. Common W. Bur. Soils. No. 7. 36 pp.

Campbell, P. G. C., A. Lewis, P. M. Chapman, A. A. Crowder, W. K. Fletcher, B. Imber, S. N. Luoma, P. M. Stokes & M. Winfrey, 1988. Biologically-available metals in sediments. NRCC No. 27694, Ottawa. 296 pp.

Catsiki, A.-V. & A. Arnoux, 1987. Etude de la variabilité des teneurs en Hg, Cu, Zn et Pb de trois espèces de mollusques de l'Etang de Berre (France). Mar. Envir. Res. 21: 175–187.

Crowder, A. & M. Bristow, 1986. Aquatic macrophytes in the Bay of Quinte, 1972-82. In C. K. Minns, D. A. Hurley & K. H. Nicholls (eds.). Can. Spec. Publ. Fish. Aquat. Sci. 86: 114–127.

Ewell, W. S., J. W. Gorsuch, R. O. Kringle, K. A. Robbillard & R. C. Spiegel, 1986. Simultaneous evaluation of the acute effects of chemicals on seven aquatic species. Envir. Toxicol. Chem. 5: 831–840.

Fantin, A. M. B., A. Franchini, E. Ottaviani & L. Benedetti, 1985. Effects of pollution on some freshwater species. II: Bioaccumulation and toxic effects of experimental lead pollution on the ganglia in *Viviparus ater* (Mollusca, Gastropoda). Bas. Appl. Histochem. 29: 377–387.

Fox, M. E. & S. R. Joshi, 1984. The fate of pentachlorophenol in the Bay of Quinte, Lake Ontario. J. Great Lakes Res. 10: 190–196.

Frank, R., R. L. Thomas, M. V. H. Holdrinet & V. Damiani, 1980. PCB residues in sediments collected from the Bay of Quinte, Lake Ontario 1972–73. J. Great Lakes Res. 6 (4): 371–376.

Hoppe, R. T., L. M. Smith & D. B. Wester, 1986. Foods of wintering diving ducks in South Carolina. J. Field Ornithol. 57: 126–134.

Mayers, R. A., A. W. McIntosh & V. L. Anderson, 1977. Uptake of cadmium & lead by a rooted aquatic macrophyte (*Elodea canadensis*). Ecology 58: 1176–1180.

McCullough, G. B., 1981. Migrant waterfowl utilization of the Lake Erie shore, Ontario, near the Nanticoke industrial development. J. Great Lakes Res. 7: 117–122.

Minns, C. K., D. A. Hurley & K. H. Nicholls (eds.), 1986. Project Quinte: Point source phosphorus control and ecosystem response in the Bay of Quinte, Lake Ontario. 270 pp. Can. Spec. Publ. Fish. Aquat. Sci. 86.

Mudroch, A. & J. A. Capobianco, 1979. Effects of mine effluent on uptake of Co, Ni, Cu, As, Zn, Cd, Cr & P by aquatic macrophytes. Hydrobiologia 64 (3): 223–231.

Mudroch, A. & J. A. Capobianco, 1980. Impact of past mining activities on aquatic sediments in Moira River basin, Ontario. J. Great Lakes Res. 6 (2): 121–128.

Mudroch, A., L. Sarazin & T. Lomas, 1988. Summary of surface & background concentrations of selected elements in the Great Lakes sediments (report). J. Great Lakes Res. 14 (2): 241–251.

Nebeker, A. V., A. Stinchfield, C. Savonen & G. A. Chapman, 1986. Effects of copper, nickel and zinc on three species of Oregon freshwater snails. Environ. Toxicol. Chem. 5: 807–811.

Newman, M. C. & A. W. McIntosh, 1982. The influence of lead in components of a freshwater ecosystem on molluscan tissue lead concentrations. Aquat. Toxicol. 2: 1–19.

Ontario Ministry of the Environment, 1983. Analytical methods for environmental samples. Ontario Ministry of the Environment, Rexdale, Ont.

Paehlke, R., L. Maynes & V. McCulloch, 1982. A legacy of arsenic. Case study: the Moira River. Alternatives 10 (2/3): 12–14.

Ryan, D. E., D. C. Stuart & A. Chattopadhyay, 1978. Rapid multi-elemental neutron activation analysis with a Slowpoke reactor. Anal. Chim. Acta. 100: 87.

Sheldon, S. P., 1987. The effects of herbivorous snails on submerged macrophyte communities in Minnesota lakes. Ecol. 68: 1920–1931.

Sly, P. G., 1986. Review of postglacial environmental changes & cultural impacts in the Bay of Quinte. In C. K. Minns, D. A. Hurley & K. H. Nicholls (eds.), Spec. Publ. Fish. Aquat. Sci. 86: 7–26.

Stagraphics Statistical Graphics System, 1986. Statistical Graphics Corp. (STSC Inc.), Rockville, Maryland.

Williamson, P., 1978. Opposite effects of age and weight on cadmium concentrations of a gastropod mollusc. Ambio. 8: 80–81.

Hydrobiologia **188/189**: 345–351, 1989.
M. Munawar, G. Dixon, C. I. Mayfield, T. Reynoldson and M. H. Sadar (eds)
Environmental Bioassay Techniques and their Application.
© 1989 *Kluwer Academic Publishers. Printed in Belgium.*

Use of aquatic macrophytes as a bioassay method to assess relative toxicity, uptake kinetics and accumulated forms of trace metals

Steven Smith & Michael K. H. Kwan
Biosphere Sciences Division, King's College London, Kensington Campus, Campden Hill Road, London W8 7AH, UK

Key words: Lemna, bioassay, cadmium, thallium, toxicity, accumulation

Abstract

Floating aquatic macrophytes such as the Lemnaceae have many attributes which commend their use in laboratory and field investigations to assess both the toxicity of substances and the quality of freshwater systems. As well as their more well known advantages of small size, relative structural simplicity, rapid growth and vegetative reproduction and genetically homogenous populations, they are also excellent accumulators of a number of metallic elements. This raises the possibility of the use of these aquatic macrophytes in water quality monitoring and also as laboratory bioassays for toxicity and uptake studies. Results are presented of a study of the comparative toxicity, uptake kinetics and accumulated forms of thallium and cadmium in the duckweed, *Lemna minor* and the role of this methodology in water quality monitoring and hazard evaluation are discussed.

Introduction

Specific aspects of the basic relationship between exposure, tissue concentration and toxicity have been examined in *Lemna minor* exposed to thallium and cadmium. These two elements form an interesting contrast in that they are both highly toxic entities but with very different chemical properties, uptake kinetics and very probably modes of toxic action. The objective is to demonstrate that an understanding of the uptake mechanisms, the distribution and accumulated forms of trace metals in plants are integral components of bioassay procedures for metal toxicity assessment and biological monitoring.

Methods

Healthy *Lemna minor* cultures, growing in quarter strength Steinberg medium with a turnover rate of 4–5 days have been subject to metal stress in a semi-static testing regime for periods of 10 days. Changes in the rate of growth were assessed in terms of the whole plant responses; frond production and increases in surface area and fresh weight. In addition, chlorophyll content and ultrastructural changes were monitored to ascertain the relative sensitivities of the various responses. Apart from the qualitative assessment of ultrastructural changes, the other responses have been compared on the basis of 'relative growth rate' i.e.

346

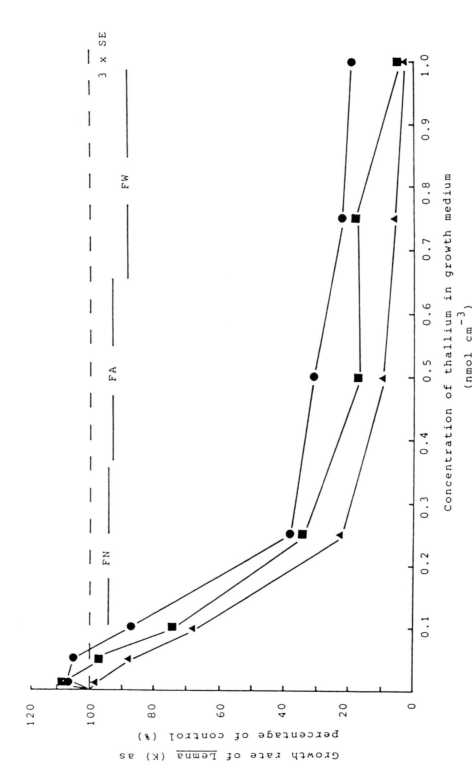

Fig. 1. Comparison of the relative increase in ▲, total frond area (FA); ■, frond fresh weight (FW); and ●, frond number (FN), in *Lemna minor* exposed to different concentrations of thallium for a period of 10 days. Lines labelled, FN, FA and FW indicate the statistical threshold, i.e. 3 × SE, three times the pooled standard error. (Reproduced with permission from Kwan & Smith, 1988).

the exponential increase in growth per unit time, expressed as a percentage of the respective control values; each replicate consisted of 10 fronds and five replicates were used for each metal exposure. Hence, from this it has been possible to estimate the Effective Concentration i.e. the concentration resulting in a 50% reduction of growth, and the Threshold Concentration, which is a statistical indication of the onset of toxicity and is measured as the growth reduction in excess of three times the pooled experimental standard error. Flowthrough, continuous exposure systems were developed to study metal uptake kinetics as a function of time and concentration and metal efflux. Short-term 24 hour metal exposures were used to examine factors influencing uptake and the accumulated forms of the metals. A full description of all the methods adopted in the study have been reported in Kwan & Smith (1988) and Kwan (1988).

Results

Whilst focusing the study on thallium and cadmium, it is nevertheless useful to interpret the toxic responses relatively to other elements; and in a suite of elements that included Cr, Ba, Cd, Cu and Tl, EC_{50} values for frond number production were found to be 75, 59, 1.7, 1.3 and 0.2 nmol ml^{-1} and for threshold concentrations 2.5, 0.02, 0.6, 0.3, 0.08 nmol ml^{-1} respectively. Thus confirming the extremely toxic nature of Tl, although Cu was found to be more toxic than Cd. One point worthy of note is that, in the actual environment the bioavailability of many metals is affected by a number of factors e.g. pH and the nature and the concentration of ligands but the ion activity of thallium is little affected by these factors.

Since highly significant curvilinear relationships have been established between metal concentration in the growth medium and levels accumulated in the plant tissues, the various growth indices may be expressed with respect to one or the other (Kwan & Smith, 1988; Kwan, 1988). The response of the growth parameters and changes in chlorophyll content to thallium

Table 1. Effective Concentration and Threshold Value for Each of the Growth Parameters in Terms of Metal Concentration in *Lemna* Tissues (μmol g^{-1})

Growth Parameter	Effective concentration		Threshold concentration	
	Thallium	Cadmium	Thallium	Cadmium
Frond area	2.18	15.90	0.96	9.40
Frond number	2.70	21.00	1.44	9.35
Fresh weight	2.50	19.00	1.38	10.50
Chlorophyll a	3.16	23.40	2.08	11.00
Chlorophyll b	4.40	35.80	2.45	13.90

and cadmium stress followed the same overall pattern (Fig. 1) and clearly the changes were proportional to metal exposure, however as can be judged from the EC_{50} and threshold values in Table 1 there were differences in sensitivity between the parameters. For thallium, 50% reduction in surface area and threshold levels were attained at significantly lower exposures than the equivalent reductions in fresh weight and frond number (Student's t test $p < 0.05$ and $p < 0.01$ respectively) and in the case of cadmium a significant difference was found between the EC_{50} values for surface area and frond number (Student's t test $p < 0.01$). The onset of thallium stress is marked by an enhanced production of fronds and this coincides with the surface area threshold, hence at the initial thallium exposures the most obvious response is an increased production of smaller fronds. The whole plant growth parameters appear to be more sensitive indices of toxicity than changes in total chlorophyll content. Ultrastructural changes (Kwan, 1988) also support this observation as marked visible deformation of chloroplasts were only observed at exposures in excess of threshold values.

Evaluation of the uptake kinetics of the Tl and Cd, both as a function of time and concentration, has demonstrated the differences in the uptake mechanisms involved and the spatial location of the metals in the plant tissues (Kwan, 1988). The salient features of the study are summarised as follows:
1) Continuous exposure to 0.2 nmol cm^{-3} Tl and

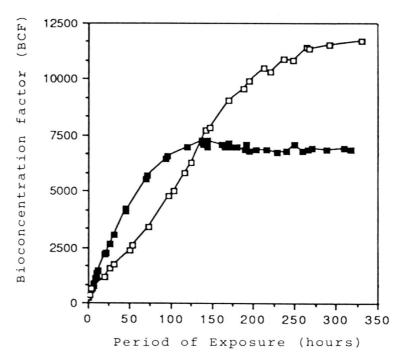

Fig. 2a. Uptake of thallium (■), and cadmium (□) by *Lemna minor* as a function of time.

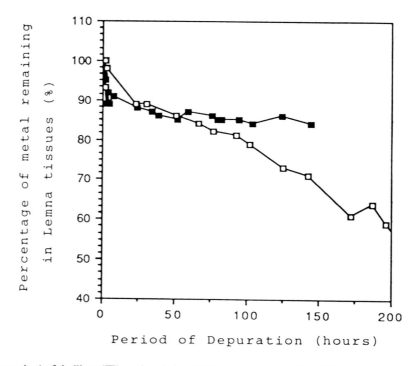

Fig. 2b. Efflux (depuration) of thallium (■), and cadmium (□) in metal-free medium following long-term exposure to either Tl or Cd.

1.0 nmol cm^{-3} Cd in flowthrough systems for up to 300 hours, revealed two distinct patterns (Fig. 2a); attainment of a well-defined thallium steady-state after 140 hours continuous exposure, whereas for cadmium a steady-state was only achieved after some 260 hours exposure. At this stage, the cadmium bioconcentration factor (BCF, 12 600) was twice that of thallium (BCF, 7000).

2) Depuration (i.e. the rate of efflux from plant to metal-free medium; figure 2b) of cadmium was essentially a continuous process and after 240 hours, only 47% of the steady-state concentration remained. Over the first three hours some 10% of thallium was relatively rapidly effluxed from the *Lemna* tissues but thereafter the rate of efflux decreased and after 48 hours no further depuration was observed and after 140 hours over 80% of the steady-state concentration remained in the plant tissues.

3) Set 24 hour exposures to a range of thallium and cadmium concentrations revealed uptake profiles characteristic of multiphasic systems; interpretation of the trends according to Nissen (1973) indicated that thallium uptake involved as many as four phases in much the same manner as potassium uptake whilst Cd uptake was less complex and consisted of two phases.

4) Immersion in boiling water, extended periods of darkness and sodium azide treatment, all inhibited uptake to a similar extent and the overall conclusion is that 90% of Tl uptake is active but that of Cd appears to involve both active and passive processes, for example, Na N$_3$ treatment resulted in a 55% reduction in cadmium uptake.

Buffer extractions (20 nM MES-Tris, pH 7.0) of plant homogenates labelled with ^{204}Tl and ^{109}Cd showed that as much as 86% of the thallium taken up by the plant may be classified as aqueous soluble and of this 84% appeared to be associated with species less than 10 000 daltons (ultrafiltration of soluble fraction using a membrane with a cut-off of 10 000 daltons); a sizeable proportion of the accumulated Cd was retained in the insoluble fraction (37%) and of the soluble component over

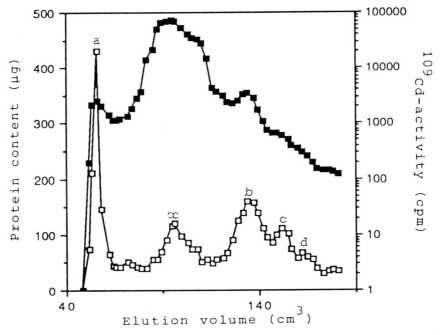

Fig. 3a. Sephadex G50 elution profile of buffer soluble extracts of *Lemna minor* grown in the presence of 1 nmol cm^{-3} Cd (labelled with ^{109}Cd) for 12 days: □, protein content; ■, ^{109}Cd-activity.

350

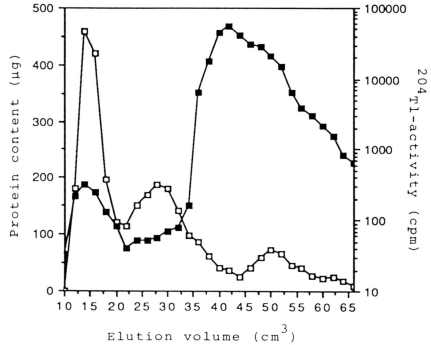

Fig. 3b. Sephadex G25 elution profile of buffer-soluble extracts of *Lemna minor* grown in the presence of 0.2 nmol cm^{-3} Tl (labelled with ^{204}Tl) for 12 days: □, protein content; ■, ^{204}Tl-activity.

50% was associated with species greater than 10 000 daltons. Enzyme digestion of fronds using cellulase, pronase and pectinase showed that much of the Cd was associated with pectins and to a lesser extent proteins and celluloses.

A very interesting pattern has emerged from applying Gel Filtration Chromatography to aqueous extracts of *Lemna* (Sephadex G100, 150 000 – 4000 daltons; G50, 30 000 – 1500 daltons; G25, 5000 – 1000 daltons) the distribution of radio activity was compared with the total protein (Lowry test) distribution in the separated fractions (Fig. 3a & b). A close correspondence was found between the cadmium and protein distributions: five distinct protein fractions were identified; a > 150 000, x = 7000, b = 1800, c and d < 1500 daltons and peaks a, x and b, in particular, coincided with a peak for cadmium activity (Fig. 3a). Clearly, the majority of the element was associated with the 7000 daltons protein fraction and it is interesting to note that this size range is similar to the metallothioneins found in animal

tissues. On the other hand, thallium shows little association with proteins; evidence of complexing with a protein fraction in the size range of > 150 000 was observed in the G100 column but as the G25 column demonstrates (Fig. 3b) the major peak of activity occurs at an elution volume corresponding to a molecular size of < 1000 and the position of this peak was very similar to that obtained from injecting with a thallous ion standard. Hence much of the thallium in the soluble fraction behaves in a similar fashion to the free metal ion.

Concluding remarks

All in all, thallium is near to one order of magnitude more toxic than cadmium but at equimolar exposures *Lemna* accumulates considerably more cadmium than thallium. It is thought that these differences may be explained in terms of the physicochemical properties of the two elements and hence in the dynamics of the metals in the plant tissues. Thus, thallium is a 'mobile' element with little affinity for organic ligands and its ion activity is relatively unaffected by changes in pH. Thallium in the cell behaves in an analogous fashion to potassium, even to the point that the cell vacuole is a likely site of accumulation. Cadmium, on the other hand, has a strong affinity for organic ligands and a major proportion of the element in *Lemna* tissues is complexed with the pectins of the cell wall and even in the soluble phase much of the Cd is bound to protein fractions.

Attention has been drawn to the role of bioassay procedures and the *Lemna* system described has applications for both single species toxicity prediction as well as a monitoring species.

However it is strongly contended that in developing these aspects, the overall relationship between exposure, tissue concentration and toxicity in the manner described is of fundamental importance.

References

Kwan, K. H. M., 1988. Comparative study on the whole plant toxicity, uptake kinetics and accumulated forms of thallium and cadmium in *Lemna minor L*. Ph. D. Thesis, University of London. 402 pp.

Kwan, K. H. M. & S. Smith, 1988. The effect of thallium on the growth of *Lemna minor* and plant tissue concentrations in relation to both exposure and toxicity. Envir. Pollut. 52 (3): 203–219.

Nissen, P. (1973). Multiphasic uptake in plants II. Mineral cations, chloride, and boric acid. Physiol. Pl. 29: 298–354.

Hydrobiologia **188/189**: 353–359, 1989.
M. Munawar, G. Dixon, C. I. Mayfield, T. Reynoldson and M. H. Sadar (eds)
Environmental Bioassay Techniques and their Application.
© 1989 *Kluwer Academic Publishers. Printed in Belgium.*

Bioassays with a floating aquatic plant (*Lemna minor*) for effects of sprayed and dissolved glyphosate

W. Lyle Lockhart, Brian N. Billeck & Chris L. Baron
Canada Department of Fisheries and Oceans, Central and Arctic Region, Freshwater Institute, 501 University Crescent, Winnipeg, Manitoba, Canada R3T 2N6

Key words: duckweed, glyphosate, *Lemna*, phytotoxicity, bioassay, herbicide

Abstract

Macrophytes in forested areas and in prairie wetlands furnish critical habitat for aquatic communities and for several species of birds and mammals. North American agriculture relies heavily on herbicides and these compounds are detected routinely in surface waters of Western Canada. The question is whether these residues have biological meaning. There is surprisingly little literature on the responses of macrophytes to herbicides, or indeed to other chemicals. Previously we have used common duckweed in efforts to detect effects of herbicides and other chemicals. Duckweed clones were developed from local collections and grown axenically. In this study the plants were exposed to glyphosate herbicide either by dissolving formulated Roundup® (Monsanto Canada Inc.) in the culture media or by spraying of the cultures in a laboratory spray chamber. Plant growth was monitored by counting the fronds present on several occasions over a 2-week period following treatment and by taking wet and dry weights of plants after the final counting period. Plant growth, as measured by increased numbers of fronds or increased wet or dry weights was relatively insensitive to glyphosate dissolved in the culture medium. However, the plants were killed by application of glyphosate as a spray.

Introduction

Duckweeds have been used as convenient bioassay organisms for the detection of phytotoxicity since the 1930s. They were among the species used to define the effects of the earliest phenoxy herbicides on plants (Blackman & Robertson-Cunninghame, 1954, 1955). Phenoxy herbicides have since become distributed widely throughout watersheds in central Canada (Gummer, 1980). A recent survey of two small rivers draining agricultural watersheds in western Manitoba revealed the presence of several other herbicides (trifluralin, triallate, bromoxynil, 2,4-D, 2,4,5-T

diclofop and dicamba) in addition to phenoxy compounds (Muir & Grift, 1987). While herbicides may seem to be the most obvious application for duckweed bioassays, these plants also respond to other materials such as heavy metals (Nasu & Kugimoto, 1981) and surfactants (Bishop & Perry, 1981).

Glyphosate is being applied in agriculture for weed control and also in forestry for 'conifer release' – the control of young deciduous trees to allow the free growth of conifers in plots being managed for the production of conifers. It is virtually impossible to spray any material over large areas by aircraft, particularly in forestry, without contamination of surface waters within those areas, and so emergent aquatic vegetation might be exposed both by contact with spray droplets and by contact with herbicide-contaminated water. Duckweeds in laboratory cultures were known to be sensitive to glyphosate in the water, however, the toxicity was reduced by addition of bentonite clay to the water (Hartman & Martin, 1984, 1985). Field trials in which glyphosate was sprayed onto duckweed, however, failed to control the plants (Thayer & Haller, 1985). The difference between bioassay experience and field experience prompted us to question whether the plants might respond differently to glyphosate applied as a surface spray than to glyphosate presented dissolved in the water.

Materials and methods

Duckweed cultures

Stock cultures of a clone common duckweed were cultured from an original collection of plants taken from Lake-of-the-Woods, northwestern Ontario. The clone was obtained by bleaching as described by Hillman (1961), and was maintained in axenic culture using Stewart's (1972) medium with asparagine at 132.1 mg l^{-1} as the source of nitrogen. Plants were grown in 250-ml Erlenmeyer flasks with 100 ml of medium per flasks, in a controlled environment room at 25 °C. Light was provided by General Electric Gro & Sho® lights at about 60 mE m^{-2} s^{-1} with a photoperiod of 16 h light and 8 h dark, essentially as described by Lockhart & Blouw (1980).

Exposures to dissolved glyphosate

Glyphosate (N-(phosphonomethyl)glycine) was used as the formulated commercial herbicide Roundup® (Monsanto Canada Inc.), supplied at 356 g l^{-1} of the isopropylamine salt of glyphosate. Plant cultures were treated with glyphosate by dissolving the herbicide in the culture medium. For exposures to dissolved glyphosate the concentrations used ranged from 10^{-7} M to 10^{-3} M of glyphosate as the isopropylamine salt. Exposures were conducted in 125-mL Erlenmeyer flasks to which sterilized culture medium and glyphosate were added before the addition of 10 representative fronds of duckweed from the stock clone. Five flasks were set up for each exposure concentration, and there were also five untreated controls.

Exposures to sprayed glyphosate

For application of a spray to the surface of the plants a laboratory sprayer was used consisting of a spray nozzle which moved along a rigid track at a rate and pressure selected by the operator. The sprayer was calibrated using dyes to allow delivery of any desired quantity of material to the surface of the cultures. The application rate used was that recommended for field control of annual weeds up to 15 cm high, namely 2.25 l of formulated Roundup® per ha, or approximately 800 g of active ingredient per hectare (Monsanto Canada Inc.). This rate of application to the surface of the dishes would produce, on complete mixing, a concentration of about 2.34×10^{-5} M glyphosate in the culture medium. Exposures were conducted by placing sufficient duckweed fronds in deep Petri dishes with a surface area of 0.00785 m^2 and containing 117.8 ml of culture medium. After spraying, the exposed plants were allowed to stand in the spray apparatus for 6, 12, or 24 hr without disturbance, and then 10 apparently normal fronds were selected from each dish and transferred to 125-ml Erlenmeyer flasks where they were grown in clean culture medium. (There was no visible injury to plants at the time the samples of 10 fronds were removed). The standing times in the apparatus were required; preliminary experiments showed that removal of the sprayed cultures immediately after spraying resulted in loss of toxicity, presumably by washing glyphosate deposits off the plants into the medium.

Controls were treated in the same way as sprayed cultures except that the dish covers were left in place. There were 10 replicates for each exposure.

Plant responses to treatments

The numbers of fronds were counted several times over a 2-week period following exposure, and growth curves were plotted. Following the final count, cultures were drained, blotted, weighed, then air dried to constant weight at 95 °C and re-weighed.

Statistical procedures

Data describing plant growth response to dissolved glyphosate were fitted to the following model, at each level of glyphosate concentration.

$$Ln \text{ (frond number)} = \beta_0 + \beta_1(\text{time}) + \beta_2(\text{time})^2 + E_i$$

Plant growth response to fixed glyphosate concentration with variable exposure time data were also fitted to this model, and considered at each level of exposure time. Plant 'doubling times' were estimated as a function of the coefficients of the above model, and were calculated as of the ninth day after treatment started. The physical interpretation of the model parameters may be described as follows: β_0 is the intercept, and is proximate to the natural log of the frond number at the outset of the experiment; β_1 is the instantaneous rate of culture growth at time zero; β_2 is the deceleration of growth over time, and E_i is the error component.

Plant weight response to dissolved, as well as sprayed, glyphosate was fitted to a simple linear model with exposure time as the independent variable and plant weight the dependent variable. Results based on plant dry weight coincided with those based on plant wet weight.

Analysis of variance, regression, and covariance analysis were used to analyze the data, employing contrasts and simultaneous multiple comparisons (Duncan) to determine differences between treatment levels (SAS, 1986).

Results and discussion

Plant growth response to dissolved glyphosate

The mean (geometric) numbers of fronds produced in cultures exposed to glyphosate dissolved in the culture medium are plotted in Fig. 1. Coefficients for the model describing the growth curves are given in Table 1, and these show that the model provided a good description of the growth, as judged by r^2 values exceeding 0.97. There was no inhibition of frond production at concentrations of 10^{-4} M or lower, however, culture growth was eliminated at 10^{-3} M, hence the threshold for phytotoxicity was somewhere in the range between 10^{-4} M and 10^{-3} M. Indeed there was a small but significant enhancement of frond production at 10^{-4} M and 10^{-5} M, presumably

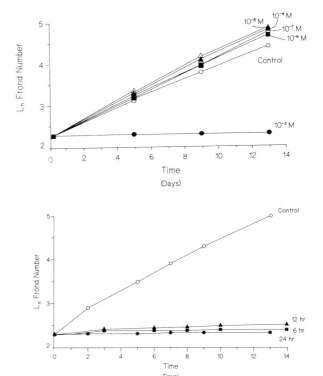

Fig. 1. Numbers of fronds present in duckweed cultures exposed to several concentrations of glyphosate dissolved in the culture medium (above) and exposed to a surface spray at 800 g active ingredient per ha and allowed to stand undisturbed for varying periods in the spray chamber (below).

Table 1. Coefficients for growth equation[1] for different concentrations of glyphosate dissolved in the culture medium.

Glyphosate concentration (M)	β_0	β_1		β_2	r^2	Doubling time (days)
0 (Control)	2.29	0.197	b	-0.00211	0.979	4.7
10^{-7} M	2.26	0.239	ab	-0.00350	0.994	4.3
10^{-6} M	2.26	0.229	ab	-0.00314	0.992	4.4
10^{-5} M	2.26	0.265	a	-0.00483	0.993	4.4
10^{-4} M	2.26	0.249	a	-0.00397	0.986	4.3
10^{-3} M	2.30	0.003	–	-0.00012	ns[2]	–

Slopes β_1 followed by the same letter failed to differ at the $\alpha = 0.01$ probability level.

[1] L_n (frond number) $= \beta_0 + \beta_1(\text{time}) + \beta_2(\text{time})^2 + E_i$

[2] Since there was no growth at 10^{-3} M glyphosate there was no significant relationship between time and frond number, so that r^2 and doubling times were not meaningful. Doubling times for other exposures were calculated arbitrarily at day 9.

illustrating the phenomenon of hormesis (Stebbing, 1982).

In these exposures the plants were somewhat less sensitive than in studies reported previously with glyphosate. Jaworski (1972) found that glyphosate at 10^{-3} M inhibited completely the growth of *Lemna gibba*, that growth was significantly (75%) inhibited at 10^{-4} M, and essentially unaffected at 10^{-5} M. Cooley & Foy (1986) reported that 10^{-4} M glyphosate reduced growth of *Lemna gibba* to about one third of untreated controls while 10^{-5} M had no effect. Hartman & Martin (1984) used a bioassay technique similar to that reported here and found growth of *Lemna minor* reduced by about one third at only 2 mg l^{-1} (approximately 10^{-5} M) with growth almost eliminated at 10 mg l^{-1}. While the plants used here were relatively insensitive to glyphosate, they are not insensitive to other herbicides. For example, the cultures used here were comparable to those from another laboratory in their sensitivity to terbutryn. We obtained a growth reduction at a starting concentration of terbutryn of 3.16×10^{-8} M (about 8 µg l^{-1}, Lockhart *et al.*, 1983), and Bahadir & Pfister (1985) reported growth reduction at a concentration which started at 30 µg l^{-1} and fell to 10 µg l^{-1} over a 1-week exposure period. Richardson (1985) has reviewed a variety of bioassays with glyphosate.

Recent study of the dissipation of glyphosate residues from forest ponds treated by several passes of an aircraft emitting glyphosate at a rate of 0.89 kg ha^{-1} has shown that maximum residues were under 200 mg l^{-1}, (Goldsborough, 1989) and so the duckweed cultures would not be sensitive to these concentrations of dissolved glyphosate. An earlier study of glyphosate sprayed at 3.3 kg ha^{-1} over a forest stream resulted in peak residues in the first 3 hr after spraying of only about 2.7 mg l^{-1} (Newton *et al.*, 1984).

Plant weight responses to dissolved glyphosate

The effect of dissolved glyphosate on frond growth in terms of wet and dry weights is shown in Table 2. The same conclusions follow from frond weights as from growth in frond numbers. Concentration of glyphosate at 10^{-4} M or lower did not reduce wet or dry weights below controls, while 10^{-3} M obviously did.

The low toxicity of glyphosate in the water is not surprising in view of the high water-solubility of this herbicide (1.2%, Weed Science Society, 1983). The tendency of a compound to accumulate in duckweed is related to its octanol/water partition coefficient (Lockhart *et al.*, 1983) which is inversely proportional to its water solubility (Chiou *et al.*, 1977). Given its high water solubility, glyphosate would have little tendency to

Table 2. Mean wet and dry weights of duckweed fronds after 21 days in cultures exposed to a range of starting concentrations of glyphosate dissolved in the culture medium.

Glyphosate (M)	Wet Weight (mg) $\pm$ S.D.		Dry Weight (mg) $\pm$ S.D.	
0	236 $\pm$ 33	a	23.9 $\pm$ 5.5	c
10^{-7} M	253 $\pm$ 18[1]	a	22.8 $\pm$ 1.6[1]	c
10^{-6} M	265 $\pm$ 22	a	22.3 $\pm$ 0.7[1]	c
10^{-5} M	280 $\pm$ 34	a	25.4 $\pm$ 1.1	c
10^{-4} M	256 $\pm$ 13	a	22.4 $\pm$ 2.4	c
10^{-3} M	7.6 $\pm$ 1.9	b	0.3 $\pm$ 0.1	d

[1] Mean of four cultures.

Figures are arithmetic means of 5 cultures except as noted.

Means followed by different letters are statistically different at the $\alpha = 0.01$ probability level using Duncan's test.

partition from the water to the plants, and it would also wash off the plants easily prior to penetration. Indeed users are warned that rainfall occurring 6 hr after treatments of weeds with glyphosate may reduce its effectiveness, and heavy rainfall within 2 hr may wash glyphosate off leaves (Weed Science Society, 1983).

Plant growth response to sprayed glyphosate

The application of glyphosate as a spray to the surface of cultures was more toxic to the plants than was adding glyphosate to the culture medium. Figure 1 shows the effect on frond production of spraying commercial Roundup® at the rate recommended for control of weeds up to 15 cm high (2.25 l formulated product per hectare = 801 g isopropylamine salt of glyphosate ha^{-1}). Whether the plants were left in the exposure chamber for 6, 12, or 24 hr made little difference; in all cases the subsequent growth of the cultures was essentially zero, while untreated plants (which were left standing covered in the spray apparatus 24 hr after treatment) grew normally. The theoretical concentration in the water if all the glyphosate deposited were mixed throughout the volume of medium in the dish was 2.3×10^{-5} M. This concentration was too low to cause measurable reductions in culture growth (Fig. 1). The plants must have experienced an exposure greater than that available through mixing of glyphosate with the water. Presumably the plants intercepted droplets of spray and were exposed through deposit contact, perhaps with some additional exposure through the water.

Plant weight responses to sprayed glyphosate

The spray deposit effectively eliminated plant weight increases (Table 3) at all three standing periods just as it inhibited frond production (Fig. 1). These results confirm the susceptibility of the plants to sprayed glyphosate. In early experiments with the track sprayer, plants were removed from the apparatus immediately after spraying, and the spray had no effect, presumably because even gentle movement of the dishes caused some mixing of plants and medium and washed glyphosate deposits off plants surfaces into the medium where there was insufficient herbicide to affect growth. This suggests that glyphosate would be relatively ineffective on duckweed in field settings where mild wave action would have the same effect as our early removal of dishes from the sprayer. In fact, glyphosate at application rates up to 6.7 kg ha^{-1} was ineffective in controlling growth of duckweed in outdoor experimental trials (Thayer & Haller, 1985). Apparently 6 hr of undisturbed standing after spraying was long enough to allow sufficient glyphosate to penetrate the plants and exert its phytotoxic action.

Table 3. Wet and dry weights of duckweed fronds after 13 or 14 days in cultures exposed to glyphosate applied as a surface spray of commercial Roundup® at 800 gha^{-1}.

Treatment	Standing time (hr)	Wet Weight (mg) ± S.D.		Dry Weight (mg) ± S.D.	
Unsprayed control	24	225 ± 31	a	14.5 ± 1.9	c
Sprayed	6	12.1 ± 1.5	b	2.3 ± 0.3	d
Sprayed	12	17.6 ± 8.1	b	2.4 ± 0.8	d
Sprayed	24	11.0 ± 1.4[1]	b	1.9 ± 0.4[1]	d

[1] Mean of nine cultures.
Cultures were allowed to stand undisturbed in the spray apparatus for 6, 12, or 24 hr after spraying before sub-cultures were taken for measurement of effects over a 13- or 14-day observation period in untreated medium.
Figures are arithmetic means of 10 cultures except as noted, with standard deviations.
Means followed by different letters are statistically different at the $\alpha = 0.01$ probability level using Duncan's test.

Conclusions

Glyphosate is inherently phytotoxic to duckweed, but the plants are relatively insensitive to glyphosate in the water, probably because glyphosate would have little tendency to partition from water to plants and hence the dosage experienced by the plants would be less than for a more hydrophobic compound. Following spray applications deposit contact may pose a greater risk to emergent aquatic vegetation than contamination of the water. Non-target emergent vegetation would be expected to suffer damage if the deposit were not washed off within some time less than six hours. With sprayed glyphosate phytotoxicity may be expressed more appropriately in terms of spray rates than in conventional units of concentrations in the water.

Acknowledgements

The Department of Plant Science, University of Manitoba, kindly allowed the use of its laboratory-scale track sprayer for the spray exposures.

References

Bahadir, M. & G. Pfister, 1985. A comparative study of pesticide formulations for application in running waters. *Ecotoxicol. Envir. Saf.* 10: 585–590.

Bishop, W. & R. Perry, 1981. Development and evaluation of a flow-through growth inhibition test with duckweed (*Lemna minor*). In: D. Branson & K. Dickson (Eds.), *Aquatic Toxicology and Hazard Assessment: Fourth Conference.* American Society for Testing and Materials, Philadelphia. Special Technical Publication 737. pp. 421–435.

Blackman, G. & R. Robertson-Cuninghame, 1954. Interactions in the physiological effects of growth substances on plant development. *J. Exp. Bot.* 54: 184–203.

Blackman, G. & R. Robertson-Cuninghame, 1955. Interrelationships between light intensity, temperature, and the physiological effects of 2 : 4-dichlorophenoxyacetic acid on the growth of *Lemna minor. J. Exp. Bot.* 6: 156–176.

Chiou, C., V. F. D. Schmedding & R. Kohnert, 1977. Partition coefficient and bioaccumulation of selected organic chemicals. *Envir. Sci. Technol.* 11: 475–478.

Cooley, W. & C. Foy, 1986. Effects of SC-0224 and glyphosate on inflated duckweed (*Lemna gibba*) growth and EPSP-synthetase activity from *Klebsiella pneumoniae. Pestic. Biochem. Physiol.* 26: 365–374.

Goldsborough, L. & A. Beck, 1989. Rapid dissipation of glyphosate in small forest ponds. *Arch. Envir. Contam. Toxicol.*, in press.

Gummer, W., 1980. Pesticide monitoring in the prairies of Western Canada. In: B. Afghan & D. Mackay (Eds.). *Hydrocarbons and Halogenated Hydrocarbons in the Aquatic Environment*, pp. 345–372. Plenum Press, New York.

Hartman, W. & D. Martin, 1984. Effect of suspended bentonite clay on the acute toxicity of glyphosate to *Daphnia pulex* and *Lemna minor. Bull. Envir. Contam. Toxicol.* 33: 355–361.

Hartman, W. & D. Martin, 1985. Effects of four agricultural pesticides on *Daphnia pulex, Lemna minor* and *Potomogeton pectinatus. Bull. Envir. Contam. Toxicol.* 35: 646–651.

Hillman, W., 1961. The *Lemnaceae*, or duckweeds, a review of the descriptive and experimental literature. *Bot. Rev.* 27: 221–287.

Jaworski, E., 1972. Mode of action of N-phosphonomethyl-glycine: Inhibition of aromatic amino acid biosynthesis. *J. Agric. Food Chem.* 20: 1195–1198.

Lockhart, W., B. Billeck, B. deMarch & D. Muir, 1983. Uptake and toxicity of organic compounds: Studies with an aquatic macrophyte (*Lemna minor*). In: W. Bishop, R. Cardwell, and B. Heidolph (Eds.), *Aquatic Toxicology and Hazard Assessment: Sixth Symposium*, American Society for Testing and Materials, Philadelphia. ASTM STP 802. pp. 460–468.

Lockhart, W. & A. Blouw, 1980. Phytotoxicity tests using *Lemna minor*. In: E. Scherer (Ed.). *Toxicity Test for Freshwater Organisms*, Can. Dep. Fish. Oceans, Winnipeg. Spec. Pub. Fish. Aquat. Sci. No. 44. pp. 119–130.

Monsanto Canada Inc. *Roundup Liquid Herbicide by Monsanto.* 26 pp.

Muir, D. & N. Grift, 1987. Herbicide levels in rivers draining two prairie agricultural watersheds (1984). *J. Envir. Sci. Health,* B22: 259–284.

Nasu Y. & M. Kugimoto, 1981. *Lemna* (duckweed) as an indicator of water pollution. I. the sensitivity of *Lemna paucicostata* to heavy metals. *Arch. Envir. Contam. Toxicol.* 10: 159–169.

Newton, M., K. Howard, B. Kelpsas, R. Danhaus, C. Lottman & S. Dubelman, 1984. Fate of glyphosate in an oregon forest ecosystem. *J. Agric. Food Chem.* 32: 1144–1151.

Richardson, W., 1985. Bioassays for glyphosate. In: E. Grossbard & D. Atkinson (Eds.). *The Herbicide Glyphosate*, pp. 286–298. Butterworths.

SAS Institute Inc., Cary, North Carolina, 1986. *SAS Systems for Linear Models.*

Stebbing, A., 1982. Hormesis – the stimulation of growth by low levels of inhibitors. *Sci. Tot. Envir.* 22: 213–234.

Stewart, G., 1972. The regulation of nitrite reductase level in *Lemna minor* L. *J. Exp. Bot.* 23: 171–183.

Thayer D. & W. Haller, 1985. Effect of herbicides on floating aquatic plants. *J. Aquat. Pl. Man.* 23: 94–95.

Weed Science Society of America, Champaign, Illinois, 1983. *Herbicide Handbook,* fifth edition. pp. 258–263.

Hydrobiologia **188/189**: 361–366, 1989.
M. Munawar, G. Dixon, C. I. Mayfield, T. Reynoldson and M. H. Sadar (eds)
Environmental Bioassay Techniques and their Application.
© 1989 *Kluwer Academic Publishers. Printed in Belgium.*

Phytomonitoring of pulverized fuel ash leachates by the duckweed *Lemna minor*

H. A. Jenner & J. P. M. Janssen-Mommen
*N.V. Kema, Environmental Research Department, P.O. Box 9035, Utrechtseweg 310, Arnhem 6800 ET,
The Netherlands*

Key words: *Lemna minor*, pulverized fuel ash, leaching, metals, accumulation, toxic effects

Abstract

The duckweed *Lemna minor* is one of the smallest vascular plants with a known strong capacity for metal accumulation. *L. minor* is proposed as a phytomonitor for coal ash drainage systems and for bio-assay studies directed to complexation and speciation. The duration of the experiment can be restricted to fourteen days; it is then possible to determine accurate data of differences in growth of the clone forming plant by using image processing techniques. Leaching of pulverized fuel ash (PFA) with acetic acid according to EPA instruction resulted in effects attributed to the acetic acid itself rather than to the metals in solution. Toxic effects of both leachates, 'natural' and 'artificial', are discussed. The order of toxicity of metals studied so far in separate metal experiments is Cd > Cu > Zn > As(Arsenite) > Se(Selenite) > Ge > B > Mo.

Introduction

The forecast for coal combustion in the Netherlands is for about 15 million tons a year in the year 2000, which will result in an annual amount of about 1.5 million tons of pulverized fuel ash (PFA). Compared with 1985, at 0.5 million tons of PFA, the production will triple in 15 years' time (Meij *et al.*, 1986). The deposition of PFA in disposal piles arouses concern in the Netherlands about long-term environmental effects due to elevated metal concentrations in PFA and the possibly elevated metal concentrations in leachate or runoff water (Bolt & Snel, 1986; Cherry & Guthrie, 1977; Cherry *et al.*, 1984). Nriagu (1988) describes the dangers of low-level elevated metal concentrations for higher animals and human populations. He discusses possible links with

cardiovascular diseases, and allergies. Sublethal effects can be expected for (aquatic) organisms lower in the food-chain.

Phytotoxicity research is still a minor component of all the tests and bio-assays commonly used in toxicity studies (Wang, 1984). The criticism of phytomonitoring with *L. minor* is focussed on lower sensitivity compared with aquatic animals, so that duckweed studies seem to be redundant. Bishop & Perry (1981) also concluded that the ecological significance of the *L. minor* testing was questionable compared with the testing of daphnias or fish. However, duckweed forms an essential component in shallow stagnant waters. The bio-assay studies of Ray & White (1976), Guthrie (1979) and Nasu & Kugimoto (1981), utilizing higher plants and especially aquatic plants, such as the duckweed *L. minor*, highlighted the advantages of aquatic plants. The importance of metal speciation and related toxic effects is discussed by Wang (1986, 1987), who used duckweed for toxicity tests of Cr, Ba, Cd and Ni in natural water samples.

Reproduction of *L. minor* is usually vegetative. Each mother frond produces two daughter fronds in two separate envelopes. However, flowering also occurs. Duckweed is clone-forming, which means that the starting material can be genetically equal. It can be disinfected and grown in a liquid medium as well as on agar, autotrophically or heterotrophically (Landolt, 1957; Hillman, 1961). From an ecological point of view it is an important species with a global distribution, eaten by several species of wildfowl, and, for instance, muskrats.

The emphasis in our study is on the toxic effects of PFA leachates prepared artificially and by leaching procedures on growth and accumulation of heavy metals. The artificially prepared leachate mixture is used to avoid natural differences in metal concentrations in coal combustion (Srivastava *et al.*, 1986).

Methods

All experiments are performed with a duckweed strain obtained from the Agricultural University of Wageningen. For the aseptic culture a specially designed cabinet was used. In the cabinet the illumination was provided by a cool white fluorescent lamp (16 h light; 8 h dark) with a light intensity of 6000 lux. The temperature was maintained at 22 ± 1.5 °C. High Petri dishes of glass were used for the stock culture and disposable Petri dishes, rinsed with approx. $0.1N$ HNO_3 were used for the experiments. Disinfection of the duckweed was effected by immersing the fronds in 70% ethanol, followed by 1% NaOCl and rinsing with sterile water (Bowker, 1980). The growth medium used (pH 5) was a Gorham medium (Gorham, 1950), modified by Rombach (1976), see Table 1.

The incubation time for all experiments was fourteen days, in which each experiment was inoculated with 4 triplets of *L. minor* on day one.

Table 1. Gorham medium (1950) modified by Rombach (1976).

Element	mg l^{-1}	
$Ca(NO_3)_2 \cdot 4H_2O$	500	
$MgSO_4 \cdot 7H_2O$	250	
KH_2PO_4	136	
H_3BO_3	2.86	
$MnCl_2 \cdot 4H_2O$	1.81	
$ZnSO_4 \cdot 7H_2O$	0.22	
$(NH_4)Mo_7O_{24} \cdot 4H_2O$	0.18	
$CuSO_4 \cdot 5H_2O$	0.07	
$Co(NO_3)_2 \cdot 6H_2O$	0.08	
NH_4VO_3	0.01	
Fe^{3+} EDTA – solution	5	ml
pure water (milli-Q) to make up to 1 litre.		

Fe^{3+} EDTA – solution:
in 1 litre of pure water (Milli-Q)
– 934 mg $FeCl_3 \cdot 6H_2O$
– 800 mg Na_2EDTA (Titriplex II)

The effects on growth were measured by counting the number of fronds and using image processing techniques to determine the percentage of the water surface covered by duckweed as a measure for biomass. EDTA is necessary (metal complexation) for optimal growth. With the image processing technique it was found that 2.5 μM EDTA resulted in a higher percentage covering than the prescribed 10 μM EDTA according to Gorham. However, the number of fronds was equal at either EDTA concentration, indicating that the fronds in 2.5 μM EDTA were larger. In all other experiments 2.5 μM EDTA was used. All experiments were carried out in triplicate. The effects measured were calculated against the controls within each experiment.

The leachates of PFA were prepared according to EPA instructions, i.e. solid/liquid of 1:20; stirring for 24 h.; a static pH 5 (acetic acid); filtering at 45 μm (U.S. EPA, 1980). This 'natural' type of leachate is used against the 'standard' of an artificially composed leachate, in accordance with (cascade technique) the mean metal concentration values in PFA in the year 1987. The destruction techniques of PFA are in accordance with ASTM D3683 (ANSI/ASTM; 1978) and NEN 6465 (NNI; 1981). The duckweed was dried for 24 h at 60 °C and destroyed in teflon-coated pressure bombs with 1 : 1 diluted HNO_3. The results are expressed on a wet weight basis. The metal concentrations in PFA, leachate and duckweed were determined by means of ICP–AES and AAS.

Results

The results of both destruction methods for determining the total metal concentrations in PFA in accordance with NEN & ASTM are listed in Table 2. The differences in micro-element composition clearly show how cautiously a proposed method has to be used. The cation concentrations in the ASTM procedure (Cu, Cr, Ni, Zn) are higher, indicating a better destruction of the matrix. Hence, only part of these cation concentrations can contribute to biological processes or is bio-available to the plant.

The results of the leaching experiments in accordance with the EPA standard procedure are presented in Table 3. The differences between the percentage of leaching of cations (Cr, Cu, Ni) and anions (As, Mo, Se) are striking. For Se, however, actual concentrations are low. This metal analysis table has served as the composition list for the artificial 'standard' leachate.

The effects in percentages of the addition of 'natural' leachate on the growth of L. minor are presented in Fig. 1. Noteworthy is the difference in effect due to the acid used to maintain a static

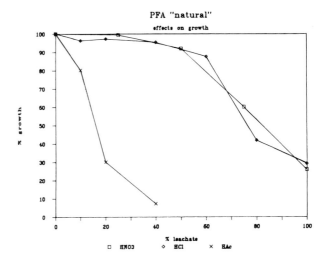

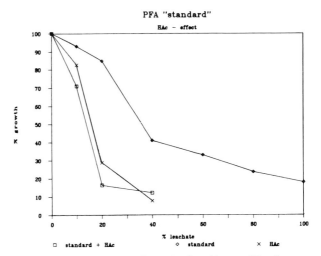

Fig. 1. Effects on growth of 'standard' and 'natural' leachates with and without acetic acid.

Table 2. Comparison between different methods of PFA element analysis for some micro-elements ($\mu g\ g^{-1}$).

Element	Dutch stand. NEN 6465	ANSI–ASTM 3683
As	38	45
Cu	100	173
Cr	60	153
Ni	38	136
Pb	50	52
Sb	8	13
Se	15	15
Zn	80	133

pH of 5 during the prescribed 24 hours stirring. Only the leachate prepared with acetic acid seems to cause a dramatic decrease in growth. This phenomenon was studied in a number of experiments with the 'standard' and 'natural' leachates with and without acetic acid (Fig. 1). The decrease in growth is caused by the toxic action of the acetic acid. Effects on growth actually start at a 60% addition. For the 'standard' leachate a more rapid decline is found with clear effects at as little as 20% addition, showing that some components in the 'standard' are missing, or that the 'natural'

Table 3. Element concentration in PFA and leachate (acetic acid) calculated as a mean value of coal combustion. Situation in 1987.

Element	PFA		Leachate	
	%	$\mu g\, g^{-1}$	$mg\, l^{-1}$	%
Al	15.2		0.33	0.004
Ca	1.3		129	20
Fe	4.8		0.012	0.0005
Mg	0.6		9.7	3.5
P	0.18		2.9	32
As		58	0.46	16
B		166	0.9	11
Cr		150	0.02	0.23
Cu		156	0.11	1.4
Mn		667	0.2	0.6
Mo		25	0.36	29
Ni		142	0.09	1.3
Se		17	0.39	46
Zn		150	0.21	2.8

leachate is less toxic because of some chemical interactions.

The accumulation of macro- and micro-elements by *L. minor* is presented in Table 4.

Anions as well as cations are accumulated, but especially cations. The accumulation of micro-

Table 4. Macro/micro-elements in *Lemna minor* ($\mu g\, g^{-1}$ ww.) grown on medium with 10% leachate (acetic acid). CF: concentration factor.

Element	Control		10% PFA leachate	
	$\mu g\, g^{-1}$	CF	$\mu g\, g^{-1}$	CF
Al	9.8	–	13.5	409
Ca	790	9.7	764	8.6
Fe	16.7	118	6.4	49
Mg	196	8	239	10.4
P	1275	41	1100	39
As	–	–	2.6	57
B	59	116	58	105
Cu	2.5	139	9.4	347
Mn	192	384	248	527
Ni	1.4	–	3.1	344
Se	–	–	0.7	18
Zn	57	1130	67	1022

Table 5. Element concentration, inducing a growth reduction of 50% after 14 days. The CF data are calculated from the EC_{50} concentrations.

Element	$mg\, l^{-1}$ (EC_{50})	CF (EC_{50})
Cu	0.13	300
Cd	0.09	825
Zn	0.33	820
Se		
SeO_2	2.75	13
Na_2SeO_3	1.78	20
Na_2SeO_4	>5	8
Ge	6.4	57
B	8.5	18
Mo	64	9
As		
$NaAsO_2$	0.82	210
Na_2HAsO_4	30	12

elements was studied in other experiments (to be published elsewhere) in which the attention was focussed on the metal speciation of As and Se. In these experiments methods were similar to the PFA leaching study. The EC_{50} values (= concentration resulting in growth reduction of 50%) for the different metals are outlined in Table 5. The concentration factor (CF) is expressed as the ratio of the concentration accumulated in the duckweed and the concentration in the medium after 14 days incubation time. Amongst the elements given in Table 5 Cd, Zn and Cu show a relatively high CF combined with a low EC_{50} value compared with the other elements. The difference in toxicity of As(III) and As(V) was a factor 35.

Discussion

The results in the study of metal accumulation with *Lemna minor* show the ease of applying duckweed as a bioassay. It should be fitted in a larger test scheme complementary to (in-)vertebrate tests. A prerequisite for future applicability of *L. minor* as phytomonitor for metal pollution is the simultaneous occurrence of high accumulation and relatively strong effect-resistance for metals, see Tables 4 and 5. Effects on growth by the 'natural' leachate occur at addition percentages of 60%; however, at additions of 10% elevated metal concentrations can already be detected. The use of a 'standard' leachate addition as a reference toxicant led to effects at 20% percent addition. This subject needs more attention.

A great deal of research has been focussed on the leaching techniques and the prediction of heavy metal concentrations in leachates (Van der Sloot *et al.*, 1982, 1985) with cascade, column and 24-hour stirring tests. Direct use of leachates in bioassays, as demonstrated with PFA leachate prepared in accordance with the EPA standard, has to be used with care. The supposed effects on growth of a combination of heavy metals appear to be attributed to the acetic acid. Acetic acid is a known phytotoxic for barley seedlings (Lynch, 1977). Instead of acetic acid, HNO_3 will be a better acid.

Considering the rate of metal leaching of PFA the anions, and especially the Se-anion, show a high washout of about 50% within 24 hours. This is caused by rather poor bonding, due to condensation of the anions (As, Se, Mo) on the PFA particles, compared with the cations, which can also be seen in the data of the two destruction methods for element analysis. (Table 2).

As opposed to the low leachability, accumulation of cations by duckweed is high, especially for the elements Cu and Zn (Table 4).

The data of EC_{50} values on growth by individual elements show a well-known order.

One should be cautious in interpreting the apparently low Se concentration data, because Se accumulates rapidly in the food chain (Lemly, 1985) and teratogenic effects on higher animals (fish) are found at concentrations lower than 15 μg/l as Se (Gillespie & Bauman, 1986). Considering the differences in effects by selenite and selenate more research on speciation is required. In the case of arsenite and arsenate, the difference in toxicity is far more pronounced, but arsenite is rapidly oxidized to arsenate (if it exists at all), and besides that As is not accumulated in the food chain and therefore poses no threat in ecotoxicology compared with Se.

The results presented in this paper point to the need for a better understanding of that part of metals that will leach from disposal piles at a certain location and subsequently that part that will be bio-available for the biota. Hence, speciation and bio-assay studies are necessary. A simple determination of total metal concentrations in PFA by means of more or less complete destruction methods, as used in legislation so far, overestimates possible toxicological effects and consequently devaluates PFA potentials for recycling purposes.

Acknowledgements

Comments and suggestions of two anonymous referees were quite useful in the revision of the manuscript.

References

ANSI/ASTM 1978 ASTM D 3683–78 In: Annual Book of Standards Vol. 05.05 – Am. Soc. Test. Mat. (Philadelphia), pp. 3.

Bishop, W. E. & R. L. Perry, 1981. Development and evaluation of a flow-through growth inhibition test with duckweed (*Lemna minor*). In: D. R. Branson & K. L. Dickson (Eds.). Aquatic Toxicology and Hazard Assessment: Fourth Conference, ASTM STP 737, pp. 421–435. Am. Soc. Test. Mat., 1981.

Bolt, N. & A. Snel, 1986. Environmental aspects of fly-ash application in The Netherlands. Kema Scientific & Technical Reports 4: 125–140.

Bowker, D. W., A. N. Duffield & P. Denny, 1980. Methods for the isolation, sterilization and cultivation of *Lemnaceae*. Freshwat. Biol. 10: 385–388.

Cherry, D. S. & R. K. Guthrie, 1977. Toxic metals in surface waters from coal ash. Wat. Resour. Bull. 13: 1227–1236.

Cherry, D. S., R. K. Guthrie, E. M. Davis, & R. S. Harvey, 1984. Coal ash basin effects (particulates, metals, acidic pH) upon aquatic biota: an eight-year evaluation. Wat. Resour. Bull. 20: 535–544.

Gillespie, R. B. & P. C. Bauman, 1986. Effects of high tissue concentrations of selenium on reproduction by bluegills. Trans. Am. Fish. Soc. 115: 208–213.

Gorham, P. R., 1950. Heterothrophic nutrition of seed plants with particular reference to *Lemna minor* L. Can. J. Res., Sec.C 28: 356–381.

Guthrie, R. K., 1979. The uptake of chemical elements from coal ash and settling basin effluent by primary producers. Sci. Total Envir. 12: 217–222.

L. I. H. E.
THE MARKLAND LIBRARY
STAND PARK RD., LIVERPOOL, L16 9JD

366

Hillman, W. S., 1961. The *Lemnaceae*, or duckweeds. A review of the descriptive and experimental literature. Bot. Rev. 27: 221–287.

Landolt, E., 1957. Physiologische und Okologische Untersuchungen an *Lemnaceen*. Berichte Schweiz. Bot. Ges. 67: 271–411.

Lemly, A. D., 1985. Toxicology of selenium in a freshwater reservoir: implications for environmental hazard evaluation and safety. Ecotox. Envir. Safety 10: 314–338.

Lynch, J. M., 1977. Phytotoxicity of acetic acid produced in the anaerobic decomposition of wheat straw. J. Appl. Bact. 42: 81–87.

Meij, R., L. H. J. M. Janssen & J. Van der Kooij, 1986. Air pollutant emissions from coal-fired power stations. Kema Scientific & Technical Reports 4: 51–69.

Nasu, Y. & M. Kugimoto, 1981. *Lemna* (duckweed) as an indicator of water pollution. I. The sensitivity of *Lemna paucicostata* to heavy metals. Arch. Envir. Contam. Toxicol. 10: 159–169.

NNI (Ned. Norm. Inst.), 1981 NEN 6465. Sample preparation of sludge, water containing sludge and air dust for the determination of elements by atomic absorption spectrometry. Destruction with nitric acid and hydrochloric acid-NNI (Delft, The Netherlands), pp. 4.

Nriagu, J. O., 1988. A silent epidemic of environmental metal poisoning? Envir. Pollut. 50: 139–161.

Ray, S. & W. White, 1976. Selected aquatic plants as indicator species for heavy metal pollution. J. Envir. Sci. Health A11: 717–725.

Rombach, J., 1976. Effects of light and phytochrome in heterotrophic growth of *Lemna minor* L. Comm. Agricultural University Wageningen The Netherl. 76–1: pp. 114.

U.S. Environmental Protection Agency, 1980. Fed. Regist. 45: 33127–33128.

Van der Sloot, H. A., J. Wijkstra, C. A. Van Stigt & D. Hoede, 1985. Leaching of trace elements from coal ash and coal ash products. In: Duedall, I. W., D. R. Kester & P. K. Park (eds.): Wastes in the ocean 4: Energy wastes in the ocean. pp. 468–495. John Wiley and Sons Inc. N.Y.

Van der Sloot, H. A., J. Wijkstra, A. Van Dalen, H. A. Das Slanina, J., Dekkers, J. J., Wals, G. D., 1982. Leaching of trace elements from coal solid waste. Neth. Energy Res. Found. ECN Report 82–120. Petten, The Netherlands.

Srivastava, V. K., P. K. Srivastava, R. Kumar & U. K. Misra, 1986. Seasonal variations of metals in coal fly ash. Envir. Pollut. Ser. B. 11: 83–89.

Wang, W., 1984. Uses of aquatic plants in ecotoxicology. Environment Internat. 10: 3 pp.

Wang, W., 1986. The effect of river water on phytotoxicity of Ba, Cd and Cr. Environ. Pollut. Ser.B. 11: 193–204.

Wang, W., 1987. Toxicity of nickel to common duckweed (*Lemna minor*). Envir. Toxicol. Chem. 6: 961–967.

Hydrobiologia **188/189**: 367–375, 1989.
M. Munawar, G. Dixon, C. I. Mayfield, T. Reynoldson and M. H. Sadar (eds)
Environmental Bioassay Techniques and their Application.
© 1989 *Kluwer Academic Publishers. Printed in Belgium.*

Sensitive bioassays for determining residues of sulfonylurea herbicides in soil and their availability to crop plants

A. Rahman
Plant Science Group, Ruakura Agricultural Centre, Ministry of Agriculture and Fisheries, Hamilton, New Zealand

Key words: herbicides, bioassays, phytotoxicity, persistence, herbicide residues, biological activity, soil residual activity, sulfonylurea herbicides

Abstract

Sulfonylurea herbicides are potent inhibitors of plant growth and are extremely active against a wide spectrum of weeds. They are used at very low rates (10–50 g ai/ha) and cause rapid inhibition of root and shoot growth of young plants. Routine chemical assays for detecting low levels of these compounds are difficult and there is need to develop sensitive bioassay methods for detecting their extremely low residue levels in the soil.

This paper describes a simple pot bioassay method with a self watering system using turnip (*Brassica rapa*) seedlings as test plants for quantitative determination of sulfonylurea herbicides. Results are presented with six of these compounds whose activity was investigated in widely differing substrates. The potential availability to plants was calculated from the dose-response curves in different substrates. The dose-response relationship has been described by a specifically developed computer model. Details are also given of a direct seeded bioassay method with controlled watering system using several test species for detection of sulfonylurea herbicides. The potential uses and practical applications of both techniques are discussed.

Introduction

The sulfonylurea herbicides have recently emerged as a major new class of herbicide and an important advance in chemical weed control technology. They are highly active against a wide spectrum of broadleaf and grass weeds. While seed germination is not usually affected, subsequent root and shoot growth are rapidly and severely inhibited in sensitive seedlings. With their unprecedented herbicidal activity, field use rates have dramatically fallen resulting in application rates of grams rather than kilograms per hectare.

The need for such broad spectrum, low dosage compounds with greater crop selectivity is an important factor contributing to the rapid success of this group of chemicals. World-wide, eight sulfonylurea herbicides have been commercialized by 1987 and by the mid 1990's it is anticipated that this number could more than double.

So far the sulfonylurea herbicides have been developed for use on wheat and other small grain cereals, rice, soybeans, and for non-selective industrial weed control. Their selectivity is based on the ability of crop plants to metabolize the herbicides to non phytotoxic products (Sweetser et al., 1982). Genetic and biochemical studies on their mode of action have demonstrated that they act in plants by blocking the production of some essential amino acids through inhibition of the enzyme acetolactate synthase (Ray, 1985).

Although some members of this group have a very short half life in soil, eg. DPX-L5300 and thiameturon-methyl (Brown et al., 1987; Rahman et al., 1988; Sionis et al., 1985), others such as chlorsulfuron and triasulfuron are known to persist much longer in the soil (eg. James et al., 1988; Peterson & Arnold, 1985). Chemical hydrolysis and microbial breakdown are the principal modes of their degradation which is highly dependent on soil pH, soil texture, moisture, and temperature, and can vary widely between different soils (Beyer et al., 1987). The time for which these herbicides persist in the soil is of particular importance as this has already been shown to have serious implications for the safety of following crops (Duffy et al., 1987; Nicholls et al., 1987).

To predict the impact of residual levels of compounds on a rotational crop, the challenge is to develop a practical assay method which is sensitive enough to detect the extremely low residues in the soil and then to correlate this with wide range of sensitivities of the crops in the field. Although detection limits of 0.2 μg/kg have been demonstrated using chemical assays, conducting routine analyses at these levels is extremely difficult (Duffy et al., 1987). Several laboratory and greenhouse plant bioassays have been used to study the behavior of sulfonylurea herbicides, particularly of chlorsulfuron (Beyer et al., 1988). Some of these are not sensitive enough for compounds like thiameturon-methyl which have low activity through soil and undergo rapid breakdown, some are very time consuming and/or laborious, while others need very precise operating conditions.

This paper provides information on two bioassay procedures for measuring the biological activity of sulfonylurea herbicides. The first method is a simple pot bioassay method with a self watering system using seedling transplants. The second method uses controlled subsurface watering with seeds planted directly into the treated soil in pots. In addition to details of the two bioassay procedures, the paper gives results on studies with a number of commercially available sulfonylurea materials in widely differing substrates.

Materials and methods

Seedling transplant bioassay method

A simple pot bioassay method with a self-watering system as described by Stalder & Pestemer (1980) has been used successfully for determining the availability of several photosynthetic inhibitor type herbicides in soils. With some modifications, this method has been further developed for measuring the biological activity and plant availability of sulfonylurea herbicides.

Turnip (Brassica rapa ssp. rapa) was found the most suitable test species for all sulfonylurea herbicides used in this study. Seeds were sown uniformly in vermiculite in seed pans. After 7-8 days' growth, seven seedlings were transplanted into plastic pots (7 cm) filled with soil or vermiculite treated the day before with a known concentration of the herbicide (by applying the chemical with a pipette and then thoroughly mixing throughout). The pots were attached to Petri dishes to produce a self-watering arrangement through a glass-fibre wick. The dishes were kept filled with nutrient solution and the soil water content during the bioassay period was 100% of water holding capacity or greater. All treatments were replicated four times and the pots were kept

Table 1. Sulfonylurea herbicides and their concentrations used in the seedling transplant bioassay method.

Common name	Trade name (% a.i.)		Concentrations (μg a.i./kg soil)
Sulfometuron	Oust	(75%)	0.010– 40
Chlorsulfuron	Glean	(75%)	0.005– 15
Metsulfuron methyl	Escort	(60%)	0.005– 40
Triasulfuron	Logran	(20%)	0.005– 30
Thiameturon methyl	Harmony	(75%)	0.100–100
DPX-L5300	Granstar	(75%)	0.100–100

Table 2. Some physical and chemical characteristics of the soils used.

Soil characteristics	BBA loamy sand	Wendhausen clay loam	Horotiu sandy loam	Hamilton clay loam
Organic C (%)	1.27	1.85	9.26	4.15
Sand (%)	54.3	25.8	58.2	29.1
Silt (%)	34.7	31.2	19.7	30.6
Clay (%)	10.5	42.1	15.6	37.2
pH	6.7	7.6	5.6	5.3
Max water holding capacity (ml/100 g)	20.7	63.7	79.0	–*
Field capacity (%)	–	–	42.3	36.8
Bulk density (kg/l)	1.34	1.40	0.70	0.91

* not determined

in a greenhouse with day temperatures of 22–26 °C and night temperatures down to 16 °C. Artificial light was used to provide a photoperiod of 16 hours. Each experiment was conducted 3 or 4 times, with the repeat experiments being conducted in different seasons.

The sulfonylurea compounds and their concentrations used in these experiments are listed in Table 1. A range of concentrations was chosen for each compound to encompass the whole dose-response relationship for each substrate. A non sorptive substrate 'vermiculite' and three widely differing soils (viz. BBA, Wendhausen and Horotiu) were used for all experiments. Some physical and chemical properties of these soils are given in Table 2. To achieve the same volume of each soil within the pot, 70 g of oven-dry soil was weighed in for BBA and Wendhausen soils while only 50 g of oven-dry soil was used for the Horotiu soil because of its low bulk density. For estimating the availability of herbicides to plants, vermiculite (mesh size 2-4 mm) was used as a standard material.

Evaluations of the results were made by adoption of logistic S-curves on the plant fresh weight data. For this, a computer programme was developed in which the dose-response curve is described by four parameters (θ) as shown in Fig. 1.

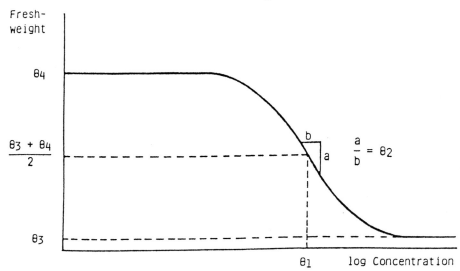

Fig. 1. Description of the dose-response curve calculated by the 'logistic' computer model.

These parameters are:

$\theta 1$: Position parameter – the dose which is half way between the maximum weight ($\theta 4$) and the minimum weight ($\theta 3$).

$\theta 2$: Slope – slope within the linear part of the curve in the range of ED_{50}.

$\theta 3$: Minimum weight – the minimum fresh weight of transplanted seedlings, which cannot be reduced further by higher rates of the herbicide.

$\theta 4$: Maximum weight – the actual weight of plants in untreated soil or a higher value estimated by the model where stimulation occurs due to the herbicide.

All the θ values are estimated by the computer model itself. However, if necessary, a value can be inserted within the programme, eg. the minimum weight, if it is not reached by the concentrations used within the experiment.

In addition to the logistic evaluations, the probit analysis method (Nyffeler *et al.*, 1982; Pestemer, 1983) was used to compare the applicability of two statistical methods for measuring the availability to plants in the range of ED_{50} and the dose-response relationships. The availability to plants in this case was calculated by comparison of ED_{50} values in vermiculite and in a given soil using the logistic model. As vermiculite is a non sorptive substrate, differences in the activity of a herbicide between vermiculite and a soil could be attributed to adsorption of the herbicide by the soil.

Direct seeded bioassay method

Although similar in approach to the seedling transplant method, this method differs in several respects. Firstly the seeds of bioassay species are planted directly into the treated soil as opposed to the seedling transplants. Secondly the pots are sub-irrigated and the soil moisture is maintained between 80% and 100% of field capacity which is much lower than the maximum water holding capacity. Thirdly the herbicide is sprayed through a greenhouse sprayer on an area basis and concentration is calculated on g/ha basis which allows a direct comparison for bioassays of soil samples collected from the treated fields. The method can however be used also for mixing concentration on a weight basis (e.g. mg or μg/kg) as described by Rahman (1977).

Two sulfonylurea compounds viz., DPX-L5300 and triasulfuron were used in the experiments reported here and their activity was evaluated in two soils viz. Horotiu sandy loam and Hamilton clay loam, the properties of which are listed in Table 2. A range of concentrations between 0.5 and 500 g a.i./ha (normal field use rates vary between 10 and 40 g a.i./ha) was used, with each concentration individually made up in 1 litre volumetric flasks and applied with a moving belt greenhouse sprayer delivering 300 litres/ha at 200 kPa to the soil which was in 5-cm deep trays. After a short time to allow the treated soil to 'dry', the soil was thoroughly mixed to incorporate the herbicide and then used to plant the test species. Four pots of each species were used for each rate of the herbicide.

The four test species used for this bioassay method were mustard (*Brassica nigra*), lentil (*Lens culinaris*), subterranean clover (*Trifolium subterraneum*) and annual ryegrass (*Lolium multiflorum*). They were chosen because of their ease of growing, range of sensitivity or the species likely to be planted in rotation in the New Zealand farming system. Twenty seeds of each test species were planted at 1 cm depth in a 12-cm diameter plastic pot. All pots were sub-irrigated to maintain the soil between 80% and 100% field capacity. The plants were grown in the greenhouse for 4–5 weeks before assessment and harvesting for dry matter weight determinations. The dose-response relationship was evaluated by subjecting the dry matter shoot weight data to the probit analysis method.

Results and discussion

Seedling transplant bioassay method

All the sulfonylurea herbicides tested caused very quick growth inhibition, with visual symptoms

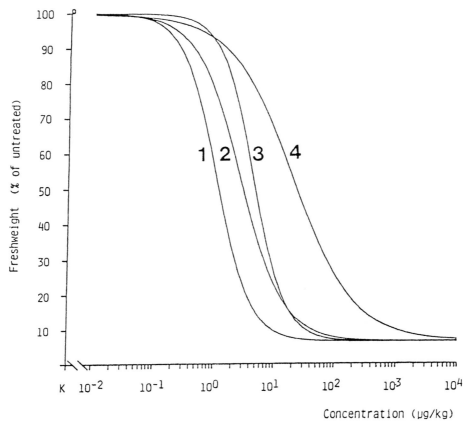

Fig. 2. Dose-response relationship of DPX-L5300 from the 'logistic' model in vermiculite and three soils. 1 = Vermiculite; 2 = Wendhausen-soil; 3 = BBA-soil; 4 = Horotiu-soil

often occurring within 1 to 2 days. In many instances the visual effects were obvious in plants but no reduction could be recorded in fresh weight at the end of the experiment. The effects were first noticed as a reduction in the number of lobes on the leaves. This effect was in inverse proportion to the herbicide concentration and continued until the leaves were reduced to small stunted knobs around the growing point. Considerably enlarged size of cotyledons was also noted in many

Table 3. ED_{50} values – μg a.i./kg – (with standard deviations) of six sulfonylurea herbicides in different substrates.

Herbicide	Substrates (soils)			
	Vermiculite	BBA	Wendhausen	Horotiu
Sulfometuron	0.12 ± 0.03	0.19 ± 0.04	0.17 ± 0.04	0.47 ± 0.16
Chlorsulfuron	0.20 ± 0.04	0.30 ± 0.10	0.58 ± 0.09	0.61 ± 0.11
Metsulfuron methyl	0.35 ± 0.15	1.05 ± 0.01	0.43 ± 0.12	2.27 ± 0.53
Triasulfuron	0.51 ± 0.17	1.64 ± 0.17	0.57 ± 0.14	3.65 ± 2.00
Thiameturon methyl	0.82 ± 0.18	1.90 ± 0.43	6.52 ± 1.00	6.98 ± 1.39
DPX-L5300	1.98 ± 1.12	4.49 ± 0.84	2.97 ± 0.76	21.69 ± 5.35

372

Table 4. Calculated plant availability of six sulfonylurea herbicides in three soils (taking availability in vermiculite as 100%).

Herbicide	% availability		
	BBA	Wendhausen	Horotiu
Sulfometuron	63.2	70.6	35.7
Chlorsulfuron	66.7	34.5	45.9
Metsulfuron methyl	33.0	81.4	22.2
Triasulfuron	31.0	90.0	14.0
Thiameturon methyl	43.2	15.8	16.5
DPX-L5300	44.1	66.7	9.1

instances. At very high rates the seedlings did not grow past the cotyledon stage, which sometimes turned yellow or reddish and decreased in size.

Using the DPX-L5300 data as an example, the typical dose-response curves for vermiculite and the three soils are shown in Fig. 2. By using the computer 'logistic' model it is possible to calculate any ED-value along the curves, including the 95% confidence limits. The most reliable quantitative detection of the compounds with this method is provided between ED_{30} and ED_{70} values, because outside this range the deviation increases markedly. In vermiculite the ED_{30} was 0.06 $\mu g/kg$ for sulfometuron, 0.10 for chlorsulfuron, 0.16 for metsulfuron methyl, 0.28 for triasulfuron, 0.46 for thiameturon methyl and 1.03 $\mu g/kg$ for DPX-L5300. However, the dose-response relationship is best described by all evaluation methods (graphically, probit or logistic methods) in the range of the ED_{50} value (Fig. 3). A comparison of ED_{50}

values for different herbicides and soils is therefore a suitable means for determining the relative activity.

The ED_{50} values calculated by the 'logistic' model from curves similar to those of Fig. 2 are summarized in Table 3. These data show that with turnips as test species, sulfometuron was the most active of the sulfonylurea compounds tested, followed by chlorsulfuron, in vermiculite as well as in the three soils. At the other end of the scale, thiameturon methyl and DPX-L5300 had much higher ED_{50} values in all substrates, exhibiting their lower activity. Both these herbicides are known to have very low activity through the soil (Beyer *et al.*, 1988; Ferguson *et al.*, 1985, Rahman *et al.*, 1988). However, their ED_{50} values are also high in vermiculite, suggesting that their concentration in the soil solution could decrease rapidly during the period of the experiment because of their short persistence and fast degradation.

The ED_{50} values for all compounds were considerably higher in the Horotiu sandy loam soil compared to the other two soils. This may be explained in part by the high organic carbon content of the Horotiu soil (Table 2) which has been shown to reduce the phytotoxicity of the sulfonylurea compounds (Beyer *et al.*, 1988; Mersie & Foy, 1985). The lower activity in the Horotiu soil is also evident from the calculated plant availability data presented in Table 4 which indicate the different adsorption behavior of the soils used. The percentage available part of the total amount was the least for all compounds in the Horotiu soil.

Table 5. ED_{50} values (g a.i./ha) for four bioassay species in two soils using direct seeded bioassay method.

Bioassay species	DPX-L5300*		Triasulfuron	
	H.S.L.**	H.C.L.**	H.S.L.	H.C.L.
Mustard	20	320	1.4 ± 0.23	2.2 ± 0.37
Lentil	60	380	2.8 ± 0.63	3.8 ± 0.77
Sub clover	100	440	2.9 ± 0.85	3.4 ± 0.94
Annual ryegrass	120	>500	2.9 ± 0.71	3.6 ± 0.82

* ED_{50} values for DPX-L5300 are approximate only.
** H.S.L. – Horotiu sandy loam; H.C.L. – Hamilton clay loam.

The pH value of the soil has been shown to have a large influence on the degradation of sulfonylurea herbicides (Beyer et al., 1987; 1988). However, during the short experimental period of 10 days this effect of pH may not be apparent for persistent compounds like sulfometuron. But for compounds like DPX-L5300 which undergo a rapid chemical hydrolysis, the effect of low pH in the Horotiu soil could be significant even in this short time. This may be the reason for the extremely low plant availability figure of 9.1% for DPX-L5300 in the Horotiu soil (Table 4).

Direct seeded bioassay method

The most common effect of both sulfonylurea herbicides on all the test species was stunting.

However, close inspection revealed that the overall effect of stunting was due to different responses in different species. For instance, with lentils a reduction occurred in growth of the main stem resulting in branching, giving stunted and distorted plants. With mustard the effects were first noticed as a reduction in the number of lobes on the leaves and continued until the leaves were reduced to stunted knobs. For both subterranean clover and annual ryegrass, the effect was a reduction in leaf size; the leaves appeared normal in other respects.

The effects of known concentrations of DPX-L5300 and triasulfuron on the four bioassay species are presented in the form of ED_{50} values (i.e. the concentration required for a 50% reduction in the growth of the species compared with the untreated control) in Table 5. These figures

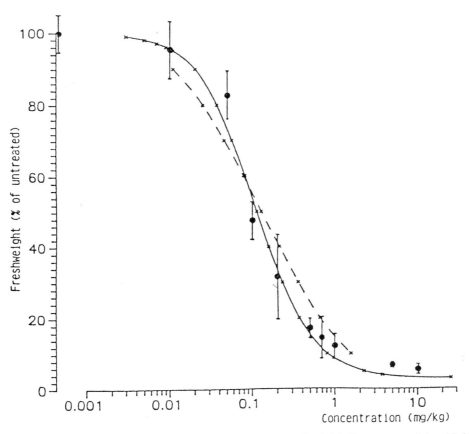

Fig. 3. Comparison of two mathematical methods (– – – = probit; and ——— = 'logistic'; ● = average data with 95% confidence limits) for calculating the dose-response relationship of sulfometuron in vermiculite.

are based on the averages of several series of 'standards' planted at different times of the year. The ED_{50} values from individual experiments showed small seasonal variations.

Data presented in Table 5 show that mustard was more susceptible to both herbicides than the other bioassay species. The level of biological activity of both compounds was much lower in the Hamilton clay loam soil than in the Horotiu sandy loam soil. The magnitude of this difference varied with the bioassay species, but the trend was similar. The visual data on percent growth reduction were supported well by the actual dry matter yields from various treatments.

As was the case with the seedling transplant bioassay technique, the influence of soil characteristics on the biological activity of sulfonylurea herbicides is demonstrated clearly also with this technique. These experiments have been conducted with many other New Zealand soils and the results have confirmed the strong influence of soil type on the ED_{50} value of these herbicides.

Comparison of the two bioassay techniques

It is evident from the results presented here that the two bioassay techniques have different sensitivity levels and probably have applicability to different situations. The seedling transplant technique is more sensitive and is able to detect lower quantities of certain sulfonylurea herbicides. Using seedlings with well developed roots, the plants come quickly into contact with the herbicide. Also because of the self watering system, the soil moisture is well over the field capacity level allowing more herbicide to come into the soil solution and thus be more available to bioassay plants. Because of the short growing time in the greenhouse for the seedling transplant technique the affected seedlings are not able to recover from injury in cases where the activity of a herbicide is diminishing quickly.

The seedling transplant technique is labour intensive, but requires a very short experimental period (10 to 11 days). If seedlings are selected carefully to maintain uniformity, reliable results could be achieved quickly with this method, allowing for better use of time and facilities. The direct seeded technique is however more useful for bioassay of field samples because it allows direct comparison of herbicide rates on an area basis. It is particularly relevant where large number of field samples are to be bioassayed, and for compounds with longer soil persistence e.g. chlorsulfuron and triasulfuron. Due to the longer growing time required for this technique there could be problems in assessing the residual activity of compounds like DPX-L5300 and thiameturon methyl which have short persistence in soil and their concentration could be decreasing rapidly (as a result of their fast degradation) during the experimental period.

The visual symptoms which appear at an early stage can often be very helpful in both techniques for noting the presence of very small quantities of herbicides which do not lead to ultimate reduction in plant dry weight.

Conclusions and practical applications

The quantitative estimation of small amounts of herbicide residues which have biological activity through the soil (but which cannot be readily measured by instrumental methods) is of practical use for selecting succeeding crops in the rotation, or for determining potential ecotoxicological side effects. For this purpose the evaluation of ED_{50} values is not as important as the range where no effect can be measured. This so called 'no observable effect level' (NOEL) cannot be easily calculated from the data through statistical methods because this value can vary widely due to dependence on factors such as replications. In most cases it is more appropriate to define a level of injury which is acceptable in the field, eg. the no response category (up to 10% damage) as defined by Pestemer *et al.* (1980).

Both bioassay techniques described here can be very useful for estimating herbicide disappearance rates quantitatively under field conditions. They are especially relevant for herbicides such as those belonging to the sulfonylurea group which are used at extremely low rates in the field and for which the resulting minute residues cannot be measured through chemical assay techniques. These techniques can also be used to give prognoses of the residue behavior and their effects on succeeding crops such as those already reported by James *et al.* (1988) and Rahman *et al.* (1988). As the use of sulfonylurea herbicides becomes more widespread, these techniques would prove very useful in measuring their persistence and their biologically active residues remaining in the soil, in formulating safe crop rotations, and in evaluating their potential ecotoxicological side effects.

Acknowledgements

A part of this research was conducted at the Federal Biological Research Centre for Agriculture and Forestry in Braunschweig, Federal Republic of Germany. Sincere appreciation is expressed to Prof. Dr. W. Pestemer and Ms P. Guenther for their help and support. Thanks are also due to T. K. James and J. Mortimer for their technical assistance with research conducted in New Zealand.

References

Beyer, E. M., H. M. Brown & M. J. Duffy, 1987. Sulfonylurea herbicide – soil relations. Proc. 1987 Br. Crop Prot. Conf. – Weeds 531–540.

Beyer, E. M., M. J. Duffy, J. V. Hay & D. D. Schlueter, 1988. Sulfonylurea herbicides. In: P. C. Kearney & D. D. Kaufman (eds.), Herbicides: Chemistry, Degradation and Mode of Action. Marcel Dekker Inc., New York. 3: 117–189.

Brown, H. M., M. M. Joshi & A. Van, 1987. Rapid soil microbial degradation of DPX-M6316. Weed Sci. Soc. Am. Abstr. 27: 75.

Duffy, M. J., M. K. Hanafey, D. M. Linn, M. H. Russell & C. J. Peter, 1987. Predicting sulfonylurea herbicide behaviour under field conditions. Proc. 1987 Br. Crop Prot. Conf. – Weeds 541–547.

Ferguson, D. T., S. E. Schehl, L. H. Hagemann, G. E. Lepone & G. A. Carraro, 1985. DPX-L5300 – A new cereal herbicide. Proc. 1985 Br. Crop Prot. Conf. – Weeds 43–48.

James, T. K., A. Rahman & T. M. Patterson, 1988. Residual activity of triasulfuron in several New Zealand soils. Proc. N.Z. Weed and Pest Cont. Conf. 41: 21–25.

Mersie, W. & C. L. Foy, 1985. Phytotoxicity and adsorption of chlorsulfuron as affected by soil properties. Weed Sci. 33: 564–568.

Nicholls, P. H., A. A. Evans & A. Walker, 1987. The behaviour of chlorsulfuron and metsulfuron in soils in relation to incidents of injury to sugar beet. Proc. 1987 Br. Crop Prot. Conf. – Weeds 549–556.

Nyffeler, A., H. R. Gerber, K. Hurle, W. Pestemer & R. R. Schmidt, 1982. Collaborative studies of dose-response curves obtained with different bioassay methods for soil-applied herbicides. Weed Res. 22: 213–222.

Pestemer, W., 1983. Methodenvergleich zur Bestimmung der Pflanzenverfugbarkeit von Bodenherbiziden. Berichte des Fachgebietes Herbologie (Hohenheim). 24: 85–96.

Pestemer, W., L. Stalder & B. Eckert, 1980. Availability to plants of herbicide residues in soil. Part 2. Data for use in vegetable crops. Weed Res. 20: 349–353.

Peterson, M. A. & W. E. Arnold, 1985. Response of rotational crops to soil residues of chlorsulfuron. Weed Sci. 34: 131–136.

Rahman, A., 1977. Persistence of terbacil and trifluralin under different soil and climatic conditions. Weed Res. 17: 145–152.

Rahman, A., T. K. James & J. Mortimer, 1988. Persistence and mobility of the sulfonylurea herbicide DPX-L5300 in a New Zealand volcanic soil. Mededelingen van Rijksfaculteit Landbouwwetenschappen Gent. 53: 1463–1469.

Ray, T. B., 1985. The site of action of the sulfonylurea herbicides. Proc. 1985 Br. Crop Prot. Conf. – Weeds 131–138.

Sionis, S. D., H. G. Drobny, P. Lefebvre & M. E. Upstone, 1985. DPX-M6316 – A new sulfonylurea cereal herbicide. Proc. 1985 Br. Crop Prot. Conf. – Weeds 49–54.

Stalder, L. & W. Pestemer, 1980. Availability of herbicide residues in soil. Part 1. A rapid method for estimating potentially available residues of herbicides. Weed Res. 20: 341–347.

Sweetser, P. B., G. S. Schow & J. M. Hutchison, 1982. Metabolism of chlorsulfuron by plants: Biological basis for selectivity of a new herbicide for cereals. Pestic. Biochem. Physiol. 17: 18–23.

Hydrobiologia **188/189**: 377–383, 1989.
M. Munawar, G. Dixon, C. I. Mayfield, T. Reynoldson and M. H. Sadar (eds)
Environmental Bioassay Techniques and their Application.
© *1989 Kluwer Academic Publishers. Printed in Belgium.*

Root and shoot elongation as an assessment of heavy metal toxicity and 'Zn Equivalent Value' of edible crops

Y.H. Cheung[1,3], M.H. Wong[1] & N.F.Y. Tam[2]
[1] *Department of Biology, Hong Kong Baptist College, 224 Waterloo Road, Kowloon, Hong Kong (* author for correspondence)*; [2] *Department of Applied Biology and Chemical Technology, Hong Kong Polytechnic, Hung Hom, Kowloon*; [3] *Environmental Protection Department, 11/F., Empire Centre, Tsim Sha Tsui East, Kowloon, Hong Kong*

Key words: seed development, edible crops, root growth inhibition, heavy metal toxicity, Zinc Equivalent Value

Abstract

Seeds of thirteen edible plant species were tested for their response to heavy metals during their early development. It was found that a short-term root elongation test of six days could be used to evaluate the degree of toxicity of aqueous samples containing heavy metals. Shoot elongation was found to be less sensitive to metals than root elongation.

The seeds were sown in pots containing freshwater sand to which known concentrations of metal solutions were added. The relative toxicity of the three metals, copper, nickel and zinc, followed the pattern of $Ni > Cu > Zn$.

Results on the relative toxicity of Zn : Cu : Ni to various plant species indicated that the ratios were species-specific. The Zn equivalent concept of Zn : Cu : Ni = 1 : 2 : 8 could not be applied to all the plant species tested.

The root growth of seeds of *Brassica parachinensis* (flowering Chinese cabbage) placed on filter papers in petri dishes to which metal solutions were added were tested. The sensitivity ranking of the metals tested was found to be as follows: $Ni > Cd > Cu > Al > Fe > Zn > Pb > Mn > Ag$. There was no significant difference ($p > 0.05$) in percentage reduction in root elongation among the four different repeated trials.

Introduction

Analytical chemical methods are usually used to detect the existence of toxic metals and their concentrations in sites of suspected pollution. However, precise chemical measurements are unlikely to be wholly reliable guides to actual toxicity under field conditions (Martin, 1982). On the contrary, living organisms in ecosystems react 'ecotoxicologically' to contamination levels. Chemical dosage can be translated into a quantifiable biological response, i.e. a bioassay, which in turn may be a convenient substitute for chemical measurement (Goodman *et al.*, 1974).

Root elongation has commonly been used in measurement of metal tolerance in plants, e.g. lead tolerance (Wilkins, 1957). Since then, the root elongation test has been extensively used as a bioassay method in detecting phytotoxic effects of various substances, e.g. comparative toxicity of several heavy metals on *Lolium pyrenne* (Wong & Bradshaw, 1982), roadside dust on *Brassica chinensis*, *B. parachinensis* and *Daucus carota* (Wong & Lau, 1983), refuse compost on *B. parachinensis* (Wong, 1985) and effluent from a molasses distillery on *Oryza sativa* (Behera & Misra, 1982).

In the era of rapid expansion of sludge utilization on land, there was concern that uncontrolled application of sewage sludge could cause toxicity of the soil due to the presence of harmful quantities of heavy metals, especially copper, nickel and zinc. The 'Zinc Equivalent Value' developed then was based on results from previous pot experiments which showed that copper and nickel were twice and eight times respectively as toxic to young barley seedlings as zinc. The value was used to indicate the combined toxicity of the three metals contained in sewage sludge (Chumbley, 1971).

The objectives of the present study include the following:
1. To test the effectiveness of shoot and root elongation of 13 edible crops in response to toxicity of Cu, Ni and Zn.
2. To verify the validity of 'Zinc Equivalent Value' (Zn : Cu : Ni = 1 : 2 : 8) by testing the relative responses of eight edible crops.
3. To study the comparative toxicity of nine heavy metals (Ag, Al, Cd, Cu, Fe, Mn, Ni, Pb and Zn) on root growth of *Brassica parachinensis* (flowering Chinese cabbage).
4. To study the reproducibility of the root elongation tests among four repeated trials.

Materials and methods

1. Root elongation of 13 edible crops in response to Cu, Ni and Zn

Seeds of 13 edible crops under six categories: leafy plant, root plant, cereal, legume, fruit plant and grass were used as testing materials (Table 1). Plastic containers (diameter: 9 cm, height: 13 cm) containing freshwater sand, with a pH of 6.5 and undetectable levels of salinity and Cu, Ni and Zn, were used as the growth medium. Seeds were first pregerminated in petri dishes lined with moisten-

Table 1. The 13 edible crops used as testing materials

Categories	Test plants
Leafy plant	*Brassica parachinensis* (flowering Chinese cabbage)
	Brassica chinensis (Chinese white cabbage)
	Spinancia oleracea (spinach)
	Lactuca sativa (lettuce)
Root plant	*Raphanus sativus* (Chinese radish)
	Daucus carota (carrot)
Cereal	*Hordeum vulgare* (barley)
	Oryza sativa (rice)
	Zea mays (maize)
Legume	*Phaseolus mungo* (mungo bean)
	Glycine max (soy bean)
Fruit plant	*Lycopersecum esculentum* (tomato)
Grass	*Lolium pyrenne* (ryegrass)

ed filter papers at 25 °C for 24 h in the dark. Only those with a single white root tip of 2 mm length were chosen for the experiment. Six pregerminated seeds of each edible crop being tested were transplanted into a container, about 1 cm beneath the sand surface. At the beginning of the experiment, solutions of Cu, Ni and Zn sulphates (30, 20 and 75 mg l^{-1}, respectively) with pH adjusted to 7.0 $\pm$ 0.5, were added to each container while the control group received distilled water only. There were triplicate samples for each treatment.

All the containers were placed in random order in a growth chamber with a 16-h light/8-h dark cycle, temperature of 23 $\pm$ 2 °C and relative humidity of 60-65% for a period of 12 days. The root and shoot lengths of the seedlings were measured every day during the growth period.

2. 'Zinc Equivalent Values' of 8 edible crops

In order to study the relative effects of Ni, Cu and Zn on the early development of edible crops, eight out of the 13 species were selected for further study according to the following criteria: fast growth rate, high sensitivity, regular growth pattern, short germination period, economic importance and availability of seed sources (OECD,

1981). *Brassica chinensis, Brassica parachinensis* (leafy plant), *Raphanus sativa, Daucus carota* (root plant), *Hordeum vulgare, Oryza sativa* (cereal), *Phaseolus mungo* and *Glycine max* (legume) were chosen.

Seeds of the eight selected species were subjected to a series concentrations of Ni, 2 to 20 mg l^{-1} for leafy plants and 3 to 30 mg l^{-1} for the other 3 groups; Cu, 3 to 30 mg l^{-1} for leafy and root plants and 4 to 40 mg l^{-1} for cereals and legumes; and Zn, 6 to 60 mg l^{-1} for all the four groups. Distilled water alone was added to the control groups. The experimental conditions were the same as for the first experiment. The 144-h EC50 values (median effective concentration reducing root length by 50% of control at Day 6) were calculated. The relative toxicity of Zn:Cu:Ni was computed by comparing their EC50 values according to the following formula: Zn:Cu:Ni = EC50 of Zn/EC50 of Zn : EC50 of Zn/EC50 of Cu : EC50 of Zn/EC50 of Ni.

3. Comparative toxicity of 9 heavy metals on root growth of B. parachinensis

Seeds of *B. parachinensis* were subjected to a series of concentrations of sulphate salts of nine heavy metals (Al, Cd, Cu, Fe, Pb, Mn, Ni, Ag, and Zn). Petri dishes lined with filter papers (Whatman No. 42) were used instead of plastic pots filled with freshwater sand. The other growth conditions were the same as for the previous experiments and the EC50 values at Day 6 were calculated.

4. Reproducibility of the root elongation tests

This study consisted of four repeated trials of an identical experiment. In each trial, seeds of *B. parachinensis* were subjected to 12 different treatments (4 different concentrations of Zn, Cu and Ni, respectively) in petri dishes lined with filter papers. All the growth conditions were the same as for the third experiment. Percentage reduction of root growth was calculated from each trial. One-way Analysis of Variance was used to test for

significant difference (at the 95% level of significance) among the percentage root reduction in the four trials. F values were calculated for each treatment condition and compared with the critical value at the 95% level of significance.

Results

1. Root elongation of 13 edible crops in response to Cu, Ni and Zn

Shoot and root lengths of the edible crops were plotted against time in days (Fig. 1). In general, all

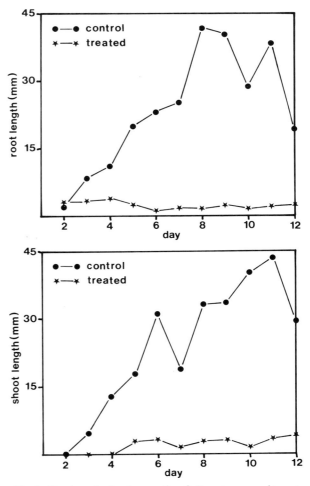

Fig. 1. Root and shoot growth of *Brassica parachinensis* seedlings in a medium containing 30 mg l^{-1} Cu, 20 mg l^{-1} Ni and 75 mg l^{-1} Zn compared to those in control without metals.

Table 2. Percentge reductions of root and shoot growth of seeds of 13 species at Day 6 grown in a medium containing 30 mg l^{-1} Cu, 20 mg l^{-1} Ni and 75 mg l^{-1} Zn compared to growth in controls without metals.

Seed	% root reduction	% shoot reduction
B. parachinensis	94.91	89.29
B. chinensis	94.35	80.35
S. oleracea	91.92	76.34
L. sativa	90.03	100.00
R. sativus	96.13	92.34
D. carota	93.74	92.00
O. sativa	97.90	42.65
H. vulgare	95.23	83.27
Z. mays	*	*
G. max	86.88	62.50
P. mungo	95.59	91.24
L. esculentum	95.82	85.26
L. pyrenne	100.00	100.00

* fungal infection occurred and the seeds were considered dead.

the 13 plant species were sensitive to the metal solutions. 'The reason for the shortening of root or shoots after 9 to 11 days was possibly due to the variability in the response of the seedlings to the treated conditions as different pots were measured during the 12-day period.' Lengths of roots and shoots of the seedlings to which distilled water alone was added increased constantly during the growth period while those treated with solutions containing mixtures of Cu, Ni and Zn had depressed root and shoot growth. Fig. 1 shows the growth curves of shoot and root growth of *B. parachinensis* in a medium containing 30 mg l^{-1} Cu, 20 mg l^{-1} Ni and 75 mg l^{-1} Zn compared to the control groups without metals.

For most of the plant species tested, root length at Day 6 provided an early indication of the toxic effect exerted by heavy metals. Table 2 summarizes the percentage reductions of shoot and root lengths of all of the 13 plant species on Day 6. Both shoot and root lengths of the seedlings, especially the latter, were significantly ($p < 0.05$) reduced when compared with the control. Among the 13 edible crops, five species were excluded from the 'Zn Equivalent Value' test. Ryegrass produced very fine and fragile roots, germination of maize required a rather long period, growth of lettuce was slow and both tomato and spinach had poor root growth.

2. 'Zinc Equivalent Values' for eight edible crops

Table 3 shows the EC50 values on Day 6 for the eight selected crop species determined from ranges of concentrations of Cu, Ni and Zn solutions. In general, the relative toxicity of Cu, Ni and Zn to the tested species followed the pattern of Ni > Cu > Zn with the sole exception of the toxicity to barley which was Cu > Ni > Zn. By comparing the EC50 values, a sensitivity ranking

Table 3. 144-h EC50 values of Ni, Cu & Zn and their relative toxicity on eight selected species.

Categories	Test plants	144-h EC50 (mg l^{-1})			Ratio (Zn : Cu : Ni)
		Zn	Cu	Ni	
Leafy plants	*B. parachinensis*	18.0	5.0	1.2	1 : 2.60 : 15.00
	B. chinensis	4.5	4.0	2.0	1 : 1.13 : 2.25
Root plants	*R. sativus*	14.5	7.1	2.6	1 : 2.04 : 5.58
	D. carota	19.8	14.5	5.2	1 : 1.37 : 3.81
Cereals	*O. sativa*	42.6	6.1	3.0	1 : 7.10 : 14.20
	H. vulgare	26.6	12.6	24.0	1 : 2.10 : 1.10
Legumes	*P. mungo*	48.8	10.3	7.5	1 : 3.00 : 3.88
	G. max	33.0	11.0	8.5	1 : 4.76 : 6.50

Table 4. 144-h EC50 values of various heavy metals applied singly on root growth of *B. prachinensis*

Metal	EC50	
	(mg l^{-1})	μM
Ag	80.00	27.68
Al	25.00	1.97
Cd	3.50	0.511
Cu	3.50	1.39
Fe	12.40	2.49
Mn	60.00	19.50
Ni	1.60	0.33
Pb	44.00	15.43
Zn	23.00	9.31

of different plant species, in the decreasing order were:

Ni: *B. parachinensis* > *B. chinensis* > *R. sativa* > *O. sativa* > *D. carota* > *G. max* > *P. mungo* > *H. vulgare*

Cu: *B. parachinensis* > *B. chinensis* > *O. sativa* > *R. sativa* > *D. carota* > *G. max* > *P. mungo* > *H. vulgare*

Zn: *B. chinensis* > *R. sativa* > *B. parachinensis* > *D. carota* > *H. vulgare* > *P. mungo* > *O. sativa* > *G. max*

The results of the relative toxicity ratios of Ni, Cu and Zn for the eight selected plant species are shown in Table 3. Although the relative toxicity of Zn : Cu : Ni of *Brassica parachinensis* and *Raphanus sativus* were close to the proposed 1 : 2 : 8 ratios as in the Zn Equivalent Value, these ratios were not observed for other test species. Therefore, the proposed 1 : 2 : 8 ratios are not applicable to a wide variety of plant species in nature.

3. Comparative toxicity of 9 heavy metals on B. parachinensis

The 144-h EC50 values of *B. parachinensis* subjected to a range of concentrations of nine heavy metals are shown in Table 4. The order of metal toxicity, in descending order was: Ni > Cd > Cu > Al > Fe > Zn > Pb > Mn > Ag.

4. Reproducibility test

The percentages of root reduction of *B. parachinensis* subjected to different concentrations of Zn, Cu and Ni, respectively, in four trials are listed in Table 5. Results indicated that the calculated F

Table 5. Reproducibility of the root elongation test among four trials using *Brassica parachinensis*.

Metal	Concentration mg l^{-1}	Percentage root reduction[1]				F value[2]
		Trial				
		1	2	3	4	
Zn	14.2	25.1	28.8	36.4	26.6	1.04
	25.3	45.7	44.1	56.5	47.7	0.55
	45.0	79.6	78.6	72.9	83.4	2.82
	142.3	97.1	96.8	95.6	93.2	15.35
Cu	3.6	61.9	59.4	59.8	61.0	0.03
	6.3	80.1	80.2	85.1	84.1	0.24
	11.2	86.2	89.3	90.8	89.8	2.01
	35.6	94.1	94.2	97.2	92.8	1.23
Ni	1.8	39.4	30.1	22.4	38.3	0.94
	3.2	59.5	58.0	63.6	66.2	1.24
	5.6	73.5	77.0	73.9	72.4	0.87
	17.8	89.2	88.2	87.2	84.7	1.54

Note 1: the results were obtained from a comparison of the treated seeds with control seeds grown without added metals
Note 2: F value was obtained by One-way Analysis of Variance and compared with the critical value of 3.24

values were smaller than the critical F value in 11 out of 12 treatments among the four trials. This indicated that there was no significant difference at the 95% level between the percentage root reduction among the four trials in 11 out of 12 treatments. Elongation in the fourth treatment of Zn, at 142.3 mg l^{-1}, differed from that in the other three concentrations because of the low variability.

Discussion

It is commonly known that certain trace metals in very low concentrations may inhibit root elongation and delay seed germination in a large variety of plants (Adriano et al., 1973; Wollan et al., 1978; Wong & Bradshaw, 1982). In the present study, pregerminated seeds were used for testing metal toxicity. Root growth seems to be a better indicator as it was more sensitive to metal solutions than shoot growth according to the present as well as previous work (Wong & Bradshaw, 1982).

The following sensitivity ranking was observed for the four categories of plants in decreasing order of sensitivity: leafy plant > root plant = cereal > legume. The present study also revealed that seeds of root plants, cereals and legumes, which contained high food reserves, had lower sensitivity than seeds of leafy plants which had lower food reserves. The order of metal toxicity in terms of root growth inhibition was established as Ni > Cd > Cu > Al > Fe > Zn > Pb > Mn > Ag. This was largely in agreement with other studies, e.g. on chlorosis produced in oats, Triticum spp. (Hunter & Vergnano, 1952); and on root inhibition of ryegrass, Lolium perenne (Wong & Bradshaw, 1982).

'Zn Equivalent Value' was developed for the control of Zn, Cu and Ni in sewage sludge spread on agricultural land (Chumbley, 1971). Combined phytotoxic effects of Zn, Cu and Ni on plants were regarded as being additive on young barley seedlings (Beckett & Davis, 1978) and ryegrass (Davis & Carlton-Smith, 1984). However, subsequent field and pot trials with sludge-treated soil indicated that the concept did not apply uniformly over a broad spectrum of plant species (Council for Agricultural Science and Technology, 1976; Coker et al., 1982; Johnson et al., 1983). The present experiment involves a simpler method by testing pregerminated seeds in metal solutions rather than in field and pot trials. It also indicated that the sensitivity of edible crops available in the subtropical region to toxic metals is also species-specific. Among the eight crop species tested, only B. parachinensis and R. sativus had Zn : Cu : Ni ratios close to the proposed 1 : 2 : 8 ratios of the 'Zn Equivalent Value'.

The simple method used in the present study seems to be reliable and reproducible. The use of root elongation of B. parachinensis, one of the most important vegetables available locally seems to provide a relatively fast (six days) and accurate assessment of metal toxicity. It is envisaged that part of the sewage sludge generated by the existing as well as future sewage treatment plants in Hong Kong will be used for agricultural application. The bioassay method adopted in the present study will therefore be useful as a convenient tool to evaluate the phytotoxicity of heavy metals contained in the sewage sludge.

Acknowledgements

We would like to thank Drs. Diane Malley and Uwe Borgmann of Fisheries and Oceans Canada for the constructive review of the manuscript and their suggestions.

References

Adriano, D. C., A. C. Chang, P. E. Pratt & R. Sharpless, 1973. Effects of soil application of dairy manure on germination and emergence of some selected crops. J. envir. Qual. 2: 396–399.

Beckett, P. H. T. & R. D. Davies, 1978. The additivity of the toxic effects of Cu, Ni and Zn in young barley. New Phytol. 81: 155–173.

Behera, B. K. & B. N. Misra, 1982. Analysis of the effect of industrial effluent on growth and development of rice seedlings. Envir. Res. 28: 10–20.

Chumbley, C. G., 1971. Permissible levels of toxic metals in sewage used on agricultural land. Agricultural Development and Advisory Service Advisory, Paper No. 10, Pinner, Ministry of Agriculture, Fisheries and Food. U.K. 12 pp.

Coker, E. G, R. D. Davis, J. E. Hall & C. H. Carlton-Smith, 1982. Field experiments on the use of consolidated sewage sludge for land reclamation effects on crop yield and composition and soil conditions, 1976-81. Technical Report: TR183. Stevenage, Water Research Centre. U.K. 83 pp.

Council for Agricultural Science and Technology, 1976. Application of sewage sludge to cropland appraisal of potential hazards of the heavy metals to plants and animals. Report PB-264 015. US Department of Commerce, National Technical Information Service. Springfield, U.S.A. 43 pp.

Davis, R. D. & C. H. Carlton-Smith, 1984. An investigation into the phytotoxicity of zinc, copper and nickel using sewage sludge of controlled metal content. Envir. Pollut. B8: 163–185.

Goodman, G. T., S. Smith, G. D. R. Parry & M. J. Inskil, 1974 (Abstract) The Use of Moss-bags as Deposition Guages for Airborne Metals. 41st Annual Conference of Natural Society of Clean Air, Oct. 1974. Cardiff, England.

Hunter, J. G. & O. Vergnano, 1952. Nickel toxicity in plants. Ann. appl. Biol. 34: 279.

Johnstons, N. B., P. H. T. Beckett & C. J. Waters, 1983. Limits of zinc and copper toxicity from digested sludge applied to agricultural land. In: Davies, R. D., Hucker, G. and Hermite, P. L. (eds.), Environmental Effects of Organic and Inorganic Contaminants in Sewage Sludge. pp. 75–81. Dordrecht, D. Reidel, The Netherlands.

Martin, M. H. & P. J. Coughtrey, 1982. Biological Monitoring of Heavy Metal Pollution. Applied Science Publishers, London. 478 pp.

OECD, 1981. OECD Guidelines for Testing of Chemicals. Organisation for Economic Cooperation and Development, Paris.

Wilkins, D., 1957. A technique for the measurement of lead tolerane in plants. Nature 180: 37.

Wollan, E., R. D. Davis & S. Jenner, 1978. Effects of sewage sludge on seed germination. Envir. Pollut. 17: 195–205.

Wong, M. H., 1985. Phytotoxicity of refuse compost during the process of maturation. Envir. Pollut A37: 159–174.

Wong, M. H. & A. D. Bradshaw, 1982. A comparison of toxicity of heavy metals by using root elongation of ryegrass, Lolium purenne. New Phytol. 91: 255–261.

Wong, M. H., & W. M. Lau, 1983. Effects of roadside soil extracts on seed germination and root elongation of edible crops. Environ. Pollut. A31: 203–215.

Hydrobiologia **188/189**: 385–395, 1989.
M. Munawar, G. Dixon, C. I. Mayfield, T. Reynoldson and M. H. Sadar (eds)
Environmental Bioassay Techniques and their Application.
© 1989 *Kluwer Academic Publishers. Printed in Belgium.*

Effects of acidity on acute toxicity of aluminium-waste and aluminium-contaminated soil

N. F. Y. Tam,[1] Y. S. Wong[1] & M. H. Wong[2]
[1] *Department of Applied Biology and Chemical Technology, Hong Kong Polytechnic, Hung Hom, Kowloon, Hong Kong;* [2] *Department of Biology, Hong Kong Baptist College, Kowloon, Hong Kong*

Key words: acidity, acute toxicity, heavy metals extractability, aluminium-waste, bioassays

Abstract

The total heavy metal concentrations of Al-waste and Al-contaminated soil were many times higher than that found in the control soil, which might pose toxic effects on nearby ecosystems under acidic condition. The present study aimed to detect the amount of Al, Cu, Zn, Mn, Pb, Ni and Cd extracted by distilled water and ammonium acetate at pH 3.8, 4.8, 5.8, 6.8 and 7.8. The acute toxicities of water extracts were assessed by two bioassays. Results showed that concentrations of heavy metals, especially Al, were the highest in extracts from Al-contaminated soil extracted with NH_4OAc, followed by Al-waste. The control displayed relatively low levels of metals. More heavy metals were extracted at acidic pH than at neutral pH. Distilled water extracts exhibited lower levels of metals than those extracted with NH_4OAc. The first bioassay, in terms of seed germination and root elongation of *B. parachinensis*, indicated that the germination rates were seriously retarded by Al-waste even at neutral pH. Less than 25% seeds were germinated in Al-waste and the toxic effect was more obvious at pH 3.8. Root growth in Al-waste and Al-soil was slower than in the control and no seedling in Al-waste had roots longer than 2 cm at the end of this study. In the second bioassay, the photosynthetic rate of *Chlorella pyrenoidosa* was significantly inhibited by Al-waste when compared with the control, although the pH effect was not clear. This study revealed that the metal availability was pH dependent and their toxicity could be rapidly assessed by two simple bioassays.

Introduction

The aluminium industry, especially Al-fabrication facilities, has grown rapidly in recent years. During the fabrication process, Al-waste in the form of particulates, is a potential source of heavy metal contamination to surrounding soil and water systems. High concentrations of Al in the soil solution exert toxic effects on plants, especially on root growth and development. Typical symptoms are restricted and abnormal root growth, swollen and brown-coloured root tips, fewer lateral roots,

short stunted curved side roots, leaf margin chlorosis, and defoliation (Uhlen, 1985; Thornton *et al.*, 1987; Baligar *et al.*, 1987). Nowadays, heavy metal toxicity especially Al is of more serious concern because large areas of the earth's surface are subjected to 'acid rain' (Likens *et al.*, 1979). Acid precipitation causes gradual acidification in soil which enhances leaching and solubility of heavy metals, particularly Al. The elevated Al ion contents not only contaminates the soil but also causes Al migration into ground and surface waters (Cronan & Schofield, 1979). Aluminium toxicity to aquatic organisms in acidic waters has been reported (Baker & Schofield, 1982; Salder & Lynam, 1987). Although sophisticated chemical techniques have been developed to analyze the total concentrations of heavy metals in the environment, the figures obtained do not necessarily reflect their availabilities for biological uptake, and consequently do not provide meaningful information concerning their potential toxicity. During the past decades, a relatively large number of biological methods for evaluating heavy metal toxicity on organisms have been developed. Among them, algal bioassay seemed to be the most commonly used technique. However, whether it is preferable to use algae instead of higher plants as assay organisms may be questionable. Therefore, the aims of this study were to (1) detect the amounts of Al and other heavy metals in Al-waste and Al-contaminated soil collected from an Al-fabrication plant; (2) examine the effects of two reagents, ammonium acetate (NH_4OAc) and distilled water on extracting metals from Al-waste and Al-soil; (3) measure the metal availability under different degrees of acidity; and (4) evaluate and compare the acute toxicity of water extracts using two bioassay techniques.

Materials and methods

Sample collection and heavy metal analysis

The aluminium waste and its contaminated surface soil (0–2 cm) were collected from an Al-fabrication factory in Hong-Kong. Surface soil (0–2 cm) from a clean rural area nearby was also sampled and used as control. These samples had pHs of 6.3, 7.2 and 7.3 and organic matter contents of 0.79%, 7.95% and 6.40%, for Al-waste, Al-soil and control soil, respectively. These samples were air-dried, passed through a 2 mm sieve and digested by mixed acids containing conc. HNO_3: conc. H_2SO_4: 60% $HClO_4$ (5.0 : 0.5 : 1.0 v/v) in a digestion assembly. The filtered digests were analyzed for Al, Cu, Zn, Fe, Ni, Mn, Pb and Cd by inductively-coupled plasma (ICP) spectrophotometry (Instrumental Laboratory Plasma 300). The lower detection limits of the ICP technique for the metals studied are as follows: 10, 2, 25, 1.5, 7, 2, 1 and 2 $\mu g/l$ for Al, Zn, Pb, Cd, Ni, Fe, Mn and Cu respectively. All determinations were in triplicate.

Preparation of rainwater extracts

Rainwater with an average pH around 5 was collected during the summer of 1987. Air-dried samples of Al-waste (Aw), Al-contaminated soil (As) and the control soil (Cs) were shaken with rainwater (1 : 10 w/v) for 1 hour in a horizontal shaker. The resulting filtrate in each case represented the corresponding rainwater extract. The pH (by pH meter) and heavy metal content (by the ICP method) of the extracts were then measured.

Assessment of acute toxicity of rainwater extracts using two bioassay techniques

The acute toxicity of rainwater extracts from Al-waste, Al-soil and control soil was assessed by the following two bioassay methods:
1. Seed germination and root elongation of *Brassica parachinensis* (Flowering Chinese Cabbage): Thirty seeds of *B. parachinensis* were placed on a filter paper in each of the 20 petri dishes. Five ml of various rainwater extracts were applied once every other day. Distilled water was used as a control. There were four replicates for each treatment. All of the seeds were placed in a growth chamber with a light/dark cycle of 16/8 hour (light intensity of 4,000 $\pm$ 75 lux) at a temperature of 20 $\pm$ 2 °C. The number of germinated seeds (with emerged embryo at least 2 mm long) and the length of main root (radicle) were recorded every two or three days for two weeks.
2. Photosynthetic rates of *Chlorella pyrenoidosa* (a unicellular green alga): Pure culture stock of *C. pyrenoidosa* in Bristol medium was centrifuged, washed in carbonate-bicarbonate buffer solution (pH 9) and resuspended in

10 ml rainwater extract contained in a 50 ml conical flask. The initial density of algal cells in each extract was maintained at 5×10^7 cells ml^{-1}. The algal cultures were then incubated in the same growth chamber as described earlier. At the 3rd- and 24th-hour intervals, 3 ml of each cell culture were centrifuged and the cell residues were resuspended in 3 ml buffer solution in a Gilson flask. The photosynthetic rate of algal suspension (μl O_2 evolved min^{-1}) was then determined using a Gilson respirometer equipped with a shaking water bath set at 30 °C and light intensity of 10 000 lux. All measurements were in quadruplicate.

pH effects on metal extractability and acute toxicity of distilled water extracts

Al-waste, Al-contaminated soil and control soil were mixed with 1 M NH_4OAc solution of different pH in the ratio of 1 : 10 w/v. The pH values of NH_4OAc solution were adjusted to pH = 3.8, 4.8, 5.8, 6.8 and 7.8 using diluted H_2SO_4 and NaOH. The mixtures were shaken for 1 hour and then filtered. The pH and heavy metal contents of these NH_4OAc extracts were determined using the same method described above. Extracts of Al-waste, Al-soil and control soil were prepared in the same way as the NH_4OAc extracts using distilled water of different pHs. Besides determining the pH and heavy metal contents in these

water extracts, the acute toxicities of these extracts were also assessed using the same bioassay techniques described above.

Results and discussion

Total and rainwater-extractable heavy metal contents in Al-waste (Aw) and Al-contaminated soil (As)

Table 1 shows that total heavy metal concentrations of both Al-waste and Al-soil were ten to hundreds fold higher than that found in the control soil. For instance, total Al contents in Al-waste were over 20 times more than that found in Al-soil which in turn had twice the amount found in control soil. Al-waste and Al-soil also contained higher concentrations of Pb, Cd, Cu and Ni than control soil. The high heavy metal content in the Al-soil was due to the dissolution, leaching and oxidation of heavy metals which originated from the deposited Al-waste. It was observed that a lot of Al-waste in the form of fine particulates was scattered on the soil surface and became a significant source of heavy metal contamination to the soil system. Despite the elevated total concentrations of heavy metals in Al-waste and Al-soil, most of them were not extracted by pH 5 rainwater. This was because the majority of metals in Al-waste was in the elemental state and chemically inactive, whereas in Al-soil, the heavy metals were tightly bound within the lattice structure of clay minerals and/or the inorganic and

Table 1. Total and rainwater (pH = 5) extractable concentrations of heavy metals ($mg\,l^{-1}$) in Al-waste and soil samples.

Metal	Al-Waste		Al-contaminated soil		Control soil	
	Tot.	Ext.	Tot.	Ext.	Tot.	Ext.
Al	823869 ± 7112[a]	0.04 ± 0.23[b]	34378 ± 5180[c]	0.12 ± 0.06[d]	16846 ± 451[e]	0.05 ± 0.02[b]
Cu	2078.9 ± 41.0[a]	0.28 ± 0.01[b]	2468.5 ± 160.9[c]	0.08 ± 0.02[d]	8.1 ± 0.9[e]	nd
Zn	681.9 ± 169.6[a]	2.33 ± 0.28[b]	4590.3 ± 403.4[c]	0.35 ± 0.01[d]	134.7 ± 21.5[e]	nd
Ni	134.1 ± 17.1[a]	nd	165.8 ± 10.6[a]	nd	14.2 ± 2.4[b]	nd
Pb	1270.9 ± 120.1[a]	nd	280.5 ± 16.2[b]	nd	46.3 ± 6.3[c]	nd
Cd	153.8 ± 23.5[a]	nd	22.6 ± 2.1[b]	nd	12.2 ± 1.4[c]	nd

nd: not detected; Tot.: total concentration; Ext.: extractable concentration.
Values of the same metal with the same superscript are not significantly different at the $P \leq 0.05$ level according to ANOVA-test.

388

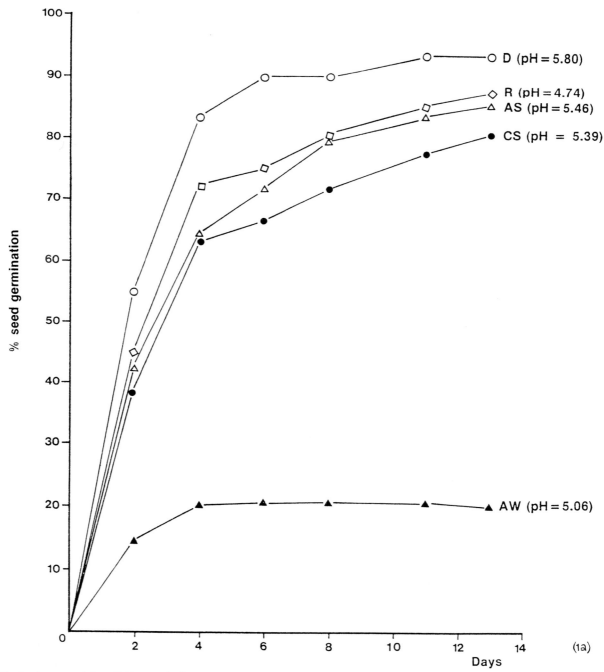

Fig. 1. Acute toxicity of Al-waste/soil extracts on seed germination and root elongation of *B. parachinensis*. (Fig. 1a: seed germination; 1b: root length; D: distilled water; R: rainwaters; As: aluminium-contaminated soil; Cs: control soil; Aw: aluminium waste).

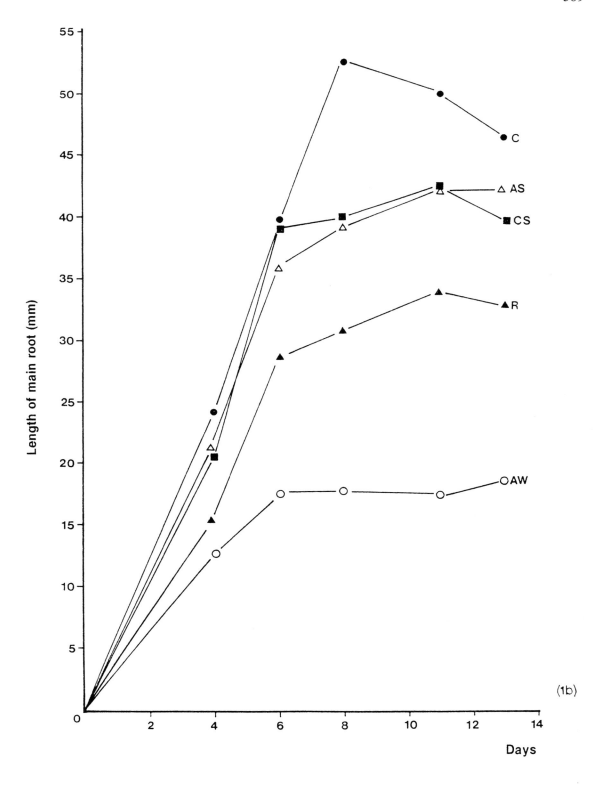

(1b)

organic complexes of the soil. Nevertheless, the concentrations of Cu and Zn in the rainwater extract of Al-waste seemed to be significantly higher than those in the control soil extract (Table 1).

Acute toxicities of rainwater extracts from Al-waste and Al-contaminated soil

Seed germination of *B. parachinensis* was severely suppressed by the rainwater extract of Al-waste and less than 20% of seeds were germinated throughout the experiment (Fig. 1a). For the Al-soil extract, germination of *B. parachinensis* did not differ from that of the control soil. Similar results were observed when root elongation was used as the bioassay parameter. Growth of the main root was greatly retarded by Al-waste rainwater extract (Fig. 1b). Besides shortening the root length, the root system in Al-waste extract was usually malformed with short, stunted, brownish and curved side roots. This indicated that the rainwater extract of Al-waste was toxic to seed germination and root elongation which might be related to the high concentrations of heavy metals. When comparing the root length of germinated seeds in Al-soil and control soil extracts, no significant differences ($P \leq 0.05$) were found. However, their lengths were even greater than those recorded in rainwater. This suggested that the amounts of heavy metals available in Al-soil extracts had not reached toxic levels. Moreover, nutrients released in this extract may have produced stimulatory effects on plant root development.

Figure 2 depicts the photosynthetic rates of *C. pyrenoidosa* in different rainwater-extracts and shows that there was no significant difference ($P \leq 0.05$) in the volume of O_2 evolved among all treatments after 3 hours of incubation. Each culture had a photosynthetic rate equivalent to 2.5 μl O_2 evolved min^{-1}, the same as that measured for cells cultivated in Bristol medium. However, when the cells were incubated for 24 hours, the photosynthetic rate of *C. pyrenoidosa* in Al-waste extract was drastically reduced to 1.8 μl O_2

evolved min^{-1}. Cells in Al-soil extract had a similar photosynthetic rate as that found in the control soil extract and Bristol medium. This confirmed the toxicity of Al-waste rainwater extract and its toxicity increased with incubation time.

Effects of extractions and the extractant pH on metal extractability

Table 2 illustrates that significantly higher ($P \leq 0.05$) heavy metal contents were found in Al-waste and Al-soil extracts than in the control soil extract. The highest concentrations of Al, Cu, Pb, Mn and Zn were extracted from Al-contaminated soil, followed by Al-waste. The control soil extracts contained a very low level of heavy metals. This might be explained by the fact that the Al-waste consisted of fine particulates; and that its heavy metal content was less chemically active and so not readily extracted by NH_4OAc. On the contrary, most of the metals in Al-soil existed as oxides or hydroxides or chemical complexes. The organic matter, clay particles and minerals in the contaminated soil may have also enhanced metal exchangeability (Logan *et al.*, 1985). Table 2 further reveals that the acidity of NH_4OAc did not change during the extraction procedure, i.e. the resulting extracts had the same pH as the original NH_4OAc. In all samples, more metals were extracted in acidic NH_4OAc solution. Ammonium acetate with pH 3.8 and 4.8 extracted the largest quantities of heavy metals especially Al, Zn, Mn and Fe, followed by extraction with pH 5.8 NH_4OAc. Relatively low levels of metals were released in extracts at pH 6.8 and 7.8. It has been found that the quantity of metals extracted from soils by NH_4OAc, particularly in the case of Al, is a continuous function of the pH of NH_4OAc; and that it decreases as the pH increases from 4 to 7 (McLean *et al.*, 1959; Logan *et al.*, 1985).

When comparing the amount of heavy metals extracted by NH_4OAc and by distilled water of the same pH, it is clear that the former extracted more heavy metals. In fact, NH_4OAc is commonly considered as a superior and strong

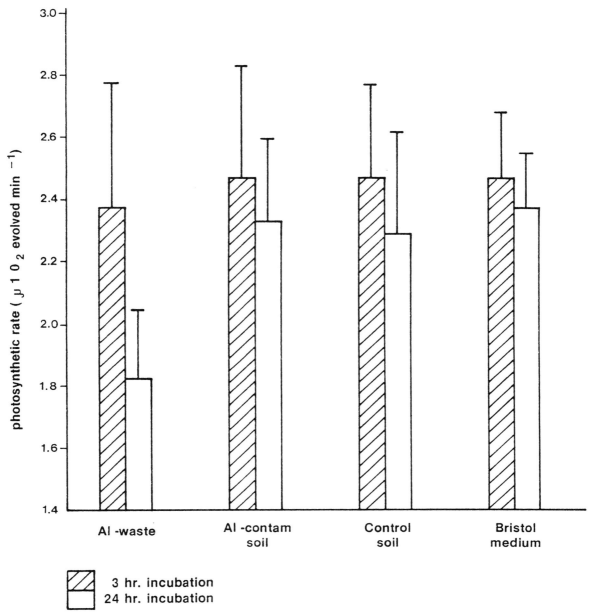

Fig. 2. Acute toxicity of Al-waste/soil extracts on photosynthetic rate of *C. pyrenoidosa*.

extractant for plant-available metals (McLean *et al.*, 1959; Lee & Sharp, 1985). When using distilled water instead of NH_4OAc as extractant, the pH values of the resultant extracts were significantly different ($P \leq 0.05$) from the original values. Table 3 shows that the pH values of all extracts were around 6 to 7, although some

extracts resulted from distilled water of pH 3.8 or 4.8. Similar findings were reported by Tyler (1978) who indicated that the pH of percolate from polluted soil leached with pH 3.2 artificial rainwater increased to about 6. This indicated that acids (H^+ ions) in the leaching solution (acidic distilled water in this experiment) were

Table 2. pH and heavy metal contents (mg l^{-1}) of soil/waste solutions extracted by NH$_4$OAc of different pH.

	pH of NH$_4$OAc				
	3.8	4.8	5.8	6.8	7.8
pH of extracts					
Aw	3.88 ± 0.04	4.58 ± 0	5.80 ± 0	6.58 ± 0	7.80 ± 0
As	3.90 ± 0.14	4.90 ± 0	5.90 ± 0	7.18 ± 0.04	7.88 ± 0.04
Cs	3.90 ± 0.07	4.86 ± 0.06	5.98 ± 0.04	7.40 ± 0	7.93 ± 0.04
Al conc.					
Aw	17.47 ± 2.04	2.88 ± 0.86	0.07 ± 0.08	0.10 ± 0.05	0.09 ± 0.03
As	346.65 ± 21.28	224.20 ± 33.52	1.20 ± 0.38	0.22 ± 0.08	0.15 ± 0.04
Cs	12.57 ± 1.06	nd	nd	nd	nd
Cu conc.					
Aw	9.06 ± 0.11	5.51 ± 1.19	4.21 ± 0.06	3.11 ± 0.31	4.35 ± 0.08
As	13.78 ± 1.14	18.46 ± 1.97	15.49 ± 0.22	9.25 ± 0.32	15.56 ± 0.47
Cs	0.18 ± 0.01	0.08 ± 0.01	0.03 ± 0	nd	nd
Zn conc.					
Aw	34.31 ± 0.04	21.82 ± 12.86	15.17 ± 0.24	11.42 ± 1.49	12.99 ± 0.06
As	164.75 ± 3.46	105.64 ± 4.99	59.15 ± 1.20	18.28 ± 0.03	15.37 ± 0.12
Cs	0.99 ± 0.11	0.74 ± 0.47	0.32 ± 0.01	0.10 ± 0.01	nd
Pb conc.					
Aw	2.75 ± 0.67	0.87 ± 0.37	0.57 ± 0.08	nd	nd
As	3.92 ± 0.65	3.99 ± 0.40	4.20 ± 0.33	0.99 ± 0	nd
Cs	1.46 ± 0.28	0.76 ± 0.27	0.13 ± 0.04	nd	nd
Mn conc.					
Aw	0.95 ± 0.08	0.39 ± 0.25	0.25 ± 0.01	0.14 ± 0.18	0.19 ± 0.01
As	13.31 ± 1.24	6.29 ± 0.19	1.39 ± 0.11	0.67 ± 0.01	0.39 ± 0.01
Cs	1.44 ± 0.25	0.60 ± 0.16	0.21 ± 0.03	nd	nd
Fe conc.					
Aw	8.52 ± 2.01	2.25 ± 0.48	0.08 ± 0	nd	nd
As	70.22 ± 0.70	17.99 ± 2.58	0.19 ± 0	0.03 ± 0	0.02 ± 0
Cs	6.84 ± 1.06	0.54 ± 0.23	nd	nd	nd

nd: not detected; Ni and Cd are not detected; Aw: Aluminium-waste; As: Aluminium-contaminated soil; Cs: control soil.

efficiently neutralized by other base cations on the soil exchange complex. It has been found that the hydrogen ions (protons) of acid precipitation were buffered in the soil due mainly to dissolution of hydroxyl-Al compounds (AlOOH) producing either Al ions (Al^{3+}) or a new solid phase in the form of AlOHSO$_4$ (Ulrich *et al.*, 1980). It has also been found that neutralization of acid rain in the Falls Brook Watershed system is accomplished by a two-step chemical reaction. Initially, rapid but incomplete neutralization occurs by the dissolution of the reactive Al phase in the soil zone, followed by slow neutralization by basic cations such as Si and Ca released by the decomposition of silicate minerals (Johnson *et al.*, 1981). In addition, as distilled water has no buffering capacity, its pH can be easily changed even though it is acidic. Being a buffer by itself, NH$_4$OAc can maintain its pH values during this extraction process and therefore can exert a pH effect on

Table 3. pH and heavy metal contents (mg l^{-1}) of soil/waste solutions extracted by distilled water of different pH.

	pH of distilled water				
	3.8	4.8	5.8	6.8	7.8
pH of extracts					
Aw	6.55 ± 0.07	6.60 ± 0.14	6.00 ± 0	6.43 ± 0.32	6.83 ± 0.04
As	6.70 ± 0	6.63 ± 0.18	6.68 ± 0.04	6.65 ± 0.07	6.83 ± 0.04
Cs	6.58 ± 0.11	6.50 ± 0	6.60 ± 0	6.60 ± 0	6.65 ± 0.07
Al conc.					
Aw	0.43 ± 0.08	0.22 ± 0.28	0.28 ± 0.25	0.24 ± 0.15	0.10 ± 0.01
As	0.52 ± 0.04	0.25 ± 0.06	0.23 ± 0.07	0.78 ± 0.07	0.65 ± 0.07
Cs	nd	nd	nd	nd	nd
Cu conc.					
Aw	0.27 ± 0.05	0.24 ± 0.04	0.22 ± 0.04	0.30 ± 0.03	0.28 ± 0.04
As	nd	nd	nd	nd	nd
Cs	nd	nd	nd	nd	nd
Zn conc.					
Aw	1.61 ± 0.12	1.39 ± 0.02	1.39 ± 0.12	1.07 ± 0.03	1.01 ± 0.03
As	nd	nd	nd	nd	nd
Cs	nd	nd	nd	nd	nd

nd: not detected; Pb, Ni, Cd, Fe and Mn are not detected; Aw: Aluminium-waste; As: Aluminium-contaminated soil; Cs: Control soil.

extraction. In the present study, high concentrations of Al and Zn were found only in Al-waste and Al-soil extracts produced from distilled water with pH 3.8 and 4.8, respectively. Nevertheless, the buffering capacity of soil or waste would become saturated if the environment were to be subjected to long-term acid precipitation. Consequently, the soil solution produced would have a low pH and high concentrations of heavy metals (Tyler, 1978). In that case, the toxicity of these extracts would be greatly enhanced.

Acute toxicity of distilled water extracts

Percentages of seed germination at days 4 and 10 together with the length of the main root at day 10 are presented in Table 4. Al-waste water-extracts at all pH values produced the lowest levels of seed germination and root growth. Morphological abnormalities including swollen and brown-colored root tips, fewer lateral roots and curved side roots were observed. These are the typical symptoms of Al toxicity (Thornton *et al.*, 1987; Baligar *et al.*, 1987). It has been observed that Al-damage to the root system starts at Al concentrations between 1–2 mg l^{-1} (Ulrich *et al.*, 1980). In the present study, the fact that some seeds germinated in Al-waste water-extracts had a smaller amount of shoot growth although their roots were damaged, may be due to the fact that shoot growth is less sensitive to Al than root growth (Thornton *et al.*, 1987).

These results suggest that the high concentration of Al in Al-waste extracts might be responsible for their inhibitory effect on plant development. In addition to Al, other heavy metals such as Zn might also exert toxic effects (Wong & Bradshaw, 1982). The low nutrient and organic matter contents in the Al-waste extracts may also enhance the toxicity of Al and other heavy metals to plants. It has been reported that root damages caused by Al might in turn reduce the plant's ability to take up many nutrient ions especially

Table 4. Acute toxicity of Al-waste and Al-contaminated soil solutions extracted by distilled water of different pH – Results of two bioassay techniques.

	pH of distilled water				
	3.8	4.8	5.8	6.8	7.8
Percentages of seed germination (%) at day 4					
Aw	17 ± 5	17 ± 5	20 ± 5	33 ± 5	20 ± 0
As	48 ± 12	62 ± 7	63 ± 9	73 ± 0	72 ± 2
Cs	45 ± 12	63 ± 5	52 ± 12	60 ± 0	60 ± 14
Percentages of seed germination (%) at day 10					
Aw	18 ± 7	25 ± 2	23 ± 9	35 ± 7	25 ± 7
As	68 ± 12	80 ± 5	85 ± 7	90 ± 0	83 ± 5
Cs	53 ± 0	68 ± 7	72 ± 2	78 ± 2	65 ± 12
Length of main root (mm) at day 10					
Aw	15.75 ± 1.72	13.13 ± 1.59	9.24 ± 0.79	11.57 ± 2.73	12.5 ± 2.12
As	23.58 ± 0.74	23.87 ± 0.55	31.91 ± 5.53	27.91 ± 4.53	29.33 ± 2.02
Cs	34.53 ± 1.46	43.27 ± 7.21	36.83 ± 14.61	31.22 ± 2.64	28.95 ± 1.23
Photosynthetic rate of chlorella (μl O_2 envolved min^{-1}) at 24 hr. incubation					
Aw	2.01 ± 0.14	1.69 ± 0.01	1.79 ± 0.30	1.65 ± 0.06	1.97 ± 0.38
As	2.62 ± 0.13	1.99 ± 0.26	2.41 ± 0.01	2.19 ± 0.03	2.44 ± 0.26
Cs	2.38 ± 0.04	2.46 ± 0.50	2.33 ± 0.12	1.81 ± 0.19	2.49 ± 0.12

Aw: Aluminium-waste; As: Aluminium-contaminated soil; Cs: Control soil.
Values are mean and standard deviations of four replicates.

those of P and Ca, rendering the plant deficient in P (Uhlen, 1985). Al-contaminated soil extracts seemed to support the same amount of seed germination as the control soil extracts, but roots in these extracts were slightly shorter than those in the control (Table 4). These effects are not restricted to higher plants such as *B. parachinensis*.

Al-waste extracts were also toxic to the unicellular green alga at all pH levels. As indicated in Table 4, the photosynthetic rates of *C. pyrenoidosa* were significantly ($P \leq 0.05$) reduced by Al-waste extracts after 24-hour incubation. It has been established that heavy metals reduce the photosynthetic activity of algae by causing structural damage to chloroplasts (Shubert, 1984). The effect of pH was not significant in all extracts because of the buffering effect, and all extracts tended to have a neutral pH and similar concentrations of heavy metals except at pH 3.8.

The two bioassays demonstrated that all extracts of Al-waste were toxic and inhibited plant growth and their physiological activities, although some extracts contained only a small amount of available heavy metals. This suggested that chemical determination of heavy metals alone may not reflect the actual hazard of the solution. Moreover, bioassays are more favourable than chemical methods. The two bioassays used in this study are standard procedures recommended by many authors. Algal strains of Chlorococeals such as *Chlorella* are sensitive organisms belonging to the first link in the aquatic food chains. Therefore, they give a rapid and good estimation of the potential hazard of any waste or chemical released to aquatic systems (Shubert, 1984). Besides algal bioassays, assays using higher plants are also used in hazard assessment. It has been claimed that root elongation is a valid and sensitive plant response to environmental toxicity (Ratsch, 1983). Furthermore, Wong &

Bradshaw (1982) used root elongation of ryegrass to determine phytotoxicity of the heavy metals and found that root elongation was invariably more responsible to toxicity than was shoot growth. Results obtained from this study demonstrated that the toxicity indicated by both bioassays were comparable although the test involving higher plants took longer to complete.

References

Baker, J. P. & C. L. Schofield, 1982. Aluminium toxicity to fish in acidic waters. Wat. Air Soil Pollut. 1–3: 289–309.

Baligar, V. C., R. L. Wright, T. B. Kinraide, C. D. Foy & J. H. Elgin Jr., 1987. Aluminium effects on growth, mineral uptake, and efficiency ratios in red clover cultivars. Agron. J. 76 (6): 1038–1044.

Cronan, C. S. & C. L. Schofield, 1979. Aluminium leaching response to acid precipitation: Effects on high-elevation watersheds in the Northeast. Sci. 204: 304–306.

Johnson, N. M., C. T. Driscoll, S. Eaton, G. E. Likens & W. H. McDowell, 1981. 'Acid rain', dissolved aluminium and chemical weathering at the Hubbard Brook Experimental Forest, New Hampshire. Geochimica et. Cosmochimica Acta 45: 1421–1437.

Lee, R. & G. S. Sharp, 1985. Extraction of 'available' aluminium from podzolized New Zealand soils of high aluminium status. Commun. Soil Sci. Plant Anal. 16: 261–274.

Likens, G. E., R. F. Wright, J. N. Galloway & T. J. Butler, 1979. Acid rain. Sci. Am. 241: 43–50.

Logan, K. A. B., M. J. S. Floate & A. D. Fronside, 1985. The determination of exchangeable acidity and exchangeable aluminium in hill soils. Part 2: Exchangeable aluminium. Commun. Soil Sci. Plant Anal. 16: 309–314.

McLean, E. O., M. R. Heddleson & G. J. Post, 1959. Aluminium in soils: III. A comparison of extraction methods in soils and clays. Proc. Soil Sci. Soc. Am. 23: 289–293.

Ratsch, H. C., 1983. Interlaboratory Root Elongation Testing of Toxic Substances of Selected Plant Species. U.S. EPA-600/3-83-051, U.S. Environ. Prot. Agency, Washington D.C. 46 pp.

Salder, K. & S. Lynam, 1987. Some effects on the growth of brown trout from exposure to aluminium at different pH levels. J. Fish. Biol. 31: 209–219.

Shubert, L. E. (ed.), 1984. Algae as Ecological Indicators. Academic Press, London. 435 pp.

Thornton, F. C., M. Schaedle & D. J. Raynal, 1987. Effects of aluminium on red spruce seedlings in solution culture. Environ. Exp. Bot 27 (4): 489–498.

Tyler, G., 1978. Leaching rates of heavy metal ions in forest soil. Wat. Air Soil Pollut. 9: 137–148.

Uhlen, G., 1985. The toxic effect of aluminium on barley plants in relation to ionic composition of the nutrient solution. II. Soil solution studies. Acta Agric. Scand. 35: 271–277.

Ulrich, B., R. Mayer, P. K. Khanna, 1980. Chemical changes due to acid precipitation in a Loess-derived soil in central Europe. Soil Sci. 130: 193–199.

Wong, M. H. & A. D. Bradshaw, 1982. A comparison on the toxicity of heavy metals, using root elongation of rye grass, *Lolium perenne*. New Phytol. 91: 255–261.

Hydrobiologia **188/189**: 397–402, 1989.
M. Munawar, G. Dixon, C. I. Mayfield, T. Reynoldson and M. H. Sadar (eds)
Environmental Bioassay Techniques and their Application.
© 1989 *Kluwer Academic Publishers. Printed in Belgium.*

Do bioassays adequately predict ecological effects of pollutants?

John S. Gray
Department of Marine Zoology and Chemistry, University of Oslo, P.B. 1064, 0316 Blindern, Oslo 3, Norway

Abstract

With some notable exceptions, such as the echinoderm and oyster larvae tests, the species traditionally used in bioassays are not sufficiently sensitive to detect subtle ecological effects of pollutants. It is suggested that by using ecological criteria, species can be identified from any pollution gradient that are sensitive to subtle effects of pollution. Examples are given using gradients of oil, sewage and titanium dioxide pollution, showing how ecologically sensitive species for use in laboratory bioassays can be selected objectively.

Many marine molluscs show microgrowth bands, which can be used as *in situ* field bioassays. An example is given using the bivalve *Cerastoderma edule*.

Introduction

The main goals of the use of bioassays in pollution research have been defined by Cairns & Pratt (1989) in this volume as: to rank hazards, to set discharge limits, to predict the environmental consequences of discharges, to protect important species and to protect ecosystem structures and functions. Whilst the first two goals are discussed in detail in this volume little attention has been given to ecological aspects such as the capability of bioassays of predicting the ecological consequences of discharges. Here I concentrate on two main points, the selection of organism used in bioassay tests and whether or not responses shown on the common organisms used in bioassay tests predict ecological effects in the environment.

Much effort has been concentrated on selection of organisms that can be used in bioassay tests. In the past the primary criterion has been their robustness in laboratory cultures rather than their sensitivity to chemical pollutants or their ability to indicate subtle ecological consequences of the effects of pollutants in the field. There are notable

exceptions. A number of highly sensitive bioassays have been developed based on echinoderm larvae (Kobayashi, 1971) which have been shown to be 2–3 orders of magnitude more sensitive than *Artemia* (Connor, 1972). An oyster larvae technique has been developed which is also highly sensitive, (Woelke, 1967, 1968, 1972).

Such techniques are the exception. The dominant organisms used in bioassays today (see papers in this volume) are crustaceans of the genus *Daphnia* (a cladoceran) as a freshwater assay organism and *Artemia* (an anostracan) as a marine assay organism. Yet if such species dominate in nature what does this mean in an ecological context? Cladocerans have a high food requirement and are, energetically, especially suited to growth in eutrophic waters. Therefore, if species of *Daphnia* dominate it often indicates that conditions are meso- or eu-trophic rather than indicating that conditions are just beginning to change from oligotrophy, (Harris, 1986, p. 221). If *Artemia* dominates in nature the habitat will be an extremely saline lake! Thus in ecological contexts dominance by either of these species does not indicate subtle effects of pollution, but rather extreme effects. The type of species that are required in bioassay tests are not species that by their presence in high numbers indicate environmental extremes, but should rather be species that are sufficiently sensitive to indicate the first stage of decline of a system.

Thus, I believe that the selection of organisms for use in bioassay techniques, such as those above, has been based on incorrect ecological arguments. Rather than selecting species which are sufficiently robust to withstand laboratory conditions one should use ecological criteria to select organisms that can indicate the first stages of decline of an ecosystem. The following strategy is suggested as being applicable to many ecological situations and can lead to the objective selection of organisms that can be used in laboratory bioassays.

Methods

Objective selection of sensitive organisms for use in bioassays

As an example of how sensitive organisms can be selected objectively I use data from transects along increasing gradients of pollution, at an oil platform in the North Sea, a titanium dioxide dump site in Norway and at a sewage dump site in the Clyde estuary, Scotland. If one plots the distribution of the organisms as individuals per species grouped into geometric classes, (the lognormal distribution) one obtains the plots in Fig. 1. It is clear that at the polluted sites the assemblages are greatly disturbed when compared with the control sites; rather than the smooth curve shown at the control sites the disturbed sites show a series of disjunct groups of species with few rare species represented by 1 or 2–3 individuals per species and by a great increase in the abundance of a few common species. It is usually these common species that are used as indicators of pollution-induced disturbance in an ecological context. For example the polychaete *Capitella capitata* is a species which increases greatly in abundance on approaching the actual

pollution source. For example, at the Statfjord oil platform it is present in geometric class XII (2048–5095 individuals per sampling unit) at site 0.5 only and at the sewage dump site it is present in geometric class XIII (5096–10191 individuals per sampling unit) at the dump site, O. *C. capitata* has been suggested as a universal indicator of pollution, but it's presence in high numbers illustrates an endpoint in the pollution gradient rather than the first indication of change. The plots in Fig. 1. clearly show the gradient of disturbance.

If instead of recording changes in abundance of the common species one takes the moderately common species (in geometric abundance classes between 8 and 63 individuals per sample) at the unpolluted sites and plots abundance changes of these species along the gradient, the plots in Fig. 2 are obtained.

Figure 2 shows that a selection of only 6 species from the total number of species (over 80 at the sewage dump site) gives clear trends, with increases in abundance approaching the dump site followed by a large decrease in abundance at the site itself. The trends shown by these 6 species are similar to those shown using the full data set with plots of total number of species and diversity, (Gray & Pearson, 1982). Thus a group of species has been selected objectively, which are sensitive to the initial stages of pollution-induced change.

This approach has been used on a number of other case-histories involving organic enrichment in Europe and has been shown to apply fairly generally, (Pearson, Gray & Johannessen, 1983). The species selected using this method then, are the species that should ideally be used in bioassay tests as they are sensitive species rather than endpoint species. However, are the species specific to a given area and do the species selected show similar responses over a wide range of pollutants?

Pearson, Gray & Johannessen, (1983) have shown that over the European N. Atlantic the suite of species responding to organic enrichment gradients is relatively small. Thus on a regional scale a small group of species has been identified which is common over a large area. Table 1 shows some of the species which were extracted using

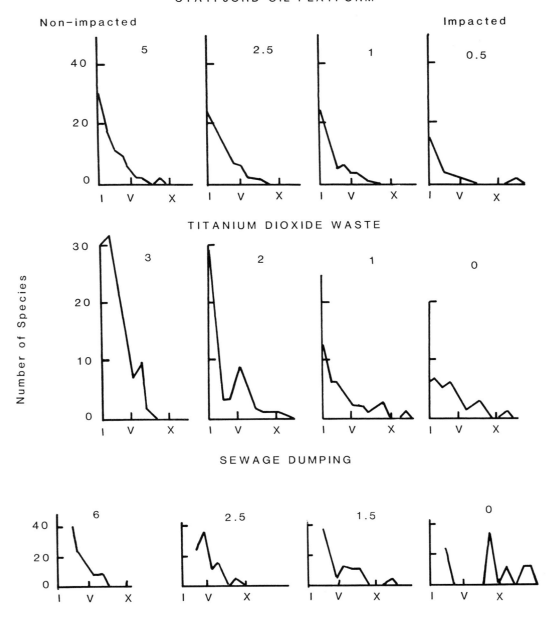

Fig. 1. Distribution of individuals among species for macrofauna of sediments subjected to various pollutants. In each case the gradient increases from left to right. Individuals per species in x2 geometric classes (I = 1 individ. per species, II = 2–3 individ. per species, III = 4–7 individ. per species, IV = 8–15 individ. per species, V = 16–31 individ. per species etc.) Distances from the discharge point are shown in km.
a) Statfjord oil platform, (Data from Olsgaard unpublished).
b) Titanium dioxide dumping in Norway, (Data from Olsgaard unpublished).
c) Sewage dumping ground in the Clyde estuary, Scotland (from Pearson *et al.*, 1983).

400

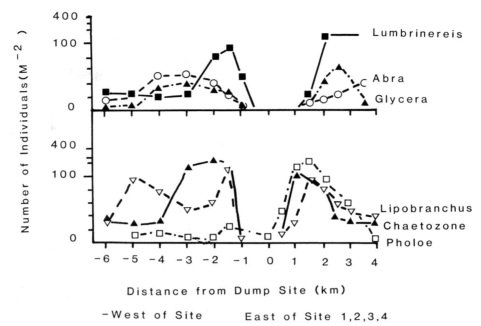

Fig. 2. Objectively selected species in geometric groups V and VI (8–63 individuals per species) from the Clyde dump site, (from Pearson *et al.*, 1983).

the above method, on the pollution gradients shown in Fig. 1. Clearly a number of species show similar responses to the different pollutants.

The group of ecologically sensitive species will, however, probably vary from geographical region to geographical region. It seems to me inappropriate to try and standardize on universal bioassay organisms 'if the goal is to predict ecological consequences'. Clearly for assessing the relative toxicity of different chemicals there is a clear case for universal standard bioassay organisms.

It must be said, however, that few of the species in Table 1 have been studied in detail and little is known about their natural history and none are routinely in culture. It is however, likely that such species will prove to be far more relevant for bioassay work, in an ecological context, than many of those used presently. More efforts should be devoted to culturing such potentially ecologically sentinel species.

One group of species that are particularly useful in monitoring, are the bivalve molluscs, (which

Table .1 Occurrence of selected moderately common species in stressed marine environments.

Species	Area				
	Loch eil (pulp mill)	Clyde (sewage)	N. Sea (oil)	Oslofjord (sewage)	Norway (TiO$_2$ waste)
Lumbrinereis spp		+		+	+
Pholoe minuta	+	+	+	+	+
Glycera spp	+	+	+	+	+
Goniada maculata	+			+	+
Diplocirrus glaucus	+			+	
Anaitides spp	+		+	+	
Chaetozone setosa		+		+	+

have often been identified using the above techniques, e.g. *Abra* in Fig. 2). Many bivalves living in the tidal and just subtidal habitat show microgrowth patterns in the shell which can be used to record the spatial and temporal influences of environmental factors including pollutant effects. Thus, a novel bioassay technique can be foreseen, where the animal is left *in situ* to monitor its own environment.

Bivalve microgrowth bands in environmental monitoring

Controversy exists as to whether the microgrowth bands are laid down in response to tidal or diurnal stimuli. Using the bivalve *Cerastoderma edule* taken from the intertidal sediment in Trondheimsfjord, Norway a sequential series of photographs showing microgrowth bands over many months were obtained, (Fig. 3). The photographs were then placed on a digitizing board so that the width of each band could easily be recorded on a computer. The raw data was then subjected to time series analyses to ascertain the dominant periods shown in the microgrowth patterns, (for details see Lønne & Gray, 1988). The analyses showed that it was tides and not diurnal patterns that initiated the microgrowth bands. The shell thus encapsulated a history of the local environment.

From such an analysis one can envisage that *C. edule* can be used to monitor retrospectively the spatial and temporal effects of a pollutant event such as an oil spill. The assumption is that growth rate will be reduced and thus by sampling over a spatial gradient the spatial extent of the reduction in growth can be plotted. By taking individuals at different time intervals after the oil has been removed it should be possible to measure the time period over which growth rate has been reduced, assuming that the organisms have survived the spill. This then is a sensitive, sub-lethal bioassay tool.

Our results show that variation between individuals from the same locality is fairly low and that a sample of 5 individuals of the same approximate size gives a measure of the natural varia-

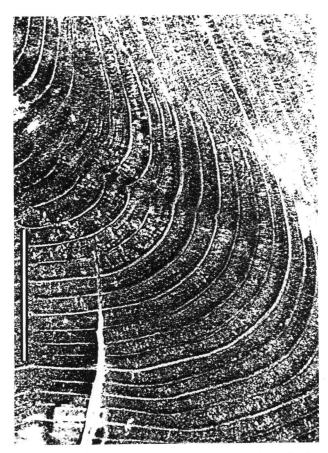

Fig. 3. Microgrowth band in *C. edule* from Trondheimsfjord, Norway, (from Lønne & Gray, 1988). See text for fuller explanation.

bility. From our studies (Lønne & Gray, 1988) many marine molluscs show suitable microgrowth bands, such as *Patella vulgata*, *Littorina littorea*, the periostracum of *Mytilus edulis*, *Mya arenaria* and probably many more. Species such as *M. arenaria* are long-lived and it is possible to reconstruct growth patterns over many years. Some marine organisms have growth records that stretch over thousands of years, e.g. the coral *Porites* (Isdale, 1985) where records of long-term hydrographic events as El Nino and the Southern Oscillation are clearly seen. *Porites* also records rainfall on the coast of Queensland in the form of fluorescent bands in the skeleton. In the future with the development of sophisticated microchemical techniques it may well prove possible to isolate pollutants from within specific bands from known dates and so build up field dose-response relationships.

References

Cairns, J. Jr. & J. R. Pratt, 1989. The scientific basis of bio-assays. In: M. Munawar, G. Dixon, C. I. Mayfield, T. Reynoldson & M. H. Sadar (eds.), Environmental Bio-assay Techniques and their Application. Hydrobiologia 188/189: 5–20.

Connor, P. M., 1972. Acute toxicity of heavy metals to some marine larvae. Mar. Pollut. Bull. 3: 190–192.

Gray, J. S. & T. H. Pearson, 1982. Objective selection of sensitive species indicative of pollution-induced change in benthic communities. I. Comparative methodology. Mar. Ecol. Progr. Ser. 9: 111–119.

Harris, G. P., 1986. Phytoplankton Ecology. Chapman & Hall N. Y., 384 pp.

Isdale, P., 1984. Fluorescent bands in massive corals record centuries of coastal rainfall, Nature, Lond. 310: 578–579.

Kobayashi, N., 1971. Fertilised sea urchin eggs as an indicatory material for marine pollution bioassay, preliminary experiments. Publs. Seto. Mar. Biol. Lab. 18: 379–406.

Lønne, O. J. & J. S. Gray, 1988. Influence of tides on micro-growth bands in Cerastoderma edule from Norway. Mar. Ecol. Progr. Ser. 42: 1–7.

Pearson, T. H., J. S. Gray & P. Johannessen, 1983. Objective selection of sensitive species indicative of pollution-induced change in benthic communities. 2. Data analyses. Mar. Ecol. Progr. Ser. 12: 237–255.

Woelke, C. E., 1967. Measurement of water quality with the Pacific oyster embryo bioassay. Wat. Qual. Crit. 416: 112–120.

Woelke, C. E., 1968. Application of shellfish bioassay results to the Puget Sound pulp mill pollution problem. N. W. Sci. 42: 125–133.

Woelke, C. E., 1972. Development of a receiving water quality bioassay criterion based on the 48 hour Pacific oyster (Crassostrea gigas) embryo. Washington Department of Fisheries, Technical Report 9: 1–93.

Hydrobiologia **188/189**: 403–406, 1989.
M. Munawar, G. Dixon, C. I. Mayfield, T. Reynoldson and M. H. Sadar (eds)
Environmental Bioassay Techniques and their Application.
© 1989 *Kluwer Academic Publishers. Printed in Belgium.*

The *Daphnia* bioassay: a critique

Donald J. Baird, Ian Barber, Mairead Bradley, Peter Calow & Amadeu M.V.M. Soares
Department of Animal and Plant Sciences, University of Sheffield, Sheffield S10 2TN, U.K.

Key words: Daphnia, bioassay, effects of food concentration, genetic variation, maternal effects, laboratory culture, genotype-environment effects

Abstract

Daphnia magna is used widely as a standard ecotoxicological indicator organism, and protocols exist for its use in assessing the toxicity of substances under acute and chronic experimental conditions. Problems exist in repeatability of such bioassays between laboratories. Sources of variation are identified using a simple quantitative genetics model. Presenting specific examples, we conclude that these problems are tractable, but only if the genotype and culture conditions prior to and during tests are strictly controlled.

Introduction

One of the essential properties of a laboratory bioassay is that it should be repeatable. This is true whether single-species or multi-species bioassays are being considered. In this paper we specifically evaluate problems which cause inconsistencies among single-species bioassays that use the freshwater cladoceran *Daphnia magna* Straus, but emphasise that the conclusions drawn are applicable to all types of ecotoxicological tests, regardless of species or compound being considered.

D. magna is used in aquatic toxicology primarily for its ease of culture, its high sensitivity to toxicants and its clonal method of reproduction (Berge, 1978; Adema, 1978). Two types of bioassays are commonly performed with it: 48-hour acute tests using neonates that are < 24 hours old, and 21-day chronic life-cycle tests that are run from birth to ca. the tenth instar. In the former, the toxicological effect is death; in the latter it is the inhibition of 'normal reproduction'.

The problem

Within the European Community (EC), bioassays are periodically checked by the 'ring test' procedure in which different laboratories follow similar protocols to assess reference toxins. The results provide a measure of interlaboratory consistency in the performance of a standard bioassay.

Such a ring-test, involving a 21-day life-cycle test with sodium bromide as reference toxin, has recently been carried out using *D. magna* (Cabridenc, 1986). The criterion used was the no-effect concentration, defined as the concentration immediately below that in which a significant effect on reproduction occurred. Of 37 participating laboratories, 22 returned results which satisfied validity criteria specified in the protocol. Most tests which failed validity criteria did so because of low fecundity in the control treatments (see also below). The final results (Table 1) exhibit the maximum possible variability in response. We believe that this extreme inconsistency illustrates a serious problem that can only be solved by a

404

Table 1. Results of the 1986 EC ring test on the chronic effects of sodium bromide on *D. magna* fecundity (After Cabridenc, 1986).

Laboratories participating:	$n = 37$
Valid test results obtained:	$n = 22$
NaBr concentration range:	3–$117 \, mg \cdot L^{-1}$
Range of results obtained:	<3–$>117 \, mg \cdot L^{-1}$
(NOEC on fecundity – see text)	

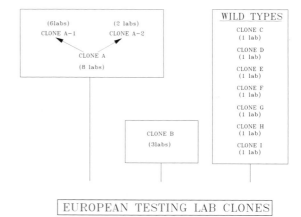

Fig. 1. Results of the EC interlaboratory genetic typing exercise. 27 laboratory clones typed at six enzyme loci (PGM, PGI, MPI, GOT, MDH-2, ES-1) from a sample of 30–40 animals drawn randomly from laboratory stock cultures.

systematic examination of each of the variability components of the toxicological response.

The model

We adopt a standard quantitative genetics approach to this problem (Falconer, 1981), viz.:

$$V_P = V_G + V_E + V_{GE} \qquad (1)$$

This equation partitions the total variability in stress-response of a group of test animals (V_P) into three subcomponents:
(1) V_G, variability due to genetic heterogeneity
(2) V_E, the variability due to environmental heterogeneity.
(3) V_{GE}, the variability due to genotype by environment interaction.

Using this approach, we not only make suggestions for improvement of culture systems and future test protocols, but also comment on the suitability of clonal organisms for use in ecotoxicology.

Genetic variability

Genetic typing of the clones used by the EC ring-test laboratories has been carried out (Bradley *et al.*, in prep.) using standard electrophoretic techniques (Hebert & Beaton, 1986). The details will be published elsewhere, but a summary is given briefly in Fig. 1. Genetic differences between laboratory stock clones appear to have arisen for two reasons. Firstly, a number of labo-

ratories obtained stocks directly from local wild populations (clones C–I), and, not surprisingly, these have proved genetically unique. Secondly, some genetic heterogeneity has been due to clonal divergence (clones A-1 and A-2 being direct descendants of clone A). Here, a number of laboratories received their stocks from a central source that at some point gave rise to two subclones, either due to mutation or sexual recombination. The former seems rather more likely than the latter as the source of clonal divergence, since the typing study failed to reveal any within-laboratory genetic variation in stocks – each consisted of a pure clone.

In a standardised series of acute toxicity tests (48 hr LC_{50} tests following a standard protocol (OECD, 1981)) using cadmium chloride (concentrations measured using graphite furnace atomic absorption spectrophotometry), we have found (Baird *et al.*, in prep.) the range in response of the ring-test laboratory clones ranged from 0.8 ± 1.1 ppb to 25.8 ± 1.1 ppb (mean $\pm$ s.e.). This clearly indicates that differences in genotype do reflect significant differences in toxicological response. Hence the results obtained from *D. magna* bioassays may depend on the clone chosen to run the test.

Environmental variability

We can recognise two separate components of environmental variability: the conditions experienced prenatally in the mother ('maternal effects', Falconer, 1981), and conditions experienced by individuals after birth. In practice these two components refer respectively to conditions in stock culture and during the bioassay itself.

Since there is as yet no standardisation between different laboratories in culture methodology, it is likely that maternal effects could be an important source of variability in bioassay results. It is well-known that maternal reserves play an important role in juvenile growth for *Daphnia* (Tessier *et al.*, 1983), but of greater importance is whether maternal provisioning influences the toxicological response of offspring. To examine this, mothers from a single clone were raised under constant culture conditions (following guidelines in Goulden *et al.*, 1982), but at two different ration levels (0.05 and 0.5 mgC L^{-1} *Chlorella vulgaris* in axenic culture), and their offspring (<24 hr old) were compared in a standard 48 hr LC$_{50}$ test using 3,4-dichloroaniline. The results (Fig. 2) indicate that offspring from high-ration mothers were approximately twice as susceptible to 3,4-dichloroaniline as those from low-ration mothers (LC$_{50}$: 104 ± 4.2 ppb (0.5 mg C$\cdot$L^{-1}) versus 195 ± 6.3

(0.05 mg C$\cdot$L^{-1}). Maternal effects are therefore of toxicological importance, particularly in neonate 48 hr acute tests, and worthy of more detailed investigation.

The second component of environmental variability relates to conditions in the bioassay itself. The EC ring-test provides a good example of this. From these data, it is clear that as ration level increases, fecundity increases (Fig. 3). In the EC ring-test, one of the validity criteria specified was a minimum fecundity of 70 neonates per animal in the control treatment. Many laboratories failed this criterion. From the same relationship of fecundity and ration level for the control animals in the EC ring-test (Fig. 4), two conclusions emerge: firstly, the specified ration level of 1.0 mg C$\cdot$L^{-1} was not always given (actual range 0.1–100 mg C$\cdot$L^{-1}), and secondly, that in those tests that were run at the correct ration level, fecundity ranged from 9 to 129 neonates per adult female. Of the variability in fecundity, almost 64% can be explained by a failure to adhere to the specified ration level. Hence it is not surprising that the ring-test results showed so much variability.

Genotype-environment interactions

Given that clones of *D. magna* exhibit different tolerances to a specified toxicant (see above), it

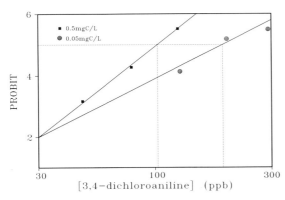

Fig. 2. The influence of maternal ration level on 48 hr acute response (measured as LC$_{50}$ – dashed lines) of <24 hr old neonates to 3,4-dichloroaniline.

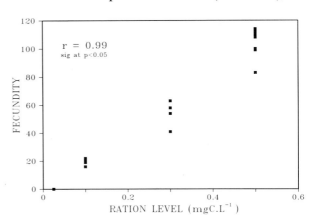

Fig. 3. Fecundity of *D. magna* over 21 days in unstressed conditions at different ration levels. (Food = *Chlorella vulgaris* maintained in axenic culture).

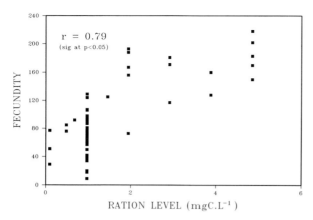

Fig. 4. Fecundity of *D. magna* in control (i.e. unstressed) treatments in the 1986 EC ring test as a function of ration. N.B. One point (ration = 100 mgC $\cdot$ L^{-1}, fecundity = 0) has been omitted to improve clarity.

would be useful to know if the rank order of tolerance among clones reared under similar conditions is the same for different classes of toxins, and to what extent interclonal differences in stress-tolerance reflect differing degrees of adaptation to the laboratory environment. These questions address the problem of subtle interactive effects between genotype and environment about which we know little but which certainly merit further investigation.

It follows that there are two ways in which two laboratories may obtain differing results in a standard test even when adopting constant culture conditions: by culturing different genotypes under the same constant environmental conditions, or by culturing the same genotype under different environmental conditions, irrespective of the constancy of those environments. Since neither genotype nor culture environment are at present specified for laboratories running standard *D. magna* bioassays, between-laboratory variability is not surprising.

Summary

i) To improve consistency in the *D. magna* bioassay among testing labs, both genotype and culture conditions must be specified. Test protocols must be unambiguous; if tests are carried out under conditions that deviate from the protocol the results must be discarded.

ii) Further research into each of the components of equation (1) should identify which clone/s is suitable for testing purposes, and should lead to the formulation of 'good laboratory practice' guidelines.

iii) While in principle it should be straightforward to minimise V_G and V_E, the interactive effects arising from V_{GE} are likely to be subtle and merit more detailed investigation.

Acknowledgements

Portions of this work were funded by NERC grant TSF/86/AeT/4, EC contracts CCAM/87/319 and B/86000160 and a NATO grant 3/A/87/PO.

References

Adema, D. M. M., 1978. *Daphnia magna* as a test animal in acute and chronic toxicity tests. Hydrobiol. 59: 125–134.

Berge, W. F. ten, 1978. Breeding *Daphnia magna*. Hydrobiol. 59: 121–123.

Cabridenc, R., 1986. Exercice d'intercalibration concernant une méthode de détermination de l'ecotoxicité à moyen terme des substances chimiques vis-à-vis des daphnies. Unpublished EC Report. Contract W/63/476 (214). Ref. I.R.C.H.A.D. 8523. Vert-le-Petit, France. 20 pp.

Falconer, D. S., 1981. Introduction to quantitative genetics. Longman. Harlow, Essex. 340 pp.

Goulden, C. E., R. M. Comotto, J. A. Hendrickson Jr., L. L. Horning & K. L. Johnson, 1982. Procedures and recommendations for the culture and use of *Daphnia* in bioassay studies. Pages 139–160 in Proceedings of the American Society for Testing and Materials Fifth Toxicology Symposium, Special Technical Publication Number 7668, Philadelphia, Pennsylvania, USA.

Hebert, P. D. N. & M. Beaton, 1986. Cellulose-acetate gel electrophoresis. Unpublished ms. Univ. Windsor, Ontario. 34 pp.

OECD., 1981. *Daphnia* sp. 14 day reproduction test (including acute immobilisation test). OECD guidelines for testing of Chemicals, no. 202. ISBN 92-64-1221-4. Paris. 15 pp.

Tessier, A. J., L. L. Henry, C. E. Goulden & M. W. Durand, 1983. Starvation in *Daphnia*: Energy reserves and reproductive allocation. Limnol. Oceanogr. 28: 667–676.

Hydrobiologia **188/189**: 407–410, 1989.
M. Munawar, G. Dixon, C. I. Mayfield, T. Reynoldson and M. H. Sadar (eds)
Environmental Bioassay Techniques and their Application.
© 1989 *Kluwer Academic Publishers. Printed in Belgium.*

Life-tables of *Daphnia obtusa* (Kurz) surviving exposure to toxic concentrations of chromium

Lidia Coniglio & Renato Baudo
CNR-Istituto Italiano di Idrobiologia, I-28048 Verbania Pallanza, Italy

Key words: Hexavalent chromium, *Daphnia obtusa*, toxicity testing, life-tables

Abstract

The life-tables of *Daphnia obtusa* surviving a 48 hours exposure to various hexavalent chromium concentrations have been used to assess whether or not the toxicant had affected demographic parameters. The statistical comparisons lead to the conclusion that even such a short exposure to a relatively low level of chromium reduces the life-span, delays the time of first reproduction, shortens the reproduction period, and decreases the brood size.

Introduction

Environmental pollution by toxic elements and compounds often takes place through intermittent waste chemical discharges, so that in the receiving ecosystem the toxicant concentrations are likely to show wide variations: most of the time the pollutant level is below the threshold value for the resident organisms, however they are occasionally subjected to dangerous levels.

Such short-term exposure to a toxicant may induce delayed mortality and affect the whole ecosystem by modifying the size of the populations, thus changing the relationships amongst the various trophic levels. In addition to this effect, it is reasonable to expect that even individuals surviving exposure to the toxicant may respond in some negative way.

To test this hypothesis, *Daphnia obtusa* cohorts were briefly exposed (48 hours) to seven different concentrations of hexavalent chromium (20 to 140 μg Cr l^{-1}); the surviving animals were then transferred into clean water to grow in optimal conditions until the death of the last individual. The resulting life-tables were then used to assess if the toxicant might in some way affect demo-graphic parameters such as survivorship curves, age specific mortality, age specific fertility and fecundity, survivorship – weighted fertility and fecundity, generation length, mean mortality rate, mean expectation of life of the cohort at birth, intrinsic rate of natural increase and finite rate of increase, net reproduction rate.

Thus the test differs from those of previous Authors (Allan & Daniels, 1982; Daniels & Allan, 1981; Walton *et al.*, 1982; Connell & Airey, 1982; Nebeker, 1982; Gentile *et al.*, 1983; Thain, 1984; Gersich *et al.*, 1985), which evaluated the influence of chronic exposure to toxicants on the life cycle of many aquatic organisms, because in this case the object of the study are the animals surviving an acute stress and then reared in a clean environment.

In addition, although the most widely used cladoceran in toxicity testing is *Daphnia magna* (Berger, 1929; Adema, 1978; Müller, 1980; Buikema *et al.*, 1982; Baudo, 1987), we preferred *Daphnia obtusa* as a test organism, because it is present in Italian lakes of widely different trophic levels.

The choice of chromium as a pollutant was motivated by its use as a reference toxicant in many official toxicity tests. In addition, the hexavalent form is not easily lost by adsorption on the vessel walls or particulate matter, thus a constant concentration can be easily maintained during the period of exposure (Guglielmucci et al., 1981).

Material and methods

Daphnia obtusa cohorts used in this research were obtained from a clone grown according to the methodologies described by Hrbàckovà (1974) & Peters (1987).

Toxicity testing consisted of exposing the cladocerans to seven different concentrations of hexavalent chromium (20 to 140 μg Cr 1^{-1}, as $K_2Cr_2O_7$). For each concentration and the control (Lake Maggiore water, filtered on 0.45 μm and UV sterilized), three groups of ten individuals less than 24 h old were used; the procedure is similar to that proposed by Dave et al. (1981), and Buikema et al. (1982). The animals were not fed during the 48 h of exposure to the toxicant.

Visual observations of the moving animals were made after 3, 6, 12, 24 and 48 h, and the lethal concentration for 50% of the individuals was calculated by probit analysis (Finney, 1971).

The surviving animals were then transferred into clean water (one individual per vial containing 70 ml of filtered and sterilized Lake Maggiore water, changed daily), and grown in optimal conditions (constant temperature of 20 ± 1 °C, photoperiod of 12 h light and 12 h dark, daily food supply of 10 cal 1^{-1} as suspension of Scenedesmus sp. cells).

From the resulting life-tables, the demographic parameters such as survivorship curves, age specific mortality, age specific fertility and fecundity, survivorship-weighted fertility and fecundity, generation length, mean mortality rate, mean expectation of life of the cohort at birth, intrinsic rate of natural increase and finite rate of increase, and net reproduction rate, were calculated as suggested by Seber (1973) and Margalef (1974).

Results

The calculated 48-h LC50 is 61 μg Cr 1^{-1}; this means that the sensitivity of Daphnia obtusa to chromium is in the range displayed by other daphnids, such as D. hyalina (Badouin & Scoppa, 1974), D. pulex (Cairns et al., 1978), and D. magna (Cairns et al., 1978; Müller, 1980).

The results of life-tables were summarized in Table 1. No data are recorded for the two highest concentrations because all animals died during 48 h exposure. For each remaining treatment the first line reports the number of survivors after the toxicity test (1_0); e_0 indicates mean expectation of

Table 1. Demographic and reproductive parameters of Daphnia obtusa: 1_0 is the initial number of individuals surviving in each cohorts (three replicates) after 48 hr exposure, all others are mean values of the three replicates. e_0 – mean expectation of life (days) of the cohort at birth; Σ 1x mx fec – total number of eggs produced; Σ 1x mx fer – total number of newborns; R_0 fec – average number of eggs produced by a female in its life span; R_0 fer – average number of newborns produced by a female in its life span; 'a' – intrinsic rate of natural increase; the last line shows the percentage of eggs which did not develop into a newborn.

	Control	Cr Conc. (μg/l)				
		20	40	60	80	100
1_0–1	10	9	8	3	2	1
–2	10	9	10	2	5	1
–3	10	9	10	7	4	0
e_0	45	51	48	41	52	27
Σ 1x mx fec	467	425	409	146	193	29
Σ 1x mx fer	303	286	286	106	157	28
R_0 fec	47	47	44	38	52	29
R_0 fer	30	32	31	27	42	28
'a'	0.180	0.182	0.182	0.164	0.179	0.130
% egg not developed	35	33	30	28	20	3

life of the cohort at birth; Σ lx mx fec is the total number of eggs produced; Σ lx mx fer is the total number of newborns; R_0 fec and R_0 fer are, respectively, the average number of eggs and newborns produced by a female in its life span; 'a' is the intrinsic rate of natural increase; the last line shows the percentage of eggs which did not develop into a newborn.

The statistical comparison (Student's-t) reveals a significant ($P \leq 0.05$) lowering of the mean expectation of life, number of eggs, and newborns, for the individuals exposed to concentrations of 60 and 100 $\mu g\,Cr\,l^{-1}$. Animals surviving the 80 μg $Cr\,l^{-1}$ treatment proved to be even more vigorous than the control. In this case, the toxicant very likely spared the strongest among the individuals, and this is reflected in the fecundity and fertility data as well as in the e_0 values.

The percentage of non-developing eggs decreases with increasing concentrations of chromium. A possible explanation for this is that the mothers that survive are better adapted for reproduction.

In contrast, no differences were detected for the genetically determined demographic parameters such as mean generation length and intrinsic rate of natural increase.

Discussion

The statistical comparisons lead to the conclusion that even such a short exposure to a relatively low level of chromium reduces the life-span, delays the time of first reproduction, shortens the reproduction period, and decreases the brood size.

The treatment did not appear to affect the mean generation length and the intrinsic rate of population increase, which reflects the theoretical physiological capacity of the species to grow in an unlimited environment; however, the actual population growth is significantly lowered. This is due to the fact that proportionally fewer females are left from the beginning of the life tables with the increase in chromium concentration.

Moreover, as shown in Fig. 1, the theoretical

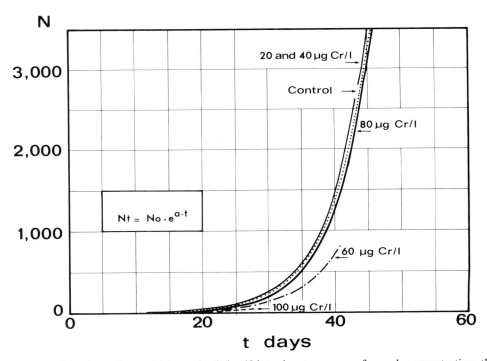

Fig. 1. Growth curves of *Daphnia obtusa* which survived the 48 h toxicant exposure: for each concentration, the number of individuals is calculated by using its own intrinsic rate of natural increase a and for a period of time corresponding to the relative mean expectation of life at birth e_0.

growth of a population, starting from one animal with its own intrinsic rate of natural increase and lasting for the experimentally measured mean expectation of life, will yield a significantly different final number of individuals when exposed to increasing concentrations of chromium.

This conclusion is of some importance especially if viewed in relation to the actual standards proposed for the protection of the aquatic environment from chromium pollution: in fact, the values reported as 'safe' for the aquatic environments range from $5 \mu g$ Cr 1^{-1} (Mance *et al.*, 1984 to $40 \mu g 1^{-1}$ (Taylor *et al.*, 1979). According to our data, if *Daphnia obtusa* is exposed to such concentrations for 48 h, and then the unpolluted conditions are restored, after a time equivalent to the mean expectation of life we could expect a mortality of 0 and 29%, respectively.

The study demonstrates the need for improved knowledge of the sublethal effects of the toxicants: in our view, the simple approach we followed can be of some help in indicating the long-lasting effects of short-term exposure to environmental toxicants that otherwise may be overlooked.

References

Adema D. M. M., 1978. *Daphnia magna* as a test animal in acute and chronic toxicity tests. Hydrobiologia 59: 125–134.

Allan J. D. & R. E. Daniels, 1982. Life table evaluation of chronic exposure of *Eurytemora affinis* (Copepoda) to kepone. Mar. Biol. 66: 179–184.

Badouin, M. F. & P. Scoppa, 1974. Acute toxicity of various metals to freshwater zooplankton. Envir. Contam. Toxicol. Bull. 12: 745–751.

Baudo, R., 1987. Ecotoxicological testing with *Daphnia magna*. Mem. Ist. Ital. Idrobiol. 45: 465–482.

Berger, E., 1929. Unterschiedliche Wirkungen gleicher Ionin und Ioningenische auf verschiedene Terarten, Pflugers Archiv für die gesamte Physiologie des Menschen und der Tiere 223: 1–39.

Buikema, A. L., Jr., B. R. Niederlehner & J. Cairns, Jr., 1982. Biological monitoring. Part IV – Toxicity testing. Wat. Res. 16: 239–262.

Cairns, J., Jr., A. L. Buikema, Jr., A. G. Heath & B. C. Parker, 1978. Effects of temperature on aquatic organism sensitivity selected chemicals. Virginia Water Resources Research Center, Virginia Polytechnic Institute and State University, Bulletin 106. 88 pp.

Connell, A. D. & D. D. Airey, 1982. The chronic effects of fluoride on the estuarine amphipods *Grandidierella lutosa* and *G. lignorum*. Wat. Res. 16: 1313–1317.

Daniels, R. E. & J. D. Allan, 1981. Life table evaluation of chronic exposure to a pesticide. Can. J. Fish. Aquat. Sci. 38: 485–494.

Dave, G., K. Andersson, R. Berglind & B. Hasselrot, 1981. Toxicity of eight solvent extraction chemicals and of cadmium to water fleas, *Daphnia magna*, rainbow trout, *Salmo gairdneri*, and zebrafish, *Brachydanio rerio*. Comp. Biochem. Physiol. 69 C: 83–98.

Finney, D. J., 1971. Probits analysis. Cambridge Univ. Press, London. 333 pp.

Gentile, J. H., S. M. Gentile & G. Hoffman, 1983. The effects of chronic mercury exposure on survival, reproduction and population dynamics of *Mysidopsis bahia*. Envir. Toxicol. Chem. 2: 61–68.

Guglielmucci, G., R. Baudo, G. Galanti & P. G. Varini, 1981. Toxicity of chemical forms of copper and chromium. Part II. Acute mortality of *Daphnia magna* Straus. Mem. Ist. Ital. Idrobiol. 39: 263–280.

Gersich, F. M. *et al.*, 1985. The sensitivity of chronic endpoints used in *Daphnia magna* Straus life-cycle tests. In: R. C. Bohner & D. J. Hansen (Eds.). Aquatic toxicology and hazard assessment. 8th Symposium, Philadelphia, ASTM STP Publ. 891: 245–252.

Hrbàckovà, M., 1974. The size of primiparae and neonates of *Daphnia hyalina* Leydig (Crustacea: Cladocera) under natural and enriched food conditions. Vest. Cs. Spol. Zool. Tom. 38: 2, 98–105.

Mance, G., V. M. Brown, J. Gardiner & J. Yates, 1984. Proposed environmental quality standards for list II substances in water. Chromium. WRC Tech. Rep. TR 207, 49 pp.

Margalef, R., 1974. Ecologia. Ediciones Omega, S.A. Barcelona: 951 pp.

Müller, H. G., 1980. Experiences with test systems using *Daphnia magna*. Ecotoxicol. Envir. Safety 4: 21–25.

Nebeker, A. V., 1982. Evaluation of a *Daphnia magna* renewal life-cycle test method with silver and endosulfan. Wat. Res. 16: 739–744.

Peters, R. H., 1987. *Daphnia* culture. Mem. Ist. Ital. Idrobiol. 45: 483–495.

Seber, G. A. F., 1973. The estimation of animal abundance and related parameters. Griffin, London, 506 pp.

Taylor, M., S. W. Reeder & A. Demayo, 1979. Guidelines for Surface Water Quality. Vol. 1. Inorganic Chemical Substances. Chromium. Inland Water Directorate, Water Quality Branch, Ottawa, Canada, 9 pp.

Thain, J. E., 1984. Effects of mercury on the prosobranch mollusc *Crepidula fornicata*: Acute lethal toxicity and effects on growth and reproduction of chronic exposure. Mar. Envir. Res. 12: 285–309.

Walton, W. E., S. M. Compton, J. D. Alla & R. E. Daniels, 1982. The effect of acid stress on survivorship and reproduction of *Daphnia pulex* (Crustacea: Cladocera). Can. J. Zool. 60: 573–579.

Hydrobiologia **188/189**: 411–413, 1989.
M. Munawar, G. Dixon, C. I. Mayfield, T. Reynoldson and M. H. Sadar (eds)
Environmental Bioassay Techniques and their Application.
© 1989 *Kluwer Academic Publishers. Printed in Belgium.*

Toxicity of the new pyrethroid insecticide, deltamethrin, to *Daphnia magna*

Ruiquin Xiu, Yongxiang Xu and Shirong Gao
Institute of Environmental Health and Engineering, Chinese Academy of Preventive Medicine, 29 Nan Wei Road, Beijing, China

Key words: deltamethrin, *Daphnia magna*, pesticide, toxicity

Abstract

The toxicity of deltamethrin, a synthetic pyrethroid insecticide, was determined under standardized conditions (ISO, 1982) in neonates and juveniles of *Daphnia magna*. Neonates (6 to 24 h old) were more sensitive than juveniles (48 to 72 h old). The 24- and 48-h EC_{50}s (immobilization) in neonates were 0.113 and 0.031 $\mu g \, l^{-1}$, respectively. The toxicity of deltamethrin was highly toxic. The 96-h EC_{50} was in the ppt ($\mu g \, l^{-1}$) range. Toxicity tests with *Daphnia* may be used to detect toxic residues in water and sediment in areas treated with deltamethrin and other highly toxic pyrethroid pesticides.

Introduction

Deltamethrin is a synthetic pyrethroid insecticide manufactured by Roussel Uclaf, Paris, France (trade name: Decis). The recommended application dose for agricultural use is 5 to 17.5 g of active ingredient per hectare, which is about 200 times lower than the dose for DDT. The acute oral toxicity for the rat is 135 mg kg^{-1} b.w., or about the same as for DDT (Worthing, 1983).

Daphnia sp. and other crustaceans are more closely related to insects than vertebrates, including fish. Therefore, highly selective insecticides like deltamethrin may pose a greater hazard to arthropods than to vertebrates. *Daphnia* has also been used extensively to test the toxicity of chemicals to aquatic organisms (Anderson, 1980; EPA, 1973; ISO, 1982). Toxicity tests with *Daphnia* sp. make it possible to compare the toxicity of compounds such as deltamethrin with that of thousands of other chemicals.

The purpose of the present study was to determine the toxicity of deltamethrin to *Daphnia magna* under standardized conditions (ISO, 1982), and to examine its possible cumulative action. Furthermore, the effect of size (age) on sensitivity to deltamethrin was examined.

Materials and methods

Daphnia magna sp. has been cultured continuously in our laboratory since 1962. They were fed algae (*Chlorella* sp. and *Scenedesmus* sp.) prior to exposure and the culture media was replaced three times weekly, but not during exposure. In this study we used neonates (6 to 24 h old) and juveniles (48 or 72 h old) to start exposures.

The insecticide deltamethrin's chemical name is (s)-α-cyano-m-phenoxybenzyl (1R,3R)-3-(2,2-dibromovinyl)-2,2-dimethylcyclo-propane-carboxylate ($C_{22}H_{19}BrNO_3$). It is soluble in acetone, ethanol, dioxane and most aromatic solvents, and is soluble in water < 0.002 ppm. The compound was purchased from Roussel Uclaf (France).

Test solutions of deltamethrin were made from

a stock solution of 1000 mg l^{-1} concentration of deltamethrin with 0.4 mg l^{-1} acetone and reconstituted in water.

Concentrations between 0.0001 and 3200 μg l^{-1} (logarithmic scale) were tested. Exposures were in 300 ml glass beakers containing 200 ml of test solution and covered with watch glasses. All concentrations were tested in triplicate with 20 Daphnids per beaker and replicated three times. The test solutions were not aerated but changed daily. Dissolved oxygen, temperature and pH were determined at the start of each day. Ambient conditions during exposure were as follows: hardness 250 mg l^{-1} expressed as $CaCO_3$; dissolved oxygen 4 to 6 mg l^{-1}; pH 7.8 to 8.0; temperature 20 ± 0.5 °C; the photo period was 16 h light and 8 h dark for all tests. Dead and live animals were recorded at 24, 48 and 96 h with a binocular dissecting microscope. Animals that showed any swimming motion were considered alive. Animals without a heart beat were considered dead, as were animals lying on the bottom of a beaker with an irregular or slight heart beat. Test animals that were not able to swim within 15 seconds of stimulation with water from the handling pipette were considered immobile. We estimated both the LC$_{50}$ (median lethal concentration) and EC$_{50}$ (median effective concentration) values on log probit paper using the above criteria for death, survival and immobility. These values are presented together with the highest concentration causing 0% mortality and the lowest concentration causing 100% mortality. All values are based on nominal concentrations calculated from added amounts of the toxicants.

Results

The results are summarized in Tables 1 and 2. Neonates were more sensitive than juveniles, especially during short-term exposures. This higher mortality in neonates results from their higher metabolic rate and molting frequency. Also the lower biomass to volume (toxicant) ratio in tests with neonates compared to juveniles may have an effect. However using the 96 h LC$_{50}$ as a

Table 1. Toxicity of deltamethrin to neonates and juveniles of *Daphnia magna*.

Period of Exposure (h)	Neonates (6–24 h old)		Juveniles (48–72 h old)	
	EC$_{50}$[a] (μg l^{-1})	LC$_{50}$[b] (μg l^{-1})	EC$_{50}$ (μg l^{-1})	LC$_{50}$ (μg l^{-1})
24	0.113	0.133	0.290	520
48	0.031	0.038	0.029	0.037
96	0.003	0.010	0.018	0.021

[a] Effective concentration for immobilization of 50% of the experimental animals.
[b] Lethal concentration (stops the heart) for 50% of the experimental animals.

Table 2. Concentrations (μg l^{-1}) of deltamethrin causing effects in all (LC$_{100}$ and EC$_{100}$) or none (LC$_0$ and EC$_0$) of neonates and juveniles of *Daphnia magna* exposed to deltamethrin.

Item	Neonates			Juveniles		
	24 h	48 h	96 h	24 h	48 h	96 h
LC$_{100}$	1.0	0.32	0.10	3200	3.2	0.32
LC$_0$	0.01	0.001	0.001	10.0	0.001	0.0001
EC$_{100}$	1.0	0.32	0.10	3.2	0.32	0.32
EC$_0$	0.01	0.001	0.0001	0.01	0.0001	0.0001

measurement of toxicity there was only a factor of two difference between tests on the two life history stages (0.010 μg l^{-1} for neonates and 0.020 μg l^{-1} for juveniles).

Our results (Table 1) also showed that LC$_{50}$'s and EC$_{50}$'s decreased with exposure time. The ratio between the 24 and 96 h LC$_{50}$'s was 13.3 for neonates. For juveniles this ratio was almost 25 000. The latter ratio is due to the very high concentrations required to kill (stop the heart) juveniles. Immobilization in juveniles occurred at only slightly higher concentrations than in neonates.

Effects of deltamethrin were seen over a very broad range of concentrations. Concentrations required to kill or immobilize all Daphnids (LC$_{100}$ or EC$_{100}$) were often two or even three orders of magnitude higher than those that did not kill or immobilize any of them (LC$_0$ or EC$_0$). Even after

96 h this difference persisted. The 96 h LC100 being $0.10 \, \mu g \, l^{-1}$ in neonates and $0.32 \, \mu g \, l^{-1}$ in juveniles, and the EC0 being as low as $0.0001 \, \mu g \, l^{-1}$.

Discussion

Acute toxicity tests with *Daphnia magna* sp. are usually terminated after 24 or 48 h (EPA, 1973; ISO, 1982). Our tests were extended to 96 h without any excessive immobility in controls ($<10\%$). Our 24 h EC_{50} of $0.133 \, \mu g \, l^{-1}$ in *Daphnia magna* sp. neonates is close to the concentration required to immobilize chironomid larvae (*Chironomus tentans*, fourth instar) of $0.20–0.22 \, \mu g \, l^{-1}$ in 24 h tests with sediment (Muir *et al.*, 1985). With fish, the 96 h LC_{50}'s for deltamethrin are generally around $1 \, \mu g \, l^{-1}$ (L'Hotellier & Vincent, 1986). Other reports, reviewed by Khan (1983), have showed that deltamethrin and some other synthetic pyrethroids are acutely toxic to some crustacea at the low ppb or even ppt range.

In comparison with these studies, our results with *Daphnia magna* sp. indicate that this animal is among the more sensitive to the toxic effects of deltamethrin. Because of the high toxicity of pyrethroids and the difficulties involved in chemical analysis at low concentrations of these compounds, bioassays with sensitive organisms can become a cost-efficient alternative to chemical analysis in the monitoring of pyrethroid residues in aquatic environments. The information on the behaviour and the fate of synthetic pyrethroids in aquatic ecosystems is limited. Adsorption appears to be a significant mode of dissipation from water, but in sediments, synthetic pyrethroids can persist for a long time (Khan, 1983). Bioassays with sediment and water may be a more relevant bioassay protocol for assessing the toxicity of pyrethroid insecticide.

In conclusion, deltamethrin is extremely toxic to *Daphnia magna* sp. and there is a great difference in susceptibility among individuals. Rune & Goran (1984) reported 48 h EC_{50} values of pp'-DDT and PCP in *Daphnia magna* sp. were $0.68 \, \mu g \, l^{-1}$ and $0.44 \, mg \, l^{-1}$, respectively. In the present study our tests have shown that the 48 h EC_{50} value of deltamethrin in *Daphnia magna* sp. was an order of magnitude lower at $0.031 \, \mu g \, l^{-1}$. A safe concentration for deltamethrin in aquatic systems would therefore be $0.0001 \, \mu g \, l^{-1}$ to avoid water pollution.

Acknowledgements

This research was supported by the National Natural Science Foundation of China. The experiments were done in China and the paper was written at the Department of Zoophysiology, University of Goteborg, Sweden. The authors are greatly indebted to Dr. Dave Goran for critical reading of the manuscript.

References

Anderson, B. G., 1980. Aquatic invertebrates in tolerance investigations – from Aristotle to Naumann. In: A. L. Builema, Jr. & J. Cairns, Jr., Eds., Aquatic Invertebrates Bioassays, ASTM STP 715, American Society for Testing and Materials, (Philadelphia, USA) pp. 3–35.

EPA (Environmental Protection Agency), 1983. Water quality criteria, 1972. National Academy of Sciences, U.S. EPA Research Series, (Washington, D.C. USA) pp. 1–6.

ISO (International Organization for Standardization), 1982. Water quality – determination of the inhibition of the mobility of *Daphnia magna* Straus (*Cladocera, Crustacea*). ISO 6341-1982, pp. 1–10.

Khan, N. Y., 1983. An assessment of the hazard of synthetic pyrethroid insecticides to fish and fish habitat. In: J. Miyamoto, & P. C. Kearney, (Eds.), Pesticide Chemistry: Human Welfare and the Environment, Vol. 3, Pergamon Press, N.Y., pp. 437–450.

L'Hotellier, M. & P. Vincent, 1986. Assessment of the impact of deltamethrin on aquatic species. Proc. Br. Crop. Prot. Conf. – Pests and Diseases, Vol. 3, pp. 1109–1116.

Muir, D. C. G., G. P. Rawn, B. E. Townsend, W. L. Lockhart & R. Greenhalgh, 1985. Bioconcentration of cypermethrin, deltamethrin, fenvalerate and permethrin by *Chironomus tentans* larvae in sediment and water. Envir. Toxicol. Chem. 4: 51–61.

Rune, B. & D. Goran, 1984. Acute toxicity of chromate, DDT, PCP, TPBS and zinc to *Daphnia magna* cultured in hard and soft water. Bull. Envir. Contam. Toxicol. 33: 63–68.

Worthing, C. R., (Ed.) 1983. The pesticide manual, 7th ed., The British Crop Prot. Council, Croydon, G.B., p. 161.

Hydrobiologia **188/189**: 415–424, 1989.
M. Munawar, G. Dixon, C. I. Mayfield, T. Reynoldson and M. H. Sadar (eds)
Environmental Bioassay Techniques and their Application.
© 1989 *Kluwer Academic Publishers. Printed in Belgium.*

Herbicide effects on planktonic systems of different complexity

Winfried Lampert, Walter Fleckner, Eckart Pott, Ursula Schober & Karl-Ulrich Störkel
Max-Planck-Institut für Limnologie, Postfach 165, 2320 Plön, FRG

Key words: herbicide, atrazine, bioassay, complexity, community response

Abstract

Bioassays of different complexity were compared with respect to their capability to predict the environmental impact of the herbicide atrazine in aquatic systems. Acute toxicity tests with *Daphnia* did not yield meaningful results. Sublethal tests with *Daphnia* (feeding inhibition, reduction of growth and reproduction) were more sensitive, but effective concentrations of atrazine were still rather high (2 mg/L). A relatively complicated 'artificial food chain' system that incorporated direct and indirect effects on *Daphnia* yielded significant reduction of daphnid population growth at 0.1 mg/L. Enclosure experiments with natural communities were by far the most sensitive tools. Community responses could be measured at concentrations as low as 1 μg/L and 0.1 μg atrazine/L. At the lowest concentration, however, communities recovered after three weeks. We conclude that in complex systems indirect effects can be more important than direct effects, so that, contrary to the conditions in simple tests, non-target organisms may be the better indicators of herbicide stress to natural communities.

Introduction

The potential environmental impact of chemicals on aquatic organisms is evaluated by means of tests of various complexities. Single species tests or combinations of species of the same trophic level may be used to estimate an EC_{50} for target or non-target organisms. Interactions between trophic levels may be included by using combinations of producers and consumers. Finally, generic microcosms or outdoor systems such as ponds may be used as test units. A question of great interest is how complex a test system must be in order to give a realistic assessment of hazardous effects in natural communities.

For the herbicide atrazine, Larsen *et al.* (1986) recently compared the sensitivity of different systems. They constructed dose-response curves and estimated the EC_{50} values in single-species tests, generic microcosms and experimental ponds. As

atrazine is a herbicide that inhibits photosynthesis, Larsen *et al.* (1986) chose ^{14}C uptake as the response parameter. Hence they measured the response of organisms closely related to the targets of the herbicide. Effects on non-target organisms, however, are of equal interest if the environmental impact is to be evaluated. Interactions between target and non-target organisms may also produce highly significant indirect effects in the natural community.

Daphnia is frequently used as a test organism to study effects of toxicants on non-target organisms. Test for acute toxicity as well as for chronic effects have been standardized. In the case of a herbicide such as atrazine that inhibits photosynthesis, however, *Daphnia* may be expected to be rather insensitive because it is not the target organism. But this does not mean that *Daphnia* cannot be affected by the herbicide under field conditions where indirect effects may be important.

We test here two hypotheses:

(1) Do indirect effects enhance the sensitivity of a test, making it more powerful for the evaluation of possible hazards of the herbicide in natural communities?

(2) Due to indirect effects in a natural community, can responses of non-target organisms to stress from a herbicide be detected at lower concentrations than those of the photosynthetic target organisms?

The first hypothesis is rather trivial. Indirect effects in the community can result from trophic interdependencies among its members. Barring unexpected toxic effects to a non-target herbivore, the indirect effects will be the result of effects on its algal food (i.e. a target organism).

The second hypothesis is not so self-evident. It must be considered, however, that zooplankton in a natural community are often food limited (Lampert, 1985) and may thus be sensitive to herbicides which affect their food resources. Growth and reproduction result from the difference between food assimilation (which in turn depends on the concentration of edible food) and metabolic losses. A slight reduction of the food concentration may push the energy balance of the zooplankton below the maintenance threshold (Lampert & Schober, 1980). This may drive the zooplankton population to extinction, although the algal biomass has been only moderately reduced.

This paper compares the minimum concentrations of atrazine that cause detectable effects in test systems of different complexity. The simplest system involves only direct acute mortality of *Daphnia* in the absence of algae. Sublethal effects are measured in the presence of algae, but are still direct effects on the daphnids as the food particles are frequently renewed and algal growth is not of interest. A two-step food chain system allows the herbicide to act on both algae and animals, thus it involves direct and indirect effects on *Daphnia*. Finally, daphnids are considered as parts of a natural community (including phytoplankton, bacteria, competing zooplankton and invertebrate predators) in plastic enclosures suspended in a lake. This system is most complex. Reduced primary production influences daphnids not only by altering the food supply but also by changing the environmental conditions (e.g. oxygen and pH). As they were conducted with varying communities at different seasons, the latter tests come closest to the natural conditions. Hence our experiments allow us to compare the effectiveness of the laboratory tests for estimating the environmental impact of the test herbicide.

Material and methods

General

Our comparison comprises laboratory tests with individuals and populations and field tests with natural communities. Laboratory tests were carried out with an obligately parthenogenetic strain of *Daphnia pulicaria* Forbes and with *D. magna* Straus. Both stocks have been kept in the laboratory for several years at 20 °C under dim continuous light. The green alga *Scenedesmus acutus* Meyen was used as exclusive food for the daphnids in all laboratory experiments. Algae were grown with CHU 12 media in chemostats as described in Lampert (1976). We used membrane-filtered (0.45 μm) lake water (either from Lake Constance or from a pond in the Frankfurt Botanical Gardens) as test water. All experiments were run at 20 °C in a temperature controlled water bath.

Community responses were measured in plastic enclosures with natural plankton communities incubated in a lake during different seasons.

Atrazine (2-chloro-4-ethylamino-6-isopropylamino-s-triazine) was provided by Ciba Geigy, Basel. A solvent was not used with the atrazine. All chemical and biological analyses were carried out according to routine protocols. Particulate organic carbon was determined in an infrared carbon analyzer described by Krambeck *et al.* (1981).

Acute toxicity

Adult *D. pulicaria* were exposed to varying concentrations of atrazine in air-saturated filtered lake water in test tubes of 100 ml. Five daphnids were placed into each test tube with 10 replicates per concentration. Numbers of dead (immobile) animals were recorded after 24 and 48 hours.

Growth and reproduction

Individual *D. pulicaria* (one day old at start of the experiment) were kept in glass beakers of 100 ml in *Scenedesmus* suspensions of 1 mg carbon/L with varying concentrations of atrazine (10 replicates per concentration). Each day the medium was changed and the number of offspring was recorded (see Schober & Lampert, 1977). We performed two series of experiments. The first ran for 28 days, the second was continued until all test animals had died (72 days). At the termination of experiment one, length, egg number, dry weight and carbon content per individual were measured for all daphnids.

Feeding inhibition

Adult *D. pulicaria* were fed ^{14}C-labeled *Scenedesmus acutus* to measure their individual feeding rates in the presence of differing concentrations of atrazine (Pott, 1982). Each atrazine treatment was run in parallel with a control without the herbicide. Results are presented relative to the controls. Daphnids were exposed to the experimental atrazine concentration for varying periods prior to the feeding trial.

'Food chain' system

A two-step system was designed to test the combined effects of atrazine on algae and *Daphnia* (Störkel & Lampert, 1981). The system is described in Fig. 1. *Scenedesmus* were grown in two chemostats of identical size and flow rate. Peris-

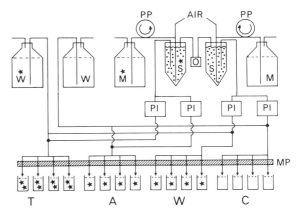

Fig. 1. Schematic representation of the 'food chain' bioassay for the separate test of atrazine effects on *Daphnia* via water and food (Störkel & Lampert, 1981). S, chemostat (*Scenedesmus*); M, algal nutrient medium; W, filtered water; PP, peristaltic pump; PI, piston pump; MP, multichannel peristaltic pump; O, overflow. Groups of beakers at the bottom contain *Daphnia* populations. Asterisk denotes presence of atrazine. Treatments labeled by letters: C, controls; W, water effect; A, algal effect; T, combined effect.

taltic pumps withdrew identical small volumes of algal suspension from the chemostats. Algal subsamples were diluted with membrane-filtered pond water from two reservoirs, then distributed into 2-liter glass beakers equipped with an overflow. Atrazine was added to the nutrient solution of one chemostat and to one reservoir with dilution water. At the beginning of an experiment, each beaker was inoculated with 10 neonate females of *D. magna*.

After the experiment had run for three weeks, the number of animals per beaker, the dry weight of the population and the number of eggs were recorded. Hence, a population response was measured in this case. The design provided four different treatments simultaneously:
(1) Controls with no atrazine.
(2) Atrazine in chemostat; effect of inhibited algal growth.
(3) Atrazine in dilution water; direct effect on *Daphnia*.
(4) Atrazine in both chemostat and dilution water; combined effects of the herbicide on *Daphnia*.

Enclosure experiments

Experiments with natural communities were performed in plastic bags mounted in an aluminum frame floating in Schöhsee, a moderately eutrophic lake at Plön, northern W. Germany. The enclosures had a diameter of 1 m and were about 2.70 m long, holding 1.70 m³ of water. Four bags were filled with water from the lake screened through a 100 μm mesh to exclude large zooplankton. The bags were inoculated with equal amounts of zooplankton collected with a plankton net. Two bags received atrazine; the other two served as controls. Preliminary experiments had shown that all four bags developed very similar communities when left untreated. Experiments were typically run for no longer than three weeks to avoid extensive growth of periphyton on the walls of the bags. Water samples for chemical and biological analysis were taken at regular intervals. Zooplankton was sampled by vertical hauls of a plankton net (100 μm). Atrazine concentrations in the bags were monitored by gas chromatography (Fleckner, 1989).

Results

Table 1 summarizes the effective concentrations of atrazine in our different test setups. The LC_{50} (48 h) is about 10 mg/L (Pott, 1980), a very high concentration considering that it is near the maximum solubility of atrazine in water. Sublethal effects can be detected at concentrations of about one order of magnitude lower. A 50% feeding inhibition of *Daphnia* is measured between 1 and 3 mg atrazine/L (Pott, 1982), varying with the prehistory of the daphnids. Animals adapted to low levels of the herbicide for several hours are less sensitive. The reduced feeding rate is reflected in growth and reproduction of *Daphnia*. Significantly ($p < 0.05$) reduced reproduction, both at the individual and at the population level, and significantly reduced growth was found at 2 mg atrazine/L (Schober & Lampert, 1977).

Atrazine concentrations of another order of magnitude lower caused significant effects in the 'food chain' system (Fig. 2) (Störkel, 1983). The final dry weight of *Daphnia* in the controls varied around 20 mg/L. Significant reductions of daphnid biomass occurred at concentrations of 0.3 and 0.1 mg atrazine/L and in one of the two experiments at 0.05 mg atrazine/L. Hence the limit of detectable effects in this system lies between 0.1 and 0.05 mg atrazine/L. Significant reductions were seen only in the treatments where algae were affected by atrazine [algal (A) and combined (T) effects]. If the herbicide was only in the water, the daphnid biomass after three weeks was sometimes greater than in the controls at the higher atrazine concentrations. This was a consequence of food limitation of *Daphnia* in the system. In the beginning of an experiment, daphnids grow until they overshoot the carrying capacity of the system (algal input to the beakers is constant). Then the population begins to oscillate. The control populations were already in the downward phase of the population curve after three weeks.

Table 1. Effective concentrations of atrazine in bioassays of different complexities. Sources: 1, Pott (1980); 2, Schober & Lampert (1977); 3, Störkel & Lampert (1981); 4, Fleckner (1989)

Assay	Parameter measured	Effective concentration (mg/L)	Source
Daphnia; acute toxicity	LC_{50} (48 h)	10	1
Daphnia; feeding inhibition	50% inhibition	1–3	1
Daphnia; growth, reproduction	sign. ($p < 0.05$) reduction	2	2
Daphnia in 'food chain'	Population dry weight, sign. reduction	0.05–0.1	3
Enclosures with natural communities	Community response	0.001 (0.0001)	4

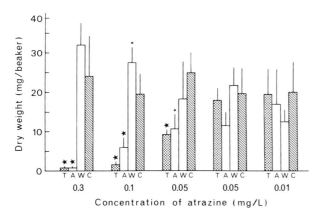

Fig. 2. Biomass of *Daphnia* populations in the 'food chain' bioassay after three weeks in experiments with different atrazine concentrations. Letters denote treatments specified in Fig. 1 (C, controls; W, water effect; A, algal effect; T, combined effect). Vertical bars = S.D. (n = 4). Asterisk indicates mean significantly different from control (Tukey's test; large symbol, p < 0.01; small symbol, p < 0.05).

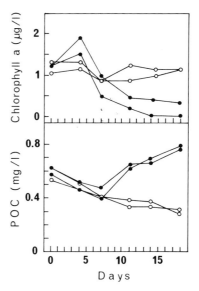

Fig. 3. Enclosure experiment with 1 μg atrazine/L. Time course of chlorophyll *a* and particulate organic carbon (POC). Open circles: control bags; full circles: bags with atrazine.

Atrazine in the water did not affect the carrying capacity, but retarded the development of the daphnid population, so that they were near their maxima when they were harvested. The duration of the test is, therefore, of critical importance. For our system, it should have been either shorter than three weeks or considerably longer.

Community effects in enclosures were monitored at the lowest atrazine levels. We found similar patterns in all experiments between 100 and 1 μg atrazine/L. After a lag time, primary production decreased in the treatments, but not in the controls. Chlorophyll *a* decreased as algae died, while a considerable growth of bacteria occurred. Oxygen concentration decreased in consequence. Reduced food availability and poor food conditions (visible in a drop in fecundity; cf. also chlorophyll levels near zero in Fig. 3) led to considerable changes in zooplankton abundance and composition, cladocerans being more affected than copepods. Atrazine proved to be very persistent in the system. About 90% of the atrazine added to the enclosures was still present in the water after 18 days at the 1 μg/L level. Losses were even lower (about 5%) at the higher concentrations. We do not present the results at

100 μg/L and 10 μg/L atrazine, as they were qualitatively identical to those at 1 μg/L, but were even stronger.

Figures 3–9 show that bags treated in the same ways usually behaved similarly. The lag time was about 7 days at 1 μg atrazine/L. After this initial period, chlorophyll *a* decreased in the herbicide treated bags while it remained constant in the controls (Fig. 3). In contrast, particulate organic carbon decreased slowly in the controls, probably due to sedimentation and grazing, but increased in the herbicide treatments, because bacterial production and detritus formation from decaying phytoplankton and zooplankton were high. Oxygen concentration in the controls was always around saturation or slightly above (Fig. 4). In the atrazine bags, however, oxygen decreased rapidly after the seventh day due to decreased photosynthesis and increased bacterial oxygen consumption. Bacterial and blue-green filaments occurred at that time. All zooplankton were negatively affected (Fig. 5), though they responded differently. *Daphnia* decreased to very low levels in the treatments, but increased slightly in the con-

420

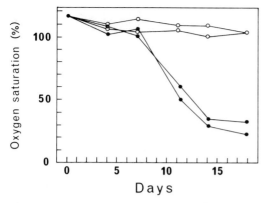

Fig. 4. Enclosure experiment with 1 μg atrazine/L. Time course of oxygen saturation. Open circles: control bags; full circles: bags with atrazine.

trols. The decrease was due to a failure of reproduction. Near the end of the experiments, daphnids carried no more eggs. *Bosmina* was rare in the beginning of the experiment and stayed rare in the atrazine bags, but increased greatly in the controls. Cyclopoids and nauplii decreased at a lower rate than *Daphnia; Cyclops* seemed to be particularly resistant. These responses were also observed in the other series. Cladocerans evidently suffered from malnutrition in the atrazine treat-

ments. The decline in algal food density and particle quality were probably both important. Moreover, the increasing abundance of flakes and filaments may have interfered with food collection. Cyclopoids may have been less affected as they are raptorial feeders that prey on small zooplankton. Besides food shortage, oxygen concentration of less than 2 mg/L may have had a negative effect on all zooplankton.

Tests with 0.1 μg atrazine/L were performed twice during different seasons in 1982, once in June/July at high water temperature (W) and once in September when the water temperature was lower (C). When the cold series was terminated, bags were left in the lake for another 17 days and sampled again at day 42. A third series was performed in August/September 1984. These series differ somewhat from the experiments at higher atrazine concentrations. At this low concentration, atrazine was lost from the system rather quickly and was not detectable after 10 days. In 1982, the response to the herbicide was faster in the warm than in the cold series, but otherwise the results were similar. No differences between treatments and controls occurred in the third series. We, therefore, present only data for 1982.

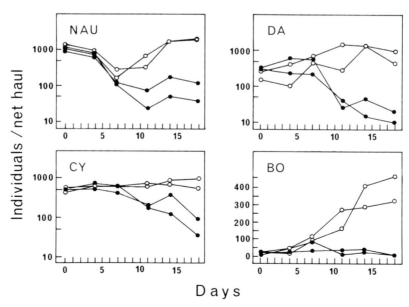

Fig. 5. Enclosure experiment with 1 μg atrazine/L. Time course of zooplankton (NAU, nauplii; DA, *Daphnia*; CY, *Cyclops*; BO, *Bosmina*). Open circles: control bags; full circles: bags with atrazine.

Algal biomass did not differ between treatments and controls in the warm series (Fig. 6). Chlorophyll *a* was slightly lower in the atrazine bags in the cold series, but returned to the same level as in the controls after 25 days. This increase continued, so that after 42 days algal biomass in the atrazine bags was nearly 5 times as high as in the controls. Atrazine evidently caused an algal bloom. Primary production was more affected by 0.1 µg/L atrazine than was chlorophyll *a*, but the recovery of the algal community was also visible here after about two weeks (Fig. 7). At this time, atrazine concentrations seem to have been too low to further inhibit photosynthesis. Oxygen concentration was reduced in the herbicide treatments, but not so severely as at higher concentrations of atrazine (Fig. 8). Recovery of the oxygen concentration occurred after two and three weeks, respectively. Oxygen saturation had returned to nearly 100% at day 42 in the cold series. *Daphnia* in the herbicide treatments were nearly eliminated, although the conditions were not as unfavorable as in the experiment at 1 µg atrazine/L (Fig. 9). However, they started to recover when the conditions improved at the end of the cold series.

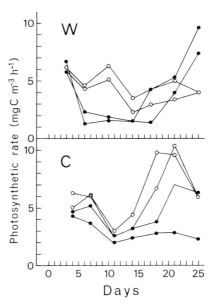

Fig. 7. Enclosure experiment with 0.1 µg atrazine/L. (W, warm series; C, cold series). Time course of primary production. Open circles: control bags; full circles: bags with atrazine.

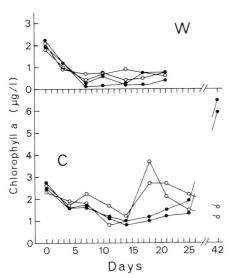

Fig. 6. Enclosure experiment with 0.1 µg atrazine/L. (W, warm series; C, cold series). Time course of chlorophyll *a*. Open circles: control bags; full circles: bags with atrazine.

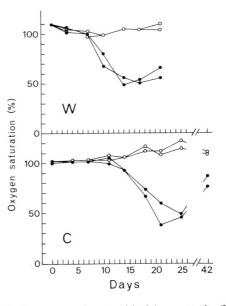

Fig. 8. Enclosure experiment with 0.1 µg atrazine/L. (W, warm series; C, cold series). Time course of oxygen saturation. Open circles: control bags; full circles: bags with atrazine.

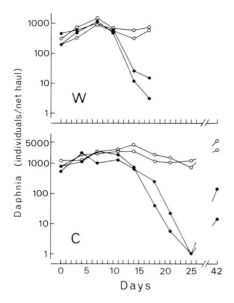

Fig. 9. Enclosure experiment with 0.1 μg atrazine/L. (W, warm series; C, cold series). Time course of *Daphnia* abundance. Open circles: control bags; full circles: bags with atrazine.

Discussion

The excessive use of atrazine has resulted in serious contaminations of surface water and ground water. Concentrations of about 2 μg/L have been detected in the U.S.A. and in the Federal Republic of Germany (e.g. Wilson *et al.*, 1987; Oehmichen & Haberer, 1986). Depending on season and land use, peak concentrations of 74 μg/L have been found in Canada (Frank *et al.*, 1987). Our results suggest that routine toxicological tests are probably not suitable for the assessment of the ecological impact of this herbicide in an aquatic ecosystem.

Comparing our own data (Table 1), our first hypothesis is strongly supported: more complex systems are more sensitive. Although a sensitive algal species may show responses in a single species test at concentrations as low as in our community tests, routine protocols will probably be not so effective. Direct effects on target organisms in single species tests (*Chlorella*) have been reported occasionally at concentrations as low as 1 μg atrazine/L (e.g. Torres & O'Flaherty,

1976), but the effective concentration is often much higher (Butler, 1977). This suggests that direct effects on algae are dependent on algal species and culture conditions. Acute toxicity to *Daphnia* as a non-target organism has no ecological meaning in the case of atrazine. Sublethal tests like feeding inhibition or the measurement of growth and reproduction are an order of magnitude more sensitive than tests of acute toxicity, but still cannot be used directly to evaluate the ecological hazard of atrazine.

Our 'artificial food chain' system includes both target and non-target organisms. It is again more sensitive than the direct tests, although it has some shortcomings. Chemostat systems are not ideal for measuring algal growth inhibition, as the effect of the herbicide may be balanced by reduced nutrient limitation of the slower growing algae. Figure 2 demonstrates that the direct effect of atrazine via the water is low, as would be expected from the simpler tests. The reduction of *Daphnia* biomass is caused by the reduced food availability.

Generic microcosms (e.g. Taub & Crow, 1981) are the next step in complexity. They indicate herbicide effects at 0.05 mg atrazine/L and above (Brockway *et al.*, 1984; Stay *et al.*, 1985), when responses of target organisms (photosynthesis) are measured. Larsen *et al.* (1986) calculated an EC_{50} of 0.1–0.16 mg atrazine/L for microcosms. It is interesting to note that the target organisms (algae) in the microcosms are not more sensitive than the non-target organisms (*Daphnia*) in our two-species 'food chain' system.

Effects of atrazine on natural communities in Kansas ponds have been measured by DeNoyelles *et al.* (1982). They concentrated on target responses and found effects at concentrations of 20 μg atrazine/L in the field, while 1 μg/L was effective in laboratory experiments. 20 μg/L was also the lowest atrazine concentration that affected the aquatic insect community in the experimental ponds (Dewey, 1986). However, as this was the lowest concentration tested, even lower levels may have been effective. Our enclosure communities responded at considerably lower levels (1 μg/L). 0.1 μg/L was probably the limit for our system, because communities recovered from the initial effects of the herbicide in two of three series, but no effect was found in the third series. Thus, climatic conditions and the initial plankton composition were important for the results at this low level of atrazine.

The differences between the two 'natural' systems are not easy to explain. Our bag system did not include the bottom of the lake as the pond system of DeNoyelles *et al.* (1982), but adsorption of atrazine to the sediment is unlikely to be important as the herbicide persisted in the pond water as well as in our bags. Our enclosures, however, may exaggerate the herbicide effects, because exchange with the surrounding water is limited. For example, the gas exchange is reduced although the bags are open to the atmosphere. This may enhance the decrease of oxygen, a problem that would probably not occur in a shallow pond. Indeed, Dewey (1986) did not observe oxygen depletion after the herbicide application. Life cycles of organisms in the bags are isolated from the rest of the lake, so that recolonization is not possible. Zooplankton may play a more important role in the bags than in the lake because fish predators are excluded. All these factors may magnify the indirect effects of the herbicide. Selection for atrazine resistant phytoplankton may also explain the difference. If atrazine is frequently used in the surroundings of the Kansas ponds, one might expect that the algal community is already adapted to low concentrations of the herbicide. Schöhsee has a very small catchment area, and use of atrazine is probably negligible, because no corn or other treated crops are grown in the basin. Thus the phytoplankton in the lake may be more susceptible.

Larsen *et al.* (1986) concluded that target organisms react more sensitively in complex systems than in pure cultures, but Plumley & Davis (1980) found the opposite. Our data suggest that in complex systems non-target organisms may be even better indicators of herbicide stress than algae. Daphnids show low sensitivity to atrazine when direct effects are measured, but are as sensitive as target organisms in the moderately complex 'food chain' assay. In the very complex natural system, daphnids seem to be even more sensitive than phytoplankton. In the experiments with 0.1 μg atrazine/L, phytoplankton biomass did not differ between treatments and controls (Fig. 6), but *Daphnia* showed a clear response (Fig. 9). This does not mean that the algae did not respond to the herbicide. Algal biomass decreased in all bags. In the control bags, however, it was reduced by intensive grazing of *Daphnia*. Photosynthetic rate was a more sensitive indicator than algal biomass (Figs. 7 and 8). Small effects on the algal resource may have been enhanced in the herbivores. It is interesting to note that in the pond experiments of Dewey (1986) predatory insects were less affected than herbivores, so that the composition of the fauna changed. This is consistent with our observations that filter-feeding cladocerans were more affected than predatory or omnivorous *Cyclops*. The basic principle may be the same in benthic and pelagic systems.

We found community responses to atrazine at unexpectedly low levels. Concentrations around 1 μg/L are frequently found in surface waters in agricultural areas. Hence, lake communities in these areas may already be modified by the chemical (Shapiro, 1980). The induction of mass development of phytoplankton in the low atrazine treatments (Fig. 6) after the significant reduction of grazers indicates how the balance between trophic levels can be disturbed by minor pulses of a toxicant, resulting in instability and unexpected events. Dewey (1986) concluded that indirect effects of the toxicant in a complex community may be more important than direct effects. Our comparison confirms this statement. As a consequence, complex systems are required if the ecological impact of a toxicant is to be evaluated.

Acknowledgements

Part of this work was done at the Limnological Institute of the University of Konstanz and at the Zoological Institute of the University of Frankfurt. Financial support was provided by Deutsche Forschungsgemeinschaft and Bundesminister für Forschung und Technologie. We are especially grateful to Barbara E. Taylor for linguistic help.

References

Brockway, D. L., P. D. Smith & F. E. Stancil, 1984. Fate and effect of atrazine in small aquatic microcosms. Bull. Envir. Contam. Toxicol. 32: 345–353.

Butler, G. L., 1977. Algae and pesticides. Residue Rev. 66: 19–62.

DeNoyelles, F., Jr., W. D. Kettle & D. E. Sinn, 1982. The responses of plankton communities to atrazine, the most heavily used pesticide in the United States. Ecology 63: 1285–1293.

Dewey, S. L., 1986. Effects of the herbicide atrazine on aquatic insect community structure and emergence. Ecology 67: 148–162.

Fleckner, W., 1989. Ökotoxikologische Untersuchungen mit Herbiziden in eingeschlossenen Wasserkörpern. Direkte und indirekte Wirkungen von 2,4-D und Atrazin auf Planktonbiocoenosen und physikalisch-chemische Parameter. Diss. Univ. Kiel, 240 pp.

Frank, R., B. D. Ripley, H. E. Braun, B. S. Clegg, R. Johnston & T. J. O'Neill, 1987. Survey of farm wells for pesticide residues, Southern Ontario, Canada, 1981–1982, 1984. Arch. Envir. Contam. Toxicol. 16: 1–8.

Krambeck, H.-J., W. Lampert & H. Brede, 1981. Messung geringer Mengen von partikulärem Kohlenstoff in natürlichen Gewässern. GIT Fachz. Labor. 25: 1009–1012.

Lampert, W., 1976. A directly coupled, artificial two-step food chain for long-term experiments with filter-feeders at constant food concentrations. Mar. Biol. 37: 349–355.

Lampert, W. (ed.), 1985. Food limitation and the structure of zooplankton communities. Arch. Hydrobiol. Beih. Ergebn. Limnol. 21: 1–497.

Lampert, W. & U. Schober, 1980. The importance of 'threshold' food concentrations. In W. C. Kerfoot (ed.), Evolution and ecology of zooplankton communities, University Press of New England, Hanover, New Hampshire, pp. 264–267.

Larsen, D. P., F. DeNoyelles, Jr., F. Stay & T. Shiroyama, 1986. Comparisons of single-species, microcosm and experimental pond responses to atrazine exposure. Envir. Toxicol. Chem. 5: 179–190.

Oehmichen, U. & K. Haberer, 1986. Stickstoffherbizide im Rhein. Vom Wasser 66: 225–241.

Plumley, F. G. & D. E. Davis, 1980. The effect of a photosynthesis inhibitor atrazine, on salt marsh edaphic algae, in culture, microecosystems, and in the field. Estuaries 3: 271–277.

Pott, E., 1980. Die Hemmung der Futteraufnahme von Daphnia pulex – eine neue limnotoxikologische Meßgröße. Z. Wasser Abwasser Forsch. 13: 52–54.

Pott, E., 1982. Experimentelle Untersuchungen zur Wirkung von Herbiziden auf die Futteraufnahme von Daphnia pulicaria Forbes. Arch. Hydrobiol. Suppl. 59: 330–358.

Schober, U. & W. Lampert, 1977. Effects of sublethal concentrations of the herbicide Atrazin on growth and reproduction of Daphnia pulex. Bull. Envir. Contam. Toxicol. 17: 269–277.

Shapiro, J., 1980. The importance of trophic-level interactions to the abundance and species composition of algae in lakes. Develop. Hydrobiol. 2: 105–116.

Stay, F. S., D. P. Larsen, A. Katko & C. M. Rohm, 1985. Effects of atrazine on community level responses in Taub microcosms. In T. P. Boyle (ed.), Validation and predictability of laboratory methods for assessing the fate and effects of contaminants in aquatic ecosystems. ASTM STP 865, American Society for Testing and Materials, Philadelphia, pp. 75–90.

Störkel, K.-U., 1983. Toxikologische Untersuchungen an einer zweistufigen Labornahrungskette. Diss. Univ. Frankfurt/Main, 105 pp.

Störkel, K.-U. & W. Lampert, 1981. Sublethale Schadstoffwirkungen auf ein Modell einer limnischen Nahrungskette. Verh. Ges. Ökol. 9: 255–260.

Taub, F. B. & M. E. Crow, 1981. Synthesizing aquatic microcosms. In J. P. Giesy (ed.), Microcosms in ecological research, CONF 781101, Technical Information Service, Springfield, VA, pp. 69–104.

Torres, A. M. R. & L. M. O'Flaherty, 1976. Influence of pesticides on Chlorella, Chlorococcum, Stigeoclonium (Chlorophyceae), Tribonema, Vaucheria (Xanthophyceae) and Oscillatoria (Cyanophyceae). Phycologia 15: 25–36.

Wilson, M. P., E. P. Savage, D. D. Adrian, M. J. Aaronson, T. J. Keefe, D. H. Hamar & J. T. Tessari, 1987. Groundwater transport of the herbicide atrazine, Weld County, Colorado. Bull. Envir. Contam. Toxicol. 39: 807–814.

Hydrobiologia **188/189**: 425–531, 1989.
M. Munawar, G. Dixon, C. I. Mayfield, T. Reynoldson and M. H. Sadar (eds)
Environmental Bioassay Techniques and their Application.
© 1989 *Kluwer Academic Publishers. Printed in Belgium.*

A new standardized sediment bioassay protocol using the amphipod *Hyalella azteca* (Saussure)

U. Borgmann & M. Munawar
Great Lakes Laboratory for Fisheries and Aquatic Sciences, Department of Fisheries and Oceans,
Burlington, Ontario, Canada L7R 4A6

Key words: amphipods, *Hyalella*, sediment bioassay

Abstract

A new standardized bioassay procedure for testing the chronic toxicity of sediments to *Hyalella azteca* was developed. Tests were initiated with 0–1 wk old amphipods exposed to sediments from Hamilton Harbour, Toronto Harbour, and Lake Ontario for 4 to 8 weeks. Both survival and growth were significantly reduced in the Hamilton Harbour sediments relative to those from the lake after 4 weeks. Exposures of 8 weeks resulted in greater variability; survival of amphipods in sediments from one of the harbour stations, and growth in sediments from both harbour stations with surviving young were not statistically different from survival and growth in lake sediments. Growth and survival in lake sediments were comparable to cultures grown with cotton gauze and no sediment after 4 weeks, but survival was poorer by week 8. Replication was good in 12 out of 13 tests done in duplicate; the difference in survival between replicates averaged 2.2 animals (20 amphipods/replicate, 4 week exposure). We propose that 4 week exposures of young (0–1 wk old) *Hyalella* would provide a suitable standardized chronic toxicity test for sediments. A detailed protocol on the methodology is presented.

Introduction

Until recently, bulk chemical characterization of sediment formed the basis of sediment guidelines used for the evaluation of dredging/disposal activities. However, such guidelines lack toxicological information essential for assessing the impact of sediment-bound contaminants on biota (Munawar *et al.*, 1984; Munawar & Munawar, 1987). The Water Quality Board of the International Joint Commission has recognized this problem (Thomas *et al.*, 1987; I. J. C., 1987). One of the recent reports developed by the Sediment Subcommittee and its Assessment Work Group to the Water Quality Board recommended a battery

of laboratory bioassays for sediment bioassessment (I. J. C., 1988). The battery included benthic invertebrate bioassays using *Hyalella azteca*, *Chironomus tentans*, and *Hexagenia limbata*. However, currently no standardized methodology is available to conduct such bioassays.

Amphipods are abundant and ecologically important components of the food webs of many freshwater ecosystems. *Pontoporeia hoyi* makes up the bulk of the benthic biomass in the deeper waters of the St. Lawrence Great Lakes (Cook & Johnson, 1974), whereas *Hyalella azteca* (Saussure) and several species of *Gammarus* often dominate the biomass of nearshore or shallow areas (Barton & Hynes, 1976; Clemens, 1950;

Winnell & Jude, 1987). These amphipods are an important component of the diet of many species of fish (Brandt, 1986; Dryer *et al.*, 1965; Johnson & McNeil, 1986; Wojcik *et al.*, 1986) and are sometimes heavily contaminated with persistent organic compounds such as PCB's and DDT (Borgmann & Whittle, 1983; Fox *et al.*, 1983; Whittle & Fitzsimons, 1983). These attributes, along with their benthic habit and extreme sensitivity to toxic substances (Arthur & Eaton, 1971; Macek *et al.*, 1976a, 1976b), make them valuable organisms for sediment toxicity bioassays.

Not all species of amphipods are equally easy to use in bioassays. *Pontoporeia* is a deep, cold-water genus with a long life cycle. Chronic bioassays lasting over 1 yr have been performed with this organism (Sundelin, 1983), but most researchers would prefer tests of shorter duration. *Gammarus lacustris* is a cool-water organism which requires a period of short days and long nights (a simulated winter) in order to reproduce (de March, 1982). *G. fasciatus* and *H. azteca* are warm-water organisms which breed and grow quickly at high temperatures (25 °C) under long photoperiods. Chronic bioassays can be conducted more quickly with these amphipods. *G. fasciatus*, however, is cannibalistic (Clemens, 1950), which can create complications in bioassays. Of the three species we have tested in our laboratory (*H. azteca*, *G. fasciatus* and *Crangonyx gracilis*), we have found *Hyalella* to be the easiest to culture and use in toxicity tests for waterborne chemicals (Borgmann *et al.*, 1989). This organism is also very widespread, occurring throughout most of North America (Bousfield, 1958).

Hyalella azteca has been used previously, or suggested for use, in sediment bioassays. Nebeker *et al.* (1984; 1986) demonstrated the effect of sediment on the acute toxicity of cadmium to *Hyalella* and suggested its use in chronic bioassays. Their suggested chronic test was not precisely defined, such as protocols used for the crustacean zooplankter *Daphnia magna* or other cladocera. For example, tests with *Hyalella* call for the use of animals of a given size, either juvenile or adult. The standard *Daphnia* tests (e.g. Biesinger & Christensen, 1972) employ 0 to 24 h old young for both acute and chronic tests. To the best of our knowledge, detailed standardized test procedures for chronic sediment bioassays using *Hyalella*, and examples of their use, are not currently available in the published literature. We describe here a standardized procedure for conducting chronic sediment bioassays using 0 to 1 wk old *Hyalella* and give some preliminary results with sediments from Hamilton and Toronto Harbours.

Methods

The procedure used to obtain young *Hyalella* is described in Borgmann *et al.* (1989). Amphipod cultures were maintained in straight-sided glass jars with pieces of cotton gauze. The gauze was of the surgical bandage type available in drug stores and is essential if large numbers of young are desired. No sediment or other substrate is required. Amphipods were fed Tetra-Min® fish food flakes several times per week. The water was changed in the jars and the young were removed every week, even if no young were required. The age of the young obtained was then always 0 to 1 week. The separated young amphipods were kept in jars with 1 l of water and 5 by 10 cm piece of gauze and 20 mg of Tetra-Min for about 2 days before being used in bioassays. This ensured that animals which would normally die within a day or two due to rough handling during collection were not used in the bioassays. It also provided us with 2 days advance notice of the number of young animals available before bioassays were set up.

Sediment samples were collected by Shipek from stations 4, 5, and 6 in Hamilton Harbour (Fig. 1) on May 18 and from Lake Ontario at an offshore station (43° 32′ 58″ N, 79° 07′ 53″ W) on May 5, 1988. These were used in the first test of the bioassay procedure. Further samples were collected from Lake Ontario and the Toronto Harbour stations at Ashbridges Bay, stations 419, 911, 739 and 910 (Fig. 1) on August 17, and from stations 1, 2, and 3 in Hamilton Harbour on August 18, 1988, and used in the second experiment. In addition, sediment collected from station 419 on March 16 and May 25, 1988 was also

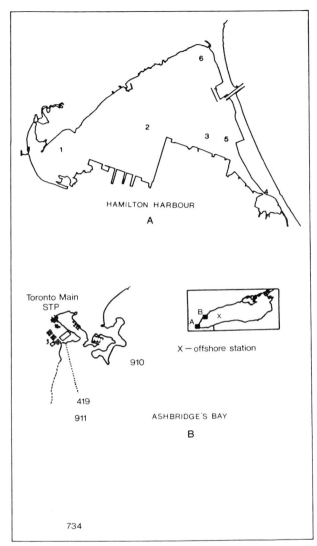

Fig. 1. Maps of Hamilton Harbour and Toronto Harbour (Ashbridges Bay) showing sediment sampling stations and an offshore station in Lake Ontario used as a control. Station 419 coincides with the sewage effluent discharge (pipe shown as dotted line) from the Toronto sewage treatment plant (STP).

give a total volume of 1.5 l and the jars were covered with a sheet of plexiglass. A tygon air line was passed through a hole in the plexiglass and an aquarium airstone attached to the end was suspended several cm above the sediment. Air was bubbled through the air line at a sufficiently slow rate so as to gently keep the water oxygenated but not to resuspend the sediment. The sediments were allowed to settle for several days and the water was given enough time to become oxygenated before amphipods were added. This bioassay is restricted to oxygenated sediments because *Hyalella* cannot withstand extended periods of anoxia. Twenty young *Hyalella* were added to each jar. Twenty mg of Tetra-Min, sifted through a 500 μm mesh screen, was added as food twice a week. The jars were incubated at room temperature (20–22 °C) under fluorescent lights with a photoperiod of 16 h light : 8 h dark. Distilled water was added as needed to keep the water level constant. After 4 weeks, and again after 8 weeks, the contents of two jars with sediment from each location were sifted through a 275 μm nylon screen and the surviving amphipods were sorted, counted, and weighed on a micro-balance.

Results and discussion

Tests were conducted in duplicate for both 4 and 8 week duration in the first experiment to determine the optimum exposure time. Survival in sediments from stations 4 and 6 (Fig. 2) were significantly lower ($P < 0.01$, X^2 or G test with Yates' correction for small sample sizes, Sokal and Rohlf 1969) than in Lake Ontario 'control' sediments after both 4 and 8 weeks. Survival in sediments from station 5 was lower at week 4 ($P < 0.05$), but not significantly different by week 8. The mean body sizes of *Hyalella* grown in sediments from station 5 and 6 (Fig. 3) were both significantly smaller than those grown in Lake Ontario sediments ($P < 0.05$, ANOVA) after 4 weeks, but not statistically different by week 8. It appears that the toxic effects of the sediments are manifest within the first 4 weeks, and that longer

tested in the second experiment. Sediments were stored refrigerated in polyethylene bags until used. Sediment from each site was sifted through a 275 μm nylon screen and added to straight-sided pyrex screwtop jars (15 cm diameter, approx. 2.5 l capacity) giving a sediment layer of 1 to 1.5 cm depth. Sifting was done under water in the test jars. Dechlorinated tap water was then added to

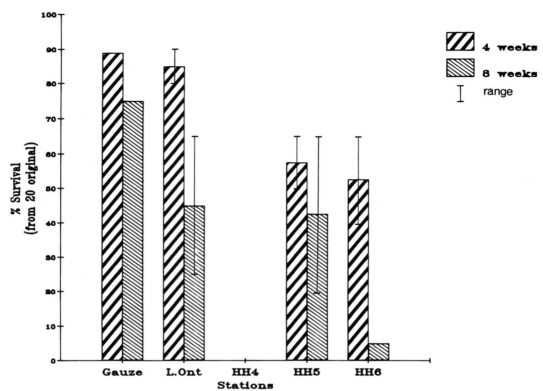

Fig. 2. Percent survival of *Hyalella azteca* after 4 or 8 weeks exposure to sediments from Lake Ontario or 3 stations from Hamilton Harbour. Also shown is the equivalent survival obtained using cotton gauze as a substrate instead of sediments. Vertical bars indicate range for duplicate jars. All animals exposed to sediments from station 4 died within 4 weeks.

exposures increase variability, possibly due to containment effects, making it more difficult to detect toxic effects statistically.

As an additional check on the sediment bioassay method we compared growth and survival with that previously reported for experiments using cotton gauze as substrate instead of sediments (Borgmann *et al.*, 1989). Growth in the Lake Ontario sediment was as good as in cultures with gauze, as was survival to week 4 (Fig. 2 and 3). Survival to week 8, however, was lower, and the amphipods did not reproduce. An incubation period of 8 weeks allowed reproduction to occur in experiments with gauze. However, the water was changed completely and uneaten food and waste products were removed every week. Perhaps this was responsible for the better reproduction with gauze. A complete water change cannot be done in the sediment assay without rough handling of the amphipods and it is there-

fore probably better to limit the exposure period to 4 weeks. Fortunately, tests of 4 to 6 weeks duration using *Hyalella* appear to be just as sensitive as longer experiments for at least some toxicants. With cadmium and pentachlorophenol, for example, reproduction in experiments with gauze as the substrate is not seriously impaired unless the toxicant concentration is high enough to also cause mortality within 4 to 6 weeks (Borgmann *et al.*, 1989).

The second bioassay was conducted for only 4 weeks using sediments collected from Hamilton Harbour, Toronto Harbour, and Lake Ontario (Table 1). Survival and growth were lower for the control (Lake Ontario) station than in the first experiment, but the differences were not statistically different. Survival was significantly reduced, relative to Lake Ontario, at stations 1 and 2 In Hamilton Harbour, but not at station 3. Growth was not significantly reduced at any of the

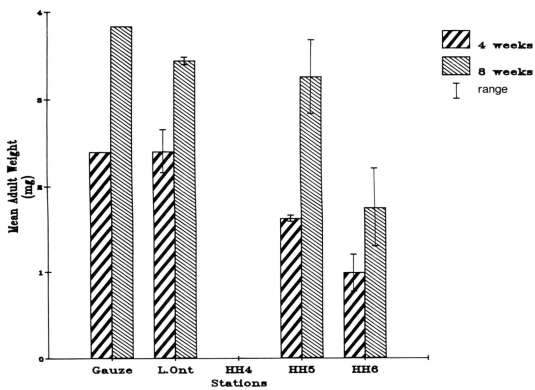

Fig. 3. Mean wet weight of *Hyalella azteca* after 4 or 8 weeks exposure to sediments from Lake Ontario or 3 stations from Hamilton Harbour. Also shown is the equivalent weight obtained using cotton gauze as a substrate instead of sediments. Vertical bars indicate range for duplicate jars. A weight was not obtainable for amphipods exposed to sediments from station 4 because there were no survivors.

stations. Interestingly, the Hamilton Harbour station giving the best survival in the second experiment (station 3) was nearest to station 5 (Fig. 1), which gave the best survival in the first experiment.

Survival of *Hyalella* exposed to sediments from station 419 (Fig. 1) collected in August, near the Toronto sewage treatment plant outflow, was nil. Survival for the nearby station 911 was extremely low, but survival at stations 734 and 910, some distance removed rom the sewage outflow, was good (Table 1). However, survival in sediments collected from station 419 earlier in the season was also good, suggesting that either sediment toxicity was reduced due to storage and delayed testing, or that sediment toxicity varies due to fluctuations in effluent discharges, at least at the station nearest the sewage outflow. Growth for surviving *Hyalella* at all the Toronto stations was

better than that of the control, possibly because of a better nutritional quality of the sediment. Final body weight of *Hyalella* exposed to sediments from stations 734 and 910 were, in fact, similar to that of the Lake Ontario 'control' from experiment 1 (Fig. 2), suggesting that sediment collected from the Lake Ontario station in August was slightly toxic, nutritionally deficient, or in some other way inhibited growth, relative to that collected in May. This would make sense if sediments from the deeper waters of Lake Ontario support better amphipod growth following the spring phytoplankton bloom.

Replicability in general appears to be good for this bioassay procedure. Performing the tests in duplicate gave sufficient data to demonstrate statistical differences among several of the sediments, as indicated above. As an additional check on replicability, we used the G-test (with Yates'

Table 1. Survival and mean weight of *Hyalella* exposed to sediments from Hamilton and Toronto Harbours for 4 weeks. All sediments were collected August 16–17, 1988, except for station 419, as indicated. Tests were run in duplicate with 20 animals/replicate except station 911, which had only a single replicate.

Station	Survival (%) Mean (range)	Weight (mg) Mean (range)
Lake Ontario	72.5 (60–85)	1.42 (1.18–1.66)
Hamilton Harbour 1	42.5 (30–55)*	1.96 (1.81–2.12)
Hamilton Harbour 2	40 (40–40)**	1.31 (1.26–1.36)
Hamilton Harbour 3	55 (20–90)#	1.38 (1.03–1.72)
Toronto 419: March	82.5 (80–85)	1.87 (1.74–2.00)
May	72.5 (65–80)	1.34 (1.29–1.39)
August	0 (0–0)**	–
Toronto 911	5 (5)**	2.50 (2.50)
Toronto 734	70 (65–75)	2.55 (2.20–2.90)
Toronto 910	90 (90–90)	2.85 (2.77–2.93)*

* Significantly different from Lake Ontario at $P < 0.05$.
** Significantly different from Lake Ontario at $P < 0.01$.
# Replicates not consistent, see text.

correction) to test for statistically significant differences in survival between replicates. For the 13 tests done in duplicate, 12 showed no significant differences. In one test, Hamilton Harbour station 3, the survival in one of the replicates (18/20) was much greater than in the other (4/20). Data from this station are, therefore, questionable. For the other 12 tests the average difference in survival between the replicates was 2.2 animals (range 0–5). It is, therefore, sufficient to perform this assay in duplicate, although a larger number of replicates may be used if space and materials are not limiting.

The number of tests done with sediments at the present time from each of the stations is not sufficient to clearly define the seasonal and spatial patterns of sediment toxicity in the two Harbours, but the results are consistent with known contaminant inputs to these Harbours and demonstrate the usefulness of the 4 week sediment bioassay technique with 0–1 week old *Hyalella*. Furthermore, the discovery that cotton gauze greatly improves reproduction in *Hyalella* cultured without sediments provides us with large numbers of young amphipods in a sediment-free medium which can then be used for the sediment assay (Borgmann *et al.*, 1989). The assay is sensitive, has good replicability, and provides data on both chronic mortality and growth. This newly standardized technique is now being applied for the bioassessment of sediments from various areas of concern in the Great Lakes and St. Lawrence River.

Acknowledgements

We thank L. H. McCarthy for collecting the sediments, S. Nielson for conducting the assays, and W. P. Norwood for supplying the young amphipods and for laboratory assistance.

References

Arthur, J. W. & J. P. Eaton, 1971. Chloramine toxicity to the amphipod *Gammarus pseudolimnaeus* and the fathead minnow (*Pimephales promelas*). J. Fish. Res. Bd. Can. 28: 1841–1845.

Barton, D. R. & H. B. N. Hynes, 1976. The distribution of amphipoda and isopoda on the exposed shores of the Great Lakes. J. Great Lakes Res. 2: 207–214.

Biesinger, K. E. & G. M. Christensen, 1972. Effects of various metals on survival, growth, reproduction, and metabolism of *Daphnia magna*. J. Fish. Res. Bd. Can. 29: 1691–1700.

Borgmann, U., K. M. Ralph & W. P. Norwood, 1989. Toxicity test procedures for *Hyalella azteca*, and chronic toxicity of cadmium and pentachlorophenol to *H. azteca*, *Gammarus fasciatus*, and *Daphnia magna*. Arch. Envir. Contam. Toxicol. 18: in press.

Borgmann, U. & D. M. Whittle, 1983. Particle-size-conversion efficiency and contaminant concentrations in Lake Ontario biota. Can. J. Fish. Aquat. Sci. 40: 328–336.

Bousfield, E. L., 1958. Fresh-water amphipod crustaceans of glaciated North America. Can. Field-Naturalist 72: 55–113.

Brandt, S. B., 1986. Ontogenetic shifts in habitat, diet, and diel-feeding periodicity of slimy sculpin in Lake Ontario. Trans. Am. Fish. Soc. 115: 711–715.

Clemens, H. P., 1950. Life cycle and ecology of *Gammarus fasciatus* Say. Ohio State University, Franz Theodore Stone Institute of Hydrobiology, Put-in-Bay, Ohio, Contrib No 12: 63 pp.

Cook, D. G. & M. G. Johnson, 1974. Benthic macroinvertebrates of the St. Lawrence Great Lakes. J. Fish. Res. Bd. Can. 31: 763–782.

de March, B. G. E., 1982. Decreased day length and light intensity as factors inducing reproduction in *Gammarus lacustris lacustris* Sars. Can. J. Zool. 60: 2962–2965.

Dryer, W. R., L. F. Erkkila & C. L. Tetzloff, 1965. Food of lake trout in Lake Superior. Trans. Am. Fish. Soc. 94: 169–176.

Fox, M. E., J. H. Carey & B. G. Oliver, 1983. Compartmental distribution of organochlorine contaminants in the Niagara River and the western basin of Lake Ontario. J. Great Lakes Res. 9: 287–294.

International Joint Commission, 1987. Guidance on characterization of toxic substance problems in Areas of Concern in the Great Lakes Basin. Report to the Great Lakes Water Quality Board. Windsor, Ontario. 177 pp.

International Joint Commission, 1988. Procedures for the assessment of contaminated sediment problems in the Great Lakes. Report from the Sediment Subcommittee and its Assessment Work Group to the Water Quality Board. Report December, 1988, Windsor, Ontario. 140 pp.

Johnson, M. G. & O. C. McNeil, 1986. Changes in abundance and species composition in benthic macroinvertebrate communities of the Bay of Quinte, 1966-84. Can. Spec. Publ. Fish. Aquat. Sci. 86: 177–189.

Macek, K. J., K. S. Buxton, S. K. Derr, J. W. Dean & S. Sauter, 1976a. Chronic toxicity of lindane to selected aquatic invertebrates and fishes. US Environmental Protection Agency, Duluth, Minnesota, Ecol. Res. Ser. EPA-600/3-76-046. 50 pp.

Macek, K. J., K. S. Buxton, S. Sauter, S. Gnilka & J. W. Dean, 1976b. Chronic toxicity of atrazine to selected aquatic invertebrates and fishes. US Environmental Protection Agency, Duluth, Minnesota, Ecol. Res. Ser. EPA-600/3-76-047. 50 pp.

Munawar, M & I. F. Munawar, 1987. Phytoplankton bioassays for evaluating toxicity of *in situ* sediment contaminants. Hydrobiologia 149: 87–105.

Munawar, M., R. L. Thomas, H. Shear, P. McKee & A. Mudroch, 1984. An overview of sediment-associated contaminants and their bioassessment. Can. Tech. Rep. Fish Aquat. Sci. 1253: 136 pp.

Nebeker, A. V., M. A. Cairns, J. H. Gakstatter, K. W. Malueg, G. S. Schuytema & D. K. Krawczyk, 1984. Biological methods for determining toxicity of contaminated freshwater sediments to invertebrates. Envir. Toxicol. Chem. 3: 617–630.

Nebeker, A. V., S. T. Onjukka, M. A. Cairns & D. K. Krawczyk, 1986. Survival of *Daphina magna* and *Hyalella azteca* in cadmium-spiked water and sediment. Envir. Toxicol. Chem. 5: 933–938.

Sokal, R. R. & F. J. Rohlf, 1969. Biometry. W. H. Freeman & Co., San Fransico. 776 pp.

Sundelin, B. 1983. Effects of cadmium on *Pontoporeia affinis* (Crustacea: Amphipoda) in laboratory soft-bottom microcosms. Mar. Biol. 74: 203–212.

Thomas, R. L., E. Evans, A. Hamilton, M. Munawar, T. Reynoldson & M. H. Sadar (eds). 1987. Ecological effects of *in situ* sediment contaminants. Hydrobiologia. 149: 272 pp.

Whittle, D. M. & J. D. Fitzsimons, 1983. The influence of the Niagara River on contaminant burdens of Lake Ontario biota. J. Great Lakes Res. 9: 295–302.

Winnell, M. H. & D. J. Jude, 1987. Benthic community structure and composition among rocky habitats in the Great Lakes and Keuka Lake, New York. J. Great Lakes Res. 13: 3–17.

Wojcik, J. A., M. S. Evans & D. J. Jude, 1986. Food of deepwater sculpin, *Myoxocephalus thompsoni*, from southeastern Lake Michigan. J. Great Lakes Res. 12: 225–231.

Hydrobiologia **188/189**: 433–443, 1989.
M. Munawar, G. Dixon, C. I. Mayfield, T. Reynoldson and M. H. Sadar (eds)
Environmental Bioassay Techniques and their Application.
© 1989 *Kluwer Academic Publishers. Printed in Belgium.*

The valve movement response of mussels: a tool in biological monitoring

Kees J. M. Kramer,[1] Henk A. Jenner[2] & Dick de Zwart[3]
[1] *Laboratory for Marine Research, MT-TNO, P.O. Box, 57, Den Helder, The Netherlands;* [2]*Joint Laboratories and Consulting Services of the Dutch Electricity Supply Companies (KEMA), P.O. Box 9035, 6800 ET Arnhem, The Netherlands;* [3]*National Institute of Public Health and Environmental Protection (RIVM), P.O. Box 1, 3720 BA Bilthoven, The Netherlands*

Key words: biological monitoring, mussels, early warning system

Abstract

Biological sensors are becoming more important to monitor the quality of the aquatic environment. In this paper the valve movement response of freshwater (*Dreissena polymorpha*) and marine (*Mytilus edulis*) mussels is presented as a tool in monitoring studies. Examples of various methods for data storage and data treatment are presented, elucidating easier operation and lower detection limits. Several applications are mentioned, including an early warning system based on this valve movement response of mussels.

Introduction

Monitoring of the quality status of natural waters has traditionally been carried out with physico-chemical techniques. For many years now, attempts have been made to include the biological response of organisms in monitoring systems for the detection of pollution in the aquatic environment (Cairns, 1979; Bayne *et al.*, 1985; Gruber & Diamond, 1988). Bivalves agree very well with the requirements that should be met when selecting a suitable monitoring organism. They are sedentary, abundant and available throughout the year. Furthermore the organisms should have a manageable size and be hardy enough to be handled in the laboratory. (Phillips, 1977, 1980).

The concentration of pollutants in molluscs can serve as an indicator for the level of pollution. Mussels (e.g. the blue mussel, *Mytilus edulis*) have been widely adopted in chemical monitoring and surveillance programmes (Goldberg *et al.*, 1978; NAS, 1980; Widdows *et al.*, 1981; de Kock,

1986). Their ability to accumulate toxicants to a level representative for the integrated environmental conditions, makes them suitable for the characterization of specific ecosystems. In the passive form of chemical biomonitoring local populations are sampled for chemical analysis, whereas the active version involves the exposure of organisms translocated from reference sites. The disadvantage of these bioaccumulation studies is, that the equilibrium concentration is usually obtained only after several weeks of exposure. This makes them unsuitable as an early warning system.

In contrast to the former method, physiological and behavioural changes or reactions are usually fast and thus potentially suitable for a fast response in continuous biological monitoring.

Several biological monitoring systems, using fish, have been developed in recent decades, which are based on the change in rheotaxis, respiration, gill activity or electrical field alterations (Juhnke & Besch, 1971; Poels, 1977;

Gruber & Diamond, 1988; Geller, 1984). The change in activity of *Daphnia sp* (Knie, 1982), the metabolic luminescent activity of bacteria in test systems like 'Microtox' (Beckman) and physiological effects of bivalves (Slooff *et al.*, 1983) have also been applied.

Following the criteria for optimal functioning of organisms in a biological warning system, bivalves offer many possibilities with respect to continuous and automatic detection, reliable and fast response, handling and data interpretation, (Koeman *et al.*, 1979; Cairns, 1979).

If we limit ourselves to the studies of the bivalve molluscs, physiological parameters in particular have been used to detect environmental changes, caused not only by pollutants, but also due to natural variation (Widdows, 1973; Davenport, 1979; Akberali & Trueman, 1985), e.g. the heart rhythm of *Scrobicularia plana* (Akberali & Black, 1980), reproduction and larval development, respiration, heart rhythm, pumping or filtration rate, shell growth and valve movement of *Mytilus edulis* (Abel, 1976; Manley & Davenport, 1979; Sabourin & Tullis, 1981; Manley, 1983; Slooff *et al.*, 1983; Manley *et al.*, 1984; Bayne *et al.*, 1985), the activity of *Anodonta cygnea* (Barnes, 1955; Salanki & Lukacsovics, 1967; Salanki & Varanka, 1976) and the burrowing activity of *Venerupis decussata* (Stephenson & Taylor, 1975). The valve movement of both freshwater and marine mussels has attracted much attention. The method is based on the fact that most mussels have their shells open for respiration and feeding most of the time. It has been shown that they close their shells under stress for an extended period of time. This valve movement detection method was used to study both natural changes in the environment and the effect of pollutants, e.g. the effects of temperature (Hiscock, 1950), light (Bennett, 1954; Ameyaw-Akumfi & Naylor, 1987), tidal movements, salinity (Davenport, 1979, 1981; Akberali & Davenport, 1982), food quantity and quality (Higgins, 1980) and a series of toxicants like trace metals (Salanki & Varanka, 1976; Davenport, 1977; Manley & Davenport, 1979; Kramer *et al.*, 1989), pesticides (Salanki & Varanka, 1978) and other trace organics (Sabourin & Tullis, 1981; Slooff *et al.*, 1983).

Since Marceau (1909) first used sooted glass and a mechanical system to record the valve movement of mussels, several comparable systems have been described (Salanki & Balla, 1964). Other investigators used strain gauges for the detection of the valve movement (Djangmah *et al.*, 1979; Higgins, 1980). Schuring & Geense (1972) were the first to use electromagnetic induction to measure the valve displacement. This system was improved into a High Frequency (HF) electromagnetic induction system, EMIS (Noppert, 1987; Jenner *et al.*, 1989). In this paper we will discuss the response of the zebra mussel and the blue mussel to several toxicants, the various possibilities of data collection and data interpretation as recorded with this new system. Examples will illustrate the various approaches that have been followed.

Materials and methods

Monitoring apparatus

The electromagnetic induction system basically contains a high frequency oscillator, two tiny coils and an amplifier, with a power supply. One coil acts as transmitter of a magnetic field generated by a HF oscillating current (500 kHz). The other coil will, depending on the distance from the transmitting coil, intercept part of this magnetic field and a current will be induced which is proportional to the distance between the shell halves (Schuring & Geense, 1972; Jenner *et al.*, 1989).

The received signal (continuous scanning) is amplified and corrected for maximum opening of the valves. The minimum span between the transmitting and receiving coils is set to about 20 mm, by appropriate placement of the coils on the shells. Each amplifying circuit (one for each mussel) is trimmed for optimal sensitivity by zero setting when the valves are closed and maximum amplification when fully open. The resolution of the device is dependent on the recording method used; an 8 bit A/D converter results in a resolution of better than 0.02 mm.

The coils are manufactured by winding copper

wire (0.08 mm diameter; 155 windings) around a ferrite core (3 mm × 1.5 mm diameter). The coils are each fixed to a thin poly-urethane coaxial cable (ca. 2 m, 1 mm diameter) (Capable BV, Breda, NL) and encased in plexiglass. The dimensions of the encapsulated coils is approximately 14 × 9 × 5 mm (*l* × *w* × *h*). The coils are glued to opposite shell valves of the mussel with a non-toxic two-component acrylic resin (Unifast, GC Dental Industrial Corp.). More technical details and electronic schemes are given in Jenner *et al.* (1989).

Our design using coaxial cables is not restricted to mussels with a byssus, fixed to a solid substrate. The method can well be applied to free-living, infaunal mussels, and allows sufficient movement at the sediment surface.

To prevent the mussel from moving through the tank, which may cause interference between the different signals, the mussels are usually glued to perspex plates attached to the tank. So far we did not observe any negative effects of the EMIS system, either on free moving or on (artificially) attached test mussels. A dampproof plastic box

contains the necessary electronic parts for the signal generation and reception of 5 mussels. Output is directed either to a chart recorder or, after A/D conversion of the signal, to a computer. Measurements were generally carried out with 10 mussels simultaneously. A schematic diagram is presented in Fig. 1.

The (prototype) early warning system consisted of a waterproof PVC housing containing a dedicated microcomputer for evaluation of the mussel response (see later) and a battery operated power supply. Eight mussels were glued to the outside of the container, the total system was submerged in the river. An output line was connected to a printer on shore (De Zwart & Slooff, 1987).

Experimental set-up

The experiments were performed with two different bivalve species: the freshwater zebra mussel, *Dreissena polymorpha* (3–3.5 cm) and the marine blue mussel, *Mytilus edulis* (3.5–4 cm). Three types of experiments were performed, two in a laboratory system:

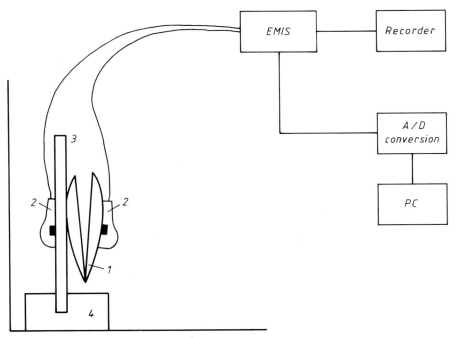

Fig. 1. Schematic presentation of the electromagnetic induction system (EMIS) based valve movement detection system. 1: Mussel; 2: coils; 3: perspex substrate; 4: support.

– continuous, single toxicant exposure of *Dreissena*;
– single toxicant exposure of *Mytilus*, with additions of separate spikes; and one in the natural environment:
– the use of *Dreissena* in an early warning system.

In the continuous toxicant exposure experiment ten organisms (*Dreissena polymorpha*) were kept in flow-through tanks ($2 \times 0.25 \times 0.25$ m, $l \times w \times h$), using river Rhine water at a flow rate of 1 l/min. The temperature was 15-18 °C. The water passed over a series of partitions (lamella separator) which resulted in a major reduction of the silt content; however, algae remained available. The experiments lasted 24 hours. Toxic chemicals (trace metals or hypochlorite) were added continuously using a peristaltic pump (Microperpex, LKB) from newly prepared stock solutions. Addition of Tributyltinoxide (TBTO) from a stock solution is hampered because of its adsorption to container walls and tubing. Therefore, inflow water was passed over different sizes (for different concentrations) of TBTO-impregnated, neoprene rubber sheets (NoFoul, Goodrich). This proved to be an efficient continuous TBTO source.

Duplicate groups of five *Mytilus edulis* were kept in Dutch coastal seawater, in static systems (16 l) for about 20 hours. The behaviour of the mussels was compared before and after addition of the toxicants (trace metals, hypochlorite, oil or TBTO), which were performed in one spike, using a timer and a peristaltic pump (Gilson). To minimize external effects upon the experiments (light, vibration, etc.) the additions were carried out during the night.

Actual pollutant concentrations were checked by chemical analysis, using standard methods for trace metal analysis (atomic absorption spectrometry or anodic stripping voltammetry). TBTO was analysed by AAS after extraction in hexane. Concentrations of hypochlorite were based on amperometric measurement of the total residual oxidants.

For the early warning system experiment zebra mussels, collected in a lake near Stockholm, were acclimated to water of the Göta Alv river for a period of three days. The early warning system was tested in the period August 23–28, 1985 in the plume of the regulation tunnel entering the Göta Alv river at the Trollhätan Olidenhalan power station (Sweden) (De Zwart & Slooff, 1987).

Results and discussion

Valve movement response to pollutants

A characteristic response of *Dreissena* to constant levels of chlorine (added as hypochlorite) is presented in Fig. 2. It shows a distinct increase of the time closed, with intermittent periods of activity. The length of these active periods decreases with the chlorine concentration until virtually no activity is recorded at a concentration level of 500 µg/l. Slooff *et al.* (1983) gave detection limits for several organic compounds of the valve movement response of *Dreissena*. The average of three measurements ranged from 105 mg/l for chloroform to 0.11 mg/l for γ-Hexachlorocyclohexane.

In the real time recording of the valve movement response of *Mytilus* in Fig. 3 an example of a mussel in the blank seawater is given. It will be clear that most of the time the mussel stays open and closes only for short periods. The time between these short enclosure events may vary substantially between mussels. Addition of 37.5 µg/l copper (as copper sulphate) results in a closure response within minutes, while at an addition of 20 µg/l Cu the closure response is delayed and is less clearly visible. This is confirmed by earlier work of e.g. Manley & Davenport (1979). A more detailed study on the effect of various copper species is given by Kramer *et al.* (1989). Addition of lower copper concentrations are less easy to detect, as the mussels instead of closing completely, only reduce the maximum 'open' reading. Application of a computer programme for data interpretation (see later) results in a detection limit of about 1–10 µg/l copper. Other estimated sensitivities are given in Table 1.

Interestingly both *Mytilus* and *Dreissena* have

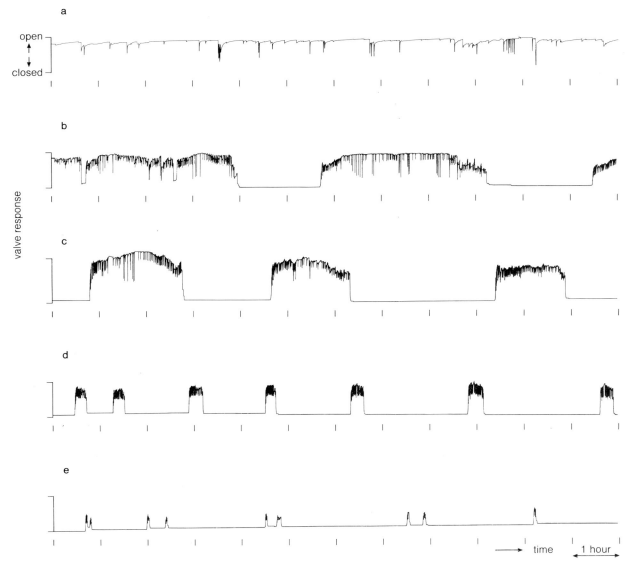

Fig. 2. Real time recording of the valve movement of *Dreissena polymorpha* under a continuous stress of chlorine. Chlorine additions (in µg/l): A:blank; B: 37; C: 55; D: 180; E: 550.

comparable low detection limits for the typical antifouling agents copper, hypochlorite and TBTO.

Data collection

The collection of the data and the kind of the information stored will be dependent on the objective of the experiment or the monitoring purpose, the storage device and the method of data treatment. Considering the kinetics of the valve movement, it is reasonable for short-term experiments to store and interpret the data in (pseudo) real-time. However, if monitoring will take weeks or even months, data storage and certainly data treatment becomes a serious problem. Therefore we distinguish:

438

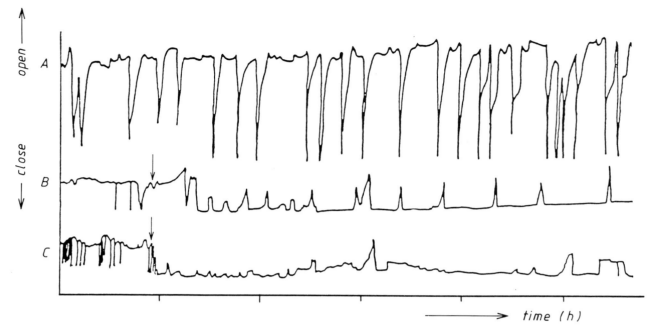

Fig. 3. Real time recording of the valve movement response of *Mytilus edulis* to the additions of one spike of 20 μg/l (B) and 37.5 μg/l (C) copper (see arrows), and in a blank situation.

– Real-time data collection

In real-time collection a continuous recording is made, either on a chart recorder or on magnetic tape. The graphs in Fig. 2 are made this way. On chart recorders the number of pens is usually

Table 1. Estimated detection ranges for several toxicants as detected by the valve closure response of *Mytilus edulis* (M) and *Dreissena polymorpha* (D) using the EMIS.

Compound	Estimated detections limit (μg/l)	Organisms
Copper	< 10	MD
Cadmium	< 100	MD
Selenium	< 100	D
Zinc	< 500	MD
Lead	< 500	MD
TBTO	< 10	MD
Chlorine	< 10	MD
Dispersed crude oil	< 6000	M

* A positive response was recorded when seven or more mussels out of ten reacted (a closing response or a change in activity, while validity was checked by analysis of variance).

limited, reducing the number of mussels that can be followed simultaneously.

– Pseudo real-time data collection

Application of an A/D conversion (datalogger), offers the possibility to sample the analogue signal at frequent intervals. Five second intervals results in a picture (Fig. 4a) that can hardly be distinguished from a real-time graph. Increase of the sampling intervals tends to reduce the sharp peaks recorded, but it takes relatively long (4 min intervals), before the information stored becomes difficult to interpret (Fig. 4b–e).

– Activity registration

It is commonly observed that mussels do not change their valve opening position for some time. To reduce the stored data volume, in this case data are collected only when changes in the signal are observed. Therefore, after an evaluation procedure, only the change from a constant value to a more positive (or a negative) value will be recorded vs time. As no values are recorded when there is no change in the signal, no 'real' time

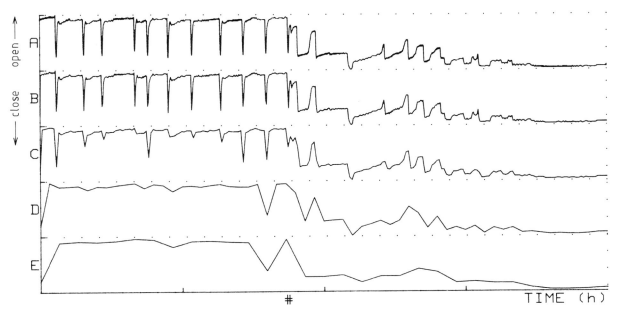

Fig. 4. Pseudo real-time recording of the valve movement response of *Mytilus edulis* as a function of sampling intervals: every 5 s (A), 20 s (B), 60 s (C), 4 min (D) and 8 min (E). A spike of 50 $\mu g/l$ copper was added at '#'.

graph can be (re)constructed from the stored data. This method results, for the mussels described, in a data reduction of about 97%, as compared with the 5 s sampling interval.

– Dedicated Early Warning system

Based on the fact that it is very unusual that several mussels are closed for a prolonged period, it is possible to construct a dedicated Early Warning system that only reacts if a preset alarm criterion has been attained. This means that if no criterion is reached almost no data have to be stored. A prototype of the early warning system was tried in the Göta River. The alarm criterion which was preset at: 'if 6 out of 8 mussels are closed for more than 5 min' made the system not sensitive enough (De Zwart & Slooff, 1987). However, from the printout of the individual mussel behaviour, information on the quality of the aquatic system could be obtained. In Fig. 5 an example is given, where the vertical axis represents the percentage of time that the eight mussels are closed for more than 5 min. In the beginning of the test some of the mussels were still responding to handling. After this first 8 hours

there was no closure for any extended time. Suddenly a closure response was recorded, indicating a discharge into the river, as was also found by other biomonitor systems applied (De Zwart & Slooff, 1987). The chemical nature of the cause of the stress was not determined. After half a day the stress disappeared.

Data treatment

One must distinguish between visual interpretation of chart recorder paper and the data treatment of digitized data. Visual inspection is virtually impossible if large data-sets are involved. For drastic changes in the environmental conditions within short periods of time a strip chart recorder is still very useful.

Once data are stored digitally it is obvious that data treatment should be performed by computer. We have developed three methods to interpret valve movement response data:

– Average valve reading

In this case the data-set is scanned for minimal and maximal reading of each mussel recorded,

440

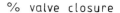

% valve closure

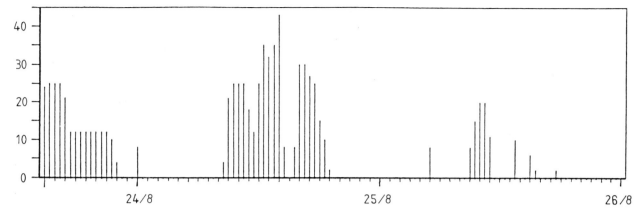

Fig. 5. Average percentage of time that *Dreissena* valves in the Early Warning system were closed for longer than 5 min during 30 min intervals (after: De Zwart & Slooff, 1987).

and set to 0 and 100% respectively. All data are thus normalized and mussels can be compared easily. Only the average reading per hour (or half hour) is recorded. Following a given mussel in time, a decrease in average value will indicate the detection of an effect. In Table 2 an example, resembling data also presented in Fig. 4, is given. About the first 2 hours represent the blank period, then copper is added to the system. A closing response can be clearly observed. It is interesting to observe that the average reading method is rather insensitive to lower sampling frequencies (Table 2). Some deviation from the 5 s sampling interval averages is visible only at relatively large sampling intervals. Two-way analysis of variance may be used to determine the confidence level or reliability of any observed effect.

– Activity interpretation

If the activity of the mussels as valve closure response parameter has been stored, it is easy to calculate the activity per unit of time (usually per h). The activity interpretation as a function of chlorine concentration that was calculated for the data set presented in Fig. 2 is given in Table 3. The ratio of the time closed/open clearly indicates that under natural conditions the mussels are closed for only a limited time (a maximum of 0.5-1 h per day was observed for *Dreissena*).

The activity per unit of time during the 'open' periods shows a marked increase in activity under stress conditions. At the lower additions of chlorine the mussels tend to be more open. Inclusion of the 'closed' periods in the calculations, makes the results less clear.

Table 2. Calculated average valve reading (per half hour) as function of the sampling frequency of the analogous signal, as graphically presented in Fig. 4. After 2 h (#) 50 µg/l copper was added.

Average reading intervals:		1	2	3	4	5	6	7	8	9
Sampling intervals (s):	5	87	87	86	68	24	26	10	4	9
	60	87	89	87	66	23	25	10	4	9
	180	89	91	90	59	22	26	10	4	9
	300	90	90	89	61	26	26	11	4	9
	600	87	93	91	49	26	24	8	4	10
	1200	84	93	90	61	22	26	15	5	10

Table 3. Calculated activity interpretation of the response of *Dreissena polymorpha* under continuous stress of several concentrations of chlorine (the data are graphically presented in Fig. 2).

Chlorine conc. ($\mu g/l$):	0	37	55	180	550
Ratio					
t(closed)/t(open)	0.06	0.5	1.3	5	18
Activity (per half h)					
open period:	4	47	47	97	62
total period:	4	31	20	16	3
Subsequent hourly activity (per h)					
hour 1	4	72	28	–	–
hour 2	8	80	93	–	–
hour 3	2	110	74	–	–
hour 4	16	102	0	–	–
hour 5	10	0	40	–	–

Mytilus showed much less changes in activity during our experiments. If plotted against time, a closure response will be indicated by a drastic decrease in hourly activity (e.g, the activity drop after 4 hours of exposure to chlorine), showing the possibilities of this method for monitoring functions.

– Alarm criterion

If mussels are to be used in a sentinel function of the environmental condition, the above-mentioned data treatments can be used to build an early warning system. It seems more appropriate to construct a dedicated system, however. If the only task of the system is to exhibit an alarm function there is no need for complicated data treatment. This is not only too time consuming (a delay in the alarm) but also requires substantial computer back-up. By looking only for the present alarm criterion, e.g. 'if 6 out of 8 mussels are closed for more than 4 minutes', the alarm function can be triggered within this short period.

Applications

A number of applications for the use of the valve closure response can be developed, especially for environments where steep gradients in the toxicological conditions are to be expected.

Some, mainly biological problems have to be sorted out, like the effect of adaptation, seasonal variation, reproduction, etc. Also a better idea of combined toxicity and the detection limits of various other chemicals should be tested. Since both freshwater and marine mussels seem to react comparably, the proposed bio-sensor can be applied anywhere in the aquatic environment. Possible applications include:
- effluent monitoring (discharge pipes);
- general water quality monitoring (rivers, coastal environments);
- monitoring of water inlets (drinking water, aquaculture);
- early warning system (alarm function, triggering a water sampler for chemical proof);
- toxicity testing;
- physiological and behavioural studies.

Conclusion

The described method has several advantages over the other methods:
- the electronic interface facilitates automated data collection and data interpretation;
- since transmitting and receiving coils attached to the mussel are quite small, and the connecting wires are thin and supple, burrowing bivalves are free to move to some extent;
- the small size and rigidity of the system allows its use both under laboratory and (semi) field conditions, the latter being essential for the application in an Early Warning system.

Acknowledgements

We are indebted to F. Noppert, J. Verburgh, T. Sikking and E. Foekema for their assistance in the various experiments performed as well as for their critical discussions on the subject.

References

Abel, P. D., 1976. Effect of some pollutants on the filtration rate of *Mytilus*. Mar. Poll. Bull. 7: 228–231.

442

Akberali, H. B. & J. E. Black, 1980. Behavioural responses of the bivalve *Scrobicularia plana* (da Costa) subjected to short-term copper (Cu II) concentrations. Mar. Environm. Res. 4: 97–107.

Akberali, H. B. & J. Davenport, 1982. The detection of salinity changes by the marine bivalve molluscs *Scrobicularia plana* (da Costa) and *Mytilus edulis* (L.). J. Exp. Mar. Biol. Ecol. 58: 59–71.

Akberali, H. B. & E. R. Trueman, 1985. Effects of the environmental stress on marine molluscs. Adv. Mar. Biol. 22: 102–198.

Ameyaw-Akumfi, C. & E. Naylor, 1987. Temporal patterns of shell-gape in *Mytilus edulis*. Mar. Biol. 95: 237–242.

Barnes, G. E., 1955. The behaviour of *Anodonta cygnea L.*, and its neurophysiological basis. J. Exp. Biol., 32: 158–174.

Bennett, M. F., 1954. The rhythmic activity of the quahog, *Venus mercenaria*, and its modification by light. Bio. Bull. Mar. Biol. Lab. 107: 174–191.

Bayne, B. L., D. R. Dixon, A. Ivanovici, D. R. Livingstone, D. M. Lowe, M. N. Moore, A. R. D. Stebbing & J. Widdows, 1985. The effects of stress and pollution on marine animals. Preager Scientific, New York, 384 pp.

Cairns, J., 1979. Biological monitoring – concept and scope. In: J. Cairns, G. P. Patil & W. E. Waters (eds.). Environmental biomonitoring, assessment, prediction and management. Int. Co-op Publ. House, Burtonsville, Ma, pp. 3–20.

Davenport, J., 1977. A study of the effect of copper applied continuously and discontinuously to specimens of *Mytilus edulis* (L) exposed to steady and fluctuating salinity levels. J. Mar. Biol. Ass. UK. 57: 63–74.

Davenport, J., 1979. The isolation response of mussels (*Mytilus edulis L.*) exposed to falling sea-water concentrations. J. Mar. Biol. Ass. UK. 59: 123–132.

Davenport, J., 1981. The opening response of mussels (*Mytilus edulis*) exposed to rising sea-water concentrations. J. Mar. Biol. Ass. U.K. 61: 667–678.

De Kock, W. C., 1986. Monitoring bio-available marine contaminants with mussels (*Mytilus edulis*) in the Netherlands. In: C. J. M. Kramer & G. P. Hekstra (eds). Monitoring in the marine environment. Part II. Environm. Mon. Assessm. 7: 209–220.

De Zwart, D. & W. Slooff, 1987. Continuous effluent biomonitoring with an early warning system. In: Bengston, Norberg-King & Mount (Eds.). Effluent and ambient toxicity testing in the Göta Alv and Viskan Rivers, Sweden. Naturvardsverket Report 3275, 40 pp.

Djangmah, J. S., S. E. Shumway & J. Davenport, 1979. Effects of fluctuating salinity on the behaviour of the west African blood clam *Anadara senilus* and on the osmotic and ionic concentrations of the haemolymph. Mar. Biol. 50: 209–213.

Geller, W., 1984. A toxicity warning monitor using weakly electric fish, *Gnathonemus petrsi*. Wat. Res. 18: 1285–1290.

Goldberg, E. D., V. T. Bowen, J. W. Farrington, G. Harvey, J. H. Martin, P. L. Parker, R. W. Risebrough, W.

Robertson, E. Schneider & E. Gamble, 1978. The mussel watch. Environm. Conserv. 5: 101–125.

Gruber, D. S. & J. M. Diamond, 1988. Automated biomonitoring – living sensors as environmental monitors. Ellis Horwood, Chichester, 208 pp.

Higgins, P. J., 1980. Effects of food availability on the valve movements and feeding behaviour of juvenile *Crassostrea virginica* (Gmelin). I. Valve movements and periodic activity. J. Exp. Mar. Biol. Ecol., 45: 229–244.

Hiscock, I. D., 1950. Shell movements of the freshwater mussel *Hyridella australis* Lam. (*Lamellibranchiata*). Aus. J. Mar. Freshwat. Res. 1: 260–268.

Jenner, H. A., F. Noppert & T. Sikking, 1989. A new system for the detection of valve movement response of bivalves. KEMA scientific and technical report 1989-7-2. ISSN-0167-8590, KEMA, Arnhem, Netherlands. 7: 91–98.

Juhnke, I. & W. K., 1971. Eine neue Testmethode zur Früherkennung akut toxischer Inhaltstoffen im Wasser. Gewaesser und Abwasser 50/51: 107–114.

Knie, J., 1982. Der Daphnientest. Decheniana, 26: 82–86.

Koeman, J. H., C. L. M. Poels & W. Slooff, 1978. Continuous biomonitoring systems for detection of toxic levels of water pollutants. In: O. Hutzinger, L. H. van Lelyveld & B. C. J. Zoeteman (eds.). Aquatic pollutants: transformation and biological effects. Pergamon, London, pp. 339–347.

Kramer, K. J. M., J. J. Verburgh & E. M. Foekema, 1989. Response of *Mytilus edulis* to various chemical species of copper. Submitted to Marine Chemistry.

Manley, A. R. & J. Davenport, 1979. Behavioural responses of some marine bivalves to heightened seawater copper concentrations. Bull. Envir. Contam. Toxicol. 22: 739–744.

Manley, A. R., 1983. The effects of copper on the behaviour, respiration, filtration and ventilation activity of *Mytilus edulis*. J. Mar. Biol. Ass. U.K. 63: 205–222.

Manley, A. R., L. L. D. Gruffydd & P. C. Almada-Villela, 1984. The effect of copper and zinc on the shell growth of *Mytilus edulis* measured by a laser diffraction technique. J. Mar. Biol. Ass. U.K. 64: 417–427.

Marceau, F., 1909. Recherche sur la morphologie, et l'histologie, et la physiologie comparées des muscles adducteurs des mollusques acephales. Arch. Zool. Exp. Gén. (Ser 5), 2: 295–469.

NAS, 1980. The international mussel watch. National Academy of Sciences, Washington DC, 148 pp.

Noppert, F., 1987. Unio, een systeem voor het automatische regristatie en verwerken van klepbewegingsgedrag van bivalven. KEMA report 00610-MOA-1637 (in Dutch). 37 pp.

Phillips, D. J. H., 1977. The use of biological indicator organisms to monitor trace metal pollution in marine and estuarine environments – a review. Envir. Pollut. 13: 281–317.

Phillips, D. H. J., 1980. Quantitative aquatic biological indicators. Pollution monitoring series. Applied Science Publ. London, 488 pp.

Poels, C. L. M., 1977. An automatic system for rapid detection of acute high concentrations of toxic substances in surface water using trout. In: Cairns, Dickson & Westlake (Eds). Biological monitoring of water and effluent quality. pp. 85–95, ASTM, STP607.

Sabourin, T. D. & R. E. Tullis, 1981. Effect of three aromatic hydrocarbons on respiration and heart rates of the mussel, *Mytilus californianus*. Bull. Envir. Contam. Toxicol. 26: 729–736.

Salanki, J. & L. Balla, 1964. Ink-Lever equipment for continuous recording of activity in mussels. Annal. Biol. Tihany. 31: 117–121.

Salanki, J. & F. Lukacsovics, 1967. Filtration and O_2 consumption related to the periodic activity of freshwater mussel (*Anodonta cygnea L.*). Annal. Biol. Tihany, 34: 85–98.

Salanki, J. & L. Varanka, 1976. Effect of copper and lead compounds on the activity of the fresh-water mussel. Annal. Biol. Tihany, 43: 21–27.

Salanki, J. & L. Varanka, 1978. Effect of some insecticides on the periodic activity of the fresh-water mussel (*Anodonta cygnea L.*). Acta. Biol. Acad. Sci. Hung. 29: 173–180.

Schuring, B. J. & M. J. Geense, 1972. Een electronische schakeling voor het registreren van openingshoek van de mossel *Mytilus edulis L.* TNO-Rapport CL 72/47 (in Dutch). 8 pp.

Slooff, W., D. de Zwart & J. M. Marquenie, 1983. Detection limits of a biological monitoring system for chemical water pollution based on mussel activity. Bull. Envir. Contam. Toxicol. 30: 400–405.

Stephenson, R. R. & D. Taylor, 1975. The influence of EDTA on the mortality and burrowing activity of the clam (*Venerupis decussata*) exposed to sub lethal concentrations of copper. Bull. Envir. Contam. Toxicol. 14: 304–308.

Widdows, J., D. K. Phelps & W. Galloway, 1981. Measurement of physiological condition of mussels transplanted along a pollution gradient in Narragansett Bay. Mar. Environm. Res. 4: 181–194.

Widdows, J., 1973. Effect of temperature and food on the heart beat, ventilation rate and oxygen uptake of *Mytilus edulis*. Mar. Biol. 20: 269–276.

Widdows, J., 1985. Physiological responses to pollution. Mar. Pollut. Bull. 16: 129–134.

Hydrobiologia **188/189**: 445–454, 1989.
M. Munawar, G. Dixon, C. I. Mayfield, T. Reynoldson and M. H. Sadar (eds)
Environmental Bioassay Techniques and their Application.
© 1989 *Kluwer Academic Publishers. Printed in Belgium.*

Physiological background for using freshwater mussels in monitoring copper and lead pollution

J. Salánki & Katalin V.-Balogh
Balaton Limnological Research Institute of the Hungarian Academy of Sciences, H-8237 Tihany, Hungary

Key words: *Anodonta cygnea* L., filtration activity, heavy metal accumulation, depuration, copper, lead

Abstract

In studying the effect of copper ($10 \pm 0.57\ \mu g$ Cu l^{-1} and $100 \pm 3.01\ \mu g$ Cu l^{-1}) and lead ($50 \pm 1.12\ \mu g$ Pb l^{-1} and $500 \pm 12.5\ \mu g$ Pb l^{-1}) on the filtration activity of *Anodonta cygnea* L. it was found that both heavy metals resulted in significant shortening of the active periods, but little change occurred in the length of the rest periods. The concentrations of copper and lead were measured in the gill, foot, mantle, adductor muscle and kidney for 840 hours of exposure to $10.9 \pm 5\ \mu g$ Cu l^{-1} and $57.0 \pm 19\ \mu g$ Pb l^{-1} as well as during subsequent depuration. Uptake was observed after 72 hours of exposure. The highest copper concentration ($59.1 \pm 16.2\ \mu g$ Cu g^{-1}) was measured at 672 h in the mantle, and the highest lead value ($143 \pm 26.1\ \mu g$ Pb^{-1}) was obtained in the kidney. Depuration of copper was fastest from the foot, and from the adductor muscle for lead. The gill had the longest half-depuration time (> 840 h for copper and > 672 h for lead).

Introduction

It is well known that bivalve molluscs are able to concentrate heavy metals such as Hg, Cd, Zn, Cu, Pb and others in their tissues (Brooks & Rumsby, 1965; Pringle *et al.*, 1968), therefore they can be used in monitoring heavy-metal pollution of the environment. Among marine species *Mytilus* meets most of the requirements necessary for a monitoring system, and it is widely used as a biological indicator of metal pollution (Coleman *et al.*, 1986; Ritz *et al.*, 1982). *Mytilus* has been used in the 'mussel watch' program for indicating additional types of pollution (Farrington *et al.*, 1983; Goldberg *et al.*, 1978).

Anodonta and *Unio* species are freshwater bi-

valves which are very suitable for monitoring heavy-metal contamination. However, not much information is available concerning their physiological responses to pollutants, and the kinetics of uptake and depuration of heavy metals. Nevertheless, there have been papers published presenting data concerning the effects of Hg and Cd on *Anodonta cygnea* L. and kinetics of their uptake (Salánki & V.-Balogh, 1985; V.-Balogh & Salánki, 1984).

In the present study the freshwater bivalve *Anodonta cygnea* L. has been subjected to Cu and Pb treatments to measure the effects of these metals upon filtration activity and to determine the characteristics of the metal uptake and depuration.

Materials and methods

In the investigations specimens of adult (11.8 ± 1.3 cm) *Anodonta cygnea* L. collected from fish ponds located at the eastern part of Hungary were used. Before the experiments the animals were kept for four weeks in an aquarium supplied with Balaton-water rich in phytoplankton. No additional food was provided.

Two separate series of experiments were carried out (both with Cu and Pb): (a) we investigated the effect of the metals on valve activity of the animals; (b) we studied the uptake and release of the metals in different tissues. In the experiments Cu^{2+} was added as $CuCl_2 \cdot 2H_2O$ and Pb^{2+} as $PbCl_2$ via water.

Series a. The animals were placed separately in plexiglass tanks (3 l each) with running lakewater. The position and movement of the shells were recorded on a mussel actograph (Salánki & Balla, 1964). Filtering takes place when the shells are open during which period fast pumping movements occur (active period), while a persistent (longer than 60 min) closed position marks a rest period (Salánki & Lukacsovics, 1967). The shell movements were continuously recorded for 168 hours before the experiments began, during exposure to heavy metals (240 hours) and during the purging period (168 hours). Metal concentrations during exposure were 10 ± 0.57 and 100 ± 3.01 $\mu g \cdot l^{-1}$ for copper, and 50 ± 1.12 and 500 ± 12.5 $\mu g \cdot l^{-1}$ for lead. The duration of the active and that of the rest periods were measured and an average was calculated from ten parallel experiments. The duration of consecutive active and rest periods was compared before, during and after exposure.

Series b. One hundred animals were placed in a glass aquarium containing 100 l water equipped with a perfusion system which assured a total change of the water within 8 h. Water temperature varied between 12 and 20 °C in accordance with changes in the lake's temperature. During exposure, the copper and lead concentrations were 10.9 ± 5 $\mu g \cdot l^{-1}$ and 57.0 ± 19 $\mu g \cdot l^{-1}$, respectively.

Exposure to metals took place for 840 h during which the concentrations of Cu and Pb were measured after 1, 4, 9, 24, 72, 168, 336, 504, 672, and 840 h. During the depuration period which followed, organ concentrations were measured after 48, 168, 336, 504, 672, 840 (only copper) h.

Both copper and lead concentrations were analyzed in the gills, foot (including viscera), mantle, adductor muscle and kidney. Samples were prepared for analysis by wet digestion according to the method of Krishnamurty *et al.* (1976) as described earlier (Salánki *et al.*, 1982).

Copper and lead concentrations were measured using a Zeiss AAS1 type AA spectrophotometer, by direct flame atomization in an airacetylene flame.

The concentrations given in the figures are mean values (± standard error of mean) of three replicate samples.

Half-depuration time ($T_{1/2}$) was defined as the time necessary to reach the half-depuration concentration (HDC) using the formula HDC = $(C_e - C_c) \cdot 2^{-1}$, where C_e = the metal concentration in the organs at the end of exposure, C_c = the metal concentration in the organs of the control animals.

Results

Change in filtering activity.

During exposure to copper and lead the activity of the mussels changed depending on the metal concentration. Before Cu-treatment the length of the active periods varied between 10-20 h, but occasionally there were also shorter periods (5-6 h). After the animals were exposed to 10 ± 0.57 μg Cu l^{-1}, the duration of the active period gradually decreased from 20 h to 8 h. Exposure of the mussels to 100 ± 3.01 μg Cu l^{-1} caused a sudden decrease in the duration of the active period to about 1 h, and this remained so during subsequent exposure. When copper exposure ceased, duration of the active periods suddenly increased to 30-40 h, then decreased again to control values (Fig. 1).

The duration of the rest period before copper treatment, was 10-18 h long and this did not change markedly with either concentration. Nevertheless, with $10 \pm 0.57\,\mu g\,Cu\,l^{-1}$ there was a cyclic shortening and lengthening of the rest period. When the copper treatment was terminated, an increase in the duration of the rest period to 30 h was observed, followed by a decrease to the level of the control values (Fig. 1).

Before the application of $50 \pm 1.12\,\mu g\,Pb\,l^{-1}$ the length of the active periods varied between 30-60 h. During exposure the duration of the active period decreased significantly to about 7 h. During purging there was a slight increase in the active period, but this did not reach the level of the control values even after one week.

During the same experiment, the rest period varied in the control between 6-18 h (average 10 h). It was stable at about 6 h at the beginning of the exposure, before increasing to produce the average control value. A shortening of the rest periods was observable during purging (Fig. 2).

Before exposure to $500 \pm 12.5\,\mu g\,Pb\,l^{-1}$, the length of the active periods varied in the control between 6-23 h, but during exposure it decreased gradually to 4-6 h. The duration of the rest periods was also shortened at this higher lead-concentration. When the lead treatment was stopped, there was an immediate increase in the duration of both the active and the rest periods. The length of these periods then returned to low levels. Control values did not return to previous levels even after one week of purging (Fig. 2).

Uptake and depuration of copper

Exposing the mussels to $10.9 \pm 5\,\mu g\,Cu\,l^{-1}$, the concentration of this metal increased in almost all organs within 1 h, but dropped below the control value after 4 h. A significant uptake could be measured only after 72 h, but different accumulation patterns were found in different organs.

The concentration of copper in the gill increased linearly up to 672 h, reaching a value of $42.8 \pm 2.33\,\mu g\,Cu\,g^{-1}$ before stabilizing. There was strong binding of copper by the gill since no

release of copper was evident after the bivalves were placed in metal-free water. Moreover, half-depuration was not achieved after 840 h of purging (Fig. 3A).

Similarly, copper concentrations increased linearly up to 672 h in the mantle, reaching a value of $59.1 \pm 16.2\,\mu g\,Cu\,g^{-1}$. A linear depuration of copper was also observed, with $T_{1/2}$ being reached in 420 h (Fig. 3B). In the foot, the concentration of copper increased linearly for the entire period of exposure (840 h), then a linear depuration pattern was observed. The $T_{1/2}$ for the foot was 195 h (Fig. 3C).

Accumulation in the adductor muscle proceeded slowly. The elevation of Cu concentrations was measured after 336 h, then saturation occurred between $15\text{-}20\,\mu g\,Cu\,g^{-1}$. There was slow depuration from this muscle, and half-depuration was not achieved during the 840 h purging period (Fig. 3D).

Only a slight concentration increase (from 15 to $27\,\mu g\,Cu\,g^{-1}$) was observed in the kidney which was reduced by half within 48 h (Fig. 3E).

Uptake and depuration of lead

No significant uptake was measured during the first 72 h of exposure to $57.0 \pm 19\,\mu g\,Pb\,l^{-1}$. Considering individual organs, two accumulation patterns could be distinguished: a linear pattern for the kidney and a logarithmical type for all the other organs.

The concentration of lead increased during the 672 h exposure period to $62.7 \pm 6.67\,\mu g\,g^{-1}$ in the gill; and this organ did not release Pb during 672 h of depuration (Fig. 4A). The foot (Fig. 4B), the mantle (Fig. 4C) and the adductor muscle (Fig. 4D) showed the same type of saturation uptake for lead, and reached $24.9 \pm 3.0\,\mu g\,Pb\,g^{-1}$, $49.3 \pm 19.4\,\mu g\,Pb\,g^{-1}$, and $38.3 \pm 2.64\,\mu g\,Pb\,g^{-1}$, respectively.

The depuration of lead was fastest from the adductor muscle ($T_{1/2} = 100$ h), followed by the mantle ($T_{1/2} = 145$ h), and the foot ($T_{1/2} = 672$ h). However, control levels were not achieved by any of them during the experimental period.

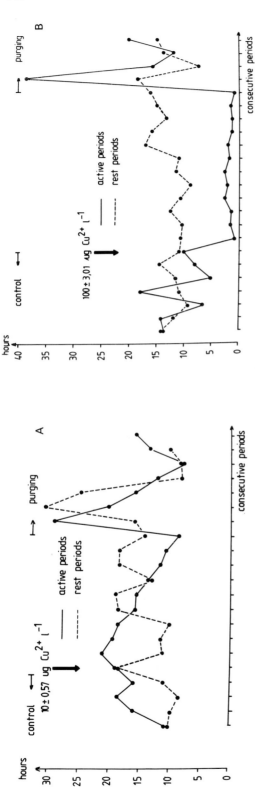

Fig. 1. Change of the duration of consecutive active and rest periods of the mussel *Anodonta cygnea* L. in control, under 240 h exposure to $10 \pm 0.57~\mu g~Cu~l^{-1}$ (A), $100 \pm 3.01~\mu g~Cu~l^{-1}$ (B) and during purging

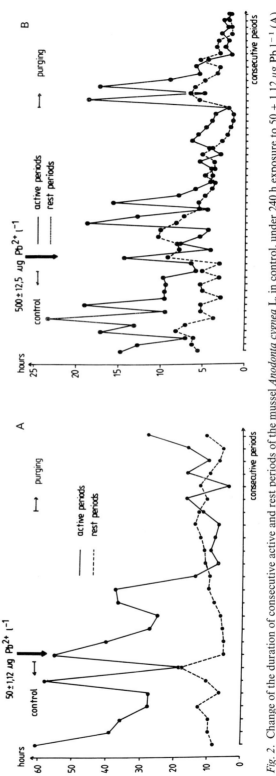

Fig. 2. Change of the duration of consecutive active and rest periods of the mussel *Anodonta cygnea* L. in control, under 240 h exposure to $50 \pm 1.12~\mu g~Pb~l^{-1}$ (A), $500 \pm 12.5~\mu g~Pb~l^{-1}$ (B) and during purging

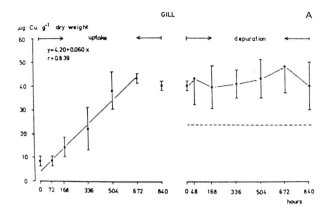

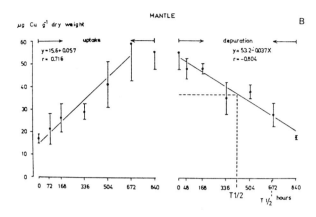

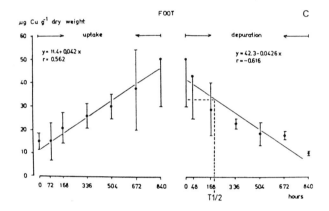

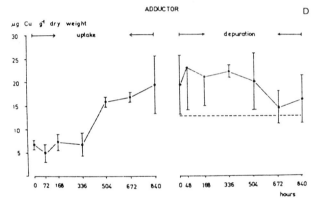

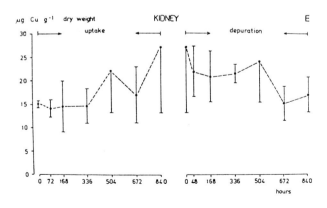

Fig. 3. Change of copper concentration in the gill (A), mantle (B), foot (C), adductor muscle (D) and kidney (E) of the mussel *Anodonta cygnea* L. during 840 h exposure to $10.9 \pm 5 \ \mu g$ Cu l^{-1} and 840 h depuration period (dotted line represents theoretical half-depuration concentrations)

Table 1. Concentration of copper in the organs of *Anodonta cygnea* L. in control and after 840 h exposure (μg Cu g^{-1} dry weight, mean $\pm$ SEM)

Organ	Control concentration (C_c)	Concentration after exposing to copper ($10.9 \pm 5 \ \mu g \ l^{-1}$) ($C_e$)	$C_e \cdot C_c^{-1}$	P
Gill	8.42 $\pm$ 2.32	40.0 $\pm$ 1.70	4.75	<0.001
Foot	14.9 $\pm$ 3.45	50.1 $\pm$ 19.2	3.36	<0.001
Adductor muscle	6.75 $\pm$ 1.02	19.3 $\pm$ 6.65	2.86	<0.001
Mantle	16.9 $\pm$ 1.73	55.4 $\pm$ 7.45	3.28	<0.001
Kidney	14.8 $\pm$ 0.784	26.9 $\pm$ 13.9	1.82	>0.05

The kidney showed linear uptake and depuration characteristics. The highest lead concentration, $143 \pm 26.1 \ \mu g \ g^{-1}$, was measured after 672 h exposure. Binding of lead was weak, as shown after 672 h depuration, when the concentration reached the initial, $29.9 \pm 8.06 \ \mu g \ Pb \ g^{-1}$ value ($T_{1/2} = 295$ h) (Fig. 4E).

Rate of bioconcentration

At the end of the exposure period, metal concentrations in the mussels were significantly higher in all organs as compared to their pre-exposure concentrations. The rate of copper bioconcentration was organ specific (Table 1), and ranged from 1.82 (kidney) to 4.75 (gill). The value of 1.82 was not significant. The bioconcentration rate of lead (Table 2) ranged between 2.35 (gill) and 4.18 (kidney).

Discussion

Measurement of metal concentrations in living organisms is an acceptable approach for detecting the level of pollution in the environment. Although this simple method is accurate for practical purposes, nevertheless, it does not offer a possibility for the more generalized evaluation of the data obtained. The combined methods of analyzing both the behavioural effects of pollutants and checking their uptake and release kinetics may increase the usefulness of biomonitoring organisms.

Mussels are remarkable because they can change their filtration rate under the effect of heavy metals and other toxicants (Abel, 1976; Salánki & Varanka, 1976). Since the activity of the animals corresponding to filtration activity can be recorded both in laboratory conditions as well as in the natural environment (Véró &

Table 2. Concentration of lead in the organs of *Anodonta cygnea* L. in control and after 840 h exposure (μg Pb g^{-1} dry weight, mean $\pm$ SEM)

Organ	Control concentration (C_c)	Concentration after exposing to lead ($57 \pm 19 \ \mu g \ l^{-1}$) ($C_e$)	$C_e \cdot C_c^{-1}$	P
Gill	26.50 $\pm$ 4.00	62.2 $\pm$ 9.14	2.35	<0.001
Foot	8.68 $\pm$ 4.46	24.9 $\pm$ 3.00	2.87	<0.001
Adductor muscle	10.8 $\pm$ 9.28	38.3 $\pm$ 2.64	3.55	<0.001
Mantle	15.7 $\pm$ 1.42	49.3 $\pm$ 19.4	3.14	<0.001
Kidney	29.9 $\pm$ 8.06	125 $\pm$ 14.5	4.18	<0.001

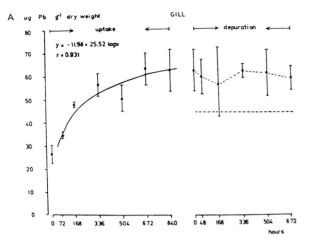

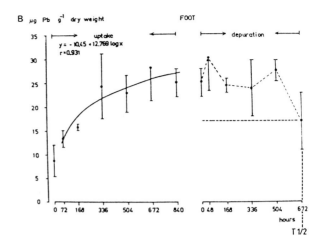

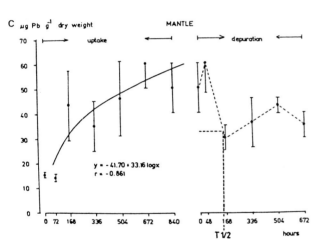

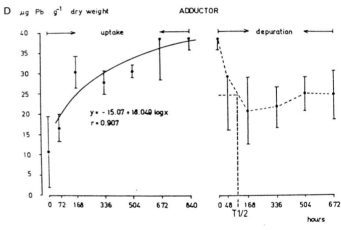

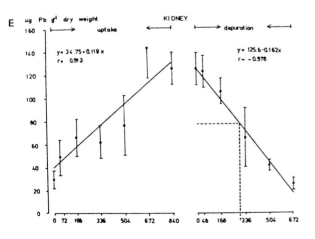

Fig. 4. Change of lead concentration in the gill (A), foot (B), mantle (C), adductor muscle (D) and kidney (E) of the mussel *Anodonta cygnea* L. during 840 h exposure to $57.6 \pm 19 \ \mu g$ Pb l^{-1} and 672 h depuration period (dotted line represents theoretical half-depuration concentrations)

Salánki, 1969), monitoring of the activity is a suitable method of predicting the biological effect of many water-soluble chemicals and hence of pollutants.

Our results show that under exposure to copper and lead the activity of the mussels changed depending on the metal concentrations. The filtering activity was reduced in both cases by shortening the duration of active periods. The rate of decrease was two and greater than ten times for copper at 10 and 100 μg Cu l^{-1}, respectively, while lead was six and ten times at 50 and 500 μg Pb l^{-1}, respectively. The decrease in duration of the active periods was gradual at both lead concentrations and at the lower copper concentration; but immediate at 100 μg·l^{-1} copper concentration.

Davenport and Manley (1978) investigated the valve closure mechanisms of *Mytilus edulis* on exposure to copper sulphate. Our findings are in good accordance with their results. They also suggested that mussels would be reasonably efficient indicators of copper pollution up to concentrations of 160-200 μg·l^{-1}.

In our earlier investigations we showed that other heavy metals such as Hg and Cd not only reduced the active periods in *A. cygnea*, but also caused the elongation of the rest periods (V.-Balogh & Salánki, 1984). In the present experiments we found that under the effects of copper and lower concentration of lead only the active periods became shorter, the rest periods did not change. This suggests that different mechanisms are involved in the regulation of activity and rest.

On the basis of our results, it is also suggested that there are differences between the effects of copper and of lead on the mechanisms which regulate activity in *A. cygnea*. In contrast to lead exposure, the duration of the active periods reached the control values after stoppage of the copper treatment, suggesting that mussels are more tolerant to copper fluctuations than to lead fluctuations.

Although filtration activity of the mussels was re-established within 168 hours after copper exposure, the concentrations of copper did not decrease to the control level. Following lead treat-

ment, neither the activity nor the lead concentration of organs was restored even after 168 hours of purging. This residual metal content can have a marked effect on an additional bioaccumulation.

In spite of changes in the filtration activity there was no significant copper and lead uptake by the freshwater mussel in the first 72 h of exposure. This suggests that on the basis of bioaccumulation one cannot detect copper or lead pollution in less than three days using mussels.

After 72 h exposure, the concentration of Cu increased significantly in all organs, except the kidney. The highest concentration rate occurred in the gill. Tallandini *et al.* (1986) found similar results while studying the uptake and distribution of Cu in various tissues of the freshwater bivalves, *A. cygnea* and *Unio elongatulus*. They reported that although copper uptake was not very marked, the highest uptake was exhibited by the gill in both organisms.

Lead uptake also remained in the range of a 2-4 fold increase and the highest lead bioconcentration rate occurred in the kidney, and the lowest in the gill. The accumulation pattern in the kidney differed from that in other organs, since there was a linear uptake. Schulz-Baldes (1974) described a similar linear uptake pattern for lead in *M. edulis*. In his experiments, however, the lead concentration increased in a linear way in all organs.

The degree of time-integration capacity of an indicator organism is an important factor in its usefulness for monitoring heavy metal contaminants (Phillips, 1980). Beside the concentration ability, the storage ability is also important. The rate of elimination and the storage ability may be characterized by half-depuration time. If $T_{1/2}$ is long, the animal or its tissues can reflect the average concentration of metals in the environment for a longer time.

The half-depuration times for different organs of *A. cygnea* treated with copper and lead show that the gill has the most prolonged retention of both metals. In our earlier investigations we found slow depuration in the gill for mercury ($T_{1/2} > 840$ h), and for cadmium ($T_{1/2}$ between 504 and 672 h), (Salánki & V.-Balogh, 1985). Consequently, the gill represents the best biomonitoring organ in the mussel, while the time lapse between two samplings should not be less than 672 h for a reliable, permanent indication of copper and lead pollution.

References

Abel, P. D., 1976. Effect of some pollutants on the filtration rate of *Mytilus*. Mar. Pollut. Bull. 7: 228–231.

Brooks, R. R. & M. G. Rumsby, 1965. The biogeochemistry of trace element uptake by some New Zealand bivalves. Limnol. Oceanogr. 10: 521–528.

Coleman, N., T. F. Mann, M. Mobley & N. Hickman, 1986. *Mytilus edulis planulatus*: an 'integrator' of cadmium pollution? Mar. Biol. 92: 1–5.

Davenport, J. & A. Manley, 1978. The detection of heightened sea-water copper concentrations by the mussel *Mytilus edulis*. J. mar. biol. Ass., UK. 58: 843–850.

Farrington, J. W., E. D. Goldberg, R. W. Risebrough, J. H. Martin & V. T. Bowen, 1983. U.S. 'mussel watch' 1976-1978: an overview of the trace-metal, DDE, PCB, hydrocarbon, and artificial radionuclide data. Envir. Sci. Technol. 17: 490–496.

Goldberg, E. D., V. T. Bowen, J. W. Farrington, G. Harvey, J. H. Martin, P. L. Parker, R. W. Risebrough, W. Robertson, E. Schneider & E. Gamble, 1978. The mussel watch. Envir. Conserv. 5C: 101–125.

Krishnamurty, K. V., E. Shpirt & M. M. Reddy, 1976. Trace metal extraction of soils and sediments by nitric acid – hydrogen peroxide. Atom. Absorp. Newslett. 15: 68–70.

Phillips, D. J. H., 1980. Quantitative aquatic biological indicators. Pollution Monitoring Series (Adv. Ed. Mellanby, K.). Applied Science Publishers LTD, London, pp. 488.

Pringle, B. H., D. E. Hissong, E. L. Katz & S. T. Mulawka, 1968. Trace metal accumulation by estuarine molluscs. J. Sanit. Engng. Div. Am. Soc. Civ. Engrs. 94: 455–475.

Ritz, D. A., R. Swain & N. G. Elliott, 1982. Use of the mussel *Mytilus edulis planulatus* (Lamarck) in monitoring heavy metal levels in seawater. Aust. J. mar. Freshwat. Res. 33: 491–506.

Salánki, J. & L. Balla, 1964. Ink-lever equipment for continuous recording of activity in mussels (Mussel – actograph). Annal. Biol. Tihany 31: 117–121.

Salánki, J. & F. Lukacsovics, 1967. Filtration and O_2 consumption related to the periodic activity of freshwater mussel (*Anodonta cygnea*). Annal. Biol. Tihany 34: 85–98.

Salánki, J. & I. Varanka, 1976. Effect of copper and lead compounds on the activity of the fresh-water mussel. Annal. Biol. Tihany 43: 21–27.

Salánki, J., Katalin V.-Balogh & Erzsébet Berta, 1982. Heavy metals in animals of Lake Balaton. Wat. Res. 16: 1147–1152.

Salánki, J. & Katalin V.-Balogh, 1985. Uptake and release of mercury and cadmium in various organs of mussels (*Anodonta cygnea* L.). In: Heavy metals in water organisms (Ed by Salánki, J.) Akadémiai Kiadó Budapest, Symposia Biologica Hungarica 29: 325–342.

Schulz-Baldes, M., 1974. Lead uptake from sea water and food, and lead loss in the common mussel *Mytilus edulis*. Mar. Biol. 25: 177–193.

Tallandini, L., A. Cassini, N. Favero & V. Albergoni, 1986. Regulation and subcellular distribution of copper in the freshwater molluscs *Anodonta cygnea* (L.) and *Unio elongatulus* (Pf.). Comp. Biochem. Physiol. 84C: 43–49.

V.-Balogh, Katalin & J. Salánki, 1984. The dynamics of mercury and cadmium uptake into different organs of *Anodonta cygnea* L.. Wat. Res. 18: 1381–1387.

Véró, M. & J. Salánki, 1969. Inductive attenuator for continuous registration of rhythmic and periodic activity of mussels in their natural environment. Med. Biol. Engng. 7: 235–237.

Hydrobiologia **188/189**: 455–461, 1989.
M. Munawar, G. Dixon, C. I. Mayfield, T. Reynoldson and M. H. Sadar (eds)
Environmental Bioassay Techniques and their Application.
© 1989 *Kluwer Academic Publishers. Printed in Belgium.*

The application of combined tissue residue chemistry and physiological measurements of mussels (*Mytilus edulis*) for the assessment of environmental pollution

John Widdows & Peter Donkin
Plymouth Marine Laboratory, Prospect Place, West Hoe, Plymouth PL1 3DH, England

Key words: Pollution, *Mytilus edulis*, QSARs, Growth

Abstract

The rationale for the use of combined tissue residue chemistry and physiological energetics measurements of *Mytilus edulis* in the assessment and monitoring of environmental pollution is outlined. Laboratory derived relationships between the concentration of toxicants in tissues and sublethal responses (eg. feeding, respiration and growth rate) provide a toxicological database for the interpretation of physiological responses measured in the field. The role of quantitative structure-activity relationships (QSAR's) in establishing tissue concentration-effect relationships for organic contaminants is discussed. The application of this approach is illustrated with reference to two field studies, a monitoring programme in the Shetlands and a practical biological effects workshop in Oslo.

Introduction

Bioassay techniques for assessing and monitoring environmental pollution should ideally fulfil the following criteria:
- The biological effects measurements should be responsive to environmental levels of pollutants, should reflect a quantitative and predictable relationship with toxic contaminants (ie. pollutants), and provide a measure of spatial and temporal changes in environmental quality.
- The bioassay, in conjunction with appropriate chemical measurements of pollutants, should provide a means of identifying the cause(s) of measured biological effects.
- The biological response should have ecological significance and be shown, or convincingly argued, to reflect a deleterious effect on growth, reproduction or survival of the individual, the population and ultimately the community.

The objectives of this paper are (a) to describe the rationale behind our approach, which uses a combination of tissue residue chemistry and physiological measurements of mussels (*Mytilus edulis*) to assess and monitor environmental pollution, and (b) to illustrate its field application and the extent to which it fulfils the criteria stated above.

Environmental-Toxicological framework

The approach is based on a combination of tissue residue chemistry and sublethal physiological energetic measurements. The main attributes of this approach are outlined below:
- Mussels have a wide geographic distribution

and are important members of coastal and estuarine communities.

- Chemical contaminants are accumulated by the body tissues of mussels and there is minimal metabolic transformation of organic toxicants; thus body burdens reflect levels of environmental contamination (Farrington *et al.*, 1983).
- Sublethal physiological responses, such as rates of feeding, respiration and excretion, are sensitive to many pollutants and reflect major mechanisms of toxicity (e.g., narcotic and neurotoxic effects on ciliary feeding, uncoupling of the respiration chain; Widdows & Johnson, 1988; Donkin *et al.*, 1989).
- Integration of these responses by means of physiological energetics provides a rapid quantitative assessment of energy available for growth and reproduction (termed scope for growth; SFG) as well as insight into the underlying causes of changes in growth rate.
- Feeding (= clearance) rate and SFG are inversely related to the tissue accumulation of toxicants over a wide range of concentrations. These tissue concentration-response relationships facilitate the establishment of cause-effect relationships (Widdows *et al.*, 1981; Widdows *et al.*, 1987; Widdows & Johnson, 1988).
- Physiological measurements can be made on mussels in the laboratory and the field, thus enabling laboratory derived tissue concentration-response relationships to be used to interpret responses measured in the field.
- There is a comprehensive literature describing the physiological responses and adaption of mussels to a wide range of extrinsic and intrinsic factors (Bayne, 1976; Widdows, 1985a).

In order to begin to identify the cause(s) of observed effects in the field situation, it is necessary to establish by laboratory experimentation an appropriate toxicological database describing the relationships between tissue concentrations of specific contaminants, or groups of contaminants, and the effects on physiological energetics. While it may be feasible to measure the tissue concentration-effect relationships for most of the

environmentally important metals, it is unrealistic to determine the sublethal effects of every single organic contaminant; within oil alone there are many thousands of individual compounds. An alternative is to use a Quantitative-Structure Activity Relationship (QSAR) approach, which facilitates prediction of the toxicological properties of organic compounds from their physico-chemical/structural properties (reviews by Hermens, 1986; Kaiser, 1987; Turner *et al.*, 1987).

Our QSAR approach, however, is distinct from that adopted by other aquatic toxicologists in that sensitive physiological responses are measured and related to toxicant concentrations in the tissues, in addition to water. This contrasts with the more common lethal (LC_{50}) or whole body (e.g., growth, EC_{50}) effects QSARs, in which toxic effects are related to aqueous toxicant concentrations only. Consequently, laboratory derived tissue concentration-effect relationships, based on more sensitive responses, are more easily related to environmental situations where animals are typically exposed to sublethal contaminant levels at varying concentrations in the water column, which are accumulated and time-integrated by the body tissues.

A number of important features concerning QSARs and the sublethal toxic effects of a range of hydrophobic organic compounds in *Mytilus edulis* are illustrated in Fig. 1 (from Donkin *et al.*, 1989).

1) There is an inverse linear relationship between the hydrophobicity of organic compounds (measured in terms of the log K_{ow}, the octanol-water partition coefficient) and the log concentration in the water required to induce a 50% reduction in the feeding (= clearance) rate of *Mytilus edulis*.

2) The log bioconcentration factor (BFC) increases with log K_{ow}.

3) Changes in bioconcentration account for most of the differences in the water-concentration based expression of toxicity. Consequently, hydrophobic organic compounds, such as aliphatic and aromatic hydrocarbons with log K_{ow}'s of <5, have equal toxicity when ex-

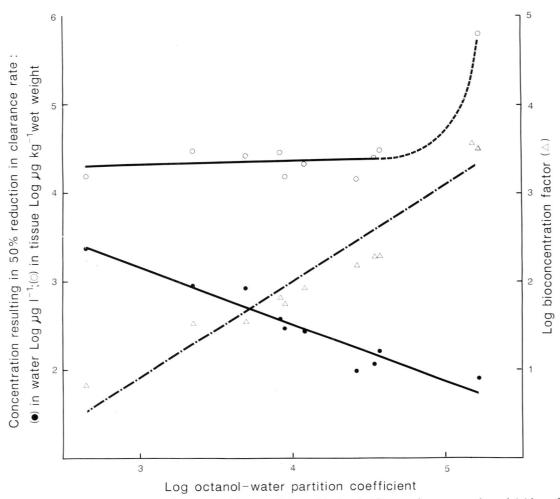

Fig. 1. Relationship between log octanol-water partition coefficient of hydrophobic organic compounds and (a) log of their bioconcentration factor into the mussel (*Mytilus edulis*) (△) (b) the concentration of these compounds in the water (●) and (c) the concentrations in the mussel tissues (○) which reduce clearance (= feeding) rate by 50% (from Donkin *et al.*, 1989). The compounds are (from left to right): toluene, naphthalene, propylbenzene, acenaphthene, biphenyl, 1-chloronaphthalene, dibenzo-thiophene, n-octane, phenanthrene, pyrene and fluoranthene.

pressed on the basis of tissue concentration. This implies a common mechanism of toxicity and suggests that effects on clearance rate of the many related hydrocarbon compounds present in oil are likely to be additive (Könemann, 1980).

4) There is a molecular weight 'cut-off' in the QSAR line occurring at a log K_{ow} of 5; a well known phenomenon associated with other compounds inducing narcosis. Compounds with a log $K_{ow} > 5$ are accumulated in the tissues but induce little or no effect on feeding rate. This 'cut-off' identifies the molecular/structural range over which compounds inhibit clearance rate and demonstrates that while many organic compounds may be detected in mussels sampled from contaminated environments, not all will produce adverse effects.

QSARs need to be established for other classes of organic compounds with different physicochemical properties and for biological responses reflecting different mechanisms of toxicity. Any

458

group forming a different QSAR is indicative of a different mode of toxicity (Hermens, 1986; Turner *et al.*, 1987).

Once established, a QSAR line then enables the toxicity of related compounds and mixtures to be predicted. Furthermore, QSARs can be used to compare the sensitivity of different organisms (Sloof *et al.*, 1983) to classes of toxicants, so extrapolation from biological effects on mussels to effects on other species becomes feasible.

Examples of field application

The results of a field monitoring programme at Sullom Voe in the Shetlands (for details see Widdows *et al.*, 1987a) and a practical Biological Effects Workshop in Oslo (Widdows & Johnson,

1988; Bayne *et al.*, 1988) are two examples used to demonstrate the application of this toxicological approach to the assessment of marine pollution.

Sullom Voe monitoring programme

The measurement of SFG (for details see Widdows, 1985; Widdows & Johnson, 1988) and the concentration of two- and three-ring aromatic hydrocarbons (a major toxic component of oil) accumulated in the tissues of *Mytilus edulis* during the period from 1982 to 1985 has provided an assessment of spatial and temporal impact of oil pollution in the vicinity of the Sullom Voe oil terminal (Widdows *et al.*, 1987a). Mussels living near the tanker loading area (sites 4 and 5; Fig. 2)

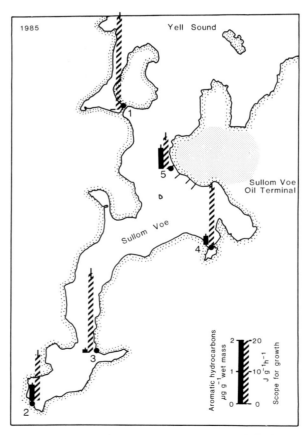

Fig. 2. Spatial variation in the tissue concentration of two- and three-ring aromatic hydrocarbons and the scope for growth of mussels (*Mytilus edulis*) sampled in the vicinity of the Sullom Voe oil terminal, Shetland, in 1985.

were consistently the most contaminated with hydrocarbon concentrations typically 10-fold higher than levels at the reference site (1) and there was a concomitant decline in the SFG of mussels at these contaminated sites. The relationship between SFG and the log concentration of two- and three-ring aromatic hydrocarbons (Shetland data 1982-1985, $r = -0.72$) indicates that an order of magnitude increase in the tissue concentration can account for an approximately 50% reduction in the growth potential of *Mytilus edulis*.

The temporal variations in SFG and hydrocarbon residues in the tissues of *Mytilus edulis* sampled from sites near the oil terminal indicate that there has been no gradual deterioration in environmental quality and no gradual increase in the level of hydrocarbon contamination within Sullom Voe since the oil terminal became operational. The recorded year-to-year variations in hydrocarbon levels in the tissues of mussels is closely correlated ($r = 0.8$) with the number and size of small oil spillages during the months preceding the annual field sampling (in July) at Sullom Voe, indicating that the system can readily recover from any transient increase in oil inputs.

For purposes of environmental management, it is necessary not only to quantify the degree of environmental contamination and the resultant deleterious biological effects, but also to address such questions as – 'how bad or good are the conditions?' and 'at what level of impact should managers start to be concerned and take remedial action?'. One approach is to present the data in a wider geographic and ecological context. Figure 3 shows a synthesis of data derived from the Sullom Voe study and from mesocosm oil experiments (Widdows *et al.*, 1987b) carried out at Solbergstrand (Southern Oslofjord, Norway). The semi-logarithmic relationship ($r = -0.87$) illustrates that, (a) there is an inverse relationship between SFG and the log concentration of aromatic hydrocarbons in the tissues of *Mytilus edulis* over three orders of magnitude and without an apparent threshold effect; this suggests a simple mode of toxicity based on loading of body tissues and the absence of any significant physiological adaption, (b) mussels from the 'reference'

site near Sullom Voe have a very low level of hydrocarbon contamination compared to other sites studied, and (c) mussels at the tanker loading jetties nearest the source of pollution are moderately contaminated, but to no greater extent than sites such as Solbergstrand, on the Oslofjord (ca 35 km south of Oslo) in a region that is not in the vicinity of any point sources of industrial or urban pollution. In addition, the Solbergstrand mesocosm studies have highlighted the important relationships between the physiological energetic responses of individuals and the long-term effects on reproduction and survival of the population and community (Widdows *et al.*, 1987b).

Laboratory (Widdows *et al.*, 1982; Donkin *et al.*, 1989), mesocosm (Widdows *et al.*, 1987b) and field studies (Widdows *et al.*, 1987a) have consistently demonstrated an inverse relationship between SFG and the log concentration of aromatic hydrocarbons (within the range log K_{ow} 2–5) in the tissues of *Mytilus edulis*. Consequently, if an observed reduction in SFG is greater than predicted from the measured body burdens of aromatic hydrocarbons (i.e. there is an increase in the slope of the relationship) this would indicate action from additional toxicants and that further chemical analysis of body tissues should be carried out. Such a situation arose in the Shetland monitoring programme in 1984 and 1985, where some SFG values were significantly lower than predicted (Fig. 3). This observation was confirmed in subsequent years, particularly as a result of increased sampling intensity around the tanker loading area, and led to the analysis of stored tissue samples for tributyltin (TBT), an antifouling agent used on ships. Preliminary evidence (unpublished data) suggests that levels of TBT in the tissues may largely explain the difference between observed effects of SFG and those predicted from the two- and three-ring aromatic hydrocarbon concentrations.

Biological effects workshop

The results of a practical workshop provide a second example of the application of this combined tissue residue chemistry and physiological

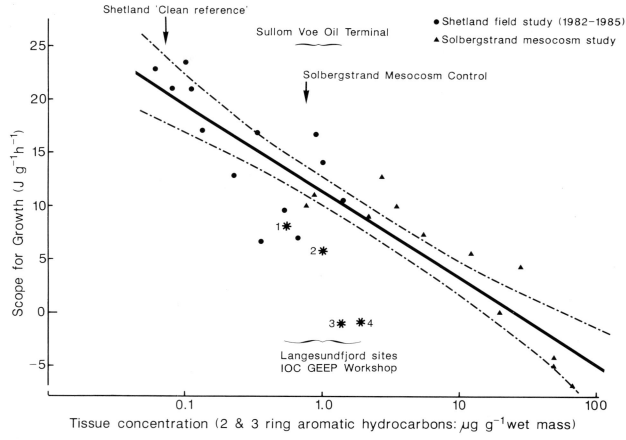

Fig. 3. Relationship between scope for growth and the log concentration of two and three ring aromatic hydrocarbons in the tissues of *Mytilus edulis*. SGF is standardised for 1 g mussel and a common ration level (0.4 mg particulate organic matter 1^{-1}), and all measurements were carried out during the spring and summer period of active growth.

(●) Data from Shetland field study (1982–1985).
(△) Data from Solbergstrand mesocosm study.
(*) Data from IOC Biological Effects Workshop.

energetic approach to the assessment of environmental pollution. At this workshop a wide selection of biological effects measurements were carried out 'blind' on samples collected from four sites along a pollution gradient in Langesundfjord (reviewed by Bayne *et al.*, 1988). The clearance rate and scope for growth of mussels showed a significant decline along the pollution gradient (Widdows & Johnson, 1988); the measured SFG values have been plotted on Fig. 3 in relation to the concentration of two- and three-ring aromatic hydrocarbons in the tissues. It is apparent, following the standardization and comparison of data, that the measured SFG of mussels at all four sites

were lower than would be predicted from the tissue accumulation of two- and three-ring aromatic hydrocarbons, particularly at the most polluted sites (sites 3 and 4; Fig. 3). The reduction in clearance rate and SFG was also greater than could be explained by the other contaminants analysed as part of the workshop (i.e. selected aromatic hydrocarbons, PCBs and selected metals; Widdows & Johnson, 1988). Furthermore, it is known that other toxic contaminants are entering Langesundfjord as a result of industrial inputs (e.g., Hg, chlorobenzenes, octachlorostyrene; Skei, 1987). The contribution that these and other unidentified toxicants make towards the total body burden is unknown.

Recently, the mussel samples from the four sites in Langesundfjord were analyzed for alkyltins (D. Page, unpublished data). Preliminary data suggest that TBT levels ($\sim 1 \mu g\, g^{-1}$ dry mass) may make a significant contribution to the overall reduction in SFG, but the relatively small TBT gradient between sites 1 to 4 does not explain the marked reduction in SFG at sites 3 and 4.

In the light of apparent discrepancies between predicted and observed effects on physiological energetics of mussels, the analysis of stored tissues for additional toxic contaminants would be profitable. If an appropriate tissue concentration-effect relationship was not available, this would then need to be established in toxicological studies.

Conclusions

The environmental toxicological approach outlined in this paper, based on the coupling of tissue residue chemistry and physiological energetic measurements, has been shown to be capable of quantifying the sublethal effects of environmental pollution. In addition, it has the potential for use in a diagnostic manner by providing a toxicological interpretation of tissue residue data and a means of identifying and establishing the relative significance of the cause(s) of sublethal effects recorded in field studies.

References

Bayne, B. L., 1976. Marine mussels, their ecology and physiology. Cambridge University Press, Cambridge: 506 pp.

Bayne, B. L., K. R. Clarke & J. Gray (eds.), 1988. Biological Effects of Pollutants: Results of a practical workshop. Mar. Ecol. Prog. Ser. 46 (Special Volume). 278 pp.

Donkin, P., J. Widdows, S. V. Evans, C. M. Worrall & M. Carr, 1989. Quantitative structure-activity relationships for the effect of hydrophobic organic chemicals on rate of feeding by mussels (Mytilus edulis). Aquatic. Toxicol. 14: 185–222.

Farrington, J. W., E. D. Goldberg, R. W. Riseborough, J. H. Martin & V. T. Bowen, 1983. US 'mussel watch' 1976-1978: an overview of the trace-metal, DDE, PCB, hydrocarbon and artificial radionuclide data. Envir. Sci. Technol. 17: 490–496.

Hermens, J. L. M., 1986. Quantitative structure-activity relationships in aquatic toxicology. Pestic. Sci. 17: 287–296.

Kaiser, K. L. E. (Ed.) 1987. QSAR in environmental toxicology II. D. Reidel Publishers, Dordrect: 465 pp.

Köneman, H., 1980. Structure-activity relationships and additivity in fish toxicities of environmental pollutants. Ecotoxicol. Envir. Saf. 4: 415–421.

Skei, J., 1981. The entrapment of pollutants in Norwegian fjord sediments – a beneficial situation for the North Sea. Spec. Publ. int. Ass. Sediment 5: 461–468.

Sloof, W., J. W. Canton & J. L. M. Hermens, 1983. Comparison of the susceptibility of 22 freshwater species to 15 chemical compounds. I. (Sub) acute toxicity tests. Aquat. Toxicol. 4: 113–128.

Turner, L., F. Choplin, P. Dugard, J. Hermens, R. Jaeckh, M. Marsmann & D. Roberts, 1987. Structure-activity relationships in toxicology and ecotoxicology: An assessment. Toxic. in vitro. 1: 143–171.

Widdows, J., 1985. Physiological measurements. In: B. L. Bayne et al., (eds.), The effects of stress and pollution on marine animals. Praeger Press, New York: pp. 3–45.

Widdows, J., D. K. Phelps & W. Galloway, 1981. Measurement of physiological condition of mussels transplanted along a pollution gradient in Narrangansett Bay. Mar. Envir. Res. 4, 181–194.

Widdows, J., T. Bakke, B. L. Bayne, P. Donkin, D. R. Livingstone, D. M. Lowe, M. N. Moore, S. V. Evans & S. L. Moore, 1982. Responses of Mytilus edulis on exposure to the water-accommodated fraction of North Sea oil. Mar. Biol. 67: 15–31.

Widdows, J., P. Donkin, P. N. Salkeld & S. V. Evans, 1987a. Measurement of scope for growth and tissue hydrocarbon concentrations of mussels (Mytilus edulis) at sites in the vicinity of the Sullom Voe oil terminal: – a case study. In: J. Kuiper & W. J. Van den Brink (eds.), Fate and effects of oil in marine ecosystems. Martinus Nijhoff, Dordrecht: pp. 269–277.

Widdows, J., P. Donkin & S. V. Evans, 1987b. Physiological responses of Mytilus edulis during chronic oil exposure and recovery. Mar. Envir. Res. 23: 15–32.

Widdows, J. & D. Johnson, 1988. Physiological energetics of Mytilus edulis: Scope for Growth. Mar. Ecol. Prog. Ser. 46: 113–121.

Hydrobiologia **188/189**: 463–476, 1989.
M. Munawar, G. Dixon, C. I. Mayfield, T. Reynoldson and M. H. Sadar (eds)
Environmental Bioassay Techniques and their Application.
© *1989 Kluwer Academic Publishers. Printed in Belgium.*

The biological assessment of contaminated sediments – the Detroit River example

Trefor B. Reynoldson[1] and Michael A. Zarull[2]
[1] *National Water Research Institute, Environment Canada, Canada Centre for Inland Waters, P.O. Box 5050, Burlington, Ontario, Canada L7R 4A6*; [2] *International Joint Commission, 100 Ouellette Avenue, Windsor, Ontario, Canada N9A 6T3*

Key words: sediments, contaminants, bioassay, benthos, assessment

Abstract

Contaminated sediments have been found in almost all water bodies which have at some time received, or are presently receiving, waste inputs from urban and industrial sources. In the Laurentian Great Lakes, sediments are classified as contaminated from bulk chemical analysis. The chemical criteria used to evaluate these results are somewhat arbitrary and only partially consider biological impacts. The absence of adequate linkage among sediment contamination, bioavailability, effects on organisms, populations, and ultimately ecosystem health, represents a major barrier to the restoration and protection of aquatic ecosystems.

An integrated strategy for the assessment and delineation of contaminated sediments is proposed which provides a comprehensive evaluation of impact, as well as a cost-effective sampling and testing program. The strategy incorporates the triad approach and is to be executed in two stages. Both stages use physical, chemical and biological information; however, the second stage requires more sampling and analyses to specify the severity and extent of the associated problems. To illustrate the type of output anticipated if the strategy is used, data assembled from the Detroit River are presented. They demonstrate that combined analysis of physical, chemical and biological data can be used to link cause and effect between sediment contaminants and benthic communities.

Introduction

Sediments contaminated with nutrients, metals, organics and oxygen demanding substances can be found in freshwater and marine systems throughout the world. While some of these contaminants are present in elevated concentrations as a result of natural processes, many are due to anthropogenic activities. In addition many of these compounds do not occur naturally. Sediments with elevated levels of contaminants have been identified in many of the nearshore, embay-ment, and tributary mouth areas in the Laurentian Great Lakes. These sediments have been designated as contaminated almost exclusively on the basis of bulk chemical analyses. In several cases, despite this designation, contaminants are present at levels lower than historic background concentrations found in sediment cores taken from the nearby, open lake depositional basins (Reynoldson *et al.*, 1988). In addition, measuring the concentrations of various chemicals present in the sediments does not address the ultimate concern; namely, whether the contaminants present are exerting biological stress and/or are being bio-accumulated. A series of bioassessment techniques, along with appropriate criteria, are necessary to identify the types of stress being exerted, their severity, and the bioavailability of the contaminants present.

Bioassessment approaches

Chapman & Long (1983) suggested that adequate assessments of sediment quality should involve three components; concentrations of toxic chemicals (bulk chemistry); measures of the toxicity of environmental samples (functional bioassessment); and measurements of the species composition and densities of the resident biota (structural bioassessment). Together, these three categories of measurement have been termed the sediment quality triad (Chapman & Long, 1983). Those authors tested this approach in Puget Sound, Washington and found good general correspondence between the three measures of sediment quality; however, chemical data alone were not always reliable indicators of biological effects (Long & Chapman, 1985).

There are many bioassessment techniques available, although not all have been used extensively in freshwater. However, a fundamental distinction can be made between structural and functional approaches to bioassessment. Much of the earliest work was of the structural or taxonomic type, where changes in community structure and indicator species were related to anthropogenic and naturally induced stresses (Warren, 1971; Hynes, 1960). Community structure information, which provides evidence of impact at one or more trophic levels within the ecosystem, can be easily and rather inexpensively obtained (Reynoldson, 1984). Sediment-related biological stress can best be demonstrated by examining the macrozoobenthic community (IJC, 1987). While structural information is necessary, desirable and readily attained, its specificity is often insufficient to influence a remedial action or regulatory change, since cause and effect cannot be easily inferred from structural data alone.

Functional tests or bioassays can provide this necessary information, since they are specific, and are usually capable of discriminating the root cause of structural changes as well as quantifying dose-response relationships.

Functional biological monitoring can be broadly defined as the measurement of any rate process or response of the ecosystem, and can utilize both taxonomic and non-taxonomic parameters (Mathews *et al.*, 1982). Such taxonomic tests include measures of species colonization or emigration rates and the rate of re-establishment of equilibrium densities following perturbation (Cairns *et al.*, 1979). The more frequently used non-taxonomic tests include acute (short-term) and chronic (longer term) measures of behavior, biochemical-physiological changes, bioaccumulation, genotoxicity (mutagenesis, carcinogenesis, teratogenesis), reproductive impairment and death. While many of these tests can be conducted *in situ*, the need for strict control over exposure conditions means that most bioassays are conducted in the laboratory. In addition, most of these tests are performed using single species which may not be indigenous to the area under investigation. Therefore, although functional tests can provide quantifiable relationships between contaminants and organisms and isolate sediment as a contributing factor, their lack of environmental realism and failure to incorporate ecosystem complexity limits their investigative utility and possibly their ability to establish actual impairment (Monk, 1983). The obvious solution to this dilemma is to use both structural and functional methods to examine stress.

Strategy

An ideal sediment assessment strategy should: (1) integrate physical data along with chemical and biological data, to provide an accurate assessment of the specific problems; (2) utilize the results from each technique to reduce subsequent sampling requirements and, therefore costs; (3) provide adequate proof of the linkage between the contaminated sediments and the problem (i.e.

cause-effect relationship); (4) quantify problem severity, thereby enabling inter-comparisons between and within areas of investigation; (5) consider the impacts or effects on different species and different trophic levels – since biological impairment may occur in both the water column (if resuspension occurs) and the sediments, and there is no such thing as the universal,

most sensitive species (Cairns, 1986). The proposed strategy (Fig. 1) consists of an initial assessment (Stage I), for areas with limited, antiquated, or no prior information, followed by a detailed investigation (Stage II). The initial assessment involves a limited physical description of the area under consideration, of the sediments and some chemical analyses. In addition

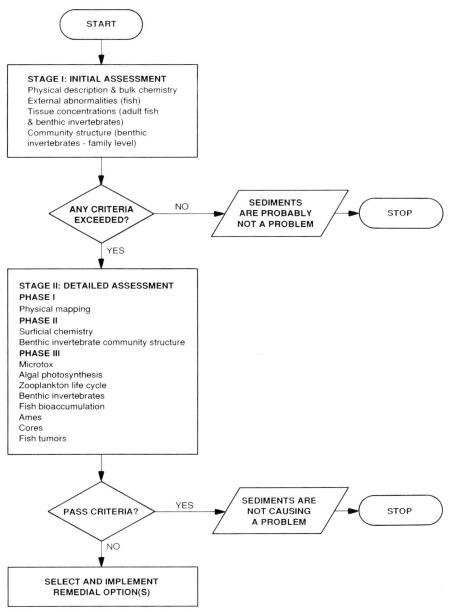

Fig. 1. Sediment assessment flowchart.

the resident benthic community structure is examined along with contaminant body burdens. Body burdens are also measured in one fish species. Available data may preclude the need for a Stage I investigation, if so, or the results of the Stage I tests suggest a sediment related problem, then a detailed assessment (Stage II) should be initiated. The detailed assessment is composed of four phases, which together define the sediment problem in the most cost effective manner. These phases do not represent rigid bounds, but rather logical groupings of work.

Stage I: Initial assessment

The initial assessment presents an opportunity to screen for potential sediment problems or to confirm the persistence of problem conditions where older or limited data are already available. The data collected, in this stage, represents the minimum necessary to assess potential sediment problems and all of the tests should be done. Should further, more detailed sampling be required, these data will further assist the investigator with sampling design.

A few sediment sampling sites should be selected based on previous sample collections, bathymetric information and/or known or suspected discharge locations. Samples must also be taken from depositional areas. Since physical, chemical and bioassessment measures are to be made on the same samples, a single field excursion should suffice.

Grain size, water content, redox potential (Eh) and pH should be measured since these factors affect contaminant concentrations and benthic community structure. In addition, surficial sediments should be analyzed for nutrients, metals, total organic carbon, persistent organics, and oxygen consuming contaminants (IJC, 1987, 1988).

Many hazardous contaminants are capable of being transferred from the sediments to the biota and accumulating at higher concentrations in subsequent trophic levels. Therefore, body burdens of hazardous contaminants should be examined in both indigenous macrozoobenthic invertebrates and adult demersal fish. The benthos, being more sedentary, provide a more direct measure of the specific relationship between localized sediment contaminant concentration and bioavailability. The fish provide a larger spatial and temporal integration of contaminants (IJC, 1987, 1988).

All of the fish collected for body burden analysis, should be examined for external abnormalities. In addition, a subsample of any fish with external deformities should be retained to allow complete histopathological analysis if a detailed assessment is necessary. The subsample should represent a cross-section of external deformity types, so that potential correlations between external abnormalities and histopathologic results can be examined in different fish species. These correlations may allow the expensive and time consuming histopathologic examination to be circumvented in future assessments.

It is recommended that a preliminary estimate of community structure impairment be made at the same sampling sites. Since this is an initial assessment only, it is not recommended that a quantitative study be undertaken; instead, qualitative estimates should suffice to identify the existence of a stressed community. Sample replication should be minimal, and taxonomic detailing to the family level at this stage will be sufficient to identify severe impact.

Criteria for initial assessment
The concentrations of metals in sediments should not exceed the background levels found in the nearest open lake depositional basin. In the Great Lakes, the background concentrations have conventionally been taken as those occurring below the *Ambrosia* (ragweed) pollen layer, which is associated with deforestation during early settlement (Thomas, 1981). Other regions may have other such convenient markers. Sediment concentrations of hazardous persistent organics should be less than detection levels using the best scientific methodology available. In the absence of existing objectives, the concentrations of hazardous persistent organics in fish or benthos should also be non-detectable using the best

scientific methodology. Concentrations of metals should be assessed on a case-by-case basis recognizing the physiological requirements for certain metals by each species.

The absence of a healthy benthic community represents an additional assessment criterion. This condition is defined by: the absence of clean water organisms such as amphipods or mayflies; a community dominated by oligochaetes; or the complete absence of invertebrates. These conclusions should be supported by evidence that the conditions are not due to a major perturbation such as dredging and/or substrate modification.

The presence of one or more external abnormalities on fish are often indicative of anthropogenically induced stress or damage. In the Great Lakes, external abnormalities such as stubbed barbels, skin discolouration (melanoma) and skin tumors are found to be highly correlated with liver cancer in brown bullheads (*Ictalurus nebulosus*; Smith *et al.*, 1988). Therefore, external abnormalities with a strongly suspected contaminant etiology should be absent (IJC, 1987, 1988).

If any of these criteria are violated, then a more detailed investigation of the sediments is required.

Stage II: Detailed assessment

The objectives of the detailed assessment are to determine the nature and severity of the sediment problem(s), and to determine the spatial and temporal extent of the contamination. The procedures are phased, but ALL ARE REQUIRED, and emphasis is placed on the use of physical information to assist the investigator in reducing sampling and testing requirements in subsequent phases. These phases do not represent rigid bounds, but rather logical groupings of work with the final design being the responsibility of the investigator. For example, in geographically small areas which require few sampling stations, Phases 2 and 3 could be combined. This modification would eliminate the need for further field collections to be made.

Phase I

The first step is the development of a three dimensional map of the physical composition of the sediments throughout the study area. The depositional areas should be precisely determined since more effort will be focused in these areas. Methods for obtaining this information are outlined in IJC (1987, 1988). Sampling results should be examined using cluster analysis (or a similar technique) which is simple to perform with available computer packages and provides good synthesis of complex data matrices. The resultant maps will define homogeneous areas for further examination in subsequent phases.

Phase 2

A second field collection will analyze benthic community structure and coincident surficial chemistry (total organic carbon, redox potential, pH, metals and persistent organics). A sampling grid based on the previous mapping of homogeneous zones of sediment type should be employed. Stratified random sampling technique should be used with more effort expended in the depositional areas and those areas with fine grained sediments.

Since the main objective of community structure assessment is to examine subtle distinctions in stress response, more detailed taxonomic data are required. In many nearshore areas, the community may be dominated by oligochaetes or chironomids, or both. While this may be indicative of impairment, there are considerable variations in environmental tolerance within the Oligochaeta and Chironomidae. Therefore species level identification can provide invaluable information for distinguishing and mapping zones of sediment impact. In addition to examining the results for indications of adverse environmental conditions (or impacts), chemical, physical and community structure data should be combined to test for further homogeneity using multivariate analyses.

Phase 3

For this phase, an additional field excursion is required for the collection of surficial sediments, sediment cores and indigenous fish. Information collected previously on impact and areas of homogeneity should allow sediment sampling sites to be reduced to an absolute minimum.

While effects on communities and ecosystems can best be studied by directly looking at changes in these systems, such methods can only be used after impacts have manifested themselves. Therefore, bioassays form an important part of a comprehensive approach to contaminant hazard assessment. Where possible, bioassays should be performed with sensitive, indigenous species so that results can be directly related to the infauna. Also, tests which examine the same effect in two or three species are better since this ensures a comprehensive data set, especially for areas with moderate contamination that may be at or near the toxicity threshold for some species. Sublethal effects should provide the major testing focus, and in particular, emphasis should be given to reproductive impairment.

Sediment bioassays which examine biochemical and physiological effects, bioaccumulation, genotoxicity, reproductive alteration and lethality (both acute and chronic) were reviewed. Tests which evaluate these effects, had standardized protocols, had extensive use, provided comparative data and were relatively simple to perform, were selected (IJC, 1988). The following bioassays also examine the effects on multiple trophic levels (bacteria, phytoplankton, zooplankton, benthic invertebrates and fish), and their exposure to dissolved as well as bound contaminants are recommended.

Microtox (Bulich & Isenberg, 1980; Bulich, 1984).

Algal Photosynthesis Bioassay (Ross *et al.*, 1987; Munawar *et al.*, 1983).

Zooplankton Life Cycle Test (Nebeker *et al.*, 1984).

Benthic Invertebrate Bioassay using the mayfly *Hexagenia limbata* or the midge larva *Chironomus tentans* (Mosher *et al.*, 1982; Malueg *et al.*, 1983; Nebeker *et al.*, 1984).

Fish and Benthic Invertebrate Bioaccumulation (Mac *et al.*, 1984).

Ames Test (Ames *et al.*, 1975).

The continuum of effects, or differences in organism response is documented in all of these bioassays, rather than the pass/fail results of a single endpoint. For example, the percent reduction in egg production and hatching of *Daphnia magna* for each sample (compared to a control or reference sediment) is measured. In this way, the relative toxicity of each sample can be examined test by test, as well as the combined test results for a geographic area. Controls should be established from samples taken well outside each area of study. Particular attention should be paid to obtaining sediments which are physically similar so that effects due to substrate differences are minimized or eliminated.

The data generated should be used in two ways: statistically significant departure of test results from control for any of the tests implies that some form of corrective action will likely be necessary; and plots of individual results should be overlapped to produce a 'toxicity map' which will rank, by severity, each area tested.

The ultimate goal of this assessment is to provide sufficient information to affect some remediation of the contaminated sediments. Furthermore decisions on remediation will require information on historic trends to allow appropriate decisions to be made. Therefore sediment cores must be collected to ascertain the vertical extent and temporal trend of contamination. Chemical analyses of the dated core sections should be conducted, based on the surficial chemical results, to identify the volume or total mass of contaminated sediments. These chemical measurements should be further supported with a few bioassays which should also be chosen on the basis of the surficial test results.

Histopathologic examinations should be conducted on indigenous, demersal adult fish since many of the observed abnormalities, particularly liver tumors, have an etiology linked with contaminants (Couch & Harshbarger, 1985; Smith *et al.*, 1988).

Criteria

Existing data are insufficient to develop precise criteria which dictate the action threshold for the remediation of contaminated sediments. Therefore, it is recommended that data necessary to establish action threshold criteria for community health, bioaccumulation, and chronic toxicity be collected.

In the absence of bioassay specific criteria, it is recommended that any effect which is significantly different from a control, at the 95% probability level, should be sufficient to establish test failure. Controls may consist of either, a standardised reference sediment, a sediment in the study area known from previous work to be uncontaminated or a sediment of similar particle size and organic content, collected from a nearby area. This latter choice is the least preferred as there may be other factors about that sediment that do not make it an appropriate control, there may however be no alternative. Mortality, or other relevant changes in the control should not exceed 10% to validate the test. The incidence of liver tumors should not exceed 2% in brown bullheads (Smith *et al.*, 1988). It is suggested that statistically significant departures from a normal or expected community (as described in Phase 2) be considered unacceptable.

The Detroit River

To illustrate the anticipated results from the proposed sediment assessment protocol, existing data sets in the Great Lakes basin were sought which had a combination of sediment chemistry, physics, benthic invertebrate community structure and sediment bioassays, for a retrospective analysis. The most comprehensive data were those from the Detroit River (Fig. 2). This river connects Lake St. Clair to Lake Erie, and is one of the most industrialized regions in the Great Lakes. Results were provided by the Ontario Ministry of the Environment from 1980 sampling of surficial sediments and benthic invertebrates from the same stations. Sediment bioassays were not conducted at that time, and although Giesy *et al.* (1988) performed assays in this region, differences in site location made them incompatible for this analysis.

Phase 1
The first step in the proposed strategy requires a detailed physical mapping of the sediments to

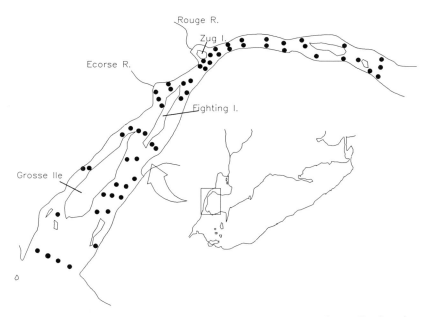

Fig. 2. Lake Erie and the Detroit River, showing the study area and sampling locations.

Table. 1. Physical and chemical sediment characteristics and biological variables analyzed in the Detroit River, 1980, and used in cluster analysis.

Physical	Chemical	Biological
Gravel (> 2.0 mm)	Mercury	Tubificidae
Sand (> 1.0 mm)	Cadmium	Naididae
Sand (> 500 μ)	Zinc	Lumbriculidae
Sand (> 250 μ)	Chromium	Enchytraeidae
Sand (> 125 μ)	Lead	Hirudinoidea
Silt (> 63 μ)	Nickel	Nematoda
Clay (> 4 μ)	Iron	Coelenterata
Fine Clay (< 4 μ)	Loss on Ignition (LOI)	Turbellaria
	Total Phophorus	Sphaeriidae
	Total Nitrogen	Gastropoda
	DDT & metabolites	Unionidae
	Lindane	Amphipoda
	PCB	Isopoda
	HCB	Decapoda
	Dieldrin	Hydracarina
	Endrin	Chironomidae
	Thiodene	Coleoptera
	Chlordane	Odonata
		Trichoptera
		Ephemeroptera

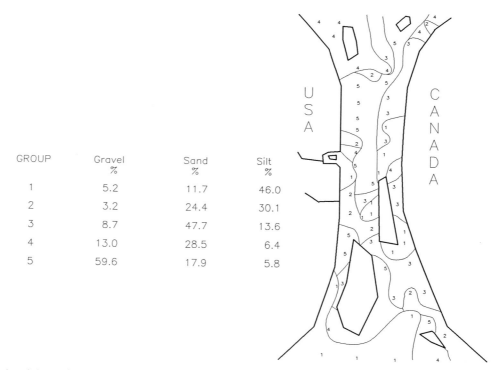

GROUP	Gravel %	Sand %	Silt %
1	5.2	11.7	46.0
2	3.2	24.4	30.1
3	8.7	47.7	13.6
4	13.0	28.5	6.4
5	59.6	17.9	5.8

Fig. 3. Results of cluster formation with physical variables and the mean value for those variables most significant in establishing site clusters.

Table. 2. Ranking of univariate correlations between environmental variables and benthic invertebrate site groups formed from cluster analysis.

Groups formed with Tubificidae			Groups formed without Tubificidae		
Variable	P	F ratio	Variable	P	F ratio
HCB	**0.000**	**273.632**			
TP	**0.000**	**78.346**			
Hg	**0.000**	**15.842**			
Cr	**0.000**	**14.134**			
Zn	**0.000**	**12.195**			
Ni	**0.000**	**10.269**			
LBNC	*0.001*	*13.466*	*Sand (>250 μ)*	*0.019*	*3.224*
Pb	*0.002*	*4.990*	*Sand (>1 mm)*	*0.027*	*2.967*
Cd	*0.002*	*4.839*	*Sand (>500 μ)*	*0.047*	*2.579*
Cu	*0.003*	*4.664*			
LOI	*0.039*	*2.710*			
Silt	0.139	1.815	LOI	0.081	2.202
Sand (>125 μ)	0.309	1.231	Silt	0.112	1.970
LCHLOR	0.384	1.193	Gravel	0.149	1.768
Sand (>1 mm)	0.468	0.904	Clay	0.265	1.346
PCB	0.521	0.871	TP	0.386	1.058
Gravel	0.643	0.630	Sand (>125 μ)	0.432	0.969
Sand (>500 μ)	0.706	0.542	Zn	0.520	0.817
Sand (>250 μ)	0.743	0.490	Ni	0.573	0.734
Clay	0.756	0.472	Cr	0.610	0.679
TN	0.904	0.257	Pb	0.655	0.613
			Cu	0.691	0.562
			Cd	0.733	0.503
			TN	0.871	0.308
			Hg	0.974	0.121

also significant ($P < 0.050$) in affecting the benthic community, as expressed by their correlation with the five benthic groups. These included three other metals, Pb, Cd, and Cu. Analysis showed that particle size only correlates for the coarse to fine sands in the sites grouped without the tubificids (Table 2). It is noteworthy that these are not the variables identified as being most important in defining the clusters for sediment physics (Fig. 3) or sediment chemistry (Fig. 4).

To determine the ability of the physio-chemical variables to predict benthic invertebrate community structure multiple discriminant analysis (MDA) was used. The MGLH module of SYSTAT was used to determine the discriminant functions and classification probabilities, and actual group membership tabulated against predicted. Analyses were done modeling two sets of variables against benthic invertebrate site groups. First using the six highly significant variables ($P < 0.001$), and second with the eleven significant variables ($P < 0.050$) identified in Table 2. The use of more variables resulted in sites being correctly assigned to groups for all stations, for which complete data were available. When only the six variables were used, site membership in the most impacted groups was accurately predicted but not so for less impacted sites (Table 3).

Finally, this type of information can be used to determine acceptable levels to which remediation should be undertaken. The mean values of the

Table. 3. The prediction of stations to correct groups using MDA with environmental variables.

Six Variables (P < 0.001)				Eleven Variables (P < 0.050)			
GP	OBS	PRED	CORRECT %	GP	OBS	PRED	CORRECT %
1	2	2	100	1	2	2	100
2	2	2	100	2	1	1	100
3	3	2	67	3	2	2	100
4	5	2	40	4	3	3	100
5	23	16	70	5	12	12	100

Table. 4. Mean values for environmental variables at site groups formed from benthic invertebrate data.

GROUP	1	2	3	4	5
n	2	2	7	5	43
VARIABLE					
HCB (ng·g)	5.500	196.00	5.670[2]	15.800	6.609[4]
TP (mg·g)	3.25	3.00	0.65	0.87	0.45
Hg (μg·g)	0.32	4.50	0.32	0.27	0.25
Cr (μg·g)	225.00	192.00	41.14	39.56	29.60
Zn (μg·g)	1465.00	975.00	199.86	165.40	150.21
Ni (μg·g)	118.50	73.50	24.14	30.00	22.30
LBHC (ng·g)	5.50	35.00[1]	18.00[3]	11.67	3.10[5]
Pb (μg·g)	300.00	404.50	70.66	43.606	68.73
Cd (μg·g)	7.00	9.85	2.73	1.50	1.18
Cu (μg·g)	239.50	171.00	49.04	32.40	39.87
LOI (%)	5.30	4.00	3.01	3.90	2.36

[1] n = 1
[2] n = 3
[3] n = 4
[4] n = 23
[5] n = 21

eleven variables shown to have impacted the benthic community have been determined for the five site groups defined by that community. Clearly, levels of contaminants in the Group 1 and 2 sites are unacceptable as demonstrated through their biological impact. Similarly levels at Group 5 sites are presumed acceptable based on these data. It is suggested, therefore, that an interim goal for sediment remediation in the Detroit River should be the concentrations observed in Group 5 sites (Table 4).

Summary

The strategy outlined here for sediment assessment uses a combination of sediment chemistry and physics together with *in situ* and laboratory biological assessment to define and bound the extent of sediment contamination. Furthermore, it expresses the impact of chemical contamination in a biologically meaningful manner and demonstrates the linkages between levels of contaminants and community response. Even with these data, which were not collected for the prime purpose of sediment assessment, and without supporting bioassays, the power of this information and analysis has been demonstrated. Definite boundaries have been placed on an impacted area, the biological effects of sediment contamination are demonstrated, six major and five minor variables have clearly been identified in a specific area of the river, and finally objective levels for clean up can be defined. This type of information is essential to provide a distinct and understandable framework for the assessment and remediation of polluted areas in the Great Lakes and other freshwater systems.

Acknowledgements

We wish to acknowledge the work and criticisms of our fellow members of the Great Lakes Water Quality Board Sediment Subcommittee (IJC) and its Assessment Work Group, in developing this strategy. We also wish to acknowledge Mr. Y. Hamdy and Mr. S. Thornley, from the Ontario Ministry of the Environment, for providing us with the Detroit River data and Mr Ron Dermott and Dr Doug Haffner for their review comments. Finally we wish to thank Ms. Mary Ann Morin for her perseverance and skill throughout the preparation of this paper.

References

Ames, B. N., J. McCann & E. Yamasacki, 1975. Methods for detecting carcinogens and mutagens with Salmonella/mammalian mutagenicity test. Mutation Res. 31: 347–364.

Bulich, A. A., 1984. A bacterial toxicity test with general environmental applications. In; D. Liu & B. Dutka (Eds.). Toxicity Screening Procedures Using Bacterial Systems. Marcel Dekker, New York. pp. 55–64.

Bulich, A. A. & D. L. Isenberg, 1980. Use of the luminescent bacterial system for the rapid assessment of aquatic toxicity. Adv. Instrum. 80: 35–40.

Cairns, J. Jr., 1986. The myth of the most sensitive species. BioScience 36 (10): 670–672.

Cairns, J. Jr., D. L. Kuhn & J. L. Plafkin, 1979. Protozoan colonization of artificial substrates. In; R. L. Wetzel (Ed.) Methods and Measurements of Attached Microcommunities: A Review. American Society for Testing and Materials, Philadelphia, PA. pp. 34–37.

Caspers, H., 1980. The relationship of saprobial conditions to massive populations of Tubificids. In; Brinkhurst, R. O. & Cook, D. G. (Eds.) Aquatic Oligochaete Biology, pp. 503–505.

Chapman, P. M., L. M. Churchland, P. A. Thomson & E. Michnowsky, 1980. Heavy metal studies with Oligochaetes. In; Brinkhurst, R. O. & Cook, D. G. (Eds.) Aquatic Oligochaete Biology, pp. 433–455.

Chapman, P. M. & E. R. Long, 1983. The use of bioassays as part of a comprehensive approach to marine pollution assessment. Mar. Pollut. Bull. 14: 81–84.

Couch, J. A. & J. C. Harshbarger, 1985. Effects of carcinogenic agents on animals: an environmental and experimental overview. Envir. Carcinogen. Rev. 3 (1): 63–105.

Giesy, J. P., R. L. Graney, J. L. Newsted, C. J. Rosiu, A. Benda, R. G. Kreis, Jr. & F. Horvath, 1988. Comparison of three sediment bioassay methods using Detroit River Sediments. Envir. Toxicol. Chem. 7: 483–498.

Hynes, H. B. N., 1960. The Biology of Polluted Waters. Liverpool Univ. Press. 202 pp.

International Joint Commission (IJC), 1987. Guidance on characterization of toxic substances problems in areas of concern in the Great Lakes Basin. A Report from the Surveillance Work Group, Windsor, Ontario, March 1987. 177 pp.

International Joint Commission (IJC), 1988. Procedures for the assessment of contaminated sediments problems in the Great Lakes. A Report by the Sediment Subcommittee, Windsor, Ontario. 140 pp.

Long, E. R. & P. M. Chapman, 1985. A sediment quality triad: measures of sediment contamination, toxicity and infaunal community composition in Puget Sound. Mar. Pollut. Bull. 16 (10): 405–415.

Mac, M. J., C. C. Edsall, R. J. Hesselberg & R. E. Sayers, Jr., 1984. Flow through bioassay for measuring bioaccumulation of toxic substances from sediment. EPA DW-930095-01-0. U.S. Environmental Protection Agency, Chicago, Illinois 26 pp.

Malueg, K. W., G. S. Schuytema, J. H. Gakstatter & D. F. Krauczyk, 1983. Effect of Hexagenia on Daphnia response in sediment toxicity tests. Envir. Toxicol. Chem. 2: 73–82.

Mathews, R. A., A. L. Buikema Jr., J. Cairns Jr. & J. H. Rodgers Jr., 1982. Biological Monitoring. Part IIA – Receiving system functional methods, relationships and indices. Wat. Res. 16: 129–139.

Milbrink, G., 1980. Oligochaete communities in pollution biology: The European situation with special reference to lakes in Scandinavia. In; Brinkhurst, R. O. & Cook, D. G. (Eds.) Aquatic Oligochaete Biology, pp 433–455.

Milbrink, G., 1983. An improved environmental index based on the relative abundance of Oligochaete species. Hydrobiologia 102: 89–97.

Monk, D. C., 1983. The uses and abuses of ecotoxicology. Mar. Pollut. Bull. 14 (8): 284–288.

Mosher, R. G., R. A. Kimerle & W. J. Adams, 1982. MIC environmental assessment method for conducting 14 day water exposure partial life cycle toxicity tests with the

476

midge Chironomus tentans. MIC Environmental Sciences Report No. ES 82-M-11. 11 pp.

Munawar, M., A. Mudroch, I. F. Munawar & R. L. Thomas, 1983. The impact of sediment-associated contaminants from the Niagara River mouth on various size assemblages of phytoplankton. J. Great Lakes Res. 9 (2): 303–313.

Nebeker, A. V., M. A. Cairns, J. H. Gakstatter, K. W. Malueg, G. S. Schuytema & D. F. Krawczyk, 1984. Biological methods for determining toxicity of contaminated freshwater sediments to invertebrates. Envir. Toxicol. Chem. 3: 617–630.

Reynoldson, T. B., 1984. The utility of benthic invertebrates in water quality monitoring. Wat. Qual. Bull. 10 (1): 21–28.

Reynoldson, T. B., A. Mudroch & C. J. Edwards, 1988. An overview of contaminated sediments in the Great Lakes, with special reference to the International Workshop held at Aberystwyth, Wales, U.K. Report to the Great Lakes Science Advisory Board, IJC, Windsor, Ontario 41 pp.

Ross, P. E., V. Jarry & H. Sloterdijk, 1987. A rapid bioassay using the green alga Selenastrum capricornutum to screen for toxicity in St. Lawrence River sediment elutriates. American Society Testing Materials STP NO 988.

Smith, S. B., M. J. Mac, A. E. MacCubbin & J. C. Harshbarger, 1988. (Abstract). External abnormalities and incidence of tumors in fish collected from three Great Lakes Areas of Concern. A paper presented at the 31st Conference on Great Lakes Research. McMaster University, Hamilton, Ontario, May 17–20, 1988.

Thomas, R. L., 1981. Sediments of the North American Great Lakes. Verh. Int. Ver. Limnol. 21: 1666–1680.

Warren, C. E., 1971. Biology and Water Pollution Control. W. B. Saunders Co. 434 pp.

Wentsel, R., A. McIntosh & V. Anderson, 1977. Sediment contamination and benthic macroinvertebrate distribution in a metal-impacted lake. Envir. Pollut. 14: 187–193.

Hydrobiologia **188/189**: 477–485, 1989.
M. Munawar, G. Dixon, C. I. Mayfield, T. Reynoldson and M. H. Sadar (eds)
Environmental Bioassay Techniques and their Application.
© 1989 *Kluwer Academic Publishers. Printed in Belgium.*

British Crown Copyright.

A method for studying the impact of polluted marine sediments on intertidal colonising organisms; tests with diesel-based drilling mud and tributyltin antifouling paint

Peter Matthiessen & John E. Thain
Ministry of Agriculture, Fisheries and Food, Directorate of Fisheries Research, Fisheries Laboratory, Remembrance Avenue, Burnham-on-Crouch, Essex CMO 8HA, U.K.

Key words: sediment, toxicity, diesel, tributyltin, colonisation, bioassay

Abstract

This paper describes a novel sediment bioassay which can be used in intertidal mud or sand, thereby exposing a contaminated sediment to a large range of naturally colonising fauna. Natural sediment, in which invertebrates had been killed by freezing, was mixed with diesel-based drilling mud (nominally 1000 mg kg^{-1} dry wt as diesel-based-mud equivalents) or particulate tributyltin (TBT) copolymer antifouling paint (nominally 0.1, 1.0 and 10 mg TBT kg^{-1} dry wt). The contaminated sediments were then re-laid intertidally in trenches lined with polythene mesh.

All treatments except 0.1 mg TBT kg^{-1} impaired the casting activity of the polychaete, *Arenicola marina*. Populations of the polychaete, *Scoloplos armiger*, and the amphipod, *Urothoe poseidonis*, were reduced in all contaminated treatments, and a dose-response effect of TBT was demonstrated. No clear effects on other groups (e.g. molluscs) were seen.

The results showed that this is a useful technique, although further development is required before it can be used routinely.

Introduction

Although freshwater, and especially marine sediments, are often the ultimate sink for hydrophobic and persistent pollutants, it is sometimes assumed that the disappearance of a xenobiotic substance from the water column is a sign that all impacted aquatic communities will recover. This is reflected in the multiplicity of toxicity tests and bioassays that are available for free-swimming organisms, and the relative dearth of methods for examining effects on benthic fauna. Nevertheless, natural benthic in-faunal populations are effective indicators of pollutant impact, and they have been shown to be sensitive to both organic enrichment and toxic contamination of fine particulate matter (Bilyard, 1987). It has been recommended, however, that relatively cheap sediment bioassays should replace expensive benthic monitoring (Long & Chapman, 1985).

Interest in sediment bioassays is also increasing because many chemicals, while in theory toxic to free-swimming fauna, are in reality efficiently scavenged from the water column by adsorption to particulates, yet their subsequent toxicity to sediment-dwellers is almost unknown. An exception to this is permethrin, a synthetic pyrethroid insecticide, which has been shown to exert persistent toxicity to burrowing larvae of the ephemeropteran, *Hexagenia rigida*, when adsorbed on mud in the laboratory (Friesen *et al.*, 1983). A number of other attempts have been made to study the toxicity of contaminated sediments in the laboratory (e.g. Keilty *et al.*, 1988; Alden *et al.*, 1988; McLeese *et al.*, 1982) and a useful review of marine sediment bioassays has been made by Swartz (1987).

Recently, methods have been developed which expose sterilised contaminated sediments to natural waters. Recolonisation may then occur by settlement of the dispersal stages of benthic organisms. Measurements of colonisation will then give an indication of the life-supporting capabilities of the sediment, and the sensitivity of such tests may be greater than LC_{50} measurements because they integrate avoidance behaviour and other sub-lethal effects with lethality. Examples of this type of test are described by Blackman *et al.* (1988a and 1988b) where natural estuarine water was pumped into the laboratory and passed over autoclaved sediment containing high concentrations of oil-based drilling mud. However, this technique failed to mimic completely results obtained under more natural conditions (Boesch *et al.*, 1981) and it is doubtful whether any type of laboratory test can be made to reproduce successfully the conditions, especially water movements, in sediments *in situ*.

An alternative approach is therefore to study colonisation of *in situ* sediments after experimental contamination. This approach has been pioneered by Tagatz (Tagatz & Deans, 1983; Tagatz *et al.*, 1987) and others (Arnoux *et al.*, 1985) and involves the placement of sediment-filled containers in inter- or sub-tidal substrates and monitoring of natural colonisation. A similar method has also been used to compare recruitment of benthic fauna in polluted and unpolluted mudflats (Thrush & Roper, 1988).

While passive advection of organisms is probably the main route of colonisation on intertidal sand flats (Savidge & Taghon, 1988), active crawling and burrowing are also significant for some species. A disadvantage of most colonisation bioassays has been the use of solid-sided exposure boxes which impede interstitial water movement and discourage migration of adult bur-

rowing organisms from the adjacent substrate. The problem has been avoided in a study of colonisation by benthic meiofauna (Decker & Fleeger, 1984) which used mesh-sided boxes, but this approach has not been used for macrofauna.

This paper describes a marine sediment bioassay that measures both passive and active colonisation and has been tested on intertidal muddy-sand flats using diesel-based drilling mud and tributyltin antifouling paint as model toxicants.

Materials and methods

The experiments were carried out on the Maplin Sands (National Grid Reference TQ 964860), a SE-facing area of intertidal muddy-sand flats on the coast of Essex UK, which supports extensive beds of eelgrass (*Zostera marina*). Physical characteristics of the sediment (moisture content, density, grain size, organic carbon content) were measured. Approximately 1 m³ of surface sediment (top 20 cm) was transferred to the laboratory in polythene sacks and taken through two 48 h cycles of deep-freezing ($-20\ °C$) and thawing to kill resident macrofauna. A 0.2 m³ sample of the sediment was then mixed in a cement mixer for 1 h with British National Oil Corporation diesel-based drilling mud (DBM) (1000 mg kg^{-1} dry wt. as DBM equivalents – nominal). This DBM was identical with that used by Blackman *et al.* (1988a). Three further 0.2 m³ sediment samples were each mixed with a tributyltin (TBT) copolymer antifouling paint to produce nominal TBT concentrations of 0.1, 1.0 and 10 mg kg^{-1} dry wt respectively. The paint was added as the washings from a dried paint film which had been abraded in water with a scouring device similar to those used for scrubbing-down yachts. The TBT treatments thus mimicked the type of sediment contamination found near boatyards. The remaining 0.2 m³ of sediment was also mixed for 1 h and used as the control treatment.

The contaminated sediments, and uncontaminated control sediment, were relaid approximately 500 m seaward of the sea-wall at the

Maplin Sands site in April 1987, each in a 3 m long × 30 cm wide × 20 cm deep trench lined with polythene mesh (5 mm apertures). The mesh was held in position with steel stakes driven into the underlying sand. The trench ends were closed off with vertical slates, also held down with stakes. On four dates between April and September 1987, the trenches were re-visited and 4 replicate sub-samples (approximately 6 l each – equivalent to 0.03 m²) were removed from one end of each trench. The sub-samples were taken by isolating a slice of sediment at the trench-end with an additional slate and excavating with a large spatula. The hole was backfilled with uncontaminated sediment and a slate finally left in position to mark the new trench-end. Some background samples were also taken from the surrounding untouched sediments. After return to the laboratory in polythene bags, the sub-samples were weighed and then sieved to 0.5 mm and the retained macrofauna preserved in 5% formol-saline containing rose-bengal stain. Each sieved sample was ultimately spread out on a white tray, and all organisms were identified and counted. Finally, the number of lugworm, *Arenicola marina*, casts visible on the surface of each trench was recorded during each visit.

On each visit, further samples of sediment were also taken from a range of depths in the experimental trenches and frozen in polythene bags to await analysis for diesel or TBT. Samples were analysed for diesel using the fluorescence spectrometric method at 230/330 nm described by Blackman *et al.*, (1988a), except that the results were standardised against the drilling mud rather than against diesel alone (R. J. Law – pers. com.). The samples were analysed for TBT by extraction with sodium hydroxide and methanol, formation of the hydride using sodium borohydride, and analysis by GC/FPD (M. J. Waldock, pers. comm.).

Results

The main physical characteristics of the Maplin sediment are given in Table 1. It falls into the category of a very fine muddy sand.

Diesel concentrations, expressed in DBM equivalents for 4 depth bands (Fig. 1a), show that concentrations in the surface 0–3 cm dropped rapidly to near-background levels within 30 days. However, in deeper bands, diesel concentrations dropped to a mean of approximately 60% of the initial concentration within that time, and only gradually declined thereafter. Degradation of the normal alkanes (of the oil) only occurred during the latter half of the experiment, but was apparent at all depths. C17/pristane and C18/phytane ratios were 2.40 and 2.35 on day-0 and unchanged by day-86, but had reduced to 1.08 and 1.03 by day-164. This equates to 55% removal of the C17 and C18 n – alkanes by bacterial action.

Fewer data are available concerning TBT concentrations in the sand (Fig. 1b) but it appears that here too, contamination rapidly disappeared from the surface layer and only slowly declined in deeper layers. In all samples, the proportion of dibutyltin (DBT) to TBT lay between 0 and 0.13, and monobutyltin was not detected. It is not possible to say whether bacterial degradation of TBT was occurring.

The fact that the diesel only degraded slowly, but nevertheless disappeared rapidly from the surface layer provides partial confirmation of visual observations that the surface 0–3 cm was mobile, while deeper layers appeared more stable. Visual observations also showed that the aerobic/anaerobic boundary lay between approximately 5

Table 1. Physical characteristics of the Maplin Sands sediment.

	Mean	Range	s.d.	n
% moisture	24.1	21.4–29.3	1.8	30
density (g cm⁻³)	1.97	–	–	5
% organic carbon	1.92	1.84–2.00	0.06	8
grain size (μm) (modal size class)	–	90–125	–	8
grain size (μm) (range of median values)	–	105–109	–	8
% silt (<63 μm)	11.9	8.5–14.4	1.7	8

480

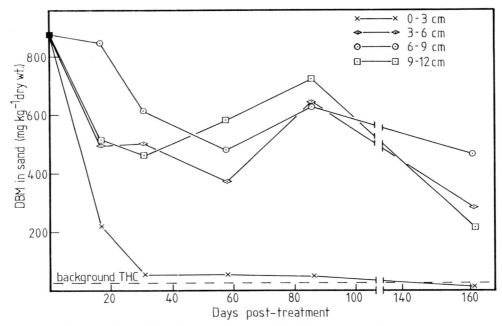

Fig. *1a*. Concentrations of diesel-based drilling mud (DBM) at 4 depths during the experiment.

and 10 cm below the surface, so the rapid surface disappearance of diesel is unlikely to have been related to differential degradation rates in the aerobic and anaerobic layers. It is more likely that

the contaminated surface particles were replaced by clean advecting sediment.

Samples taken from a range of depths in undisturbed natural Maplin Sands sediments near the

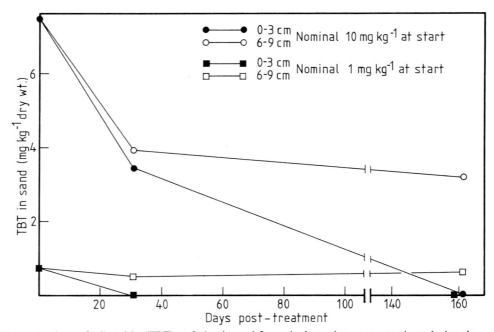

Fig. *1b*. Concentrations of tributyltin (TBT) at 2 depths and 2 nominal starting concentrations during the experiment.

experiment revealed that the majority of macrobenthic fauna lived in the top 5 cm. Figure 2 shows the vertical distribution of the five most common species on 7/5/87, expressed both as numbers per kilogram wet wt of sediment and as numbers per square metre. As the areas sampled in the experimental treatments were not measured accurately, all species density values (apart from lugworms) are reported below as numbers per kilogram averaged over the entire 20 cm depth of the experiment.

The species found throughout the duration of the experiment are listed in Table 2. There were no major differences between the species represented in the various treatments and it is likely that the small apparent differences were artefacts of the small number of replicate samples. All species found in the surrounding untouched sediment were subsequently recorded in the experimental sediments. As several of the rarer species were probably under-sampled, no attempt has been made to calculate species diversity in the various treatments.

The sediment sample volumes were too small to effectively include the larger fauna such as the lugworm, *Arenicola marina*, but the numbers of surface casts gave an indication of the abundance of this species. Figure 3 shows the density of *A. marina* casts in the various treatments, and although each point represents a single count, it is reasonable to conclude that all treatments except the nominal 0.1 mg TBT kg^{-1} effectively prevented most casting. Casting in the nominal 1.0 mg TBT kg^{-1} treatment, however, was showing evidence of recovery by the end of the experiment. Judging by the size of the casts, the affected animals were mainly adults living in the contaminated zone below 3 cm, but there was no indication whether they were merely avoiding the higher treatments or being poisoned by them.

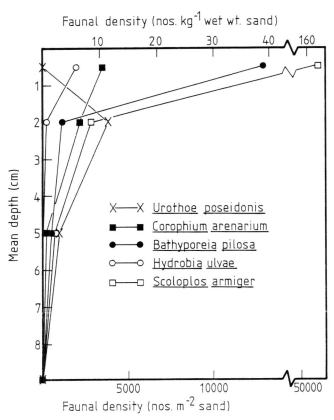

Fig. 2. Faunal distribution by depth in undisturbed natural sediment near the experimental trenches on 7/5/87.

Table 2. Species found in the various treatments. Those found at the highest densities are marked with an asterisk.

	Control	0.1 mg kg⁻¹ TBT	1.0 mg kg⁻¹ TBT	10 mg kg⁻¹ TBT	1000 mg kg⁻¹ DBM
CRUSTACEA					
*Bathyporeia pilosa**	+	+	+	+	+
*Corophium arenarium**	+	+	+	+	+
*Urothoe poseidonis**	+	+	+	+	+
*Gammarus locusta**	+	+	+	+	+
Stenothoe marina			+		
Hyale nilsonni	+				
Idotea linearis		+		+	+
Pseudocuma longicornis		+		+	+
Leander sp.	+	+	+		+
Crangon crangon	+	+	+	+	+
juvenile Brachyuran	+	+	+	+	+
Carcinus maenas	+	+	+	+	+
larval Decapod	+	+	+		
harpacticoid Copepods	+	+	+	+	+
ANNELIDA					
*Scoloplos armiger**	+	+	+	+	+
Nereis virens	+	+	+		
Nereis sp	+	+	+	+	+
*Capitella capitata**	+	+	+	+	+
Eteone longa	+				+
Pygospio elegans					+
Spio martinensis					+
Capitomastus minimus					+
Phyllodoce maculata	+	+	+		
Arenicola marina	+	+	+	+	+
Tubificoides benedeni	+		+	+	+
MOLLUSCA					
Macoma balthica	+	+	+	+	+
*Hydrobia ulvae**	+	+	+	+	+
Mytilidae juv.	+	+	+	+	+
*Cerastoderma edule**	+	+	+	+	+
Mysella bidentata			+		
PISCES					
Pomatoschistus minutus		+	+	+	+
Unidentified nematodes*	+	+	+	+	+
Total species/groups	24	24	25	21	26

Not all species were present in sufficient numbers in the samples to allow meaningful comparisons of relative density. Figure 4 presents data for two species, the small burrowing polychaete, *Scoloplos armiger* (Orbiniidae), and the burrowing amphipod, *Urothoe poseidonis* (Haustoriidae), which showed reasonably clear concentration-related effects. Significant differences (Mann-Whitney test, $p < 0.05$) from the controls were consistently found for *S. armiger* in the 10 mg TBT kg⁻¹ treatment only, but this species nevertheless showed apparently concentration-related

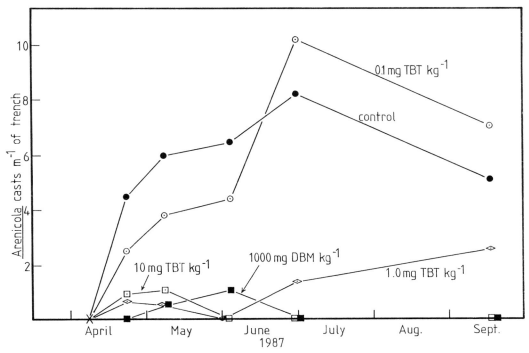

Fig. 3. Density of *Arenicola marina* casts in the various treatments.

population declines on all sampling dates. The DBM treatment caused significant reductions in *S. armiger* during the second half of the experi-

ment. *U. poseidonis* populations were generally sparse except on day-164, when numbers in the control were significantly higher than those in the

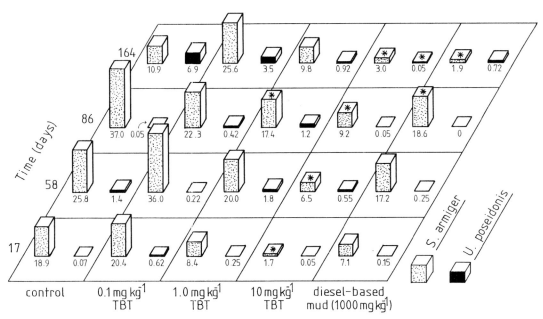

Fig. 4. Mean density (nos kg^{-1}) ($n = 4$) of *Scoloplos armiger* and *Urothoe poseidonis* in the various treatments. Values significantly different from the controls are indicated by an asterisk ($p < 0.05$, Mann Whitney test).

highest TBT treatment. Again, there was an apparently concentration-related effect within the TBT treatments. The populations of the burrowing polychaete, *Capitella capitata*, also showed a decline with increasing TBT levels on day-164, but this was not statistically significant.

There were no consistent treatment-related effects at any time in other common species. For example, populations of neither the amphipod *Corophium arenarium* (Corophiidae) nor the molluscs *Cerastoderma edule* (Bivalvia) and *Hydrobia ulvae* (Gastropoda) were apparently depressed by any treatment. Similar results were obtained for the other common species.

Discussion

Perhaps the most striking feature of these results is that very few species appeared to have been affected by the treatments even though major effects were observed in two polychaetes and an amphipod. It is possible that treatment-related effects occurred in some of the rarer species and that these effects were not seen due to the relatively insensitive sampling procedure. Nevertheless, the lack of impact on some of the more common species must be addressed. One explanation for this phenomenon is the mobility of the sediment which caused a rapid reduction in surficial contaminant concentrations; therefore surface-dwelling animals which dominated the community (e.g. *H. ulvae*, *Bathyporeia pilosa*, *S. arenarium*) were not exposed to harmful concentrations. Some evidence for this is provided in Fig. 2 which shows that *U. poseidonis*, one of the affected species, was absent at the surface but was the most common species in the deeper layers. It is interesting to note that Blackman *et al.* (1988a) observed surface recolonisation in their DBM treatment when the surficial diesel concentration had dropped, despite negligible loss of diesel from deeper layers.

Another effect related to sediment mobility, that is a feature of intertidal sand flats, is the continual passive advection of surface-dwellers (especially amphipods) into vacant habitats (Savidge & Taghon, 1988). This may have tended to swamp treatment-related effects in surface animal populations by rapidly substituting new organisms for those which would eventually have been affected by the treatment under more stable conditions. This is an artefact of the relatively small treatment area, and could perhaps be eliminated by laying the experiment in a more sheltered position.

A third possibility is that feeding behaviour influenced exposure. The affected polychaetes (including *C. capitata*) all feed non-selectively on fine sedimentary material (Fauchald & Jumars, 1979) and therefore would be expected to receive greater exposure than bivalves such as *C. edule* which filter water from above the sediment surface. Furthermore, amphipods such as *B. pilosa* selectively feed on detritus and avoid wholesale ingestion of particulates (Nicolaisen & Kanneworff, 1969).

Lastly, it is well known that pesticides such as TBT have a wide spectrum of activity against aquatic organisms (Waldock *et al.*, 1987) and it is therefore not unexpected that some groups should have remained apparently unaffected by the treatments. Toxicity data reviewed by Waldock *et al.* (1987), and based almost exclusively on tests in which marine animals were exposed to TBT as a solution in clean water, paradoxically reveal that most molluscs are highly susceptible to TBT under these conditions while crustaceans and polychaetes are much less affected. On the other hand, Salazar & Salazar (1985) failed to demonstrate mortality in 10- and 20-day laboratory toxicity tests using crustaceans, polychaetes and bivalves exposed to mud naturally contaminated with 0.78 mg TBT kg^{-1}. It therefore seems unlikely that these groups were being killed in the two least-contaminated treatments of the Maplin experiment. These observations highlight the drawback of traditional tests when attempting to predict the ecological effects of contaminated sediments, and suggest that the colonisation bioassay is primarily measuring a sublethal parameter such as avoidance behaviour. Laboratory tests to investigate this question are currently in progress.

The methods described in this paper have proved the value of the mesh-sided exposure trench which allows colonisation by large burrowing organisms as well as by swimming forms. There is a particular need to optimise the sampling strategy and minimise the effects of surface sediment mobility. An ideal bioassay would allow unrestricted access by swimming, crawling and burrowing organisms, but would minimise the diluting effect of large quantities of advected sediment and accompanying fauna. An improved method is at present being tested using a contaminated dredge spoil as the experimental toxicant.

Acknowledgements

The assistance of many Burnham Fisheries Laboratory staff has been essential to the success of this work. We particularly thank Mr R. J. Law and Dr M. J. Waldock for conducting the chemical analyses; Dr H. L. Rees and Mr D. A. Sheahan for assistance with identification of fauna; and Mr R. G. Lees and Mr D. S. Limpenny for conducting the sediment particle size analysis. We also thank the Proof and Experimental Establishment at Shoeburyness for permission to conduct the experiments on their firing range.

References

Alden, R. W., A. J. Butt & R. J. Young, 1988. Toxicity testing of sublethal effects of dredged materials. Arch. Envir. Contam. Toxicol. 17: 381–389.

Arnoux, A., G. Stora & C. Diana, 1985. *In situ* experimental study of the evolution and recolonisation of polluted sediments. Mar. Pollut. Bull. 16: 313–318.

Bilyard, G. R., 1987. The value of benthic infauna in marine pollution monitoring studies. Mar. Pollut. Bull. 18: 581–585.

Blackman, R. A. A., T. W. Fileman, R. J. Law & J. E. Thain, 1988a. The effects of oil-based drill-muds in sediments on the settlement and development of biota in a 200-day tank test. Oil. Chem. Pollut. 4: 1–19.

Blackman, R. A. A., R. J. Law & J. E. Thain, 1988b. The effects of new oil-based drill-muds in sediments on the settlement and development of biota in an improved tank test. Oil. Chem. Pollut. (in press).

Boesch, D. F., E. M. Burreson, L. C. Shaffner, H. I. Kator, C. L. Smith, D. M. Alongi, M. A. Bowen & P. O. DeFur, 1981. Experimental Colonization of Crude Oil Contaminated Sediment by Benthos on the Middle Atlantic Continental Shelf. USDI Bureau of Land Management Final Report No. BLM/YL/SR-81/01. Virginia Institute of Marine Science, Gloucester Point, 258 pp.

Decker, C. J. & J. W. Fleeger, 1984. The effect of crude oil on the colonization of meiofauna into salt marsh sediments. Hydrobiologia 118: 49–58.

Fauchald, K & P. A. Jumars, 1979. The diet of worms. Oceanogr. Mar. Biol. Ann. Rev. 17: 193–284.

Friesen, M. K., T. D. Galloway & J. F. Flannagan, 1983. Toxicity of the insecticide permethrin in water and sediment to nymphs of the burrowing mayfly *Hexagenia rigida* (Ephemeroptera; Ephemeridae). Can. Ent. 115: 1007–1014.

Keilty, T. J., D. S. White & P. F. Landrum, 1988. Short-term lethality and sediment avoidance assays with endrin-contaminated sediment and two oligochaetes from Lake Michigan. Arch. Envir. Contam. Toxicol. 17: 95–101.

Long, E. R. & P. M. Chapman, 1985. A sediment quality triad: measures of sediment contamination, toxicity, and infaunal community composition in Puget Sound. Mar. Pollut. Bull. 16: 405–415.

McLeese, D. W., L. E. Burridge & J. van Dinter, 1982. Toxicities of five organochlorine compounds in water and sediment to *Nereis virens*. Bull. Envir. Contam. Toxicol. 28: 216–220.

Nicolaisen, W. & E. Kanneworff, 1969. On the burrowing and feeding habits of the amphipods *Bathyporeia pilosa* Lindström and *Bathyporeia sarsi* (Walker). Ophelia 6: 231–250.

Salazar, M. H. & S. M. Salazar, 1985. Ecological Evaluation of Organotin – Contaminated Sediment. Naval Ocean Systems Center, San Diego, Tech. Rep. no. 1050, 22 pp.

Savidge, W. B. & G. L. Taghon, 1988. Passive and active components of colonization following two types of disturbance on intertidal sandflat. J. Exp. Mar. Biol. Ecol. 115: 137–155.

Swartz, R. C., 1987. Toxicological methods for determining the effects of contaminated sediment on marine organisms. In K. L. Dickson, A. W. Maki & W. A. Brungs (eds). Fate and Effects of Sediment–Bound Chemicals in Aquatic Systems. Pergamon Press, New York: 183–198.

Tagatz, M. E. & C. H. Deans, 1983. Comparison of field- and laboratory-developed estuarine benthic communities for toxicant exposure studies. Wat. Air Soil Pollut. 20: 199–209.

Tagatz, M. E., R. S. Stanley, G. R. Plaia & C. H. Deans, 1987. Responses of estuarine macrofauna colonising sediments contaminated with fenvalerate. Envir. Toxicol. Chem. 6: 21–25.

Thrush, S. F. & D. S. Roper, 1988. Merits of macrofaunal colonization of intertidal mudflats for pollution monitoring: preliminary study. J. Exp. Mar. Biol. Ecol. 116: 219–233.

Waldock, M. J., J. E. Thain & M. E. Waite, 1987. The distribution and potential toxic effects of TBT in UK estuaries during 1986. Appl. Organometal. Chem. 1: 287–301.

Hydrobiologia **188/189**: 487–496, 1989.
M. Munawar, G. Dixon, C. I. Mayfield, T. Reynoldson and M. H. Sadar (eds)
Environmental Bioassay Techniques and their Application.
© 1989 *Kluwer Academic Publishers. Printed in Belgium.*

The effect of heavy metal speciation in sediment on bioavailability to tubificid worms

Alistair M. Gunn, David T. E. Hunt & D. Alan Winnard
Water Research Centre, Medmenham Laboratory, P.O. Box 16, Marlow, Bucks, SL7 2HD. U.K.

Key words: heavy metals, sediment, speciation, bioavailability, tubificid worms

Abstract

The bioavailability of heavy metals in sediment to freshwater tubificid worms was compared with measures of chemical extractability using a sequential extraction procedure. In order to provide a range of test sediments of different quality, various mineral phases were prepared, in which the metals were spiked by adsorption or coprecipitation and these were then mixed with a bulk base sediment in known proportions. Results indicated good correlation between worm metal burden and metal mobilised from the sediments in the first ('exchangeable') sequential extraction step for Cd, Cu and Pb. Of the other metals tested, Zn levels in the worms were found to be constant, suggesting regulation, and Ni uptake was too small for accurate measurement. In general, metals spiked to the sediment directly, or adsorbed on the clay mineral phase were found to be much more available than those bound to sewage sludge, carbonate or hydrous ferric oxide phases.

Introduction

Heavy metals are of concern as contaminants of water bodies because of their persistence and high toxicity to many aquatic organisms. They may be present in both water and sediment in a wide variety of physico-chemical forms or 'species', and the speciation of a metal has a profound influence on its transport, fate and effects in the aquatic environment (Leppard, 1983). In particular, a knowledge of chemical speciation is of fundamental importance to an understanding of heavy metal bioavailability and toxicity, and to the rational setting of Environmental Quality Standards to protect aquatic life.

Computer modelling using available thermodynamic data can provide valuable information on the speciation of heavy metals in water (e.g. French & Hunt, 1986) but the shortage of reliable data for certain important species (e.g. complexes of heavy metals with humic substances) limits the value of this approach, making it best suited to use in laboratory studies of speciation/toxicity relationships, under carefully controlled conditions of water quality (Stumm & Morgan, 1981).

On the other hand, the separations achieved using the numerous analytical procedures for heavy metal speciation, in both the aqueous phase (Neubecker & Allen, 1983) and the sedimentary phase (Salomons & Förstner, 1984; Förstner, 1985), are almost exclusively operational in character. It must therefore be demonstrated that the results obtained are capable of predicting effects of interest – such as bioavailability and toxicity – if they are to be of practical value.

Two recent reviews of the relationships between speciation and toxicity (O'Donnel *et al.*, 1985; Hunt, 1987) have both concluded that far more is known about the bioavailability of metals from water than is known about their bioavailability from sediment. Although there are as yet no widely-accepted and reliable analytical procedures for assessing the bioavailability of dissolved metal, the free metal ion activity has in most cases proved to be the best predictor of metal uptake and toxicity. The essential problem is to obtain valid and reliable measures of bioavailable metal, given that the free metal ion activity cannot at present be measured satisfactorily for all metals in natural waters.

By contrast, there remains considerable uncertainty about the nature of the more bioavailable forms of heavy metals in sediments; the problem is made more complex by the very varied nature of the routes by which benthic organisms may take up metal from the sediment and its interstitial water (Hunt, 1987). Some progress has been made, however.

Luoma & Jenne (1976) prepared three cadmium-enriched artificial substrates (hydrous iron oxide, organically-coated hydrous iron oxide and homogenised detrital matter) and found that the clam, *Macoma balthica*, took up cadmium only from the uncoated hydrous iron oxide. The same authors later reported (1977) another study with clams – of the uptake of cobalt, silver and zinc from five different substrates: calcite, manganese and iron oxides, biogenic calcium carbonate and detritus. They found that the uptake of the metals from the different substrates was significantly correlated with the liquid/solid distribution coefficient, which they also measured; the more the metals tended to partition into a particular substrate from the water, the less available was the metal content of the substrate for uptake. Oakley *et al.* (1983), in experiments with the deposit feeding polychaete *Abarenicola pacifica* also demonstrated uptake of cadmium from ferric hydroxide and bentonite clay.

Luoma & Bryan (1979) studied the uptake of zinc by two bivalves (*Macoma balthica* again, and *Scrobicularia plana*) in San Francisco Bay and English estuaries. They found that the uptake correlated best with the fraction of the zinc extracted from the sediments by ammonium acetate, and

concluded that, in San Francisco Bay, zinc uptake was controlled by the ratio of humic substances to total organic carbon. Both Hall & Bindra (1979) and Diks & Allen (1983) observed that copper accumulation by benthic organisms correlated best with the copper present in an easily-reducible fraction.

It should be noted, however, that a number of these studies (Luoma & Bryan, 1979; Diks & Allen, 1983) have also found significant correlations between uptake and some dilute-acid-extractable fraction, and there is likely often to be a considerable degree of cross-correlation between different extracted metal fractions – a point made recently by Tessier *et al.* (1984). These workers studied the relationship between metal uptake and speciation for a freshwater bivalve (*Elliptio complanata*), finding that several 'readily-extractable' metal fractions were good predictors of uptake, because they were strongly inter-correlated.

Several studies have made use of artificial substrates to assess the bioavailability of different metal forms. Similarly, artificially enriched phases have been used to evaluate, in chemical terms, the separations achieved by sequential extraction procedures in speciation analysis (Rapin & Förstner, 1983; Meguellati *et al.*, 1983). The work reported here has combined these two approaches, by amending a natural sediment with specific, metal-enriched phases (clay mineral, hydrous ferric oxide, carbonate and sewage sludge), and conducting both sequential chemical extractions and uptake studies upon the resulting materials. In this way, it has sought to obtain further information on the bioavailability of metal in different sedimentary phases, and to identify that fraction of a sequential extraction scheme which provides the best predictor of metal uptake.

Materials and methods

Preparation of test materials

A bulk sample of sediment, sufficient for all the proposed tests, was collected from a site on the Severn. Only the top or oxic layer (1–2 cm) was collected and this was sieved to 62 μm using polyester sieves and stored as a slurry at 4 °C.

Model geochemical phases were spiked with appropriate loadings of the metals of interest Pb, Cd, Zn (as nitrate) and Cu, Ni (as sulphate) by adsorption or coprecipitation of the aqueous metal salt with a suspension of the solid phase:

(i) Clay mineral phase with adsorbed metals was prepared by addition of solutions of the metal salts to a stirred aqueous suspension of kaolin, the pH was then adjusted to 8 with sodium hydroxide.

(ii) Carbonate phase with coprecipitated metals was prepared by addition of an equivalent amount of sodium carbonate to a calcium chloride solution containing the metal salts.

(iii) Hydrous ferric oxide phase, again with coprecipitated metals, was prepared by slow addition of sufficient sodium hydroxide to a well stirred solution of ferric chloride/metal salts to give a final pH of 8.

(iv) In addition to these model phases, a sample of digested sewage sludge with elevated metal levels was obtained from Beckton sewage treatment works (Thames Water).

(v) Finally, a portion of the bulk sediment was spiked by direct adsorption of the metal ions from solution; again the slurry was stirred during the addition and the pH adjusted to 8 with sodium hydroxide.

All the phases were then thoroughly washed in a synthetic freshwater of medium hardness (100 mg Ca/l, 50 mg Mg/l, 16 mg Na/l, 4 mg K/l; pH 8) by repeated centrifuging and resuspension. Table 1 gives the nitric acid extractable metal concentrations found in the prepared phases. Prior to the

Table 1. Acid extractable metal concentrations in spiked mineral phases (μg/g dry wt)

	Zn	Cu	Ni	Pb	Cd
Kaolin	3300	350	250	1300	3.5
Carbonate	2800	300	260	750	9.0
Ferric hydroxide	4600	470	380	1900	12.5
Sludge	4200	680	94	1200	18.0

uptake experiments each of the phases was mixed with the bulk sediment in varying proportions to give a series of test sediment/phase mixtures of different metal content. The maximum phase contents (dry weight) were 11% kaolin, 14% carbonate, 8% ferric hydroxide, and 7% sludge, in order to minimise as far as possible effects on the bulk matrix.

Chemical extraction procedure

A volume of slurry containing 1 gram dry weight of sediment was pipetted into a centrifuge tube. After centrifuging, the aqueous layer was removed and the retained solids extracted sequentially by the following procedure (based largely on the schemes developed by Tessier *et al.* (1979), and Rapin & Förstner (1983)):

(i) Exchangeable fraction – shaken at room temperature with 30 ml of 1 M ammonium acetate/0.25 M calcium chloride at pH 6 (acetic acid) for 3 hours.

(ii) Carbonate fraction – shaken at room temperature for 5 hours with 20 ml of 1 M sodium acetate adjusted to pH 5 with acetic acid.

(iii) Reducible fraction – digested at 96 °C with occasional agitation for 6 hours with 30 ml 0.1 M hydroxylamine hydrochloride in 25% (v/v) acetic acid.

(iv) Organic fraction – extracted at 85 °C for 2 hours with 5 ml of 0.02 M nitric acid and 5 ml 30% hydrogen peroxide (to pH 2 with nitric acid), then continued extraction for a further 3 hours after addition of another 5 ml of 30% hydrogen peroxide. After cooling, additional 30 minute room temperature extraction with 10 ml of 3.2 M ammonium acetate in 20% (v/v) nitric acid.

(v) Acid-extractable (residual) fraction – digestion for 4 hours at 120 °C with 20 ml of concentrated nitric acid.

After each stage, the residues were separated from the extracts by centrifugation and washed with de-ionised water. The combined extracts and washings were then acidified and made up to volume prior to analysis (50 mls for stages (i) to (iv), 200 mls for stage (v)).

Cd, Pb, Ni and Cu were determined by graphite furnace atomic absorption spectrometry using a 3 point standard addition calibration for each sample. Zn was determined by flame atomic absorption after initial checks on recoveries in the various extractant matrices. Using these techniques the analytical detection limits were 0.1, 1, 1, 1 and 2 $\mu g/l$ for Cd, Pb, Ni, Cu and Zn respectively. Results are expressed in $\mu g/g$ (dry weight).

Bioaccumulation test procedure

Freshwater tubificid worms were selected as the test organism. These were sufficiently small to allow a reasonable number to be accomodated in a limited volume of sediment, thus averaging out effects of individual variability due to age, size, breeding status etc. In waters of medium hardness they have been shown to be tolerant to high concentrations of heavy metals (Brković-Popović & Popović, 1977), indicating that bioaccumulation rates in the test would not be likely to be influenced by toxic effects. Furthermore, they are readily available from aquarist dealers and a source of worms of low initial metal concentration was established.

Test vessels (250 ml flat-bottomed Drechsel bottles) were filled to a depth of 1 cm with freshly prepared sediment/phase mixtures and synthetic freshwater was then added without disturbing the sediment layer, thus avoiding fractionation during resettlement. Each test was conducted in duplicate and, prior to introduction of the worms, the vessels were left for 4 days, with aeration, to allow any immediate redistribution of metal between geochemical phases to occur. Following this 20–30 worms were added to each vessel and aeration was maintained throughout an exposure period of 14 days at 15 °C. At the end of this time, the worms were removed by careful sieving and held in clean water for 2 days in order to depurate gut contents. They were then dried, weighed and digested in nitric acid prior to analysis by flame or

furnace atomic absorption spectrometry as previously described for sediment extracts (worm digests contained ca. 300 mg/l solids).

Results and discussion

Choice of 'exchangeable' extraction reagent

The suitability of a variety of extraction reagents for 'exchangeable' metals was examined by comparison of metal recoveries obtained on mineral phases and directly spiked sediment. Those tested included: 1 M potassium fluoride at pH7; 1 M barium chloride at pH7; 1 M magnesium chloride at pH7; 1 M calcium acetate at pH 6, 6.5 and 7; 1 M ammonium acetate at pH 6, 6.5 and 7; and 1 M ammonium acetate/0.25 M calcium chloride at pH6. The potassium fluoride, barium chloride and magnesium chloride reagents all gave poor recovery of metals adsorbed on kaolin (intended as a model exchangeable phase) but of the three magnesium chloride appeared marginally the strongest. For the ammonium and calcium acetate reagents, recoveries from spiked kaolin and sediment increased, as expected, with decreasing pH; at pH6, ammonium acetate gave quantitative recoveries from kaolin but also partially attacked carbonate phase, resulting in pH drift and appreciable metal extraction. Other reagents also failed to maintain pH in carbonate and sediment extractions, thus leading to possible readsorption or precipitation of released metal. The calcium acetate reagent held pH very well and extracted metals from carbonate to a limited extent only. On the other hand, it proved less efficient than ammonium acetate for kaolin extraction and for this reason the mixed ammonium acetate/calcium chloride reagent was investigated. Figure 1 shows metal recoveries for this reagent from kaolin, carbonate and spiked sediment phases compared with those for the commonly used pH7 magnesium chloride and ammonium acetate reagents, and pH7 calcium acetate. For kaolin, the mixed reagent demonstrated essentially quantitative recovery of all metals but it also extracted important amounts of Cu, Zn and Ni from car-

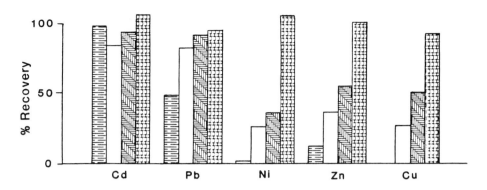

(a) Extraction from kaolin

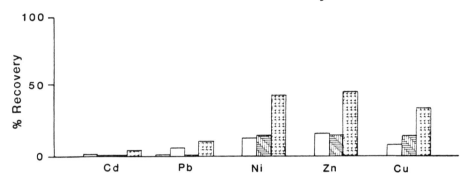

(b) Extraction from CaCO₃

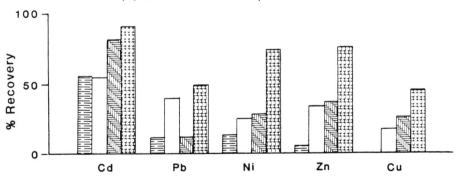

(c) Extraction from spiked sediment

1M Magnesium chloride, pH7
1M Calcium acetate, pH7
1M Ammonium acetate, pH7✳
1M Ammonium acetate/0.25M Calcium chloride, pH6

✳ pH drifted to 8.2 on carbonate

Fig. 1. Selection of 'exchangeable' reagent.

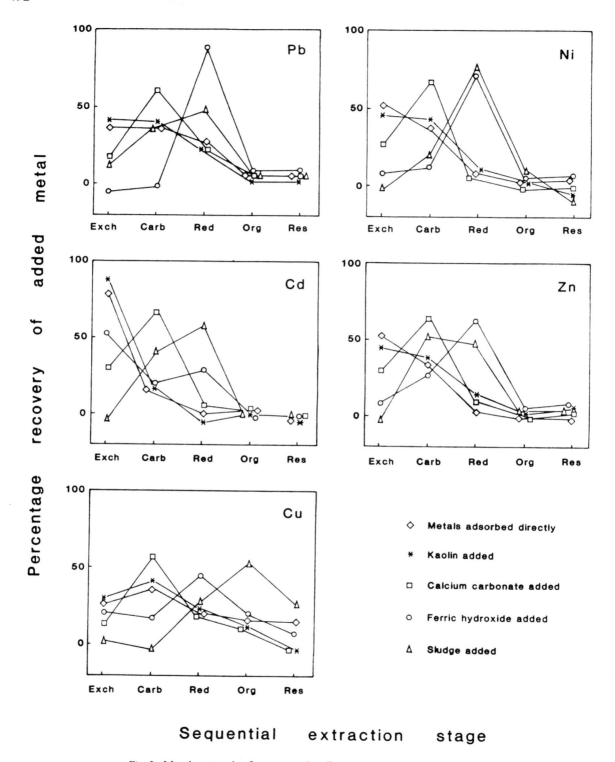

Fig. 2. Metal recoveries from treated sediments by sequential extractions.

bonate phase. Nevertheless this reagent was selected for use in the present application as it was considered better to use a stronger extractant for potential bioavailability estimation.

Sequential extractions on test sediment/mineral phase mixtures

Metal recoveries for all the treated sediments in each fraction of the sequential extraction procedure are shown in Fig. 2. It is noticeable that in

kaolin spiked sediments the recoveries in the 'exchangeable' fraction are generally lower than were observed for the pure spiked kaolin phase with the same mixed reagent (Fig. 1). This suggests either redistribution to other binding sites in the sediment after mixing, or migration during the extraction procedure. However, the results demonstrate substantial metal enrichment in the target sites after spiking, thus confirming that the intended variation of metal speciation in the test sediments had been achieved.

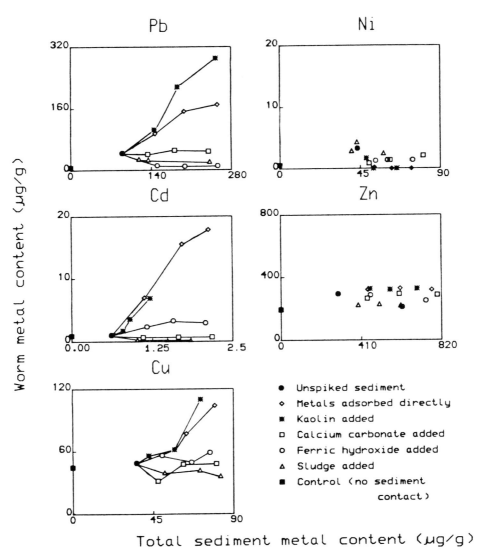

Fig. 3. Metal uptake by tubificid worms against total metal in sediments

Metal availability to tubificid worms

Results for worm uptake against total metal in sediment (sum of sequential extractions) were found to demonstrate poor correlation as is shown in Fig. 3. It can be seen, however, that increasing proportions of certain spiked mineral phases in the bulk sediment do give rise to corresponding increases in worm metal concentrations. For sediments spiked by direct adsorption or with kaolin phase, similar patterns were observed for Pb, Cu and particularly for Cd, where a bioaccumulation factor of ∼10 was found. Sediments spiked with hydrous ferric oxide, calcium carbonate or sewage sludge showed very little accumulation, which is in broad agreement with the findings of Tessier *et al.* (1984) who suggested a protective or competitive effect due to iron and to some extent organic matter. Of the remaining metals, Ni showed very

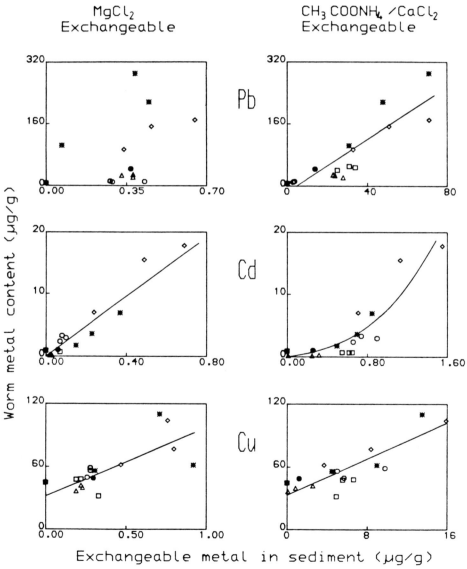

Fig. 4. Metal uptake by tubificid worms against exchangeable metal in sediment (for key to characters see Fig. 3).

little uptake and no apparent trends, while no measureable change was observed in Zn content, suggesting possible regulation of its uptake.

Overall, the results highlight the problems in relating bioavailability to total metal concentrations in sediments and reinforce the need for concerted investigation of chemical speciation. When data for the individual sequential extractions of all the test sediment/mineral phase mixtures were examined, good correlations with biological data were only found for the 'exchangeable' fractions. Results for two exchangeable reagents of different strengths are compared in Fig. 4. Extraction with magnesium chloride at pH7 ('weaker') proved to be an extremely good predictor of cadmium availability with a correlation coefficient of 0.97, whereas Pb and Cu gave values of 0.39 and 0.75 respectively. The pH6 ammonium acetate/calcium chloride reagent gave more consistent data with correlation coefficients of 0.86, 0.88 and 0.87 for Cd, Pb and Cu, and the greater extraction power of this reagent is reflected in the much higher metal recoveries. However, the data indicate its inability to adequately predict the apparent difference in availability of Pb from kaolin and directly spiked test sediments. Significant effects of this type would limit the reliability of a single reagent as a chemical surrogate of bioavailability.

At the conclusion of the experiment metal concentrations were determined in filtered samples of the overlying water and were found to be low in relation to those in the solid phase (maximum 1.4% of total in the case of Cd). Correlations of worm metal content vs. aqueous phase concentrations were relatively poor ($r = 0.59$, 0.28 and 0.21 for Cd, Pb and Cu, respectively), but it must be noted that no attempt was made to investigate aqueous phase metal speciation, which would itself be expected to have a profound effect on availability of metal by this route. Indeed, it has been suggested (Jenne et al., 1986) that metal availability to many benthic organisms might be largely controlled by the free metal ion activity in the sediment pore water, although further research is still required in this area.

Conclusions

The bioavailability of sediment metals to tubificid worms is strongly dependent on the chemical forms in which they are present and the exchangeable metal fraction of a sequential extraction scheme provides the best available predictor of uptake. The study demonstrates the importance of undertaking concerted chemical and biological tests to judge the validity of any procedure for estimation of bioavailability. A major advantage of the spiked model phase approach adopted in this study is that it overcomes the tendency, if only real sediments are used, for a high degree of intercorrelation to occur between different metal fractions and obviates the need to employ large numbers of samples from widely differing sites.

References

Brković-Popović, I. & M. Popović, 1977. Effects of heavy metals on survival and respiration rate of tubificid worms: Part 1 – Effects on survival. Envir. Pollut. 13: 65–72.

Diks, D. M. & H. E. Allen, 1983. Correlation of copper distribution in a freshwater sediment system to bioavailability. Bull. Envir. Contam. Toxicol. 30: 37–43.

Förstner, U., 1985. Chemical forms and reactivities of metals in sediments. In R. Leschber, R. D. Davis & P. L'Hermite (eds.), Chemical methods for assessing bioavailable metals in sludges and soils. Elsevier, London: 1–30.

French, P. & D. T. E. Hunt, 1986. Thermodynamic calculations of dissolved trace metal speciation in river, estuary and seawaters. WRc Technical Report TR 249, Water Research Centre, Medmenham, U.K., 38 pp.

Hall, K. J. & K. S. Bindra, 1979. Geochemistry of selected metals in sediments and factors affecting organism concentration. In Management and control of heavy metals in the environment. London, CEP Consultants: 337–340.

Hunt, D. T. E., 1987. Trace metal speciations and toxicity to aquatic organisms – A review. WRc Technical Report TR 247, Water Research Centre, Medmenham, U.K., 51 pp.

Jenne, E. A., D. M. Ditoro, H. E. Allen & L. S. Zarba, 1986. An activity-based model for developing sediment criteria for metals: Part 1. A new approach. In J. N. Lester, R. Perry & R. M. Sterritt, Eds. Proceedings of the International Conference on Chemicals in the Environment, London: 560–568.

Leppard, G. G., (ed.), 1983. Trace element speciation in surface waters and its ecological implications, Plenum Press, N.Y., 320 pp.

Luoma, S. N. & G. W. Bryan, 1979. Trace metal bioavailability: Modelling chemical and biological interactions of sediment-bound zinc. In: E. A. Jenne, ed., Chemical modelling in aqueous systems, ACS Symposium series No. 93, American Chemical Society, Washington: 577–609.
sediment – bound zinc. In: E. A. Jenne, ed., chemical modelling in aqueous systems, ACS Symposium series No. 93, American Chemical Society, Washington: 577–609.

Luoma, S. N. & E. A. Jenne, 1976. Factors affecting the availability of sediment-bound cadmium to the deposit-feeding clam, *Macoma balthica*, in: C. E. Cushing, Ed., Radioecology and energy resources, Dowden, Hutchinson & Ross, Stroudsberg, Pa.: 283–290.

Luoma, S. N. & E. A. Jenne, 1977. The availability of sediment-bound cobalt, silver and zinc to a deposit-feeding clam, in: R. E. Wildung & H. Drucker, eds., Biological implications of metals in the environment, NTIS, Springfield, Va: 213–230.

Meguellati, N., D. Robbe, P. Marchandise & M. Astruc, 1983. A new chemical extraction procedure in the fractionation of heavy metals in sediments – Interpretation. Heavy metals in the environment, Heidelberg, CEP Consultants: 1090–1093.

Neubecker, T. A. & H. E. Allen, 1983. The measurement of complexation capacity and conditional stability constants for liquids in natural waters. Wat. Res. 17: 1–14.

Oakley, S. M., K. J. Williamson & P. O. Nelson, 1983. Accumulation of cadmium by *Abarenicola pacifica*. Sci. Tot. Env. 28: 105–118.

O'Donnel, J. R., B. M. Kaplan & H. E. Allen, 1985. Bioavailability of trace metals in natural waters. In: R. D. Cardwell, R. Purdy & R. C. Bahner, eds., Aquatic Toxicology and Hazard Assessment: Seventh symposium. ASTM, Philadelphia: 485–500.

Rapin, F. & U. Förstner, 1983. Sequential leaching techniques for particulate metal speciation: The selectivity of various extractants. Heavy metals in the environment, Heidelberg, CEP Consultants: 1074–1077.

Salomons, W. & U. Förstner, 1984. Metals in the Hydrocycle. Springer-Verlag, Berlin, 349 pp.

Stumm, W. & J. J. Morgan, 1981. Aquatic Chemistry, 2nd edition, Wiley, N.Y., 780 pp.

Tessier, A., P. G. C. Campbell & M. Bisson, 1979. Sequential extraction procedure for the speciation of particulate trace metals. Analyt. Chem. 51: 844–851.

Tessier, A., P. G. C. Campbell, J. C. Auslair & M. Bisson, 1984. Relationships between the partitioning of trace metals in sediments and their accumulation in the tissues of the freshwater mollus, *Elliptio complanata* in a mining area. Can. J. Fish. Aquat. Sci. 41: 1463–1472.

Hydrobiologia **188/189**: 497–506, 1989.
M. Munawar, G. Dixon, C. I. Mayfield, T. Reynoldson and M. H. Sadar (eds)
Environmental Bioassay Techniques and their Application.
© *1989 Kluwer Academic Publishers. Printed in Belgium.*

Metal accumulation by chironomid larvae: the effects of age and body weight on metal body burdens

Gail Krantzberg
Ontario Ministry of the Environment, Water Resources Branch, 1 St. Clair Ave. W., Toronto, Ontario, M4V 1P5

Key words: bioaccumulation, chironomid(s), body burdens(s), body size, biomonitoring, metals

Abstract

Age and body weight affected the extent of metal retention in larval chironomids. Elements differed with respect to age- and size-dependent metal uptake. The slopes of the regressions of metal burdens against age and size varied depending on the range in body size considered. Among fourth instar larvae, younger chironomids had higher concentrations of Cd, Mn, Ca, Ni, Fe, and Cu than older instars. When all instars were included, only concentrations of Cd and Ni were greater in young as compared to older larvae. Concentrations of Fe, Mn, Ca, and Ni were greater in larger chironomids than smaller larvae of similar age, but the effect of body weight on metal content was significant only when a wide range in biomass was considered. For the design of biological monitoring programs that use information on tissue residues of contaminants to assess contaminant bioavailability, individuals of different ages and sizes should be collected from each site in order to validate intersite comparisons. The advantages of considering metal burdens in addition to metal concentrations are emphasized.

Introduction

Benthic macroinvertebrates are known to accumulate metals and studies have been directed toward identifying organisms that can serve as reliable monitors of environmental contamination. Even though biological monitoring programs are currently being implemented, the relationships between physicochemical properties that modify metal bioavailability to benthic macroinvertebrates, and the physiological factors that determine metal uptake, elimination and regulation by the biota remain poorly understood. Body size has been considered as one factor that can affect concentrations of metals in tissues of animals (Williamson, 1979a; Strong & Luoma, 1981). The existence of age- and size-dependent metal accumulation by aquatic biota has been docu-

mented by several authors, particularly for marine and estuarine molluscs and gastropods (Strong & Luoma, 1981; Boyden, 1977). Metal concentrations in tissues may change while metal burdens remain constant, or concentrations can remain constant while burdens change. In this context, an organism's metal burden is defined as the total metal content of an individual and is the sum of surface adsorbed and internally incorporated metal. If total metal content remains constant while body weight increases or decreases, measurements of concentration alone will mask the dynamics of metal uptake, storage, and elimination.

When biomass fluctuations influence metal concentrations, spatial and temporal differences in metal concentrations among organisms are difficult to interpret in terms of environmental quality. This is of particular concern when uniformly sized individuals are not available, as is commonly the case in field monitoring conditions. In addition, the relationship between body size and metal concentrations may not be the same for different populations (Strong & Luoma, 1981).

Because size can influence metal concentrations in benthic organisms, selecting organisms on the basis of a standardized biomass may not be justifiable. Individuals of a specified weight will not necessarily reflect environmental conditions as well as those of a greater or lesser biomass. Understanding how metal concentrations vary with body size, then, is critical for establishing acceptable standard sampling protocols for temporal and intersite assessments.

Seasonal fluctuations in biomass are often responsible for observed changes in metal concentrations in tissues (Cain & Luoma, 1986). Phillips (1976) studied seasonal variation of Zn, Cd, and Cu concentrations in the common mussel *Mytilus edulis* and attributed low levels during particular seasons to increases in organism weight. Zinc and Cu burdens remained constant when normalized for biomass fluctuations. Seasonal fluctuation in the soft tissue weight of *Mytilus edulis* was the principal factor responsible for the variation observed in Pb, Cd, Zn, and Cu (Amiard *et al.*, 1986), Pb and Cd (Farrington *et al.*, 1983), and Pb and Cu concentrations (Ritz *et al.*, 1982). Seasonal changes in Zn, Mn, Mg, Ca, and Cu in the digestive gland of the slug *Arion ater* L. were correlated with fluctuations in gland biomass (Ireland, 1984). For the marine oyster *Crassostrea virginica*, Cd concentration (Zaroogian, 1980) and Zn, Cd, Cu, and Ag concentrations (Phelps *et al.*, 1985) were negatively correlated with body weight. Similarly, concentrations of Cu, Fe, Pb, and Zn in amphipods were significantly affected by body size, with smaller individuals having the greatest concentrations (Rainbow & Moore, 1986).

Boyden (1974, 1977) examined the relationship between body size and metal concentration in shellfish, and concluded that by determining how metal burden or content varied with body weight, samples that differed due to biomass could be compared. Williamson (1979a, 1979b) applied metal content data to consider the separate effects of age and weight on metal burdens for the land snail *Cepaea hortensis*. Cadmium content and concentration increased with age and weight and, at a given age, larger animals had lower concentrations than smaller organisms. For snails having comparable body weights, younger individuals had lower Cd burdens than did older ones.

In the present investigation I considered the effects of age and weight on metal body burdens (or metal contents) for a single community of the freshwater larvae of the insect, *Chironomus* (Diptera: Chironomidae). This cosmopolitan genus is a particularly important member of benthic communities in acidic and circumneutral softwater lakes of the Canadian Shield. In south-central Ontario, *Chironomus* is generally univoltine. Eggs are deposited in the spring and larvae develop through 4 successive instars, overwinter as fourth instar larvae, and emerge early next spring. By sampling from early spring to late autumn individuals that differed in age and weight were collected. I also artificially increased the longevity of fourth instar larvae in temperature controlled chambers, thus inducing fourth instar age classes that were up to two years old. Such prolonging of the fourth instar stage is commonly exhibited by arctic and subarctic species. By comparing larvae of similar ages but different weights, and of similar weights but different ages, I could determine the separate effects of age and size on metal accumulation in chironomids. If biological monitoring programs are to be effective, identifying the factors that govern the extent of contaminant retention by the biota is vital.

Materials and methods

Chironomid larvae were collected by Ekman grab at a depth of 10–13 m from an oligotrophic acidic lake (pH c.a. 5.5) on the Canadian Shield (Plastic Lake). A summary of lake characteristics is provided in Krantzberg & Stokes (1988). Larvae were separated from sediments by sieving through a 200 μm 'Nytex' nylon screen and were placed in polyethylene bags containing a small amount of whole sediment and lake water. Organisms were placed in coolers and returned to the laboratory for sorting and culturing. By sampling larvae from May to October, 1984–1986 it was possible to compare several generations of larvae that were of similar age, since significant differences in metal content among years could mask age or weight effects.

In preparation for the assembly of laboratory cultures, sediments were collected by Ekman grab. The cultures were constructed by placing 1.5 L (10 cm depth) of unsieved lake sediment and 5 L of lake water into 7 L polyethylene basins of surface area 225 cm^2. Approximately 200 larvae were added to each basin. The cultures were covered, aerated, and maintained in darkness in a 'CANVIRON' (environmentally-controlled) chamber at 6 °C. All chironomids remained as fourth instar larvae and none completed their development to adulthood. Cultures were harvested as necessary to collect individuals of ages 11, 16, 18, 21, and 25 months. After harvest, larvae were held for 24 hours in lake water to purge their gut contents. Water was changed periodically and faeces removed in order to prevent coprophagy. Chironomids were dried at 60 °C on polyethylene weigh boats, weighed, digested to dryness in 3 mls hot, analytical grade HNO_3, and the residue was resuspended in dilute HNO_3. I analyzed for Al, Mn, Ca, Zn, Ni, Fe, and Cu by Inductively-Coupled Argon Emission Spectrophotometry, and for Pb and Cd by flameless Atomic Absorption Spectrophotometry. National Bureau of Standards Oyster Tissue and National Research Council of Canada Lobster Hepatopancrease were used as checks on the accuracy of the determinations. Recoveries were within 7% of certified values.

The relationships among metal burdens, age, and weight were compared for each metal by calculating simple linear regressions for log transformed data to examine the separate effects of age and weight on metal content in chironomid larvae. Forward-entry stepwise multiple regressions were also calculated to evaluate the age and weight dependence of metal bioaccumulation. The variability in metal content for larvae of similar age but different weight and of similar age but collected in different years was examined by Analyses of Variance. To determine whether metal accumulation by chironomids grown in culture was similar to that of field populations, I compared 11 month old larvae collected from the field with 11 month old larvae reared in the laboratory from 4 month old chironomids. Metal uptake and/or loss by larvae older than 11 months could not be compared with field populations. Log transformation of the data was necessary to achieve a normal distribution.

Results

When considering chironomids of similar age, but collected in different years, metal burdens did not differ significantly ($p > 0.05$) for any of the age classes examined (Table 1). I therefore considered all individuals of a given age class to be representative of that age class, and not representative of the sampling year. The 11 month old chironomids collected from the field did not differ in their body burdens from 11 month old chironomids raised in culture from 4 month old field populations ($p > 0.05$). For the first 7 month incubation period, then, it appeared that metal accumulation in laboratory and field populations were comparable. In addition, 11 month old larvae reared in sediments collected at the time they were 4 months old (Sept-Oct) did not differ in their metal burdens from 11 month old larvae collected directly from the field (May-June, Table 1). A decrease in the mean weight of fourth instar larvae with age was observed (Fig. 1).

Mean metal concentrations varied substantially when all age and size classes were pooled (Fig. 2). For most metals, much of this variation was explained by both age and weight (Table 2). By examining the slopes of the log-log plots, information can be obtained on how metal concentrations differed between small and large larvae. Log-log plots of metal content against with slopes significantly less than 1 would reveal that metal concentrations are greater in smaller as compared

Table 1. Variation in metal burdens in chironomids due to sampling year and laboratory incubation. All values are nmoles per individual (standard error). (JL = July, Au = August, SE = September, OC = October, M = May). N = number of pooled samples analyzed (50–200 individuals per sample). For each age group, mean values for different years were not significantly different as tested by ANOVA (p > 0.05). Metal burdens for 11 month old larvae collected in the field were not significantly different from those cultured in the laboratory.

| | Field collections | | | | | | | | | Laboratory culture | | | | | |
| | 2 months | | | 3 months | | | 4 months | | | 11 months | | | 11 months | | |
	JL/84 N = 3	JL/85 N = 4	JL/86 N = 5	AU/84 N = 3	AU/85 N = 4	AU/86 N = 3	SE/84 N = 8	OC/85 N = 9	OC/85 N = 10	M/84 N = 6	M/85 N = 6	M/86 N = 6	M/85 N = 3	M/86 N = 3	M/87 N = 3
Pb	10 (1)	11 (2)	11 (1)	27 (1)	26 (2)	28 (2)	29 (2)	29 (1)	28 (2)	62 (7)	70 (9)	65 (7)	66 (4)	68 (6)	63 (6)
Cd	.75 (.06)	.74 (.04)	.69 (.07)	2.0 (.1)	2.2 (.2)	2.0 (.2)	2.8 (.6)	3.0 (.6)	2.4 (.7)	5.2 (.7)	5.0 (.2)	5.0 (.6)	5.0 (.6)	4.9 (.3)	5.2 (.4)
Al	128 (15)	143 (17)	144 (17)	300 (14)	305 (17)	289 (18)	490 (26)	483 (29)	480 (28)	1349 (60)	1402 (57)	1384 (53)	1360 (56)	1378 (50)	1403 (61)
Mn	9 (1)	9 (1)	10 (1)	33 (3)	35 (2)	33 (2)	44 (2)	45 (4)	44 (2)	75 (6)	80 (7)	80 (6)	77 (4)	79 (3)	78 (3)
Ca	159 (7)	150 (8)	157 (7)	916 (26)	918 (22)	934 (41)	1429 (40)	1406 (44)	1420 (39)	3709 (106)	3685 (133)	3839 (135)	3745 (63)	3712 (79)	3787 (87)
Zn	41 (3)	41 (3)	40 (2)	86 (7)	90 (5)	93 (8)	140 (5)	138 (4)	141 (5)	429 (16)	433 (21)	437 (19)	430 (10)	428 (8)	439 (18)
Ni	5.0 (.5)	5.3 (.4)	5.4 (.6)	24 (2)	26 (4)	25 (2)	29 (4)	28 (4)	25 (4)	39 (4)	44 (5)	46 (6)	45 (4)	47 (4)	45 (5)
Fe	211 (8)	217 (10)	218 (10)	670 (31)	665 (37)	649 (35)	1225 (101)	1293 (95)	1321 (113)	2265 (72)	2241 (83)	2205 (69)	2222 (57)	2260 (61)	2209 (68)
Cu	10 (3)	12 (2)	12 (2)	26 (4)	26 (4)	28 (3)	40 (4)	43 (5)	38 (4)	85 (7)	89 (6)	90 (7)	90 (8)	90 (10)	87 (6)

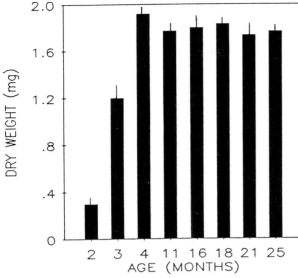

Fig. 1. The relationship between age (months) and dry weight (mg) for chironomid larvae collected from Plastic Lake (<11 months old) and cultured in the laboratory (>4 months old). Bars represent one standard deviation.

to larger organisms, while slopes close to 1 indicated that metal concentrations are not size-dependent. Slopes significantly greater than 1 reveal that metal concentrations increase with size, and that metals are accumulated more rapidly than tissue.

For a given age, larger individuals had higher concentrations of Fe and Ca than smaller ones, and for a given weight, older larvae had higher concentrations of Fe and Ca than younger larvae (Table 2). The regression of Cu and Mn on weight and age generated slopes close to 1, indicating that individuals of all ages and sizes had similar concentrations of Cu and Mn (Table 2). For a given weight, young larvae had higher Zn and Ni concentrations than older larvae, and at a given age, larger individuals had higher Zn and Ni concentrations than smaller ones. The relationship between Al and body size was not significant, but the regression of Al on age demonstrated that older organisms had higher Al concentrations than younger ones of similar size (Table 2). Lead burdens increased with age, although concentrations were similar for young and old larvae of a given size, the slope of the regression not differing

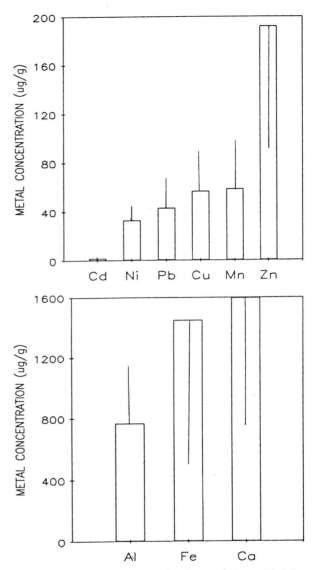

Fig. 2. Mean metal concentration ($\mu g \cdot g^{-1}$ dry weight) for all chironomid larvae. Vertical lines represent one standard deviation.

significantly from 1. Lead was not significantly related to changes in weight. Cadmium content but not concentration increased with age and with weight, and slopes were significantly less than 1.00 (Table 2) such that smaller or younger individuals had relatively higher Cd concentrations than larger or older larvae.

A broad range in weight was important for achieving a significant relationship between body

Table 2. Regressions of metal burdens (ng) against age (months) or weight (mg) for larval chironomids from Plastic Lake. Results of the regressions of metal burdens on age for fourth instar larvae, alone, are also provided. No regressions of metal burdens on weight in fourth instar chironomids were significant. Regressions follow the model: Log Y = X^m + Log b

All instars	r^2	Fourth instars	r^2	All instars	r^2
Pb $=$ age $^{.98}$ + 0.85	0.94	Pb $=$ age $^{.93}$ + 0.90	0.93	not significant	
Cd $=$ age $^{.67}$ − 0.67	0.77	Cd $=$ age $^{.36}$ + 0.26	0.85	Cd $=$ weight $^{.85}$ + 0.38	0.79
Al $=$ age $^{1.16}$ + 1.92	0.97	Al $=$ age $^{.99}$ + 2.10	0.98	not significant	
Mn $=$ age $^{.99}$ + 0.93	0.87	Mn $=$ age $^{.78}$ + 1.16	0.84	Mn $=$ weight $^{1.05}$ + 1.37	0.61
Ca $=$ age $^{1.16}$ + 2.28	0.81	Ca $=$ age $^{.72}$ + 2.75	0.95	Ca $=$ weight $^{1.42}$ + 3.08	0.76
Zn $=$ age $^{.99}$ + 0.67	0.89	Zn $=$ age $^{.81}$ + 1.71	0.94	Zn $=$ weight $^{1.03}$ + 1.46	0.93
Ni $=$ age $^{.89}$ + 0.77	0.79	Ni $=$ age $^{.62}$ + 1.05	0.88	Ni $=$ weight $^{1.08}$ + 1.38	0.74
Fe $=$ age $^{1.11}$ + 2.26	0.88	Fe $=$ age $^{.80}$ + 2.06	0.89	Fe $=$ weight $^{1.20}$ + 3.05	0.65
Cu $=$ age $^{1.03}$ + 0.94	0.94	Cu $=$ age $^{.84}$ + 1.12	0.96	Cu $=$ weight $^{1.01}$ + 1.65	0.57

burden and weight. The multiple regressions of Cd, Ca, and Zn versus age and weight for the whole data set and for fourth instar larvae only show that the significant relationship between body weight and Cd burden disappears when the early instars are removed (Table 3). There was a 10-fold difference in biomass between second and fourth instar larve, compared with a 1.2-fold difference in biomass among fourth instar larvae.

Discussion

Metal concentrations and burdens can become uncoupled when fluctuations in concentrations do not reflect similar gains or losses in metal content. To examine age and size effects on metal accumulation by freshwater chironomids, I have expressed metals in tissues in terms of body burdens in order to facilitate intersample comparisons. To illustrate the relationship between body burdens and concentrations, I discuss both values.

Table 3. Multiple regression of log age (A) and log weight (W) versus log metal. %A and %W indicate the percent of variance explained by log age and log weight, respectively. n = 135 and 113 for all instars and fourth instars only, respectively.

Metal	Data range	Model	%A	%W	r^2
Lead	all instars	log Pb $= 0.91 + 0.21A + 0.88W$	93	2	95
	fourth instar	log Pb $= 0.04 + 1.07A + 2.88W$	93	4	97
Cadmium	all instars	log Cd $= 0.08 + 0.41A + 0.55W$	76	21	97
	fourth instar	log Cd $= 0.53 + 0.32A$	87	0	87
Aluminum	all instars	log Al $= 2.00 + 1.02A + 0.28W$	97	2	99
	fourth instar	log Al $= 1.99 + 1.01A$	99	0	99
Manganese	all instars	log Mn $= 1.06 + 0.77A + 0.47W$	87	8	95
	fourth instar	log Mn $= 0.68 + 0.86A + 1.61W$	84	1	85
Calcium	all instars	log Ca $= 2.51 + 0.75A + 0.85W$	81	18	99
	fourth instar	log Ca $= 3.11 + 0.66A$	96	0	96
Zinc	all instars	log ZN $= 1.56 + 0.86A + 0.36W$	93	5	98
	fourth instar	log Zn $= 2.15 + 0.73A$	94	0	94
Nickel	all instars	log Ni $= 0.95 + 0.58A + 0.64W$	79	17	96
	fourth instar	log Ni $= 0.30 + 0.75A + 2.54W$	88	6	94
Iron	all instars	log Fe $= 2.42 + 0.84A + 0.57W$	88	9	97
	fourth instar	log Fe $= 1.58 + 0.97A + 3.44W$	89	7	96
Copper	all instars	log Cu $= 1.00 + 0.86A + 0.36W$	94	5	99
	fourth instar	log Cu $= 0.54 + 0.93A + 1.93W$	97	2	99

Several parameters that modify metal accumulation may be age and/or size dependent. The contribution of metals adsorbed to the surface to the total metal burden is principally subjected to allometric constraints, while the absorbed portion may also be responsive to metabolic control. Metabolic activity, feeding rates, and food preferences vary between young and old or small and large individuals. Exposure to and incorporation of metals, then, can change throughout the life of an individual. Dallinger and Wieser (1977) showed that the assimilation efficiency and metal accumulation in terrestrial isopods were dependent upon metabolic rate. Williamson (1979a, b) attributed higher concentrations of Pb, Cd, and Zn in small versus large land snails (*Cepea hortensis*) to the effects of metabolic activity and feeding rates on metal ingestion. Metal accumulation by the polychaete *Capitella capitata* was significantly influenced by food quality (Windom *et al.*, 1982). Metal accumulation varied with N content of detritus. Food quality and metabolic demands for nutrients, then, can alter the rates of metal accumulation, even when total metal content of the food source remains constant.

While accumulation patterns can be influenced by metabolic activity, they are clearly metal and organism dependent. Fowler & Benayoun (1974) showed that ^{109}Cd uptake by the snail *Lysmata seticaudata* increased with temperature, although temperature did not affect ^{109}Cd retention in *Mytilus galloprovincialis*. The uptake dynamics of ^{75}Se were reversed, with no temperature effect for snails, and a positive temperature dependence for the mussel. Thus, general increases or decreases in metabolic rates associated with age or size are not sufficient to explain the variation observed among metals. The effects of age or size on metal accumulation may operate at more subtle levels. For example, the energy required for metal transport across membranes or to specific storage sites may differ among elements as a result of their physical and chemical properties, or importance in metabolic processes.

Age- and weight-dependent relationships exist for chironomids and differ among the elements examined. While the direct causes for the observed slopes must remain speculative, some functional significance may be presented to account for the observed relationships.

Iron concentrations were greater in older and larger individuals. The chironomids I studied produce haemoglobin and are known to be tolerant of low O_2 conditions. Early instars are typically planktonic and become progressively more benthic with age. The likelihood that larvae will encounter conditions of low O_2 availability, then, increases with time and it is possible that chironomids can increase their rate of haemoglobin synthesis with age to enhance their ability to tolerate short periods of anoxia. Similarly, as body size increases, the surface area: volume ratio decreases, and O_2 absorption across surfaces could also decrease. If increased Fe accumulation is, in part, related to increased haemoglobin production, then larger individuals would be expected to produce disproportionately more haemoglobin to compensate for allometric changes in size.

Calcium concentrations also increased with age and weight and may be related in a variety of ways to changing metabolic demands for this element. One such function could be the amelioration of the potential toxicity of other trace elements. Many invertebrates produce intracellular Ca-P granules and concretions that, in some organisms, are involved in metal accumulation (Doyle *et al.*, 1978 in Fowler *et al.*, 1984). Metal sequestering and accumulation into such intracellular packages has been reported for many invertebrates (Fowler *et al.*, 1984; Brown, 1982; George & Pirie, 1979).

Zinc and Ni concentrations were lower in older than in younger larvae, while Cu and Mn concentrations remained constant with age. These different patterns may reflect different metabolic demands for these elements, or differences in storage and elimination capacities that are metal specific. The ability to regulate essential elements has been reported for a wide range of organisms, with Zn and Mn regulation in annelids (Roberts & Johnson, 1978; Ireland, 1975), amphipods (Icely & Nott, 1980), and Cu and Zn in other invertebrates (Anderson *et al.*, 1978; Amiard-Triquet *et al.*, 1983). In the present study, the concentrations of Zn, Ni, Cu and Mn remained constant with body size and suggests that these elements may be under homeostatic control.

Aluminum concentrations were higher in small than in larger larvae. Like Fe and Ca, Al concentrations in older larvae were higher than in young larvae of similar size. No essential function has been attributed to Al, and Al toxicity has been considered as having a role in altering community composition in acidifying water bodies (Grahn, 1980; Nilssen, 1982; Havas, 1985; Neville, 1985; Cummins, 1986). It is unlikely, then, that changes in Al concentrations with age represent changes in metabolic demand for this element. Rather, the ability to eliminate Al, or the exposure to bio-available Al may change with time. Alternately, older larvae could be exposed to higher concentrations of bioavailable Al, particularly as they progress from second to fourth instars, and develop a more benthic lifestyle.

Cadmium concentrations were greater in small and young larvae than in older or larger individuals, so that with time or with increasing biomass, tissue was added more rapidly than was Cd. This type of relationship was found for Cu and Ag in some populations of bivalves by Strong & Luoma (1981) and for Cd in unionid clams (Hemelraad *et al.*, 1986). Body size may be important in metal accumulation when metals are mainly adsorbed to the surface. As Boyden (1977) stated, if metal concentrations were highly dependent on surface adsorption, one would predict that slopes of metal content versus size would be close to 0.67. That is, with increasing body size there is a decrease in the surface area:volume ratio, and the relative contribution of surface adsorbed metal to the total body content should become less important. Conversely, slopes close to 1 imply metabolic regulation of metal content. From a previous study (Krantzberg & Stokes, 1988) it is known that only 25% of the total Cd burden was adsorbed to the exoskeleton, and it is apparent that adsorptive phenomenon are not sufficient to explain the observed relationship. Physiological mechanisms must exert some control on Cd dynamics.

Observations on the age-dependence of metal accumulation changed when considering only fourth instar larvae (Table 1, Figs. 4 and 5). Except for Al, older individuals had lower concentrations of all elements than younger individuals of similar weight. This would be expected if metabolic activity influenced the rate of metal accumulation. Older age classes reared in the laboratory had lower mean biomass, suggesting that metabolic activity declined. If metabolic rates declined, ingestion and the metabolism of metal-bearing detritus should also decrease, and reduce the extent of metal bioaccumulation.

Implications for biological monitoring

Slopes of age and size regressions with metal content vary among elements and are dependent upon the range in age and size considered. There is ample evidence that slopes vary among populations (Boyden, 1977; Williamson, 1980; Strong & Luoma, 1981), so that comparisons of body burdens among individuals of different populations, normalized for age or weight, can only be justified if the slopes are shown to be similar. While the ideal biological monitor would be one that exhibits no age or size dependence for metal accumulation, it is highly improbable that such an organism exists. To adequately evaluate environmental contamination, several sampling strategies have been proposed. Phillips (1980) recommended the deliberate selection of a large range of sizes from all locations followed by bulk or individual analyses. In the former case, the results would represent a generalized population exposure, while the latter would permit weight normalization based on regression analysis. It is emphasized that age-dependence should be examined concurrently and that metal content be determined in addition to metal concentration. Where metabolic activity modifies metal accumulation, it may also be desirable to normalize body burdens against the burdens of conservative cellular components, such as proteins. Zaroogian (1980) recommended that metal content be normalized against N, which is directly related to protein content.

The choice of a sampling strategy for assessing

environmental quality will, of course, depend upon the objective of the study. Strong & Luoma (1981) suggested that to assess chronic contamination among sites, selection should be of older and larger individuals if age and weight are positively correlated. These individuals would be more likely to integrate exposure to pollutants through time than would younger individuals. To fulfill the objective of assessing short-term fluctuations in metal inputs, several age-size classes should be collected, if rates of metal accumulation vary with rates of growth.

In order to develop a more comprehensive understanding of metal contamination and accumulation by the biota, sampling a broad range of age and size classes will provide substantially more information than will the collection of individuals of uniform age and size. This approach will lead to the development of biomonitoring protocols that are more likely to be successful in tracking spacial and temporal changes in environmental quality.

For chironomid larvae, seasonal changes in metal burdens may reflect differences due to age-dependent metal retention as well as seasonal fluctuations in environmental quality. Differences in biomass among individuals of similar age, however, are likely to be small enough to avert influencing the relationship between metal concentrations in tissues and those in the environment.

Additional time and effort should be incorporated into sampling strategies in order to account for the demonstrated effects of size and age on metal bioaccumulation. This approach, which considers biotic variability, will enhance the development of effective and predictive biomonitoring programs.

Acknowledgements

This work was conducted at the Instiute for Environmental Studies and the Department of Botany at the University of Toronto, with funding by research grants from the Canadian Wildlife Service and the Wildlife Toxicology Fund. I thank Dr. S. N. Luoma, Dr. P. M. Stokes and an anonymus reviewer for their comments on the manuscript.

References

Amiard, J. C., C. Amiard-Triquet, B. Berthet & C. Metayer, 1986. Contribution to the ecotoxilogical study of cadmium, lead, copper and zinc in the mussel *Mytilus edulis*. Mar. Biol. 90: 425–431.

Amiard-Triquet, C., J. C. Amiard, J. M. Robert, C. Metayer, J. Marchand & J. L. Martin, 1983. Etude comparative de l'accumulation biologique de quelques oligo-elements metalliques dans l'estuaire interne de la Loire et les zones neritiques voisins (Baie de Bourgneuf). Cah. Biol. mar. 24: 105–118.

Anderson, R. V., W. S. Vinikour & J. E. Brower, 1978. The distribution of cadmium, copper, lead, and zinc in the biota of two freshwater sites with different trace metal inputs. Holarct. Ecol. 1: 377–384.

Boyden, C. R., 1974. Trace element content and body size in molluscs. Nature, London 251: 311–314.

Boyden, C. R., 1977. Effect of size upon metal content of shellfish. J. Mar. Biol. Ass. U.K. 57: 675–714.

Brown, B. E., 1982. The form and function of metal-containing 'granules' in invertebrate tissues. Biol. Rev. 57: 621–672.

Cain, D. J. & S. N. Luoma, 1986. Effect of seasonally changing tissue weight on trace metal concentrations in the bivalve *Macoma balthica* in San Francisco Bay. Mar. Ecol. Prog. Ser. 28: 209–217.

Cummins, C. P., 1986. Effects of Al and low pH on growth and development in *Rana temporaria* tadpoles. Oecol. 69: 248–252.

Dallinger, R. & W. Wieser, 1977. The flow of copper through a terrestrial food chain. I. Copper and nutrition in isopods. Oecol. 30: 253–264.

Farrington, J. W., E. D. Goldberg, R. W. Risebrough, J. H. Martin & V. T. Bowen, 1983. US 'mussel watch' 1976–1978: an overview of the trace-metal DDE, PCB, hydrocarbon, and artificial radionuclide data. Envir. Sci. Tech. 17: 490–496.

Fowler, B., C. Czop, D. Elliot, P. Mistry & C. Chignell, 1984. Studies on the binding site of the Cd-binding protein (CdBP) from the american oyster (*Crassostrea virginica*). Mar. Envir. Res. 14: 451–452.

Fowler, S. W. & G. Benayoun, 1976. Influence of environmental factor on selenium flux in two marine invertebrates. Mar. Biol. 37: 59–68.

George, S. S. & B. J. S. Pirie, 1980. Metabolism of zinc in the mussel *Mytilus edulis* (L.) a combined ultrastructural and biochemical study. J. Mar. Biol. Ass. U.K. 60: 575–590.

Geode, A. A., 1985. Mercury, selenium, arsenic, and zinc in waders from the Dutch Wadden Sea. Envir. Pollut. A 37: 287–309.

506

Grahn, O., 1980. Fishkills in two moderately acid lakes due to high aluminum concentration. In: D. Drablos & A. Tollan (Eds.). Ecological Impact of Acid Precipitation. Proc. Int. Conf., Sandefjord, Norway. March 11–14, 1980. SNSF Project. pp. 310–311.

Havas, M., 1985. Aluminum bioaccumulation and toxicity to *Daphnia magna* in soft water at low pH. Can. J. Fish. Aquat. Sci. 42: 1741–1748.

Hemelraad, J., D. A. Holwerda & D. I. Zandee, 1986. Cadmium kinetics in freshwater clams. 1. The pattern of cadmium accumulation in *Anodonta cygnae*. Arch. Envir. Contam. Toxicol. 15: 1–7.

Icely, J. D. & J. A. Nott, 1980. Accumulation of copper within the 'hepatopancreatic' caeca of *Corophium volutate* (Crustacea: Amphipoda). Mar. Biol. 57: 193–199.

Ireland, W. P., 1974. Variations in the zinc, copper, manganese, and lead content of *Balanus balanoides* in Cardigan Bay, Wales. Envir. Pollut. 7: 65–75.

Ireland, W. P., 1975. Metal content of *Dendrobaena rubida* (Oligochaeta) in a base metal mining area. Oikos 26: 74–79.

Ireland, W. P., 1977. Lead retention in toads *Xenopus laevis* fed increasing levels of lead-contaminated earthworms. Envir. Pollut. 12: 85–92.

Ireland, W. P., 1984. Seasonal changes in zinc, manganese, mercury, copper, and calcium content in the digestive gland of the slug *Arion ater* L. Comp. Biochem. Physiol. 78A: 855–858.

Krantzberg, G. & P. M. Stokes, 1980. The importance of surface adsorption and pH in metal accumulation by chironomids. Envir. Toxicol. Chem. 7: 653–670.

Neville, C., 1985. Studies on the mechanisms of the toxicity of acid and aluminum to juvenile rainbow trout in a low ion environment. Int. Symp. Acidic Precipitation, Muskoka, Ontario, 15–20 Sept. Abs., p. 143.

Nilssen, J. P., 1982. Acidification in southern Norway: seasonal variation of aluminum in lake waters. Hydrobiol. 94: 217–221.

Phelps, H. L., D. A. Wright & J. A. Milursky, 1985. Factors affecting trace metal accumulation by estuarine oysters, *Crassostrea virginia*. Mar. Ecol. Prog. Ser. 22: 187–198.

Phillips, D. J. H., 1976. The common mussel *Mytilus edulis* as an indicator of pollution by zinc, cadmium, lead, and copper. 1. Effects of environmental variables on uptake of metals. Mar. Biol. 38: 59–69.

Phillips, D. J. H., 1980. Quantitative Aquatic Biological Indicators: Their use to monitor trace metal and organochlorine pollution. Applied. Sci. Publ. Ltd., London, 488 pp.

Rainbow, P. S. & P. G. Moore, 1986. Comparative metal analysis in amphipod crustaceans. Hydrobiol. 141: 273–289.

Rainbow, P. S. & P. G. Moore, 1986. Comparative metal analysis in amphipod crustaceans. Hydrobiol. 141: 273–289.

Ritz, D. A., R. Swain & N. G. Elliot, 1982. Use of the mussel *Mytilus edulis planulatus* (Lamarck) in monitoring heavy metal levels in seawater. Aust. J. Mar. Freshwat. Res. 33: 491–506.

Roberts, R. D. & M. S. Johnson, 1978. Dispersal of heavy metals from abandoned mine workings and their transference through terrestrial food chains. Envir. Pollut. 16: 293–310.

Strong, C. R. & S. N. Luoma, 1981. Variations in the correlation of body size with concentrations of copper and silver in the bivalve *Macoma balthica*. Can. J. Fish. Aquat. Sci. 38: 1059–1064.

Wieser, W., R. Dallinger & G. Busch, 1977. The flow of copper through a terrestrial food chain. II. Factors influencing the copper content of isopods. Oecol. 30: 265–273.

Williamson, P., 1979a. Opposite effects of age and wight on cadmium concentrations of a gastropod mollusc. Ambio 8: 30–31.

Williamson, P., 1979b. Comparison of metal levels in invertebrate detritivores and their natural diets. Concentration factors reassessed. Oecol. 44: 75–79.

Williamson, P., 1980. Factors affecting body burdens of lead, zinc, and cadmium in a roadside population of the snail *Cepaea hortensis* Miller. Oecol. 44: 213–220.

Windom, H. L., K. T. Tenore & D. L. Rice, 1982. Metal accumulation by the polychaete *Capitella capitata*: Influences of metal content and nutritional quality of detritus. Can. J. Fish. Aquat. Sci. 39: 191–196.

Zaroogian, G. E., 1980. *Crassostrea virginica* as an indicator of cadmium pollution. Mar. Biol. 58: 275–284.

Hydrobiologia **188/189**: 507–516, 1989.
M. Munawar, G. Dixon, C. I. Mayfield, T. Reynoldson and M. H. Sadar (eds)
Environmental Bioassay Techniques and their Application.
© 1989 *Kluwer Academic Publishers. Printed in Belgium.*

A terrestrial micro-ecosystem for measuring effects of pollutants on isopod-mediated litter decomposition

J. van Wensem
Biologisch Laboratorium, Vrije Universiteit, de Boelelaan 1087, 1081 HV Amsterdam, the Netherlands

Key words: decomposition, micro-ecosystem, microflora, *Porcellio scaber*, respiration, soil fauna

Abstract

Soil fauna-mediated litter decomposition was simulated in micro-ecosystems, using oak and poplar litter, with the isopod *Porcellio scaber* (Crustacea) as a representative soil fauna species. The aim was to identify the conditions under which decomposition rate and influence of isopods could be established accurately, yet keeping the system complex enough to be relevant for the field situation. The results showed that it was possible to divide total carbon dioxide production of a micro-ecosystem into three components: microbial respiration, isopod respiration and extra microbial, isopod-mediated respiration. The isopods stimulated the microbial respiration by 10–50%, depending on litter type and experimental design. It appears that this stimulation is achieved in two ways: 1. by increasing litter moisture content and 2. by production of faecal pellets. It is concluded that the micro-ecosystem described in this paper combines ecological relevance with reproducibility and simplicity of operation and thus provides a useful tool for measuring effects of soil contaminants.

Introduction

The accumulated dead organic matter in an ecosystem represents a valuable store of essential elements. Decomposition is a process which leads to the release of these immobilized elements, partly in the form of usable nutrients for the primary production of an ecosystem. Therefore, decomposition is indispensable for ecosystem function. In terrestrial ecosystems, the process involves a complex of biochemical reactions by which, under aerobic conditions, various substrates (cellulose, lignin, phenols, sugars and waxes) are degraded to simple compounds, such as carbon dioxide, ammonium and water. Simultaneously, new substances are synthesized by polymerization, leading to the production of stable humic compounds.

Decomposition is concentrated in the upper layers of the soil, where dead organic matter is accumulated. The thickness of the litter layer is determined by the balance between primary production of the system and the rate of decomposition, the rate being primarily determined by litter quality, the decomposer organisms and abiotic factors (Swift *et al.*, 1979). Litter quality is dependent on the amount of decomposition resistant compounds, such as lignin, and toxic compounds, such as phenols and tannins.

The soil microflora makes the greatest contribution to decomposition. Fungi and bacteria function by secreting extracellular enzymes, which break down complex polymers to smaller molecules, which can then be absorbed and utilized intracellularly.

Soil fauna can accelerate decomposition by

increased mineralization of elements. This stimulation is considered to be a result of the way in which soil fauna treats the litter, both chemically and physically. This treatment can be divided into three major categories (Visser, 1985):

1. comminution (faecal pellet production), mixing and channeling of litter and soil
2. grazing on the microflora
3. dispersal of microbial propagules.

Comminution is the reduction in particle size of the organic 'resource' (Swift et al., 1979). From a variety of micro-arthropod exclusion studies, Seastedt (1984) estimated that the average rate of litter decay, as a result of soil faunal activities, increased by 23%. The direct contribution of soil fauna to decomposition, using respiration as an indicator is, at most, 10% of the total respiration of a litter layer (Petersen & Luxton, 1982).

Pollutants affect the decomposition process. This is observed at heavily contaminated sites, e.g. in the vicinity of smelters, where litter is accumulated and decomposition is reduced (Coughtrey et al., 1979; Beyer et al., 1987). Killham & Wainwright (1981) have suggested that the decrease in decomposition rate may be a result of scarcity of soil arthropods at the polluted site. In general, however, it is not clear precisely how pollutants affect decomposition. Accumulation of litter, caused by pollutants, is a result of an imbalance between primary production and decomposition, only becoming visible with longer periods of disturbed decomposition.

Laboratory experiments, in which effects of chemicals on the decomposition process may be demonstrated in a relatively short time, can indicate the potentially toxic effects of the chemical in the field. To this end a micro-ecosystem (MES) was designed to simulate soil fauna-mediated litter decomposition. In the experiments reported here, it was examined how the decomposition process behaves in the MES in a non-polluted situation. The aim was to identify conditions under which the decomposition rate and influence of soil fauna could be established accurately, in a way that is relevant for the field situation. Several factors of importance for the process were varied:

density of isopods, ratio between litter and isopods, litter species and microbial composition in the litter.

Materials and methods

The micro-ecosystem

The MES (Fig. 1) was designed to measure the carbon dioxide production of the microflora in the litter with an infra-red gas analyser. The MES consists of a perspex cylinder, the bottom being closed with a gauze. Inside there is a tube, sealed with a frit, which is used for introducing air into the MES when measuring carbon dioxide production. The MES was filled with wet sand (chemically purified, 40–100 mesh, BDH Chemicals) to the level of the frit, oak or poplar litter being placed on the top of it. The litter was kept separated from the sand by means of a nylon gauze.

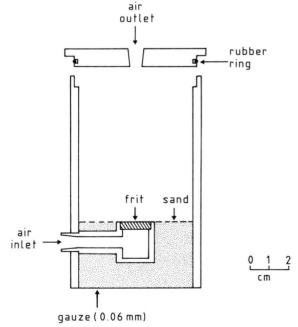

Fig. 1. Scale drawing of the micro-ecosystem. The MES is a perspex cylinder, 12 cm high and 6 cm diameter. The air for carbon dioxide measurements enters the MES via the tube at the left, passes the frit and leaves via the tube in the airtight lid. The MES is placed in a petri dish during experiments.

During the experiments the MES were placed in large, almost closed containers to prevent desiccation.

For the measurement of carbon dioxide production the MES were taken out of their containers and connected for 1 hour to the equipment, after which they were replaced. On top of the MES a lid with a gauze in the middle was placed, to provide free air exchange. During carbon dioxide measurement, this lid was replaced by the type shown in Fig. 1.

Carbon dioxide measurement

The equipment used to measure carbon dioxide production can handle twelve MES at the same time. Carbon dioxide free gas (a synthetic mix of 20% O_2 and 80% N_2) is obtained from a gas cylinder. The gas stream is divided into twelve streams, each controlled by its own flowmeter (Sho-rate, 2-1355-VMZ, Brooks instruments). After passing through the flowmeters (flow $\simeq 200$ ml min^{-1}) the gas is led through the MES. The gas streams from the twelve MES are separately connected to a gas handling unit (WA-161, ADC), which selects one gas stream to be measured by an infra-red gas analyser (IRGA, 225-MK3, ADC) and automatically switches to the next gas stream every 5 min. The other eleven streams are led into the open air. Before a gas stream of a MES is connected to the analyser, the MES has been flushed with carbon dioxide free gas for 55 min, to remove all carbon dioxide originally present, so only production is measured. The IRGA reads the carbon dioxide concentration in the gas stream for 5 min. and at the end of this period the concentration is recorded (Gila flatrecorder, Laumann). Trials with constant monitoring of the concentration showed that, once stabilized, the carbon dioxide concentration thus measured remains constant for several hours. The concentration registrated by the recorder combined with the flowrate of the gas through the MES gives the carbon dioxide production, expressed in μl per hour at 20 °C and atmospheric pressure. From the recordings obtained during the experiments, the total carbon dioxide evolution for each MES was calculated from the (bi)weekly values by numerical integration using the trapezoidal rule.

Preparation of the litter

Oak litter was collected at the Spanderswoud, a forest near Hilversum, the Netherlands, in October 1986 and April 1987. The litter layer was 10–15 cm deep. Litter from the same decomposition phase was collected; that is, fragmented but still recognizable leaves were sampled. The litter was air dried immediately after collection. Litter with particle sizes 4 to 11 mm was selected by sieving. Prior to preparation for the experiments, the air dried and sieved litter was stored at room temperature (± 20 °C, 2–4 weeks). Prior to inoculation with microflora, the litter was sterilized by autoclaving (wet, 20 min, 126 °C) and rewetted. Then a spore suspension of the fungus *Trichoderma harzianum* (3×10^5 spores g^{-1} litter) was introduced. The spore suspensions were taken from mycelium grown on malt extract agar for seven days. The fungus was obtained from the Centraalbureau voor Schimmelcultures, the Netherlands (CBS nr. 354.33).

Poplar litter was collected at the Roggebotzand, near Dronten, the Netherlands, in April 1987. The litter layer was about 3 cm deep, and only complete leaves could be found, all having lost approximately half the leaf material between the veins. This litter was treated as above with the exception of mechanical fragmentation to obtain the required particle size.

Soil fauna

To represent the soil fauna, the isopod *Porcellio scaber* Latr. (Crustacea) was used. The animal is a common saprotrophic macro-arthropod of temperate deciduous forests. The isopods were collected from the field before each experiment. They were stored at 20 °C (first exp.) or 15 °C (second exp.) and fed with oak litter. For the experiments

Table 1. Outline of experiments

a. Micro-ecosystem contents

Exp. number	Litter	Weight	Microflora	Number of isopods
I	oak	6 g	*T. harzianum*	0, 2, 4, 8, 16, 20
II	oak poplar	3 g 3 g	*T. harzianum* or original	0, 4, 8

b. Experimental set up

Exp. number	Nr. of replicates per treatment	Duration	Temp.
I	4	67 days	20 °C
II	6	49 days	15 °C

homogeneous groups of animals were composed, considering sex and weight (for exp. I sex ratio 1 ♀ : 1 ♂, mean weight per isopod 60 mg; exp. II sex ratio 3 ♀ : 1 ♂, mean weight per isopod 60 mg).

Experiment design

Experiment I was designed to examine the influence of isopod density on the microbial respiration (Table 1). The oak litter was sterilized, rewetted and inoculated with *Trichoderma harzianum*. The MES were filled with an equivalent of 6 g dry litter and remained undisturbed for 28 days before the isopods were introduced. Carbon dioxide production was measured in each MES twice weekly.

Experiment II was designed to examine the influence of litter type and microflora on the respiration and interaction between isopods and microflora (Table 1). Oak and poplar litter were halved. One half was treated as above (exp. 1). The other half was not sterilized, but a spore suspension of *T. harzianum* was introduced to prevent differences in taste between the treatments. Parts of the oak and poplar litter were placed on malt extract agar plates, and showed

that the sterilization treatment resulted in *T. harzianum* dominating the fungal flora, compared with the non-sterilized litter. The MES were filled with an equivalent of 3 g dry oak or poplar litter, and after 21 days the isopods were introduced. Carbon dioxide production was measured once a week during the whole experimental period. Based on the results of experiment I, the temperature was lowered to 15 °C (see results).

At the end of the experiments the moisture content of the litter, the dry weight loss, and isopod survival were determined.

Results

Number of isopods in the MES

In experiment I the isopod density was varied, to establish an optimal number of isopods for this size of MES and amount of litter. The objective being to maximize feeding activity of the isopods and thus the microbial response as measured by increased respiration.

Specific mortality of the isopods increased with the number of animals in the MES, possibly as a result of crowding (Ganter, 1984). In experiment II, with 4 or 8 isopods, mortality was very low, possibly because of better initial isopod condition, due to shorter storage time (exp. I 10 weeks, 20 °C: exp. II 1 week, 15 °C).

In experiment I, the MES design included a petri dish of water to prevent desiccation of the litter, especially during carbon dioxide measurement when dry air was led through the MES for 1 hour. This caused an unexpected phenomenon. The isopods were capable of moistening the litter in the MES by depressing the leaf fragments and intensifying the contact with the wet sand beneath the litter. There was a high correlation between number of isopods (mean of initial number and number surviving) and the moisture content of the litter at the end of the experiment (Pearson correlation coefficient 0.77, $n = 34$). Litter moisture content at the beginning of the experiment was 150% of the dry weight; at the end it ranged between 150% and 300%.

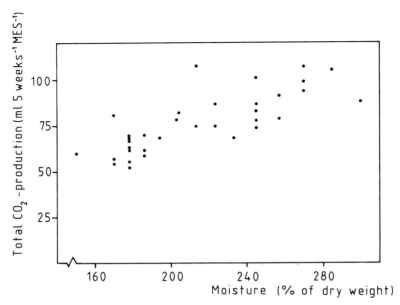

Fig. 2. Correlation between the total carbon dioxide production of a micro-ecosystem and its litter moisture content (expressed as percentage of the dry weight) at the end of the experiment. Pearson correlation coefficient 0.77; $n = 34$.

Figure 2 shows the correlation between litter moisture content and respiration of the microflora. Partial correlation (Hays, 1974) between number of isopods and respiration of the microflora, keeping moisture content constant, showed that the differences in respiration are not correlated with the number of isopods (partial correlation coefficient 0.0, $n = 34$). All differences in respiration are caused by humidity which means that the isopods have a clear influence on microbial respiration by increasing moisture content and not by comminution, which is generally thought to be their main activity in the litter (Hassal & Sutton, 1978).

The absence of a microbial response on comminution might have two reasons: 1, the over-ruling effect of litter moisture content and 2, a low activity of the isopods. Experiment II tried to avoid the influence of humidity by increasing the litter moisture content at the start of the experiment to the optimum value for microbial respiration, estimated at 270% of the dry weight. Further water supplied by the isopods should then have a less dramatic effect. This approach was successful: in two of the four treatments there

was no significant correlation between litter moisture content and carbon dioxide production. In order to increase relative isopod activity in the litter, the amount of litter in the MES was halved in experiment II. The experimental temperature was lowered to 15 °C, so that the scope for activity of *Porcellio scaber* may be assumed maximal (Newell *et al.*, 1974). In three out of four treatments, eight isopods produced a greater stimulation of microbial respiration than four isopods. It was thus concluded that eight isopods may be an optimal number for the MES when filled with 3 g (dry weight) litter.

Effects of litter type and microbial composition

Experiment II examined the influence of litter type, microflora conditioning and isopods on the microbial respiration.

As an example of the change in respiration with time, the carbon dioxide production of poplar litter for 7 weeks is shown in Fig. 3. The following pattern was observed in all experiments. Probably as a result of the pre-treatment of the litter (drying

512

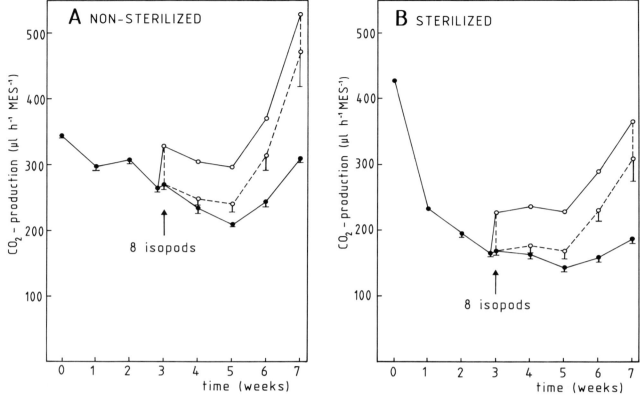

Fig. 3. Course of carbon dioxide production of micro-ecosystems in time (exp. II, mean values and standard error). A = non-sterilized poplar litter, B = sterilized poplar litter. Open circles, solid lines: total respiration of MES with isopods. Open circles, broken lines; total respiration corrected for isopod respiration (for method of calculation, see results). Closed circles, solid lines: microbial respiration of MES without isopods.

and wetting), the initial respiration is high and declines constantly. This effect is more pronounced in the sterilization treatment (Fig. 3b) and may be due to increased nutrient availability. After 3 weeks, the microbial respiration shows a gradual decrease in both treatments.

The microbial respiration in the sterilized litter (Fig. 3b) is, in spite of the higher initial respiration, lower than the respiration in the non-sterilized litter (Fig. 3a).

The isopods were introduced at week three. This resulted in an increase of the total MES respiration, which can be completely ascribed to the respiration of the isopods. Total MES respiration was corrected for this, to obtain microbial respiration only. This correction was used throughout the last 4 weeks of the experiment,

unless mortality was observed. In that case, linear interpolation was used to estimate the respiration of the isopods. The resulting values give the respiration of microflora as influenced by isopods, and allow for comparison with microbial respiration in the controls. Figure 3 shows that in both treatments the microbial respiration in MES with isopods is higher than in MES without isopods; the difference increases during the 4 weeks of isopod presence.

The increase of respiration in all MES during the last 2 weeks is a result of an accidental increase of temperature in the carbon dioxide measuring room.

To compare carbon dioxide production of the MES without effects of change in time, the respiration was integrated during the time that the

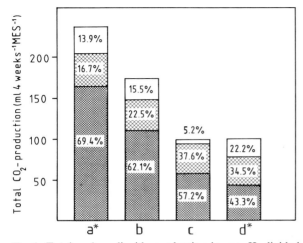

Fig. 4. Total carbon dioxide production in exp. II, divided into microflora-derived CO_2 in absence of isopods (lower part), isopod-derived CO_2 (middle part) and extra microflora-derived isopod-mediated CO_2 (upper part). a = poplar, not sterilized, b = poplar, sterilized; c = oak, not sterilized, d = oak, sterilized. * Significant correlation between carbon dioxide production and litter moisture content.

isopods were present in the MES, resulting in total carbon dioxide production.

Figure 4 shows budgets for total respiration (4

weeks) based on a comparison of MES without isopods to MES with eight isopods. Microflora in poplar litter has a higher respiration than microflora in oak litter. In all cases a stimulation of the microbial respiration by isopods was found. The way by which this stimulation was achieved may be a combination of several factors. A significant correlation between litter moisture content and total respiration of MES was found with nonsterilized poplar litter and sterilized oak litter. This suggests that in these treatments litter moisture content influences microbial respiration. One way analysis of variance, followed by the Student Newman Keuls procedure (Sokal & Rohlf, 1981), showed that eight isopods increased litter moisture content significantly with nonsterilized poplar litter (Table 2). Table 2 also shows that total respiration of MES with all litter types and treatments is significantly increased by four and eight isopods. From this it may be deduced that, although litter moisture content influenced respiration in this experiment, there is a second factor by which the isopods influence the respiration of the microflora and this might be a result of comminution.

Table 2. Results of a posteriori comparison of means, using the Student-Newman-Keuls-multiple comparison method. Within each litter type the effects of three isopod treatments were compared. Groups sharing a common letter (a, b) do not differ significantly at the 5% level. N.S. = not sterilized, S. = sterilized, s.d. = standard deviation

	CO_2-production (ml 28 days^{-1} MES^{-1})			% moisture in litter (% of dry weight)		
	Number of isopods					
	0	4	8	0	4	8
poplar, N.S.	164.6[a]	177.1[ab]	197.5[b]	317[a]	328[a]	392[b]
(s.d.)	7.7	16.0	28.1	41	40	54
poplar, S.	108.9[a]	129.0[b]	136.1[b]	290[a]	301[a]	305[a]
(s.d.)	7.2	14.5	21.3	21	33	52
oak, N.S.	56.8[a]	65.8[b]	62.0[ab]	330[a]	392[a]	339[a]
(s.d.)	4.1	5.5	4.3	75	12	25
oak, S.	43.6[a]	58.5[b]	65.9[b]	310[a]	336[a]	339[a]
(s.d.)	6.1	8.7	9.4	20	17	25

Discussion

The results demonstrate that the measurement of carbon dioxide production is a sensitive way to monitor the decomposition process in micro-ecosystems; before the litter loses a detectable amount of dry weight, it is possible to estimate decomposition by carbon dioxide evolution.

The estimated optimal isopod density is in accordance with the result of Hanlon & Anderson (1980), who found that 2–4 isopods gave a maximal microbial response in 1 g litter (dry weight).

The high microbial respiration for poplar litter, compared with oak litter, is a result of litter quality; this is determined by the C/N ratio of the litter, polyphenol, lignin and readily metabolizable components (sugars), which are all influenced by the age of the litter (Swift *et al.*, 1979). Of these variables, the polyphenol content of the leaves is probably of minor importance as most of these substances are leached out in the first phase of decomposition (Kuiters & Sarink, 1986). Compared to oak litter, freshly fallen poplar litter contains less lignin (14% versus 21%), more cellulose (43% versus 16%) and more readily metabolizable components (35% versus 22%) (Taylor & Parkinson, 1988; Waksman (1952), cited in Swift *et al.*, 1979, p. 134). However, one should keep in mind that these data have been established using different methods and might not be completely comparable.

Pattern of carbon dioxide production

Litter moisture content interfered with the sterilization treatments and therefore the present results do not allow us to conclude how sterilization influences the interaction between microflora and isopods. It appears that the sterilization treatment (heating under wet conditions) has the greater influence, probably by physically increasing nutrient availability, compared to the introduction of microflora. The system cannot be kept sterile since other fungi and bacteria are introduced with the isopods. It therefore seems unreasonable to maintain the sterilization treatment in this type of experiment.

Respiration of the microflora showed a characteristic time course: it started high and decreased rapidly, until, after some weeks, a reasonably stable level was achieved. In MES without isopods a small gradual decrease, particularly in the sterilization treatment, in microbial respiration may be observed with time. The initial rapid decrease is probably due to the exhaustion of readily available nutrients in the litter. The subsequent gradual decrease in MES without isopods is thought to be a result of the absence of disturbing factors in the litter. Physico-chemical factors, such as temperature and precipitation, especially with rhythmic changes, can increase microbial activities (Adu & Oades, 1978). A general explanation for this phenomenon may be that the microflora in the litter becomes unable to exploit new substrates due either to physical barriers or to the production of secondary, toxic metabolites. The disruptive activities of physico-chemical factors (or soil fauna) may be necessary to give a new impetus to the microflora, by removing these barriers or inhibitors. The absence of such factors in MES under laboratory conditions may explain why respiration slowly decreases.

Isopod respiration

The most uncertain variable for the total carbon dioxide production of the MES is isopod respiration. The mean respiration of the isopods was estimated as $0.12 \pm 0.03 \, \mu l \, CO_2 \, h^{-1} \, mg^{-1}$. This value is not unrealistic when compared to values for oxygen consumption of isopods, which range between $0.15 \, \mu l \, h^{-1} \, mg^{-1}$ (20 °C) and $0.21 \, \mu l \, h^{-1} \, mg^{-1}$ (16 °C) while assuming a respiratory quotient smaller than one (Wieser & Oberhauser, 1984; Phillipson & Watson, 1965). Isopod respiration was estimated once, 1 hour after introduction, and was then assumed to be a constant. This assumption may be questioned as several factors are known to influence respiration, i.e. feeding conditions, growth, routine metabolism (Kooijman, 1986), reproductive stage (Phillipson & Watson, 1965) and aggregation (Takeda, 1984). One should, therefore, use a litter type which has a high microbial respiration, thus reducing the isopods contribution to total MES respiration.

Stimulation of microbial respiration by isopods

Eight isopods stimulated microbial respiration by 10% and 50% when compared with microbial respiration in the absence of isopods. This value is in accordance with a mean stimulation of respiration by 23% found by Seastedt (1984). However, when judged by the ratio of faunal to microbial respiration in the field, i.e. 10% faunal, 90% microbial (Petersen & Luxton, 1982), the isopod density used here is too high.

It is concluded that there may be two ways by which isopods stimulate microbial respiration in the MES: one, by moistening the litter and two, by comminution. The increase in litter moisture content with isopods and thereby the increase in respiration of the microflora in this experimental design is in accordance with the results of Richardson & Morton (1986), who used MES with *Olearia* leaves and amphipods as representative soil fauna. The variation in litter moisture content in these MES could explain 84% of the variation in respiration of the microflora. Although it is reasonable to assume that soil fauna can influence moisture content and that this is important for the decomposition process, it is difficult to decide whether this effect should be considered as an artifact. If soil fauna is able to increase litter moisture content in the field, it is possible that this activity is overruled by precipitation induced drying and wetting cycles.

No effort was made to find out why comminution stimulated microbial activity, but several explanations are given by other researchers: increase of surface for microbial attack (Hanlon, 1981a, b; Hassal & Sutton, 1978), increase of leaching of nutrients (Gunnarsson, 1987) and the shift from fungi-dominated to bacteria-dominated decomposition (Hanlon & Anderson, 1980; Ineson & Anderson, 1985).

The micro-ecosystem described in this paper combines ecological relevance with reproducibility and simplicity of operation, thus providing a valuable addition to single species toxicity tests for measuring the effects of soil contaminants. Experiments using cadmium and triphenyltin hydroxide have been performed and will be reported elsewhere.

Acknowledgements

This project is financially supported by the Netherlands Organization for Applied Scientific Research, Delft. The author is indebted to Ms. D. M. M. Adema, Prof. Dr. E. N. G. Joosse, Dr. H. A. Verhoef, Dr. J. W. Vonk, Prof. Dr. S. A. L. M. Kooijman and Dr. N. M. van Straalen for the stimulation and advice given during the work; the two last persons are also thanked for critical reviewing the manuscript. Furthermore, the author is grateful to Ms. M. Aldham-Breary M. Sc. for correcting the English, to Mr. L. Sanna for drawing the figures, to the members of the photography department for reproducing the figures, to the members of the biotechnical department for designing and manufacturing the micro-ecosystems and to Ms. T. Laan for typing the manuscript.

References

Adu, K. J. & M. J. Oades, 1978. Physical factors influencing decomposition of organic materials in soil aggregates. Soil Biol. Biochem. 10: 109–115.

Beyer, W. N., W. J. Fleming & D. Swineford, 1987. Changes in litter near an aluminium reduction plant. J. Envir. Qual. 16: 246–250.

Coughtrey, P. J., C. H. Jones, M. H. Martin & S. W. Shales, 1979. Litter accumulation in woodlands contaminated by Pb, Zn, Cd and Cu. Oecologia (Berl.) 39: 51–60.

Ganter, P. F., 1984. The effects of crowding on terrestrial isopods. Ecology 65: 438–445.

Gunnarsson, T., 1987. Soil arthropods and their food choice, use and consequences. Ph. D. thesis, University of Lund, Sweden. Lund, 1987. 85 pp.

Hanlon, R. D. G., 1981a. Some factors influencing microbial growth on soil animal faeces: I. Bacterial and fungal growth on particulate oak leaf litter. Pedobiologia 21: 257–263.

Hanlon, R. D. G., 1981b. Some factors influencing microbial growth on soil animal faeces: II. Bacterial and fungal growth on soil animal faeces. Pedobiologia 21: 264–270.

Hanlon, R. D. G. & J. M. Anderson, 1980. Influence of macro-arthropod feeding activities on microflora in decomposing oak leaves. Soil Biol. Biochem. 12: 255–261.

Hassall, M. & S. L. Sutton, 1978. The role of isopods as decomposers in a dune grassland ecosystem. Sci. Proc. Roy. Dublin. Soc. 6: 235–245.

Hays, W. L., 1974. Statistics for the social sciences. Holt, Rinehart & Winston, London.

Ineson, P. & J. M. Anderson, 1985. Aerobically isolated bacteria associated with the gut and faeces of the litter feeding macro-arthropods Oniscus asellus and Glomeris marginata. Soil Biol. Biochem. 17: 843–849.

Killham, K. & M. Wainwright, 1981. Deciduous leaf litter and cellulose decomposition in soil exposed to heavy atmospheric pollution. Envir. Pollut. 26: 79–85.

Kooijman, S. A. L. M., 1986. Energy budgets can explain body size relations. J. Theor. Biol. 121: 269–282.

Kuiters, A. T. & H. M. Sarink, 1986. Leaching of phenolic compounds from leaf and needle litter of several deciduous and coniferous trees. Soil Biol. Biochem. 18: 475–480.

Newell, R. C., W. Wieser & V. I. Pye, 1974. Factors affecting oxygen consumption in the woodlouse Porcellio scaber L. Oecologia (Berl.) 16: 31–51.

Petersen, H. & Luxton, 1982. Quantitative ecology of microfungi and animals in soil and litter. Oikos 39: 287–388.

Phillipson, J. & J. Watson, 1965. Respiratory metabolism of the terrestrial isopod Oniscus asellus L. Oikos 16: 78–87.

Richardson, A. M. M. & H. P. Morton, 1986. Terrestrial amphipods (Crustacea, Amphipoda, F. talitridae) and soil respiration. Soil Biol. Biochem. 18: 197–200.

Seastedt, T. R., 1984. The role of micro arthropods in decomposition and mineralization processes. Ann. Rev. Ent. 29: 25–46.

Sokal, R. R. & F. J. Rohlf, 1981. Biometry. Freeman, San Francisco.

Swift, M. J., O. H. Heal & J. M. Anderson, 1979. Decomposition in terrestrial ecosystems. Blackwell Scientific Publications, Oxford. 372 pp.

Takeda, N., 1984. The aggregation phenomenon in terrestrial isopods. Symp. Zool. Soc. Lond. 53: 381–404.

Taylor, B. R. & D. Parkinson, 1988. Annual differences in quality of leaf litter of aspen (Populus tremuloides) affecting rates of decomposition. J. Can. Bot. 66: 1940–1947.

Visser, S., 1985. Role of the soil invertebrates in determining the composition of soil microbial communities. In Fitter, H. A. (ed.) Ecological interactions in soil. Blackwell Scientific Publications, Oxford. pp. 297–317.

Wieser, W. & C. Oberhauser, 1984. Ammonia production and oxygen consumption during the life cycle of Porcellio scaber (Isopoda, Crustacea). Pedobiologia 26: 415–419.

Hydrobiologia **188/189**: 517–523, 1989.
M. Munawar, G. Dixon, C. I. Mayfield, T. Reynoldson and M. H. Sadar (eds)
Environmental Bioassay Techniques and their Application.
© 1989 Kluwer Academic Publishers. Printed in Belgium.

Scope for growth in *Gammarus pulex*, a freshwater benthic detritivore

Caroline Naylor, Lorraine Maltby & Peter Calow
Department of Animal & Plant Sciences, University of Sheffield, Western Bank, Sheffield S10 2TN, U.K.

Key words: scope for growth, freshwater crustacean, *Gammarus pulex*, zinc, pH

Abstract

Although toxic substances affect the physiological processes of individual organisms, their ecological impacts occur at the population and community levels. However, physiological processes can often be assessed more easily and precisely than population and community ones. Here we argue that 'scope for growth', the difference between the energy input to an organism from its food and the output from respiratory metabolism, can give a good physiological measure of stress that, at least in principle, is straightforwardly related to population and community processes. We describe, in detail, how 'scope for growth' can be measured in *Gammarus pulex* (Crustacea, Amphipoda). The results indicate that both zinc and low pH can significantly reduce the scope for growth of individuals and that the most sensitive component of the energy budget is food absorption.

Introduction

The scope for growth assay has been successfully applied in the marine environment as a tool for the examination of sub-lethal, toxic effects (Bayne *et al.*, 1979; Gilfillan *et al.*, 1985; Widdows *et al.*, 1987). Here we develop a similar test for use in the freshwater environment. The species used is the freshwater amphipod, *Gammarus pulex*.

The scope for growth technique involves monitoring the energy budget of individuals and testing how the various components are affected by stress. Scope for growth can be calculated from the balanced energy equation of Winberg (1960):

$$C - F = A = R + U + P$$

or

$$C - F - R - U = P,$$

where C = energy consumed as food; F = energy lost as faeces; A = energy absorbed; R = energy metabolized (measured as respiration); U = energy lost as excretion; and P = energy available for production. The definition of scope for growth (SfG) is therefore

$$C - F - R = \text{SfG}.$$

Energy loss via excretion was assumed to be minimal and consequently not measured.

SfG can be either positive, indicating that energy is available for growth and reproduction, or zero, when energy input balances energy expenditure, or negative, when animals must use their body reserves for essential metabolism.

When SfG is used to measure the impact of a stress, the observed changes in the energy budget are assumed to be closely related to actual changes in individual growth rate and fecundity; i.e. they can be readily interpreted in terms of the response of whole individuals and the dynamics of populations to which they belong. However, SfG has two main advantages over actual measurements of growth and fecundity. Firstly, results can be obtained rapidly; in hours or days rather than in days or weeks. Secondly, by definition, SfG provides an insight into which particular components of the energy budget are affected by a stress. For example scope for growth can be reduced by: a reduction in feeding rate; a reduction in absorption efficiency; or conversely, an increase in respiration or excretion rates. Which particular effect is observed may vary for different stresses and this could then influence factors such as recovery periods and effects on population dynamics.

Gammarus pulex was selected as the test species because it is common in lotic systems and is known to be sensitive to a range of stresses. *G. pulex* is a benthic species feeding on a variety of coarse organic matter including decaying leaves.

Materials and methods

Two measurements were made: the amount of energy absorbed (food intake – faeces production) when *G. pulex* was fed a diet of alder (*Alnus glutinosa*) inoculated with *Cladosporium*; and energy lost during respiration. An outline of the method is given in Fig. 1 and details are provided below.

Animals

All animals used in these studies were collected from a small, spring-fed stream (Crags stream -NGR SK497745). In the laboratory animals were maintained in APW (artificial pond water, Table 1) under ambient temperature and lighting conditions and fed a mixed diet of elm (*Ulmus vegeta*), sycamore (*Acer pseudoplatanus*) and alder leaves for at least two weeks prior to use. Only males (7–10 mg dry weight) were used in the experiments. One week prior to a test, animals were removed from the main holding tanks and kept under experimental conditions. These were: APW; 15 °C; 12L : 12D photoperiod; and a diet of conditioned alder leaves produced according to standard techniques (see below).

Food

The food source used for the test was alder leaves inoculated with the fungus *Cladosporium*. This combination was a compromise between a food that provided optimum growth and a food that could be replicated through time.

Alder leaves were collected from the field after abscission but before leaf fall. This ensured that all leaves were at the same stage of decay. The leaves were then dried and stored. When required for use, the leaves were rehydrated and cut into discs (diameter 1.6 cm). The discs were placed in 'enriched' water ($+ CaCl_2 \cdot 2H_2O$, $MgCl_2.6H_2O$, KH_2PO_4, $(NH_4)_2HPO_4$) and autoclaved to remove any microorganisms already present. The discs were then inoculated with a standard amount of the fungus *Cladosporium* (fungi : leaves, 1 : 20), and incubated for 10 days (Fig. 1). The fungi have two functions; they are utilized as a food source by *G. pulex* (Barlocher & Kendrick, 1973) and they also facilitate the breakdown of the leaf material, making it more palatable (Kaushik & Hynes, 1971; Barlocher & Kendrick, 1975; Chamier & Willoughby, 1986).

The initial dry weight of the leaf material was obtained for leaf discs that were dried at 60 °C for

Table 1. Artificial pond water (APW)

Stock solution	g l^{-1}
1. $CaCl_2 \cdot H_2O$	58.80
2. $MgSO_4 \cdot 7H_2O$	24.65
3. $NaHCO_3$	12.95
4. KCl	1.15)

Mix 5 ml of each of the stock solutions and make up to one litre with distilled water.

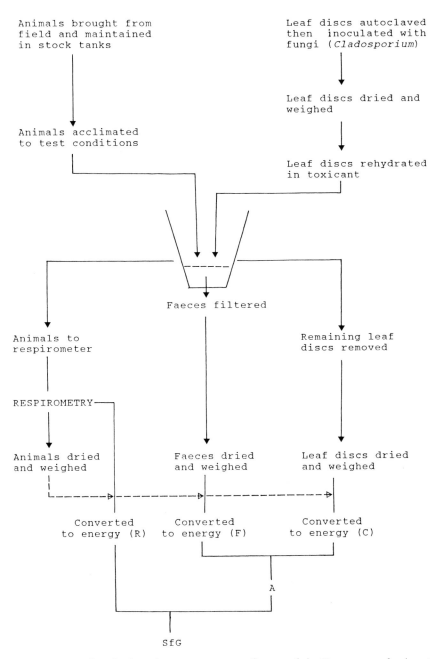

Fig. 1. Summary of method used to measure scope for growth in *Gammarus pulex* (see text).

48 hours. Before being fed to the animals, discs were rehydrated. This had two purposes; firstly, it increased palatability, and secondly, it allowed toxicants to come into equilibrium with the discs.

For example, heavy metals tend to adsorb to the leaf discs and fungi (Duddridge & Wainwright, 1980; Abel & Barlocher, 1988).

Measurement of energy absorbed

Animals (usually 20/test) were placed in individual experimental chambers (see Fig. 1) and provided with a known amount of food (L_1). Five control chambers were also used in which there were leaves but no animals. These controls were used to correct for weight lost due to leaching (C_1). After 6 days, the remaining food was removed, dried at 60 °C to constant weight and reweighed (L_2). Faecal material was filtered onto preweighed filter paper (Whatman, diameter 5.5 cm) (F_1), dried to constant weight and reweighed (F_2). Again there were 5 control filters to correct for any weight change in filters during the drying and filtering processes (C_f). The animals themselves were removed and placed in respirometry chambers.

Calculations

The correction factor for weight loss due to the leaching of soluble organics from leaf discs (C_1) was calculated from equation 1 where A_1 and A_2 refer to the initial and final weights of control leaves.

$$C_1 = \frac{(A_2/A_1)}{5}. \tag{eq. 1}$$

The amount of energy consumed in J per mg animal per day (C) (over the 6 day experimental period) is given by

$$C = \frac{[((L_1 * C_1) - L_2) * E_1]}{W * 6}, \tag{eq. 2}$$

where L_1 and L_2 refer to the initial and final weights of experiment leaf discs, E_1 is the equivalent of the food determined by bomb calorimetry (21.552 J mg^{-1}) and W is the dry weight (mg) of the animal.

The correction factor for change in weight of filters during processing (C_f) was calculated using

equation 3

$$C_f = \frac{(B_2/B_1)}{5}, \tag{eq. 3}$$

where B_1 and B_2 refer to the initial and final weight of control filters.

The amount of energy lost in faeces in J per mg animal per day (F) is given by

$$F = \frac{[(F_2 - (F_1 * C_f)) * E_f]}{W * 6}, \tag{eq. 4}$$

where F_1 and F_2 refer to the initial and final weight of experimental filter papers and E_f is the energy equivalent of the faeces obtained by bomb calorimetry (18.737 J mg^{-1}).

The amount of energy absorbed (A) in J mg^{-1} d^{-1} is

$$A = C - F. \tag{eq. 5}$$

Measurement of energy respired

After the feeding period, the animals were transferred to a flow-through respirometer (Fig. 2) (Wrona & Davies, 1984) and left for at least 15 hours before any measurements were made. This allowed the animals to acclimate to test conditions.

The oxygen content, in Torr (160 Torr = 10.2 mg l^{-1} at 15 °C), of the water leaving the respirometry chamber (PLO$_2$) was measured by slowly withdrawing 50 μl of water from the chamber and measuring its oxygen content with a Radiometer oxygen electrode. The oxygen content of the water entering the respirometry chambers (PEO$_2$) was determined from the control chamber which contained no animal. The weight-specific oxygen uptake of each animal (MO$_2$, μmol mg^{-1} h^{-1}) is given by

$$MO_2 = \frac{(PEO_2 - PLO_2) * AO_2 * Fl}{W},$$

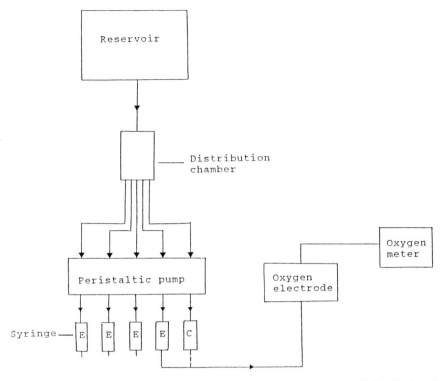

Fig. 2. Respirometer: the system consists of a reservoir of aerated APW which leads to a small distribution chamber from which a series of channels run over a peristaltic pump and into 2 ml glass syringes. These syringes act as experimental chambers (E), each one housing an individual *Gammarus pulex*. One syringe was left vacant to act as a control (C).

where AO_2 is the solubility coefficient of oxygen in water at 15 °C (= 2.01 μmol l^{-1} Torr^{-1}) and Fl is the flow rate of water through the chambers (1 hr^{-1}).

Oxygen uptake (MO_2) was corrected from a rate per hour to a rate per day ($\times 24$), and from μmol to litres ($\times 22.41 * 10^{-6}$). The conversion to energy assumes that all respiration is oxidative and that there is consequently a direct relationship between oxygen consumption and heat loss. The oxyjoule equivalent varies with the substrate being metabolized and for carbohydrate it is 21×10^3 J lo_2^{-1} (Elliott & Davison, 1975). Therefore energy respired per mg animal per day (R) is given by

$$R = MO_2 * 24 * 22.41 \times 10^{-6} * 21 \times 10^3.$$

The test has been used to evaluate the effect of two stresses, zinc ($ZnSO_4 \cdot 7H_2O$) at three con-

centrations 0.3, 0.5 and 0.7 mg l^{-1} zinc and increased acidity to pH 5 (H_2SO_4). Both food absorption and respiration were determined for each concentration.

Results

The results given here are examples from a more extensive data set that will be published elsewhere. The effect on energy absorption and energy loss via respiration, when various levels of zinc were added to the APW is shown in Fig. 3 and the effect of increasing acidity to pH 5 is shown in Fig. 4. The difference between energy absorbed and energy lost is a measure of SfG.

For zinc there was a significant decrease in energy absorbed with increasing concentration of zinc ($F_{3.50} = 10.597, p < 0.001$) but no significant change in energy loss via respiration

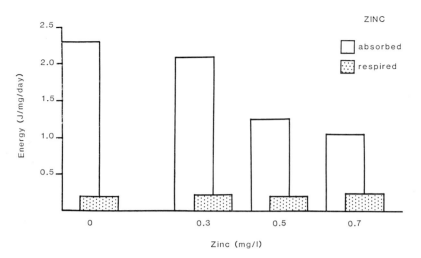

Fig. 3. The effect of various concentrations of zinc on energy absorbed and energy respired in *Gammarus pulex*. The difference between columns is a measure of scope for growth.

($F_{3.50} = 1.240$, n.s.). This resulted in a decrease in SfG ($F_{3.50} = 10.899$, $p < 0.001$). A decrease in pH resulted in a similar pattern. Energy absorbed showed a significant decrease between pH 7.6 (APW) and pH 5 ($t_{23} = 3.012$, $p < 0.001$) but the change in respiration was not significant. There was again a significant decrease in SfG ($t_{23} = 2.921$, $p < 0.001$).

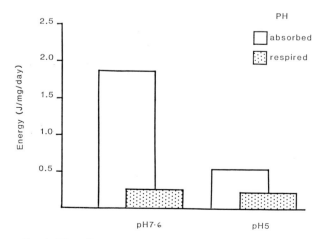

Fig. 4. The effect of reducing pH on energy absorbed and energy respired in *Gammarus pulex*.

Discussion

These preliminary data address two questions about the use of SfG in environmental monitoring: Firstly, how sensitive is the method compared with other tests (e.g. acute LC_{50} tests)? Secondly which component of the energy budget is most sensitive to toxins?

Acute tests have been carried out on large male *G. pulex* from the study population (Naylor *et al.*, unpublished). For zinc, the 24 h LC_{50} value was 8.00 ± 0.58 mg l^{-1} (mean $\pm 95\%$ C.L.). Therefore the range of zinc concentrations over which the SfG was tested (0.3–0.7 mg l^{-1}) was approximately 10–25 times less than the acute response. For pH, the 24 hr LC_{50} value was 3.82 ± 0.03. The concentration of hydrogen ions at pH 5 was approximately 15 times less than this. Hence for both acidity and zinc, SfG was detecting effects at least an order of magnitude below acute levels. The only other types of sub-lethal tests that have been carried out with *G. pulex* in the laboratory are behavioural tests, where, for example, individuals were given a choice between clean and stressed conditions (Abel & Green, 1981; Costa, 1967). Avoidance behaviour was reported for zinc at all levels tested (1 mg l^{-1} to 5 mg l^{-1}) and for pH values of less than 6.2.

For both our experiments the most sensitive component of SfG to stress was the amount of energy absorbed rather than the amount of energy metabolized. Other studies are being carried out to test the generality of this result.

The main conclusions, therefore are that the SfG test can provide a relatively rapid assay, that is at least as sensitive as more long-term chronic tests. Moreover, if the amount of food absorbed is consistently the most sensitive component of SfG to stress, then this test could be made even less demanding in time and effort by eliminating the need to measure respirometry. The feasibility of deployment in both laboratory and field situations is the subject of current work.

Acknowledgements

We would like to thank Paul Eady and Lesley Pindar for technical assistance. The project was supported by a grant from the Commission of the European Community (Contract STP-0086-1-UK(CD)).

References

Abel, P. D. & D. W. J. Green, 1981. Ecological and toxicological studies on invertebrate fauna of two rivers in the northern Pennine orefield. In Say, P. J. & B. A. Whitton (eds), Heavy Metals in Northern England: Environmental and Biological aspects. Department of Botany, University of Durham, 109–122.

Abel, T. & F. Barlocher, 1988. Uptake of cadmium by *Gammarus fossarum* (Amphipoda) from food and water. J. Appl. Ecol. 25: 223–231.

Barlocher, F. & B. Kendrick, 1973. Fungi and food preferences of *Gammarus pseudolimnaeus*. Arch. Hydrobiol 72: 501–516.

Barlocher, F. & B. Kendrick, 1975. Leaf conditioning by micro-organisms. Oecologia 20: 359–362.

Bayne, B. L., M. N. Moore, J. Widdows, D. R. Livingstone & P. Salkeld, 1979. Measurement of the responses of individuals to environmental stress and pollution: studies with bivalve molluscs. Phil. Trans. Roy. Soc., Lond. B 286: 563–581.

Chamier, A.-C. & L. G. Willoughby, 1986. The role of fungi in the diet of the amphipod *Gammarus pulex* (L.): an enzymatic study. Freshwat. Biol. 16: 197–208.

Costa, H. H., 1967. Responses of *Gammarus pulex* to modified environment (ii) Reactions to abnormal hydrogen ion concentrations. Crustaceana 13: 1–10.

Duddridge, J. E. & M. Wainwright, 1980. Heavy metal accumulation by aquatic fungi and reduction in viability of *Gammarus pulex* fed cadmium contaminated mycelium. Wat. Res. 14: 1605–1611.

Elliott, J. M. & W. Davison, 1975. Energy equivalents of oxygen consumption in animal energetics. Oecologia 19: 195–201.

Gilfillan, E. S., D. S. Page, D. Vallas, L. Gonzalez, E. Pendergast, J. C. Foster & S. A. Hanson, 1985 (Abstract). Relationship between glucose-6-phosphate dehydrogenase and aspartate amina transferase activities, scope for growth and body burdens of Ag, Cd, Cu, Cr, Pb, Zn in populations of *Mytilus edulis* from a polluted estuary. Marine Pollution & Physiology Recent Advances. Meeting Mytic Conn. U.S.A. Nov. 1–3 1983. University of South Carolina. pp. 107–124.

Kaushik, N. K. & H. B. N. Hynes, 1971. The fate of the dead leaves that fall into streams. Arch. Hydrobiol. 68: 465-515.

Naylor, C., L. Pindar & P. Calow, Inter- and intra-specific variation in sensitivity to toxins; the effects of acidity and zinc on the freshwater crustaceans *Asellus aquaticus* (L.) and *Gammarus pulex* (L.). unpublished.

Widdows, J., P. Donkin, P. N. Salked & S. V. Evans, 1987. Measurement of scope for growth and tissue hydrocarbon concentrations of mussels (*Mytilus edulis*) at sites in the vicinity of Sullen Voe oil terminal: a case study. In; Kuiper, J. & W. J. van den Brink (eds) fate and effect of oil in marine ecosystems ISBN 90-247-3489-4, Martinus Nijhoff Publishers, Dordrecht, Netherlands. pp. 269–277.

Winberg, G. G., 1960. Rate of metabolism and food requirements of fishes. Fish. Res. Bd. Can. Trans. Ser. 194: 202.

Wrona, F. J. & R. W. Davies, 1984. An improved flow-through respirometer for aquatic macroinvertebrate bioenergetic research. Can. J. Fish. Aquat. Sci. 41: 380–385.

Hydrobiologia **188/189**: 525–531, 1989.
M. Munawar, G. Dixon, C. I. Mayfield, T. Reynoldson and M. H. Sadar (eds)
Environmental Bioassay Techniques and their Application.
© 1989 *Kluwer Academic Publishers. Printed in Belgium.*

Feeding and nutritional considerations in aquatic toxicology

R. P. Lanno, B. E. Hickie & D. G. Dixon
Department of Biology, University of Waterloo, Waterloo, Ontario, N2L 3G1, Canada

Key words: aquatic toxicity testing, nutrition, feeding regime, diet

Abstract

The nutritional status of an aquatic organism, both prior to and during testing, can significantly modify the apparent toxicity of a chemical. In order to decrease the variability of toxicity test results, both within and between laboratories, feeding regimes, feed types and proximate composition of diets should be routinely reported. The advantages and metabolic effects of various feeding practices currently used in toxicity testing (fasting, starvation, feeding as a percentage of body weight, feeding *ad libitum* and pair feeding) are discussed. The disadvantages of monitoring nutritional status of test organisms are also reviewed and suggestions are offered as to how best to monitor nutritional status.

Introduction

The importance of nutrition as a factor modifying the toxicity of waterborne chemicals to aquatic organisms has been largely overlooked, even though dietary factors are known to modify the toxicity of various compounds to mammals (Boyd & Campbell, 1983) and birds (Dahlgren, 1988). Although some work describing diet-toxicant interactions in fish is appearing in the literature (Mehrle *et al.*, 1977; Dixon & Hilton, 1985), interpreting published toxicity results in the light of reported diet interactions is usually difficult. All too often toxicity results are published without a description of nutritional status of the test organisms, both before and during the test. This paper summarizes how various nutritional considerations can modify the toxicity of waterborne chemicals and outlines various feeding regimes that are available for the standardization of nutritional status in toxicity tests.

From a toxicologist's perspective, the nutritional status of an organism is concerned with the quantity and quality of the organism's diet, as well as the levels of contaminants present in the diet. The term 'feeding regime' describes the quantity of food offered to the test organisms and the manner in which it is presented (number of feedings per day). The quality of the diet refers to its proximate composition (percentage of protein, lipid, carbohydrate, water, ash), the specific composition of these various components (amino acid profiles, amount of essential fatty acids) and the presence of adequate amounts of micronutrients (vitamins and minerals).

Many effects of nutritional status on test organisms are intimately related to nutritional effects on metabolic rate. Nutritional status determines whether an organism is in a maintenance, a catabolic or an anabolic physiological state. Feeding an organism will increase its metabolic rate (heat increment of feeding) (Jobling, 1981) and hence the uptake, metabolism and depuration rates of a toxicant. As an example, the rate of uptake of [^{14}C]-benzo(a)pyrene (BaP) by fed bluegill sunfish (*Lepomis macrochirus*) was twice that of unfed fish and the depuration of [^{14}C]-radioactivity was ten times greater in fed fish (Jimenez *et al.*, 1987). Feeding greatly increased the rate of conversion of BaP to its polar metabolites and their subsequent excretion by the sunfish. The feeding of invertebrates during toxicity testing will also modify toxicity. Feeding decreased the toxicity of copper to copepods (*Acartia tonsa*) (Sosnowski *et al.*, 1979), as well as the toxicity of linear alkylbenzene sulfonate to *Daphnia magna* (Taylor, 1985).

The proximate composition of a diet will affect the proximate composition of an organism feeding upon that diet. High levels of dietary lipid will result in an increased deposition of lipid in rainbow trout (*Salmo gairdneri*) (Hilton & Slinger, 1981; Watanabe, 1982). An increase in the lipid compartment of the fish will increase the size of the compartment available for the deposition of lipophilic toxicants. Strong correlations between bioconcentration factor (BCF) and lipid content of test organisms have been noted by a number of researchers (Geyer *et al.*, 1985; Hickie & Dixon, 1989; Shubat & Curtis, 1986). The lipid content of fish can also be altered by feeding regime. Rainbow trout fed a growth ration (4% body weight/day) have been shown to have a significantly higher lipid component than trout reared on a maintenance ration (2% body weight/day) of the same diet (Shubat & Curtis, 1986).

Physiological parameters are also affected by nutritional status. Blood parameters will vary in relation to whether an animal is in a maintenance, an anabolic or a catabolic state. Short-term starvation in teleosts (48 to 72 h) can alter plasma levels of thyroid hormones (Eales *et al.*, 1981), as well as serum glucose, serum cortisol and both the level and type of serum lipid (White & Fletcher, 1986). Starvation over longer periods can alter rates of protein synthesis (Watt *et al.*, 1988), lipid metabolism (Boon & Duinker, 1985) and rates of xenobiotic transformation by liver and kidney (Andersson *et al.*, 1985). Since both the rate of

gall bladder emptying and gastric motility are related to the presence of food in the gastrointestinal tract (Holmgren *et al.*, 1983), feeding regime can potentially alter the rate of depuration of any chemical that is subject to biliary – fecal excretion. Mixed function oxidase (MFO) enzyme activity in teleosts may be altered by food deprivation (Collodi *et al.*, 1984), type of diet (commercial or synthetic) (Ankley & Blazer, 1988) and the protein level of the diet (Stott & Sinnhuber, 1978).

Pre-experimental nutritional status can also have a marked effect on toxicant impact. Although Marking *et al.* (1984) found that the type of pre-experimental diet had only a marginal effect on the acute toxicity of eleven chemicals to rainbow trout, other researchers have noted more significant effects (Hickie & Dixon, 1987; Mehrle *et al.*, 1977). If the feeding regime is held constant, then the proximate composition of the pre-experimental diet can significantly alter toxicity. Increased quantity and availability of protein will increase the tolerance of rainbow trout and bluegills to chlordane and the PCB Arochlor 1254 (Mehrle *et al.*, 1974; Mehrle *et al.*, 1977). Feeding a diet that increases lipid deposition may alter the potential for the subsequent bioaccumulation and/or toxicity of some lipophilic toxicants (Hickie & Dixon, 1987). Monitoring pre-experimental nutritional status of test organisms can also allow a researcher to monitor the health of test organisms prior to the actual experiment. If the organisms refuse their ration or exhibit poor growth, they should not be considered for use in future experiments.

The contamination of diets with toxicants can affect the response of organisms to waterborne toxicants. Relative to the waterborne contribution, the dietary input to toxicant body burden, and potentially to toxicity, is usually only important with highly lipophilic contaminants, those with a log octanol-water partition coefficient greater than 5. It should be noted that exposure to dietary toxicants can also result in the induction of detoxification and/or excretory mechanisms (eg. MFO, metallothionein) (Lanno *et al.*, 1987).

Close scrutiny of the feeding regime of test organisms can allow the monitoring of toxicant-exposure impacts on feeding behaviour. Lett *et al.* (1976) observed an initial cessation of feed intake by rainbow trout exposed to sublethal levels of waterborne copper, with a subsequent increase in feed intake to the levels of control fish over a 40-day period. Similar results were reported by Kumaraguru & Beamish (1986) for rainbow trout exposed to sublethal levels of permethrin. By quantifying feed intake and growth, the energetic status of test organisms can be monitored. Collvin (1985) observed a reduced feed conversion efficiency in perch (*Perca fluviatilis* L.) exposed to waterborne copper. He attributed this decrease to the increased metabolic energy demands of copper detoxification. Decreases in feed conversion efficiency have also been reported for larval white sucker (*Catostomus commersoni*) and young common shiners (*Notropis cornutus*) exposed to sublethal levels of cadmium, 2, 4-dichlorophenol or pentachlorophenol (Borgmann & Ralph, 1986).

Although it is obvious that nutritional status is an important modifying factor of toxicity, most suggested standard bioassay techniques for fish pay little attention to diet and feeding regime. Sprague (1973) suggests that 'Fish should not be fed during the usual acute bioassays, nor for at least a full day before testing. In tests which last longer than a week, feeding may be carried out, ...' The APHA (1980) addresses feeding of test organisms in much greater detail. Suggestions on feeding regimes prior to testing are lacking, but three different feeding regimes are proposed for special purpose growth-rate bioassays. Organisms are not to be fed during short-term bioassays, but the suggestion is made that 'When some test animals, though still alive, are dying or evidently affected after 96-hr exposure, prolong the test. If tests are continued for longer periods, feed the test organisms.' Unfortunately feeding test organisms that have been fasted for 96 hours or longer will completely change their physiological state and severely alter their response to the test toxicant.

The selection of a feeding regime for a particular testing situation will depend upon a number a factors. The feeding regime selected will vary depending on whether the test is acute or chronic and on whether it is conducted in a static or a flow-through system. One must also consider the species tested and the life stage of the organism. This is of particular importance with larval fish. Sac-fry are extremely susceptible to toxicants at the time of conversion from endogenous to exogenous feeding (Van Leeuwen *et al.*, 1985), a situation which will be intensified by the absence of food. Finally, the toxicological endpoints that one wishes to monitor should also be considered. Feeding may not be as important if lethality is being monitored over a short period of time with relatively large organisms, but may become very important if the physiological parameters of interest are affected by fasting.

Acute tests

The fundamental question with acute toxicity tests involves whether or not the organisms should be fed. Since organisms in the environment usually have food available during toxicant exposure, care must be taken in basing environmental management decisions on tests where organisms do not have the option of feeding. Traditionally, test organisms have not been fed during acute bioassays. The importance of feeding during acute tests will vary with the size of the test organism and the duration of the test. The smaller the organism and the longer the test, the more likely that the organism will go into a negative energy balance as the test progresses. As a result, the test will be performed on an organism that is in a state of physiological transition from feeding to fasting, or from fasting to starving, depending on the organism and the duration of the test. The control mortality of 2-d-old fathead minnow (*Pimephales promelas*) observed during 39 bioassays with fish that were either unfed, or fed brine shrimp nauplii (Fig. 1; Hickie, unpublished data), provides an interesting example. The mortality of fed larvae stabilized at 48 h and remained constant for the duration of the 144 h bioassays. The mortality of unfed larvae, how-

528

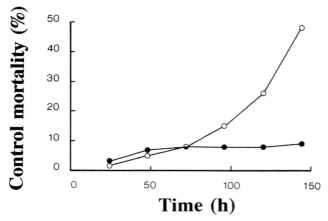

Fig. 1. The effect of feeding on mean control mortality during 39 bioassays with fed (●·······●) and unfed (○·······○) 2-d-old fathead minnow.

ever, increased over the entire course of the tests, becoming critically different from fed control larvae at 96 h. The use of unfed larvae, relative to fed larvae, led to a higher proportion of unsuitable LC50 estimates (greater than 10% control mortality). Fifty percent (³/₆) of the tests conducted using unfed larvae were unacceptable, while only 10% (³/₃₃) of the tests conducted with fed control larvae were unacceptable. Table 1 demonstrates how the feeding of larval fathead minnow affected the actual 96 h LC50 estimates. At a given age (2 or 4 d), fed larvae were approximately three times more tolerant of waterborne sodium pentachlorophenate than unfed larvae. Holdway & Dixon (1985) also showed that fed juvenile flagfish (*Jordanella floridae*) were much more tolerant of

Table 1. The effect of food availability on representative 96-h LC50s of sodium pentachlorophenate to larval fathead minnow. Each continuous-flow bioassay exposed larvae to 5 toxicant concentrations plus a control. Fed fish received brine shrimp larvae *ad libutum* three times daily. The LC50s were calculated using Spearman-Karber analysis (Hamilton *et al.*, 1977).

Age (d)	LC50 (95% confidence interval) (μg l^{-1})	
	Fed	Unfed
2	427 (388–470)	142 (101–201)
4	305 (251–371)	95 (67–135)

pulse doses of methoxychlor than starved juvenile flagfish.

Chronic tests

During chronic toxicity tests the organisms must be fed. The major question confronting the researcher involves the choice of the feeding regime appropriate to the experimental design. The four major regimes available for toxicity testing are: 1) feeding as a percentage of organismal body weight, 2) feeding *ad libitum*, 3) pair feeding, and 4) pair feeding with an *ad libitum* control.

Feeding as a percentage of body weight

Under this protocol an *a priori* decision is made, based upon available data from previous experiments or from suggested feeding tables (Hilton & Slinger, 1981), as to what percentage of body weight the organisms will be fed. The organisms are then weighed, and the daily ration calculated. If significant growth occurs during the study, it will be necessary to adjust the ration by reweighing the organisms on a regular basis throughout the exposure period. If all the food offered to the organisms is consumed, this feeding regime is quite adequate for the majority of toxicity experiments, although the danger of underfeeding still exists.

If feed intake is affected by the toxicant, however, the danger exists of developing differential growth rates, and rates of toxicant metabolism, at different exposure levels of the toxicant. This may lead to difficulty in separating feed reduction effects of the toxicant from actual effects on the bioenergetics of the test organism. There may also be a problem with unconsumed feed binding the toxicant or increasing the biological oxygen demand in the test chambers.

Feeding ad libitum

Often termed 'feeding to satiety', this feeding regime allows the test organisms to consume as much feed as they desire. Feeding *ad libitum* is easily accomplished with larger fish, but problems may arise with monitoring feed consumption of smaller test organisms such as *Daphnia* sp. or larval fish. Again, if one effect of the toxicant is to alter feed intake, then there is the danger of developing differential growth rates between treatments. Differential rates of growth and feed intake can often result in changes in physiological parameters between treatments. Interpretation of toxicant effects on these parameters becomes difficult, since it is unclear whether the effects are due to the toxicant itself or to changes in feed intake and hence metabolism.

Pair feeding

Pair feeding allows a distinction to be made between those physiological effects due to the toxicant and those due to differences in feed intake. Each treatment replicate is fed *ad libitum* and is randomly matched with a control replicate, which receives the same amount of food given to the treatment replicate. If decreased feed intake is an effect of the toxicant, then the ration given to the control replicate is reduced accordingly. As a result, feed intake as a modifying factor of toxicity is accounted for, and differences in growth and/or physiological response can be attributed more directly to toxicant level.

Pair feeding with an *ad libitum* control is simply an extension of pair feeding. This feeding regime incorporates all of the advantages of pair feeding and adds a separate *ad libitum* control which allows full expression of the phenotypic/genotypic characteristics of the species or strain of test organism. This may be an excellent means of monitoring the consistency of test organisms between batches of the same strain or species, and may greatly enhance the quality control of toxicity tests.

Conclusions

Although this paper deals with the advantages of monitoring the nutritional status of test orga-

nisms, there are certain disadvantages that should not be overlooked. The added time and effort involved in monitoring nutritional status may be troublesome, but the enhanced understanding of toxicant effects that can be obtained outweigh the additional work.

While attempts have been made to standardize diet formulations in aquatic toxicology testing (Bengtson *et al.*, 1985), we feel that this is unrealistic considering the diversity of components which can be used in feeds. Researchers should, however, routinely report the type of formulation used, along with an analysis of its proximate composition.

As low level environmental contamination increases, it will inevitably become more difficult to obtain practical diets and dietary components that are free of background xenobiotic contamination, contamination which could interact with the toxicant being studied. Great care should be taken in selecting feed components low in contaminants, and routine trace contaminant analysis of diets and feed components used in toxicity studies should be reported.

The effect of nutritional status cannot be overlooked as an important modifying factor of toxicity. Various feeding regimes are available depending upon the type of test conducted and the parameters that are being measured. If feeding regimes, along with the proximate composition of diet formulations, are routinely reported in the literature, then it may be possible both to account for more of the inter- and intra-laboratory variability in test results and to maximize the information that can be gained from each toxicity test.

Acknowledgements

The concepts put forward here were developed and refined during research supported by a Natural Sciences and Engineering Research Council of Canada Operating Grant (A8155) to D.G.D.

References

Andersson, T., U. Koivusaari & L. Forlin, 1985. Xenobiotic biotransformation in the rainbow trout liver and kidney during starvation. Comp. Biochem. Physiol. 82C: 221–225.

Ankley, G. T. & V. S. Blazer, 1988. Effects of diet on PCB-induced changes in xenobiotic metabolism in the liver of channel catfish (*Ictalurus punctatus*). Can. J. Fish. Aquat. Sci. 45: 132–137.

APHA, AWWA & WPCF, 1980. Standard methods for the examination of water and wastewater, 15th Edition, American Public Health Association, Washington, D.C., 1134 pp.

Bengtson, D. A., A. D. Beck & K. L. Simpson, 1985. Standardization of the nutrition of fish in aquatic toxicological testing. In; C. B. Cowey, A. M. Mackie & J. B. Bell (eds.) Nutrition and feeding in fish. Academic Press, London, pp. 431–446.

Boon, J. P. & J. C. Duinker, 1985. Kinetics of polychlorinated biphenyl (PCB) components in juvenile sole (*Solea solea*) in relation to concentrations in water and to lipid metabolism under conditions of starvation. Aquat. Toxicol. 7: 119–134.

Borgmann, U. & K. M. Ralph, 1986. Effects of cadmium, 2, 4-dichlorophenol, and pentachlorophenol on feeding, growth, and particle-size conversion efficiency of white sucker larvae and young common shiners. Arch. Envir. Contam. Toxicol. 15: 473–480.

Boyd, J. N. & T. C. Campbell, 1983. Impact of nutrition on detoxication. In; J. Caldwell & W. B. Jakoby. (eds) Biological basis of detoxication. Academic press, New York, pp. 287–306.

Collodi, P. M., M. S. Stekoll & S. D. Rice, 1984. Hepatic aryl hydrocarbon hydroxylase activities in coho salmon (*Oncorhynchus kisutch*) exposed to petroleum hydrocarbons. Comp. Biochem. Physiol. 79C: 337–341.

Collvin, L., 1985. The effect of copper on growth, food consumption and food conversion of perch *Perca fluviatilis* L. offered maximal food rations. Aquat. Toxicol. 6: 105–113.

Dahlgren, J., 1988. Variation in diet composition: a hazard to the reliability of the LD_{50} test. Envir. Pollut. 49: 177–181.

Dixon, D. G. & J. W. Hilton, 1985. Effects of available dietary carbohydrate and water temperature on the chronic toxicity of waterborne copper to rainbow trout (*Salmo gairdneri*). Can. J. Fish. Aquat. Sci. 42: 1007–1013.

Eales, J. G., M. Hughes & L. Uin, 1981. Effect of food intake on diel variation in plasma thyroid hormone levels in rainbow trout, *Salmo gairdneri*. Gen. Comp. Endocrinol. 45: 167–174.

Geyer, H., I. Scheunert & F. Korte, 1985. Relationship between the lipid content of fish and their bioconcentration potential of 1, 2, 4-trichlorobenzene. Chemosphere 14: 545–555.

Hamilton, R. A., R. C. Russo & R. V. Thurston, 1977. Trimmed Spearman-Karber method for estimating med-ian lethal concentrations in toxicity bioassays. Envir. Sci. Technol. 11: 714–719; correction, 12: 417 (1978).

Hickie, B. E. & D. G. Dixon, 1987. The influence of diet and preexposure on the tolerance of sodium pentachlorophenate by rainbow trout (*Salmo gairdneri* Richardson). Aquat. Toxicol. 9: 343–353.

Hickie, B. E., D. G. Dixon & J. F. Leatherland, 1989. The influence of dietary carbohydrate: lipid ratio on the chronic toxicity of sodium pentachlorophenate to rainbow trout (*Salmo gairdneri* Richardson). Fish. Physiol. Biochem. 6: 175–185.

Hilton, J. W. & S. J. Slinger, 1981. Nutrition and feeding of rainbow trout. Can. Spec. Publ. Fish. Aquat. Sci. 55. 15 pp.

Holdway, D. A. & D. G. Dixon, 1985. Acute toxicity of pulse-dosed methoxychlor to juvenile American flagfish (*Jordanella floridae* Goode and Bean) as modified by age and food availability. Aquat. Toxicol. 6: 243–250.

Holmgren, S., D. J. Grove & D. J. Fletcher, 1983. Digestion and the control of gastrointestinal motility. In; J. C. Rankin, T. J. Pitcher & R. Duggan (eds) Control processes in fish physiology. Croom Helm, London, pp. 23–40.

Jimenez, B. D., C. P. Cirmo & J. F. McCarthy, 1987. Effects of feeding and temperature on uptake, elimination and metabolism of benzo(a)pyrene in the bluegill sunfish (*Lepomis macrochirus*). Aquat. Toxicol. 10: 41–57.

Jobling, M., 1981. The influence of feeding on the metabolic rate of fishes: a short review. J. Fish. Biol. 18: 385–400.

Kumaraguru, A. K. & F. W. H. Beamish, 1986. Effect of permethrin (NRDC-143) on the bioenergetics of rainbow trout, *Salmo gairdneri*. Aquat. Toxicol. 9: 47–58.

Lanno, R. P., B. Hicks & J. W. Hilton, 1987. Histological observations on intrahepatocytic copper-containing granules in rainbow trout reared on diets containing elevated levels of copper. Aquat. Toxicol. 10: 251–263.

Lett, P. F., G. J. Farmer & F. W. H. Beamish, 1976. Effect of copper on some aspects of the bioenergetics of rainbow trout (*Salmo gairdneri*). J. Fish. Res. Bd. Can. 33: 1335–1342.

Marking, L. L., T. D. Bills & J. R. Crowther, 1984. Effects of five diets on sensitivity of rainbow trout to eleven chemicals. Prog. Fish-Cult. 46: 1–5.

Mehrle, P. M., W. W. Johnson & F. L. Mayer, 1974. Nutritional effects on chlordane toxicity in rainbow trout. Bull. Envir. Contam. Toxicol. 12: 513–517.

Mehrle, P. M., F. L. Mayer & W. W. Johnson, 1977. Diet quality in fish toxicology: effects on acute and chronic toxicity. In; F. L. Mayer & J. L. Hamelink. (eds) Aquatic Toxicology and Hazard Evaluation. American Society for Testing and Materials, Publ. No. ASTM STP 643, Philadelphia, PA, pp. 269–280.

Shubat, P. J. & L. R. Curtis, 1986. Ration and toxicant preexposure influence dieldrin accumulation by rainbow trout (*Salmo gairdneri*). Envir. Toxicol. Chem. 5: 69–77.

Sosnowski, S. L., D. J. Germond & J. H. Gentile, 1979. The effect of nutrition on the response of field populations of the calanoid copepod *Acartia tonsa* to copper. Wat. Res. 13: 449–452.

Sprague, J. B., 1973. The ABC's of pollutant bioassay using fish. In; J. Cairns, Jr. & K. L. Dickson. (eds) Biological methods for the assessment of water quality. American Society for Testing and Materials, Publ. No. ASTM STP 528, Philadelphia, PA, pp. 6–30.

Stott, W. T. & R. O. Sinnhuber, 1978. Dietary protein levels and aflatoxin B1 metabolism in rainbow trout (*Salmo gairdneri*). J. Envir. Pathol. Toxicol. 2: 379–388.

Taylor, M. T., 1985. Effect of diet on the sensitivity of *Daphnia magna* to linear alkylbenzene sulfonate. In; R. D. Cardwell, R. Purdy & R. C. Bahner. (eds) Aquatic toxicology and hazard assessment: Seventh symposium. American Society for Testing and Materials, Publ. No. ASTM STP 854, Philadelphia, PA, pp. 53–72.

Van Leeuwen, C. J., P. S. Griffioen, W. H. A. Vergouw & J. L. Mass-Diepeveen, 1985. Differences in susceptibility of early life stages of rainbow trout (*Salmo gairdneri*) to environmental pollutants. Aquat. Toxicol. 7: 59–78.

Watanabe, T., 1982. Lipid nutrition in fish. Comp. Biochem. Physiol. 73B: 3–15.

Watt, P. W., P. A. Marshall, S. P. Heap, P. T. Loughna & C. Goldspink, 1988. Protein synthesis in tissues of fed and starved carp, acclimated to different temperatures. Fish Physiol. Biochem. 4: 165–173.

White, A. & T. C. Fletcher, 1986. Serum cortisol, glucose and lipids in plaice (*Pleuronectes platessa* L.) exposed to starvation and aquarium stress. Comp. Biochem. Physiol. 84A: 649–653.

Hydrobiologia **188/189**: 533–542, 1989.
M. Munawar, G. Dixon, C. I. Mayfield, T. Reynoldson and M. H. Sadar (eds)
Environmental Bioassay Techniques and their Application.
© *1989 Kluwer Academic Publishers. Printed in Belgium.*

Hypothesis formulation and testing in aquatic bioassays: a deterministic model approach

L. S. McCarty, [1,2,3] G. W. Ozburn,[1] A. D. Smith,[1] A. Bharath,[1] D. Orr[1] & D. G. Dixon[2]
[1] *Aquatic Toxicity Research Group, Lakehead University, Thunder Bay, Ontario, Canada P7B 5E1;*
[2] *Department of Biology, University of Waterloo, Waterloo, Ontario, Canada N2L 3G1;* [3] *To whom correspondence may be addressed*

Key words: aquatic toxicity, bioconcentration, bioassay, modelling, narcotic organics, chlorobenzenes

Abstract

The paper examines the significance of toxicant kinetics information obtained from aquatic toxicity bioassays and bioconcentration tests. The data, bioconcentration kinetics and acute mortality versus exposure-duration information for juvenile American flagfish (*Jordanella floridae*) exposed to 1,4-dichlorobenzene, are interpreted in terms of a one-compartment, first-order kinetics model. The output of the model is used to formulate a testable hypothesis regarding the comparison of toxicant kinetics derived from both bioconcentration test exposures and toxicity bioassays. The model's estimates of the toxicant body burden attained at mortality are compared with theoretical and observed body burdens from literature sources. The use of a simple, deterministic residue-based, one-compartment, first-order kinetics model to evaluate existing data, as well as to formulate hypotheses to direct experimental designs, is examined.

Introduction

Aquatic toxicology has evolved largely as a descriptive discipline; investigations focus primarily on detailed descriptions of the conditions during, and the outcome of, bioassays (APHA, 1981). This stochastic approach correlates results with changes in one or more experimental parameters, generally without reference to underlying processes. Although the approach has provided some measure of success, it is ultimately handicapped by the lack of a general framework for hypothesis testing. Despite the fact that copious amounts of toxicity information have been collected on a variety of chemicals and organisms, both for scientific and regulatory purposes, the data are often of limited use in situations with different bioassay organisms or environmental conditions.

This limitation may well be related to the goal-oriented focus of standard bioassay methodologies, which emphasize specific applications rather than basic investigative science.

In order for the science of aquatic toxicology to advance, a deterministic or mechanistic approach must be adopted. This will facilitate both the correlation of the results with experimental parameters, as in stochastic modelling, and their explanation in terms of fundamental biological, physical, and chemical processes. Furthermore, the scientific method of explicit hypothesis formulation and testing will be more readily applicable to the consolidation and expansion of the knowledge base.

Some of the basic concepts of a deterministic approach are inherent in standard toxicity bioassays. However, with some notable exceptions, the concepts have not been exploited, nor the data interpreted, to any great extent (Spacie & Hamelink, 1982; Mancini, 1983; Chew & Hamilton, 1985; Connolly, 1985; Gobas & Mackay, 1987; Barber *et al.*, 1988). Bioconcentration bioassays on the other hand are fundamentally deterministic in nature; they focus on the kinetics of toxicant transfer between the environment and the exposed organism. Little work has been carried out to determine the relationships between bioconcentration and toxicity exposures and the degree of similarity of the data obtained from each. This paper will focus on examining the relationship between information obtained from toxicity and bioconcentration bioassays and determining the utility of any relationship which might be found.

The basis for our approach to this problem is previous work on quantitative structure-activity relationships (QSAR) (McCarty, 1987a, b) which suggested, at least for poorly metabolized, neutral, narcotic chemicals, that the kinetic information from both toxicity bioassays and bioconcentration bioassays is similar in nature. The differences that occur appear to be primarily a result of the difference between the biological end-points being measured: a relatively fixed body burden associated with mortality and a variable body burden associated with bioconcentration.

The general objectives of this paper are: to briefly review some of the basic concepts and assumptions involved in current aquatic bioconcentration and acute-toxicity bioassays; to compare and contrast the information which is obtained from these two basic bioassay types; to examine this information in terms of a general framework of toxicological processes; and to explore, with a specific data set, hypothesis formulation and testing based on a simple deterministic model which incorporates the concepts outlined in the previous objectives.

Background

Our first task is to define the term bioassay and to determine the exact status of bioconcentration and toxicity bioassays with reference to this definition. Finney (1978) defines a biological assay or bioassay as 'an experiment for estimating the nature, constitution, or potency of a material (or of a process) by means of the reaction that follows its application to living matter'. A typical bioassay involves a *stimulus* of a measured *dose* which is applied to a *subject* whose *response* to the stimulus is estimated as the change in some biological characteristic or state.

Finney (1978) indicated that acute, lethality bioassays can be categorized as indirect quantal bioassays. The end-point examined in bioconcentration bioassays (a non-quantal increase in the concentration of a compound in the body of an organism subsequent to exposure) is not usually considered to be a biological response in the same sense as a toxicity end-point. Nevertheless, changes in the body concentration of a compound are a response related to the nature of the compound and as such bioconcentration bioassays should be considered as direct, non-quantal (quantitative) bioassays. According to these definitions, both typical acute toxicity and bioconcentration bioassays appear to meet the criteria required to be classified as bioassays (Rand & Petrocelli, 1985). Although there are some additional differences to be examined, it is evident that these two basic aquatic protocols have some fundamental similarities which can be further exploited.

Bioconcentration and toxicity bioassays can both be placed into a general toxicological framework which will allow information to be compared and contrasted. Ariens (1980) subdivides the phenomena of toxicity into three broad components, the exposure phase, the toxokinetic phase, and the toxodynamic phase. The exposure phase describes the environmental bioavailability of a chemical toxicant to an exposed organism. The toxokinetic phase describes the uptake, distribution, metabolization, and elimination of the bioavailable portion of the chemical within the organism. The toxodynamic phase describes the time course of biological response resulting from the chemical reaching the site(s) of toxic action in the organism. The processes which are being explored in toxicity and bioconcentration bioassays will be examined with reference to these three phases of the toxicological process.

Categorizing the information obtained from bioassays is not as easy a task as it might first appear. Chemical and physical characteristics of the toxicant, such as hydrophobicity, molecular size, or molecular charge distribution, can dramatically affect kinetic rates (Gobas *et al.*, 1987). Differences in the biological characteristics of the bioassay organism, such as species, body size, sex, age, or metabolic rate, may also modify toxicant kinetics and hence toxicity (Sprague, 1969, 1970). For the purposes of this paper we will employ data where the influence of these factors has been minimized.

Toxicity bioassays

The objective of a toxicity bioassay is to determine the potency of a chemical relative to the potency of chemicals which are already known (Bliss, 1957; Filov *et al.*, 1979). On the surface this appears to be a primarily stochastic approach which focuses firmly on the toxodynamic phase of toxicity. Further examination of some of the basic assumptions made in designing toxicity bioassay protocols is, however, warranted.

The fundamental assumption of a toxicological event is the existence of a dose-response relationship; the biological response of an organism is some function of the amount of toxicant to which that organism has been subjected. In aquatic toxicology this is often referred to as the concentration-response relationship. (Rand & Petrocelli, 1985). In practical terms this means that the concentration of chemical external to the organism is used as a surrogate for the dose, the internal concentration of chemical at the site(s) of toxic action. In employing the concentration-response relationship in acute toxicity bioassays it is assumed that a steady-state equilibrium occurs, or at least could occur, between the external and internal concentrations of the chemical. At the 'threshold' concentration causing 'incipient' toxicity, when the system is most likely to be in equilibrium, the use of the external surrogate to

compare the potency of chemicals is assumed to be valid, since all modifying factors other than potency are equal (Sprague, 1969, 1970; Filov *et al.*, 1979; Cox, 1987). Although non-equilibrium circumstances can occur, bioassay protocols are designed to minimize the possibility.

The use of the concentration-response relationship also assumes that metabolic transformation of the chemicals being compared is either similar or negligible. It is theoretically possible to compensate for considerable differences in metabolism by correcting for the proportion of the external exposure concentration which is metabolized, and is hence unavailable to cause a toxic response. Unfortunately this is not usually done, since metabolic transformation rates are usually unavailable for the typically employed aquatic organisms.

A third assumption, that there is a normal distribution of the logarithm of tolerance (lognormal) in the exposed population, is commonly made to facilitate statistical analysis of toxicity bioassay data (Finney, 1978). Several types of tolerance distributions, including probit, logistic, angle, rectangular, and Weibull, are virtually indistinguishable over most (5% to 95%) of the range of toxicity (Finney, 1978; Christensen & Chen, 1985; Hanes & Wedel, 1985). Although it is an important assumption, knowledge of the exact nature of the tolerance distribution does not appear to be crucial to either the validity of the statistical procedures or the utility of the toxicokinetic information.

In summary, it is evident that although aquatic toxicity bioassays are employed primarily in a stochastic manner to compare tolerance estimates derived using a standard protocol, these protocols are essentially a method of minimizing influences on kinetics such that realistic estimates of relative toxicity (toxodynamics) can be obtained. Toxicity bioassays have very definite deterministic roots.

Bioconcentration bioassays and toxicokinetic modelling

Bioconcentration bioassays focus explicitly on the kinetic aspect of the toxicological process:

536

toxicant uptake, distribution, biotransformation, and elimination. Although a body burden or dose is usually estimated, it is of little use in a toxodynamic sense since exposure levels are deliberately chosen to produce no biological response in the test organisms at the achieved body burden. The bioavailability aspect of the toxicological process can be examined with bioconcentration bioassay protocols, but this has not been common. Some recent work has examined the effect of bioavailability on kinetics data, since uncertainty in the exposure phase can be a significant confounding factor in some bioconcentration bioassay protocols (Leversee *et al.*, 1983; McCarthy & Jimenez, 1985).

Since most bioconcentration bioassays are designed to be interpreted in terms of compartmentalized, kinetics-based models (Spacie & Hamelink, 1982) it is clear that they are inherently capable of being directly addressed with a dynamic model approach. Although a variety of models can be employed (Dedrick, 1986; Bischoff, 1986) the most common and simplest approach has been to use a one-compartment, first-order kinetics (1CFOK) model (Mancini, 1983; Spacie & Hamelink, 1982; Hawker & Connell, 1985; Niimi, 1987). Although uptake and elimination are often used in a non-specific whole-organism manner (the classic 'black box' approach), rate constants for several pathways, as well as for the impacts of growth dilution and metabolism, can be included (Fig. 1). While uptake from food can also be considered, it usually is not since the inter-

mittent nature of feeding, and hence of chemical uptake, makes it less amenable to first-order kinetics interpretation. The equations which are used to describe a 1CFOK model (Spacie & Hamelink, 1982; Mancini, 1983) are as follows:

Uptake $\quad Cf = [(Cw)\,(k1/k2)]\,e^{-(k2)(t)}$

Elimination $\quad Cf = (Co)\,[e^{-(k2)(t)}]$

where Cf is the concentration in the fish, Cw is the concentration in the water, Co is the concentration in the fish at the beginning of elimination, t is time, $k1$ is the uptake rate constant, $k2$ is the elimination rate constant, and e is the base of natural logarithms, 2.7183.

Although the simplification of a complex multicompartment system to a 1CFOK model has many advantages, there are also several disadvantages. Approximating a multi-compartment system with a single-compartment model is reasonable when the character of the compartments is similar. A complicating factor, especially for hydrophobic chemicals, is variation in lipid content both between compartments and between organisms. The obvious solution is to use a multi-compartment model. Since the basic biology needed to apply these more sophisticated techniques is unavailable in aquatic biology, two interim procedures are often employed. The first is to do comparisons only when body lipid contents are similar. The second is to report data corrected to a common lipid content.

Another problem is that the single uptake and elimination rate constants of the 1CFOK model are in fact composites of several factors which act simultaneously. For example, elimination may occur through both the gill and the gut, with the contribution of each to the total rate varying as a function temperature, body size, diet, and ration level. Similarly, although growth and metabolic detoxification may appear to be elimination in 1CFOK terms, the former acts through the production of additional tissue mass which effectively dilutes the body burden, while the latter acts by transformation of the toxicant to a less toxic metabolite.

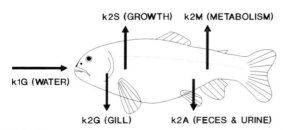

Fig. 1. Diagram of a one-compartment, first-order kinetic (1CFOK) model for the uptake and depuration of chemicals by fish. The rate constants for uptake from water by gills ($k1G$) as well as the constants for elimination by gill ($k2G$), feces and urine ($k2A$), metabolism ($k2M$), and growth dilution ($k2S$) are represented.

Until more biological details are available to properly address these modifying factors, the most useful modelling approaches appear to be those which choose organisms and chemicals for which a minimum number of processes are occurring simultaneously. It is possible to obtain from situations where the lipid contents of the test organisms are similar, where growth is minimal, where toxicant uptake from food is minimized, and where the test chemical is known to be poorly metabolized.

Body burden and biological response

Ferguson (1939) hypothesized that it is the activity of the number of molecules of a toxicant in an organism and not their actual number which is most closely related to the biological response. The concentration of a toxicant in the body of an exposed organism is therefore an approximation for the activity of the effective dose. The relationship between concentration and activity is similar to that encountered with hydrogen ion concentration and pH; a fixed number of molecules will yield different measures of activity with changes in the temperature and the ionic strength of a solution. The whole-body internal concentration of a chemical is a measure of the toxic dose more closely associated with the biological response than the external environmental concentration. It is, however, a less accurate estimate than the concentration of the toxicant bound to the receptors at the site(s) of toxic action. Despite being a much more accurate measure, the latter is considerably more difficult to estimate since the exact site(s) and mechanism(s) of toxic action are usually poorly understood. Thus the whole-body concentration, incorporating adjustments for known modifying factors, will be taken as a reasonable approximation of the internal toxic dose.

In the typical toxicity bioassay the body burden is essentially being measured indirectly by the mortality of the exposed population. This indirect method of estimating body burden will introduce error into the estimated value since, among other things, the variability associated with the toler-ance distribution of the exposed population will be included. Since body burdens are actually measured in bioconcentration bioassays, the estimates should be more accurate. For many commonly studied organic chemicals the ratio of exposure concentration to body burden is relatively constant over a range of exposure concentrations. The ratio follows the well-established positive relationship between body burden at steady-state equilibrium and the lipophilicity of the bioassay chemical (Mackay, 1982).

In toxicity bioassays with chemicals having a common mode of toxic action, using organisms of the same species and condition, the molar body burden of the toxicant should be relatively constant (Connolly, 1985). Support for this comes from QSAR work. McCarty (1987a), working with toxicity and bioconcentration data, reported that for the acute toxicity of a group of unrelated neutral, narcotic, chlorinated hydrocarbons, the body burden associated with estimates of the 96 h LC_{50} (median lethal concentration) appeared to be approximately 2 mmol l^{-1} for fish of 5% lipid content. For a group of non-halogenated alcohols and ketones the body-burden estimate was approximately 5.5 mmol l^{-1}.

Although fish have an overall density essentially equal to that of water, the average tissue density is about 1.06 to 1.09 due to the presence of the gas bladder (Bond, 1979). As such, body-burden estimates based on a mean of assayed tissue levels would be somewhat higher than estimates based on a whole-body measurement including the gas bladder. Never-the-less, we will use whole-body concentrations expressed in mmol l^{-1} and mmol kg^{-1} interchangeably. Van Hoogen & Opperhuizen (1988) measured body burdens of chlorobenzenes at the 96 h LC50 and found a body concentration of 2.5 mmol kg^{-1} for fish containing 5% lipid, which is in good agreement with the predicted value. Hodson et al. (1988), using an intraperitoneal injection technique with rainbow trout, found an LD50 (median lethal dose) for 1,4-dichlorobenzene of 10 mmol kg^{-1}. Given the substantial difference in the toxicological technique, as well as the species difference, the value is in reasonable agreement with the LC50 data.

538

From the above discussion it appears clear that bioconcentration and toxicity bioassays are alike. Both provide information on the kinetics phase of the toxicological process, bioconcentration bioassays by directly estimating rate constants and body burdens, and toxicity bioassays by using mortality-time information to indirectly estimate rate constants. This interrelationship has been commented upon previously (McCarty *et al.*, 1985; McCarty, 1986). We will examine some actual experimental results to see if they support this contention and comment upon the consequences.

Methods and results

The data set used in this discussion consists of the results from a bioconcentration bioassay and a 96-h toxicity bioassay with American flagfish (*Jordanella floridae*) exposed to the neutral organic chemical 1,4-dichlorobenzene (1,4-DCB) (Smith *et al.* manuscript; ATRG, 1987). Although the two experiments were carried out independently, they were conducted in the same laboratory with the same fish stock. Since whole-body concentrations of 1,4-DCB were not obtained during the original toxicity work, predicted and observed estimates for similar circumstances were obtained from the literature.

The general approach to examining the similarity of toxicity- and bioconcentration-derived kinetics information is relatively simple. The kinetics information obtained directly in bioconcentration bioassays is compared to kinetics information obtained indirectly in toxicity bioassays. The non-linear curve-fitting process to obtain this information from the time-toxicity relationships obtained in toxicity bioassays has been discussed (Zitko, 1979; Mancini, 1983; McCarty, 1987b). It should be noted that we have made the following assumptions in our analysis: that uptake and elimination can be reasonably described by a 1CFOK model independent of the toxic response; that the toxic response (mortality) occurs at a relatively constant body burden of the toxicant, and that there is no change in bioavailability of the toxicant,

no uptake of toxicant from the diet, negligible biodegradation, and no growth of the fish during the exposure.

Toxicity and bioconcentration data

The toxicity data for 1,4-DCB were obtained in continuous-flow bioassays with 2- to 4-month old juvenile American flagfish with an approximate mean weight of 2 g (range, 0.3 to 5 g) and an approximate mean lipid content of 10% (range, 7 to 16%) (Smith *et al.*, manuscript). The estimated 96 h LC_{50} was 2.05 mg l^{-1} (0.014 mmol^{-1}; 95% C.I. 0.0133–0.0147). Acetone, at concentrations from 79 to 196 mg l^{-1} (less than 0.02 of the 96 h LC50 for acetone) was used as a solubilizing agent for the 1,4-DCB.

The time-toxicity data were used to obtain estimates of k2, the elimination rate constant, and the factor (1/Cf) (k1/k2), the product of the inverse of the fish's whole-body toxicant concentration and the ratio of the uptake rate constant to the elimination rate constant (Fig. 2). The nonlinear routine from the PC-based statistics program SYSTAT 4.0 (Leland, 1988) was employed to carry out the curve-fitting. Since (k1/k2) is equivalent to the bioconcentration factor (Kb), (1Cf) (k1/k2) was solved (McCarty, 1987a, b) by substituting a value for k1/k2 obtained from the octanol-water partition coefficient (Kow) versus

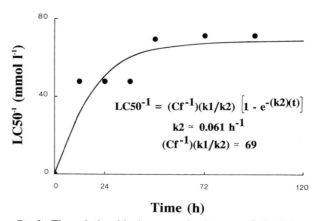

Fig. 2. The relationship between the inverse of the LC_{50} water toxicant concentration and time for flagfish exposed to 1,4-DCB. The relationship was used to obtain estimates of (Cf^{-1}) $k1/k2$) and $k2$.

Kb relationship of Mackay (1982). As the fish in this study had about twice the 5% lipid level of those on which Mackay based his relationship (McCarty, 1987a), the Kb values were adjusted upward accordingly. Hence it is possible to completely solve the equation and obtain estimates of $k1$, the uptake rate constant, and Cf, the whole-body toxicant concentration.

The bioconcentration data were obtained with fish of the same characteristics employed in the toxicity bioassay. The fish were exposed for 28 days to a mean concentration of 2.7 μg 1,4-DCB l^{-1} (0.000018 mmol l^{-1}; S.D. 0.0000016) in 79 mg acetone l^{-1} followed by a 14-day depuration period (ATRG, 1987). The exposure concentrations and body-burden estimates at different times were analyzed with the BIOFAC program, a nonlinear curve-fitting routine customized for use in analysis of bioconcentration data (Blau & Agin, 1978), to generate kinetics constants and a bioconcentration factor based on whole-body toxicant level (Fig. 3).

Hypothesis

Our working hypothesis is that kinetics are independent of the biological response being investigated. It follows that, allowing for variance due to experimental factors and organism sensitivity, the kinetics from a bioconcentration bioassay should be similar to those obtained in a toxicity bioassay. Thus we have taken the bioconcentration-based kinetics information from Fig. 3 (k1, k2), and modelled, via a 1CFOK model, the time course of toxicant concentration in the body which should occur when flagfish are exposed to the 96 h LC50 for 1,4-DCB, 0.014 mmol l^{-1} (Fig. 4). Also indicated is the approximate theoretically-determined body burden associated with the 96 h LC50, 2 mmol l^{-1} or 2 mmol kg^{-1}, assuming a fish density of 1.0 (McCarty, 1987a). Although the body-burden estimate associated with this endpoint would be more realistically represented by a range or confidence interval about the mean estimate, further refinement is currently being carried out and this value must suffice for the moment. The body-burden estimate is probably accurate to within plus or minus one-half an order of magnitude.

If the hypothesis is true, within the bounds of the assumptions we have made, toxicity-derived kinetics information, when modelled in the same way, should demonstrate two things. Firstly, the kinetic constants estimated should be of a similar value. Secondly, the body burden predicted

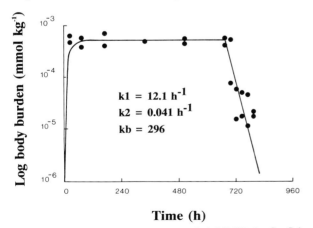

Fig. 3. The tissue concentrations of 1,4-DCB in flagfish exposed to 0.000018 mmol l^{-1} 1,4-DCB for 28 d, followed by a 14-d depuration. The relationship was analyzed (BIOFAC, Blau & Agin, 1978) to give estimates of $k1$, $k2$, and Kb.

In Fig. 3:
$$k1 = 12.1 \ h^{-1}$$
$$k2 = 0.041 \ h^{-1}$$
$$kb = 296$$

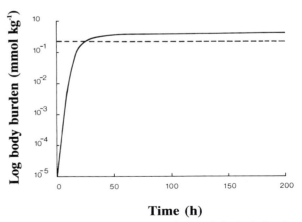

Fig. 4. The relationship between time and the body burden of 1,4-DCB in flagfish as predicted by a 1CFOK model using the kinetic information from Figure 3 ($k1$, 12.1 h^{-1}; $k2$, 0.041 h^{-1}) and an exposure concentration of 0.014 mmol l^{-1} (the 96 h LC$_{50}$). The dashed line is the theoretically determined body burden associated with the 96 h LC$_{50}$ (McCarty, 1987a).

540

should be similar to that associated with the toxic biological response in question, mortality.

Testing the hypothesis

In order to test the hypothesis, the outputs of a 1CFOK for an exposure to 0.014 mmol 1,4-DCB l^{-1} (the 96 h LC50 for flagfish) were obtained using different sets of parameters identified as BCF, observed-TOX, and calculated-TOX (Fig. 5). The BCF output, the hypothesis, is identical to the model in Fig. 4. The observed-TOX output employed both bioconcentration- and toxicity-based information. The elimination constant, k2, was derived from the non-linear curve fitting of the time-toxicity data (Fig. 2) while the BCF value was obtained from the bioconcentration results (Fig. 3). The uptake constant, k1, was derived from the toxicity-based factor (1/Cf) (k1/k2) using these two values.

The calculated-TOX output used information entirely unrelated to the bioconcentration-derived data used in the BCF output. The k2 estimate was derived from the toxicity data (Fig. 2) while the BCF parameter was calculated from Mackay's (1982) relationship for bioconcentration and Kow. A log Kow estimate for 1,4-DCB of 3.52, a

measured estimate from the MED CHEM 3.53 database (Leo & Weininger, 1988) was employed and the constant in the Mackay's (1982) equation was changed from 0.05 to 0.10 to reflect the higher lipid content of the flagfish used in this study. From inspection of Fig. 5 it can be seen that all three of the curves plotted are in very close proximity to each other. The estimated whole-body toxicant concentrations of 4.1 and 4.6 mmol l^{-1} compare favourably with the previously reported (McCarty, 1986, 1987a; Van Hoogen & Opperhuizen, 1988) estimates of toxicant body burden for acute toxicity, 2 and 2.5 mmol l^{-1}.

Discussion

Although this is not an exhaustive evaluation of the situation, it is apparent that there is reasonable agreement between the model-estimated toxicant body burden and an independent estimate of body burden for the same toxicological endpoint. A one-compartment, first-order kinetics model appears to be the deterministic model which, at least in the first approximation, is adequate for the study of toxicant kinetics in aquatic bioassays measuring both acute toxicity and bioconcentration. Since little attempt has been made to compensate for the impact of numerous factors which can modify the uptake and depuration of chemicals by fish, the observations are currently limited to the circumstances examined. However, since this deterministic model has been shown to have some validity, further study and refinement would appear to be warranted.

A particular advantage of this approach is that explicit hypothesis formulation and testing can be exploited. Several interesting questions could be addressed in this way. Firstly, currently available bioassay results could be better interpreted and interpolated. An example might be a situation where bioconcentration-based kinetics information and 96 h LC50 data was available for several related chemicals. The 1CFOK model could be used to produce hypothesized body-toxicant concentrations for increasing exposure durations at the respective LC50 concentrations in water. The

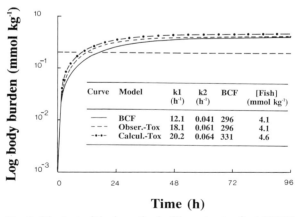

Curve	Model	k1 (h⁻¹)	k2 (h⁻¹)	BCF	[Fish] (mmol kg⁻¹)
——	BCF	12.1	0.041	296	4.1
- - -	Obser.-Tox	18.1	0.061	296	4.1
-··-	Calcul.-Tox	20.2	0.064	331	4.6

Fig. 5. The test of the hypothesis. The outputs of a 1CFOK model for a 96 h exposure of flagfish to 0.014 mmol 1,4-DCB l^{-1} (the 96 h LC_{50}) using three different sets of parameters. See the text for a full explanation. The horizontal dashed line is the theoretically determined body burden associated with the 96 h LC_{50} (McCarty, 1987a).

change in the total body burden with time for the mixture could also be obtained by summing the contribution of each chemical at each sampling time. Depending on the information available, the analysis could be expanded. Comparisons between bioconcentration-based and toxicity-based rate constants could be made. Toxicant body-burden estimates could be compared and contrasted both within the data set between chemicals and external to the data set using literature values. Inferences regarding the mode(s) of action of the various chemicals and mixtures, as well as the influence of some chemical and biological modifying factors, could be made. The information obtained could be used to design experiments to determine if kinetics rate constants, body-toxicant burdens, and biological responses occur as hypothesized for various exposure concentrations and durations, both for single chemicals and mixtures.

A second broad area of application for the concepts outlined here involves interpretation of the toxicological significance of the contaminant body burdens present in tissues of fish sampled from natural populations. Although the levels of chemicals in fish from contaminated environments are often reported, it is usually impossible to relate those levels to a toxicological response by the fish. One possible way of evaluating the degree of impact associated with these body burdens is by comparing them to the levels attained by fish during controlled laboratory exposures where toxicological end-points can be quantified. The use of kinetic body-burden modelling could facilitate this approach.

The final area of potential application involves interpretation of the impacts of pulse (intermittent) toxicant exposures (Hodson et al., 1983; Holdway & Dixon, 1985). Contaminant water quality standards are typically set as a mean aqueous concentration not to be exceeded, and are usually derived from laboratory tests where organisms are continuously exposed to fixed concentrations of the toxicant. Contaminant concentrations in aquatic ecosystems are rarely constant due to varying rates of input, dilution, and degradation. The most promising approach to under-standing and dealing with pulse and fluctuating exposure toxicity is through the use of bioconcentration models, since ultimately the level of toxicant in the organism best defines the level of exposure for non-equilibrium conditions.

Within the limitations specified, we trust that we have demonstrated both how toxicity and bioconcentration bioassay data are related and how this information might be exploited within the confines of a deterministic model based on the concepts of one-compartment, first-order kinetics.

Acknowledgements

The authors would like to acknowledge the financial support of the Ontario Ministry of the Environment. We are also grateful for the input of the reviewers, Dr. P. Hodson and Dr. A. Opperhuizen.

References

Apha, 1981. Standard Methods for the Examination of Water and Wastewater, 15th ed. American Public Health Association, Washington, DC. 1334 pp.

Ariens, E. J., 1980. Design of safer chemicals, pp. 1–44. In E. J. Ariens (ed.), Drug Design, Vol. 9. Academic Press, New York, NY.

ATRG, 1987. Aquatic Toxicity Studies of Multiple Organic Compounds. Part 1. Chlorinated Benzenes and Chlorinated Phenols. A Report to the Ontario Ministry of the Environment and Environment Canada. Aquatic Toxicity Research Group, Lakehead University, Thunder Bay, Ontario. 277 pp.

Barber, M. C., L. A. Suarez & R. R. Lassiter, 1988. Modelling bioconcentration of nonpolar organic pollutants by fish. Envir. Toxicol. Chem. 7: 545–558.

Bischoff, K. B., 1986. Physiological pharmacokinetics. Bull. Math. Biol. 48: 309–322.

Blau, G. E. & G. L. Agin, 1978. A users manual for BIOFAC: A computer program for characterizing the rates of uptake and clearance of chemicals in aquatic organisms. Dow Chemical Co., Midland, MI. 10 pp.

Bliss, C. I., 1957. Some principles of bioassay. Am. Sci. 45: 449–466.

Bond, C. F., 1979. Biology of Fishes. Saunders College Publishing, Philadelphia, PA. 514 pp.

Chew, R. D. & M. A. Hamilton, 1985. Toxicity curve estimation: fitting a compartment model to median survival times. Trans. Am. Fish. Soc. 114: 403–412.

Christensen, E. R. & C. Y. Chen, 1985. A general noninteractive multiple toxicity model including probit, logit, and Weibull transformations. Biometrics 41: 711–725.

Connolly, J. P., 1985. Predicting single-species toxicity in natural water systems. Envir. Toxicol. Chem. 4: 573–582.

Cox, C., 1987. Threshold dose-response models in toxicology. Biometrics 43: 511–523.

Dedrick, R. L., 1986. Interspecies scalling of regional drug delivery. J. Pharm. Sci. 75: 1047–1052.

Ferguson, J., 1939. The use of chemical potentials as indices of toxicity. Proc. Royal Soc. B 127: 387–404.

Filov, V., A. Golubev, E. Liublina & N. Tolokontsev, 1979. Quantitative Toxicology: Selected Topics. John Wiley & Sons, New York, NY. 462 pp.

Finney, D. J., 1987. Statistical Method in Biological Assay, 3rd ed. C. Griffin & Co. Ltd., London, Eng. 508 pp.

Gobas, F. A. P. C. & D. Mackay, 1987. Dynamics of hydrophobic organic chemical bioconcentration in fish. Envir. Toxicol. Chem. 6: 495–504.

Gobas, F. A. P. C., W. Y. Shiu & D. Mackay, 1987. Factors determining partitioning of hydrophobic organic chemicals in aquatic organisms. pp. 107–123. In; K. L. E. Kaiser (ed.), QSAR in Environmental Toxicology-II. D. Reidel Publishing Company, Dordrecht, Holland.

Hanes, B. & T. Wedel, 1985. A selected review of risk models: one hit, multihit, multistage, probit, Weibull, and pharmacokinetic. J. Am. Coll. Toxicol. 4: 271–278.

Hawker, D. W. & D. W. Connell, 1985. Relationships between partition coefficients, uptake rate constants, and the time to equilibrium for bioaccumulation. Chemosphere 14: 1205–1219.

Hodson, P. V., B. R. Blunt, U. Borgmann, C. K. Minns & S. McGaw, 1983. Effect of fluctuating lead exposures on lead accumulation by rainbow trout (*Salmo gairdneri*). Envir. Toxicol. Chem. 2: 225–238.

Hodson, P. V., D. G. Dixon & K. L. E. Kaiser, 1988. Estimating the acute toxicity of waterborne chemicals in trout from measurements of median lethal dose and the octanol-water partition coefficient. Envir. Toxicol. Chem. 7: 443–454.

Holdway, D. A. & D. G. Dixon, 1985. Acute toxicity of pulse-dosed methoxychlor to juvenile American flagfish, *Jordanella floridae* Goode and Bean as modified by age and food availability. Aquat. Tox. 6: 243–250.

Leland, L., 1988. SYSTAT: The System for Statistics. SYSTAT Inc., Evanston, IL. 822 pp.

Leo, A. & D. Weininger, 1988. MED CHEM Database, Version 3.53. Pomona Medicinal Chemistry Project, Pomona College, Claremont, CA.

Leversee, G. J., P. F. Landrum, J. P. Giesy & T. Fannin, 1983. Humic acids reduce bioaccumulation of some polycyclic aromatic hydrocarbons. Can. J. Fish. Aquat. Sci. 40: 63–69.

Mackay, D., 1982. Correlation of bioconcentration factors. Envir. Sci. Technol. 16: 274–278.

Mancini, J. L., 1983. A method for calculating the effects, on aquatic organisms, of time varying concentrations. Wat. Res. 17: 1355–1362.

McCarthy, J. F. & B. D. Jimenez, 1985. Reduction in bioavailability to bluegills of polycyclic aromatic hydrocarbons bound to dissolved humic material. Envir. Toxicol. Chem. 4: 511–521.

McCarty, L. S., P. V. Hodson, G. R. Craig & K. L. E. Kaiser, 1985. On the use of quantitative structure/activity relationships to predict the acute and chronic toxicity of chemicals to fish. Envir. Toxicol. Chem. 4: 595–606.

McCarty, L. S., 1986. The relationship between aquatic toxicity QSARs and bioconcentration for some organic chemicals. Envir. Toxicol. Chem. 5: 1071–1080.

McCarty, L. S., 1987a. Relationship between toxicity and bioconcentration for some organic chemicals. I. Examination of the relationship. In; K. L. E. Kaiser (ed.), QSAR in Environmental Toxicology-II. pp. 207–220. D. Reidel Publishing Company, Dordrecht, Holland.

McCarty, L. S., 1987b. Relationship between toxicity and bioconcentration for some organic chemicals. II. Application of the relationship. In; K. L. E. Kaiser (ed.), QSAR in Environmental Toxicology-II. pp. 221–230. D. Reidel Publishing Company, Dordrecht, Holland.

Niimi, A. J., 1987. Biological half-lives of chemicals in fishes. Rev. Envir. Contam. Toxicol. 99: 1–46.

Rand, G. M. & S. R. Petrocelli, 1985. Fundamentals of Aquatic Toxicology. Hemisphere Publishing Co., Washington, DC. 666 pp.

Smith, A. D., A. Bharath, C. Mallard, D. Orr, J. A. Sutton, J. Vukmanich & G. W. Ozburn, 1989. The acute toxicity of 10 chlorinated organic compounds to the American flagfish *Jordanella floridae* (Goode and Bean). Aquat. Toxicol. Manuscript submitted.

Spacie, A. & J. Hamelink, 1982. Alternative models for describing the bioconcentration of organics in fish. Envir. Toxicol. Chem. 1: 309–320.

Sprague, J. B., 1969. Measurement of pollutant toxicity to fish I. Bioassay methods for acute toxicity. Wat. Res.: 793–821.

Sprague, J. B., 1970. Measurement of pollutant toxicity to fish II. Utilizing and applying bioassay results. Wat. Res. 4: 3–32.

Van Hoogen, G. & A. Opperhuizen, 1988. Toxicokinetics of chlorobenzenes in fish. Envir. Toxicol. Chem. 7: 213–219.

Zitko, V., 1979. An equation of lethality curves in tests with aquatic fauna. Chemosphere 2: 47–51.

Hydrobiologia **188/189**: 543–560, 1989.
M. Munawar, G. Dixon, C. I. Mayfield, T. Reynoldson and M. H. Sadar (eds)
Environmental Bioassay Techniques and their Application.
© 1989 *Kluwer Academic Publishers. Printed in Belgium.*

In situ and laboratory studies on the behaviour and survival of Pacific salmon (genus *Oncorhynchus*)

Ian K. Birtwell & George M. Kruzynski
Department of Fisheries and Oceans, Biological Sciences Branch, West Vancouver Laboratory, West Vancouver, British Columbia, V7V 1N6, Canada

Key words: Pacific salmon, behaviour, estuaries, pulp mill effluent, hypoxic conditions

Abstract

Juvenile Pacific salmon display a marked surface water orientation during downstream migration, estuarine and nearshore coastal rearing phases. Many estuaries in British Columbia are vertically stratified with a shallow, well-defined halocline which can restrict the dispersion of wastes discharged into less saline surface waters and impose constraints upon aquatic organisms.

In situ experiments in an estuary receiving a surface discharge of treated pulp mill wastes, revealed conditions which were lethal to underyearling salmon at, and below the halocline (4.0–6.5 m depth). Behavioural bioassays determined that juvenile chinook salmon were biased towards the water surface and avoided waters at depth. Dissolved oxygen was the variable which affected this distribution most significantly. Surface waters receiving effluent from another pulp mill were lethal to juvenile salmon within 350 m, and a significant vertical avoidance response occurred within 350–950 m of the outfalls. The behavioural response was significantly correlated with *in situ* temperature, pH and colour (effluent).

As a complement to field experiments we developed a 4500 l water column simulator (WCS) to examine salmon behaviour in the laboratory. We investigated the surface water orientation behaviour of juvenile salmon in relation to variations in salinity and dissolved oxygen. Under simulated vertically stratified estuarine conditions, the fish moved freely between overlying fresh water and salt water. Induction of hypoxic conditions in fresh water elicited a downward distribution shift towards the halocline and oxygenated, but more saline, waters. Avoidance reactions (50% level) occurred consistently up to 7–8 mg·l^{-1} dissolved oxygen. Salmon continued to examine the hypoxic freshwater zone despite sub-optimal conditions.

Introduction

The continuing intrusion of man's activities into estuaries has been a major stimulus for the progression of research on the ecology of juvenile Pacific salmon (genus *Oncorhynchus*). This, in turn, has emphasized the importance of understanding how salmon utilize these transition zones so that sound fisheries and environmental management may be accomplished. There exists, however, relatively little information on the behaviour of salmon in estuaries, despite its inherent linkage to survival (Mace, 1983; Piercey *et al.*, 1985; Macdonald *et al.*, 1987).

In British Columbia, many estuaries are vertically stratified, with a well-developed halocline close to the surface. Wastes discharged into rivers or the fresh-water lens of vertically stratified estuaries may be confined there due to prevailing density gradients and thereby impose constraints

544

upon aquatic organisms such as juvenile salmon which, depending upon species and life history stage, will reside in estuaries for hours to months (Levy & Northcote, 1982). During this time these fish display a marked surface water orientation, an innate behavioural trait which we have examined in relation to the discharge of pollutants (Birtwell, 1977; Birtwell & Harbo, 1980; Birtwell *et al.*, 1983a; McGreer & Vigers, 1983), and to simulated estuarine conditions and hypoxic waters (Birtwell & Kruzynski, 1987).

The intent of this paper is to demonstrate the application of complementary bioassay techniques to investigate the behaviour of Pacific salmon in waters receiving pulp mill effluent, and their responses to hypoxia under controlled laboratory conditions within a Water Column Simulator (WCS).

Methods and materials

Field experiments

The technique employed to investigate the *in situ* behaviour of juvenile salmon has been described by Birtwell (1977), and only brief comment will be made here. Figure 1 shows the main components of the floating cage. An aluminum framework, $0.5 \times 0.5 \times 6.0$ m is intersected (at 1 m intervals) along its length by a system of perforated aluminum gates. These may be opened or closed by a lever located adjacent to styrofoam floats which hold the cage vertically in the water column. Typically ten experimental fish (hatchery or wild stock, acclimated to the unpolluted environmental conditions) are placed into each of the six compartments while the apparatus is suspended horizontally at the water surface. The gates between adjacent compartments, closed at this time, are opened once the apparatus is positioned to float vertically in the water column. In this way, the experimental fish are given access to the enclosed water column.

Concurrent determinations of selected aquatic characteristics (temperature, salinity, dissolved oxygen, pH, oxidation-reduction potential,

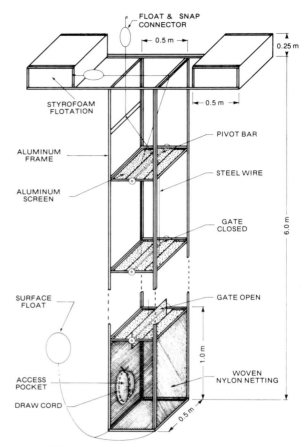

Fig. 1. Diagram of the apparatus used to examine the *in situ* vertical distribution of juvenile fish (after Birtwell, 1977).

colour) are made at specific depths adjacent to each compartment to provide vertical profiles at time intervals dictated by the requirements of the study. Upon termination of the experiment the gates are closed and the cage raised to the horizontal position so that the fish in each compartment can be examined. Using this technique, the vertical distributions of fish at 1 m depth intervals were examined at a number of polluted and unpolluted locations in relation to prevailing aquatic conditions. Replicate and coincident experimentation was possible through the use of up to 8 identical cage units.

We investigated the rapidity with which the experimental fish would move from their placement level in the cages to that which they may 'prefer'. Using marked and unmarked juvenile chum (*O. keta*) and coho (*O. kisutch*) salmon we recorded the responses of fish at a reference and a polluted (with pulp mill effluent) site, at 0.25, 0.5, 1.0, 2.0, 3.0, 6.0, 12.0 and 24.0 h. Although the fish would respond within as little as 0.25 h we preferred to conduct most investigations over 3 h for logistical purposes, and to permit at least some acclimation of the fish to their new environment. A similar, but independent assessment was also conducted by McGreer & Vigers (1983).

During studies of 24 h duration, we compared the survival and distribution of juvenile salmon at a reference site and at a polluted site (Birtwell & Harbo, 1980). Using the experimental cages as 6 discrete 1.0 m compartments (the gates remained closed) we held fish at specific depth intervals. Coincidentally, fish in other cages at the same locations were given access to the water column (gates opened) and the behavioural experiment was conducted in the usual manner.

We assessed the distribution and survival of previously dye-tagged (fluorescent grit) fish at locations near pulp mill outfalls. The movement of specifically-coloured fish between compartments and the significance of the initial placement location in relation to any subsequent mortality were examined. Because lethal conditions often existed in surface waters due to the tendency for effluent to concentrate there, we wished to determine if any of the fish initially placed higher in the water column died more rapidly than those placed in deeper compartments of the apparatus. The relevance of our approach relates to the innate behavioural traits of juvenile salmon which, at times, appear to encourage occupancy of suboptimal habitats; a topic which will be dealt with in more detail later.

Data analysis

A consistent data analysis format was employed (Birtwell, 1977, 1978; Birtwell & Harbo, 1980; McGreer *et al.*, 1982; McGreer & Vigers, 1983). Correlation analysis, Analysis of Variance, Duncan's Multiple Range Test, Student Newman-Keul's Test and Stepwise Multiple Re-gression analysis were used as required on transformed (log $\times$ + 1) or untransformed data to determine and define relationships between variables, and to separate homogeneous data sets.

Location of in situ experiments

Using the experimental cages, the behaviour of juvenile salmon was examined at 3 locations along the coast of British Columbia (Fig. 2). Concern about the adequacy of water quality in these areas to maintain populations of salmonids stimulated a number of studies; the *in situ* experiments reported here were components of these investigations. Effluent from the Port Alberni, Port Alice and Port Mellon pulp mills is discharged into Alberni and Neroutsos Inlets and Howe Sound, respectively. These inlets range in length from 25 to 40 km and are fed by rivers which give rise to vertically stratified estuaries, with a well-defined halocline close to the water surface.

Port Alberni

This pulp mill produces about 1000 ADT (air dried tonnes)$\cdot$d^{-1} and discharges 150×10^3 m$^3\cdot$d^{-1} of clarified and biologically-treated effluent from the production of newsprint, kraft paper, bleached and unbleached pulp: the effluent is of 'relatively' low acute toxicity to rainbow trout (96 h LC$_{50}$ > 80% v/v; Department of Environment, unpublished information). The effluent flows directly into the freshwater lens of the Somass River estuary, at the head of Alberni Inlet. Although dissolved oxygen levels are slightly depressed in the saline waters below the halocline due to natural phenomena, they were higher before the pulp mill commenced operations (Tully, 1949; Tully *et al.*, 1957). The hypoxic condition below the halocline is considered to be exacerbated by pulp mill effluent in the freshwater lens. Stain from the effluent effectively reduces photosynthetic activity (Parker & Sibert, 1973; Sibert & Parker, 1973), and hypoxic conditions persist in the deeper waters (often be-

546

Fig. 2. Location of three pulp mills along the coast of British Columbia where *in situ* experimentation was undertaken on the vertical distribution of juvenile salmon.

low 3 m depth). The levels of dissolved oxygen in these waters are at lethal or stressful concentrations to salmonids (Birtwell, 1978; Birtwell & Harbo, 1980). Replicate 3 h experiments were carried out with juvenile chinook (*O. tshawytscha*) salmon within cages located at 0.2, 0.7 and 1.4 km from the pulp mill outfall, along the seaward effluent dispersal path which follows the eastern shoreline of Alberni Inlet.

Port Mellon

This pulp mill produces bleached, unbleached and semi-bleached kraft (600 $ADT \cdot d^{-1}$) and $70 \times 10^3 \, m^3 \cdot d^{-1}$ effluent with a 96 h LC_{50} value $< 30\%$ v/v to rainbow trout (Department of Environment, unpublished information). At the time of our studies, the effluent was discharged without biological treatment into the brackish surface waters of Howe Sound. Experimental cages were placed at 80, 350, 950 and 1750 m from the outfalls; sites 1, 2, 3 and 4 respectively. Sites 1 and 3 were south of the outfalls, along the seaward dispersal path of the effluent, site 2 was along the same shoreline, but to the north. Site 4 acted as

a reference site, to the east, away from the typical effluent dispersion route.

Port Alice

This is a sulphite mill with a production of approximately 430 $ADT \cdot d^{-1}$. At the time of the study, effluent was discharged into the surface waters of Neroutsos Inlet at a rate of about $120 \times 10^3 \, m^3 \cdot d^{-1}$; the effluent is typically acutely toxic to rainbow trout (96 h $LC_{50} < 30\%$ v/v; Department of Environment, unpublished information). Experimental cages were located at 6 sites along the inlet encompassing a wide range of aquatic conditions. A reference site was located approximately 17 km seaward from the pulp mill (McGreer & Vigers, 1983).

Laboratory experiments

To investigate the behaviour of juvenile salmon under controlled conditions we developed a large experimental aquarium (Water Column Simulator – WCS), which is shown in Fig. 3 (from Birtwell & Kruzynski, 1987). The apparatus provides a water column ($2.4 \times 2.4 \times 0.8$ m) which can be vertically stratified, and in which dissolved oxygen, salinity, water velocity and temperature can be varied and controlled under continuous flow or recirculating conditions. A major feature of the WCS is the ability to maintain a well-defined halocline between adjacent and horizontally-flowing water masses allowing us, for example, to mimic a vertically stratified estuarine situation. Since the publication of the description of the WCS (Birtwell & Kruzynski, 1987) we have incorporated a control system which can modulate the temperature of incoming water ($\pm 0.3 \, °C$) or chill it continuously by as much as $2.5 \, °C$. This has facilitated year-round use of the equipment and effectively avoided a potential total gas pressure problem which was a concern associated with the seasonal use of the WCS prior to the installation of the temperature control system.

Lighting for behavioural experiments is pro-

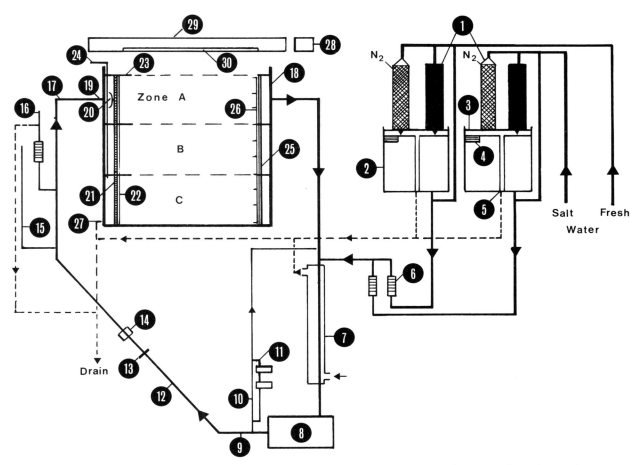

Fig. 3. Schematic showing the main components of the Water Column Simulator: 1. aeration-deaeration columns; 2. constant head reservoirs; 3. perforated lid; 4. float switch; 5. drain; 6. water flow regulators; 7. heat exchanger; 8. water pump; 9. vertical pipe; 10. bypass-recirculation loop; 11. filters; 12. main water supply line to aquarium; 13. orifice plate; 14. main control valve; 15. vertical standpipe; 16. water level control standpipe; 17. sensor location; 18. aquarium (lateral view); 19. entry location of main water supply; 20. water dispersion cone; 21. water dispersion plate; 22. nylon screen, entry side; 23. acrylic horizontal wings; 24. capillary air bleeds; 25. nylon screen, exit side; 26. manifold system with sampling ports; 27. drain valve; 28. metal halide light source with photoperiod control; 29. Light Pipe™; 30. red fluorescent lights. Water circulation loops to the middle (B) and bottom (C) zones of the aquarium are omitted for clarity (from Birtwell & Kruzynski, 1987).

vided by a metal halide light source, fitted with a shutter mechanism which can both control photoperiod and simulate dawn and dusk. The location of fish within the WCS aquarium is monitored using a high resolution video camera coupled to a high resolution closed-circuit black and white video system. The precise location of individual and groups of fish is quantified using a computerized image analysis system, in real time or from an analysis of video tapes.

The experimental protocol employed during the

examination of the behaviour of juvenile chinook salmon, prior to, and during the time of their natural migration to sea, was explained by Birtwell & Kruzynski (1987). In weekly sequential experiments (April, May and June) we used groups of 20 freshwater-reared juvenile chinook salmon. After transfer to the WCS, the fish were given a 48 h adjustment period in air-equilibrated fresh (well) water. Subsequently a vertically stratified water column was created with fresh water overlying salt water. After a further 42 h, hypoxic water was introduced into the freshwater zone progressively reducing its level of dissolved oxygen. The stability of the halocline and the sharp physical separation of fresh from salt waters (refer to Birtwell & Kruzynski, 1987) enabled us to examine the response of the juvenile chinook salmon to hypoxic conditions in the freshwater lens. After 3 h, air-equilibrated fresh water was once more admitted into the system, and in this way an asymmetric 'oxygen sag curve' was obtained. The entire experiment was recorded on video tape and the response of the experimental fish to the changing conditions was quantified.

The response of the fish to hypoxia was most sensitive during earlier experiments, when the majority of the fish occupied the freshwater layer. An experiment carried out in April exemplifies the response. Twenty juvenile chinook salmon (54.2 ± 4.3 mm; 1.41 ± 0.4 g: $\bar{x} \pm$ SD) were placed in the WCS aquarium. After 24 h in fresh water we examined their vertical distribution by recording the positions of individual fish in 3 identical horizontal zones of the water column, at 2 min intervals (n = 30), over a 1 h period around midday.

Results

In situ experimentation

Port Alberni
At each of the study sites the distribution of chinook salmon was similar and biased towards the water surface. In all experiments the greatest number of fish was in the uppermost (0–1.0 m) compartment of the apparatus, and complete avoidance was recorded when the cage penetrated the halocline into the colder, more saline, but hypoxic waters.

Data on fish abundance were related to the measured water characteristics, and through Stepwise Multiple Regression analyses it was determined that dissolved oxygen (% air saturation) was the most significant variable related to the fish distribution: log (Fish Number + 1) = 2.48 + 1.95 log dissolved O_2; n = 70, $R^2 = 0.69$.

In situ bioassays (n = 10) at six sites 150–1600 m from the pulp mill outfall revealed that the waters at 6.5 m depth were lethal to juvenile chinook salmon within 24 h. At 4.0 m depth (around the halocline) mortality was lower (22.5% in 96 h, n = 8), and it was lowest at 0.5 m (10% in 96 h, n = 24). However, on one occasion, at the site adjacent to the outfall, all the test fish died within 24 h. This was considered to be an abnormal situation related to the pulping process. Dissolved oxygen levels ranged between 5 and 90%; 5 and 100% and 65 and 160% of air saturation at depths of 6.5, 4.0 and 0.5 m respectively.

Port Mellon
More detailed studies on the responses of juvenile salmonids to the surface discharge of pulp mill effluent were undertaken in Howe Sound. Here, as in Alberni Inlet, sub-optimal conditions existed in the vicinity of the effluent outfalls. Effluent from this mill was not biologically treated, and was substantially more toxic to salmonids than that discharged at Port Alberni.

Juvenile chinook , chum, and coho salmon reacted similarly within the experimental cages. At the reference location (site 4) the majority of fish were in the uppermost compartment (0–1.0 m), whereas at sites closest to the effluent outfalls, the fish tended to avoid the upper water layers (Table 1). The fish distribution shifted downwards into waters containing lower concentrations of pulp mill effluent.

The results of experiments with juvenile chum salmon illustrate the nature of the response. Colour was used as a very approximate tracer for pulp mill effluent; higher values reflected higher concentrations of effluent (a non-linear relationship, influenced by pH). Mean colour values at each of the experimental sites are presented in Table 1. The most marked differences in colour values among sites and depths occurred at site 1; 80 m from the effluent sources. Here, colour values ranged from 7 to 1250, 0.5 to 5.5 m depth (n = 150); whereas at site 4 the range of values over the same depths was relatively narrow (6–99; n = 150). Site 1 had significantly higher

(6–99; n = 150). Site 1 had significantly higher (P < 0.05) colour values at 0.0, 0.5 and 1.5 m depth than sites more distant from the outfalls. Although sites 1 and 2 were significantly different in colour from sites 3 and 4, at depths of 2.5, 3.5 and 4.5 m, no significant differences at 5.5 m were observed indicating that the distinction between sites diminished with depth. The pattern of colour variation with depth and among sites reflects the increased concentration of pulp mill effluent in water layers closer to the surface, and at sites closest to the source.

While the separation of experimental sites using fish abundance information does not match exactly that recorded for colour data, there are some close similarities (Table 1). At site 4, 68.5% of the test fish were in the 0–1.0 m compartment

(0.5 m depth interval) and 9.7% at 1.0–2.0 m (1.5 m depth interval). Similar values were recorded at site 3, 950 m from the outfalls, but significantly different results were obtained at sites 1 and 2. The surface 0.5 m zone contained 39.5 and 22.7% of test fish at sites 2 and 1 respectively. At 1.5 m, however, these values were 15.8 and 19.9%, reflecting a downward shift in the distribution of fish in the water column.

To further investigate potential relationships, we analyzed data on fish abundance and aquatic characteristics from the 0.5 and 1.5 m zones (equivalent to 0–1, 1–2 m compartments of the apparatus). Although Stepwise Multiple Regression analysis indicated no significant relationship between the numbers of fish and the measured aquatic characteristics at 1.5 m depth,

Table 1. A summary of the results of Analysis of Variance and Student Newman-Keul's Test (P < 0.05) applied to the number of juvenile chum salmon at 6 depths within experimental cages at sites 1 to 4* in Howe Sound, and corresponding colour values. Horizontal bars depict homogeneous data subsets.

Depth (m)	n	Mean number and (%) of fish per site*				n	Mean colour units per site*			
0.0						69	4	3	2	1
							17.2	37.1	127.0	655.0
0.5	69	1	2	3	4					
		13.6	23.7	39.5	41.1	69	18.3	30.6	100.5	390.1
		(22.7)	(39.5)	(65.8)	(68.5)					
1.5	69	4	3	2	1					
		5.8	6.1	9.5	11.9	69	17.4	21.8	46.8	88.7
		(9.7)	(10.2)	(15.8)	(19.9)					
2.5	69	3	4	2	1					
		4.1	6.8	10.2	12.4	69	16.6	19.4	31.6	39.0
		(6.8)	(11.3)	(17.1)	(20.6)					
3.5	69	4	3	1	2					
		2.7	5.5	9.8	11.1	69	14.4	15.8	27.6	29.7
		(4.5)	(9.1)	(16.3)	(18.4)					
4.5	69	4	3	2	1	69	4	3	1	2
		2.1	3.2	3.6	6.3		12.7	14.4	20.8	21.4
		(3.5)	(5.5)	(5.9)	(10.5)					
5.5	69	4	3	2	1	69	4	3	2	1
		1.4	1.6	1.9	5.9		11.5	12.4	16.2	17.6
		(2.4)	(2.6)	(3.2)	(9.9)					

* Sites 1, 2, 3, and 4 –80, 350, 950 and 1750 m from pulp mill outfalls, respectively.

colour was the most significant variable at 0.5 m, giving the equation: log (Fish Number + 1) = 2.29 − 0.55 log colour (n = 150, R^2 = 47.97).

Colour, in turn, was significantly and negatively correlated with distance from the effluent discharge (log colour = 4.039−8.517 log distance, n = 150; R^2 = 66.98), and to a number of the measured variables (log colour = 13.15 + 1.30 log temperature − 2.06 log dissolved O_2 − 3.36 log pH − 2.29 log oxidation − reduction potential, n = 150; R^2 = 52.27).

Data on the presence of salmonids (chinook, coho and chum salmon) at 0.5 and 1.5 m depth intervals were combined and examined for differences between experimental sites. The results of these analyses are presented in Table 2. A separation of sites 1 and 2 from 3 and 4 was obtained for both depth intervals revealing, significantly, the downward shift in distribution of fish at sites

Table 2. A summary of the results of Analysis of Variance and Student Newman-Keul's Test (P < 0.05) applied to the number of juvenile chum, chinook and coho salmon at 0.5 and 1.5 m depths during *in situ* experimentation at sites 1 to 4 * in Howe Sound. Horizontal bars depict homogeneous data subsets.

Depth (m)	n	Mean number of fish and (%) per site*			
		1	2	3	4
0.5	148	18.03	24.31	39.28	40.86
		(30.0)	(40.5)	(65.0)	(68.0)
1.5	150	4	3	2	1
		6.36	6.89	12.11	12.36
		(10.6)	(11.5)	(20.2)	(20.6)

* Sites 1, 2, 3, and 4 −80, 350, 950 and 1750 m from pulp mill outfalls, respectively.

1 and 2 which were within 350 m of the pulp mill outfalls.

Aquatic characteristics, recorded during *in situ*

Table 3. Results of Analysis of Variance and Student Newman-Keul's Test (P < 0.05) applied to colour, temperature, dissolved oxygen and pH data, recorded adjacent to *in situ* experimental cages at sites 1 to 4 * in Howe Sound. Horizontal bars depict homogeneous data subsets.

Depth (m)	Variable	n	Values ($\bar{x}$) per site*			
0.0	colour	148	4	3	2	1
			19.1	32.9	146.1	602.6
0.5		150	19.0	32.9	126.2	363.9
1.5		150	17.9	25.1	53.5	94.0
0.0	temperature	148	3	4	2	1
	(°C)		13.4	16.6	17.1	20.1
0.5		150	14.5	16.5	16.8	17.7
1.5		150	15.7	16.3	16.3	16.6
0.0	dissolved oxygen	148	1	2	3	4
	(% air saturation)		63.4	96.7	105.2	112.7
0.5		150	89.9	98.5	104.4	110.9
1.5		150	100.3	102.6	107.8	110.0
0.0	pH	148	3	2	4	1
			7.69	8.11	8.26	8.65
0.5		150	1	2	3	4
			7.73	8.00	8.02	8.26
1.5		150	7.51	7.81	8.18	8.27

* Sites 1, 2, 3, and 4 −80, 350, 950 and 1750 m from pulp mill outfalls, respectively.

experiments with the three species of salmon, which were most significantly related to the distribution of fish, are shown in Table 3. Colour values differed significantly between sites whereas the numbers of fish at sites 1 and 2 were significantly different from those at sites 3 and 4 at 0.5 and 1.5 m depth intervals (Table 2), 950 and 1750 m from the outfalls. In general, water colour (pulp mill effluent) and temperature were higher with proximity to the discharge, but dissolved oxygen and pH were reduced (0.5 and 1.5 m depths; Table 3). Temperature at site 3 was significantly lower at 0.5 m due to the influence of cold McNair Creek waters which enter Howe Sound in the vicinity of this experimental location. The effect was noticeable only in the water layers very close to the surface, and not at 1.5 m depth. Values of pH tended to increase with distance from the outfalls at depths of 0.5 and 1.5 m but no trend was evident at the surface (Table 3). Despite the buffering capacity of sea water, pH variation was often quite dramatic close to the effluent source: pH ranged between 6.1 and 10.1 at 0.5 m depth and 3.5 and 10.8 over all the depths sampled (0.0–5.5 m) at site 1.

The relationship between the presence of salmon and aquatic characteristics was explored more fully using Stepwise Multiple Regression.

The results for each salmon species, and for the combined data set, at 0.5 and 1.5 m depth intervals are shown in Table 4. In the uppermost cage compartment (0.5 m depth), where aquatic characteristics and the distribution of fish were most varied, colour and colour plus temperature were the variables most related to the distribution of chum and chinook salmon. The distribution of coho salmon, was, however, related to a number of variables. Thus although effluent (colour) was the variable which most affected the vertical position of fish in the water column, other variables cannot, necessarily, be eliminated as influential in this regard.

During comparative 24 h survival and behavioural experiments at sites 1 and 4 we encountered unexpected mortality in the latter tests (at site 1). Since all fish which were denied the choice of position in the water column at site 1 died at depths 0–3 m but survived at 3–6 m, we anticipated 100% survival of fish given full access to the water column. This experiment demonstrated that not only were there no fish present at depths 0–3 m, but of those fish in deeper waters, some (25% on one occasion) had died. In other experiments at site 1 similar mortalities occurred over 24 h on a number of occasions, suggesting that these fish had died while attempting to occupy the

Table 4. Significant ($P < 0.05$) results of Stepwise Multiple Regression analyses relating the number of juvenile chum, coho and chinook salmon at specific zones in the water column, to aquatic variables, during *in situ* experimentation in Howe Sound.

Species	Depth	n	Regression equation (log Fish Number + 1 =)	R^2
Chum	0.5	69	2.29 − 0.55 log colour	47.97
	1.5	59	no significant relationship	
Coho	0.5	39	1.67 log diss. O_2 + 6.16 log pH − 1.01 log temp. − 2.05 log ORP − 0.2 log colour − 0.38	74.34
	1.5	39	2.72 log temp. − 2.39	17.44
Chinook	0.5	42	3.41 − 1.1 log temp. − 0.38 log colour	53.57
	1.5	42	9.53 − 4.60 log diss. O_2 + 0.5 log colour	34.86
Species Combined	0.5	168	2.63 − 1.11 log temp. + 0.84 log pH − 0.39 (log colour)	42.88
Species Combined	1.5	168	2.04 log temp. + 0.19 log colour − 0.90 log diss. O_2 − 0.56	11.24

ORP – oxidation-reduction potential.

diss. O_2 – dissolved oxygen (% air saturation).

surface (albeit lethal) waters. We considered that this mortality was a consequence of the innate behavioural response of the fish to occupy the uppermost portion of the water column, effectively jeopardizing their survival due to exposure to toxic dilutions of pulp mill effluent. Alternatively, those fish initially placed in the uppermost section of the experimental cages could have had a significantly longer exposure to the toxic waters than those which had been placed in the deeper compartments.

Using dye-tagged juvenile coho salmon, we were able to determine that the recorded mortality was not a function of initial placement position within the experimental cages. Unfortunately, we were not able to obtain significant mortalities when desired, but one experiment illustrates the response. In this instance only 5 batches of 5 fish were used in the experimental cages, and they were placed at 0–1, 1–2, 2–3, 3–4, and 4–5 m depth intervals (blue, orange, red, green and 'no colour', respectively), at site 1 and site 4. No mortalities were recorded at the reference site 4, and the typical distribution pattern revealed earlier was apparent. Here, 76% were at 0–1, 16% at 1–2, 4% at 2–3 and 3–4 m depth intervals. In contrast, at site 1, only 36% were alive and all of them were in the 0–1 m compartment. There was 12% mortality of each of the 'no colour', green, red and blue, and 16% of the orange coloured fish. Dead fish were found at 0–1, 1–2, 2–3 m (4% in each) and at 5–6 m (52%) where no fish had been placed at the start of experimentation. Colour values ranged between 30 and 325, at 0.5 m (n = 4) and 4 to 8, at 5.5 m depth (n = 4). Similar results were obtained using unmarked juvenile chum and coho salmon over the course of 8 experiments at site 1. Mortalities were recorded at 1–2 and 2–3 m depth intervals (0.3%); 4–5 m (1.3%) and at 5–6 m (37.6%).

The utility of our *in situ* cage technique was also demonstrated during a two-year study at Port Alice by McGreer & Vigers (1983). They recorded mortalities of juvenile chum salmon (in less than 4 h) at sites close to the outfalls of this sulphite pulp mill. At sites beyond those where mortalities

had occurred during 'preference' experiments, they examined the responses of fish to vertical and horizontal changes in certain aquatic characteristics. They also examined the response time of fish within the apparatus, corroborating our findings. Since effluent from the Port Alice pulp mill moves seaward, a 'zone of influence' could be defined using deviation from the surface water orientation behaviour characteristic of unpolluted environments (Fig. 4). As at Port Mellon, juvenile chum salmon avoided surface waters containing higher concentrations of pulp mill effluent. On an ebb tide, the zone of influence extended 10 km from the pulp mill (McGreer & Vigers, 1983).

Water column simulator

We examined the vertical distribution of juvenile (age 0 +) chinook salmon under fresh water and

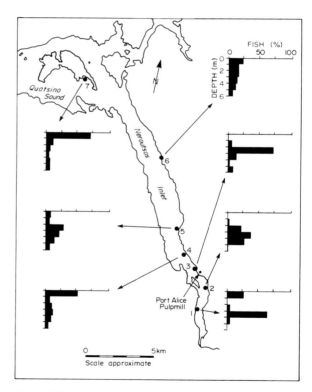

Fig. 4. The vertical distribution of juvenile chum salmon during *in situ* behaviour experiments in Neroutsos Inlet (from McGreer & Vigers, 1983).

vertically stratified estuarine conditions within the WCS. After the 48 h acclimation period in fresh water, salt water was added to the lower two-thirds of the water column. Within one minute this change stimulated exploratory behaviour by the fish. They schooled more tightly and moved to explore the lower sections of the WCS aquarium, within the brackish water. Although vertical stratification, with salt water beneath fresh water, occurred quickly (< 1 h), maximum salinity was

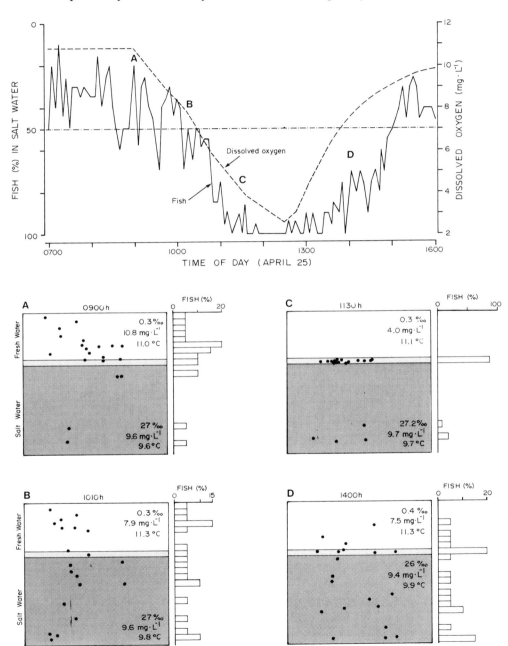

Fig. 5. The responses of 20 freshwater-reared juvenile chinook salmon to changes in dissolved oxygen in fresh water during simulated vertically stratified estuarine conditions. The precise location of individual fish in the apparatus is depicted in diagrams A, B, C, and D at specific times during the experiment.

generally reached after 3–4 h, depending upon the replenishment flows entering the apparatus. At this time we attained stable, vertically stratified conditions, which included a very well-defined halocline.

In all of the weekly experiments (April, May and June) the salmon maintained a strong surface water orientation. They would move across the halocline and freely enter both salt and freshwater zones. Although they displayed great flexibility in this regard and remained primarily biased towards the water surface, over time more fish spent longer periods in salt water.

In April, and after 24 h in fresh water, the mean number of fish ($\pm$ SD) in the top, middle and bottom zones (0–0.75; 0.75–1.5; 1.5–2.25 m water depth) was 10.5 ± 2.8, 6.1 ± 3.4 and 3.5 ± 2.2 respectively. Under vertically stratified estuarine conditions, the equivalent distribution was 12.5 ± 2.2, 2.5 ± 1.9, and 5.2 ± 1.8, demonstrating little difference in the distribution of the fish despite radically different conditions in the WCS. At 0900 h, under continuous flow stratified conditions, hypoxic water was added to the freshwater zone of air-equilibrated well water. At 1230 h air-equilibrated fresh water was readmitted to this zone. The asymmetric curve showing the changes in dissolved oxygen over time in the fresh water layer is presented in Fig. 5. The percentage of fish in salt water represents those fish at and below the halocline, at 5-min intervals. The data reveal the bias of the fish towards fresh water at this date. The response of the juvenile chinook to hypoxic conditions is revealed more precisely by the associated diagrams in Fig. 5 which depict the exact location of fish in the WCS aquarium.

With the onset of hypoxic conditions, the distribution of the fish changed, and they became more active. Figure 5A shows that initially most of the fish occupied the freshwater zone at 0900 h. However, greater activity and dispersion occurred as dissolved oxygen was lowered (Fig. 5B; 1010 h; $7.9 \text{ mg} \cdot \text{l}^{-1}$) and the fish moved away from the surface. With progressively lower levels of dissolved oxygen, greater numbers of fish moved from the freshwater zone and occupied positions at or below the halocline. A unique and reproducible phenomenon was observed during exposure of the fish to levels of dissolved oxygen below, in this experiment, $5.9 \text{ mg} \cdot \text{l}^{-1}$. The juvenile chinook tended to position themselves at the halocline – a boundary not only between salt and fresh water, but also, under these conditions, a boundary between hypoxic and normoxic waters (e.g. Fig. 5C). The fresh-water-reared juvenile chinook remained at the halocline for approximately 2 h while dissolved oxygen levels decreased from 5.9 to 2.6 and increased marginally to $2.8 \text{ mg} \cdot \text{l}^{-1}$. Only when dissolved oxygen levels were increased did the fish start to disperse. With progressive elevation in the level of dissolved oxygen fewer fish occupied the saltwater zone (Fig. 5D), and eventually their vertical distribution approximated that prior to the onset of hypoxic conditions.

There was a marked avoidance of hypoxic waters and, although we cannot accurately determine a threshold response level for 50% of the test fish, it is apparent that reduced levels of dissolved oxygen stimulated activity and the fish moved deeper in the water column. Fifty percent of the experimental fish were consistently in normoxic

Table 5. Water characteristics ($\bar{x} \pm$ S.D.) in 3 zones of the Water Column Simulator during an examination of the responses of juvenile chinook salmon to hypoxic conditions in a freshwater layer overlying salt water.

Variable	n	Zone in WCS		
		Top	Middle	Bottom
Temperature (°C)	48	11.1 ± 0.2	9.7 ± 0.2	9.7 ± 0.2
Salinity (‰)	49	0.3 ± 0.1	26.5 ± 1.8	26.8 ± 1.5
Dissolved oxygen ($\text{mg} \cdot \text{l}^{-1}$)	49	2.6 to 10.8*	9.6 ± 0.2	9.5 ± 0.1

* Range.

salt water or at the halocline at dissolved oxygen levels around 7–8 mg·l^{-1}.

Throughout this experiment the conditions within the 2 water layers of the apparatus remained relatively constant, with the exception of the variation in dissolved oxygen which occurred in the upper, freshwater zone. Table 5 presents data for dissolved oxygen, temperature and salinity within the three equivalent depth zones of the WCS aquarium.

Discussion

The preference of juvenile salmon for surface waters has been demonstrated in both *in situ* and laboratory (WCS) experiments, indicating that this behavioural trait can be modified by changes in aquatic conditions. Field experiments at Port Mellon and Port Alice, where surface waters contained higher concentrations of pulp mill effluents, elicited a retreat by the salmon to deeper and less polluted waters. On the other hand, at Port Alberni, juvenile chinook salmon avoided deeper hypoxic waters by moving upward in the water column. Such vertical movements were confirmed in controlled laboratory (WCS) experiments where juvenile chinook salmon voluntarily moved downward, away from increasingly hypoxic fresh water stratified over normoxic sea water, and upward as oxygen levels were increased.

Thus, in both laboratory and field experiments we have demonstrated two techniques which were developed to capitalize on the innate behavioural trait of salmon to occupy surface waters and move vertically in the water column. The results have allowed us to quantify the responses of the fish to aquatic characteristics and thereby corroborate the utility of the techniques for application in environmental research.

Field studies

There has been little research on the surface water orientation behaviour of juvenile Pacific salmon. Beak Consultants Ltd., (1981) used a trapping technique which sampled at discrete depth intervals in the estuary of the Fraser River. They found, as we did, that juvenile salmon were biased towards the water surface, but there was a tendency for larger (age 1 +) individuals to occupy the faster-flowing mid-channel waters, while underyearlings occupied shallower, slower-moving waters close to shore. Studies on the relationship between salmon migration and water characteristics have revealed that certain hydrographic features may be related to salmon orientation. Adult Atlantic salmon have been recorded close to the water surface (for months) in estuaries, seemingly associated with the depth of the halocline (Brawn, 1982). Furthermore, Westerberg (1984) and Döving et al. (1985) consider that the fine-scale hydrographic features (e.g. thermal microstructure) may provide a necessary reference system for successful orientation by salmon in nearshore coastal regions. Our own observations in the estuary of the Campbell River, British Columbia, also emphasize the importance and influence of microhabitat characteristics on the behaviour of juvenile salmon (Macdonald et al., 1987).

At Port Alberni, the progeny of approximately 200×10^3 spawning adult Pacific salmon, plus the release of 10×10^6 juvenile salmon from the Robertson Creek Hatchery, reside in the Somass River estuary during their migration to the Pacific Ocean via Alberni Inlet. Research on the distribution and ecology of fish in the estuary in relation to aquatic characteristics revealed marked differences in fish use of the region, primarily associated with industrial activities and urban development (Birtwell, 1978; Birtwell & Harbo, 1980; Birtwell et al., 1983a; Birtwell et al., 1984). Studies on the vertical distribution of fish in surface waters, employing a stratified gill net technique, showed that adult and sub-adult (age > 1 +) salmon were generally found in waters closer to the surface (0–2.5 m depth) rather than in deeper (2.5–5.0 m) waters at or below the halocline (Birtwell et al., 1983a). The results of these studies on wild populations, reinforced by the findings of our *in situ* cage experiments, led us to conclude that the presence of hypoxic conditions at depth (>4 m) limit both juvenile and older salmon in their use of the water column in the Somass River estuary.

Even though salmon may make temporary excursions into sub-optimal conditions at depth (as evidenced in laboratory studies with the WCS) it is unlikely that they would reside there. *In situ* bioassays demonstrated lethal conditions to juvenile chinook salmon at 6.5 and 4.0 m depths in the Somass River estuary, effectively forcing the fish to occupy surface waters containing dilutions of pulp mill effluent. While some species of salmon would exit the estuary quickly, others may reside for months (Levy & Northcote, 1982). Thus in light of recent research on the adverse effects of pulp mill effluent on 'resident' fish (Andersson *et al.*, 1988) and our own findings, concern is probably warranted over the continued health of the Somass River estuary. In addition to current concerns about sublethal effects, if effluent from the pulp mill were to become more toxic, a severe environmental problem could result for salmon which would be 'obligated' (behaviourally) to occupy sub-optimal habitats.

The results of *in situ* experiments at both Port Mellon and Port Alice, revealed that the discharge of highly toxic mill wastes into surface waters elicited a very strong response from juvenile salmon in the experimental cages. Behavioural changes were used to define a 'zone of influence' from the outfalls which, at Port Mellon extended 350–950 m. At Port Alice it extended 10 km from the mill on ebb tides (McGreer & Vigers, 1983). These results have been used, in conjunction with others that examined the environmental quality around the pulp mills (Birtwell & Harbo, 1980; Tollefson, 1982), to assist with environmental management decisions. At both locations, effluents from the mills are now discharged via deep diffusers, obviating the need to rely on initial surface water dilution, thereby reducing the immediate impact on juvenile salmon (and other organisms) which use the uppermost parts of the water column.

However, despite significant efforts to improve the water quality of Neroutsos Inlet particularly through modification of industrial activities in the pulp mill, there still persist at times, severe hypoxic conditions along the length of this 20 km-long inlet.

At Port Mellon, studies on fish distribution in the shallow intertidal zone revealed an impoverished fauna close to the pulp mill outfalls, and an absence of salmon within 400 m (Birtwell & Harbo, 1980). Similarly, at Port Alice, Poulin & Oguss (1982) also recorded an absence of juvenile chum salmon close to the pulp mill, and a reduced habitat utilization at more distant locations along Neroutsos Inlet. These studies suggest that the salmon avoided regions around the pulp mill due to the adverse environmental conditions (McGreer *et al.*, 1982). However, one cannot dismiss the potentially more serious consequences of entry into the toxic surface waters. At Port Mellon, juvenile salmon were captured in gill nets at a distance of about 80 m from the outfalls in 0–2.5 m surface waters which subsequently proved to be lethal to the fish (Department of Fisheries and Oceans, unpublished data). Furthermore during *in situ* behaviour experiments some test fish died despite having the option of using deeper waters which would have permitted survival. Even under such toxic conditions close to the water surface, the surviving fish chose to occupy the uppermost compartment of the apparatus which was immersed in the highest concentration of effluent. McGreer & Vigers (1983) also recorded mortalities during their behaviour experiments close to the Port Alice pulp mill (but over a shorter time period: <4 h vs 24 h at Port Mellon). They did not explore these findings and located their study sites more distant from the pulp mill so that sublethal behavioural effects could be studied without the complication of death of any of the test fish. It appears, therefore, that the innate surface water orientation behaviour of juvenile salmon may override some potentially adaptive avoidance responses (i.e., a vertical distribution shift), and thereby jeopardize survival. Alternatively, discrimination of potentially less harmful waters may be impaired due to the lack of suitable gradients and sensory cues (Höglund, 1961). Another factor may be the attraction of the test fish to the effluent and its constituents. It has been demonstrated that attraction to lethal levels of contaminants occurs, for example, with rainbow trout and chlorine (Hadjinicolaou & Laroche, 1988). Consequently, it may be significant that aquatic conditions which the juvenile salmon may choose to prefer could inadvertently jeopardize their survival. In this context, the paucity or absence of juvenile salmon close to the Port Alice pulp mill at the time of studies by McGreer & Vigers (1983) and Poulin & Oguss (1982), may not be solely due to avoidance of the region but also to mortality of fish that had encountered, and chosen to remain in, this degraded habitat.

It is apparent that the use of fish presence as indicative of appropriate water quality in the vicinity of effluent discharges is fraught with problems, for presence *per se* is, at the very best, only a coarse indicator of environmental health.

At Port Alberni and Port Alice (and to a much lesser extent, Port Mellon), dissolved oxygen was one of the variables most significantly associated with the distribution of fish during *in situ* experimentation. The observed behaviour is likely to have been the product of the combined effects of contaminants present in the effluent and hypoxia. Alderdice & Brett (1957) and Hicks & DeWitt (1971), found that the toxic effects of pulp mill effluent to sockeye and coho salmon were enhanced under hypoxic conditions. More recently, Sprague (1985) concluded from a review of the literature that, in general, a reduction in dissolved oxygen from 100 to 80% of air saturation will cause increased mortality of fish exposed to some contaminants.

In Neroutsos Inlet, perennial hypoxic conditions currently exist in the spring, summer and early autumn periods (Department of Fisheries and Oceans, unpublished information). *In situ* experimentation on the survival of salmon has recently revealed lethal conditions at the time of salmon spawning migrations, prompting federal government authorities to request the maintenance of an average of $5 \text{ mg} \cdot 1^{-1}$ dissolved oxygen in the top 10 m of the water column to promote a successful migration (Departments of Fisheries and Oceans and Environment, unpub-

lished information). In the light of the foregoing comments, this level may be a significant compromise over that which is optimal for the salmon.

Laboratory studies

Because of the dearth of knowledge on the interactive effects of hypoxic conditions and industrial effluent discharges, we have focussed our attention on the responses of juvenile salmon to reduced levels of dissolved oxygen through the use of the WCS. The results obtained on the responses of juvenile chinook salmon have applicability in the context of water quality objectives and the maintenance of suitable aquatic conditions. We hope to generate baseline information from which we can progress to examine the combined effects of contaminants and hypoxic conditions and quantify the degree to which fish will make use of sub-optimal habitats.

Juvenile salmon occupied the upper part of the water column and to this extent their behaviour was similar to that recorded during *in situ* experimentation. The behaviour of the fish under simulated estuarine conditions was somewhat unanticipated. As the salmon were underyearlings, reared in fresh water, we anticipated little movement of the fish between fresh and sea water in April. However, the fish would move across the distinct halocline between waters of widely differing salinities (0.3 to 26‰) but overall they spent most time in fresh water. Whether this stock of Fraser River salmon behaves similarly in the Fraser River estuary during their protracted residency (Levy & Northcote, 1982) is not known. The extensive Fraser River estuary is typically vertically stratified but the halocline is generally not as close to the surface, nor frequently as well-defined as, for example, in the estuaries within Alberni and Neroutsos Inlets, or that which was produced in the WCS. Nevertheless, very sharp gradients do occur in the Fraser River estuary between salt and fresh waters, but processes relating to such hydrographic features are not well understood (A. Ages, Department of Fisheries and Oceans, personal communication).

The juvenile chinook salmon in the WCS

tended to occupy the freshwater lens more so than saltwater layer, thus the responses of the fish to changes in dissolved oxygen were considered applicable to the majority of the population under experimentation. The downward shift in distribution of the fish in response to decreasing dissolved oxygen levels in fresh water was a highly reproducible event, and prior to moving to salt water the fish moved away from the surface of the freshwater zone.

In that the fish demonstrated behavioural flexibility in moving between salt and freshwater layers under normoxic conditions, the positioning of up to 80% of the fish at the halocline between these horizontally-flowing water layers was unanticipated. Gee *et al.* (1978) reported that *Salmonidae*, in contrast to families of 'great plains fishes' did not rise to breathe the surface film under hypoxic conditions. However, we have observed this surfacing behaviour in salmonids exposed to suspended sediments and chemical contaminants in our laboratory (unpublished data). In his review of dissolved oxygen and fish behaviour, Kramer (1987) stated that hypoxia caused vertical and horizontal habitat shifts, but made no reference to the apparent behaviour of juvenile salmon to use, or be limited by, deep water boundaries. Thus, in contrast to species which move upwards to the boundary layer where diffusion maintains dissolved oxygen levels in the surface film (Kramer, 1987), these juvenile chinook salmon moved downwards to an analogous boundary between normoxic and hypoxic waters of different salinities. Only after 2 h did the behaviour change on April 25 and the fish dispersed, especially into more saline waters. It is probable that this group of salmonids was not sufficiently adapted to move directly into sea water at this time, as would be deduced from their protracted use of the Fraser River estuary until July (Levy & Northcote, 1982). Dissolved oxygen variation across the halocline was quite abrupt with substantial differences occurring over 2 cm depth. Recordings of dissolved oxygen in water extracted at specific depths across the halocline prior to the water masses exiting the WCS aquarium (where the effects of turbulent mixing and diffusion would

probably have been greatest), revealed stable conditions in both fresh and saltwater layers (Birtwell & Kruzynski, 1987). Therefore, fish occupying the halocline region could, within a vertical plane of 6 cm, be immersed in waters of widely differing dissolved oxygen and salinity levels.

Gas transfer across the boundary between still fresh and salt waters in vertically stratified conditions is effectively inhibited due to the lack of vertical mixing (Waldichuk, 1984), however in the WCS some vertical mixing is inevitable under the continuous-flow conditions imposed. In vertically stratified estuaries where pollutants in the freshwater layers can produce hypoxic conditions, it would appear that even for freshwater-acclimated salmon an option may exist at the halocline that would assist survival. Kulakkattolickal & Kramer (1987) report the effects of an extract of *Croton tiglium* seeds on the zebrafish (*Brachydanio rerio*) given or denied access to the water surface. They concluded that aquatic surface respiration permitted the fish to reduce their ventilation rate in hypoxic waters and hence reduce uptake of toxins. It seems quite feasible that in stratified estuaries, salmonids could make use of similar adaptive movements to the halocline or the water surface to aid survival under sub-optimal conditions. However, even though surfacing behaviour may be physiologically beneficial, it has been shown to render fish more susceptible to avian predators (Birtwell *et al.*, 1983b).

In June, juvenile chinook salmon in the WCS were still biased towards the water surface but under hypoxic conditions, they did not display the strong association with the halocline that younger fish had demonstrated in April. Greater numbers of fish spent more time below the halocline, no doubt reflecting their increasing adaptability to sea water. However, in all experiments, when dissolved oxygen levels in the freshwater layer were at highly stressful or lethal levels (see Doudoroff & Shumway, 1967) some fish would make transient excursions into these waters. This suggests, again, that fish presence, *per se*, is not necessarily the most sensitive criterion to define suitable environmental conditions. Furthermore, such behaviour is quite consistent with our field observations in an unpolluted estuary (Macdonald *et al.*, 1987) and in contaminated habitats. The innate behavioural trait of juvenile salmon to occupy surface waters is extremely strong, and even under adverse (lethal) conditions, they will strive to occupy specific habitats, possibly to their detriment.

We determined that 7–8 mg·l^{-1} was the level at which 50% of the test fish responded to the hypoxic conditions by moving to or below the halocline. This level is, to our knowledge, higher than that which has previously been reported to cause avoidance reactions by salmonids in response to hypoxia (refer to Whitmore *et al.*, 1960; Bishai, 1962; Doudoroff & Shumway, 1970; Davis, 1975; EPA, 1986). However, measurable physiological responses occur when salmonid blood ceases to become fully saturated with oxygen at about 120 mm Hg; 76% of the saturation level, (D. Randall, University of British Columbia, personal communication). Such adaptations could be manifested in sensitive behavioural responses possibly mediated via pO$_2$ receptors. The fact that the responses which we observed occurred around 7–8 mg·l^{-1} (equivalent to about 101–115 mm Hg) attests to the relatively high sensitivity of this technique. It would, however, be preferable to quantify the responses of a more sensitive fraction of the population (EC$_{10}$ vs EC$_{50}$, for instance). In doing so, the results may have greater applicability for the protection of salmonids. Reliance is often placed on the response of 50% of the population, frequently for very legitimate reasons, but this could be misleading. The present lack of understanding of the complex interactions between contaminants and hypoxic conditions warrants a cautious approach to the use of receiving water dissolved oxygen criteria (Davis, 1975; EPA, 1986). It is therefore important that our research focusses not only upon sensitive responses, by sensitive individuals in populations, but also upon the ecological significance of such interactions.

Acknowledgments

We are especially grateful for the assistance from numerous staff in the Habitat Management Division, and the Salmon Habitat Section of the Department of Fisheries and Oceans. The cooperation of the pulp and paper industry facilitated *in situ* experimentation. The cooperation of Drs J. C. Davis and G. L. Greer, and assistance of R. M. Harbo during the field component is particularly appreciated. G. E. Piercey and S. Spohn helped with laboratory experimentation and together with Dr. J. S. Macdonald, B. Raymond, A. Chu, and D. Jefferies, assisted with data analysis. B. Gordon drafted the figures and Mrs. D. Price diligently typed the manuscript.

References

Alderdice, D. F. & J. R. Brett, 1957. Some effects of kraft mill effluent on young Pacific salmon. J. Fish. Res. Bd. Can. 14: 783–795.

Andersson, T., L. Förlin, J. Hardig & A. Larsson, 1988. Biochemical and physiological disturbances in fish inhabiting coastal waters polluted with bleached kraft mill effluents. Mar. Envir. Res. 24: 233–236.

Beak Consultants Ltd., 1981. Environmental impact statement of proposed improvement to the Fraser River shipping channel. Public Works Canada, Vancouver, B.C. 635 pp.

Birtwell, I. K., 1977. A field technique for studying the avoidance of fish to pollutants. In: Proc. 3rd Aquatic Toxicity Workshop, Halifax, N.S., Nov. 2–3, 1976. Envir. Prot. Serv. Tech. Rep. No. EPS-5-AR-77-1, pp. 69–86.

Birtwell, I. K., 1978. Studies on the relationship between juvenile chinook salmon and water quality in the industrialized estuary of the Somass River. p. 57–78. In: B. C. Sheppard & R. M. J Ginetz (Eds.), Proceedings of the 1977 Northwest Pacific Chinook and Coho Workshop. Fish Mar. Serv. Tech. Rep. 759. pp. 57–78.

Birtwell, I. K. & R. M. Harbo, 1980. Pulp mill impact studies at Port Alberni and Port Mellon, B.C. Trans. Tech. Sect. Can. Pulp Pap. Assoc. 81: 85–88.

Birtwell, I. K. & G. M. Kruzynski, 1987. Laboratory apparatus for studying the behaviour of organisms in vertically stratified waters. Can. J. Fish. Aquat. Sci. 44: 1343–1350.

Birtwell, I. K., S. Nelles & R. M. Harbo, 1983a. A brief investigation of fish in the surface waters of the Somass River estuary, Port Alberni, British Columbia. Can. Manuscr. Rep. Fish. Aquat. Sci. 1744: 31 pp.

Birtwell, I. K, G. L. Greer, M. D. Nassichuk & I. H. Rogers,

1983b. Studies on the impact of municipal sewage discharged into an intertidal area within the Fraser River estuary, British Columbia. Can. Tech. Rep. Fish Aquat. Sci. 1170: 55 pp.

Birtwell, I. K., M. Wood & D. K. Gordon, 1984. Fish diets and benthic invertebrates in the estuary of the Somass River, Port Alberni, British Columbia. Can. Manuscr. Rep. Fish. Aquat. Sci. 1799: 49 pp.

Bishai, H. M., 1962. Reactions of larval and young salmonids to water of low oxygen concentrations. J. Cons. Int. Explor. Mer. 27: 167–180.

Brawn, V. M., 1982. Behaviour of Atlantic Salmon (*Salmo salar*) during suspended migration in an estuary, Sheet Harbour, Nova Scotia, observed visually and by ultrasonic tracking. Can. J. Fish. Aquat. Sci. 39: 248–256.

Davis, J. C., 1975. Minimal dissolved oxygen requirements of aquatic life with emphasis on Canadian species: a review. J. Fish. Res. Bd. Can. 32: 2295–2332.

Doudoroff, P. & D. L. Shumway, 1970. Dissolved oxygen requirements of freshwater fishes. Food Agriculture Organization of the United Nations. FAO Tech. Pap. 86. Rome, Italy. 291 pp.

Döving, K. B., H. Westerberg & P. B. Johnson, 1985. Role of olfaction in the behavioural and neuronal responses of Atlantic salmon, *Salmo salar*, to hydrographic stratification. Can. J. Fish. Aquat. Sci. 42: 1658–1667.

Environmental Protection Agency, 1986. Quality criteria for water, U.S., EPA 440/5-86-001, Washington, D.C. 35 pp.

Gee, J. H., R. F. Tallman & H. J. Smart, 1978. Reactions of some great plains fishes to progressive hypoxia. Can. J. Zool. 56: 1962–1966.

Hadjinicolaou, J. & G. LaRoche, 1988. Behavioural responses to low levels of toxic substances in rainbow trout (*Salmo gairdneri*, Rich). In: W. J. Adams, G. A. Chapman and W. G. Landis (Eds.), Aquatic Toxicology and Hazard Assessment: 10th Volume, pp. 327–340. ASTM STP 971, American Society for Testing and Materials, Philadelphia.

Hicks, D. B. & J. W. De Witt, 1971. Effects of dissolved oxygen on kraft pulp mill effluent toxicity. Wat. Res. 5: 693–701.

Höglund, L. B., 1961. The reactions of fish in concentration gradients. Rep. Inst. Freshwat. Res. Drottningholm. 43: 1–147.

Kramer, D. L., 1987. Dissolved oxygen and fish behaviour. Envir. Biol. Fishes. 18: 81–92.

Kulakkattolickal, A. T. & D. L. Kramer, 1987. Effect of surface access and oxygen concentration on the toxicity of *Croton tiglium* (Euphorbiaceae) seed extract to *Brachydanio rerio* (Cyprinidae). Can. J. Fish. Aquat. Sci. 44: 1358–1361.

Levy, D. A. & T. G. Northcote, 1982. Juvenile salmon residency in a marsh area of the Fraser River estuary. Can. J. Fish. Aquat. Sci. 39: 270–276.

Macdonald, J. S., I. K. Birtwell & G. M. Kruzynski, 1987. Food and habitat utilization by juvenile salmonids in the Campbell River estuary. Can. J. Fish. Aquat. Sci. 44: 1233–1246.

Mace, P. M., 1983. Bird predation on juvenile salmonids in the Big Qualicum River, Vancouver Island. Can. Tech. Rep. Fish. Aquat. Sci. 1117: 79 pp.

McGreer, E. R. & G. A. Vigers, 1983. Development and validation of an *in situ* fish preference-avoidance technique for environmental monitoring of pulp mill effluents. In: W. E. Bishop, R. D. Cardwell and B. C. Heidolph, (Eds.), Aquatic Toxicology and Hazard Assessment: sixth symposium. pp. 519–529. ASTM STP 802, American Society for Testing and Materials, Philadelphia.

McGreer, E. R., D R. Munday & G. A. Vigers, 1982. Acute toxicity and fish avoidance behaviour studies in Neroutsos Inlet – 1980. Vol 6 of Environmental Improvement at Neroutsos Inlet, B.C. Western Forest Product Ltd., Vancouver, B.C. 25 pp.

Parker, R. R. & J. Sibert, 1973. Effect of pulp mill effluent on dissolved oxygen in a stratified estuary – II. Empirical Observations. Wat. Res. 7: 503–514.

Piercey, G. E., I. K. Birtwell, H. Herunter, M. Kotyk, G. M. Kruzynski, J. S. Macdonald & K. Seaman, 1985. Data report on physical habitat characteristics and observations of fish at two regions in the estuary of the Campbell River, British Columbia, 1984. Can. Data Rep. Fish. Aquat. Sci. 551: 107 pp.

Poulin, V. A. & E. Oguss, 1982. Migration characteristics of juvenile chum salmon in Neroutsos Inlet with particular emphasis on the effects of sulphite mill effluent on migratory behaviour, 1980. Vol. 4 of Environmental Improvement at Neroutsos Inlet, B.C. Western Forest Products Ltd., Vancouver, B.C. 45 pp.

Sibert J. & R. R. Parker, 1973. Effect of pulp mill effluent on dissolved oxygen in a stratified estuary – II. Numerical Model. Wat. Res. 7: 515–523.

Sprague, J. B., 1985. Factors that modify toxicity. Chap. 6 In: G. M. Rand & S. R. Petrocelli (Eds.). Fundamentals of Aquatic Toxicology. Hemisphere Publ. Corp. Washington, 666 pp.

Tollefson, R., 1982. Environmental Improvement at Neroutsos Inlet, B.C. Western Forest Products Limited, Vancouver, British Columbia. 84 pp.

Tully, J. P., 1949. Oceanography and prediction of pulpmill pollution in Alberni Inlet. Fish. Res. Bd. Can. Bull. No. 83. 169 pp.

Tully, J. P., H. J. Hollister, R. J. I. Fjarlie & W. Anderson, 1957. Physical and chemical data record Alberni Inlet and Harbour 1939–1941. Fish. Res. Bd. Can. File N7-1-3(1). Pacific Oceanographic Group, Nanaimo, B.C. 89 pp.

Waldichuk, M., 1984. Laboratory observations on transfer of atmospheric oxygen into stratified seawater, In: W. Brutsaert and G. H. Jirka (Eds.), Gas Transfer at Water Surfaces. pp. 547–556. D. Reidel Publ. Co. 639 pp.

Westerberg, H., 1984. The orientation of fish and the vertical stratification at fine and microstructure scales. In: J. D. McCleave, G. P. Arnold, J. J. Dodson and W. H. Neill (Eds.), Mechanisms of Migration in Fishes. pp. 179–204. Plenum Publishing Corp., New York, New York.

Whitmore, C. M., G. E. Warren & P. Doudoroff, 1960. Avoidance reactions of salmonids and centrarchid fishes to low oxygen concentrations. Trans. Am. Fish. Soc. 89: 17–26.

Hydrobiologia **188/189**: 561–566, 1989.
M. Munawar, G. Dixon, C. I. Mayfield, T. Reynoldson and M. H. Sadar (eds)
Environmental Bioassay Techniques and their Application.
© 1989 *Kluwer Academic Publishers. Printed in Belgium.*

Analysis of fish bile with HPLC – fluorescence to determine environmental exposure to benzo(a)pyrene

Eric P. Johnston[1] & Paul C. Baumann[2]
[1] *The Ohio State University, Department of Zoology, 1735 Neil Avenue, Columbus, Ohio 43210;* [2] *U.S. Fish and Wildlife Service, 2021 Coffey Road, Columbus, Ohio 43210*

Key words: benzo(a)pyrene equivalents, bile, HPLC

Abstract

Brown bullhead from the Black River, Ohio, have a high incidence of liver neoplasia which is associated with elevated concentrations of polynuclear aromatic hydrocarbons (PAHs) in the sediment. We evaluated the use of biliary concentrations of benzo(a)pyrene [B(a)P] equivalents as a means for determining PAH exposure. Bile was collected from 16 brown bullheads and 8 common carp taken from each of two Lake Erie tributaries in Ohio, the industrialized Black River and the non-industrialized Old Woman Creek. Hatchery bullhead (n = 8) were used to determine base levels of PAHs. A high performance liquid chromatography (HPLC) – fluorescence technique was used to determine the concentration of B(a)P equivalents in the bile samples. The area of all peaks fluorescing at 380/430 nm was summed to give a single value for B(a)P equivalents in each sample. Concentrations of B(a)P equivalents generally reflected concentrations of PAH in sediment where fish were collected. Bile taken from Black River carp contained the highest concentration of B(a)P equivalents and was significantly different from all other groups. The value obtained for Black River bullhead was also high and was found to be significantly different from hatchery bullhead. B(a)P equivalents varied between carp and bullhead from the same habitat possibly because of differing food habits or metabolic pathways. However, our results indicate that relative levels of B(a)P equivalents in the bile of fish correspond well to B(a)P levels in sediment and may offer a means of determining environmental exposure of fish to the parent compound.

Introduction

The Black River is an industrially polluted tributary of Central Lake Erie. A USX Steel Plant and associated coking facility are located on the lower reach of this river. Sediments contain particularly high levels of polynuclear aromatic hydrocarbons (PAHs) including the known mammalian carcinogen benzo(a)pyrene [B(a)P] (Fabacher *et al.*, 1988; West *et al.*, 1984). Examination of brown bullheads (*Ictalurus nebulosus*)

captured in the Black River has revealed high incidences of liver neoplasia, suggesting a cause-and-effect relationship between the chemically contaminated sediments and the observed neoplasms (Baumann & Harshbarger, 1985; Baumann *et al.*, 1982). Supporting this hypothesis, are laboratory experiments which demonstrate the ability of benthic fish to accumulate PAHs through sediment and diet (McCain *et al.*, 1978; Roubal *et al.*, 1977). Residue analysis of Black River fish has demonstrated the presence of

562

certain PAHs at levels greater than 5000 ppb. However, B(a)P and most of the larger 4 and 5 ring PAHs, were found at levels in fish relative to concentrations in sediment. Fish rapidly metabolize PAHs to intermediates which either bind to liver DNA or form conjugates which then pass into the bile (Varanasi *et al.*, 1986). Therefore, to obtain accurate estimates of PAH exposure, it is necessary to use analytical techniques which identify both the parent and its metabolites. Studies by Krahn *et al.*, (1984) and Maccubbin *et al.*, (1988) demonstrate the presence of complex assemblages of PAH metabolites in the bile of fish captured in contaminated waterways. These authors have shown that biliary concentrations of material fluorescing at specific PAH wavelengths may be used to monitor exposure of fish to PAH. In our study, we have further evaluated the use of biliary concentrations of material fluorescing at B(a)P wavelengths (380/430 nm) as a means for determining PAH exposure in freshwater fish populations.

Sample collection

Bile was collected from 16 brown bullheads and 8 common carp (*Cyprinus carpio*) taken from each of two study locations, the Black River and Old Woman Creek Estuary (OWC) (Fig. 1). OWC is a less contaminated Lake Erie tributary which receives agricultural runoff but no known industrial effluent. The lower reach of OWC, where collections were made, is part of the OWC State Nature Preserve. Fish were collected in fyke nets in July and August when metabolic rates would be high and after the spawning season when metabolism of steroids might interfere with metabolism of xenobiotics. Bile was collected by making an incision in the gall bladder which allowed the bile to flow into a light-proof vial. Immediately after collection, the samples were placed on dry ice. They were then transported to the laboratory where they were stored at $-10\,°C$. Bile was collected only from fish measuring 250 mm and larger. Hatchery reared brown bullheads (n = 8) were

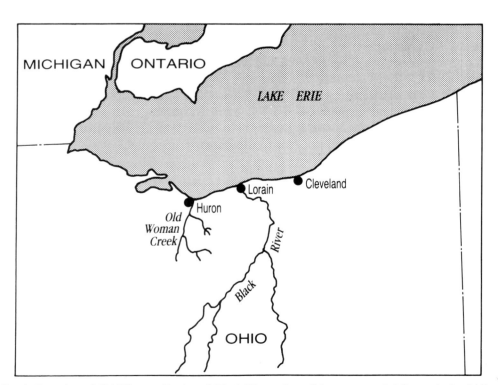

Fig. 1. Location of Old Woman Creek and Black River where fish were sampled for analysis of bile.

used to determine base levels of PAH. We also collected sediment at OWC from three locations. Since recent sediment values had been obtained from the Black River, we did not independently sample this location.

HPLC analysis of bile and sediment

Bile samples were analyzed using a reversed-phase high performance liquid chromatography (HPLC) – fluorescence technique adapted from Krahn et al. (1984). A 0.26ID-X-25 cm ODS-HC sil x-1 Vydak column was used to perform separations. Acetic acid-water (5 μl/l) (solvent A) and methanol (solvent B) formed a linear gradient as follows: 100% solvent A to 100% solvent B in 15 minutes; 15 minutes at 100% solvent B; 3 minutes to return to 100% solvent A. Bile samples (5.0 μl) were chromatographed at a flow rate of 1.2 ml/minute, with the fluorescence detector set at B(a)P wavelengths (380/430 nm, excitation/emission). A Perkin-Elmer Series 3 liquid chromatograph, Perkin-Elmer 650-10s fluorescence spectrophotometer, and a 3340A Hewlett Packard Integrator were used. The area of all peaks fluorescing at 380/430 nm was summed and converted to B(a)P equivalents (the amount of B(a)P that would be present if the area was attributed only to B(a)P using the response factor from a B(a)P standard.

We analyzed a total of 56 bile samples, each representing a different fish. Bile samples were grouped by species and location for statistical analysis. We used Tukey's studentized range test to test for significance between the groups (SAS, 1985). OWC sediment samples (n = 3) were collected at each of three sampling sites within OWC. Equal volumes of the three samples collected at each site were combined to form one composite sample which was analyzed following the procedures outlined by Black et al. (1979).

Results

PAH concentrations in OWC were lowest in sediment samples collected in the center of the estuary, while samples collected near a highway bridge and a railway bridge were higher. Samples from the latter location (Table 1) were higher than PAH sediment concentration values observed in Buckeye Lake, a reference location used by Baumann et al. (1982). The mean concentration of B(a)P equivalents for the sixteen Black River bullhead was significantly greater (P < 0.01) than the mean for the hatchery bullhead (Table 2). The mean for OWC bullheads was intermediate between that for Black River bullheads and that for hatchery bullheads, but there was no significant difference (P > 0.05) between either pair of values.

Table 1. Concentration (ng/g dry weight) of selected PAH in sediment collected in Buckeye Lake and Old Woman Creek.

PAH	Buckeye[a] Lake	Old Woman Creek	Black[b] River
Phenanthrene	40	47	52,000
Anthracene	4	13	15,000
Fluoranthene	110	222	33,000
Pyrene	72	136	24,000
Benz(a)anthracene	21	78	11,000
Chrysene	28	89	10,000
Benzofluoranthenes	36	187	15,000
Benzo(a)pyrene	14	271	8,000
Benzo(g, h, i)perylene	16	76	5,400
Indeno(1, 2, 3, c, d)pyrene	15	51	6,400

[a] Baumann, unpublished data.
[b] Fabacher et al. 1988.

Table 2. Mean concentrations (ng/µl) of B(a)P equivalents found in the bile of brown bullhead and common carp collected from the Black River (BR), Old Woman Creek Estuary (OWC) and Fender's Fish Hatchery (FH). Different superscripts indicate significant differences (P < 0.05).

Species	Location	Summed peak area	Range	Standard deviation
bullhead	BR	3.71 [b]	10.1–104.7	25.9
bullhead	OWC	2.19 [bc]	0.7– 71.7	20.3
bullhead	FH	0.21 [c]	0.0– 4.9	2.0
carp	BR	19.65 [a]	102.1–354.1	93.2
carp	OWC	0.47 [c]	1.2– 11.5	3.4

The highest mean concentration of B(a)P equivalents was found in the bile of Black River carp (Table 2). Black River carp had significantly greater (P < 0.01) B(a)P equivalent concentrations in bile than all other groups. B(a)P equivalents in bile of OWC carp were not significantly different (P > 0.05) from those of OWC bullhead or hatchery bullhead, but were significantly lower (P < 0.01) than B(a)P equivalent concentrations in Black River bullhead. Sample chromatograms for the five groups are presented in Fig. 2.

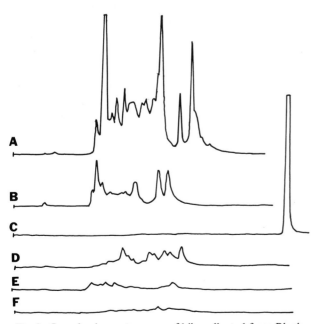

Fig. 2. Sample chromatograms of bile collected from Black River carp 96.8 ng/µl (A), Black River bullhead 24.9 ng/µl (B), benzo(a)pyrene standard, 25 ng/µl (C), Old Woman Creek bullhead 18.9 ng/µl (D), Old Woman Creek carp 1.8 ng/µl (E), and hatchery bullhead 0.5 ng/µl (F).

Discussion

Our data are consistent with the results of Krahn *et al.* (1984) and Maccubbin *et al.* (1988) in that fish from contaminated sites have greater concentrations of B(a)P equivalents in bile than fish from less contaminated sites. However, concentrations of B(a)P equivalents varied by species and did not always reflect the magnitude of differences observed in sediment PAH concentrations. Concentrations of B(a)P equivalents in carp bile did reflect the observed differences in sediment PAH, but concentrations in bullhead bile from OWC and the Black River did not. These differences may have arisen as a result of different feeding habits or habitat selection by bullhead and carp. However, large differences in feeding habits and habitat selection do not seem probable, since both species are omnivorous benthic feeders. Thus, the species differences cannot be fully explained at this time and additional study of metabolism and feeding habits are needed.

Black River sediment samples were collected in 1984 (Fabacher *et al.*, 1988) and OWC sediment samples were collected in 1985, while fish used for study were collected in 1985 and 1986. High concentrations of PAH are present in the Black River samples collected in 1984 (Table 1). Although the USX coking facility closed in 1984, concentrations of PAHs in Black River sediment should have remained high in 1985 and 1986. The closing of the coking facility, however, would cause lower PAH exposure to Black River fish collected in 1985 and 1986 than are suggested by the 1984 Black River sediment data. The highest PAH concentrations in OWC sediments are two orders of magnitude lower than PAH concentrations in the Black River. However, the concentrations of sediment PAHs found in OWC, a state nature preserve, were higher than expected, when compared to Buckeye Lake (Table 1), a heavily used recreational lake with many old septic systems and two small towns in its drainage (Baumann et al., 1982). Concentrations of B(a)P in OWC sediment collected near a railroad tressel were an order of magnitude higher than concentrations of B(a)P in Buckeye Lake sediment. Many other PAHs were found in OWC at concentrations two to three times higher than in Buckeye Lake sediment. It is possible that even higher sediment concentrations of PAHs are present in localized areas of OWC, such as sites of discarded railroad ties, where sediment samples were not taken. The closing of the coking facility at the Black River and the unexpectedly high concentrations of PAHs at certain locations in OWC may partly explain the lack of significant difference between concentrations of biliary B(a)P equivalents in OWC and Black River bullhead.

The results from common carp bile suggests that certain species may serve as better indicators of exposure than others. Concentrations of B(a)P equivalents in the bile of Black River carp were approximately six times greater than concentrations of B(a)P equivalents in the bile of Black River bullhead. English sole (Parophrys vetulus), a species exhibiting high tumor frequencies, have been shown to form greater numbers of liver adducts than starry flounder (Platichthys stellatus), a species showing much lower tumor frequencies, when both species are collected from the same contaminated sites (Varanasi et al., 1986). This suggests that species exhibiting a high frequency of tumors may retain more metabolites as adducts and less metabolites in bile than species with a low tumor frequency, even though both are exposed to identical concentrations of PAHs. Common carp and brown bullhead do exhibit different susceptibilities to hepatic tumor induction; bullheads have a high incidence of liver lesions and carp do not (Black, 1988) and (Baumann, 1988, unpublished data). This lack of

liver neoplasia in carp would be consistent with lower adduct formation, and therefore a higher concentration of PAH metabolites in bile. Different rates of adduct formation may account for some of the differences in bile metabolite levels noted between the two species.

The results of this study suggest that biliary concentrations of B(a)P equivalents may be used to successfully indicate PAH exposure of fresh-water fish. Biliary concentrations of B(a)P equivalents cannot, however, be used as a predictor of specific tumor frequencies. In addition to B(a)P and B(a)P metabolites, metabolites of pyrene and fluoranthene also fluoresce at B(a)P wavelengths (Krahn et al., 1987). Because fish are exposed to sediment containing complex mixtures of PAHs, the exact portion of biliary metabolites fluorescing at 380/430 nm which is carcinogenic cannot be determined without identification of individual chromatographic peaks.

Use of an HPLC-fluorescence method to determine PAH exposure has several advantages. The HPLC-fluorescence method is simple to perform and does not require complicated and time-consuming extractions and derivations. Biliary concentrations of B(a)P equivalents provide information that cannot be obtained from histopathological or sediment data. High concentrations of PAHs in sediment indicate a potential for exposure, while high concentrations of PAHs in bile indicate actual exposure. Because depuration of metabolic PAHs is rapid the concentration of biliary B(a)P equivalents indicates current levels of exposure. Therefore, this technique may be used to monitor rapidly changing exposure levels due to disturbances such as contaminant spills. Current exposure levels cannot be determined through histopathological data since the latency period associated with tumors in unknown.

The HPLC-fluorescence method could be used in conjunction with histopathological analyses and sediment analyses to obtain a more detailed scenario of the source of exposure, extent of exposure and the resultant pathology occurring in a fish population. Since this method is rapid and inexpensive, the HPLC-fluorescence technique could be used to prescreen waterways for the

566

purpose of evaluating PAH exposure in freshwater fish populations.

References

Baumann, P. C., W. D. Smith & M. Ribick, 1982. Hepatic tumor rates and polynuclear aromatic hydrocarbon level in two populations of brown bullhead (*Ictalurus nebulosus*). In: M. W. Cooke, A. J. Dennis and G. L. Fisher, (Eds), Polynuclear Aromatic Hydrocarbons: Sixth International Symposium of Physical and Biological Chemistry. pp. 93–102. Battelle Press, Columbus, OH.

Baumann, P. C. & J. C. Harshbarger, 1985. Frequencies of liver neoplasia in a feral fish population and associated carcinogens. Mar. Envir. Res. 17: 324–327.

Black, J. J., P. P. Dymerski & W. F. Zapisek, 1979. Routine liquid chromatographic method for assessing polynuclear aromatic hydrocarbon pollution in freshwater environments. Bull. Envir. Contam. Toxicol. 22: 278–284.

Black, J. J., 1988. Chronic effects of toxic contaminants in large lakes. In: N. Schmidthe (Ed.), Toxic Contamination in Large Lakes, Vol. 1, pp. 55–81, Lewis Publishers, Chelsea, MI.

Fabacher, D. L., C. J. Schmitt, J. M. Besser & M. J. Mac, 1988. Chemical characterization and mutagenic properties of polycyclic aromatic compounds in sediment from tributaries of the Great Lakes. Envir. Toxicol. Chem. 7: 529–543.

Krahn, M. M., M. S. Myers, D. G. Burrows & D. C. Malins, 1984. Determination of metabolites of xenobiotics in the bile of fish from polluted waterways. Xenobiotica. 14: 633–646.

Krahn, M., D. G. Burrows, W. D. Macleod & D. C. Malins, 1987. Determination of aromatic compounds in hydrolyzed bile of English sole (*Parophrys vetulus*) from polluted sites in Puget Sound, Washington. Arch. Envir. Contam. Toxicol. 16: 511–522.

Maccubbin, A. E., S. Chidambaram & J. J. Black, 1988. Metabolites of aromatic hydrocarbons in the bile of brown bullheads (*Ictalurus nebulosus*). J. Great Lakes Res. 14: 101–108.

McCain, B. B., H. O. Hodgins, W. D. Gronlund, J. W. Hawkes, D. W. Brown & M. S. Meyers, 1978. Bioavailability of crude oil from experimentally oiled sediments to English sole (*Parophrys vetulus*) and pathological consequences. J. Fish. Res. Bd. Can. 35: 657–664.

Roubal, W. T., T. K. Collier & D. C. Malins, 1977. Accumulation and metabolism of carbon-14 labeled benzene, naphthalene and anthracene by young coho salmon (*Oncorhynchus kisutch*). Arch. Envir. Contam. Toxicol. 5: 513–529.

SAS, 1985. SAS Users Guide: Statistics, Version 5. SAS Institute, Cary, North Carolina.

Varanasi, U., M. Nishimoto, W. L. Reichert & B. J. LeEberhart, 1986. Comparative metabolism of benzo(a)pyrene and covalent binding to hepatic DNA in English sole, starry flounder and rat. Cancer Res. 46: 3817–3824.

West, W. R., P. A. Smith, P. W. Stokes, G. M. Booth, T. Smith-Oliver, B. E. Butterworth & M. L. Lee, 1984. Analysis and genotoxicity of a PAC-polluted river sediment. In: M. W. Cooke & A. J. Dennis, (Eds.), Polynuclear Aromatic Hydrocarbons: Mechanisms, Methods and Metabolism. pp. 1395–1411. Battelle Press, Columbus, OH.

Hydrobiologia **188/189**: 567–572, 1989.
M. Munawar, G. Dixon, C. I. Mayfield, T. Reynoldson and M. H. Sadar (eds)
Environmental Bioassay Techniques and their Application.
© 1989 *Kluwer Academic Publishers. Printed in Belgium.*

The use of sheepshead minnow (*Cyprinodon variegatus*) and a benthic copepod (*Tisbe battagliai*) in short-term tests for estimating the chronic toxicity of industrial effluents

T. H. Hutchinson & T. D. Williams
ICI Brixham Laboratory, Freshwater Quarry, Brixham, Devon TQ5 8BA, England

Key words: effluent, chronic, sheepshead minnow, copepod

Abstract

Summary results of laboratory investigations into potential chronic effects of industrial effluent discharges are presented and discussed. The sheepshead minnow (*Cyprinodon variegatus*) and the benthic copepod (*Tisbe battagliai*) were selected as test species. Toxicity tests were conducted on newly hatched (approximately 24 hours old) sheepshead minnow larvae. Survival and growth (as dry weight) effects were measured over 7 days. Two different stages of the copepod life cycle were tested: effects on adult female survival and reproduction were measured over 9 days, and naupliar survival over 7 days. The results were incorporated into an existing monitoring programme of the effluent disposal area in the North Sea. Predicted effluent dilutions in the disposal area would exceed one million times within 8 hours. This dilution is 18 and 100 times greater than the 7 day lowest no observed effect concentration (NOEC) for copepod and sheepshead minnow respectively.

Introduction

Interest is increasing in the potential long-term biological effects resulting from the discharge of effluents into the aquatic environment. A limit to this approach has been the cost and length of time required to generate sublethal toxicity data. Since effluents vary considerably over time, frequent measurements of toxicity are essential. Therefore, tests for estimating the chronic toxicity of industrial effluents should be of short duration. The sensitivity of fish and invertebrate larval stages to a range of single toxicants has been widely reported (McKim, 1977, 1985; Patin, 1982) and therefore tests should also incorporate sensitive early life history stages.

Short-term methods have been developed to estimate the chronic toxicity of effluents to aquatic

fauna (Norberg & Mount, 1985; Horning & Weber, 1985). This paper describes the application of short-term (7–9 day) saltwater test methods using fish larvae and benthic copepods in a site specific study.

The methods were selected in order to estimate the chronic toxicity of effluent byproducts from the manufacture of methyl methacrylate (Perspex). Byproduct acid (BPA) from the manufacture of methyl methacrylate is discharged into an approved disposal area in the North Sea. Some of this effluent is used to manufacture ammonium sulphate, a process that produces an ammonium sulphate purge liquor (PL) effluent which is also discharged to sea.

The composition of these effluents is shown in Table 1. The effluents contain acid and ammonia, together with a number of organic components

Table 1. Composition of methyl methacrylate derived wastes (%).

Constituent	Byproduct acid (BPA)	Ammonium sulphate purge liquor (PL)
Ammonium sulphate	20–30	5–15
Sulphuric acid	35–42	–
Ammonium acetone disulphonate	2–4	30–50
Water-soluble organics	2–4	
Water	to 100	to 100

which are difficult to characterize. The concern that the disposal operations might result in a local accumulation of potentially harmful organics in the disposal area required studies to establish possible biological effects. Laboratory results are discussed in relation to measured and predicted effluent concentrations in the disposal area.

Materials and methods

Fish larvae

The sheepshead minnow (*Cyprinodon variegatus*) was selected as the test species since it is widely distributed, is an important link in the food chain, is widely reported in the literature, and can be readily cultured in the laboratory (Ward & Parrish, 1980). Larvae were produced following induced spawning of adult fish using human chorionic gonadotropin injections. Toxicity tests commenced within approximately 24 hours of the larvae hatching. Byproduct acid (BPA) and ammonium sulphate purge liquor (PL) effluent samples were held at reduced temperature (4 °C) before testing began. Test solutions were made up directly on a % volume : volume (v/v) basis with dilution water drawn from the local seawater supply (approximately 34.5‰) acclimated to test temperature (25 ± 2 °C).

A non-aerated, static-renewal test system was used, adapted from the method described in detail by Norberg & Mount (1985). Duplicate test vessels with a 500 ml working volume were used per test concentration. Ten larvae were allocated to each test vessel and the vessels covered during the exposure period to prevent contamination or loss of test solution. Larvae were fed twice daily (generally before and after test solution renewal) with 24 hour old brine shrimp (*Artemia*) nauplii in order to maintain a constant supply of food.

Test material and control solutions were renewed daily, with pH, dissolved oxygen and temperature being measured before and after the renewal process. Mortalities were recorded daily and dead larvae discarded. At the end of the 7 day exposure period the surviving larvae in each test vessel were rinsed with distilled water, dried overnight at 105 °C and the pooled larvae weighed to 0.01 mg. Mean dry weights were adjusted for the number of larvae surviving.

The 7 day survival data were transformed by converting the square root of the proportion of surviving larvae to arc sine (a variance-stabilizing transformation). Along with the dry weight of larvae alive at the end of the tests, these data were subjected to one way analysis of variance (ANOVA, $P < 0.05$). If the results were significantly different, Dunnett's procedure was used to compare concentration means to control means ($P < 0.05$). Full details of the statistical methods are described in Horning & Weber (1985).

Copepods

The harpacticoid copepod (*Tisbe battagliai*) was also selected as a test species. Marine benthic copepods of the genus *Tisbe* (*Harpacticoida*) are particularly suitable for studying toxicological effects: they have a wide geographic distribution, possess short lifecycles, and require minimal space and equipment for testing. They are comparatively easy to culture in the laboratory (Battaglia, 1970) and animals at all developmental stages can be harvested from cultures at any time of the year.

Tisbe battagliai, a sibling species of the holothuriae group, has been found in shallow waters in coastal regions of Europe and the Atlantic coast of the USA (Volkmann-Rocco, 1972). During its life cycle *Tisbe* passes through six naupliar and six copepodid stages, the last of which is the adult. At 20 °C the development from nauplius to adult takes about 10 days and adult females release their first brood of offspring after approximately 14 days.

Harpacticoid copepods are important components of the marine meiobenthos and constitute an important food resource for larval and juvenile fish (Hicks & Coull, 1983).

The stock cultures were maintained in a filtered, natural seawater, identical to the test dilution water, at a temperature of 20 ± 1 °C, and fed a diet of algae (*Isochrysis galbana*). Toxicity tests were conducted on two different stages of the copepod life cycle: adult females carrying egg sacs and newborn offspring (nauplii). This provided a comparison of sensitivity between the two different life history stages. Newborn (<24 hours) nauplii were obtained by isolating ovigerous females from a mass culture 24 hours before an experiment so that nauplii collected the next day would be nearly the same age (<24 hours). Adult females were obtained from a mass culture derived initially from a cohort of nauplii collected during 24 hours. After approximately 11 days females carrying their first egg sac were visible and these were selected as test animals.

A stock solution of the PL effluent was prepared in dilution water. Test solutions were prepared by the addition of appropriate volumes of stock solution to dilution water. Experiments were conducted at 20 ± 1 °C and a photoperiod of 16L : 8D provided.

Reproduction experiments were initiated by introducing females at the start of their adult reproductive period, ie. when they produced their first egg sac. Males were excluded since a single copulation enables females to produce fertile eggs throughout their reproductive period.

At the start of the test, one ovigerous female was randomly assigned to each test chamber. The test containers were multi-chambered disposable polystyrene tissue culture cells with close-fitting lids. Each test chamber contained 5 ml of test solution.

Observations (by microscope) were made daily for mortality and the presence of offspring in each test chamber.

The test solutions were prepared on the day of use and were renewed every 48 hr. On each renewal day surviving females were transferred to the new solutions and fed 1×10^5 cells/ml *Isochrysis galbana*. The number of offspring produced by each female was counted after each renewal and the end of the test. Temperature was measured daily. The duration of the experiment was 9 days.

Further evidence of biological effects was sought in studies of naupliar survival after 7 days. Groups of 20 nauplii were exposed to concentrations of PL effluent under semi-static conditions in test chambers containing 5 ml of test solution. At the start of the test, 5 nauplii were randomly added to each of 4 test chambers giving a total of 20 nauplii per concentration. The test solutions were prepared on the day of use and partially renewed every 48 hr. On each renewal day 80% (4 ml) of the test solution was replaced with freshly prepared test solution containing 1×10^5 cells/ml *Isochrysis galbana*. Mortality was recorded daily for each test chamber.

Statistically significant ($P < 0.05$) effects on survival of the adult females were calculated using a contingency table (exact test) procedure. The reproduction data were then examined using Dunnett's procedure following one way analysis of variance to identify significant differences from the control. Two approaches were used for analyzing the reproduction data. The mean number of offspring produced among all the animals exposed was calculated. If an animal died after producing young, the total number of offspring produced up to the time of death was used. The reproduction data were also analyzed after making an adjustment for mortality. Those animals that died during the experiment were not included in the analyses.

The data for naupliar survival did not meet the assumptions for normality therefore a non-parametric test, Steel's Many-One Rank Test was used to determine statistically significant effects on naupliar survival. Full details of the statistical methods are described in Horning & Weber (1985).

Results

Fish larvae

The larval sheepshead minnow toxicity test method was applied successfully for both BPA and PL effluents. A summary of the results is presented here. The measured parameter ranges for BPA were: dissolved oxygen 3.0 to 6.4 mg l^{-1}; pH 5.80 to 8.40 and temperature 23.2 to 25.8 °C. Similarly, for PL the values were: dissolved oxygen 3.0 to 6.6 mg l^{-1}; pH 7.54 to 8.31 and temperature 24.0 to 25.0 °C. No aeration was introduced in order to avoid the potential loss of volatile test materials.

The 7 day survival and growth (dry weight) data are shown in Table 2. Data are pooled from both test vessels for each treatment. Both effluents tested showed a similar toxicity to larval sheepshead minnow during the 7 day exposure period. From the calculated NOEC values growth was more sensitive than survival when BPA was tested, while the reverse was true when larvae were exposed to PL. However, the most sensitive NOEC was calculated to be 0.01% v/v for both effluents.

Copepods

The effect of PL effluent on the survival and reproduction of adult females is shown in Table 3. The results provide a comparison of 5 and 9 day exposures.

All females in the control produced 2 broods of offspring after 5 days and a minimum of 3 broods after 9 days. After 5 days there was a significant effect on reproduction at the maximum effluent concentration tested which was 0.0056%. Similar results were obtained after 9 days. The results were not changed after adjusting the reproduction data for the adult mortality noted.

The mortality of the nauplius stage after 7 days is shown in Table 4. After 7 days mortality of the nauplii at effluent concentrations $> 0.0018\%$ was significantly greater than the control.

Naupliar survival appeared to be a more sensitive indicator of effluent toxicity in this study than adult reproduction. The NOECs for naupliar survival and adult reproduction were 0.0018% and 0.0032% respectively.

The temperature throughout the experiments ranged from 19.5 to 20.7 °C, the pH from 7.66 to 8.30 and dissolved oxygen concentration from 6.4 to 7.6 mg l^{-1}.

Table 2. 7 day % survival and growth of sheepshead minnow larvae exposed to BPA and PL effluents.[a]

Concn (% v/v)	BPA effluent		PL effluent	
	% survival	Mean dry weight (mg)	% survival	Mean dry weight (mg)
Control	95	1.346 (0.064)	100	1.609 (0.522)
0.00125	100	1.156 (0.016)	nt	nt
0.0025	95	1.146 (0.203)	100	1.450 (0.277)
0.005	95	1.209 (0.086)	100	1.285 (0.139)
0.01	95	1.199 (0.117)	95	2.376 (0.696)
0.02	95	1.003 (0.111)*	90*	1.548 (0.059)
0.04	nt	nt	0*	–

* Significant difference from the control (P < 0.05).
[a] Dry weight data are summarised as the means of 2 replicates per treatment (± standard deviations).
nt Not tested.

Table 3. Results of the 9 day reproduction test with adult *Tisbe battagliai* exposed to PL effluent.[a]

Concn (% effluent)	5 days		9 days	
	% mortality	No of offspring produced	% mortality	No of offspring produced
Control	0	50.5 (8.4)	0	80.0 (19.0)
0.00056	0	41.0 (12.6)	0	66.3 (17.6)
0.0010	0	43.6 (7.0)	0	82.1 (19.4)
0.0018	0	52.3 (10.4)	0	101.9 (33.2)
0.0032	0	43.8 (9.8)	0	68.3 (16.6)
0.0056	10	30.8 (17.5)*	20	47.7 (29.5)*

* Significant difference from the control (P < 0.05).
[a] Reproduction results are summarized as the means of 10 replicates per treatment (± standard deviations) not adjusted for mortality.
One female was exposed per replicate.

Table 4. Results of the 7 day mortality test with naupliar *Tisbe batagliai* exposed to PL effluent.[a]

Concn (% effluent)	Total number tested	Total number of mortalities	% mortality
Control	20	0	0
0.0010	20	0	0
0.0018	20	0	0
0.0032	20	6	30*
0.0056	20	18	90*
0.010	20	20	100*
0.018	20	20	100*

* Significant difference from the control ($P < 0.05$).

[a] Mortality data are summarized as the means of 4 replicates per treatment.

5 nauplii were exposed per replicate.

Discussion

Both effluents showed a similar toxicity to sheepshead minnow larvae during the 7 day exposure period. Generally, % survival was inversely related to effluent test concentration. However, the relationship between larval dry weight and effluent test concentration was less clear. In particular, the mean dry weight of larvae exposed to 0.01% v/v of PL (2.376 mg) was greater than that of the control larvae, and was markedly greater than other mean larval dry weight values. The reason for this atypical value is unclear. The reasons for the differential sensitivity of survival and growth between the two effluents are not known but may include a temporary delay in early growth, chemically-induced growth stimulation or the influence of density on larvae surviving exposure (Woltering, 1984). The most sensitive NOEC for both effluents was calculated to be 0.01% v/v (equivalent to a dilution of 1×10^4).

Both stages of the copepod life cycle were more sensitive to PL effluent than larval sheepshead minnow. BPA was not tested using *Tisbe*.

Naupliar survival appeared to be a more sensitive indicator of effluent toxicity than adult reproduction. The 7 day NOEC for naupliar survival was 0.0018% (equivalent to a dilution of 5.6×10^4). These results demonstrate the impor-

tance of including critical early life history stages in toxicity assessments. Predictive results based on adult stages alone may underestimate the effect on more sensitive early developmental stages. Survivorship data alone require the least effort to collect. Due to the greater sensitivity of the nauplius stage further biomonitoring work on PL effluent should focus on naupliar survival and development.

The static culture renewal system used in copepod tests was a simple system requiring little maintenance. In order to reduce the level of effort required the test solutions were renewed every 48 h. In view of the unstable nature of effluents the test methods could be improved by renewing the test solutions daily.

Because the animal is easy to culture in the laboratory and tests were performed with simple laboratory equipment, the tests are economic to perform. The culture and toxicity tests were carried out using natural seawater, however *Tisbe* species have also been cultured in synthetic seawater (Verriopoulos & Dimas, 1988). This would enable a wide range of laboratories to participate in toxicity tests with these marine species.

The cost and difficulty in performing chronic biomonitoring tests will have a major effect on the number of effluent samples tested and their testing frequency. Short-term sublethal tests provide a cost effective means of biomonitoring effluents. However, to be of most value they should first be validated against longer term chronic experiments.

The lifespan of *Tisbe battagliai* is short (approx 31–36 days at 20 °C) therefore life cycle toxicity data can be produced with reduced expenditures of time and cost compared to other marine species. Further work has been undertaken with *Tisbe battagliai* to validate the short-term (7–9 d) estimates of chronic toxicity against longer term sublethal experiments.

The most sensitive NOECs for copepod and larval sheepshead minnow were related to dilution studies in the disposal area (Lewis, 1984). Within 200 seconds and one hour of discharge dilutions of 1×10^4 and in excess of 4×10^4 were achieved respectively. Predicted effluent dilutions in the wake of the disposal vessel would exceed one million times within 8 hours. This is 18 and 100 times more dilute than the 7-day lowest no observed effect concentration (NOEC) for copepod and larval sheepshead minnow respectively.

The larval fish and copepod bioassays were found to be relatively rapid and cost-effective means of estimating the chronic toxicity of effluents. The methods proved useful in the situation described where effluent sample volumes were small and flow-through testing was impractical. The methods are being further evaluated using reference toxicants to compare species sensitivity.

Additionally, a biomonitoring programme using these tests is being undertaken to provide a broad base of estimated chronic values for important industrial effluents.

References

Battaglia, B., 1970. Cultivation of marine copepods for genetic and evolutionary research. Helgolander Wiss. Meeresunters. 20, 385–392.

Hicks, G. R. & B. C. Coull, 1983. The ecology of marine meiobenthic harpacticoid copepods. Oceanogr. Mar. Biol. A. Rev. 21: 67–175.

Horning, W. B. & C. I. Weber, 1985. Short term methods for estimating the chronic toxicity of effluents and receiving waters to freshwater organisms. US EPA/600/4-85/014.

Lewis, R. E., 1984. Studies of dilution in the wake of the methyl methacrylate dumping vessel. ICI Brixham Laboratory report BL/B/2529.

McKim, J. M., 1977. Evaluation of tests with early lifestages of fish for predicting long term toxicity. J. Fish Res. Bd. Can. 34: 1148–1154.

McKim, J. M., 1985. Early life stage toxicity tests. In: G. M. Rand & S. R. Petrocelli, (eds), Fundamentals of Aquatic Toxicology. Hemisphere Pub. Co., Washington, D. C., pp. 58–95.

Norberg, T. J. & D. I. Mount, 1985. A new fathead minnow (*Pimephales promelas*) subchronic toxicity test. Envir. Toxicol. Chem. 4: 711–718.

Patin, S. A., 1982. Pollution and the biological resources of the oceans. Butterworth Scientific Chapt. 8: 124–129.

Ward, G. S. & P. R. Parrish, 1980. Evaluation of early life-stage toxicity tests with embryos and juveniles of sheepshead minnows (*Cyprinodon variegatus*). Aquat. Toxicol. ASTM STP 707: 243–247.

Woltering, D. M., 1984. The growth response in fish chronic and early life-stage toxicity tests: a critical review. Aquat. Toxicol. 5: 1–21.

Verriopoulos, G. & S. Dimas, 1988. Combined toxicity of copper, cadmium, zinc, lead, nickel and chrome to the copepod *Tisbe holothuriae*. Bull. Envir. Contam. Toxicol. 41: 378–384.

Volkmann-Rocco, B., 1972. *Tisbe battagliai* n. sp., a sibling species of *Tisbe holothuriae* Humes (Copepoda Harpacticoida). Arch. Oceanogr. Limnol. 17: 259–273.

Hydrobiologia **188/189**: 573–576, 1989.
M. Munawar, G. Dixon, C. I. Mayfield, T. Reynoldson and M. H. Sadar (eds)
Environmental Bioassay Techniques and their Application.
© 1989 *Kluwer Academic Publishers. Printed in Belgium.*

Preliminary investigations on the influence of suspended sediments on the bioaccumulation of two chlorobenzenes by the guppy (*Poecilia reticulata*)

S. Marca Schrap[1,2] & Antoon Opperhuizen[2,3]
[1] *Laboratory of Environmental and Toxicological Chemistry, University of Amsterdam, Nieuwe Achtergracht 166, 1018 WV Amsterdam, The Netherlands;* [2] *Department of Basic Veterinary Sciences, Environmental Toxicology Section. University of Utrecht, P.O. Box 80176, 3508 TD Utrecht, The Netherlands;* [3] *to whom correspondence should be addressed.*

Key words: bioavailability, bioaccumulation, hydrophobicity, fish, chlorobenzenes

Abstract

In this study the uptake by guppies (*Poecilia reticulata*) of 1,2,3-trichlorobenzene (TCB) and hexachlorobenzene (HCB) from sediment-free water was compared with uptake in the presence of suspended sediment. The results show that the influence of suspended sediment on the uptake of chlorobenzenes varies with test compound. For TCB uptake was not influenced by the presence of suspended sediment. This is probably due to the large amount of the chemical which is dissolved relative to the amount which is present in the sorbed state. For the more hydrophobic HCB, the concentration found in the fish from the system with suspended sediment was significantly higher than in fish from the control experiment.

Introduction

The uptake of persistent chemicals by fish from water has been investigated in many studies. In such experiments, uptake rates and bioconcentration factors are measured in fish/water systems. However, the presence of particulate matter (for instance sediment) in water has been shown to influence bioaccumulation. A reduction of bioconcentration factors for fish by the presence of particulate matter has been demonstrated for Kelthane (Eaton *et al.*, 1983), lindane, a trichlorobiphenyl and a trichlorophenol (Carlberg, 1986) and polycyclic aromatic hydrocarbons (McCarthy, 1985).

In this study the uptake of two chlorobenzenes by guppies in the presence of suspended sediment was compared with the uptake in a system without sediment.

Materials and methods

Reagent grade 1,2,3-trichlorobenzene (TCB) and hexachlorobenzene (HCB) (purity $> 97\%$) were used as the test compounds. Laboratory bred male guppies (*Poecilia reticulata*) (weight, approximately 150 mg; length, 15 to 23 mm) were used in the experiments. The sediment, a natural sediment from the Oostvaardersplassen (The Netherlands) was wet sieved (1 mm) and stored in water at 5 °C until use. It had a dry fraction of about 40%, and an organic carbon content of 3% of the dry weight. Before the experiments the sediment was oven dried at 200 °C.

574

The experiments were carried out in 3.5 l vessels as shown in Fig. 1. Each vessel and pump constituted a closed system with essentially no head space for air. Fish and water(/sediment) samples were taken from the top of the vessel. Three simultaneous experiments were carried out. In experiment A 30 fish and 2450 mg (dry weight) of sediment were present, in experiment B only 2450 mg sediment was added and in experiment C only 30 fish were present. Experiment B was used as a control for the loss of the chemicals by volatilization or degradation by microorganisms. During the experiments the water was circulated to keep the sediment in suspension. Fish were fed (Tetramin fish food) four times during the experimental period. Water (2 : 1 Amsterdam tap water and distilled water) was saturated with the chlorobenzenes using a generator column (Opperhuizen *et al.*, 1988), and was added to the three 3.5 l vessels. After dilution the concentrations in water were 125 μg/l for TCB and 1.5 μg/l for HCB. To avoid lethal toxic effects to the fish, the concentrations of the test compounds in the water were chosen so that during the uptake period the sum of the concentrations of the chlorobenzenes in the fish would not exceed 2 μmol per gram of fish (Van Hoogen & Opperhuizen, 1988).

To reach an equilibrium between the concentrations of the chlorobenzenes in the water and on the sediment, sediment was added to the systems 48 hours before the fish. As indicated by previous sorption experiments (unpublished data of the authors), equilibrium will be reached in 48 hours for these chlorobenzenes. The total amount of chlorobenzenes was the same in each system whether sediment was present or not. The uptake of the two chlorobenzenes by the fish from experiments A and C were compared over 290 hours. Twelve fish were sampled from each of systems A and C, and 14 water or water/sediment suspension samples were taken from all three systems.

Fish samples (2 fish per sample) were homogenized in a mortar and extracted with distilled hexane by heating under reflux for 90 minutes. After centrifugation the hexane layer was concentrated under a gentle nitrogen stream and analyzed on a Tracor 550 gas chromatograph with a [63]Ni electron capture detector (2.4 m column packed with 2% Dexil 300 Chromosorb W HP 80–100 mesh; measured detection limit, $\pm$ 100 pg for both chlorobenzenes). Water samples from system C (10 ml) were extracted with 4 ml distilled hexane and analyzed after concentration of the hexane layer. Sediment-suspension samples (10 ml) were taken from systems A and B and extracted as described for the water samples.

Results and discussion

Analysis of fish samples showed that there was no detectable background contamination by the chlorobenzenes. In addition, less than 10 pg/g of TCB and less than 1 pg/g of HCB were found as background contamination in the sediment. Before the experiments started, a blank experiment (without chemicals) was carried out. Fifteen fish were placed in a sediment suspension of 700 mg/l for 17 days. During the exposure the oxygen concentration was consistently > 7 mg/l. Only one mortality occurred, on day 17. Otherwise the fish showed no reduction in activity and

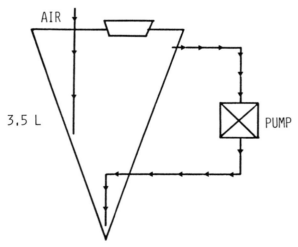

AIR

3,5 L

PUMP

Fig. 1. A diagram of the experimental system. All parts except for the pump were made of glass. The gap on the top, for taking fish and water samples, was closed by a stopper. By circulating the water, the sediment was kept in suspension.

only a slight color change. During the actual experimental period fish from experiment C had low activity, and after four days three fish had died. There were no mortalities in the system where sediment was present (experiment A). In Fig. 2 the concentration of TCB in water is shown. The concentration of HCB in the water and the sediment suspension was below the detection limit in all three test systems. This is probably due to the small sample amounts of 10 ml. Since there was no decrease in the concentration of TCB in the system with only sediment (experiment B) (Fig. 2), it can be concluded that there were no losses of the compound by volatilization or degradation by microorganisms. The uptake of TCB by the fish was not influenced by the presence of sediment in the system (Fig. 3), while the concentration of HCB in the fish from the system with sediment was significantly higher than that in the fish from the system without sediment (Fig. 4).

Black & McCarthy (1988) concluded from their experiments with trout and 2,2',5,5'-tetrachlorobiphenyl and benzo(a)pyrene, that compounds sorbed to humic acids are not available for uptake by fish. Also Eaton *et al.* (1983),

McCarthy (1985) and Carlberg (1986) found a decrease of uptake and bioconcentration factors for different compounds when clay or humics were present. An enhanced concentration for HCB in fish, as was found in the described experiments, is in disagreement with these observations. However, from experiments of Kuehl (1987) it is clear that sediment can be found in the intestinal tract of carp. Depending on the sediment and on the test compound, the presence of sediment in the intestinal tract of fish may influence the analytical results. For the described experiments it is not known if sediment was ingested by the fish. Furthermore, even if sediment was in the intestinal tract of the fish, it remains unclear whether or not the chemicals sorbed on the sediment would be taken up by the fish. Additional research is required.

In conclusion, it is evident that the system outlined here is suitable for measurement of the uptake of hydrophobic chemicals in the presence of suspended sediments. In addition, the concentration of HCB in fish can be influenced by the presence of suspended sediment. This is not the case for TCB.

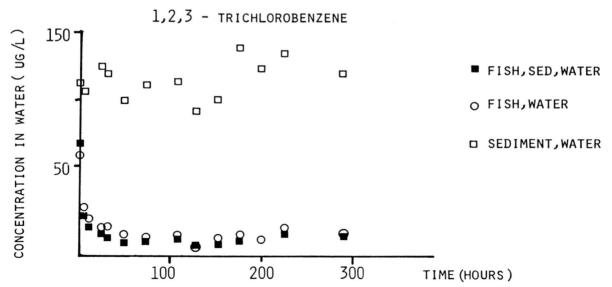

Fig. 2. The concentration of 1,2,3-trichlorobenzene in the water or sediment suspension in all three experimental systems during the uptake period.

576

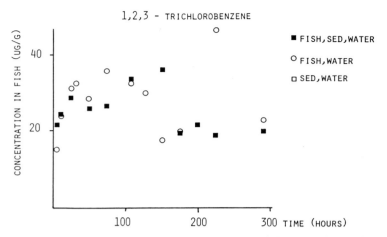

Fig. 3. The concentration of 1,2,3-trichlorobenzene in the fish during the uptake period.

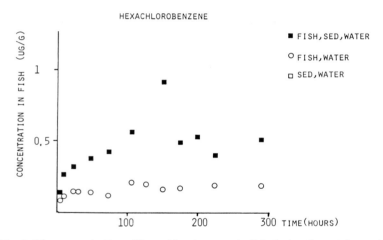

Fig. 4. The concentration of hexachlorobenzene in fish during the uptake period.

Acknowledgements

This work was supported by the Institute for Inland Water Management and Waste Water Treatment, Ministry of Transport and Public Works, The Netherlands.

References

Black, M. C. & J. F. McCarthy, 1988. Dissolved organic macromolecules reduce the uptake of hydrophobic organic contaminants by the gills of rainbow trout (*Salmo gairdneri*). Envir. Toxicol. Chem. 7: 593–600.

Carlberg, G. E., K. Martinsen, A. Kringstad, E. Gjessing, M. Grande, T. Kallqvist & J. U. Skare, 1986. Influence of aquatic humus on the bioavailability of chlorinated micropollutants in Atlantic salmon. Arch. Envir. Contam. Toxicol. 15: 543–548.

Eaton, J. G., V. R. Mattson, L. H. Mueller & D. K. Tanner, 1983. Effects of suspended clay of bioconcentration of Kelthane in fathead minnows. Arch. Envir. Contam. Toxicol. 12: 439–445.

Kuehl, D. W., P. M. Cook, A. R. Batternman, D. Lothenbach & B. C. Butterworth, 1987. Bioavailability of polychlorinated dibenzo-p-dioxins and dibenzofurans from contaminated Wisconsin river sediment to carp. Chemosphere 16: 667–679.

McCarthy, J. F. & B. D. Jimenez, 1985. Reduction in bioavailability to bluegills of polycyclic aromatic hydrocarbons bound to dissolved humic material. Envir. Toxicol. Chem. 4: 511–521.

Opperhuizen, A., F. A. P. C. Gobas, J. M. D. Van der Steen & O. Hutzinger, 1988. Aqueous solubility of polychlorinated biphenyls related to molecular structure. Envir. Sci. Technol., 22: 638–646.

Van Hoogen, G. & A. Opperhuizen, 1988. Toxicokinetics of chlorobenzenes in fish. Envir. Toxicol. Chem. 7: 213–219.

Hydrobiologia **188/189**: 577–585 1989.
M. Munawar, G. Dixon, C. I. Mayfield, T. Reynoldson and M. H. Sadar (eds)
Environmental Bioassay Techniques and their Application.
© *1989 Kluwer Academic Publishers. Printed in Belgium.*

Identification of developmental toxicants using the Frog Embryo Teratogenesis Assay-*Xenopus* (FETAX)

John A. Bantle [1], Douglas J. Fort & Brenda L. James
Department of Zoology, Oklahoma State University, Stillwater, Oklahoma, 74078 USA; [1]*To whom correspondence should be addressed*

Key words: Xenopus, FETAX, teratogenicity, developmental toxicant, groundwater

Abstract

Because growth and development are processes sensitive to the action of many chemicals, bioassays that screen for developmental toxicants may be more indicative of chronic effects than acute toxicity assays. FETAX is a 96 h whole embryo static renewal test employing the embryos of the frog *Xenopus laevis*. Endpoints are mortality, malformation and growth. Because of the frog's fecundity, its extensive use in basic research and the ability to obtain embryos year-round, it is an ideal organism to use in screening for developmental toxicants. By validating using known mammalian teratogens and the use of rat liver microsomes to stimulate mammalian metabolism, we have extended the use of the system for the prescreening of human developmental toxicants. In past validation work, we have correctly identified the teratogenicity of 15 to 17 compounds used in validation for a predictive accuracy of approximately 88%. In the present study, the ability of FETAX to detect developmental toxicants in groundwater samples taken from an industrial waste dump was evaluated. FETAX showed that it was sensitive enough to detect developmental toxicants in samples without prior concentration. In some samples, less than half the LC50 concentration was required to cause significant malformation. In some cases, a dose-response curve was not obtainable but the test results nonetheless indicated some developmental toxicity. The results of this study indicate that it is necessary to routinely screen for developmental toxicants when establishing water quality criteria for the preservation of species and for human health.

Introduction

The Frog Embryo Teratogenesis Assay- *Xenopus* (FETAX) is a 96 h static renewal test for the presence of developmental toxicants in the environment (Dumont *et al.*, 1983). The test has mortality, malformation and growth inhibition as its primary endpoints. It is designed for both the prescreening of samples for possible human developmental toxicity as well as the preservation of other aquatic species.

A number of validation studies using compounds of known mammalian developmental toxicity have suggested that FETAX will have a predictive accuracy of greater than 85% (Bantle *et al.*, 1988; Sabourin & Faulk, 1987; Dawson & Bantle, 1987; Courchesne & Bantle, 1985). For proteratogens and compounds that rapidly lose their teratogenicity following metabolism, an *in vitro* metabolic activation system using Aroclor 1254-induced rat liver microsomes (Bantle & Dawson, 1988; Fort *et al.*, 1988) has been developed. This system should help reduce the number of false positives and false negatives in future testing.

In two previous studies, FETAX detected developmental toxicants in both surface waters and sediment samples (Dawson *et al.*, 1985; Dawson *et al.*, 1988). These developmental toxicants proved to be heavy metals. Zinc was responsible for most of the effects seen. In the present study, we used FETAX to evaluate the developmental toxicity of ground water contaminated by an industrial waste site containing waste organics. FETAX was sensitive enough to detect these compounds although a complete dose-response curve was not always obtained. The results indicate that not only is FETAX an acceptable test for developmental toxicants but also that it is necessary to screen for developmental toxicants.

Methods and materials

Water samples

Samples were taken from water wells bored around an industrial waste site containing, among other wastes, organic solvents. Well samples were received on 12/12/87. Samples were not opened until sufficient numbers of embryos were obtained to carry out the assays. Two 500 ml bottles were used to store each sample. Only one bottle was opened at a time. After first warming to room temperature, a small aliquot of the sample(s) being tested was removed and pH, dissolved oxygen, hardness, and alkalinity measurements were taken. Larger samples were removed for each series of experiments. The pH of each concentration was adjusted to within a range of 7.2-7.8. Every 24 h during the 96 h test, fresh additions of sample were made to each test dish after pH adjustment (static renewal protocol). Thus, the storage bottle was opened four times during each experiment.

Samples of the distilled water used to make the diluent, the $ZnSO_4$ reference toxicant and FETAX solution were taken for standard metal analysis by atomic absorption.

Animal care

The *Xenopus* adults were obtained from *Xenopus* I (Ann Arbor, MI) and maintained in glass aquaria and/or fiberglass raceways in dechlorinated tap water. This water was filtered through activated carbon-zeolite and aerated for 48 h prior to use. It was periodically tested to ensure that the pH, dissolved oxygen content, hardness, and heavy metal content were at acceptable levels (Peltier & Weber, 1985). Adult frogs were fed beef liver and lung supplemented with liquid vitamins (Polyvisol®).

Assay procedure

FETAX solution (Dawson and Bantle, 1987) was used for all testing both as a defined test medium for controls and as a diluent. FETAX solution is composed of 625 mg NaCl, 96 mg $NaHCO_3$, 30 mg KCl, 15 mg $CaCl_2$, 60 mg $CaSO_4$, $2H_2O$, and 75 mg $MgSO_4$ per L of deionized, distilled water. The pH was 7.9 after mixing and gentle aeration. No additional buffer was required to maintain the pH at 7.6 to 7.9.

Breeding tanks and all glassware were washed in dilute HCl, and then rinsed thoroughly in deionized water. The tanks were filled with FETAX solution and aerated for a short time before introducing the animals.

To induce mating, the male and female received 500 and 1000 IU, respectively, of human chorionic gonadotropin (Sigma®, St. Louis, MO.) via injection into the dorsal lymph sac. Amplexus normally ensued within 2 to 6 h and the deposition of eggs took place from 9 to 12 h after injection.

After breeding the adults and fecal material were removed from the tank and the embryos collected in 60 mm plastic Petri dishes. The jelly coating surrounding the embryos was removed by gentle swirling for 2 to 3 min in a 2% w/v cysteine solution, prepared in FETAX solution. The pH of the cysteine solution was adjusted to 8.1 with NaOH.

Following the removal of the jelly coat, abnormally cleaving embryos and necrotic eggs were removed from the collection of embryos. A second selection ensured that only normally developing embryos (at blastula) were used in the tests. For each test concentration two sets of 20 embryos each were placed in plastic Petri dishes containing a total of 8 ml of solution. At least four sets of 20 embryos were used as controls for each test. The control solutions contained 8 ml FETAX solution. The dishes were covered to minimize evaporation.

Four water samples were tested using FETAX solution as a diluent. Water samples were kept tightly stoppered, in the dark, and at 4 °C until use. The water samples were labelled Sample A, B, C, and D. $ZnSO_4$ served as a positive control while FETAX solution was the negative control. In a previous series of experiments, ground water taken from wells just removed from the study site showed no effect on FETAX endpoints so we did not include a negative groundwater control in this study. Range finder tests were performed immediately but definitive experiments had to be performed using water from the opened sample bottles after several weeks of storage. The use of duplicate sample bottles minimized head space in the bottle. The samples had a large amount of sediment present which we did not pipet into the sample dishes.

The embryos were maintained in the test dishes at 23-24 °C for 96 h. At 24, 48, and 72 h the solutions were changed. At the time of daily test solution changes, dead embryos were removed and the number dead recorded. Determination of gross structural malformations and mortality were made using a dissecting microscope. All embryos were scored by the same individual. Death at 24 h (stages 26,27) was determined by skin pigmentation, structural integrity, and irritability of the embryos, while at 48 (stages 37-39), 72 (stage 42) and 96 h (stage 46) the absence of a heartbeat in the transparent embryos was also used as an indicator of death. Staging was according to Nieuwkoop & Faber (1975).

At 96 h, dead embryos were removed from the dishes and the surviving embryos were fixed in formalin (0.5 to 0.75% w/v). The numbers of dead and malformed embryos were then noted and recorded. Malformed embryos that died prior to 96 h were not included in the number malformed. Probit analysis was performed using the computer program provided by Peltier & Weber (1985). Embryos surviving to fixation were then individually measured (head-tail length) using a Radio Shack® digitizer and Model 16 microcomputer.

Results

Water quality

Each water sample was tested for pH, dissolved oxygen, hardness and alkalinity immediately after the initial opening of the sample bottle. Samples A, B, C, and D all had pH values between 5.2 and 6.1 (Table 1). These values would have been too low to conduct FETAX without a pH effect. For this reason all samples were adjusted to between pH 7.2 and 7.8 immediately prior to testing. All samples had relatively high oxygen contents after initial bottle opening as opposed to some previously tested samples from the same waste site. The O_2 levels, ranging from 7.3 to 8.8 mg/L, were all high enough to support life. Very often we see precipitates from when deoxygenated samples from this waste site were diluted with oxygenated FETAX solution. We did not observe any such precipitation in this particular study. Samples B, C, and D had hardness values higher than FETAX solution, and sample A had a hardness value lower than FETAX solution. Sample B and D hardness readings were three to eight times higher than FETAX solution which may have affected test results somewhat. The alkalinity of sample B was approximately 2.5 times higher than that of FETAX solution but it was not considered to be so high as to affect the test.

Table 2 shows the metal analysis of deionized, distilled H_2O, FETAX solution and $ZnSO_4$ in FETAX solution. Observed values were within the expected range except for chromium which was high. These high chromium readings may have been artifactual, however. In later definitive

Table 1. Initial water quality parameters.

Sample	pH (SU)	Conductivity (μS)	Dissolved oxygen (mg/L)	Hardness (mg/L CaCO3)	Alkalinity (mg/L CaCO3)
FETAX solution	8.1	1750	8.6	112	68
Sample A	5.2		7.3	68	16
Sample B	6.1		8.8	820	164
Sample C	5.5		8.7	148	36
Sample D	5.6		7.9	328	72
ZnSO$_4$ and FETAX solution	6.8	1580	9.3	144	62

Measurements on site samples were performed immediately after opening the bottles. Samples were stored at 5°C prior to analysis.

Table 2. Metals content* analysis of double distilled water, FETAX solution, and ZnSO$_4$ reference toxicant in FETAX solution.

Sample	Na	Ca	Mg	K	Fe	Pb	Zn	Cu	Cr	Ni	Cd
Double distilled H$_2$O	<0.5	<0.5	<0.5	<0.5	0.18	<0.005	<0.01	<0.04	0.031	0.1	<0.001
ZnSO$_4$**	259.0	18.4	14.1	13.7	0.25	<0.005	19.22	<0.04	0.035	<0.1	<0.001
FETAX solution	263.7	20.7	14.2	14.8	0.24	<0.006	<0.01	<0.04	0.033	<0.1	<0.001

* Values are mg/L, as determined by atomic absorption analysis.
** ZnSO$_4$ was made up as a 20 mg/L solution.

experiments we were able to obtain distilled water that read low in chromium (10 fold lower) and the rate of control malformation more typical of FETAX (about 7.5%). The 20 mg/L stock ZnSO$_4$ solution contained only 19.2 (96.1%) mg/L Zinc so values reported in all subsequent tables for Zinc should be reduced 3.9% for actual concentrations. We did not perform a metal analysis on any of the test samples.

FETAX assay results

General plan of testing

Two range finding and a single definitive test were run for each sample. Table 3 shows only the results from the definitive tests. In all cases the definitive tests gave results consistent with the range finding tests indicating not only the repeatability of FETAX but also that there were no pronounced changes in the water samples. By

Table 3. 96 h LC50, EC50 (malformation), and Teratogenicity Indicies.

Sample	96 h LC50 (%)	96 h EC50 (malformation) (%)	TI LC50/EC50
Sample A	>100	>100	–
Sample B	36.3 (29.3–44.8)	21.2 (19.4–23.1)	1.7
Sample C	79.8 (75.5–83.2)	35.9 (33.2–38.8)	2.2
Sample D	>100 (57.1–64.3)	60.6	>1.7
ZnSO$_4$	– (4.0–5.8)	4.8	–

working quickly, minimizing headspace in the sample bottles and refrigerating the samples at 4 °C, we were able to ensure repeatable results as evidenced by the closeness of results between range finding and definitive tests.

Sample A

Sample A exhibited little developmental toxicity. Mortality and malformation rates remained much the same despite increasing concentrations of sample water (Table 3; Fig. 1A). Range finder tests suggested the use of concentrations ranging from 37.5-100%. The controls were adequate as control mortality was 2.5% and the malformation rate was 12.7%. Growth was only affected in the highest concentrations of the A water sample (Fig. 2).

Sample B

Results from the range finding tests suggested that the definitive experiment be set up with concentrations between 18.8-87.5% (Fig. 1B). There was no reduction in toxicity from range finding experiments as the 100% kill occurred at only 43.8% concentration and 100% malformation was at 31.3%. The 96 h LC50 was 36.6% and the 96 h EC50 (malformation) was 21.2% for a TI of 1.7 (Table 3). The teratogenic risk for this sample

must be considered slight as our criteria suggest that sample TIs less than 1.5 convey little teratogenic risk. Figure 1B does show, however, that the dose-response curve for malformation is clearly separate from the mortality curve. Embryolethality is the main concern with this sample. Fig. 2 shows that growth reduction was very sharp with this sample and an overall growth reduction of 40% was observed (60% of control length).

Sample C

Fig. 1C shows the results of the definitive experiment which was set up from 27-81.3% concentration based on the range finder. Some slight loss of toxicity and teratogenicity was evident as we failed to obtain a total kill at the highest concentration and we only obtained 100% malformation at 62.5%. The 96 h LC50 was 79.8% and the 96 h EC50 (malformation) was 35.9% for a TI of 2.2 (Table 3). This sample was not only toxic but also posed a high teratogenic risk. Growth was significantly affected at only 25% concentration and

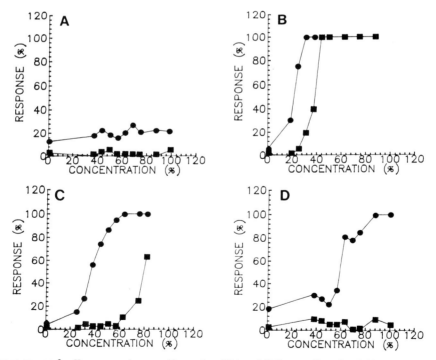

Fig. 1. Dose-response curves for *Xenopus* embryo malformation (●) and (■) mortality after 96 h exposure to site samples A-D. All dilutions were prepared in FETAX solution.

582

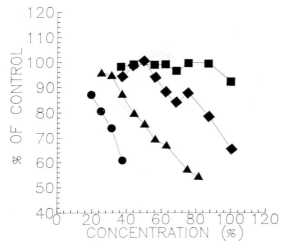

Fig. 2. Representative growth curves for *Xenopus* embryos after 96 h exposure to site samples (■) A; (◆) B; (▲)C; and (●) D.

Fig. 2 shows that C caused embryos at 81.3% concentration to be only 55% of control length. Even further reduction in size could have occurred had there been higher concentrations.

Sample D

Sample D proved highly teratogenic but relatively non-lethal. Based on the results of the range finding experiments, we set up the definitive experiment from 37.5-100% concentration (Fig. 1D). Mortality in the control was only 2.5% but malformation was higher at 18%. Despite the high background, an excellent dose-response was observed for malformation. A TI higher than 1.8 indicated a clear teratogenic risk and only 56.3% concentration was able to significantly inhibit growth (Fig. 2). Figure 2 shows a clear reduction of head-tail length as the concentration increased. Overall length was 65% of controls at 100% concentration.

Zinc sulfate controls

In order for the screening test to be valid, it was important to show that FETAX responded in a reliable manner. A reference standard is one method whereby variability between experimental runs can be accounted for. We performed two

$ZnSO_4$ positive controls for malformation during the definitive part of the test and we present the results of a typical run in Fig. 3 and Table 3. Table 3 shows that the 96 h EC50 values was 4.8 mg/L (Second $ZnSO_4$ control was 5.0 mg/L) and this was within the expected range of 4 to 6 mg/L. Fig. 3 shows that mortality did not occur in this range as expected. Measured zinc concentrations were 3.9% less than shown. The dose-response curves exhibited no significant mortality but showed a smooth increase in malformation with increasing zinc concentrations although a high rate of malformation was initially seen in Fig. 3. Over the concentration range employed, zinc primarily caused malformation of the gut but other malformations also occurred. In the concentration ranges employed, embryonic·growth was not affected significantly. We concluded that FETAX was both repetitive and reliable and that genetic variation was slight.

Order of observed toxicity

The order of embryotoxicity from low to high was A < D < C < B. However, the order of malformation risk from low to high was A < B < C < D. The A sample showed the least

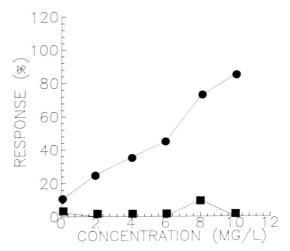

Fig. 3. Dose-response curves for *Xenopus* embryo (●) malformation and (■) mortality after 96 h exposure to $ZnSo_4$. Dilutions were prepared in FETAX solution.

effects as the 100% concentration exhibited only a 21% malformation rate (FETAX solution control was 12.7%) and a 5% death rate. In contrast Sample D at 100% concentration caused a 100% net malformation rate but only 5% death rate. There was a high risk of being born alive but malformed with this sample. Samples B and C showed intermediate risk levels. With regard to the growth inhibition endpoint, the order was A < D < C < B.

Types of malformations observed

Figure 4A shows a typical control larva at the end of the 96 h test. The Stage 46 larva is about to commence feeding and all major organ systems have been formed although histogenesis is still ongoing. The eye is darkly pigmented and the gut is tightly coiled. The tail is straight and the actual head-tail length is about 9.7 mm. Figure 4B shows three larvae that had been exposed for 96 h to another water sample from this same waste site. All surviving larva at this concentration showed identical malformations. The most striking feature was the complete lack of pigmentation especially in the eye region. The larvae were 20% shorter than controls but the picture was enlarged to control size to facilitate observation. Note the poor coiling of the gut, the abnormal mouth region and the pericardial edema. The tail fin is also malformed but not kinked.

Virtually all organ systems for all test samples were malformed, but the most common malformation was moderate to severe loose gut coiling. Head-eye malformations and tail kinking were the next most common malformations observed. We did observe a moderate amount of tissue edema. Although tissue edema is not strictly classified as a malformation, we have generally observed that other associated structures were deformed.

At times, abnormal embryos exhibiting multiple deformities were seen in the control groups. However, they always occurred at a much higher frequency in the sample water in the second or definitive test. Malformations observed included abnormal head, eyes, heart, spine and gut. Edema

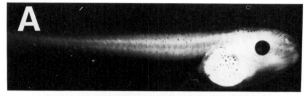

Fig. 4. Control embryo exposed to FETAX solution for 96 h. 4B-Embryos exposed to 25% of an environmental sample from the same waste site for 96 h but different wells. Notice the lack of pigment, abnormal gut development, edema, and craniofacial malformations.

about the head and eyes was evident. Pericardial edema was also seen in controls. Sample B caused severe malformation primarily of the embryonic gut. Severe head and tail malformations were the next most commonly observed malformations in tests with B. The malformations observed in sample C were less severe but still serious. At concentrations below 50%, gut malformations predominated but above 50%, head-eye malformations became very prevalent. Sample D was more teratogenic than toxic and the order of malformations was loose gut coiling > tail kinking > head malformations. As in sample C loose gut coiling occurred first in the lower concentrations and the other malformations were observed with loose gut coiling at the higher concentrations of test sample. Sample A was the least toxic and teratogenic sample tested and the malformations observed were not as severe as in the other samples. The order of

584

malformations were loose gut coiling > edema > tail kinking.

Discussion

The importance of developmental toxicity end-points

We feel that it is most important to measure end-points such as mutagenicity, carcinogenicity and developmental toxicity in water samples. The end-points may relate to chronic effects not seen in ordinary acute toxicity assays. Even though FETAX may be less sensitive than some other bioassays we predict that if the embryos of these organisms were tested, they would be more sensitive than the adults, perhaps by the same factor observed in FETAX. Once FETAX detects a developmental toxicant, a selected battery of other developmental toxicity tests could be used to confirm the result and prioritize samples for mammalian testing. In general, the TI is the most important indicator of teratogenic risk. The TI is a measure of the teratogenicity of a sample and is derived by dividing the 96 h LC50 by the 96 h EC50 (malformation). Any value over 1.5 represents sufficient separation of the two dose-response curves such that the toxicant poses a significant teratogenic risk. Another indicator is the severity of the malformation. Teratogens cause severe malformations while nonteratogens generally kill but do not malform. Teratogens often, although not always, cause growth inhibition at concentrations far lower than that needed to induce malformations.

Sample A was not teratogenic because there was not enough developmental toxicity to cause a significant malformation and mortality rate from which a TI could be derived. Sample D was highly teratogenic as its 96 h EC50 (malformation) was 60.6% and even 100% concentration caused little death (Fig. 3). Its TI was greater than 1.8 and it would require that the sample be concentrated and tested again in order for the exact TI to be determined. Sample C exhibited both toxicity and teratogenicity as the TI was 2.2 indicating a moderate teratogenic risk. The most toxic of the

samples was B with a 96 h LC50 of 21.2%, but even this sample exhibited a TI of 1.7 (Table 3). All the samples exhibited developmental toxicity except for A. Because of these results and our previous results using Tar Creek surface waters and sediments (Dawson et al., 1985 & Dawson et al., 1988), we feel that testing for developmental toxicity using a screen such as FETAX should be routine.

Developmental toxicity of the four water samples

Based on all the above criteria, all samples except A were teratogenic and further testing should be performed to establish the risk to human populations. A exhibited only slight developmental toxicity. The three end-points measured in FETAX are death, malformation and growth inhibition. In human terms, these are spontaneous abortion, birth defects and low birth weight, respectively. Each end-point is therefore meaningful although it is presently impossible to predict the dosage required to cause human effects.

Caution is also warranted regarding the presence of proteratogens and the loss of teratogenicity by metabolic activation. We tested only for direct acting teratogens. Xenopus embryos do not have an active metabolic activation system. An in vitro metabolic activation system for FETAX using Aroclor 1254 induced rat liver microsomes has been developed (Fort et al., 1988) but is has not yet been extensively validated. When available, this addition to the assay will make it more relevant to potential human hazard identification although it is not necessary in order to predict the hazard to other aquatic organisms.

The present study shows that the four water samples demonstrated a rather complete range teratogenicity, toxicity and growth inhibiting effects. These effects remained relatively constant upon sample aging suggesting that the samples did not undergo significant chemical changes upon storage. The FETAX results indicate that a definite hazard exists in these samples for embryos and larvae of aquatic organisms. Further testing using mammals may elucidate the potential human hazard of these samples.

In some cases such as the mortality curve for sample D, FETAX failed to provide complete dose-response information. In such cases it is necessary to evaluate the sample in light of the malformation and growth inhibition end-points. *Xenopus* embryos are more tolerant than fathead minnow embryos (Dawson *et al.*, 1988) and FETAX may generate a number of results such as those seen in Fig. 1D. In such cases, the data is interpretable but the sample may require concentration and retesting in order to obtain definitive results. Although slightly more tolerant than fathead minnow embryos, *Xenopus* embryos develop more quickly and are available in far higher numbers thereby facilitating statistical analysis.

$ZnSO_4$ was used as a positive control for malformation and concentrations between 2–10 mg/L were used. Malformation increased in typical dose-response fashion and the 96 h EC50 (malformation) was within the expected limits. This control accounts for variability in test system components, genetic variability and differences in operator technique.

In summary, there is a need to test specifically for developmental toxicants in the aquatic environment. FETAX provides a 96 h whole embryo bioassay that can be used in initial hazard assessment. By validation with known human teratogens and by adding an *in vitro* metabolic activation system, the results can be used in human hazard identification although retesting with mammalian systems is recommended. FETAX is slightly tolerant compared to fathead minnow but its sensitivity proved adequate for these environmental samples.

Acknowledgements

This work was supported by a Reproductive Hazards in the Workplace Grant 15-30 from the March of Dimes Birth Defects Foundation and an OSU University Center for Water Research Grant. DJF is a recipient of a Presidential Fellowship in Water Quality.

References

Bantle, J. A. & D. A. Dawson, 1988. Uninduced Rat Liver Microsomes as a Metabolic Activation System for the Frog Embryo Teratogenesis Assay-*Xenopus* (FETAX). In: W. J. Adams, G. A. Chapman & W. G. Landis, (Eds) Proceedings of the 10th Aquatic Toxicology Symposium ASTM STP-971., pp. 316–326.

Bantle, J. A., D. J. Fort & D. A. Dawson, 1988. Bridging the gap from short-term teratogenesis assays to human health hazard assessment by understanding common modes of teratogenic action. In: W. J. Adams, G. A. Chapman & W. G. Landis, (Eds.) Proceedings of the 12th Aquatic Toxicology Symposium ASTM, In Press.

Courchesne, C. L. & J. A. Bantle, 1985. Analysis of the activity of DNA, RNA, and protein synthesis inhibitors on *Xenopus* embryo development. Teratogen. Carcinog. Mutagen. 5: 177–193.

Dawson, D. A., C. A. McCormick & J. A. Bantle, 1985. Detection of teratogenic substances in acidic mine water samples using the Frog Embryo Teratogenesis Assay-*Xenopus* (FETAX). J. Appl. Toxicol. 5: 234–244.

Dawson, D. A. & J. A. Bantle, 1987. Development of a reconstituted water medium and preliminary validation of the Frog Embryo Teratogenesis Assay-*Xenopus* (FETAX). J. Appl. Toxicol. 7: 237–244.

Dawson, D. A., E. Stebler, S. A. Burks & J. A. Bantle, 1988. Evaluation of the developmental toxicity of metal-contaminated sediments short-term fathead minnow & frog embryo-larval assays. Envir. Toxicol. Chem. 7: 27–34.

Dumont, J. N., T. W. Schultz, M. Buchanan & G. Kao, 1983. Frog Embryo Teratogenesis Assay: *Xenopus* (FETAX)-A short-term assay applicable to complex environmental mixtures. In: M. D. Waters, S. S. Sandhu; J. Lewtas, L. Claxton, N. Chernoff & S. Nesnow (Eds.) Symposium on the Application of Short-term Bioassays in the Analysis of Complex Environmental Mixtures III. pp. 393–405. Plenum Press, New York.

Fort, D. J., D. A. Dawson & J. A. Bantle, 1988. Development of a metabolic activation system for the Frog Embryo Teratogenesis Assay-*Xenopus* (FETAX). Teratogen. Carcinog. Mutagen. 8: 251–263.

Nieuwkoop, P. D. & J. Faber, 1975. Normal Tables of *Xenopus laevis* (Daudin). 2nd ed. Amsterdam: North Holland, 243 pp.

Peltier, W. H. & C. I. Weber, 1985. United States Environmental Protection Agency, 'Methods for Acute Toxicity Tests With Fish, Macroinvertebrates and Amphibians', EPA-600/4-85-013, pp. 1–78.

Sabourin, T. D. & R. T. Faulk, 1987. Comparative evaluation of a short-term test for developmental effects using frog embryos. Branbury Report 26: Develop. Toxicol.: Mechanisms and Risk. Cold Spring Harbor Laboratory pp. 203–223.

Hydrobiologia **188/189**: 587–594, 1989.
M. Munawar, G. Dixon, C. I. Mayfield, T. Reynoldson and M. H. Sadar (eds)
Environmental Bioassay Techniques and their Application.
© *1989 Kluwer Academic Publishers. Printed in Belgium.*

Cellular and biochemical indicators assessing the quality of a marine environment

Jocelyne Pellerin-Massicotte [1], Emilien Pelletier [2] & Mariane Paquet [2]
[1] *Université du Québec à Rimouski, 300 Allée des Ursulines, Rimouski, Qué,. Canada G5L 3A1;*
[2] *INRS-Océanologie, 310 Allée des Ursulines, Rimouski, Qué, Canada, G5L 3A1*

Key words: environnement, bioindicateur, lysosomes, malate déshydrogénase, pollution

Abstract

Environmental stressors as well as the direct or combined effects of pollutants could be harmful to the populations living in a marine environment and the reproductive and nutritive processes could be impaired in a deteriorating environment. Sublethal effects of pollutants were studied in the blue mussel *Mytilus edulis* L., a good bioaccumulator of contaminants. Blue mussels of 3.5 cm were sampled on a rocky substrate at Pointe-Mitis (48° 40′ N, 68° 02′ W) along the coast of the St-Lawrence estuary. Mussels were placed in experimental tanks, fed, supplemented with mineral salts and continuous sea water flow and kept 72 h before the exposure to 0.01 μg l^{-1} and 0.3 μg l^{-1} methylmercury hydroxide in the presence or absence of selenium, at a concentration of 125 μg l^{-1}, a possible antagonist of methylmercury. The contamination protocol was performed during 45 days and a 14 day period of recuperation was allowed. The stress caused by the transplantation of mussels in the laboratory tanks and/or by the presence of pollutants was evaluated by a general indicator of stress developed in our laboratory, the measure of the lysosomal membrane fragility (LMF) of the digestive gland, according to the method developed by Moore (1976). The effects of contamination on metabolism were measured by the study of the variations of the malate dehydrogenase activity (MDH), a key enzyme of the aerobic metabolism. The first days of the contamination period led to an increased metabolism in the mantle and to a detoxifying mechanism in the hepatopancreas. At days 22 and 29 of the experiment, the affinity of the MDH was greatly decreased with both concentrations of methylmercury and selenium, suggesting a competitive inhibition of the enzymatic activity by the pollutants. LMF increased as the mussels were kept longer in the tanks. Methylmercury increased the stress undergone by the mussels. LMF gives information about the degree of stress of the organism while the biochemical indicator informs about the metabolic effects of sublethal concentrations of pollutants.

Introduction

La qualité de l'environnement marin se doit d'être une préoccupation pour les individus et pour les gouvernements. Plusieurs programmes de vérification des sources de pollution en milieu marin ont déjà été élaborés et l'utilisation de biotests a conduit au classement des sources de pollution par leur degré de toxicité. Ces biotests, cependant, ne permettent de déterminer que les effets aigus des contaminants sur les processus vitaux des organismes (Kolber *et coll.*, 1983; Conway, 1982).

588

Les effets des polluants peuvent être plus subtils et ne pas affecter de façon visible un écosystème, mais plutôt affecter les processus physiologiques de base des individus pour diminuer l'effort de reproduction d'une population (Phillips, 1986). L'évaluation de ces effets subtils et sous-létaux n'est pas facile. L'approche qui privilégie l'utilisation de bivalves marins comme bioindicateurs de pollution n'est valable que si les niveaux de pollution sont suffisants pour que la bioaccumulation soit proportionnelle à la concentration ambiante de polluants (Konasewich et coll., 1986; Pellerin-Massicotte et coll., 1988). Si l'on veut plutôt évaluer la qualité d'un environnement marin par le biais de l'impact physiologique de concentrations sous-létales de polluants, il s'agit alors de choisir une mesure physiologique et/ou biochimique et/ou cytochimique qui sera susceptible d'évaluer avec précision et justesse les différents stress environnementaux (Bayne et coll., 1985).

Ces dernières années nous avons développé un indice biochimique capable de renseigner sur la performance biologique de l'organisme en présence de méthylmercure par la mesure de la variation de l'activité de la malate déshydrogénase chez *Mytilus edulis L.* et un indice cytochimique, par la mesure de la déstabilisation de la membrane lysosomale en réponse à un stress. Ce dernier indice est général et réflète autant la présence de

stress environnementaux, thermiques (Moore, 1976), nutritionnels (Bayne et coll., 1976), de salinité (Bayne et coll., 1979) que celle d'hydrocarbures (Pellerin-Massicotte et coll., 1988).

Nos principaux objectifs de recherche sont donc de détecter des niveaux de pollution sous-létale pour un environnement marin, par le développement de différents indicateurs complémentaires, cellulaires et biochimiques chez les invertébrés marins. L'objectif spécifique de ce travail est de vérifier si l'utilisation de la mesure de la déstabilisation de la membrane lysosomale s'avère pertinente pour une pollution sous-létale au méthylmercure. Des comparaisons des niveaux de déstabilisation seront faites avec les variations de l'activité de la malate déshydrogénase.

Matériel et méthodes

Le protocole expérimental a déjà été décrit par Pelletier (1986) et Pellerin-Massicotte et Pelletier (1987a). Brièvement, le processus de contamination est réalisé grâce à l'appareillage illustré à la Fig. 1. L'eau de mer est pompée dans les bassins contenant les moules à partir d'un bac à décantation et les solutions mères de sélénium (Na_2SO_3) et de méthylmercure (($[CH_3Hg]_3O)OH$), adsorbées sur de fines parti-

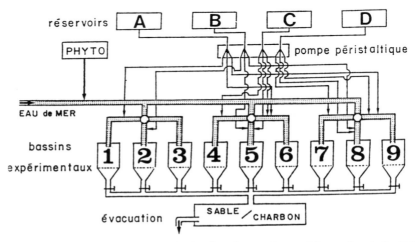

Fig. 1. Appareillage expérimental pour réaliser la contamination des moules en laboratoire. A, B, C, D sont les réservoirs contenant les polluants. Les bassins contenant les moules sont numérotés de 1 à 9.

cules de TiO$_2$, sont gardées dans des réservoirs de verre de 13 l et mélangées ultérieurement avec de l'eau de mer à l'aide d'une pompe péristaltique. Les effluents contaminés sont par la suite filtrés sur un mélange de sable-charbon. La photopériode a été réglée à 12 h.

Des moules bleues (*Mytilus edulis L.*; 3.5 ± .3 cm), ont été échantillonnées sur un substrat rocheux à Pointe-Mitis (48° 40′ N, 68° 02′ W), le long du littoral de l'estuaire du Saint-Laurent. Cent cinquante moules ont été placées dans chaque bassin, pour un total de 4 bassins, pour ensuite être mises en présence d'eau de mer (témoins), de 0.01 μg l^{-1} de méthylmercure, 0.3 μg l^{-1} de méthylmercure et 125 μg l^{-1} de sélénium. Une période de récupération de 14 jours a suivi la période de contamination.

Pendant toute la durée du protocole expérimental, les moules ont été nourries avec *Phaeodactylum tricornutum*; le niveau de polluant et la nourriture ont été maintenus constants par le remplacement des moules prélevées aux jours 0, 1, 3, 22, 29, 45 et 59 par d'autres moules mises dans les bassins dans un filet de Nytex®. Les jours d'échantillonnage, les moules ont été pesées, les manteaux et les muscles adducteurs postérieurs ont été disséqués et pesés. La malate déshydrogénase étant présente dans les fractions tissulaires mitochondriale et cytosolique avec des paramètres cinétiques différents, la séparation et la purification partielle de ces isoenzymes a été réalisée par centrifugation différentielle en utilisant du tampon Tris et différentes concentrations de sucrose (Scopes, 1982); une fois obtenues, ces fractions des homogénats tissulaires ont été gardées − 70 °C jusqu'à la réalisation des analyses enzymatiques. Les protéines totales solubles ont été mesurées par la méthode de Bradford (Scopes, 1982), à l'aide des réactifs achetés chez BIO-RAD avec l'albumine bovine comme référence. L'activité de la malate déshydrogénase a été évaluée en suivant l'oxydation de la NADH à 340 nm avec un spectrophotomètre Perkin-Elmer Coleman 575. L'oxaloacétate a été utilisé comme substrat avec environ 5 à 20 μg de protéines par essai enzymatique, pour un volume total de 3.1 ml. La déstabilisation de la membrane lysosomale a été mesurée d'après la méthode décrite par Moore (1976). Cette méthode cytochimique se base sur le temps de préincubation à pH 4.5 nécessaire pour donner un maximum d'intensité de coloration pour l'enzyme N-acétyl-glucosaminidase.

Méthodes analytiques

Les tissus ont été digérés dans de l'acide nitrique concentré et le mercure total a été analysé en duplicata par absorption atomique sans flammes à l'aide d'un spectrophotomètre Perkin-Elmer modèle H6-3, selon la méthode déjà décrite par Pelletier (1986).

Analyses statistiques et calculs spécifiques

Les calculs de l'activité catalytique de la malate déshydrogénase ont été réalisés selon la procédure décrite par Bergmeyer (1983). Les calculs conduisant à l'évaluation du Km de l'enzyme ont été faits grâce au programme ENZYME, gracieusement donné par Rudolf A. Lutz, D. Sc. et David Rodbard, M.D. (Laboratory of Theoretical and Physical Biology, National Institute of Child Health and Human Development, National Institutes of Health, Bethesda, Maryland, USA). Ce programme ajuste les courbes d'activité par itérations répétitives de la méthode de régression non-linéaire. Le Km qui correspond à la concentration du substrat pour laquelle l'enzyme travaillera à la moitié de sa vitesse maximale et le Vmax qui réflète la capacité maximale de transformation du substrat par l'enzyme sont estimés ensuite par la méthode de linéarisation de Eadie. Les erreurs standards résultantes sont alors utilisées pour détecter les différences entre les courbes par le test de t de Student pour données non-pairées (Lutz *et coll.*, 1986).

Résultats

Un premier protocole de contamination à l'été 1986 utilisant 3.0 μg l^{-1} de méthylmercure a dé-

590

montré chez les moules la présence d'un processus de bioaccumulation linéaire et proportionnelle à cette concentration de polluant dans les tissus mous de *Mytilus edulis L.* (Pellerin-Massicotte & Pelletier 1987a). Cette concentration est cependant sous-létale car elle ne provoque pas la mort des organismes, ne modifie pas les poids humides de chair, du manteau et du muscle adducteur postérieur, ni de la concentration en protéines dans les tissus; cependant, des effets inhibiteurs importants après 21 jours de contamination sont observés au niveau de l'activité de la malate déshydrogénase du manteau et du muscle adducteur postérieur. Pour des concentrations en mercure de 10 et 30 fois inférieures, les niveaux de bioaccumulation du mercure atteignent un plateau entre 7 et 21 jours du protocole expérimental (Pellerin-Massicotte *et coll.*, 1988).

Il était dès lors intéressant de vérifier quelles étaient les modifications cellulaires et biochimiques causées par le méthylmercure et de vérifier la complémentarité de ces réponses. Le manteau de la moule est un site d'entreposage de matières de réserve et de gamètes et toute modification métabolique à ce niveau peut être de grande importance. La réponse de la malate déshydrogénase à des concentrations croissantes de substrat dans le manteau (voir Fig. 2) se situe à un Km calculé de 5.9 10^{-5} mole d'oxaloacétate, au jour 0 du protocole expérimental, avec une activité catalytique maximale calculée de 2700 U l^{-1}. Une journée de mise en bassins et de mise en présence des moules avec les polluants modifie significativement les paramètres cinétiques de la malate déshydrogénase (Fig. 3). Le Km des moules témoins se déplace de quatre (4) fois vers la droite tandis que la contamination par le méthylmercure et le sélénium contrebalance cette perte de l'affinité de l'enzyme par le maintien d'une valeur de Km similaire à celle observée avant la mise en bassins. De plus, cette variation de Km est proportionnelle à la concentration de polluant, la plus faible concentration ne déplaçant la courbe que légèrement vers la gauche par rapport aux moules témoins. Celles-ci, subissent de plus une déstabilisation de la membrane lysosomale d'environ 27% (Fig. 4) avec une valeur

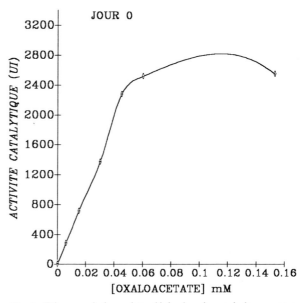

Fig. 2. Réponse de la malate déshydrogénase à des concentrations croissantes d'oxaloacétate, chez des moules témoins, avant la garde prolongée en laboratoire. L'activité catalytique retrouvée dans le manteau de la moule est exprimée en UI (capacité de transformation de 1 nmole d'oxaloacétate en une minute et par litre).

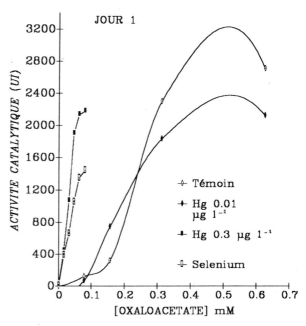

Fig. 3. Variations de l'activité de la malate déshydrogénase, au jour 1 du protocole de contamination, dans le manteau de la moule, en réponse au méthylmercure et au sélénium.

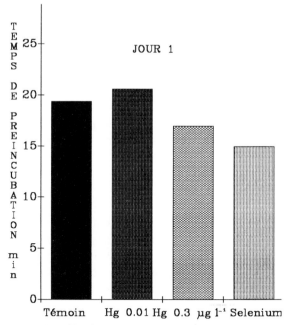

Fig. 4. Déstabilisation de la membrane lysosomale chez des moules témoins, gardées en bassin, ou des moules subissant une contamination au méthylmercure et au sélénium. La déstabilisation est exprimée en minutes de préincubation pour obtenir le maximum de coloration.

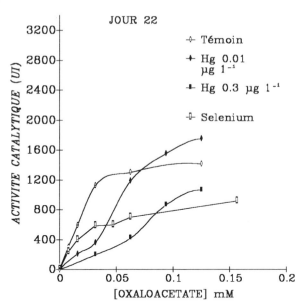

Fig. 5. Variations de l'activité de la malate déshydrogénase au jour 22 du protocole de contamination.

moyenne de 19 minutes, les valeurs normalement retrouvées chez des organismes non stressés étant de 26 minutes de temps de préincubation (Bayne *et coll.*, 1985). Contrairement aux deux autres contaminants (mercure, 18 min; sélénium, 16 min), la présence de la plus faible concentration de méthylmercure n'affecte pas de façon importante (21 min) la stabilité de la membrane lysosomale. Après 22 jours de contamination (Fig. 5), le Km de la malate déshydrogénase du manteau des moules témoins revient à une valeur comparable à celle observée avant la mise en bassin. Cependant, la présence de méthylmercure déplace le Km vers la droite, proportionnellement aux concentrations du polluant, mais sans affecter le Vmax de façon significative. La présence de sélénium, modifie la réponse maximale de l'enzyme sans toutefois modifier le Km.

Au niveau cellulaire (Fig. 6), les moules témoins subissent une déstabilisation de 50% de la membrane lysosomale par la longue période de

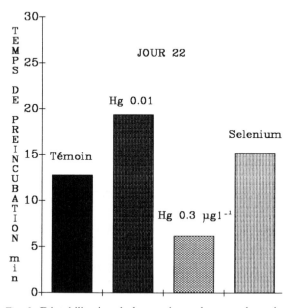

Fig. 6. Déstabilisation de la membrane lysosomale au jour 22 du protocole expérimental. Les moules témoins sont nourries et sont en présence d'un apport d'eau de mer en continu. Les moules contaminées reçoivent en surplus, les polluants mentionnés.

592

présence en bassin. La présence de sélénium et de la plus faible concentration de méthylmercure, curieusement, diminue la fragilité de la membrane lysosomale tandis que le stress occasionné par la plus forte concentration de méthylmercure, déstabilise de 75% la membrane. A la fin de la période de contamination, au jour 45 du protocole expérimental, peu importe le contaminant étudié, les valeurs de Km reviennent similaires à celles des moules témoins avant la mise en bassin et à celle des moules dans le bassin témoin. Les réponses cellulaires à la présence des contaminants, sont le reflet des différents processus d'intégration physiologique qui prennent place lors de la récupération des organismes. Elles sont alors plus difficiles à interpréter. L'observation des coupes d'hépatopancréas nous signale une modification dans la population et la proportion des différents lysosomes. Deux (2) pics sont observés, l'un démontrant une déstabilisation prononcée de la membrane et l'autre, une stabilité normale.

La période de récupération allouée de 14 jours est suffisante pour que les Kms reviennent aux valeurs normales, toutefois, les réponses maxi-males en réponse aux deux concentrations de méthylmercure sont significativement augmentées (Fig. 7) de même que les réponses cellulaires au niveau de la membrane lysosomale; celle-ci voit sa stabilité augmentée par la présence de polluants tandis que les moules du bassin témoin présentent, après 59 jours de mise en bassin, un état de stress prononcé, avec plus de 85% de déstabilisation de la membrane lysosomale.

Discussion

Ces résultats démontrent nettement que des concentrations de polluants et plus spécifiquement de métaux lourds tels le mercure sous forme organique, en concentrations sous-létales qui, de plus, ne conduisent pas à une bioaccumulation chez *Mytilus edulis* L., peuvent perturber fortement le métabolisme énergétique dans un tissu clé ainsi que la réponse d'organites spécifiques au processus de détoxification, les lysosomes.

Les résultats de cette expérience et de celle de 1986 (Pellerin-Massicotte & Pelletier, 1987b) montrent que les doses utilisées sont sous-létales, les poids de l'organisme, des tissus, les concentrations de protéines et la survie des bivalves n'ayant pas été affectés par la présence des contaminants. L'utilisation de la malate déshydrogénase comme indicateur biochimique se trouve également à être validée par la mesure des Kms qui demeure constante, d'une année à l'autre, lors de l'échantillonnage au début de la saison, et avant la ponte de ces mollusques. Cet indicateur biochimique s'avère fiable, reproductible et quantifiable. Les comparaisons entre les différents protocoles expérimentaux et entre les résultats obtenus au cours des différentes années se trouvent donc facilitées, d'autant plus que le Km calculé dans la fraction cytosolique du manteau est comparable à celui rapporté dans la littérature scientifique (Bergmeyer, 1983). La cinétique de bioaccumulation qui plafonne entre 7 et 21 jours (Pellerin-Massicotte & Pelletier, 1987b; Pellerin-Massicotte *et col.*, 1988) de mise en présence des contaminants démontre que les processus de détoxification sont mis en marche dès le début de la

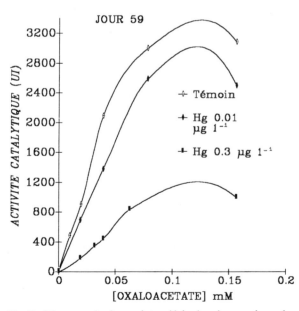

Fig. 7. Réponse de la malate déshydrogénase chez des moules témoins gardées en bassin depuis 59 jours et chez des moules ayant subi 45 jours de contamination suivis d'une période de récupération de 14 jours.

contamination. Concomitamment, la présence des polluants, et la mise en bassins, conduit à des effets dramatiques sur le manteau de la moule. Au jour 1, les réponses cellulaires et métaboliques s'inversent. Les lysosomes sont des organites subcellulaires capables de cataboliser des composés cellulaires endogènes et exogènes. Leurs fonctions sont associées à la digestion intracellulaire, la régulation des processus sécrétoires, les mécanismes de défense cellulaire, le recyclage des protéines et des organites et la séquestration des substances xénobiotiques. Différentes conditions de stress déstabilisent la membrane lysosomale, augmentent la perméabilité de la membrane aux substrats, activent les enzymes hydrolytiques et les libèrent dans le cytoplasme, causant des effets cataboliques à ce niveau. Il est donc vraisemblable que la mise en présence des polluants déclenche des mécanismes de détoxification au niveau des lysosomes, ce qui engendre, par le fait même, une augmentation du métabolisme énergétique. L'évolution du Km dans le manteau des moules témoins illustre bien la capacité de récupération et d'adaptation des moules à leur nouvel environnement. En effet, le Km de la MDH revient aux valeurs normales 22 jours après le début de la période de contamination. Par contre, la présence de méthylmercure affecte le métabolisme énergétique. Le polluant, en n'affectant que l'affinité de l'enzyme pour son substrat et non la capacité de transformation du substrat illustrée par le Vmax, démontre qu'il y a une inhibition compétitive exercée par le mercure, probablement par la liaison du polluant au niveau du site actif de l'enzyme sur les ponts disulfure. La déstabilisation de la membrane lysosomale, comme l'a déjà démontré Moore (1976), est un excellent indicateur de stress général. Il suffit de constater qu'au jour 22, la réponse cellulaire des moules témoins est indicatrice d'un stress physiologique et que, sur une aussi longue période de contamination, la présence de la plus forte concentration de mercure amplifie le stress de la mise en bassin. A ce moment, les deux indicateurs nous renseignent sur la capacité métabolique et l'état physiologique général de l'organisme. A la fin de la période de contamination, la capacité métabo-

lique de l'organisme est retrouvée, grâce possiblement, à des mécanismes d'adaptation physiologique, telle l'induction de la synthèse de nouvelles protéines enzymatiques, qui vient compenser l'inefficacité des enzymes présentes inactivées par les polluants. Cette récupération métabolique est aussi évidente 14 jours après l'enlèvement des polluants.

Conclusion

L'évaluation de l'état de santé de mollusques soumis à des conditions environnementales adverses doit tenir compte du stress subi par les organismes soit par leur manipulation, leur transplantation ou la présence des substances à étudier. Les résultats présentés ci-haut illustrent bien la nécessité de choisir des indicateurs complémentaires qui ne négligent pas les variations physiologiques de base des individus. De plus, les indicateurs reflétant les variations métaboliques pouvant résulter de la présence de contaminants, doivent avoir un niveau de sensibilité et de réponse tel qu'ils puissent être utilisables dans les écosystèmes naturels.

References

Bayne, B. L., D. R. Livingstone, M. N. Moore & J. Widdows, 1976. A cytochemical and a biochemical index of stress in *Mytilus edulis L.* Mar. Pollut. Bull. 7: 221–224.

Bayne, B. L., M. N. Moore, J. Widdows, D. R. Livingstone & P. Salkeld, 1979. Measurement of the responses of individuals to environmental stress and pollution: studies with bivalve molluscs. Phil. Trans. r. Soc. Lond. B. 286: 563–581.

Bayne, B. L., D. A. Brown, K. Burns & D. R. Dixon, 1985. The effects of stress and pollution in marine animals. Praeger Publ., New York, 384 pp.

Bermeyer, H. U., 1983. Methods of enzymatic analysis. Vol. I Fundamentals. Verlag Chemie Weinheim GmBh, 575 p.

Conway, R. A., ed., 1982. Environmental risk analysis for chemicals. Van Nostrand Reinhold Environmental Engineering Series, New York, xxiv-558 pp.

Kolber, A. R., T. K. Wong, L. D. Grant, R. S. De Woskin & T. J. Hugues, 1983. In vitro toxity testing of environmental agents. Current and future possibilities. Plenum Press, New York. xix-553 pp.

594

Konasewich, D. E., E. R. McGreer & H. Sneddon, 1986. Feasibility assessment of sediment toxicity tests suitable for OCDA application review. Report for the Regional Ocean Dumping Advisory Committee (RODAC), Environnement Canada, 79 pp.

Lutz, R. A., C. Bull & D. Rodbard, 1986. Computer analysis of enzyme-substrate-inhibitor kinetic data with automatic model selection using IBM-Pc compatible microcomputer. Enzyme 36: 197–206.

Moore, M. N., 1976. Cytochemical demonstration of latency of lysosomal hydrolases in digestive cells of the common mussel *Mytilus edulis L.* and changes induced by thermal stress. Cell Tiss. Res. 175: 279–287.

Pellerin-Massicotte, J. & E. Pelletier, 1987a. Evaluation of sublethal effects of methylmercury on malate dehydrogenase activity in the mantle of the blue mussel. Proc. of Oceans '87; Halifax: 1550–1554.

Pellerin-Massicotte, J. & E. Pelletier, 1987b. Evaluation of sublethal effects of pollutants. Can. Tech. Rep. Fish Aquat. Sci., 1575: 146–149.

Pellerin-Massicotte, J. & E. Pelletier, C. Rouleau & M. Pâquet, 1988. Mise au point d'indicateurs biochimiques et cellulaires de la qualité d'un environnement marin. Can. Tech. Rep. Fish Aquat. Sci. 1607: 113–126.

Pelletier, E., 1986. Modification de la bioaccumulation du sélénium chez *Mytilus edulis L.* en présence du mercure organique et inorganique. J. Can. des sc. hal. et aquat. 43: 203–210.

Phillips, D. J. H., 1986. Use of bio-indicators in monitoring conservative contaminants: program design imperatives. Mar. Pollut. Bull. 17: 10–17.

Scopes, R., 1982. Protein purification. Principles and practice. Springer-Verlag, New York, 282 pp.

Hydrobiologia **188/189**: 595–600, 1989.
M. Munawar, G. Dixon, C. I. Mayfield, T. Reynoldson and M. H. Sadar (eds)
Environmental Bioassay Techniques and their Application.
© 1989 *Kluwer Academic Publishers. Printed in Belgium.*

The role and application of environmental bioassay techniques in support of the impact assessment and decision-making under the Ocean Dumping Control Act * in Canada

K. L. Tay
Marine Environmental Branch, Environmental Protection, Conservation and Protection, Environment Canada, 45 Alderney Drive, 5th Floor, Dartmouth, Nova Scotia, Canada B2Y 2N6

Key words: sediment, toxicity, bioaccumulation, application, research, ocean dumping, bioassay techniques, impact assessment

Abstract

A brief description of the Canadian Ocean Dumping Control Act (ODCA) and the current ocean dumping permit process in Canada is given. Case studies are used to discuss and review the role of various experimental bioassay techniques and *in situ* bioaccumulation studies in the permit assessment and evaluation processes.

The results and progress of research work related to biotesting funded under the Ocean Dumping Control Act Research Fund (ODCARF) are discussed and reviewed with special attention to the practical uses of these studies.

Future directions for ocean dumping related research, particularly with respect to the development of appropriate sediment bioassay tools and ecotoxicity evaluation, are discussed

Introduction

Ocean dumping is regulated in Canada under the Ocean Dumping Control Act (ODCA) which was promulgated in 1975 to meet Canada's commitment to the 1972 London Dumping Convention on the prevention of marine pollution by the dumping of wastes and other matters.

The ODCA is administered by Environment Canada through a permit system in which approximately 150 Ocean Dumping Permit Applications are reviewed and processed annually. As many as 90% of these applications relate to the disposal of dredged materials originating from natural sedimentation in harbours and coastal ports where dredging is essential to safety of navigation and the local resource based economy.

The day-to-day management of ocean dumping is based upon the application of established criteria for assessing the environmental acceptability of each dumping operation. Following the general principles of the London Dumping Convention, the ODCA prohibits the dumping of potentially toxic substances in the marine ecosystem, unless evidence exists that they can be rapidly rendered harmless in the marine environment (Ocean Dumping Control Act, 1975). In general, limits of acceptability for these substances lie close to what

---

* On June 30, 1988, the Ocean Dumping Control Act (ODCA) was replaced by part VI of the new Canadian Environmental Protection Act (CEPA).

are considered to be the background levels. For example, in the cases of cadmium and mercury, the maximum acceptable limits of proposed dump materials are 0.6 mg/kg and 0.75 mg/kg respectively (Ocean Dumping Regulations, 1988). If the concentration of one of these heavy metals exceeds the permissible limits, no dumping will be allowed unless the applicant can demonstrate that the proposed materials will be rapidly rendered harmless at the proposed dump site. Toxicity levels are used to establish acceptable limits only in the case of organohalogen compounds.

At present, no toxicity tests have officially been adopted under the ODCA for permit application assessment. When required by the regulatory authority to supplement permit application information, bioassays are carried out based on the standard bioassay methods developed by Environment Canada regional laboratories. For example, in 1980, when an Ocean Dumping Permit Application for a dredging and ocean dumping operation at Dalhousie Harbour, New Brunswick, was rejected by Environment Canada based on the high concentration of cadmium, zinc and lead in the sediments, a bivalve, *Macoma*, common to both the proposed dredged and dump sites, was employed in static bioassay tests to assess the toxicity of these contaminants. Based on the results of the *Macoma* test and the other chemical and physical information, a decision was made to require the dredged material be contained in a bermed embayment (MacLaren Marex Inc., 1980; Ray *et al.*, 1981; McIver & Tay, 1986).

Bioaccumulation studies are sometimes conducted by permit applicants during pre-dumping and post-dumping monitoring programs specified in an Ocean Dumping Permit. A good example is the four-year assessment and monitoring program conducted by Public Works Canada for the Miramichi Estuary dredging project. In a bioaccumulation study of this program, Public Works Canada was able to demonstrate that the tissue cadmium concentrations in oyster, *Crassostrea virginica*, crab, *Cancer irroratus*, and clam, *Macoma balthica* increased significantly during the dredging project and subsequently returned to the levels present before the initiation of the operation (MacLaren Plansearch, 1985) (Fig. 1). Based on this finding, they concluded that the possible long term impacts of sediment cadmium contamination within the estuary appeared to be negligible after the termination of the dredging project.

The search for suitable bioassay tests and ecotoxicity methods has always been one of the top priorities of the Canadian Ocean Dumping Program. To date, approximately twenty projects funded by the Ocean Dumping Control Act Research Fund (ODCARF), a research grant managed by the Canadian Government to fund research related to ocean dumping, have been related to bioaccumulation studies or bioassay tests.

This paper will review some of the important ODCARF studies related to bioassay techniques and the future needs in ocean dumping research.

Macoma balthica as a test organism

The first toxicological test funded by ODCARF was conducted at the Atlantic Regional Laboratory of Environment Canada in 1976. In this study, a bivalve, *Macoma balthica* was chosen as the most suitable test organism for bioaccumulation study among the three species of organisms used (the other two are the polychaete, *Nereis virens* and the mud snail, *Nassarius obsoletus*) (Sharp, 1977). However, subsequent study conducted in the same laboratory using contaminated sediments collected from Halifax Harbour indicated that only PCBs were accumulated in the *Macoma* tissues (Ken Doe, personal communication). Several tests conducted during the period of 1979 to 1981 showed no uptake of either cadmium or PCBs in *Macoma* (Ken Doe & Roy Parker, personal communication). In contradiction to the above studies, McGreer *et al.* (1984) reported from the Canadian west coast that *Macoma* is successful 86% of the time in predicting bioavailabilities of cadmium, lead and PCBs. Thus, it is quite evident that *Macoma* is less than ideal as a test organism for bioaccumulation studies as results are, at times, ambiguous.

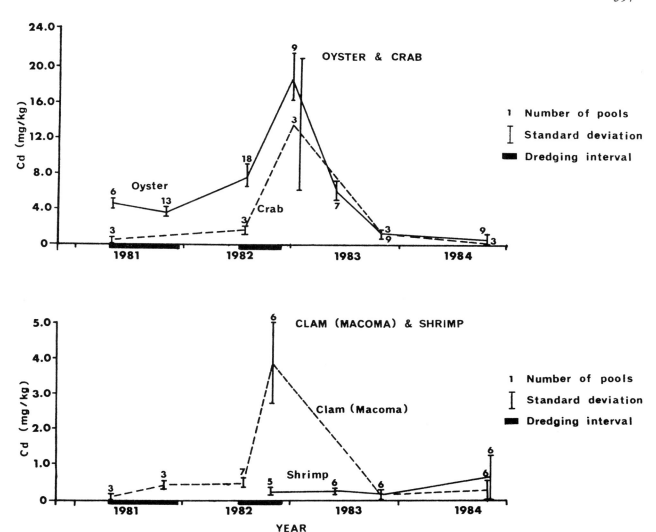

Fig. 1. Cadmium contamination of Biota (from MacLaren Plansearch, 1985).

Amphipod *Rhepoxynius abronius* lethality test

In 1987, with the support of the ODCARF, an amphipod acute lethality test was conducted in the Pacific Region following the *Rhepoxynius abronius* method currently used by the State of Washington, U.S.A., in the assessment of the ocean disposal of dredged materials.

The study was performed using sediment samples collected from six different locations within the Vancouver Harbour. Results of these tests indicated that sediments from three of the test sites are acutely toxic to the amphipods, however, the toxicity of these sediments could not be correlated to the concentration of any particular chemicals (As, Cd, Cu, Pb, Hg, Zn, TOC, TCP/PCP, PCB and PAH) within the sediment (Chapman *et al.*, 1988). This was consistent with previous results reported by Chapman & Barlow (1984) in sediments from Vancouver coastal areas. Thus, the toxicity demonstrated in this test may not be directly related to the pollutants of concern under the ODCA Regulations. Based on these results, the 'amphipod test' remains to be proven as an acceptable tool for permit assessment under the ODCA. Furthermore, no test has yet been done using the Atlantic amphipod, *Corophium volutator*.

Bioassays using natural assemblages of microorganisms in field samples

At the same time when the above technique was developed in the Pacific Region, a microbial test procedure focusing on metabolic processes of indigenous species was developed with the support of the ODCARF in Atlantic Canada.

Three possible procedures were developed in this study (Lee *et al.*, 1988):

Method for exoenzyme activity bioassay in sediment slurries.

Based on the application of the fluorescent 4-methylumbelliferone (MUF) substrates, the activities of three exoenzymes (glucosidase, phosphatase and protease) in natural sediments were used to assess the toxicity of a series of known toxic metals (Cd, Cu, Hg, Ni, Pb, and Zn). Sediments were spiked by the selected metals and the concentration of toxicant that resulted in a 50% decrease in enzyme activity was calculated as the EC_{50}. The sensitivity of this method was comparable to that reported in other sediment bioassays based on the use of colorimetric substrates. The similarity in toxicity orders and EC_{50} values

between the three enzyme systems assayed indicates that the procedure is both reliable and consistent (Table 1). While any one of the three enzymes can be used to test toxicant impacts, the use of multiple enzyme system is recommended for more accurate estimates of pollutant impact on the biota. Due to the simplicity of the analytical procedure, results from fluorometric sediment slurry assays can be obtained within 2.5 hours.

Method for exoenzyme activity bioassay in sediment cores.

While bioassays with sediment slurries can be performed rapidly, metabolic activity of bacteria in suspended sediment may not necessarily represent that found under actual conditions due to the physical disruption of the micro-environments present in the intact cores when the sediments are resuspended. Using the same principle employed in the sediment slurries method, the activities of three exoenzymes in intact sediment core samples were assessed. Preliminary results showed that EC_{50} values obtained using the sediment cores were generally lower than those obtained with sediment slurries, suggesting a greater sensitivity of bacteria in intact cores than in slurries (Table 2). However, the increased ecological validity of the sediment core bioassay over slurry bioassays is offset by the complexity of the test which increased its cost, its lack of time efficiency and increased variability in results.

Table 1. The toxicity of six metals to glucosidase, phosphatase and protease activity and the uptake of thymidine and glutamic acid in slurries of muddy sediment. All values in MG/L. (from Lee *et al.*, 1988).

Assay procedure	Ni	Hg	Cu	EC_{50} Cd	Pb	Zn
Glucosidase	310	14	5	231	407	33
Phosphatase	196	130	11	999	671	67
Protease	130	18	1	484	33	24
Thymidine	241	1	1	144	105	2
Glutamic acid	346	1	1	100	113	12

Table 2. The toxicity of metals to glucosidase activity in cores of sandy sediment. All values are mg/l. (from Lee *et al.*, 1988).

Core horizon	Ni	EC_{50} Hg	Zn
0–1 cm	38	1.3	108
1–2 cm	40	1.3	107
2–3 cm	33	0.5	77
3–4 cm	49	1.4	80
4–5 cm	48	1.3	77
5–6 cm	52	1.0	94

Method for radiolabelled organic uptake bioassay in sediments.

This method was developed to assess the effect of toxic metals on bacterial uptake of 14C-glutamic acid and 3H-thymidine. The results indicated that uptake of radiolabelled organic substrates appeared to be more sensitive than the exoenzyme bioassays for the range of metals tested (Table 1).

Future research under ODCA

This year, four of the eight research projects funded under the ODCARF concern biotesting. These projects are:

1. Heron Island Dump Site Monitoring: The new bacteria tests will be used to measure the toxicity of cadmium, lead and zinc in sediments at both dredged and dump sites and to measure the recovery rate of the benthic community at the dump site.
2. The Effects of Suspended Dredge Material on Aquaculture Organisms: The study will attempt to develop a guideline for measuring the maximum tolerance levels of aquaculture organisms and the minimum zone of influence for measurable impacts of suspended dredge material during dredging and ocean dumping operations.
3. Development of Biochemical Indices for the Evaluation of Sublethal and Chronic Toxicity of Contaminants in Marine Environment: A sea water flow-through system for biochemical indices reported by Pellerin-Massicotte and Pelletier (1987) will be modified for sediment bioassay testing.
4. Biological Testing Using Recommended Tests in Concert with Heavily Contaminated Environment to Assess the Tiered Testing Approach: This study will verify the Sediment Quality Triad proposed by Chapman (1986) by using amphipod bioassay test, sediment chemical and physical data and bioaccumulation studies on resident fish species.

Future research on sediment bioassays for use in the management of the Ocean Dumping Program in Canada should include the following:
1. Methods for short-term effect measurement and the development of a simple screening tool:
   a. Studies are needed to adapt bioassay tests (*Macoma*, amphipods, eggs and larvae stages of fish species and indigenous bacteria) to the permit application screening process.
   b. Development of new sediment bioassay techniques which are simple, cost-effective, practical and reproducible should be encouraged.
2. Methods for long-term effect measurement and dump site monitoring program:
   a. Research on the Sediment Quality Criteria reported by Chapman *et al.* (1987) should be supported.
   b. Field verification of the existing toxicity techniques, especially those related to dump site monitoring and assessment programs should be promoted.
3. Quality Control and Quality Assurance programs relating to sampling and chemical analytical methods in support of bioassays should be supported.
4. Inventory of data bases on a spectrum of test species used in established bioassays should be supported.

Conclusion

The proper application of toxicity bioassays in the evaluation of the risks and impacts of the ocean disposal of wastes depends on the specific nature of the wastes and the target environment. A single-species test, such as the *Macoma* or *Amphipod* test, might be useful in detecting the toxicity of a few selected pollutants, however, when the assessment of more complex environmental impacts caused by the dumping of multiple contaminated substances is required, a battery approach using single species lethal and sublethal bioassays supplemented with bulk sediment chemical analysis and benthic community studies is more applicable. This type of approach has successfully been used in several occasions when sediment chemical and physical data alone could not be used for decision making.

The test organisms in existing sediment bioassay tests are usually species that are alien to the benthic community at the dump site. The new microbial tests developed by Lee *et al.* (1988) have adopted a different approach by directly measuring the metabolic disturbance in a natural benthic community caused by the impacts of ocean dumping at the dump site. The use of organisms obtained from the dump site will increase the accuracy in predicting the 'true' impact of dumping wastes at sea.

While there is much value in exploring and refining our bioassay techniques, it is even more important that there is a continuing review and assessment of the present regulatory limits. Results of bioassays will provide useful feedback for the establishment of better criteria for the ODCA Permit reviewing system. Without scientifically sound regulatory limits, the usefulness of any 'good' biotesting technique will be greatly reduced.

Acknowledgements

I thank H. Hall, A. McIver and W. Barchard of the Marine Environment Branch, Environmental Protection, C & P, Environment Canada, for their valuable comments and suggestions in the preparation of this paper. Special thanks also to Dr. K. Lee of the Kenneth Lee Research Ltd., Dartmouth, N. S., Canada, for his assistance in providing data and helpful suggestions. Data and results of research summarized in this paper are based on research reports submitted under the Ocean Dumping Control Act Research Fund (ODCARF) in Canada.

References

Chapman, P. M., 1986. Sediment quality criteria from the sediment quality triad – An example. Envir. Toxicol. Chem. 5: 957–964.

Chapman, P. M. & C. T. Barlow, 1984. Sediment bioassays in various B. C. coastal areas. E. V. S. Consultants. Report submitted to Environment Canada, Pacific Region, Kapilano 100, Park Royal South, West Vancouver, British Columbia, Canada, under Ocean Dumping Control Act Research Fund. 22 pp.

Chapman, P. M., R. C. Barrick, J. M. Neff & R. C. Swartz, 1987. Four independent approaches to developing sediment quality criteria yield similar values for model contaminants. Envir. Toxicol. Chem. 6: 723–725.

Chapman, P. M., R. R. Rousseau & E. A. Power, 1988. Amphipod bioassays on sediments along the B. C. coast containing heavy metals and organic contaminants. E.V.S. Consultants. Report submitted to Environment Canada, Pacific Region, Kapilano 100, Park Royal South, West Vancouver, British Columbia, Canada, under Ocean Dumping Control Act Research Fund. 32 pp.

Lee, K., C. N. Ewing & E. M. Levy, 1988. Tests for toxicity and environmental impact assessment based on the activity of indigenous bacteria measured with fluorogenic and radiolabelled substrates. Kenneth Lee Research Ltd.. Report submitted to Environment Canada, Atlantic Region, 45 Alderney Drive, Dartmouth, Halifax Co., Nova Scotia, Canada, under Ocean Dumping Control Act Research Fund. 160 pp.

MacLaren Marex Inc., 1980. Bioavailability of Cadmium in the Restigouche Estuary, Task 1: Total Cadmium distribution. Report for Brunswick Mining and Smelting Corp. Ltd., Bathurst, Gloucester Co., New Brunswick, Canada. 23 pp.

MacLaren Plansearch Ltd., 1985. Environmental monitoring of the dredging of the Miramichi River. Volume I, Summary of 1984 monitoring. Report for Public Works Canada, Saint John, Saint John Co., New Brunswick, Canada. 182 pp.

McGreer, E. R., R. Deverall, D. R. Munday & E. Gerencher, 1984. Development and evaluation of a bioassay protocol for predicting the bioaccumulation potential of sediment-associated contaminants. E.V.S. Consultants. Report submitted to Environment Canada, Pacific Region, Kapilano 100, Park Royal South, West Vancouver, British Columbia, Canada. 57 pp.

McIver, A. & K. L. Tay, 1986. The interpretation of sediment data in support of Ocean Dumping Permit Applications and the role of ocean dumping related research in impact assessment and decision-making – An overview. In: J. S. S. Lakshminarayana. (Ed.) Proceedings of the 3rd Annual Aquatic Toxicity Workshop: Nov. 12–14, 1986, Moncton, Westmorland Co., New Brunswick, Canada. pp. 123–128.

Ocean Dumping Control Act, 1975. Canada Gazette, Part III. Queen's Printer for Canada, Ottawa, Canada. Vol. 1(9): 1–25.

Ocean Dumping Regulations, 1988. Canada Gazette, Part I. Queen's Printer for Canada, Ottawa, Canada. pp. 4903–4922.

Pellerin-Massicotte, J. & E. Pelletier, 1987. Evaluation of sublethal effects of pollutants with biochemical indicators. Oceans'87 Proceedings, Volume 5: 1550–1554.

Ray, S., D. W. McLeese & M. R. Peterson, 1981. Accumulation of copper, zinc, cadmium and lead from two contaminated sediments by three marine invertebrates. Bull. Envir. Contam. Toxicol. 26: 315–322.

Sharp, G., 1977. Studies to develop a marine benthic bioassay. Unpublished report submitted under Ocean Dumping Control Act Research Fund to Ocean Dumping Program, Environment Canada, Atlantic Region, 45 Alderney Drive, Dartmouth, Halifax Co., Nova Scotia, Canada. 36 pp.

Hydrobiologia **188/189**: 601–618, 1989.
© 1989 *Kluwer Academic Publishers. Printed in Belgium.*

In situ bioassessment of dredging and disposal activities in a contaminated ecosystem: Toronto Harbour

M. Munawar[1], W.P. Norwood[2], L.H. McCarthy[3] & C.I. Mayfield[4]
[1,2,3] *Department of Fisheries and Oceans, Great Lakes Laboratory for Fisheries and Aquatic Sciences, Ecotoxicology Division, Canada Centre for Inland Waters, Burlington, Ontario, L7R 4A6;* [4] *Biology Department, University of Waterloo, Waterloo, Ontario, Canada. N2L 3G1*

Key words: bioassessment, dredging, disposal, sediment, bioassays

Abstract

The contamination of Toronto Harbour is a very serious problem. The major sources of pollution are the Don River and sewer outflows, as well as industrial, and municipal effluents. The problem is further compounded by perturbations of the toxic sediment caused by dredging, dredge-disposal, navigation, and recreational activities. The impact of contamination and nutrient enrichment was reflected in the size-fractionated primary productivity experiments. Generally, microplankton/netplankton ($> 20 \mu$m) productivity was enhanced whereas ultraplankton ($< 20 \mu$m) productivity was inhibited. These observations are attributable to interactions between ameliorating nutrients and toxic contaminants as well as to the differential sensitivity of natural phytoplankton size assemblages to the bioavailable chemical regime. *In situ* environmental techniques applied in Toronto Harbour were effective, sensitive, and rapid, and provided a better understanding of the impact of dredging/disposal activities under natural conditions. These techniques have great potential in the assessment of the ecotoxicology of harbours and other stressed environments.

Introduction

Prior to 1974, the Keating Channel in Toronto Harbour, (Ontario, Canada), was extensively dredged at an average of 88 500 m³ of sediment annually (Dredging Subcommittee, 1983). However, the dredging of the Channel was curtailed due to lack of suitable confined disposal facilities for contaminated sediments. Only maintenance dredging was performed in the Channel mouth and harbour slips from 1974 to 1980, resulting in the removal of an average 17 923 m³ of sediment annually (Dredging Subcommittee, 1983). Since an average 50 000 m³ of sediment was transported down the Don River annually during this same

period, problems soon arose (Fricbergs, 1986). Dredging activities increased at the mouth of the Channel and in harbour slips, resulting in the dredging of 43 045 m³ in 1981 and 102 873 m³ in 1982. These sediments were disposed of in an endikement called the Leslie Street Spit, also known as the Eastern Headland (Dredging Subcommittee, 1983). However, the increased dredging did not keep pace with sediment input and silt not only filled the Keating Channel but was slowly accumulating in the harbour, docking slips, and navigation channels. Further disturbance by navigation activities and dredging throughout the harbour compounded the problem.

Dredging causes considerable disturbance of

bottom sediments, leading to resuspension of particulates, a possible source of contaminants. The impact of this disturbance from physical, chemical, and toxicological viewpoints has not been investigated in the past (Munawar *et al.*, 1984) due to lack of appropriate techniques. However, on-site monitoring methodologies have been developed by our laboratory during a study conducted at a contaminated pond in the Leslie Street Spit, Toronto (Munawar *et al.*, 1986) and in Toronto Harbour (Munawar & Thomas, 1989). These procedures have been applied to Toronto Harbour on a large scale.

Methods

Dredging

In situ impact on primary productivity

In order to study the impact of dredging in Toronto Harbour under *in situ* conditions, it was important to collect water samples before, during, and after dredging. Water samples were carefully collected from the front of a 16-foot Boston Whaler. This craft, with a very shallow draft compared to dredging barges and cargo ships, did not disturb the bottom sediments in test areas. The boat was carefully manœuvered bow first into the test site and the water sample collected using a 6 m by 3 cm tube sampler from the bow of the boat. In this way, the boat and its engine did not pass through or disturb the water to be collected. The boat was then slowly backed out of the site before the dredging operations commenced. This technique was repeated for samples collected during- and post-dredging in order to monitor the bioavailability of contaminants. Three dredging episodes were monitored within the harbour during July, 1985 (Fig. 1), and included the following:

Pre-dredging (control)
During-dredging
Post-dredging

A portion of the well-mixed water sample was preserved in Lugol's solution for identification and enumeration of phytoplankton using the Utermohl inverted microscope technique (Munawar *et al.*, 1974; Munawar & Munawar, 1978). Aliquots of 50 ml were transferred into four 125 ml polycarbonate bottles fitted with lids, to determine size-fractionated primary productivity. All bottles were spiked with 1 μCi of sodium bicarbonate ^{14}C and incubated for four hours at constant light levels (212 μE m^{-2} sec^{-1} at 400-700 nm) in a Conviron E7 Plant Growth Chamber maintained at the lake temperature measured at the time of sample collection. At the end of the four hour incubation, the entire contents of each bottle were size-fractionated by filtration through a 20 μm Nitex screen and subsequently passed through a 0.45 μm Millipore membrane filter at a vacuum pressure of 5 inches Hg, in order to estimate ultraplankton/picoplankton (< 20 μm) productivity. The material retained by the 20 μm Nitex screen was immediately backwashed and further filtered through a 0.45 μm filter for the estimation of microplankton/netplankton (> 20 μm) productivity. The filters were then acidified with 10 ml of 0.1 N HCl under the same vacuum pressure, then placed in a phase combining system (PCS) for liquid scintillation counting (Lind & Campbell, 1969). Analyses included the calculation of carbon assimilation rates, means, and standard errors (Munawar & Munawar, 1987).

Samples for size-fractionated chlorophyll *a* were collected in triplicate by filtration, as described above. Chlorophyll *a* concentrations were estimated according to the spectrophotometric method of Strickland & Parsons (1968). Production per Biomass (P/B) quotients (mg C. ass. mg^{-1} Chl. h^{-1}) using chlorophyll *a* as a crude biomass indicator were calculated (Munawar *et al.*, 1986).

The water samples were sent to the Wastewater Technology Centre, Environment Canada for chemical analyses of nutrients, trace metals, and organic substances (pesticides, PCBs, and GC-MS analyses of PAHs).

Sediment samples were also collected simultaneously by bucket from the dredging barges on site during each dredging episode. These sedi-

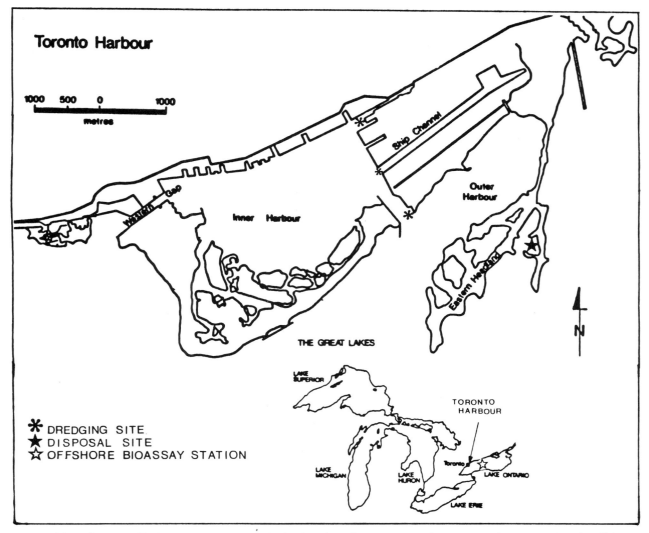

Fig. 1. Map of Toronto Harbour on Lake Ontario indicating sites of dredging and disposal operations, as well as the offshore bioassay station.

ments were stored at 4 °C in plastic bags for use in sediment elutriate bioassays. Two sub-samples (100 gm) of each sediment were set aside for geochemical analyses. Concentrations of trace elements in one sub-sample were determined by X-ray fluorescent spectrometry while the second sub-sample was analyzed for pesticides and PCBs. All chemical analyses were carried out according to the Analytical Methods Manual of the Water Quality Branch (Inland Waters Directorate, 1979).

Disposal

The dredged material collected from the dredging sites was transported by means of barges as indicated earlier and disposed of in designated confined areas (Fig. 1). There, disposal operations were monitored both biologically and chemically during the following operations:

    pre-disposal (control)
    post-disposal (10 minutes)
    post-disposal (60 minutes)

During various phases of disposal, water samples were carefully collected from the front of a 16-foot Boston Whaler as described earlier. Nutrients, trace metals, and organic analyses were performed on water samples according to the methods previously described. Similarly, the impact of disposal material and resuspended particulates were evaluated by conducting ^{14}C size-fractionated primary productivity experiments.

Sediment elutriate bioassays

The sediments collected from the barges during the three dredging episodes were used to prepare elutriates for bioassessment. The elutriates were made by mixing one volume of sediment with four volumes of 0.45 μm filtered offshore Lake Ontario water. The mixture was agitated for 30 minutes with a rotary tumbler at 5 rpm. After one hour of settling, the supernatant was collected and filtered through 0.45 μm Millipore membrane filters (U.S. E.P.A., 1977). The elutriates were sub-sampled for chemical analyses, according to the methods described earlier.

The remaining elutriate portions were kept at 4 °C until sediment elutriate bioassays were performed using ^{14}C techniques. These assays were made using natural offshore phytoplankton as a test assemblage and were spiked with 1, 5, 10, and 20 percent elutriate dosages. These samples were then compared to the control, to which no elutriate had been added. Incubation and size-fractionation procedures were similar to those described earlier (Munawar et al., 1983; 1985).

Statistical evaluation

Pairwise comparisons of production during the three sampling times for dredging or disposal were obtained using least squares to fit the regression model $Y = \beta X$ which passes through the origin. The significance of the regression was determined using an ANOVA table. A t-test was then used to determine whether the estimate of the slope differed from a slope of 1.0. A slope which was equivalent to 1.0 would indicate that algae production had not changed between the two sampling times.

Results

Dredging

In situ impact on primary productivity

Dredging Episode I (22 July 85)
Nutrient and trace metal concentrations of water samples are presented in Table 1. The data revealed decreases in NH_3, TKN, total P, alkalinity, and Al during dredging compared to pre-dredge values. Increases in turbidity, Cr (Hexa & Total), Cu, and Fe were also observed.

The species composition of commonly occuring taxa at the dredging site prior to the dredging operation is given in Table 2. The algal composition was dominated by phytoflagellates belonging to Chrysophyceae and Cryptophyceae, particularly Uroglena americana (Calkins) Lemmermann, Cryptomonas erosa Ehrenberg, and C. rostrata Skuja.

The P/B quotients for both size assemblages were significantly enhanced by dredging activity (Fig. 2). The microplankton/netplankton (> 20 μm) quotient was only enhanced slightly during dredging, but 15 minutes after dredging exhibited a significant 36 percent enhancement ($P < 0.01$) over the pre-dredging levels. The quotient for the ultraplankton (< 20 μm) was greatly affected by dredging, showing a 43 percent enhancement of productivity ($P < 0.01$) during this period. Forty percent enhancement ($P < 0.001$) was still occuring 15 minutes after termination of the dredging operation (Fig. 2).

Dredging Episode II (25 July 85)
Nutrients and trace metals found in the water are given in Table 1. The data showed decreases in Cr(Total), and Ni concentrations, whereas increases were observed in TKN, P(Total), alkalinity, turbidity, Al, Cr(Hexa), Fe, Mn, Pb, and Zn concentrations both during- and post-dredging.

Table 1. Dredging water chemistry (mg l^{-1})

	NH$_3$	TKN[1]	TP[2]	ALK	PH	Turb[3]	Al	Cr[4]	Cd	Cr[5]	Cu	Fe	Mn	Ni	Pb	Zn	Hg
Dredge I																	
Pre-dredge	0.35	2.0	0.3	101.7	7.86	4.5	0.38	0.003	<0.001	0.006	0.01	0.4	0.03	<0.01	0.004	0.02	<0.02
During-dredge	0.22	1.2	0.2	100.7	7.92	5.5	0.32	0.018	<0.001	0.022	0.03	0.42	0.03	<0.01	0.004	0.02	<0.02
Post-dredge	0.32	0.8	0.2	94.1	7.95	5.5	0.27	0.01	<0.001	0.008	0.01	0.33	0.02	<0.01	0.004	0.02	<0.02
Dredge II																	
Pre-dredge	<0.1	2.5	0.1	90.3	8.15	1.8	0.27	0.006	<0.001	0.011	<0.001	0.13	0.01	0.03	0.002	0.01	<0.02
During-dredge	<0.11	3.7	0.1	102.6	7.93	4.3	0.30	0.010	<0.001	0.009	<0.001	0.36	0.03	<0.01	0.005	0.02	<0.02
Post-dredge	<0.1	4.4	0.5	103.6	7.9	5.6	0.32	0.010	<0.001	0.011	<0.001	0.42	0.03	<0.01	0.005	0.02	<0.02
Dredge III																	
Pre-dredge	0.34	1.5	<0.1	104.0	8.0	9.0	0.41	0.008	<0.001	0.007	<0.001	0.66	0.03	<0.01	0.005	0.02	<0.02
During-dredge	0.31	1.3	<0.1	94.0	8.1	4.5	0.23	0.008	<0.001	0.006	<0.001	0.40	0.02	<0.01	0.005	0.02	<0.02
Post-dredge	0.26	0.7	<0.1	95.0	8.0	10.0	0.51	0.004	<0.001	0.004	<0.001	0.70	0.02	<0.01	0.005	0.02	<0.02
* WQA 78	0.02		0.5	–	–	–	–		0.0002	0.05	0.001	0.3	–	0.025	0.025	0.03	0.0002

* Water Quality Agreement, 1978
[1] Total Kjeldahl nitrogen
[2] Total phosphorus
[3] Turbidity
[4] Hexa Chromium
[5] Total Chromium

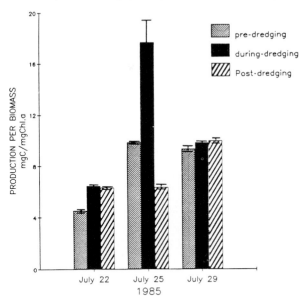

DREDGING
Ultraplankton/picoplankton <20 μm

pre-dredging
during-dredging
Post-dredging

PRODUCTION PER BIOMASS
mgC./mgChl.a

July 22   July 25   July 29
1985

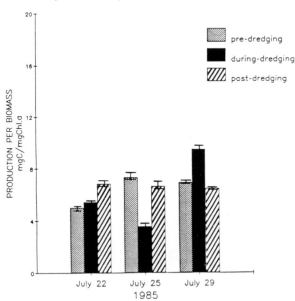

DREDGING
Microplankton/netplankton >20 μm

pre-dredging
during-dredging
post-dredging

PRODUCTION PER BIOMASS
mgC/mgChl.a

July 22   July 25   July 29
1985

Fig. 2. P/B quotients (production/biomass) for microplankton/netplankton and ultraplankton/picoplankton at the dredging site.

Table 2. Species composition – contributing at least 1% of the total biomass of Toronto Harbour prior to dredging, 1985.

Genus species:	July 22	July 25	July 29
Oscillatoria tenuis Agardh	4		5
Carteria cordiformis (Carter) Diesing			5
Chlamydomonas sp.	1	2	2
C. globosa Snow			2
Dictyosphaerium pulchellum Wood		2	
Oocystis borgei Snow			5
Pandorina morum (O. Mueller) Bory	2		
Chrysidiastrum catenatum Lauterborn		2	
Chrysochromulina parva Lackey		4	1
Chromulina sp.		1	
Dinobryon divergens Imhof		4	2
Ochromonas sp.			2
Uroglena americana (Calkins) Lemmermann	15	9	14
Melosira islandica O. Mueller	1		
Nitzschia sp.	4		
Surirella sp.			3
Tabellaria fenestrata (Lyng.) Kuetzing		2	
Cryptomonas sp.	2		
C. erosa Ehrenberg	19	20	18
C. erosa var. reflexa Marssonii	4	6	4
C. ovata Ehrenberg		2	
C. phaseolus Skuja			1
C. rostrata Skuja	11	3	3
Katablepharis ovalis Skuja	1	2	1
Rhodomonas minuta Skuja	2	3	3
Gymnodinium ordinatum Skuja		1	

Also NH_3 levels were below detection limits during this dredging episode whereas concentrations were relatively high during Episode I.

A number of organic pesticides were also found in the samples collected during dredging. Lindane was found in high concentrations (0.001 mg l^{-1}) while Dieldrin and BHC were present in concentrations of 0.006 and 0.003 mg l^{-1} respectively.

The species composition of phytoplankton in this Episode was generally similar to that observed for Episode I, although Chrysophyceae species were more frequent this time. *Uroglena americana* (Calkins) Lemmermann and *Cryptomonas erosa* Ehrenberg continued to dominate (Table 2).

The phytoplankton size assemblages exhibited a differential response to dredging operations during this Episode (Fig. 2) as opposed to Episode I. The microplankton/netplankton (> 20 μm) P/B quotient was inhibited by 51 percent ($P < 0.001$) over the pre-dredging level. However, its ratio recovered to a level only 10 percent (non-significant) below that of the pre-dredge condition, 15 minutes after the termination of the dredging operation. The ultraplankton (< 20 μm) demonstrated a 79 percent ($P < 0.001$) enhancement of P/B quotient over the pre-dredging operation (Fig. 2). However, 15 minutes after the termination of the dredging operation it showed a 36 percent ($P < 0.001$) inhibition compared to the P/B quotient observed before the commencement of dredging.

Dredging Episode III (29 July 85)

Nutrient and trace metal concentrations in water samples for this Episode are presented in Table 1. A decrease was observed in NH_3, TKN, alkalinity, Al, Cr(Hexa & Total), Fe, and Mn during – and/or post-dredging. No increase was observed except in Fe and Al, which had concentrations higher in the post-dredge sample. Throughout the three periods of dredging investigated, Fe concentrations remained above Water Quality Agreement (1978) levels, as did NH_3.

The phytoplankton species composition was once again similar to the preceding Episodes dominated by *Uroglena americana* (Calkins) Lemmermann and *Cryptomonas erosa* Ehrenberg (Table 2).

Again, as observed in Episode I, both size assemblages demonstrated an enhancement of P/B quotient during the dredging operation (Fig. 2). The microplankton/netplankton (> 20 μm) was enhanced by 37 percent ($P < 0.001$) but 6 percent inhibition ($P < 0.05$)

was recorded 15 minutes after the dredging activity. The ultraplankton (< 20 μm) showed only a slight enhancement of 5 percent and 7 percent for the during- and post-dredging operations respectively, although statistically, they were not significant.

The statistical analysis of the combined data for all three dredging Episodes was carried out to evaluate the overall impact of dredging on the natural phytoplankton productivity (Fig. 3). The statistics are summarized below:

Microplankton/netplankton (> 20 μm) – the fit of each of the regression lines was significant ($P < 0.001$). During-dredging production increased significantly ($P < 0.001$) by 21 percent over pre-dredging. Post-dredging production increased significantly ($P < 0.001$) by 18 percent over pre-dredging. Post-dredging production was not significantly different ($P < 0.05$) from during-dredging.

Ultraplankton (< 20 μm) – the fit of each of the regression lines was significant ($P < 0.001$). During-dredging production increased significantly ($P < 0.001$) by 28 percent over pre-dredging. Post-dredging production increased significantly ($P < 0.001$) by 21 percent over pre-dredging. Post-dredging production decreased significantly ($P < 0.05$) by 8 percent over during-dredging.

Disposal
In situ impact on primary productivity

Disposal Episode I (July 22, 1985)

Nutrient and trace metal concentrations are presented in Table 3. The data revealed increases in NH_3, TKN, turbidity, Al, Cu, Fe, Mn, Pb, and Zn concentrations immediately after disposal. Concentrations of Pb decreased 60 minutes after disposal and both Cr(Hexa and Total) concentrations decreased to lower than the initial levels.

The phytoplankton species composition at the disposal site prior to the disposal operation is given in Table 4. Phytoflagellates belonging to

608

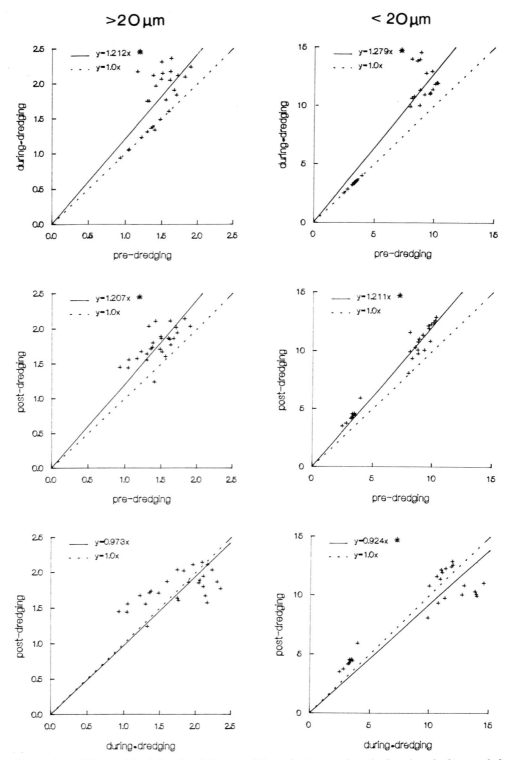

Fig. 3. Statistical analysis of the combined data for all three dredging episodes on microplankton/netplankton and ultraplankton/picoplankton.

Table 3. Disposal water chemistry (mg·l⁻¹)

	NH₃N	TKN[1]	TP[2]	ALK[3]	PH	Turb[4]	Al	Cr[5]	Cd	Cr[6]	Cu	Fe	Mn	Ni	Pb	Zn	Hg
Disposal I																	
Pre-disposal	<0.1	0.8	0.2	85.5	8.12	3.4	0.27	0.11	<0.001	0.012	<0.01	0.25	0.01	<0.01	0.003	0.01	<0.02
Post 10 min	0.28	1.4	0.2	86.5	8.03	18.0	1.0	0.014	<0.001	0.018	0.01	1.1	0.03	<0.01	0.015	0.03	<0.02
Post 90 min	0.24	1.4	0.2	87.4	8.1	7.2	0.54	0.005	<0.001	0.003	<0.01	0.48	0.02	<0.01	0.006	0.02	<0.02
Disposal II																	
Pre-disposal	<0.1	2.0	<0.1	83.6	8.3	5.5	0.32	0.018	<0.001	0.019	<0.01	0.29	0.01	<0.01	0.009	0.02	<0.02
Post 10 min	<0.1	3.1	<0.1	84.6	8.3	6.0	0.37	0.020	<0.001	0.037	<0.01	0.37	0.01	<0.01	0.005	0.02	<0.02
Post 60 min	<0.1	2.4	<0.1	81.7	8.3	4.5	0.22	0.009	<0.001	0.020	<0.01	0.23	<0.01	<0.01	0.005	0.02	<0.02
Disposal III																	
Pre-disposal	0.32	1.1	<0.1	86.0	8.2	2.5	0.12	0.005	<0.001	0.004	<0.01	0.15	<0.01	<0.01	0.002	0.01	<0.02
Post 10 min	0.47	1.0	0.5	113.0	8.0	35.0	5.0	0.009	<0.001	0.015	0.03	5.8	0.14	<0.01	0.074	0.10	<0.02
Post 60 min	0.69	1.1	<0.1	86.0	8.1	9.0	0.52	0.010	<0.001	0.008	<0.01	0.57	0.02	<0.01	0.007	0.02	<0.02
WQA*	0.02	–	0.5	–	–	–	–	–	0.002	0.05	0.005	0.3	–	0.025	0.025	0.03	0.0002

*WQA: Water Quality Agreement, 1978
[1] Total Kjeldahl nitrogen
[2] Total phosphorus
[3] Alkalinity
[4] Turbidity
[5] Hexa Chromium
[6] Total Chromium

Table 4. Species composition – contributing at least 1% of the total biomass of Toronto Harbour prior to disposal, 1985.

Genus species:	July 22	July 25	July 29
Oscillatoria minima Gickl- horn			1
Chlamydomonas sp.	2		
C. globosa Snow	2	1	2
Carteria cordiformis (Car- ter) Diesing		5	4
Scenedesmus arcuatus Lemmermann	1		
Tetraspora lacustris Lem- mermann			1
Cosmarium tenue West & West		2	
Chrysochromulina parva Lackey	9	5	6
Dinobryon divergens Imhof		1	1
Ochromonas sp.			3
Uroglena americana (Cal- kins) Lemmermann	5	2	3
Diatoma elongatum var. tenuis (Agardh) Kütz	1		1
Melosira islandica O. Mueller	4		4
Fragilaria construens (Ehr.) Grunow	1		
Cryptomonas sp.	2	3	4
C. erosa Ehrenberg	21	42	19
C. erosa var. reflexa Marssonii	9	8	6
C. ovata Ehrenberg	6	3	3
C. rostrata Skuja	6	3	1
Katablepharis ovalis Skuja	3	3	2
Rhodomonas minuta Skuja	6	5	7
Gymnodinium ordinatum Skuja	2		3

Cryptophyceae and Chrysophyceae such as *Cryptomonas erosa* Ehrenberg, *C. erosa v. reflexa* Marssonii, *C. ovata* Ehrenberg, *C. rostrata* Skuja, *Rhodomonas minuta* Skuja, *Chrysochromulina parva* Lackey, and *Uroglena americana* (Calkins) Lemmermann dominated.

During the disposal Episode I, both size assemblages showed an inhibition of P/B quotient immediately after the disposal activity (Fig. 4). The microplankton/netplankton quotient drop- ped by 27 percent ($P < 0.001$). However, 60 minutes later it had recovered and even showed a 50 percent ($P < 0.01$) enhancement above that of the pre-disposal level. The ultraplankton was inhibited by 50 percent ($P < 0.001$) immediately after disposal and remained 18 percent ($P < 0.01$) below the pre-disposal level after 60 minutes.

Disposal Episode II (July 25, 1985)
Nutrient and trace metal concentrations are presented in Table 3. Increases in TKN, turbidity, Al, Cr(Hexa and Total), and Fe concentrations immediately after disposal were recorded. Conversely, the concentration of Pb decreased. All parameters which exhibited increased concentrations returned generally to pre-disposal levels or lower after 60 minutes.

Algal composition in this episode was similar to that observed for disposal in Episode I, although *Cryptomonas erosa* Ehrenberg contributed overwhelmingly to the total biomass (Table 4).

Again, as observed in disposal Episode I, the P/B quotient of microplankton/netplankton was slightly inhibited immediately after disposal and remained low at 12 percent (non-significant) 60 minutes later. The ultraplankton responded quite differently this time and demonstrated a 61 percent enhancement ($P < 0.001$) immediately after disposal. This enhancement trend continued when a 107 percent higher P/B ($P < 0.001$) quotient than pre-disposal level was observed, 60 minutes after the disposal activity (Fig. 4).

Disposal Episode III (July 29, 1985)
Nutrient and trace metal concentrations are presented for Episode III in Table 3. The water analyses revealed significant increases immediately after disposal in total P, turbidity, Al, Cr(Hexa and Total), Cu, Fe, Mn, Pb, and Zn concentrations. Most of these concentrations decreased 60 minutes after disposal.

Once again, the species composition remained the same as observed in the preceding Episodes. The most common species were *Cryptomonas erosa* Ehrenberg, *C. erosa var. reflexa* Marssonii, *Rhodomonas minuta* Skuja, and *Chrysochromulina parva* Lackey (Table 4).

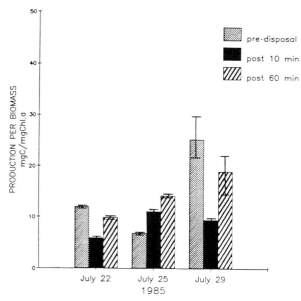

DISPOSAL
Ultraplankton/picoplankton <20 μm

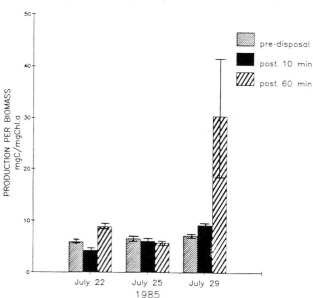

DISPOSAL
Microplankton/netplankton >20 μm

Fig. 4. P/B quotients (production/biomass) for micro-plankton/netplankton and ultraplankton/picoplankton at the disposal site.

The responses of phytoplankton size assemblages are shown in Fig. 4. The microplankton/netplankton was affected inversely compared to the first two disposal Episodes. The P/B quotient was immediately enhanced by 29 percent (non-significant) and showed an extremely high P/B level 60 minutes after the disposal operation, although statistically non-significant.

The ultraplankton was again inhibited by the disposal activity, similar to the observations of the disposal Episode I (July 22, 1985). The P/B quotient decreased by 63 percent ($P < 0.01$) immediately after the activity and continued to remain suppressed by 25 percent below the pre-disposal level, although this was non-significant statistically (Fig. 4).

Generally, the nutrients and trace metals increased following disposal activities. P/B quotients were usually inhibited after disposal of dredged material but sometimes slight enhancement of primary productivity occurred. The statistical evaluation of all three disposal Episodes is synthesized in Fig. 5 and is summarized below:

Microplankton/netplankton ($> 20\ \mu$m) – the fit of each of the regression lines was significant ($P < 0.001$). During disposal, production decreased significantly ($P < 0.001$) by 24 percent over pre-disposal. Post-disposal production decreased significantly ($P < 0.05$) by 10 percent over pre-disposal. Post-disposal production increased significantly ($P < 0.001$) by 17 percent over during-disposal.

Ultraplankton ($< 20\ \mu$m) – the fit of each of the regression lines was significant ($P < 0.001$). During-disposal production decreased significantly ($P < 0.001$) by 22 percent over pre-disposal. Post-disposal production decreased significantly ($P < 0.001$) by 14 percent over pre-disposal. Post-disposal production was not significantly different from during-disposal.

Sediment elutriate bioassays

Episode I (22 July 1985)
The sediment contained levels above Ontario Ministry of the Environment Dredging Guide-

612

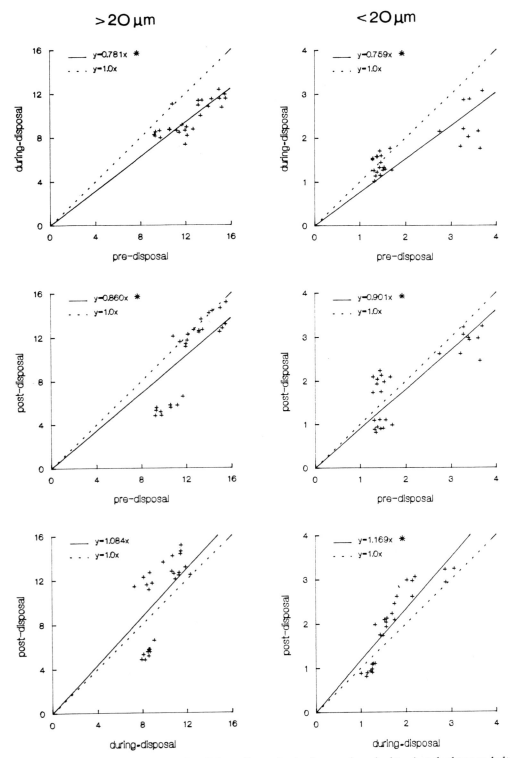

Fig. 5. Statistical analysis of the combined data for all three disposal episodes on microplankton/netplankton and ultraplankton/picoplankton.

Table 5. Sediment chemistry ($\mu g \cdot g^{-1}$).

Sediment	Date	LOI[2] (%dry)	TKN[3] (%dry)	TP[4] (%dry)	Al (%dry)	Cd	Cr	Cu	Fe (%dry)	Mn	Ni	Pb	Zn	Hg
Episode I	Jul 22–85	9.5	0.005	0.005	4.6	2.0	60	65	2.5	560	30	203	278	0.23
Episode II	Jul 25–85	7.6	0.019	0.008	4.4	<1	52	52	2.3	550	20	176	235	0.26
Episode III	Jul 29–85	9.0	0.028	0.024	4.6	2.0	60	60	2.5	560	30	212	300	0.30
Mean		8.7	0.017	0.012	4.5	1.6	57	59	2.4	557	27	197	271	0.26
*O.M.E.		6.0	0.2	0.1	–	1.0	25	25	1.0	–	25	50	100	0.30

Elutriates chemistry ($mg \cdot l^{-1}$)

Sediment elutriate	NH₃	NO₃	TKN	SRP	DOC	DIC	SiO²	ALK	Cd	Co	Cu	Fe	Mn	Ni	Pb	Zn
Episode I	0.239	0.102	24.08	0.0021	9.8	56.3	8.45	207.0	<0.001	<0.001	0.008	0.024	<0.02	0.012	0.11	0.17
Episode II	17.9	0.022	17.2	0.0015	8.9	57.2	8.16	203.0	<0.001	0.001	0.004	0.021	<0.02	0.002	<0.001	0.004
WQA[1]	0.02	–	–	–	–	–	–	0.0002	–	0.005	0.3	–	–	0.25	0.25	0.03

+ Date not available
* Ontario Ministry of Environment Dredging Guidelines (1985)
[1] Water Quality Agreement, 1978
[2] Loss on ignition (%)
[3] Total kjeldahl nitrogen
[4] Total phosphorus

614

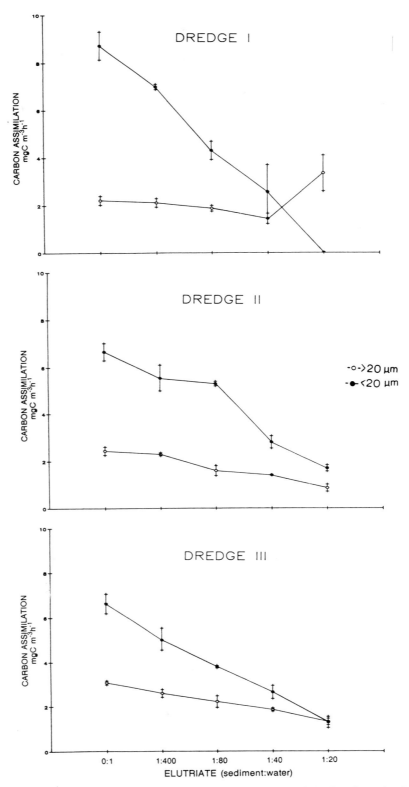

Fig. 6. Algal Fractionation Bioassays (AFB) using various concentrations of elutriate from the dredging episodes.

lines (1985) of Cr, Cu, Fe, Ni, Pb, and Zn (Table 5). Therefore, this sediment was unsuitable for offshore disposal.

The chemical analyses results of the sediment elutriate are presented in Table 5. Trace metals were found at low concentrations, similar to the concentrations found in the water samples collected during dredging activities.

The offshore ultraplankton used as a test assemblage was greatly inhibited with increasing doses of standard elutriate (Fig. 6). The 1 : 20 concentration caused 100 percent inhibition ($P < 0.001$) and even the 1 : 400 concentration caused 20 percent inhibition ($P < 0.01$). The microplankton/netplankton offshore assemblage did not show much response to elutriate additions with the exception of the 1 : 20 concentration when a 52 percent enhancement was observed which was statistically significant.

Episode II (25 July 1985)

The chemical analyses of this sediment revealed higher percent concentrations of TKN and total phosphorus than found in Episode I (Table 5). However, out of all three sediments analyzed, this had the lowest concentrations of trace metals, except for Hg. However, Cr, Cu, Fe, Pb, and Zn concentrations still exceeded Ontario Ministry of the Environment guideline (1982; 1985) levels, which would likely label this sediment unsuitable for offshore disposal.

The sediment elutriate exhibited some chemical concentrations different from those observed for Episode I (Table 5). An extremely high concentration of NH_3 was observed. Other nutrient concentrations were similar to or lower than the elutriate of Episode I concentrations. Trace metal concentrations were similar to that of Episode I.

Both the ultraplankton and microplankton/netplankton assemblages were inhibited with increasing doses of sediment elutriate (Fig. 6). The maximum inhibition observed for the $< 20 \mu m$ assemblage was 7 percent ($P < 0.001$) at the 1 : 20 elutriate concentration. The maximum inhibition observed for the $> 20 \mu m$ assemblage was 64 percent ($P < 0.001$) with the 1 : 20 elutriate concentration.

Episode III (29 July 1985)

This sediment, collected on July 29, 1985, exhibited the highest concentrations of nutrients (TKN and total P), as well as the trace metals Pb, Zn, and Hg compared to the other two episodes (Table 5). Also, all trace metal concentrations, except Hg, exceeded the Ontario Ministry of Environment (1982; 1985) sediment guidelines, thus labelling this sediment unsuitable for offshore disposal.

Results of chemical analyses for the sediment elutriate were unavailable. However, chemical analyses of the sediment revealed concentrations similar to those observed for Episodes I and II, except for TKN and total P. Therefore, it is possible that similar concentrations of trace metals and nutrients may have been eluted.

As seen with the other two elutriate bioassays, inhibition was observed with increasing doses of elutriate (Fig. 6). The $< 20 \mu m$ assemblage exhibited a maximum inhibition of its productivity by 80 percent ($P < 0.001$) at the 1 : 20 sediment to water ratio. The $> 20 \mu m$ assemblage was similarly inhibited with increasing doses of elutriate to a maximum inhibition of 57 percent ($P < 0.001$) at the 1 : 20 sediment to water concentration.

Discussion

The perturbation of the Toronto Harbour Water Front is a very complex problem resulting from the inputs from the Don River and sewer outflows, discharges of municipal and industrial effluents, commercial and recreational uses, and lack of continued maintenance. The complexity of the ecosystem is reflected in the contaminated sediments which result from a number of events.

Firstly, the duration of dredging operations at the mouth of the Keating Channel is continuous 6 hours a day, five days a week, for a minimum of two months of the year. Secondly, it takes two hours of dredging to fill one barge with sediment, during which time a plume of silt of approximately 800 m is formed. Thirdly, approximately five minutes was required to dump the sediment from the bottom of the barge at the disposal site, which would only be a maximum of 5 m deep (Fricbergs, 1986). The sediment therefore would fall through the water in a matter of seconds. The plume of silt from this operation was 50 m and remained visible for 10 to 20 minutes. In terms of duration of perturbation and affected area, dredging operations cause the largest plume over the longest time. Therefore, the major disturbance of sediments is dredging.

It is obvious that Toronto Harbour water, and especially sediment, are polluted with trace metals, nutrients, and some organic substances. Disturbance of these sediments by any action causes the release of these contaminants and make them available to biota. Sediments that are heavily contaminated with iron, cadmium, chromium, copper, nickel, lead, zinc, arsenic, and PCBs are unsuitable for offshore disposal, according to the guidelines designated by the Ontario Ministry of the Environment (1985, 1985) and Dredging Subcommittee (1983). Toronto Harbour water contained many trace metals and nutrients (copper, zinc, mercury and ammonia) whose concentrations exceeded those limits.

Results also indicated that dredging caused changes in nutrient and trace metal concentrations. Generally, a decrease in some nutrient concentrations was observed. P/B quotients were generally enhanced by dredging and sometimes regained initial levels observed prior to the commencement of dredging operations.

Sediment elutriates, made from the dredged sediments collected from the barges, severely inhibited offshore phytoplankton primary productivity. Thus, there is potential that these sediments will be toxic to natural offshore phytoplankton populations. For example, traditional EC50s were calculated for the 29 July, 1985 Episode. The EC50 for the $> 20 \mu m$ size fraction was 17.3 mg C m^{-3} h^{-1} while the $< 20 \mu m$ fraction was 7.5 mg C m^{-3} h^{-1}. Based on this data, it can be concluded that the sediment was the most toxic for ultraplankton. Computations were also made of EC25s as an early warning tool (Munawar & Thomas, 1989) and were 3.5 and 2.2 mg C m^{-3} h^{-1} respectively.

It is apparent that both dredging and disposal activities are instrumental in resuspending the bottom sediments. Consequently, changes are made to the productivity of indigenous phytoplankton in the harbour, a fundamental process in sustaining the food web. As mentioned earlier, the results indicate that the observed enhancement and inhibition of primary productivity may be the result of complex nutrient/contaminant interactions. The offshore phytoplankton, however, was severely inhibited by the addition of sediment elutriates, indicating that the nutrient/contaminant interactions were not operating under the same conditions as the *in situ* experiments. Moreover, the indigenous population may also be adapted to such perturbations in the harbour (Munawar, 1989), unlike the sensitive offshore phytoplankton whose primary productivity was extremely inhibited.

It was not possible to isolate, in the natural environment, effects of turbidity and the subsequently reduced light penetration which resulted after each dredging and disposal activity. The light penetration was measured before and after each activity, indicating an increased turbidity which could inhibit *in situ* primary productivity. However, since enhancement, rather than inhibition, was generally observed in our results, it seems that turbidity may not be a limiting factor. Furthermore, the predominant species were phytoflagellates which are mobile and can swim to the upper strata to avail themselves of the needed radiation. Moreover, our experiments were conducted in a constant light incubator equipped with a shaker which kept the test containers continuously agitated, thus avoiding the problems associated with light. Therefore, the observed results appear to be governed by nutrient and contaminant availability, and not by physical factors such as light limitations.

Enhancement of primary productivity could be problematic in terms of causing nuisance algal blooms if microplankton/netplankton growth is high. On the other hand, inhibition of ultraplankton could result in the elimination of the edible component of the plankton resource essential for zooplankton grazing (Ross & Munawar,

1981), which may result in a disruption of food web interactions.

The Toronto Harbour project has been instrumental in developing, on a large scale, *in situ* bioassessment technology and an understanding of the complexities of the harbour ecosystem. The techniques applied were successful in elucidating the impact of nutrients and contaminants. The results clearly suggest the need for such *in situ* procedures in addition to complementary laboratory assays. It is obvious that the results from the field provide realistic but different conclusions compared to the laboratory assays with sediment elutriates.

The following action plan is suggested to improve the ecosystem health of Toronto Harbour:

1. Resume dredging within the Keating Channel, which is the most efficient location for trapping silt and sediments transported down river. Maintain this dredging on a continuing basis using the best possible means for containment.
2. Dispose of the dredge material into the Eastern Headland where containment of contaminants is the most effective.
3. Clean up the major source of silt and contaminants, i.e., the Don River drainage basin, by reducing contaminated discharges into the river and reducing soil erosion from construction sites and agricultural lands in the drainage basin.
4. Clean up the other major source of contaminants, the sewer outflows.
5. The amount of sediment-bound contamination in Toronto Harbour could be reduced if:
   a. the most efficient means of dredging was utilized
   b. the Eastern Headland continued to be used for containment of contaminated sediments
   c. reduction of silting of slips and navigation channels occurred, which would limit redistribution of contaminants through sediment resuspension, and
   d. implementation of clean-up measures of the Don River drainage basin were set in place,

which would reduce the amount of sediment transported down-river as well as the levels of contaminants discharged into the river.

In conclusion, the application of an *in situ* environmental technology, such as the one conducted in the present study, appears to be a promising tool in the investigation of the ecotoxicology of harbours, and has a great potential for the improvement of the ecosystem health of the Great Lakes Areas of Concern and other stressed environments.

Acknowledgements

We would like to thank the following colleagues and personnel who assisted in various aspects of the work: I.F. Munawar, A. El-Shaarawi, M. Dutton, H.F. Nicholson, J. Milne, and L. Keeler. The assistance provided by Mr. I. Orchard (Environmental Protection Service, Toronto) and Mr. K. Fricberg (Toronto Harbour Commission) is greatly appreciated. The financial support, encouragement and interest of the following agencies who sponsored the project is acknowledged:
   Toronto Harbour Commission
   Metropolitan Toronto and Region Conservation Authority
   Environmental Protection Service, Environment Canada

References

Dredging Subcommittee, 1983. Evaluation of Dredged Material Disposal Options for two Great Lakes Harbours Using the Water Quality Board Dredging Subcommittee Guidelines. Report of The Dredging Subcommittee to the Great Lakes Water Quality Board. April. pp. iii-vi, 1 –67.

Fricbergs, K. S., 1986. Operations and management of the disposal facility – Toronto Harbour. In: A Forum to Review Confined Disposal Facilities for Dredged Materials in the Great Lakes. Report of the Dredging Subcommittee to the Great Lakes Water Quality Board. October 31, pp. 51–52.

Inland Waters Directorate, 1979. Analytical Methods Manual. Water Quality Branch, Ottawa: Environment Canada.

Lind, O. T. & R. S. Campbell, 1969. Comments on the use of liquid scintillation for routine determination of C-14 activity on production studies. Limnol. Oceanogr. 14: 787–789.

Munawar, M., 1989. Ecosystem health evaluation of Ashbridges Bay environment using a battery of tests. Fisheries and Oceans Report to MISA, Ministry of the Environment. 43 pp.

Munawar, M. & I. F. Munawar, 1978. Phytoplankton of Lake Superior 1973. J. Great Lakes Res. 4 (3-4): 415–442.

Munawar, M. & I. F. Munawar, 1987. Phytoplankton bioassays for evaluating toxicity of in situ sediment contaminants. In: R. L. Thomas, R. Evans, A. Hamilton, M. Munawar, T. Reynoldson and H. Sadar (Eds.). Ecological Effects of in situ Sediment Contaminants. Hydrobiologia 149: 87–105.

Munawar, M. & R. L. Thomas, 1989. Sediment toxicity testing in two areas of concern of the Laurentian Great Lakes: Toronto (Ontario) and Toledo (Ohio) harbours. Hydrobiologia 176/177: 397–409.

Munawar, M., P. Stadelmann & I. F. Munawar, 1974. Phytoplankton biomass, species composition and primary production at a nearshore and a mid-lake station of Lake Ontario during IFYGL. In: Proc. 17th Conf. Great Lakes Res. pp. 629–652.

Munawar, M., A. Mudroch, I. F. Munawar & R. L. Thomas, 1983. The impact of sediment-associated contaminants from the Niagara River mouth on various size assemblages of phytoplankton. J. Great Lakes Res. 9 (2): 303–313.

Munawar, M., R. L. Thomas, W. P. Norwood & S. A. Daniels, 1986. Sediment toxicity and production-biomass relationships of size-fractionated phytoplankton during on-site simulated dredging experiments in a contaminated pond. In: P. G. Sly (Ed.) Sediments and Water Interactions, Springer-Verlag, New York. pp. 407–426.

Munawar, M., R. L. Thomas, W. P. Norwood & A. Mudroch, 1985. Toxicity of Detroit River sediment-bound contaminants to ultraplankton. J. Great Lakes Res. 11 (3): 264–274.

Munawar, M., R. L. Thomas, H. Shear, P. McKee & A. Mudroch, 1984. An overview of sediment-associated contaminants and their bioassessment. Can. Tech. Rep. Fish. Aquat. Sci. 1253: i–vi, 1–136.

Ontario Ministry of the Environment (O.M.E.), 1982. Evaluating the impacts of marine construction activities on water resources (addendum 1978). In: Guidelines and Register for Evaluation of Great Lakes Dredging Projects. Report of the Dredging Subcommittee to the Water Quality Programs Committee of the Great Lakes Water Quality Board. January. pp. 92–93.

Ontario Ministry of the Environment (O.M.E.), 1985. Guidelines for the Management of Dredged Materials in Ontario. Water Resources Branch, Toronto, Ontario. 18 pp.

Ross, P. E. & M. Munawar, 1981. Preference for nannoplankton size fractions in Lake Ontario zooplankton grazing. J. Great Lakes Res. 7: 65–67.

Strickland, J. D. H. & T. R. Parsons, 1968. A practical handbook of seawater analysis. Bull. Fish. Res. Bd. Can. 167: 283–293.

U.S. Environmental Protection Agency/U.S. Corps of Engineers, 1977. Ecological evaluation of dredged material into ocean waters. Environ. Effects Lab. U.S. Army Corps of Eng. Waterways Exp. Station, Vicksburg, Mass.

Water Quality Agreement (W.Q.A.), 1978. Agreement, with annexes and terms of reference, between the United States of America and Canada. Signed at Ottawa, November 22, 1978. International Joint Commission, Windsor, Ontario.

Hydrobiologia **188/189**: 619–631, 1989.
M. Munawar, G. Dixon, C. I. Mayfield, T. Reynoldson and M. H. Sadar (eds)
Environmental Bioassay Techniques and their Application.
© 1989 *Kluwer Academic Publishers. Printed in Belgium.*

An improved elutriation technique for the bioassessment of sediment contaminants

S.A. Daniels [1]*, M. Munawar [2] & C.I. Mayfield [1]

[1] *Biology Department, University of Waterloo, Waterloo, Ontario, Canada N2L 3G1*; [2] *Great Lakes Laboratory for Fisheries and Aquatic Sciences, Fisheries and Oceans, C.C.I.W., P.O. Box 5050, 867 Lakeshore Road, Burlington, Ontario, Canada L7R 4A6*; (*Present address: Research and Applications Branch, National Water Research Institute, C.C.I.W., P.O. Box 5050, 867 Lakeshore Road, Burlington, Ontario, Canada L7R 4A6)*

Key words: sediment elutriate, rotary tumbling, phytoplankton bioassay, sediment contaminants, toxicity, carbon assimilation

Abstract

An improved method is proposed for the preparation of sediment elutriates which permits relatively realistic determination of bioavailable contaminants. It suggests the use of rotary tumbling in a cycle of 3–4 rpm to achieve sediment-water mixing. Experiments were undertaken to evaluate the mixing efficiency of the rotary tumbler as compared to that of the compressed air, wrist-action shaker, and reciprocal shaker methods. Sediment to water ratios of 0:1, 1:20, 1:10, and 1:4 were tested over 0.5, 1.0, 24, and 48-h elution periods. Elutriate evaluations were based on chemical, physico-chemical and gravimetric determinations; and also on ^{14}C-phytoplankton bioassays using *Chlorella vulgaris* (Beyerinck). Results indicated that rotary tumbling produced the most consistent bioassay-supportable data. It was also the most efficient procedure when used for 1 h with 1:4 sediment-water mixtures.

Introduction

The role of aquatic sediments in garnering large quantities of hazardous contaminants has aroused much interest and promoted scientific investigation (Lee *et al.*, 1975; Engler, 1979; Munawar *et al.*, 1983, 1986). These are contaminants which are subject to release from adsorbing sediment particles by normal mechanical and physico-chemical forces operating in the aquatic ecosystem. In addition, these substances possess the ability to penetrate the plasma membrane of aquatic organisms (Boudou *et al.*, 1983). Hence,

much research activity has focussed on the circumstances and influences affecting the release of toxicants so adsorbed, and their effects on the biota (Mac *et al.*, 1984; Seeley & Mac, 1984).

The Elutriate Test was developed as a broad-spectrum leaching procedure primarily to determine the 'solubility' of contaminants subject to release when dredged sediments were deposited in open water (Keely & Engler, 1974). The method was subsequently formalized and promulgated as the Standard Elutriate Test (USEPA, 1977). It proposed mixing sediment and site water by mechanical shaker at a rate of 100 excursions per minute, or by bubbling compressed air through an air-diffuser system accompanied by periodic stirring for 30 minutes.

Recently, there has been renewed interest in the use of interstitial (pore) water in the assessment of potential sediment toxicity (Giesy *et al.*, 1988a, b). This method is useful and may serve as a valuable complement to the elutriate technique.

Côté & Constable (1982) examined a number of options in studies of leaching techniques for evaluating the contaminant potential of solid waste. A number of solid-liquid ratios were tested using a variable speed rotary tumbler to promote interaction between the solid and the liquid phases.

Thus, the thrust of this study was firstly, the development of a procedure to produce elutriates based on the bioavailability of sediment contaminants, and secondly, to approximate the natural leaching conditions of the natural system.

Methods and materials

Eight experiments were conducted which investigated sediment-water combinations with ratios of 1 : 20, 1 : 10, 1 : 4, and a control of 0 : 1, agitated initially for periods of 1.0 h and 48 h (Table 1). Since the overall goal concerned the pattern of contaminant removal, distilled water was used as the liquid phase in most preliminary experiments. The 48-h mixing period was eventually dropped and the 0.5-h period included, since it had been the recommended mixing period for the Standard Elutriate Test (Keely & Engler, 1974; USEPA, 1977).

Sample collection

Sediment used in preliminary studies was collected by Ponar grab sampler from Triangle Pond, a settling pond on Toronto Harbour's East Headland (Munawar *et al.*, 1986). Immediate subsampling was conducted for chemical characterization, and samples stored in teflon-lined containers at 4 °C until used in elutriate preparation (usually within 72 h). Collection and handling of sediment was generally consistent with guidelines defined by the USEPA (1977).

Glass-distilled water was used to prepare elutriates in studies I, II, IV, V, and VI; while water collected at Triangle Pond provided dilution water in study III. In studies VII and VIII water collected from Toronto Harbour was used. Site water was settled for four hours, decanted, and then stored at 4 °C until used.

Elutriate preparation

Sediment-water combinations (v/v) were prepared using the method of volumetric displacement (USEPA, 1977). Studies I, II, III, and IV evaluated the 'ferris-wheel' type rotary tumbler (Côté & Constable, 1982), Rugged Rotator, Model RD 250 (Kraft Apparatus Inc., Mineola, N.Y.) which was modified by the addition of six retort arms. To each of these was attached a metal clamp capable of accomodating a 1-litre Nalgene polypropylene widemouth bottle to ensure secure attachment of each vessel (Fig. 1). The device was operated at a rate of 4 rpm, with sediment-water combinations and agitation periods as shown in Table 1. In study IV, the pH of the elutriation mixtures was adjusted to 6.0 before mixing.

Study V evaluated the air-agitation method advocated by USEPA (1977). The system consisted of a water-trap assembly through which compressed air was saturated before delivery through fritted-glass diffusers to sample mixtures (Fig. 2b). Study IV concentrated on the Burrel wrist-action shaker (Fig. 2a), used at a rate of approximately 100 excursions per minute. Studies VII and VIII compared all options by including the three methods of agitation, the sediment-water combinations detailed above, and mixing periods of 1.0 h and 24 h in study VII, and 0.5 h, 1.0 h and 24 h in study VIII. The Eberbach reciprocal shaker was substituted for the wrist-action shaker in studies VII and VIII, to accomodate 1-litre containers.

After agitation, treatment mixtures were settled for a minimum of 6.0 h at 4 °C before the liquid phase was removed by aspiration. The super-

Table 1. Summary of treatments and procedures used in elutriation studies.

Study	Elutriate preparation and treatments
I	– Agitation by ferris-wheel type rotary tumbler operated at speed of 4 rpm; Evaluation of the procedure; – S : W ratios, 0 : 1, 1 : 20, 1 : 10 & 1 : 4; Triangle Pond sed.; – Dilution water: distilled water; – Periods of agitation of 1.0 h and 48 h; – Conductivity, pH, volatile & total solids monitored.
II	– Agitation by rotary tumbler at a rate of 4 rpm; – S : W ratios, 0 : 1, 1 : 20, 1 : 10 & 1 : 4; Triangle Pond sed.; – Dilution water: distilled water; – Periods of agitation of 1.0 h, 24 h & 48 h; – Conductivity, pH, volatile & total solids monitored.
III	– Agitation by rotary tumbler at a rate of 4 rpm; – S : W ratios, 0 : 1, 1 : 20, 1 : 10 & 1 : 4; Triangle Pond sed.; – Dilution water: Triangle Pond site water; – Periods of agitation of 1.0 h & 24 h; – Conductivity, pH, volatile & total solids monitored.
IV	– Agitation by rotary tumbler at a rate of 4 rpm; – S : W ratios, 0 : 1, 1 : 20, 1 : 10 & 1 : 4; Triangle Pond sed.; – Dilution water: distilled water; – Periods of agitation of 1.0 h, 24 h & 48 h; – pH of elutriation mixture adjusted to 6.0 with acetic acid prior to agitation.
V	– Agitation by air-diffuser system; – S : W ratios, 0 : 1, 1 : 20, 1 : 10 & 1 : 4; Triangle Pond sed.; – Dilution water: distilled water; – Periods of agitation of 1.0 h, 24 h & 48 h; – Conductivity, pH, volatile & total solids monitored.
VI	– Agitation by wrist-action shaker at rate, 100 cycles/minute; – S : W ratios, 0 : 1, 1 : 20, 1 : 10 & 1 : 4; Triangle Pond sed.; – Dilution water: distilled water; – Periods of agitation of 1.0 h, 24 h & 48 h; – Conductivity, pH, volatile & total solids monitored.
VII, VIII	– Comparison of three methods of agitation with the reciprocal shaker substituted for wrist-action shaker; – S : W ratios, periods of agitation, and bioassay levels detailed in Table 4; Toronto Harbour water & sediment; – Volatile and total solids, chemical, and biological parameters monitored.

natant was partitioned by spinning in a Sorvall RCB2-B automatic refrigerated centrifuge at 10 000 rpm for 30 min before filtration through pre-washed 0.45-μm Millipore HA 47-mm membrane filters.

Elutriate sub-samples for trace metal, nutrient-major ion and trace organics were taken immediately after preparation for chemical analysis using standard procedures (Inland Waters Directorate,

1985). The remaining portion of each elutriate was reserved for use in bioassays. All samples were stored at 4 °C in the dark and used within 72 h.

Physical and chemical assessment

The pH of sediments and elutriates was monitored using a Corning 125 single-probe digital pH

Fig. 1. Rugged Rotator Rotary Tumbler

meter. Measurements of specific conductance (conductivity) were carried out using a Radiometer CDM 83 conductivity meter. Gravimetric determinations of total dissolved solids (TDS), volatile solids (VS, loss on ignition), and moisture content were performed using a Mettler digital analytical balance. Sediment and elutriates were analyzed to determine concentrations of eight trace metals using inductively-coupled argon plasma emission spectrometry (Inland Waters Directorate, 1985).

Organic parameters were measured by gas chromatography and included organochlorine pesticides (OCs) and polychlorinated biphenyls (PCBs). A Varian Vista 6000 gas chromatograph equipped with a Vista 8000 autosampler, splitless capillary injector, and a J & W DB5-30N 0.5-mm ID fused silica column was used. Sediment samples received pre-treatment on an Autoprep 1200A gel permeation chromatograph (Analytical Biochemical Laboratories Inc., Columbia, Mo.) prior to chromatographic analysis. All analyses were carried out at the Canada Centre for Inland Waters using the procedures and methods of the National Water Quality Laboratory (Inland Waters Directorate, 1985). Organic analyses were performed in the Ultra-trace Laboratory of the Great Lakes Laboratory for Fisheries and Aquatic Sciences.

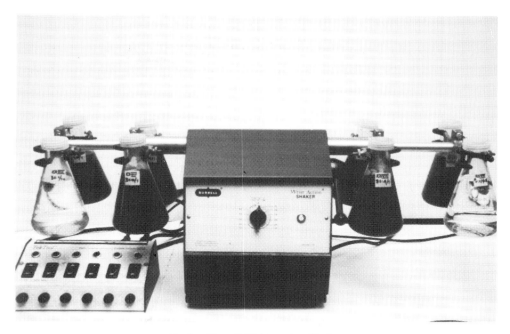

Fig. 2a. Burrell Wrist-action Shaker

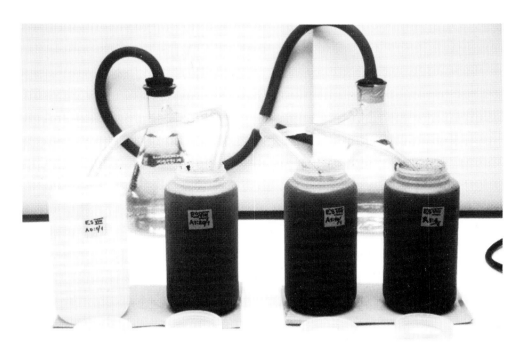

Fig. 2b. Aeration System of Agitation

Biological assessment

Axenic cultures of *Chlorella vulgaris* Bayerinck were acquired from Carolina Biological Supply Co., (Burlington, North Carolina, USA) and grown aseptically at a constant temperature of 20 °C in CHU-10 medium (Nichols, 1973) and an illumination period of 16 hours light ($\simeq 252\,\mu E\,s^{-1}\,m^{-2}$) and 8 hours dark. When appropriate densities were achieved (usually about 5×10^6 cells ml^{-1}), cultures were diluted while still in the log phase of replication. The diluent consisted of 0.45-μm filtered water collected one km offshore in Lake Ontario. This test mixture was in each instance allowed to equilibrate at 20 °C for 24 h, before use in bioassays.

The bioassays were conducted as described below:
1. Polycarbonate Erlenmeyer bottles (100 ml), were washed with detergent, and 10 percent HNO$_3$ before multiple distilled water rinses.
2. Four replicate bottles per treatment were each charged with 50 ml of the test sample. Two sets of replicates, untreated with elutriates, were used as controls – one set exposed to light during incubation, and the other incubated in the dark. The dark control monitored the level of any heterotrophic carbon assimilation occurring during incubation, and resulting values were used to correct data from all light-incubated samples.
3. Five percent and 25 percent elutriates were spiked into the bottles.
4. All bottles were then spiked with 2 μCi ^{14}C as NaHCO$_3$, well mixed and incubated for 4 h at 20 °C in a Conviron E7 Plant Growth Chamber providing constant light at 252.5 μE m^{-2} sec^{-1}.
5. At the end of the incubation period the temperature automatically dropped to 4 °C and the lights were extinguished. Before filtration, a 1-ml sub-sample was removed from each bottle for estimation of total available activity. These were each placed in separate scintillation vials and preserved with monoethanolamine. Samples were filtered through 0.45 μm Millipore filters. Filters with collected plankton were washed under negative pressure (not exceeding 17 kPa) with 0.1 N HCl to remove surface residue of

^{14}C-NaHCO$_3$. Each filter was placed in a separate scintillation vial and subsequently treated with 10 ml of the scintillation fluor PCSII (Amersham Corp.). Samples were vortex-mixed for scintillation counting by means of an LKB Wallac 1211 Rackbeta automatic liquid scintillation counter (Lind & Campbell, 1969).

Data acquired from scintillation counting were corrected for dark uptake, and assimilation coefficients (Assim. Coeff.) were computed. Assim. Coeff. is defined in this study as follows:

$$\text{Assim. Coeff.} = 100\ (\text{dpm}_{\text{Assim}}/\text{dpm}_{\text{Avail}})\quad \text{(Eq. 1)}$$

where,

dpm$_{\text{Assim}}$ = the level of ^{14}C-uptake by *Chlorella*

dpm$_{\text{Avail}}$ = the level of ^{14}C available for assimilation (Total Activity).

The experimental design applied was a $2 \times 3 \times 3 \times 4$ factorial involving the two levels of elutriate addition, three methods of agitation, the three time periods, and the four sediment-water ratios, as described above. Analysis of variance was utilized in the analysis of data from studies VII and VIII employing the F-test as a test of significance. Data assessment was focussed on interactions between level of addition (L) and ratio of sediment and water (R), because conclusions and interpretations regarding the other variables had already been facilitated by studies I to VI. Student's t-test comparisons were also applied to data enabling other intra-experimental evaluations. Data were then tabulated and respective levels of significance indicated.

Results

Constraints on space do not permit the presentation of the entire volume of data collected during this study. Hence, most of the data presented will relate to the evaluation of the rotary methods of mixing. At appropriate stages data will be presented and/or discussed comparing the other two methods, but in many instances formal results presentation will not be feasible.

Table 2. Mean PH values of elutriates prepared with distilled water and Triangle Pond sediments by rotary tumbling.

Treatment (S : W)	Period of agitation		
	1 h	24 h	48 h
0 : 1	6.71	6.55	6.42
1 : 20	8.07	8.30	8.00
1 : 10	8.11	8.25	8.06
1 : 4	8.22	8.39	8.18

Table 3. Mean conductivity of elutriates produced with DW and Triangle Pond sediments – rotary tumbling.

Treatment (S : W)	Specific conductance (μs cm^{-1})		
	1 h	24 h	48 h
0 : 1	9	3	2
1 : 20	387	734	674
1 : 10	495	739	848
1 : 4	369	636	701

pH Changes

In preliminary studies where rotary tumbling and distilled water were used, only small changes were observed when samples mixed for 24 h were compared with those treated for only 1.0 h (Table 2). A pH reduction of 0.16 was noted in the elutriate control (0 : 1), while all other treatments increased by 0.14–0.23. Similarly, a pH decrease of 0.29 in the control occurred with 48-h agitation. All other 48-h treatments also experienced pH reductions with mean differences of 0.04 for the 1 : 4 mixture, and 0.07 for the 1 : 20 sediment-water mixture (Table 2).

By adjusting the pH to 6.0 before tumbling, the pH increased in most 24-h samples compared to 1-h agitated samples. The 48-h mixing caused an overall pH reduction, with its reading of 7.46 being further reduced compared to the 24-h 1 : 4 sample. Elutriates prepared by tumbling with site water produced pH changes between 1.0 h and the 24 h samples which showed no great differences compared to the distilled water elutriates. Other methods of mixing reduced pH readings for both the 24 h- and 48-h treated samples compared to the respective 1.0 h-treatments.

Changes in conductivity

Conductivity values of rotary-tumbled elutriate controls prepared with distilled water, showed values which, as expected, were greatly reduced, 62.5 percent in the 24-h control and 76 percent in the 48-h control (Table 3). Other treatment values increased several orders of magnitude, with the difference generally greater with longer periods of agitation. When the pH of elutriate mixtures was adjusted to 6.0 before agitation, the same general trend was observed. This was also the case when site water was used to prepare elutriates, except that measured values increased almost two-fold.

Changes in dissolved solids

Elutriates prepared from Toronto Harbour sediment and site water showed only small differences in volatile solids values (loss-on-ignition) between treatment periods; but within a given mixing period, a different trend was observed (Table 4). Generally, there were small differences in elutriate volatile solids between 1 : 20 and 1 : 10 treatments, with both mixtures showing greater disparity compared to their respective 1 : 4 combinations, and to their controls in each case. The values in 1 : 4 sed : water mixtures were invariably significantly increased compared to their controls. This was true for both treatment periods. Note that rotary treatment produced values which progressively increased as the sediment constituent in the treatment combination increased.

Total dissolved solids produced by rotary-tumbling indicated that, except for an 8 percent reduction (in the 24-h control) all treatments showed increased values as the mixing period increased. Moreover, concentrations of total dissolved solids were significantly increased ($P < 0.05$–0.01) in all treatments compared to respective controls (Table 4). Mixing by com-

Table 4. Dissolved solids prepared from Toronto Harbour sediment and site water by three methods of agitation (mg/L).

Treatment (S : W)	Volatile solids		Total dissolved solids	
	1 h	24 h	1 h	24 h
A 0 : 1	39	53	97 ± 3	108 ± 2
A 1 : 20	72	82	127 ± 13	257 ± 5**
A 1 : 10	78	63	268 ± 8**	275 ± 5**
A 1 : 4	75	90	313 ± 13**	455 ± 75*
E 0 : 1	53	53	100 ± 3	100 ± 0
E 1 : 20	82	70	230 ± 7**	210 ± 7**
E 1 : 10	83	88	255 ± 15**	312 ± 12**
E 1 : 4	118	135	403 ± 13**	457 ± 55*
R 0 : 1	53	70	112 ± 14	103 ± 13
R 1 : 20	77	87	207 ± 3**	270 ± 3**
R 1 : 10	90	97	297 ± 13*	327 ± 3**
R 1 : 4	122	147	430 ± 57*	492 ± 42*

* Statistically significant at the 95% probability level compared to the control.
** Statistically significant at the 99% probability level compared to the control.
*** Statistically significant at the 99.9% probability level compared to the control.
A = Agitation by aeration.
E = Agitation by Eberbach reciprocal shaker.
R = Agitation by rotary tumbler.

pressed air also produced values which increased as the time of mixing and sediment content of mixtures increased. Use of the reciprocal shaker generated values showing no significant change in trend between 1.0-h and 24-h controls, nor between sed-water treatments except for a 9.0% reduction in the 1 : 20/24-h treatment (compared to 1 : 20/1.0) The 1 : 10/24 and 1 : 4/24 treatments increased significantly after mixing for 24 h.

Trace metal analysis

Comparative studies produced mean trace metal concentrations which showed no great differences between the three methods when mixed for 0.5 h. Values generally increased as sediment level increased. There were instances (usually involving 1 : 10 treatments) where the trend was disrupted by marked decreases. Mn values in rotary-tumbled samples increased 200 to 300 percent over corresponding data from the other two forms of mixing, but the levels of Mn in air-mixed samples were about 50 percent less than that in 0.5-h mixed samples.

Agitation for 1.0 h produced small changes in Cu concentrations where the 1 : 4 treatment values usually increased, particularly when mixed by rotary and reciprocal methods (Table 5). Zn, Fe, and Mn levels followed a generally increasing progression as the sediment content increased. Fe and Mn data also showed a general increase. This was especially true for Mn data from reciprocal mixing, though air-mixed generated values were approximately 50 percent less.

Regardless of mixing method, 24-h leaching produced no major changes in Cu values. Mn levels virtually doubled with rotary mixing, but were less elevated with reciprocal, and even less with air mixing. Trace metal concentrations of sediment used to prepare elutriates (Table 5) indicated that only very small quantities of the total constituents were actually partitioned by the various mixing procedures.

Organic analysis

The only organic parameter of consequence was total PCBs. Traces of α-BHC were detected when all three mixing methods were applied for 0.5 h and 1.0 h, and with reciprocal mixing for 24 h. PCBs were released after 0.5 h only with reciprocal shaking; but 1.0-h mixing generated greatly increased PCB partitioning with all methods (Table 6). The reciprocal and air-diffuser methods produced similar PCB values. Significantly increased concentrations resulted from 1 : 4 rotary mixing. Twenty-four-h mixing also produced PCB partitioning by all methods but at greatly reduced levels. HCB, α- and γ-Chlordane, Heptachlor Epoxide, Aldrin, and three other organochlorine parameters were also monitored, but were not detected in elutriates.

Table 5. Mean trace metal values from elutriates prepared with Toronto Harbour dilution water using three methods of agitation for 1.0 h (μg l^{-1}).

Treatment (S : W)	Cd	Co	Cu	Fe	Mn	Ni	Pb	Zn
Sediment	5.8*	12*	151*	2.8*	–	44*	489*	480*
A Q : 1/1	<1	<1	<1	<10	<1	<1	<1	
A 1 : 20/1	<1	<1	5	28	2	<1	7	
A 1 : 10/1	<1	<1	8	36	2	<1	20	
A 1 : 4/1	<1	<1	6	64	3	<1	13	
E 0 : 1/1	<1	<1	<1	10	<1	<1	<1	
E 1 : 20/1	<1	<1	4	162	2	<1	11	
E 1 : 10/1	<1	<1	5	167	2	<1	15	
E 1 : 4/1	<1	<1	15	299	2	1	20	
R 0 : 1/1	<1	<1	<1	<10	<1	<1	<1	
R 1 : 20/1	<1	<1	11	178	2	<1	9	
R 1 : 10/1	<1	<1	12	286	3	<1	14	
R 1 : 4/1	<1	<1	41	369	3	<1	24	

* Values expressed in units of mg l^{-1}.
** Values expressed in percent.
A = Agitation by aeration.
B = Agitation by Eberbach reciprocal shaker.
R = Agitation by rotary tumbler.

Table 6. Mean organic values of elutriates prepared with Toronto Harbour dilution water and sediment using three methods of agitation for 1.0 h (ng l^{-1}).

Treatment (S : W)	α-BHC	γ-BHC	Total PCB
A 0 : 1/1	<0.4	<0.4	<4
A 1 : 20/1	21.6	<0.4	381
A 1 : 10/1	<0.4	<0.4	277
A 1 : 4/1	13.1	6.5	496
E 0 : 1/1	<0.4	<0.4	<4
E 1 : 20/1	4.3	9.8	307
E 1 : 10/1	10.8	4.3	334
E 1 : 4/1	8.6	<0.4	459
R 0 : 1/1	<0.4	<0.4	<4
R 1 : 20/1	1.4	<0.4	276
R 1 : 10/1	0.9	<0.4	302
R 1 : 4/1	<0.4	<0.4	507

A = Agitation by aeration.
E = Agitation by Eberbach reciprocal shaker.
R = Agitation by rotary tumbler.

Biological assessment

In general, rotary mixing reflected significant inhibition ($P < 0.05$–0.01) of ^{14}C-carbon uptake by *Chlorella vulgaris* with the 1.0-h and the 24-h treatments. This was particularly evident with the 25 percent elutriate addition. Some samples receiving a 5.0 percent addition of 1.0-h mixed elutriate showed significant differences ($P < 0.05$) (Table 7), but no change was observed in similarly spiked 24-h samples. Sediment-water ratios, elutriate addition levels, and their interactions also played highly significant ($P < 0.001$) roles with 1.0-h mixing. While significant differences occurred among ratios ($P < 0.001$) and among levels ($P < 0.01$) in 24-h treatments, there was no significance in the relationship between sediment-water ratios and elutriate addition levels.

However, the 1 : 4 sediment-water mixture promoted significant ($P < 0.01$) phytoplankton C-assimilation in 0.5-h rotary tumbled elutriates when compared to bioassay controls (LC, light control – untreated bioassay mixture incubated in light)

Table 7. Elutriation methodology study: mean bioassay values of elutriates prepared with Toronto Harbour dilution water and sediment using three methods of agitation for 1.0 h.

Treatment (S : W)	Assim. coeff.		Probability > F		
	5% amended	25% amended	R	L	ææR × L
LC	0.141 ± 0.014	0.141 ± 0.014			
A 0 : 1/1.0	0.141 ± 0.018	0.141 ± 0.011			
A 1 : 20/1.0	0.156 ± 0.007	0.118 ± 0.003	0.30	0.003	0.03
A 1 : 10/1.0	0.146 ± 0.009	0.109 ± 0.003 [+]			
A 1 : 4/1.0	0.158 ± 0.003	–			
LC	0.150 ± 0.004	0.150 ± 0.004			
E 0 : 1/1.0	0.156 ± 0.011	0.156 ± 0.007			
E 1 : 20/1.0	0.180 ± 0.005**	0.112 ± 0.003**[+]	0.14	0.0001	0.001
E 1 : 10/1.0	0.177 ± 0.006*	0.115 ± 0.002**[++]			
E 1 : 4/1.0	0.187 ± 0.007*	0.096 ± 0.002***[+++]			
LC	0.248 ± 0.014	0.248 ± 0.014			
R 0 : 1/1.0	0.272 ± 0.006	0.272 ± 0.010			
R 1 : 20/1.0	0.255 ± 0.005	0.208 ± 0.002*[++]	0.0001	0.0001	0.0002
R 1 : 10/1.0	0.262 ± 0.190*[++]	0.190 ± 0.000*[++]			
R 1 : 4/1.0	0.239 ± 0.006[+]	0.211 ± 0.004[++]			

* Statistically significant compared to untreated phytoplankton control (LC): *, **, *** 95%, 99%, 99.9% significance respectively.

[+] Statistically significant compared to elutriate control (0 : 1): [+], [++], [+++] 95%, 99%, 99.9% significance respectively.

A = Air diffuser; E = Reciprocal shaker; R = Rotary tumbler.

ææ Interactions between ratio (R) of sed. to water, and level (L) of elutriate addition.

and also to the elutriate control, 0 : 1. The levels of significance were $P < 0.05$ and $P < 0.01$ compared to LC, and $P < 0.01$ and $P < 0.01$ compared to the 0 : 1 controls for 5.0 percent and the 25 percent elutriate additions, respectively. Significant interactions ($P < 0.001$) were recorded between sediment-water ratios.

Air mixing produced no statistically significant differences in carbon uptake with 5 percent spikes regardless of mixing period. One-hour mixed elutriates spiked at 25 percent exhibited significant interactions among levels ($P < 0.01$), interactions among ratios ($P < 0.05$), and also in the relationship between ratios and elutriate spiking levels, although changes in C-assimilation levels were slight (Table 7). The 25 percent addition of 1 : 4 24-h elutriates produced significant enhancement ($P < 0.01$) in carbon uptake compared to the control, and to elutriate (0 : 1) controls. Only

interactions between ratios showed any significance ($P < 0.01$) among the interacting variables.

Reciprocal shaking for 0.5 h caused significant reductions in carbon uptake especially at the 25 percent elutriate spiking level. These lowered values were significant compared to those of both experimental controls. Moreover, the variables sediment-water ratios, elutriate levels, as well as their interacting relationship were all statistically significant ($P < 0.001$). The 5 percent addition of 1.0-h reciprocally-shaken elutriate produced enhanced carbon uptake ($P < 0.05-0.01$), while the 25 percent addition significantly inhibited all carbon assimilation ($P < 0.05-0.01$). Only the 24-h 1 : 10 and 1 : 4 mixtures showed significant ($P < 0.05-0.01$) differences. Statistical differences among ratios, among levels, and between ratios and levels was observed with 1.0-h mixing, but no such significance was recorded with 24-h mixing.

Discussion

Distilled water was used as a leaching medium in preliminary experiments to achieve experimental conditions which permitted discrimination of treatment effects without complications from the influence of dissolved salts, buffering capacity, etc. Therefore, the pronounced elevated pH trend of elutriates as exposure time and sediment volume increased was evidently due to the calcareous nature of Triangle Pond sediment (Munawar et al., 1986). This was so because the alkaline effect was reduced when the pH had been previously adjusted to 6.0, but this measurement was generally similar to the initial pH of the leaching distilled water. The large buffering capacity of the sediment required excess H^+ to achieve a pH of 6.0. Ions were then available to neutralize some of the alkaline influences of the sediment, thus resisting large pH increases.

A reduced pH range in air-mixed samples was likely caused largely by oxidation of Ca^{2+}- and Mg^{2+}-producing insoluble oxides and carbonates. In addition, the air diffuser system was not very effective in creating sustained general turbulence when sediment volume exceeded 150 ml which explains the observed relative resistance to pH increase. The insignificant pH changes observed with the use of site-water resulted from the absence of an effective pH difference between the alkaline dilution water and the sediment.

The tendency of pH increases to be less with longer mixing periods and increasing sediment volume was due to a well recognized phenomenon. High biological and chemical activity in these sediments demonstrate a sequence of oxygen-consuming events after which nitrate and sulphate oxidation occurs, accompanied by eventual CH_4 production (Forstner et al., 1986). In addition, Calmano & Forstner (1983) reported observing a pH decrease and bacterial leaching of metals under conditions of this kind. Adams et al. (1982) reviewed this phenomenon and pointed out that in situ oxygen demand is comprised of a water oxygen demand and sediment oxygen demand.

Therefore, the reduced sediment oxygen demand would have been the principal contributor to total oxygen demand, since distilled water was used in most instances. Reduced species such as Fe^{2+}, Mn^+ and H_2S would then undergo rapid oxidation resulting in reduced availability of such metals. Thus, with prolonged mixing the pH dropped as anoxic conditions developed, and as the 48-h data indicated, the pH of elutriates either stabilized or actually decreased in some instances.

Generally, specific conductance is influenced by the same factors as pH since this measurement indicates the level of ionic activity in a sample. Thus, the trends in conductivity values generally paralleled those for pH where values actually declined. Nevertheless, large increases in pH-adjusted elutriate, and in air-mixed samples should not be overlooked. The scope of this study did not permit the pursuit of a definitive explanation, but it is reasonable to conclude that the buffering systems in both cases must have had the capacity to facilitate the release of more ions. The significance of their rather similar pH measurements (7.46 and 7.43 respectively) cannot be merely co-incidental.

Volatile solids (loss on ignition, LOI) provide a crude collective measurement of the organic constituent of a sample. Ideally, this parameter should show direct proportionality with values for total dissolved solids. This was generally so, except for a few inconsistencies which resulted most likely from the fact that the sample size was only 30 ml. Increased replication may have been instrumental in minimizing this flaw. Besides producing data with more stable trends in 1.0-h samples, the rotary tumbler showed greater effectiveness. Evidently, this was due to the ability of this method to promote complete sediment-water interaction by permitting the non-colloidal sediment particles to completely traverse the entire water column during each cycle.

This argument acquires greater credence when the pattern of trace metal partitioning is critically evaluated. The consistent and efficient performance of the tumbler is indisputable as indicated by the values for Cu, Fe, Mn, and Zn. Moreover,

630

even the inconsistent trend observed for Zn at the 0.5 and 24-h treatment periods was absent in 1-h treated samples. In addition, organic analysis did not dispute this contention.

It is interesting to note that findings of carbon uptake inhibition by trace metals, PCBs and other constituents were supported by the bioassay results. Given that the strong partitioning capabilities of the reciprocal (and wrist-action) shaking as conventionally practised, it was not surprising that significant inhibition of carbon uptake was observed for all mixing periods. It is also helpful to reiterate at this point that the levels of contaminants partitioned in this manner do not necessarily have any meaningful relationship to levels partitioned under normal lacustrine conditions.

Conclusions

Since the air-mixing method is apparently inappropriate, and the reciprocal and the wrist-action methods of mixing (at 100 rpm) tend to produce overestimates of bioavailable contaminant levels, rotary tumbling emerges as a suitable and realistic approach to sediment elutriate preparation. Therefore, the use of this method is proposed with a 1:4 sediment site-water mixture agitated in cycles of approximately 4 rpm for 1.0 h. Moreover, it is suggested that elutriates prepared in this manner would provide better data for the assessment of contaminant bioavailability in polluted sediments. Moreover, the gentle cycle of the 'ferris-wheel' type rotary tumbler, appears to more closely mimic the natural dynamics of lacustrine sediment-water interactions. The mildness of the mixing process also maintains the structural characteristics of the sediment particles, thereby minimizing the likelihood of particle damage. Resulting data may be more representative of available rather than extractable contaminants.

In summary, low-frequency rotary tumbling (a) exerts no excessive mechanical stress on sediment particles; (b) when conducted for short periods does not encourage great oxidation-reduction changes; (c) tends to produce more uniform data by effectively exposing sediment particles to the leaching medium.

Finally, it is suggested that a more accurate approach to the preparation of wet sediment-water mixtures would be to compute the sediment component using dry weight calculations, rather than the wet weight, volume:volume, relationships currently in vogue. By basing calculations on the dry weight equivalent of the sediment, more accurate extrapolation of data and comparisons between studies would be possible. During the conduct of this study, sediment was encountered which varied in moisture content from 55 percent to 70 percent. Without a dry weight correction therefore, great variation may occur from experiment to experiment where different sediments are used.

Acknowledgements

This study is a part of the PhD thesis of S. Daniels (Univ. Waterloo). The work was conducted at the Great Lakes Laboratory for Fisheries and Aquatic Sciences, Fisheries & Oceans Canada under the guidance of Drs. C. Mayfield & M. Munawar. We are indebted to successive directors of GLLFAS, Drs. R.L. Thomas & J. Cooley who generously provided excellent field and laboratory support. We thank W. Norwood, L. McCarthy, D. Myles & M. Dutton (Fisheries & Oceans) for their assistance in the field, and D. Sergeant for conducting organic analyses. Drs. P. Ross (Illinois Natural History Survey), D. Gregor (Water Quality Branch, Saskatchewan), G. Leppard (National Water Research Institute), as well as Mr. W. Norwood reviewed and provided helpful comments on the manuscript. The assistance of Mr. H.F. Nicholson in technical editing is gratefully acknowledged.

References

Adams, D. A., G. Matisoff & W. J. Snodgrass, 1982. Flux of reduced chemical constituents (Fe^{2+}, Mn^{2+}, NH_{4+} and CH_4) and sediment oxygen demand in Lake Erie. Hydrobiologia 92: 405–414.

Boudou, A., D. Georgescauld & J. P. Desmazes, 1983. Eco-toxico-logical role of the membrane barriers in transport and bioaccumulation of mercury compounds. *In*: J. Nriagu (ed.), *Aquatic toxicology*, Wiley Series in Environmental Science and Technology, Vol. 13, John Wiley & Sons, New York, pp. 177–136.

Calmano, W., W. Alf & U. Förstner, 1983. Heavy metal removal from contaminated sludges with dissolved sulfur dioxide in combination with bacterial leaching. Proc. Internatl. Conf. Heavy Metals in the Environment, Heidelberg, September 1983, vol. 2, CEP Consultants, Edinburgh, pp. 952–955.

Côté, P. L. & T. W. Constable, 1982. Evaluation of experimental conditions in batch leaching procedures. Resources and Conservation 9: 59–73.

Engler, R. M., 1979. Bioaccumulation of toxic substances from contaminated sediments by fish and benthic organisms. In: Proc. Fourth U.S.-Japan Experts' Meeting, Oct. 1978, Tokyo, Japan. Ecol. Res. Series U.S. Envir. Protect. Agency, pp. 325–354.

Förstner, U., W. Alf, W. Calmano, M. Kersten & W. Salomons, 1986. Mobility of heavy metals in dredged harbour sediments. In: P. G. Sly (ed.), Proc. Third Internat. Symp. on Interactions Between Sediment and Water. Geneva, Switzerland, August 27–31, 1984, Springer-Verlag New York Inc., pp. 371–380.

Giesy, J. P., R. L. Graney, J. L. Newsted, C. J. Rosiu, A. Benda, R. G. Kreis Jr. & F. J. Hovarth, 1988a. Comparison of three sediment bioassay methods using Detroit River sediments. Envir. Toxicol. Chem. 7: 483–498.

Giesy, J. P., C. J. Rosiu, R. L. Graney, J. L. Newsted, A. Benda, R. G. Kreis Jr. & F. J. Hovath, 1988b. Toxicity of Detroit River sediment interstitial water to the bacterium *Photobacterium phosphoreum*. J. Great Lakes Res. 14(14): 502–513.

Inland Waters Directorate, 1985. National Water Quality Laboratory Manual of Analytical Methods. Environment Canada, Canada Centre for Inland Waters, Burlington, Ontario.

Keely, J. E. & R. M. Engler, 1974. Dredged material research program. Miscellaneous Paper D-74-14, US, Army Engineer Waterways Experiment Station, Vicksburg, Mississippi.

Lee, G. F., J. M. Lopez & G. M. Mariani, 1975. Leaching and bioassay studies on the significance of heavy metals in dredged sediments. Proc. Int'l. Conference on Heavy Metals in the Environ. Vol. II, Part 2, Toronto, pp. 731–764.

Lind, O. T. & R. S. Campbell, 1969. Comments on the use of liquid scintillation for routine determination of Carbon-14 activity in production studies. Limnol. Oceanogr., 14: 787–789.

Mac, M. J., C. C. Edsall, R. J. Hesselberg & R. E. Sayers, 1984. Flow-through bioassay for measuring bioaccumulation of toxic substances from sediment. Rept. Contrib-61; EPA/905/3-84/007, 26 pp.

Munawar, M., A. Mudroch, I. F. Munawar & R. L. Thomas, 1983. The impact of sediment-associated contaminants from the Niagara River mouth on various size assemblages of phytoplankton. J. Great Lakes Res. 9: 303–313.

Munawar, M., R. L. Thomas, W. P. Norwood & S. A. Daniels, 1986. Sediment toxicity and production-biomass relationships on size fractionated phytoplankton during on-site simulated dredging experiments in a contaminated pond. In: P. G. Sly (Ed.), Sediment and Water Interactions, Springer-Verlag, New York Inc., pp. 407–426.

Nichols, H. W., 1973. Growth media – freshwater. In: J. R. Stein (ed.), Handbook of Phycological Methods. Culture Methods and Growth Measurements. Cambridge Univ. Press, pp. 9–51.

Salomans, W., N. M., de Rooij, H. J. Kerdick & J. Brill, 1987. Sediments as a source of contaminants? In: R. Thomas, R. Evans, A. Hamilton, M. Munawar, T. Reynoldson & H. Sadar (eds.), Ecological Effects of In Situ Sediment Contaminants. Hydrobiologia 149: 13–30.

Seelye, J. G. & J. Mac, 1984. Bioaccumulation of toxic substances associated with dredging and dredged material disposal. Rept. EPA-905/3-84-005, 54 pp.

USEPA, 1977. Dredged material sample collection and preparation. Appendix B, In: Ecological Evaluation of Proposed Discharge of Dredged Material into Ocean Waters, Environmental Protection Agency/Corps of Engineers Technical Committee for Dredged and Fill Material. Environmental Effects Laboratory, U.S. Army Engineer Waterways Experiment Station, Vicksburg, Mississippi.

Hydrobiologia **188/189**: 633–640, 1989.
M. Munawar, G. Dixon, C. I. Mayfield, T. Reynoldson and M. H. Sadar (eds)
Environmental Bioassay Techniques and their Application.
© *1989 Kluwer Academic Publishers. Printed in Belgium.*

Assessing the impact of episodic pollution

John Seager [1] & Lorraine Maltby [2]
[1] *Water Research Centre, Medmenham Laboratory, PO Box 16, Marlow, Buckinghamshire, SL7 2HD, UK*; [2] *Department of Animal and Plant Sciences University of Sheffield, Sheffield, S10 2TN, UK*
(Correspondence to J. Seager)

Key words: episodic pollution, fish monitor, water quality criteria

Abstract

The increased tightening of controls on industrial and municipal wastewater discharges has resulted in steady improvements in the quality of many important rivers over recent years. However, episodic pollution, particularly from farm wastes and combined sewer overflows continues to pose a major problem, and is one of the main causes of poor quality rivers today. Despite our acknowledgement of this continuing problem, very little is known of the mechanistic basis of responses and recovery of aquatic organisms and communities exposed to intermittent pulses of common pollutants. The majority of ecotoxicological studies to date have been concerned with the effects of continuous exposure. Although such studies may provide a means of predicting the impact of episodic pollution events, a more appropriate test design would be to assess toxicity under pulsed and fluctuating exposure. Studies should also include a post-exposure observation period and should consider recovery of individuals and communities. This paper reviews the results of reported studies relevant to the investigation of episodic pollution and pays particular attention to the effects of magnitude, duration and frequency of exposure. Results of field investigations using an *in situ* bioassay technique are also presented to emphasize the importance of field validation of proposed water quality criteria for intermittent pollution events.

Introduction

Increased investment in the treatment of domestic and industrial wastewater has resulted in a steady improvement in water quality of many important rivers over recent years. Contrary to this trend we have witnessed the deterioration in quality of a significant number of rivers in both urban and agricultural catchments which are affected by episodic pollution events. For example, the results of the last published survey of river quality in England and Wales (Department of the Environment, 1986) indicated that, although 90 per cent

of the river and canal lengths were of satisfactory quality, there had been an overall deterioration in some areas since 1980. Transient pollution from both combined sewer overflows in urban rivers and the discharge of farm wastes in agricultural catchments were both cited as major causes of decline in water quality.

Despite our acknowledgement of these existing problems, very little is known about how aquatic organisms and communities respond to the physical and chemical changes in receiving waters brought about by transient pollution events. Aquatic ecotoxicology has been principally concerned with assessing the effects of continuous exposure of toxins. This has provided the basis for deriving water quality criteria to be applied in controlling 'continuous' discharges. However, this approach is considered to be inappropriate for pollution control management involving intermittent discharges. It cannot be assumed that standards based on the responses of aquatic organisms in conventional continuous exposure toxicity tests may be applied in deriving criteria for the control of episodic pollution.

This paper argues that an entirely different approach is required for studying the effects of transient pollution in aquatic biota. It involves investigating the relative importance of magnitude, duration and frequency of exposure of common pollutants and establishing the mechanistic basis for responses in selected organisms both in the laboratory and the field. Reported data relevant to this approach are reviewed and the results of *in situ* bioassays are presented to emphasize the importance of field validation.

Prediction of responses of aquatic organisms to pollution episodes

The prediction of responses of aquatic organisms to pulses of pollutants involves one of two approaches; either direct experimentation under controlled conditions, or the application of dynamic models to data from more conventional continuous exposure bioassays.

Laboratory investigation of pulsed exposure

A diversity of predictive laboratory tests of varying duration and end-point are reported in the literature. The most usual experimental design is to apply a constant dose of a toxin or mixture of toxins in the water and measure either lethal or sub-lethal responses of the test organisms for a fixed period of time. However, this investigative approach is unlikely to yield predictive data appropriate to the field situation where the concentrations and toxic effect of toxins may fluctuate

markedly, either by changes in the pollution input or by natural variations in the properties of the site itself such as flow, pH or temperature.

The importance of investigating pulsed and fluctuating exposures has been illustrated by a number of laboratory studies. Thurston *et al.* (1981) examined the effect of fluctuating ammonia concentrations on both rainbow trout, *Salmo gairdneri*, and cutthroat trout, *Salmo clarki*, under a range of exposure regimes of different duration and periodicity. They found that the fish could resist short ammonia pulses at concentrations slightly above the 96 hour LC50 without apparent long-term effects provided that these pulses were followed by a recovery period. However, over a 96 hour period, fish could withstand a fixed concentration of ammonia better than fluctuations with the same mean. A similar observation was reported by Brown *et al.* (1969) in investigations of the effect of fluctuating exposures of toxicants to rainbow trout. Survival of fish exposed alternatively every 2 hours to low and high concentrations of ammonia was significantly decreased compared to fish continuously exposed to the mean concentration. However, when the same dosing regime was applied with zinc, or mixtures of zinc and ammonia, significant differences in mortality were not detected.

Acclimation of aquatic organisms to the presence of toxicants throughout continuous exposure has been widely reported in the literature. The ability of organisms suddenly exposed to pulses of pollutants for short periods of time to acclimate to these rapidly changing conditions is unlikely to be the same as under continuous exposure. This may in part explain the observations of Thurston *et al.* (1981) and Brown *et al.* (1969). However, there is another element to our understanding of how aquatic organisms respond to pulsed and fluctuating exposures and this relies on our knowledge of how chemicals continue to exert a toxic effect following exposure. The fundamental importance of observing post-exposure effects has been illustrated by several studies in which animals have been returned to 'clean' water following brief exposure to a toxicant. For example, Pascoe & Shazili (1986) reported that the LT50 of rainbow trout continuously exposed to 1.0 mg/l of cadmium was 1900 minutes. However, an exposure of only 32 minutes ultimately resulted in 50% mortality of the batch of fish. Results such as these clearly demonstrate the inadequacy of conventional continuous exposure tests which make no attempt to quantify post-exposure effects. Indices have been proposed to take account of post-exposure mortality. Abel (1980) proposed the use of the term 'medium lethal exposure time' (MLET) which is the time required to kill half the animals within a set post-exposure period. A similar index, 'median post exposure lethal time' (pe LT50) has been proposed by Pascoe & Shazili (1986) which represents the time from the end of the exposure period by which 50% of the animals are dead.

Of course, we are not only interested in post-exposure mortality. A knowledge of the processes of recovery of organisms intermittently exposed to pulses of toxicants is clearly fundamental to our understanding of how individuals, populations and communities respond to episodic pollution, yet it has received relatively little attention.

The rate of recovery of an organism following exposure to a toxicant is dependent on a large number of both intrinsic and extrinsic factors. One consideration is the mode of action of the chemical and the importance of this factor was demonstrated by Wright (1976) in examining the relative toxicities of surfactants. When 30 min EC50 values were calculated from the results of short-duration tests on Stage II nauplii of the barnacle *Elminius modestus* the relative toxicities of the different decyl surfactants were, in descending order, nonionic > anionic > cationic. However, when recovery during a 48 hour post-exposure period was included as an index of toxicity the sequence was found to be the complete reverse of that determined on the basis of 30 min EC50s. Other factors found to influence recovery rate are exposure time, exposure concentration and temperature during exposure and recovery periods (Green *et al.*, 1988) during laboratory exposures of phenol to the freshwater crustacean, *Asellus aquaticus*.

Modelling approaches to predicting the effects of pulsed and fluctuating exposures

The results of the relatively few studies which have considered the effects of pulsed and fluctuating exposures of toxicants have emphasized the real need to gain more insight into the relative importance of magnitude, duration and frequency of exposure. This will require extensive laboratory experimentation which will inevitably be labour-intensive and costly. An alternative approach which may have some merit is to use our existing knowledge of responses of aquatic organisms to pulsed and fluctuating exposures to create mathematical models. These models can then be applied to the comparative wealth of data from continuous bioassays in order to make predictions for a range of different toxicants and exposure conditions.

Mancini (1983) has presented a kinetic model to predict the effects of fluctuating exposures based on the assumption that the concentration of a chemical in an organism at any one time is a function of uptake and detoxification rates. The approach may be summarised as follows:

$$\frac{dC_N}{dt} = K_u \cdot C_w - K_r \cdot C_N \qquad (1)$$

where K_u = uptake from the water,
K_r = detoxification rate,
C_w = concentration of chemical in the water
C_N = concentration of chemical in the organism

Equation 1 may be integrated and rearranged:

$$\frac{C_D}{K_u} = \frac{C_w}{K_r} \, [1 - exp(-K_r t_D)] \qquad (2)$$

where C_D = organism related concentration at death
t_D = time of organism death

The parameters of Equation 2 may be obtained from continuous bioassay tests to predict the effects of time varying exposure. Mancini (1983) used the data of Brown *et al.* (1969) on the mortality of rainbow trout exposed to both constant and fluctuating concentrations of zinc to validate the model and it produced a very good fit. The same data set has been used in the validation of a similar modelling approach (Connolly, 1985) based on uptake and elimination of toxicants. Using parameters obtained from constant exposure toxicity tests, the survival time under fluctuating exposure was calculated and compared to the observed data, again resulting in a good correlation.

Although these kinetic modelling approaches proved reasonably accurate in predicting toxicity of zinc, they have not proved to be equally as successful when applied to data for other chemicals. Wang & Hanson (1985), for example, used a kinetic model to predict the effects of fluctuating exposure of chlorine to four species of fish. Differences between observed and expected mortality were found to vary between 8 and 300%. The success of these types of simple mathematical predictions will depend on how realistically the model parameters represent biological processes such as uptake and detoxification which will vary widely depending on the physico-chemical characteristics of the toxicant itself and the range of extrinsic factors which affect its bioavailability and its toxic effect on the organism in question. Nevertheless, such modelling approaches demonstrate potential for further development and refinement and represent a potentially useful tool for deriving standards for episodic pollution.

Field Assessment of Effects of Pollution Episodes

Field studies on the ecological effects of pollution have traditionally been concerned with making spatial comparisons between community structure at upstream 'reference' and downstream 'sites'. When considering the impact of pollution episodes on receiving water biota it is also crucially important to recognize the variation in time-scale of different effects. For example, in studies to evaluate the nature of oxygen depletion following combined sewer overflow events, Harremoes

(1982) distinguished between the 'immediate' depletion caused by the degradation of organic matter in the water column, and the 'delayed' depletion effect lasting 12 to 24 hours after the discharge caused by degradation of organic material settled out on the stream bed. Similarly, the discharge of specific toxins such as heavy metals and certain organic compounds may result in acute 'knock-down' effects on more long-term chronic effects due to accumulation of persistent chemicals in sediments and biota.

One useful way of investigating both the spatial and temporal impact of pollution episodes in the field is to use *in situ* bioassays which exploit sublethal physiological responses of aquatic organisms. Methods have been developed which allow the investigation of sensitive responses to polluting episodes of different magnitude and duration, and also permit us to examine post-exposure recovery. One approach which has shown considerable potential for the study of episodic pollution is to exploit the changes in ventilatory patterns of fish which may occur as a result of pollutant induced stress. Both coughing responses and ventilation rates of fish have been shown to be affected by the presence of a wide range of toxic substances (Cairns & van der Schalie, 1980). Ventilatory signals of fish may be conveniently measured in continuous monitoring systems and systems have been used in a variety of biomonitoring applications including protection of potable supply intakes (Evans *et al.*, 1986) and effluent toxicity screening (Hayward *et al.*, 1989).

One such system is the Water Research Centre Mark III Fish Monitor described by Evans *et al.* (1986). The monitor is based on the measurement of fish breathing frequency and operates by the detection of the small oscillating voltage produced by the musculature involved in the gill ventilation process. As the measurement of this response does not involve destruction of the organism, it is also ideally suited to investigating recovery of the individual following the passage of pollutant pulses.

The fish monitor has been used successfully in field experiments to investigate the physiological

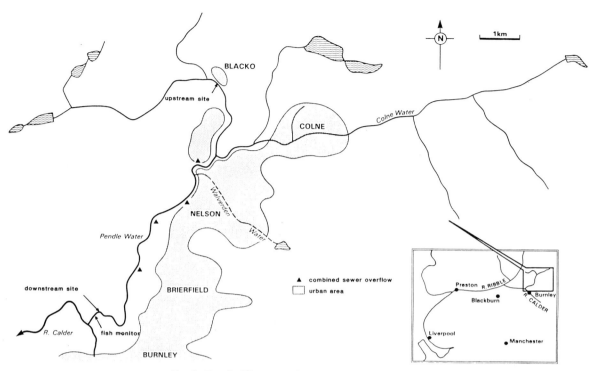

Fig. 1. Pendle Water catchment showing study sites.

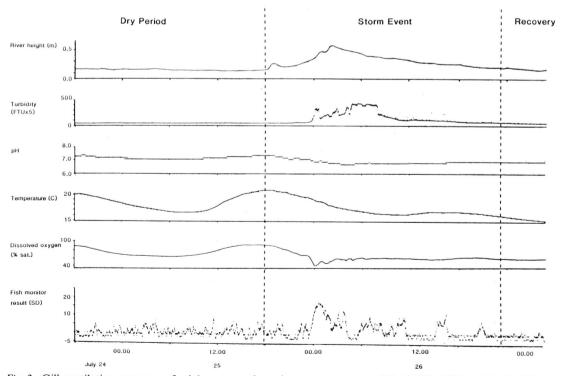

Fig. 2. Gill ventilation response of rainbow trout throughout a storm event (July 24-26, 1988) on Pendle Water.

responses of brown trout, *Salmo trutta*, to induced 12 hour episodes of low pH and elevated aluminium (Ormerod *et al.*, 1987). It has also been deployed in an investigation to assess the response of rainbow trout to pulses of pollutants caused by combined sewer overflow discharges. The study has been carried out on an urban river, Pendle Water in Lancashire, UK (Fig. 1).

The chemical and biological data output from a Fish Monitor sited on Pendle Water before, during and after a typical storm event are summarised in Fig. 2. Chemical determinands were measured continuously by pHOX DPM units and data were logged every minute. The gill ventilation rates of four fish were also recorded every minute and analysed using a statistical method based on the percentile test (Jolly, 1983). The ventilation frequency data collected for each fish are stored to establish a 'background' data set over the previous two hours. The most recent ten minutes' data for each fish provide the current 'inspection' data set. Each of the individual values of the inspection set is then compared to see if it falls outside the non-parametrically estimated 5 and 95 percentile limits of the background set. The total number of excursions outside these limits is then transformed to a value, called 'Fish Monitor Result' which relates to the binomial distribution of excursions standardized such that the value at the mean number of excursions is zero. The units are expressed as the number of standard deviations from the mean. A Fish Monitor Result of greater than 10 is generally taken to indicate that the observed excursions in ventilation rate can no longer be explained by random chance fluctuations, and that the fish is under some degree of physiological stress.

The fish responded rapidly (Fig. 2) at the onset of the storm event which was characterized by rapid increases in suspended solid loading and both 'immediate' and 'delayed' oxygen depletion similar to that observed by Harremoes (1982). The Fish Monitor thus enables interrogation of the gill ventilation response throughout an event and the subsequent recovery period, and allows us to compare this response with the continuous chemical record for each measured determinand.

Clearly, the fish exposed to the river water may be responding to a complex mixture of pollutants, many of which cannot be conveniently monitored on a continuous basis. Nevertheless, the system provides us with useful ecotoxicological information for measured parameters of interest for the purposes of validation of proposed water quality criteria. If, for a measured concentration profile of our chemical of interest throughout a storm event, we exceed the defined water quality standard yet observe no significant excursion in ventilation rate, the standard proposed may be excessively stringent and may require amendment. This decision will also necessitate consideration of how this sub-lethal physiological response relates to population and community effects. It is therefore recommended that *in situ* bioassays are deployed in association with other biological methods available for the assessment of the dynamics of fish populations and the structure and function of aquatic communities. This will allow examination of the relationship between sub-lethal responses of the individual level and population and community effects. Interpretation will also be assisted by the development of an appropriate database which allows comparison of ventilation rate responses with known ecotoxicological effects in other organisms.

Water quality criteria for episodic pollution

Water quality criteria for a wide range of determinands relating to specific water uses have been applied globally as the basis of water quality management. Criteria have traditionally been derived with the assumption that exposure of aquatic organisms to toxins will be continuous. Consequently, they have little relevance to the management of water quality where the predominant pattern of pollution is episodic.

An entirely different approach is therefore required in deriving criteria for intermittent pollution. Water quality models based on rainfall time series data are currently being developed and will allow us to predict the impact of intermittent discharges on receiving water quality in terms of

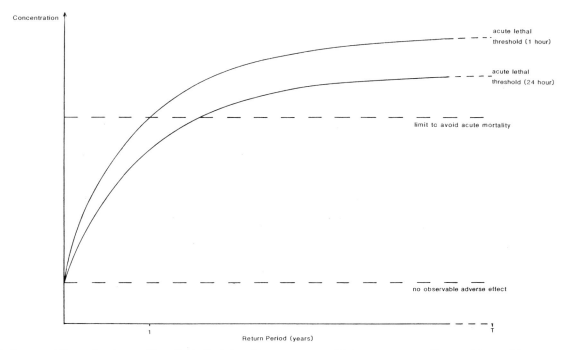

Fig. 3. Proposed form of water quality criteria for intermittent exposure of fish to common pollutants. Top and bottom curves relate to events of 1 and 24 hours duration respectively. T is the minimum acceptable return period of any event which may result in fish mortality.

magnitude, duration and frequency of events (Hvitved-Jacobsen, 1982). Criteria incorporating these three parameters have been proposed for three different fish classes (Danish Engineering Union Wastewater Committee, 1985). Although these criteria represent a conceptually important step, their biological foundation appears to be weak. They are essentially extrapolations from reported acute toxicity data coupled with a seemingly arbitrary assessment of what is an acceptable return period for a pollution event which results in fish mortality.

A model for future development of water quality criteria for intermittent pollution is proposed in Fig. 3. The model is based on a number of different premises. Firstly, it relies on the assumption that conventional criteria based on 'no observable effect levels' will be applied at the point where the return period of an event equals zero, which represents 'continuous' exposure. Secondly, it assumes that the fish population is protected from mortality caused by pollution episodes during an average year. Thirdly, it is accepted that, in rivers

seriously affected by episodic pollution, an extreme event may result in mortality of fish and that there is a threshold frequency for such an event (represented by T) which, if not exceeded, will not jeopardise the long-term viability of the population.

The form of the curves will clearly depend on the determinand in question and decisions taken on acceptable return periods of events of varying severity. The robustness of the criteria relies entirely on the available database for the reported effects of short-term intermittent exposure of aquatic organisms to individual toxins. The way forward is to improve our knowledge of the relative effects of magnitude, duration and frequency of exposure by controlled laboratory experimentation. Tentative criteria may then be formulated and then validated under field conditions using *in situ* bioassay techniques. If successful, they will be applied in conjunction with river impact models as the biological yardstick for defining acceptable quality for the required uses of the waterbody.

References

Abel, P. D., 1980. A new method for assessing the lethal impact of short-term, high-level discharges of pollutants on aquatic animals. Prog. Wat. Technol. 13: 347–352.

Brown, V. M., D. H. M. Jordan and B. A. Tiller, 1969. The acute toxicity to rainbow trout of fluctuating mixtures of ammonia, phenol and zinc. J. Fish Biol., 1: 1–9

Cairns, J. & W. H. van der Schalie, 1980. Biological Monitoring. Part I- Early warning systems. Wat. Res. 14: 1179–1196.

Connolly, J. P., 1985. Predicting single-species toxicity in natural water systems. Envir. Toxicol. Chem. 4: 573–582.

Danish Engineering Union Wastewater Committee, 1985. Forurening af vandlob fra overlobsbygvaerker. Dansk Ingeniorforening Spildevandskomiteen, Skrift nr. 22.

Department of the Environment, 1986. River Quality Survey, England and Wales. HMSO, London.

Evans, G. P., D. Johnson & C. Withell, 1986. Development of the WRc Mk III Fish Monitor: Description of the system and its response to some commonly encountered pollutants. WRc Environment Technical Report TR 233. Water Research Centre, UK.

Green, D. W. J., K. A. Williams, D. R. L. Hughes, G. A. R Shaik & D. Pascoe, 1988. Toxicity of phenol to *Asellus aquaticus* (L.) – effects of temperature and episodic exposure. Wat. Res. 22: 225–231.

Harremoes, P., 1982. Immediate and delayed oxygen depletion in rivers. Wat. Res. 16: 1093–1098.

Hayward, R. S., N. G. Reichenbach, L. A. Dixon & T. J. Wildoner Jr., 1989. Variability among bluegill ventilatory rates for effluent toxicity biomonitoring. Wat. Res (in press).

Hvitved-Jacobsen, T., 1982. The impact of combined sewer overflows on the dissolved oxygen concentration of a river. Wat. Res. 16: 1099–1105.

Jolly, P. K., 1983. Statistical methods for the detection of significant changes in fish ventilation frequency. WRc Environment Report 475-M, Water Research Centre, Medmenham Laboratory, UK.

Mancini, J. L., 1983. A method for calculating effects on aquatic organisms of time varying concentrations. Wat. Res. 17: 1355–1362.

Ormerod, S. J., N. S. Weatherley, P. French, S. Blake & W. M. Jones, 1987. The physiological response of brown trout *Salmo trutta* to induced episodes of low pH and elevated aluminium in a Welsh hill-stream. Annls Soc. r. Belg. 117 supplement 1: 435–447.

Pascoe, D. & N. A. M. Shazili, 1986. Episodic pollution – a comparison of brief and continuous exposure of rainbow trout to cadmium. Ecotox. Envir. Safety 12: 189–198.

Thurston, R. V., C. Chakoumakos & R. C. Russo, 1981. Effect of fluctuating exposures on the acute toxicity of ammonia to rainbow trout (*Salmo gairdneri*) and cutthroat trout (*S. clarki*). Wat. Res. 15: 911–917.

Wang, M. P. & S. A. Hanson, 1985. The acute toxicity of chlorine on freshwater organisms: time concentration relationships of constant and intermittent exposures. In R. C. Bahner & D. J. Hansen (eds.): Aquatic Toxicology and Hazard Assessment: Eighth Symposium. ASTM STP 891: 213–232.

Wright, A., 1976. The use of recovery as a criterion for toxicity. Bull. Envir. Contam. Toxicol. 15: 747–749.

Hydrobiologia **188/189**: 641–648, 1989.
M. Munawar, G. Dixon, C. I. Mayfield, T. Reynoldson and M. H. Sadar (eds)
Environmental Bioassay Techniques and their Application.
© 1989 *Kluwer Academic Publishers. Printed in Belgium.*

Acute toxicity of industrial and municipal effluents in the state of Maryland, USA: results from one year of toxicity testing

D. J. Fisher, C. M. Hersh, R. L. Paulson, D. T. Burton & L. W. Hall, Jr.
Aquatic Ecology Section, Environmental Sciences Group, The Applied Physics Laboratory, Johns Hopkins University, Shady Side, MD 20764 USA

Key words: regulating toxics, whole effluent testing, acute bioassay, invertebrate, vertebrate (fish), TRE (Toxicity Reduction Evaluations)

Abstract

In July, 1986 the Johns Hopkins University Applied Physics Laboratory's Aquatic Ecology Section established a bioassay facility for conducting an effluent biomonitoring program for the State of Maryland. Acute toxicity test procedures were developed and implemented for testing freshwater (*Pimephales promelas, Daphnia magna,* and *Ceriodaphnia* sp.) and estuarine (*Cyprinodon variegatus* and *Mysidopsis bahia*) invertebrates and fish. Procedures and test species are similar to those used by the U.S. Environmental Protection Agency (EPA) except that low salinity testing ($\leq 15\%_0$) is conducted with the sheepshead minnow and an estuarine mysid found in the Chesapeake Bay.

Results from the first year of the program involving acute screening bioassays of major industrial and municipal dischargers are presented. Over 90 dischargers were tested during the first full year of operation. The frequency of toxicity was 36% and 14% for industrial and municipal dischargers, respectively. Reference toxicity test results are also summarized. Results from these tests are presented by outfall type. A number of examples are presented concerning the use of these data for regulating toxic discharges by the State of Maryland's Department of the Environment. These results indicate the importance of implementing a biomonitoring program with proven, easily managed methods that can be readily understood by both regulators and permit holders.

Introduction

The control of the discharge of toxic substances into the Chesapeake Bay (USA) from point sources has become of increasing importance over the last decade with the realization that this ecosystem has been highly impacted by man's activities. Prior to 1972, there were relatively few legal tools available to control the pollution of the nation's surface waters (Williamson & Burton, 1987). In recognition of this problem, the Congress passed the Federal Water Pollution Control Act Amendments of 1972, now more commonly known as the Clean Water Act (Public Law 92-500). This Act requires that any person wishing to discharge any pollutant from a fixed location must first obtain a National Pollution Discharge Elimination System (NPDES) permit. This Act still serves as the primary legal framework for the control of surface water pollution in the United States.

In the early days of the administration of the Act, NPDES permits focused primarily on the control of conventional and known non-conventional pollutants. Conventional pollutants (e.g., high biological oxygen demand) served as general indicators of pollution and their control was desirable because these properties in water were shown to produce undesirable consequences. While controlling toxicity was not an initial focus of NPDES permits, the scientific basis for the assessment of toxicity to aquatic organisms had already be laid. (i.e., Sprague, 1973). Early toxicity studies concentrated primarily on individual toxicants under the assumption that the toxic constituents of a discharge were known, and that it would be sufficient to control each such constituent.

In 1977, the U.S. Congress amended the Federal Water Pollution Control Act (Public Law 95–217) to state that 'it is the national goal that the discharge of toxic pollutants in toxic amounts be prohibited'. The amendments reflected a growing recognition that controls on conventional and non-conventional pollutants did not always assure that discharges would not be toxic. Effluent limitations based on individual chemicals may not provide adequate protection for aquatic life when there are additive or synergistic effects between components in a complex mixture; when the toxicity of components is not known; and/or when a chemical characterization of the effluent has not been done.

As a result of the increased awareness of the value of effluent toxicity test data for toxic control in the water quality program and the NPDES permit program, the U.S. Environmental Protection Agency (EPA) issued a national policy statement entitled, 'Policy for the Development of Water Quality-Based Permit Limitations for Toxic Pollutants' (USEPA, 1984). This new Agency policy proposed the use of toxicity data to assess and control the discharge of toxic substances to the Nation's waters through the NPDES permits program. The policy stated that 'biological testing of effluents is an important aspect of the water quality-based approach for controlling toxic pollutants. Acute and short-term chronic effluent toxicity data, in conjunction with other data, can be used to establish control priorities, assess compliance with State water

quality standards, and set permit limitations to achieve those standards.' All states have water quality standards which include narrative statements prohibiting the discharge of toxic materials in toxic amounts.

The Clean Water Act allows EPA's authority to issue NPDES permits to be delegated to the states under certain conditions. All of the relevant jurisdictions with drainage into the Chesapeake Bay, except the District of Columbia, have authority to issue NPDES permits. A typical permit operates primarily by placing concentration or mass limits on the amount of a pollutant or category of pollutants which can be contained in a discharged effluent. The legal authority to impose such limitations can derive from either Federal or State law. Currently, limitations on toxic substances can be technology-based, water-quality based, or federal or state effluent standards. In addition, biologically-based toxicity limitations are beginning to be used.

It is the authors' view that the careful use of biologically-based toxicity testing in the NPDES permit program offers the potential for cost-effective improvements in the ecological well-being of the Chesapeake Bay. This is partially because of the inherent benefits of using biological systems to protect ecosystems, and in part because the other three approaches mentioned above appear to be approaching their limits of marginal utility, as discussed by Williamson & Burton (1987).

With the realization that effluent biomonitoring was an important method for regulating toxic discharges, Maryland's Department of the Environment contracted the Aquatic Ecology Section of the Johns Hopkins University's Applied Physics Laboratory (JHU/APL) to build and staff an effluent bioassay laboratory to screen NPDES-permitted discharges to the waters of the State. The laboratory was also directed to develop standard operating procedures (SOP) for conducting freshwater and saltwater bioassays of both an acute and chronic nature. The following paper presents results from the first year of testing, dealing entirely with acute screening bioassays on the major industrial and municipal discharges to State waters. The discussions and conclusions presented in this paper represent the views of the authors and are not official policy of the State of Maryland or it's Department of the Environment.

Materials and methods

Acute toxicity test procedures were developed and implemented for testing freshwater fathead minnows (*Pimephales promelas*) and daphnids (*Daphnia magna* and *Ceriodaphnia* sp.). The estuarine sheepshead minnow (*Cyprinodon variegatus*) and mysid shrimp (*Mysidopsis bahia*) are now used for testing low salinity effluents. Chronic test procedures have been developed for the same species. Initial acute procedures and test species were similar to those used by EPA (USEPA, 1985) except that low salinity testing ($\leq 15\%_0$) was conducted with the sheepshead minnow and the grass shrimp *Palaemonetes pugio* while test methods were developed for an estuarine mysid from the Chesapeake Bay (*Neomysis americana*). Most of the acute testing effort dealt with fresh water organisms since the majority of the effluents tested consisted of fresh water even if discharged to an estuarine system. The State is presently concerned with addressing the toxicity of the effluent itself, not the toxicity upon dilution in the receiving stream.

Acute tests for screening major industrial and municipal dischargers were conducted using a 48-h static renewal methodology. Effluent samples (24-h composite or grab) were taken by the State of Maryland, iced, and delivered to the bioassay facility. Samples were collected during normal operating periods of the facilities in order to get a representative toxicity value. Permittees were notified as to the time of sampling. Part of each sample was delivered to the State chemistry laboratory for analysis for NPDES-permitted constituents. The toxicity tests were initiated within 24 hours of receipt of the sample. Aeration was not used unless the dissolved oxygen concentration showed a steep downward trend in the first 4 to 8 hours of the test. (USEPA, 1985). If this occurred, all treatments, including controls,

were aerated. The effluent itself was aerated prior to test initiation if the dissolved oxygen concentration was below 6.0 mg l^{-1}. Each effluent was tested with an invertebrate and a fish species.

Each test consisted of one control treatment and a series of five effluent dilutions (100%, 56%, 32%, 18%, and 10% effluent). The test temperature in all acute tests was 20 °C, a common water temperature found in the Bay in spring and fall. The control water was the same as the test dilution water and the water used to culture the test organisms. Most of the tests were conducted with organisms cultured in the JHU/APL laboratory. The ages of the test organisms were as follows: fathead minnow and sheepshead minnow – 14 to 35 days; daphnids – < 24-h neonates; grass shrimp – juvenile. Both 48-h LC_{50} and EC_{50}s were calculated for each test but the EC_{50} was used by the State of Maryland in its subsequent regulatory actions. The toxicity data presented in this paper will be in the form of 48-h EC_{50} values expressed as % effluent. Acute $CdCl_2$ reference toxicity tests using the same culture and dilution water as the effluent tests were conducted throughout the period, initially with each test but later on a monthly basis. These toxicity data are presented as 48-h LC_{50} values based on measured cadmium concentration.

Results

A total of 91 separate outfalls were tested. Since some outfalls were tested more than once, there were 101 toxicity tests conducted on fish and the same number on invertebrates. Of the 91 effluents tested, 80 were freshwater and 11 were estuarine. There were 69 industries and 22 municipalities tested. All of the municipal effluents tested represented discharges from waste water treatment plants. The State of Maryland considers an effluent to be acutely toxic if the 48-h EC_{50} is 100% or less. Thus, any effluent affecting more than 50% of the test organisms in the 100% effluent treatment is considered to be acutely toxic. Of the 69 industrial discharges, 15 were toxic to both species tested, 5 toxic to the fish only, and 5 toxic to the invertebrates only. Thus, 36% of the industrial dischargers had effluents toxic to one or both species. On the municipal side, only 3 effluents were toxic, two to both species and one to the fish only. Only 14% of the municipal effluents were toxic to one or both species. It should be remembered though that fewer municipalities than industries were tested during this period.

There are a wide range of industries operating in the State of Maryland under the NPDES permit system. The effluents tested during the first year were divided into major groups according to products produced (Table 1). Chemical and chemical-related facilities represented the largest group of industrial dischargers, with electrical generating facilities also represented strongly. Although the number of waste water treatment plants tested was much smaller than the total number of industrial dischargers tested, they form the largest category when all the dischargers are divided into groups.

In contrast to all other groups, no electrical generating facility effluents were found to be toxic (Fig. 1). This is somewhat misleading on two accounts. First, these tests did not measure any thermal effects which these discharges may have since all the tests were conducted at a standard temperature. Second, these facilities were not tested when chlorine may have been present in the discharge. Chlorination is used extensively during the warmer months of the year (May – October) to inhibit biofouling. All of the generating facilities tested were sampled during the cooler months when chlorine is not used. Thus, until these facilities are retested during the summer months these results should be viewed with caution.

The remaining industrial discharge categories had similar frequencies of toxicity (Fig. 1). Approximately 40 to 50% of the effluents tested in each of the other industrial categories were toxic to one or both species. As stated earlier, the municipal discharges were found to be toxic much less frequently (14%). These data can be presented in a number of ways. If we consider industrial dischargers only, 36% were toxic. If we eliminate the electrical generating facilities for the reasons mentioned above, 43% of the remaining industrial discharges were toxic. Finally, if considered on the basis of the total number of tested discharges to State waters, 31% of the effluents were toxic.

These data presented above give an indication of frequency of toxicity, not of the magnitude. Results from the acute tests indicate 48-h EC_{50}s ranging from much less than 10% to 90% effluent

Table 1. An overview of testing effort with major NPDES dischargers characterized by type. Some dischargers had multiple outfalls.

Major Group (Abbreviation)	Types of Industries Tested (# tested in each subgroup)
Electrical Generating (Elec)	Coal (10) or Incineration (1)
Chemical and Related (Chem)	Inorganic Pigments and Chemicals (8), Soaps and Detergents (2), Munitions (2), Organics and Pesticides (1), Waste Disposal (1)
Water Treatment (WWTP)	Wastewater (22) and Water Filtration (3)
Primary Metals (Metal)	Blast Furnaces and Plates (3), Fabricated Metal Products (2), Aluminum (1), Stainless Steel (1)
Petroleum and Related (Petrol)	Asphalts (2), Refinery (2), Oils and Greases (1), Storage (1), Thermoplastics (1), Laminates (1)
Food and Related (Food)	Poultry and Feeds (3) Bakery (1), Meat Packing (1) Rendering (1), Sugar Refining (1)
Machinery/Construction (Machine)	Electronics (3), Insulating (2), Conveyor and Sandblasting (1), Stone and Concrete (1), Lawn and Garden (1)
Miscellaneous (Misc)	Printing (2), Tanning (1), Assembly of Cars and Trucks (1), Mining (1), Groundwater (1), Hazardous Waste Landfill (1), Drinking Water (1)

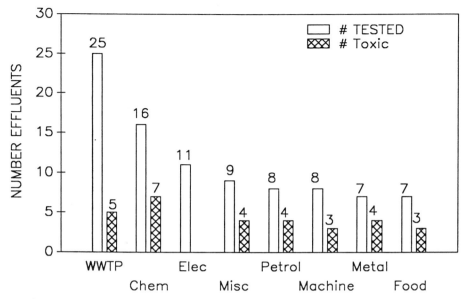

Fig. 1. Number of acutely toxic effluents from each major group described in Table 1 (See Table 1 for abbreviations).

(Fig. 2). Since the greatest dilution tested in each case contained 10% effluent, it was not possible to calculate EC_{50} values when there were greater than 50% of the organisms affected at this treatment. The range of toxicity was somewhat less for the municipal wastewater treatment plants (WWTP) than for the industrial groups, possibly reflecting either the limited number of toxic components in these effluents or better treatment systems for these components. The most toxic effluent tested was from a rendering facility which produces feather and poultry byproducts, meal, and fat. This effluent killed all the fish and 95% of the daphnids exposed at a 10% effluent treatment in less than 1 hour. This effluent was found to contain 12 mg l^{-1} dissolved NH_3.

There were a total of 28 48-h freshwater $CdCl_2$ reference toxicity tests conducted in conjunction with the above effluent tests. The results indicate that the daphnid is more sensitive to this toxicant than the fathead minnow (Fig. 3). The mean 48-h LC_{50}s were 0.090 mg l^{-1} and 0.043 mg l^{-1} for the fathead and daphnid, respectively. The organisms showed a consistent response to $CdCl_2$, with only a few instances where results were not within ± 1 standard deviation of the mean of all tests conducted. The LC_{50} values generated dur-

ing this testing are within the range acceptable by the EPA when using their acute test methods (USEPA, 1985). This indicates that the effluent toxicity results generated in the present study were generated from a healthy pool of test organisms.

Discussion

The Maryland biomonitoring protocol states that biomonitoring is required as a condition of NPDES permits whenever the potential for toxicity of the effluent is apparent. The JHU/APL effluent bioassay facility was contracted by the State as an independent agency to supply information concerning the acute and chronic toxicity of effluents. If JHU/APL tests indicate that an effluent is toxic, the permittee is notified by the State of the results. If the permittee questions the results, they can repeat the test inhouse or have the test repeated by an independent consultant. When the toxicity is verified, a toxicity reduction evaluation (TRE) plan must be submitted and approved by the State's Department of the Environment. The TRE consists of a series of toxicity tests on chemically and physically altered

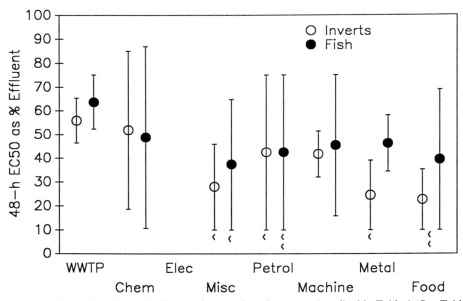

Fig. 2. Range of acute toxicity values for toxic effluents from each major group described in Table 1 (See Table 1 for abbreviations).

effluent in a effort to identify the toxic constituents. The permittee, or a consultant, must then conduct the TRE, and upon completion, must submit a plan to the State to reduce the toxicity of the effluent.

The testing effort conducted by JHU/APL indicated a number of acute toxicity problems with effluents discharged into State waters. When 36% of the industrial discharges are toxic at the point of discharge, it is an indication of a potentially serious problem. What has the State done in response to these findings? A few examples may be of interest. The rendering plant mentioned above as the most toxic effluent tested by JHU/APL was ordered by the State to submit a TRE. The plan was submitted and approved and ammonia was found to be the major toxic component. At the time of our initial testing, the effluent contained 312 mg l^{-1} ammonia nitrogen. This represents approximately 12 mg l^{-1} dissolved NH_3 at the temperature and pH of the effluent. Operational changes were made to the facility to reduce the ammonia level and a waste treatment system upgrade has been proposed. A recent retest of this effluent conducted by JHU/APL showed a marked reduction in toxi-

city, with only 10% mortality of fish in the 100% effluent treatment. This treatment killed 90% of the daphnids but there was no mortality at the 56% effluent treatment.

Other industries with toxicity problems have also been contacted by the State. A petroleum plant specializing in asphalt, flooring, and felt was found to be very toxic to daphnids (48-h EC_{50} < 10%). The State has approved a TRE plan for this facility and the work is in progress. Initial findings indicate a problem with a surfactant. A leather tanning plant was found to have an EC_{50} of < 10% with fathead minnows. This company has completed its TRE and has found that ammonia and an anionic surfactant are the toxic constituents. A plan to eliminate the toxicity is under development.

At the time of this writing, all of the major discharge permittees whose effluents have been shown to be toxic by JHU/APL testing have been contacted by the State and required to either retest their effluent or submit a TRE. Two facilities have retested their effluents and found no toxicity. The initial toxicity of one of these was found to be due to rainwater runoff from a large parking facility. A retest by our laboratory during

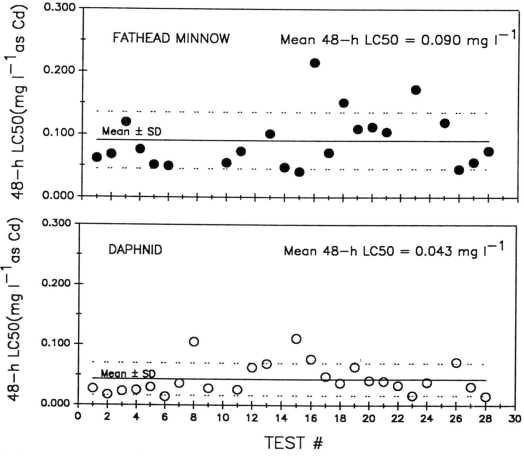

Fig. 3. Acute toxicity of CdCl₂ to fathead minnows and daphnids as determined in reference toxicity tests.

dry weather indicated no toxicity. The other effluent was from a petroleum facility which produced thermoplastic materials. Their retest indicated to them that our initial toxicity values were due to a spill which occurred at the time the effluent was sampled. Our tests conducted on this sample showed 48-h EC_{50}s of less than and much less than 10% for the daphnid and fathead minnow, respectively. We conducted a retest of this effluent at a later date with the fathead minnow and found less toxicity, with a 48-h EC_{50} of 56%. Thus, although less toxic, the facility still had problems, and has been informed so by the State.

Conclusion

These results indicate the importance of implementing a biomonitoring program with proven, easily managed methods that can be readily understood by both regulators and permit holders. Many of the papers presented at this conference dealt with new bioassay techniques and methods to determine the health of ecosystems in general. We felt it was important to present results from a biomonitoring program which measures toxicity of specific effluents at the point of discharge and which produces results which are currently used by the State of Maryland to regulate discharge of toxics to State waters. These tests are short-term, simple, and cost effective bioassay methods which can be used to regulate the discharge of toxic pollutants. While these methods may not insure complete protection of the health of the aquatic environment, they can lead to significant reductions in the discharge of toxic constituents from point sources. The important point is that the methods are available *now* and their use can have a significant impact on the discharge of toxic constituents immediately.

Acknowledgements

This work was supported by the State of Maryland's Department of the Environment. The program is administered through the Division of Toxics Programs Development under the direction of Mr. G. H. Harman. Much of the work conducted was in conjunction with effluent sampling and monitoring coordinated by Mr. J. A. Veil and Mr M. H. Knott of the Industrial Permits group and Ms. M. J. Garreis of the Municipal Permits group.

References

Sprague, J. B., 1973. The ABC's of pollutant bioassay using fish. In; J. Cairns, Jr. & K. L. Dickson (eds.). Biological Methods for the Assessment of Water Quality. ASTM Spec. Tech. Publ. 528, Amer. Soc. Testing Materials. Philadelphia. pp. 6–30.

USEPA, 1984. Policy for the Development of Water Quality-Based Permit Limitations for Toxic Pollutants. 49 Federal Register, Friday, March 9, 1984, pp. 9016–9010.

USEPA, 1985. Methods for measuring the acute toxicity of effluents to freshwater and marine organisms. 3rd. ed. Environmental Monitoring and Support Laboratory, U. S. Environmental Protection Agency, Cincinnati, Ohio. EPA 600/4–85–013, 216 pp.

Williamson, R. L., Jr. & D. T. Burton, 1987. Use of aquatic biological testing under the NPDES permit system to reduce toxic pollution of the Chesapeake Bay. In; S. K. Majumdar, L. W. Hall, Jr. & H. M. Austin (eds.). Contaminant Problems and Management of Living Chesapeake Bay Resources. The Pennsylvania Academy of Science. Easton. pp. 518–540.

Hydrobiologia **188/189**: 649–658, 1989.
M. Munawar, G. Dixon, C. I. Mayfield, T. Reynoldson and M. H. Sadar (eds)
Environmental Bioassay Techniques and their Application.
© 1989 *Kluwer Academic Publishers. Printed in Belgium.*

Nitrification rates in the lower river Rhine as a monitor for ecological recovery

Yves J. H. Botermans & Wim Admiraal
National Institute of Public Health and Environmental Protection, P.O. Box 1, 3720 BA Bilthoven (U), The Netherlands

Key words: nitrification, river Rhine, biological monitoring, seasonal dynamics, toxic substances

Abstract

The rate of *in situ* nitrification was tested as an indicator of the toxicological quality of the river Rhine. Concentration changes of ammonium ions over 85 to 133 km long reaches of three river branches downstream of the densely populated Ruhr-area (F.R.G.) were calculated from a data base for the period 1972 to 1986. Concentrations of ammonium in the river exceeded values of 1 mg N/l in winter. Because of the very high input of ammonium, bacterial nitrification dominated over other nitrogen processes. Relative rates of nitrification in the three river branches were proportional to the water temperature for the individual years. Nitrification rates in the river increased by a factor of ca. 4 during the period of 1972 to 1986. Toxic substances, whose concentrations decreased in the same period of time, were proposed as inhibitors of *in situ* nitrification rather than e.g. a low oxygen saturation of the water. The improvement of the conditions in the river, indicated by the *in situ* rate of nitrification, was also documented by data on macrofauna and fish populations.

Introduction

It is a well known fact that nitrifying bacteria are sensitive to chemical pollution (Hockenbury & Grady, 1977; Sharma & Ahlert, 1977; Neufeld *et al.*, 1986). This is of practical importance for operating wastewater treatment plants that are loaded with organic waste and chemicals. Here the oxidation of ammonium ions may be impaired by concentrations of both organic and inorganic chemicals in the parts per million range (Tomlinson *et al.*, 1966; Beg *et al.*, 1982 and 1987). Analogously, nitrification in surface waters should be a process sensitive to chemical pollution.

Rivers passing through heavily industrialized and urbanized areas contain unnaturally high nitrogen concentrations, mainly as a result of indus-

trial and urban waste (Meybeck, 1982). The Rhine receives massive discharges of nitrogen, mostly as ammonia. Ammonia (NH_4^+ and NH_3) is a threat to the ecosystem because of the toxicity of the un-ionized ammonia (Ball, 1967). Microbial nitrification may alleviate some of these problems by oxidizing un-ionized ammonia to nitrate (NO_3^-) (Suzuki *et al.*, 1974). Thus chemical inhibition of nitrification in the river may create problems (additional to those caused by the chemicals themselves) associated with the accumulation of ammonia in the water.

The present study concentrates on the Netherlands part of the river Rhine 70 km downstream of the German Ruhr district, an area with a very high population density (Fig. 1) and vast industry. Since nitrification dominates over other nitrogen processes in the Netherlands section of the Rhine,

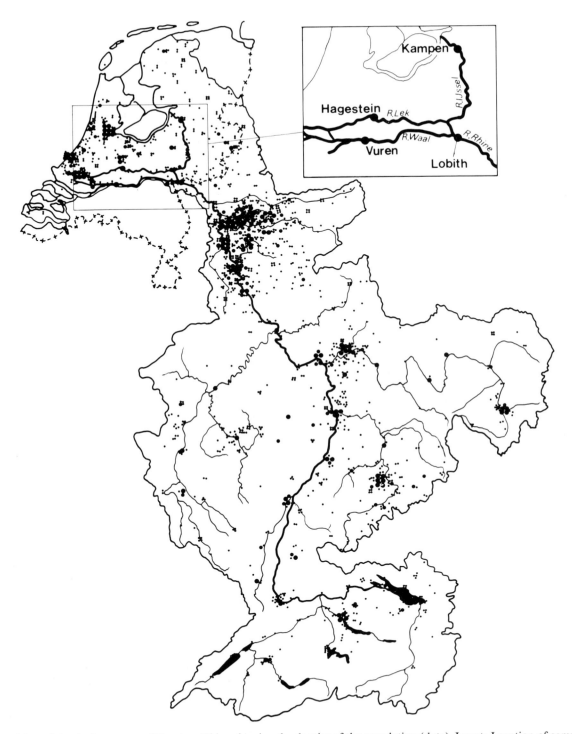

Fig. 1. Map of the drainage area of the river Rhine showing the density of the population (dots). Insert: Location of sampling stations and river sections studied.

(Admiraal & Botermans, 1989) this river section seems suitable for comparing changes in nitrification rate under the influence of a reduced discharge of toxic compounds. Furthermore, changes in nitrification rate can be compared with the observed recovery of the rivers' fauna. The nitrification rates presented here were derived from the analysis of data sets on ammonium concentrations recorded in the period from 1972 to 1986 by the Netherlands Department of Public Works (Anonymous, 1972–1986).

Methods

Ammonium concentrations were measured at four different stations on the river (Fig. 1). The photometric measurements, using the Berthelot-reaction, were done almost weekly until 1981 and at least fortnightly from 1982 on by the Netherlands State Department of Public Works (Anonymous, 1972–1986). The nitrification rates over the three reaches of the Netherlands part of the Rhine were determined by the difference in ammonium concentration between station Lobith (km 862) on one hand, and station Vuren (km 952), station Hagestein (km 947) or station Kampen (km 995) on the other hand. Pairs of samplings were only accepted if the difference in time was not more than two days. As such, the residence time of the water between two stations was considered. The three lower stations are located on the banks of, respectively, the rivers Waal, Lek and IJssel, and the three river reaches will be referred to accordingly although the reaches do not cover these river branches entirely.

Nitrification leads to decreases in ammonium concentrations between Lobith and the other stations. These decreases were used to calculate the parameters R_1 and R_2 for the nitrification rate. R_1 is based on the assumption that nitrification proceeds as a zero-order process, while R_2 is applicable to the case of a first-order decrease of ammonium concentration:

$$R_1 = \frac{[NH_4^+]_1 - [NH_4^+]_2}{[NH_4^+]_1} \cdot \frac{1}{t}$$

$$R_2 = \ln \frac{[NH_4^+]_1}{[NH_4^+]_2} \cdot \frac{1}{t}$$

$[NH_4^+]_1$ = ammonium concentration at station Lobith.

$[NH_4^+]_2$ = ammonium concentration at station Vuren, Hagestein or Kampen.

t = residence time of the water in days.

The use of the rate parameters R_1 and R_2 was based on two observations. Huang & Hopson (1974) showed that ammonium oxidation in a fixed film reactor at influent NH_4^+-N concentrations greater than 2.5 mg/l was appropriately described by zero-order kinetics. Furthermore, Kennedy & Bell (1986) propose a shift from zero- to first-order kinetics in the White River, due to the change in concentrations of ammonia from 10 mg N/l in 1982 to 1 mg N/l in 1983, assuming that Monod kinetic theory applies to substrate oxidation by nitrifying bacteria in river ecosystems.

The hydrographical characteristics of the rivers Waal, Lek and IJssel relevant to the nitrification process are given in Admiraal & Botermans (1989).

Results

At station Lobith yearly average ammonium concentrations showed a rapid decline until 1975 and a slow decrease until 1980, towards an average of less than 1 mg NH_4^+-N/l (Fig. 2A). The wide ranges shown are caused by a pronounced yearly cycle (Fig. 3). The average discharges and temperatures at station Lobith (Fig. 2B and 2C) seem to be constant, but certain years (1976) produced deviations.

Ammonium concentrations at station Vuren were similar to the concentrations at the upstream station Lobith (Fig. 3). The decrease in NH_4^+-N concentration over the Lobith to Vuren reach was analyzed using the nitrification rate parameters R_1 and R_2. R_1 (Fig. 4B) and R_2 (Fig. 4C) showed similar seasonal variations, different from the absolute concentration decrease over the river reach

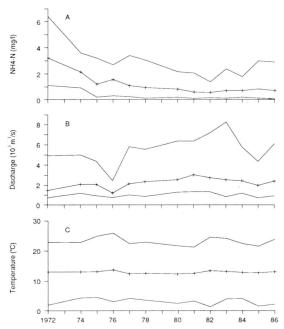

Fig. 2. Yearly averages of ammonium concentration (A), discharge (B) and temperature (C) of most years between 1972 and 1986 measured at station Lobith. Yearly minima and maxima are indicated (Anonymous 1972–1986).

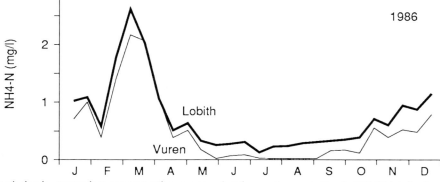

Fig. 3. Seasonal variation in ammonium concentration measured at the up-stream station Lobith (thick line) and the downstream station Vuren (thin line) in 1986.

(Fig. 4A). R_1 and R_2 were not very different in the earlier years.

R_2 values tended to increase over the years 1972 to 1986 (Fig. 5). In order to disentangle the trends in the nitrification rate we analyzed the temperature dependence of the parameter R_2 and compared it for the three river sections studied. Correlations with the water temperature, measured at station Lobith are shown in Fig. 6. For each year a significant correlation with water temperature was found. The slope of the regression line is low in the early years and showed a better response to temperature in recent years. The trend is also shown in the yearly averages of the nitrification rate parameter R_2 for the river Waal and is supported by similar calculations for the rivers IJssel and Lek (Fig. 7).

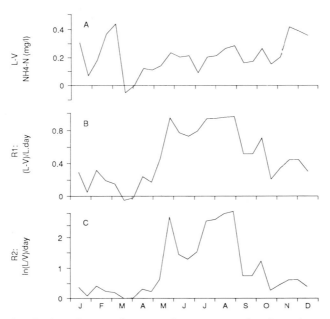

Fig. 4. Seasonal variation in the absolute decrease in ammonium concentration from the upstream station Lobith to the downstream station Vuren (river Waal) for the year 1986. The data are presented as A, the difference in NH_4-N concentration between station L (Lobith) and station V (Vuren); B, the nitrification rate parameter R_1 and; C, the nitrification rate parameter R_2.

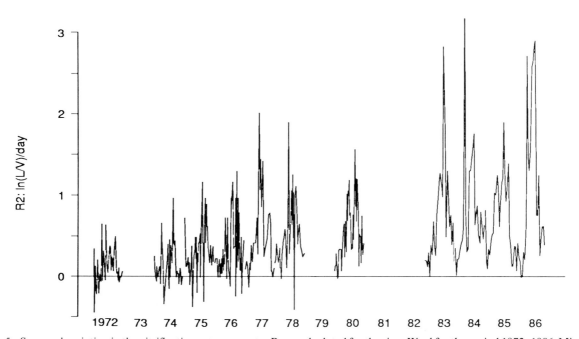

Fig. 5. Seasonal variation in the nitrification rate parameter R_2 as calculated for the river Waal for the period 1972–1986. Missing years were not considered. L = ammonium concentration at station Lobith, V = ammonium concentration at station Vuren.

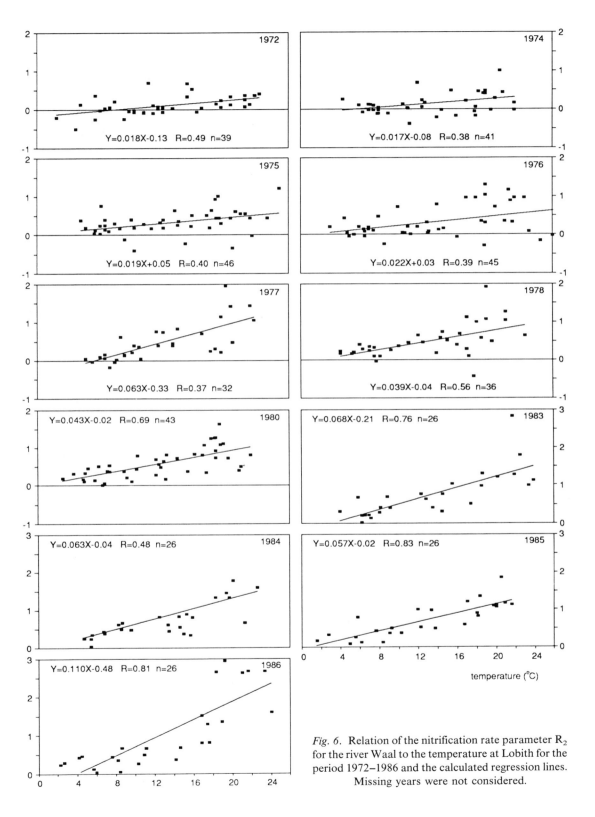

Fig. 6. Relation of the nitrification rate parameter R_2 for the river Waal to the temperature at Lobith for the period 1972–1986 and the calculated regression lines. Missing years were not considered.

Discussion

Environmental conditions in the river Rhine have changed considerably in the period from 1972 to 1986. The average oxygen concentration in the water at Lobith increased from ca. 5.5 mg/l in 1972 to ca. 8.0 mg/l in 1985 (Anonymous, 1972–1986). Oxygen concentrations in this range are not limiting for nitrification since Helder & de Vries (1983) indicated that the oxidation of nitrite is inhibited below an oxygen concentration of 3 mg/l and that the oxidation of ammonia is inhibited below an oxygen concentration of 1 mg/l. Furthermore, the river branch with the highest oxygen concentration, river Lek, did not show the highest nitrification rates (Admiraal & Botermans, 1989). Therefore the increase in oxygen over the years may have contributed to the increased nitrification rate in the river but is probably not a regulator of overriding importance.

Nitrification is usually concentrated at the interface of sediment and water (Belser, 1979; Wolter et al., 1985), or in the sediment (Curtis, 1974; Cooper, 1983). The amounts of ammonia oxidized per m^2 river bottom were 1.6, 0.6 and 0.3 mg NH_4^+-N $\cdot$ day^{-1} for the rivers Waal, Lek and IJssel respectively (Admiraal & Botermans, 1989). Nitrification rates expressed in this way did not change much in the period from 1972 to 1986, but the values indicated specific differences in the capacity of the river bottom to support intense nitrification (Fig. 7). A stable nitrification rate per m^2 of river bottom, together with a decreasing input of ammonia into the river water (Fig. 2), may also lead to an increase in the nitrification parameter R_2, thus expressing the relative change in ammonia concentration in the water. It appears that the oxidation of ammonia in Rhine water, and by implication the efficiency of microbial nitrification, has been greater in the recent past relative to the more distant past.

The installation of advanced waste-water treatment plants on the Rhine in the 1970s resulted in a significant decrease in ammonia and nitrite concentrations. It may be argued that this decrease diminished a possible substrate inhibition in populations of nitrifying bacteria in the river. Substrate inhibition is caused by un-ionized ammonia (NH_3) and un-ionized nitrous acid (HNO_2). Under average pH and temperature conditions the un-ionized ammonia concentration over the past 15 years was ca. 0.02 mg N/l with the highest value, of ca. 0.1 mg N/l, in the early years. Since nitrite in the Rhine is more than 99.9% dissociated into NO_2^- at given pH (ca. 7.8, RIWA, 1986), the HNO_2 concentrations are very low (< 1 μg/l). According to Anthonisen et al. (1976) the inhibition of *Nitrosomonas* was initiated at concentrations of NH_3 between 10 to 150 mg/l (8.2 to 123.5 mg NH_3-N/l) and that the inhibition of *Nitrobacter* begins between 0.1 and 1.0 mg NH_3/l (0.08 to 0.82 mg NH_3-N/l). In Neufeld et al. (1980) nitrification inhibition is assumed to begin at a concentration of 10 mg/l un-ionized NH_3 (8.2 mg NH_3-N/l). Anthonisen et al. (1976) also mentions an inhibition of nitrite oxidation beginning at HNO_2 concentrations between 0.22 and 2.8 mg/l (67 to 852 μg N/l). In view of the observations by Anthonisen et al. and Neufeld et al. it seems unlikely that any important substrate inhibition of nitrifiers has occurred in the Rhine.

Concentrations of toxic chemicals have declined in the river Rhine over the past 15 years (Fig. 8, RIWA, 1986). The occurrence of specific nitrification inhibitors (Hockenbury & Grady, 1977; Sharma & Ahlert, 1977; Neufeld et al., 1986) was not reported for the river Rhine, but a trend as shown in Fig. 8 can be expected for many compounds, despite the fact that the concentrations of many other compounds may not have decreased. The decrease in concentration of toxic substances could have had an important influence on the increase in the relative nitrification rate over the same period of time. The year of 1976, with an extraordinary hot and dry summer, had high water temperature and low discharge, resulting in an increase in the concentration of heavy metals, and presumably other toxic chemicals, possibly inhibiting nitrification. If this resulted in increased nitrification inhibition, a rapid recovery of nitrification could be possible with the return of more normal weather conditions in the next year. An indication for this could be the

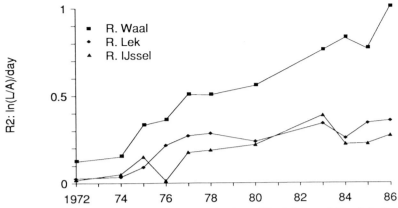

Fig. 7. Yearly averages of the nitrification rate parameter R_2 for the rivers Waal, Lek and IJssel during the period 1972–1986. Missing years were not considered. L = ammonium concentration at station Lobith, A = ammonium concentration at station Vuren, Hagestein or Kampen (for the rivers Waal, Lek and IJssel respectively).

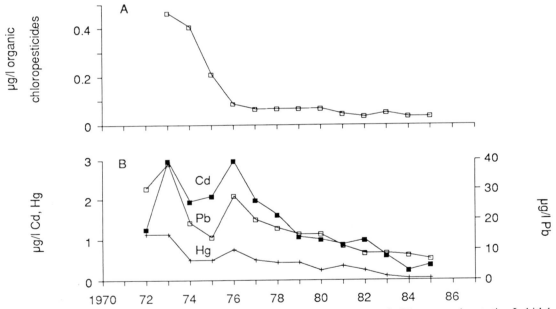

Fig. 8. Yearly average concentrations of organochlorine pesticides (A) and heavy metals (B) measured at station Lobith between 1972 and 1985 (RIWA 1986).

decrease in 1976 and the rapid recovery of nitrification in the river IJssel (Fig. 7), although this effect was not seen in the other two river branches. Müller & Kirchesch (1985) reported, for the German section of the river Rhine, a 10% inhibition of all biological processes for the section from Ludwigshafen to Cologne in 1973.

After the installation of waste water treatment plants, no inhibition has been reported.

Improvement of the condition of the Rhine is also indicated by biological parameters other than nitrification (Fig. 9). Between 1975 and 1980 the macro-invertebrates of the river IJssel have increased both in density and in number of species

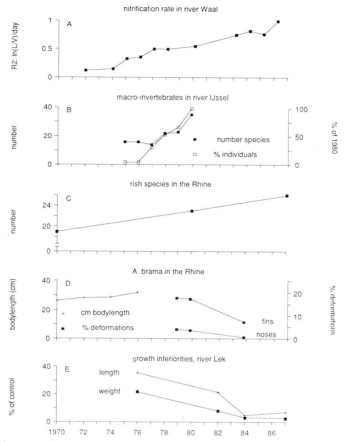

Fig. 9. Improvements over the years shown by biological data derived from the literature against the calculated nitrification rate for the river Waal. (A): Nitrification rate in the river Waal (cf Fig. 7). (B): Macro-invertebrates: Number of species except Chironomids (filled squares) and individuals as a percentage of the number in 1980 (open squares) on stones in the river IJssel (van Urk, 1981). (C): Number of fish species in the Rhine, data from Hadderingh *et al.* (1983) and van de Velde (1988). (D): Bodylength of 5-year old Bream (*Abramis brama*), and incidence of deformations of fins and noses, from Slooff *et al.* (1985). (E): Length and weight inferiorities of trout (*Salmo gairdneri*) incubated during 6–18 months in water from the river Lek (van der Gaag, 1987), expressed as a percentage of an unpolluted control.

(Fig. 9B, van Urk, 1981). From 1970 to 1987 the number of fish species increased (Fig. 9C), Hadderingh *et al.*, 1983, van de Velde, 1988) and their condition improved. The body length of the bream (*Abramis brama*) increased and the frequency of deformations of fins and noses decreased (Fig. 9D, Slooff *et al.*, 1985). Also, the growth rate of young trout (*Salmo gairdneri*) incubated in water from the river Lek (Fig. 9E, van der Gaag *et al.*, 1987) was enhanced.

These observations on the fauna and on nitrification indicate a similar recovery of the ecosystem in response to the decrease in toxicity. Therefore, changes in nitrification can be used as a sensitive indicator for ecosystem recovery in rivers flowing through urbanized areas. Under certain conditions nitrification rates calculated from simple chemical measurements could be used for routine monitoring of river ecosystems.

References

Admiraal, W. & Y. J. H. Botermans, 1989. Comparison of nitrification rates in three branches of the lower river Rhine. Biogeochemistry, in press.

658

Anonymous, 1972–1986. Quarterly reports of water quality. Department of Public Works. Lelystad. The Netherlands.

Anthonisen, A. C., R. C. Loehr, T. B. S. Prakasam & E. G. Srinath, 1976. Inhibition of nitrification by ammonia and nitrous acid. J. Wat. Pollut. Cont. Fed. 48: 835–852.

Ball, I. R., 1967. The relative susceptibilities of some species of freshwater fish to poisons-I. Ammonia. Wat. Res. 1: 767–775.

Beg, S. A., R. H. Siddiqi & S. Ilias, 1982. Inhibition of nitrification by arsenic, chromium, and fluoride. J. Wat. Pollut. Cont. Fed. 54: 482–488.

Beg, S. A. & M. M. Hassan, 1987. Effects of inhibitors on nitrification in a packed-bed biological flow reactor. Wat. Res. 21: 191–198.

Belser, L. W., 1979. Population ecology of nitrifying bacteria. Ann. Rev. Microbiol. 33: 309–333.

Cooper, A. B., 1983. Population ecology of nitrifiers in a stream receiving geothermal inputs of ammonium. Appl. Envir. Microbiol. 45: 1170–1177.

Curtis, E. J. C., K. Durrant & M. M. I. Harman, 1975. Nitrification in rivers in the Trent basin. Wat. Res. 9: 255–268.

Gaag, M. A. van der, 1987. Onderzoek naar semi-chronische toxiciteit van water uit de Lek in winterperiode 1986/1987. KIWA swo 87–307.

Hadderingh, R. H., G. H. F. M. van Aerssen, L. Groeneveld, H. A. Jenner & J. W. van der Stoep, 1983. Fish impingement at power stations situated along the rivers Rhine and Meuse in the Netherlands. Hydrobiol. Bull. 17: 129–141.

Helder, W. & R. T. P. de Vries, 1983. Estuarine nitrite maxima and nitrifying bacteria (Ems-Dollard estuary). Neth. J. Sea Res. 17: 1–18.

Hockenbury, M. R. & Grady C. P. L., 1977. Inhibition of nitrification-effects of selected organic compounds. J. Wat. Pollut. Cont. Fed. 49: 768–777.

Huang, C. S. & N. E. Hopson, 1974. Nitrification rate in biological processes. J. Sanit. Eng. Div. Proc. Am. Soc. Civ. Eng. 100: 409–422.

Kennedy, M. S. & J. M. Bell, 1986. The effects of advanced wastewater treatment on river water quality. J. Wat. Pollut. Cont. Fed. 58: 1138–1144.

Meybeck, M., 1982. Carbon, nitrogen, and phosphorus transport by world rivers. Am. J. Sci. 282: 401–450.

Müller, D. & V. Kirchesch, 1985. On nitrification in the river Rhine. Verh. Int. Ver. Limnol. 22: 2754–2760.

Neufeld, R. D., A. J. Hill & D. O. Adekoya, 1980. Phenol and free ammonia inhibition to Nitrosomonas activity. Wat. Res. 14: 1695–1703.

Neufeld, R. D., J. Greenfield & B. Rieder, 1986. Temperature, cyanide and phenolic nitrification inhibition. Wat. Res. 20: 633–642.

RIWA, 1986. De samenstelling van het Rijnwater in 1984 en 1985. Report of the Werkgroep Waterkwaliteit; Samenwerkende Rijn- en Maaswaterleidingbedrijven. 193 pp. Amsterdam.

Sharma, B. & R. C. Ahlert, 1977. Nitrification and nitrogen removal. Wat. Res. 11: 897–925.

Slooff, W., J. H. Canton & M. A. van der Gaag, 1985. Betekenis van tien jaar ecotoxicologisch onderzoek aan Rijnwater. H₂O, 18: 119–121 & 126.

Suzuki, I., U. Dular & S. C. Kwok, 1974. Ammonia or ammonium ion as substrate for oxidation by Nitrosomonas europaea cells and extracts. J. Bact. 120: 556–558.

Tomlinson, T. G., A. G. Boon & C. N. A. Trotman, 1966. Inhibition of nitrification in the activated sludge process of sewage disposal. J. Appl. Bact. 29: 266–291.

Urk, G. van, 1981. Veranderingen in de macro-invertebratenfauna van de IJssel. H₂O 14: 494–499.

Velde, van de, 1988 (Abstract) Seasonal fluctuations in the occurrence of fish and macro invertebrates in the major Rhine branch, the river Waal, as studied by sampling at cooling water intakes of a power plant. Presentation at the congress 'Biology of the Rhine', Darmstadt, Fed. Rep. Germany.

Wolter, K., H.-D. Knauth, H.-H. Kock & F. Schroeder, 1985. Nitrification and nitrate reduction in water and sediment of river Elbe. Vom Wasser 65: 63–80.

Hydrobiologia **188/189**: 659–679, 1989.
M. Munawar, G. Dixon, C. I. Mayfield, T. Reynoldson and M. H. Sadar (eds)
Environmental Bioassay Techniques and their Application.
© 1989 *Kluwer Academic Publishers. Printed in Belgium.*

Conditional stability constants and binding capacities for copper (II) by ultrafilterable material isolated from six surface waters of Wyoming, USA

John P. Giesy[1] & James J. Alberts[2]
[1] *Department of Fisheries and Wildlife, Pesticide Research Center and Center for Environmental Toxicology, Michigan State University, East Lansing, Michigan 48824, USA; [2] University of Georgia, Marine Institute, Sapelo Island, Georgia 31327, USA*

Key words: metals, chelation, model, geochemistry, speciation, prediction, site-saturation

Abstract

Ultrafilterable material < 0.15 μm was collected from six Wyoming surface waters, for which the chemical limnology had also been determined. The material was separated into four nominal size-fractions and the binding capacity of each for copper was determined by a hyperbolic, site-saturation model. The conditional, overall, thermodynamic stability constants ($\overline{K}'$) for binding of copper were determined by two discrete models: The one- and two-component Scatchard functions, and two continuous multiligand models; The one- and three-component Gaussian Scatchard functions. The accuracy of the stability constants to predict the speciation of Cu(II) in the titration of the isolated fractions and of whole waters was evaluated by comparing the predictions of the thermodynamic, geochemical simulation model, GEOCHEM to those measured by selective ion electrode. The Cu-binding capacity of material retained by ultrafilters was positively correlated with the hardness and alkalinity of the surface waters, from which they were isolated as well as the percent ash content of the ultrafilter-retained material. The magnitude of the conditional stability constant (K_i') decreased as the ratio of the total Cu concentration to total concentration of Cu-binding sites increased. The cumulative frequency distribution of K_i' was log-normally distributed. All four of the models used to estimate the conditional stability constants gave reasonable prediction of the speciation of Cu for both fractionated and whole waters but, depending on the situation, the one- or two-component Scatchard estimate generally gave the best predictions of the proportion of copper, which would be expected to be bound.

Introduction

There are a number of compounds in surface waters which can bind to metals (Giesy, 1983a). Important, refractory components of the organic carbon which can bind to metals are the humic and fulvic acids (Giesy *et al.*, 1986b). Humic and fulvic materials are ubiquitous in surface waters (Buffle & Deladoey, 1982), and they can account for a significant portion of the metal binding capacity of surface waters (Alberts & Giesy, 1983; McKnight *et al.*, 1983). This is especially

true in soft waters, where metal contamination is more likely to be of concern (Giesy, 1983a; Giesy *et al.*, 1983). The mechanism of binding to metals is poorly described (Christman & Gjessing, 1983). The heterogeneous nature of humic and fulvic materials makes it impossible to assign a molecular weight to these compounds (Buffle & Deladoey, 1982). Also, the heterogeneous nature of these materials makes it difficult to define the types of metal binding sites (Buffle, 1980; Bunzl *et al.*, 1976; Gamble *et al.*, 1980). Therefore, the exact number of sites (stoichiometry) and strength of affinity for metals (stability or formation constants) cannot be known, and direct application of theoretical thermodynamics is not appropriate.

With this being the case, is it impossible to use thermodynamic models to predict the relative concentrations of different forms of metals in surface waters, in which these humic and fulvic materials contribute a significant proportion of the total metal-binding ligand concentration? We think not. While the natural systems of interest are not as amenable to traditional, theoretical, thermodynamic relationships as model compounds such as EDTA, we feel that thermodynamic models can be used to study and predict the interactions of metals with both organic and inorganic ligands in fresh waters.

The thermodynamic simulation models, which are used to predict the relative proportion of metals existing as organic and inorganic complexes, ion pairs, and the free or hydrated ions predict equilibrium conditions based on competitive interactions between anions and cations. To make predictions based on these models, one must have estimates of the number of each type of site available to bind competitively and the relative strength of each type of site for all of the metal-ligand combinations. This information, in conjunction with the molar concentrations of metals as well as organic and inorganic ligands, is necessary to stimulate and predict the relative proportions of metal species in aqueous solutions.

The stability constants (K_{stab}) for many inorganic complexes are fairly well known. However, even for these, the values reported in the literature can vary by as much as 100 to 1000 times. The heterogeneous structure of humic and fulvic materials results in an apparent range of strengths of binding, rather than a discrete binding site. Because a greater proportion of the thermodynamically stronger sites would have a tendency to be filled at equilibrium than the weaker sites, one observes a continuum of different strength sites as one adds metal ions to a solution containing humic or fulvic materials. For this reason, the stability constant is not constant, but rather varies as a function of the ratio between the concentration of a metal and the number of

ligand sites to which it can bind (Perdue & Lytle, 1983; Perdue et al., 1984). Estimates of this conditional stability constant are what is necessary for use in thermodynamic, geochemical simulation models to predict the speciation of metals in surface waters. Because the value of the apparent, conditional stability constant is dependent on the metal to ligand ratio (MLR), theoretically, no single stability constant can be determined or assigned. One solution to this problem would be to use the stability constant derived for the appropriate MLR. Alternatively, one could select a statistical model, which describes the K_{stab} across a greater range of MLR.

A number of discrete and continuous models have been proposed to describe the different sites available for binding of metals to humic and fulvic substances (Giesy et al., 1986a & b; McKnight et al., 1983; Buffle, 1980; Gamble et al., 1980; Sanders & Bloomfield, 1980; Bloom & McBride, 1979; Chopping & Kullberg, 1978; Zunino & Martin, 1977; Buffle et al., 1977; Zunino et al., 1977; Bunzl et al., 1976). Sposito (1981) has suggested the use of a quasiparticle model of metal speciation to represent a simplification of the complex metal-humic interactions in surface waters. A quasiparticle model is a mathematical description of an aqueous solution in which the actual assembly of organic compounds is replaced by an assembly of hypothetical, identical macromolecules, whose mole mass is the number-average mole mass of the actual mixture and whose metal-complexation reactions closely mimic those of the real system. To use such a model, one must derive an appropriate mathematical description of the frequency descriptions of the different types of sites. An example of these theoretical constructs is a discrete model such as the multi-component Scatchard model (Giesy et al., 1986a; Alberts & Giesy, 1983; Giesy, 1980; Buffle, 1980; Mantoura et al., 1978). Alternatively, one could use a statistical model to describe the entire range of apparent binding sites by the use of a continuous frequency distribution.

The studies presented here were conducted to investigate the copper (Cu^{++}) binding properties of materials extracted by centrifugation and ultrafiltration from surface waters in the vicinity of Laramie, Wyoming, USA. The one-component and two-component Scatchard, discrete models and the one-component and three-component Gaussian-Scatchard, continuous, multiligand models were examined. Their relative predictive power by using these four estimators of the stability constants in thermodynamic geochemical simulation models to predict the proportion of free Cu^{+2} present during titrations of solutions containing ultrafilterable material with Cu^{+2} was determined. These predictions were then compared to measurements of free Cu^{+2} obtained with a copper selective ion electrode during actual titrations in the laboratory. The variation in binding capacity and stability constants of ultra-filterable materials isolated from six surface waters was also investigated. Finally, the stability constants and binding capacities determined in this study were used with the thermodynamic geochemical simulation model, GEOCHEM, and the results were compared to the simulations of the speciation of Cu, which was measured by selective ion electrode or ion exchange techniques for the natural waters. This last comparison was made to evaluate the utility of thermodynamic models to predict speciation observed in surface waters, where the heterogeneous humic-type materials constitute a significant proportion of, but not all of the metal binding ligands. In addition, the generality of K_{stab} among locations within a geographic region that exhibited different inorganic water chemistries was assessed.

Materials and methods

Sample collection

Surface water samples (approximately 50 liters) were collected during July of 1978 at six locations in Albany County, Wyoming, USA. Three of the locations, Lodgepole (41° 12′ N; 106° 10′ W), Rag Top Mountain (41° 15′ N; 105° 23′ W), and North McKechnie (41° 18′ N, 105° 19′ W), are ponds on the eastern slope of the Laramie

Mountains. Hunting Camp Pond (41° 10′ N, 106° 26′ W) is on the western slope of the Medicine Bow Mountains, while MSU Pond (41° 12′ N, 10° 10′ W) is on the eastern slope. Alsop Lake (41° 27′ N, 106° 45′ W) is in the Laramie Basin, which lies between the Laramie and Medicine Bow Mountains. Samples were transported to the University of Wyoming, Laramie, Wyoming, where they were processed within eight hours of collection. Samples for metals analyses were fixed with 10% HNO_3, which had been distilled in Teflon®. Samples were stored at ~2 °C in acid-washed polyethylene bottles.

Some parameters were measured in the field: these included redox potential, pH, temperature, dissolved oxygen, and conductivity, which were measured with a Model 800 Surveyor System (Hydrolabs Corp.). Alkalinity and hardness were determined by standard colorimetric titration (American Public Health Association, 1976). Water samples were separated into several fractions. An unfractionated water sample was used to determine total concentrations of constituents or to investigate the Cu-binding properties of whole waters. Particulate matter (FI) had a nominal diameter of $>0.15 \, \mu m$ and was removed by centrifugation in a Sorvall SS-1 titanium centrifuge head (Sorvall; Norwalk, Connecticut) equipped with a KSB continuous-flow attachment. Particulates with a Svedberg coefficient of 20 000 S were sedimented simultaneously by using a rotor speed of 7 000 rpm and a flow rate of 242 ml/min (Perhac, 1972). For particles with a specific gravity of 2.65 g/cc, such as kaolinite (Grimm, 1968), this corresponds to a particle diameter of approximately 0.15 μm.

Centrifuged water samples were ultrafiltered sequentially through two hollow fiber bundles (Amicon Corp., Bedford, MA). The fraction retained by a XM-100 bundle (F II, $<0.15 \, \mu m$ but >120 Å, nominal molecular weight, NMW, $>100 000$) and the fraction retained by a UM-05 bundle (F III, <120 Å but >15 Å, NMW $<100 000$ but >500) were used in subsequent studies. The water which passed through the smallest ultrafilter comprised the final fraction

(FIV, < 15 Å, NMW < 500) which contained potential complexing ions including inorganic species. The material isolated, concentrated, and fractionated by ultrafiltration, was that material which is often referred to as dissolved organic carbon and has been characterized as having properties similar to materials referred to as humic or fulvic substances. While this material did have organic carbon associated with it, there was a considerable inorganic fraction to this material, so we have referred to the material as that which was isolated by ultrafiltration. Aliquants of the fractions were reserved for the determination of copper binding, total Cu, dissolved organic carbon (DOC), equivalent weights, and ash content of DOC.

Trace metal concentrations associated with each fraction were determined by atomic absorption (Perkin-Elmer Model 306 Spectrometer, with a model HGA-2100 flameless atomizer). Interferences were evaluated by the use of internal standards and blanks (Giesy & Briese, 1977). Reportable limits of 10 standard deviation units are used throughout.

DOC samples were diluted when necessary and measured with a Beckmann Model 915 Total Organic Carbon Analyzer. The exchangeable proton concentrations per gram of organic matter were estimated by the $Ba(OH)_2$ precipitation technique (Schnitzer & Khan, 1972; Alberts & Giesy, 1983). Subsamples of ultrafilter fractions were dried at 85 °C and the ash-free dry weight determined by pyrolysis at 400 °C for 4 hr.

Concentrations of reactive phosphate (PO_4^{-3}), chloride (Cl^-), sulfate (SO_4^{-2}) and DOC were determined in the laboratory. PO_4^{-3} was determined by the ascorbic acid technique (APHA, 1976). Chloride was quantified by the mercuric nitrate-diphenyl carbazone method (APHA, 1976). SO_4^{-2} concentration was measured by the 2-aminoperimidine method on samples of centrifuged water from which organic carbon had been removed by XAD resins and diluted to the optimum sensitivity range (Baumann, 1976).

Titrations

Fifty-milliliter volumes of solution containing a known concentration of ultrafilter isolate (reported as mg/l organic carbon) (Table 1) were adjusted to a pH of 6.5 with NaOH and HNO_3 and allowed to equilibrate for three hours. The ionic strength was adjusted to 0.1 M with sodium perchlorate. Titrations were done with either 1.6×10^{-3} or 3.1×10^{-3} M $Cu(NO_3)_2$. The pH was held at 6.5 during the titrations by additions of 1×10^{-4} N NaOH, by the titrimeter. Titration was done by 0.01 ml additions from a Metrohm Dosimatt E535 Automatic Titrimeter. Copper activities were measured by an Orion Model 94–29 selective ion electrode. The electrode along with a glass combination pH electrode and reference electrode were interfaced to a Brinkman Model 102 meter (Giesy et al., 1983a). Titrations were done under a Faraday cage to minimize electrical and light interferences. Standard Cu solutions were measured with the electrodes at the same pH and ionic strength as the samples. Electrode response was linear between 7.9×10^{-7} and 1.6×10^{-5} M. Equilibration time was approximately 10 minutes with vigorous stirring between additions. Pilot studies indicated that when batch additions were made and allowed to stand in a sealed polyurethane flask for as long as 4 hr, the measured free Cu did not change significantly from the value determined after approxi-

Table .1 Concentrations of DOC (mgC/L) in ultrafilater fractions* which were titrated with Cu.

Location	Fraction	
	F II	F III
MSU	7.97	48.87
Lodgepole	6.76	14.20
Hunting Camp	12.92	13.73
Alsop	15.01	17.78
Rag Top Mt.	7.47	7.20
N. McKechnie	5.48	17.89

* Nominal Moleculat Weights (NMW) and estimates of spherical diameters of particles are given in the Experiment Station.

mately 10 minutes. This indicates that the equilibration time was of sufficient duration to give accurate results.

Reagents were prepared from analytical grade materials. Standards and Cu titration solutions were made from certified solutions of $Cu(NO_3)_2$ (Fisher Scientific, Fair Lawn, N.J.). Trace metal contamination was removed from NaOH and perchlorate by passing the solutions through a chelex-100 chelating resin and HNO_3 was redistilled in a Teflon® distillation unit. The relative concentrations of copper existing as cationic and anionic forms in surface waters were determined by selective removal of cations by a column containing strongly acidic nuclear sulfonic cationic exchange resin (Dowex®-50w-X8, 50-100 mesh, Na^+ form). Anions were recovered by a strongly basic trimethyl amine anionic exchange resin (Dowex® 1-X8, 100–200 mesh, OH^- form). Three bed volumes of sample were allowed to pass through the columns (flow rate = 2.5 ml/min) before an aliquant was collected for metal analysis by AA (Giesy, 1983). The resins were cleaned and then checked for contaminant effects with suitable blanks.

Statistics

Maximum binding capacities or the number of Cu binding sites were estimated by fitting the titration data to a hyperbolic, site-saturation function (Fig. 1). In this function, the dependent variable was the amount of bound Cu (MB). When the bound Cu was plotted as a function of the total Cu concentration, the maximum bound Cu (CM) was estimated. The function was fit by Marquardt, non-linear, least squares (Proc NLIN; Helwig & Council, 1979). Unless otherwise specified, two-way comparisons between means were made by Student's T-Test ($\alpha = 0.05$; $\beta = 0.20$). Multiple comparisons were made by ANOVA and Tukey's W-Test ($\alpha = 0.05$; $\beta = 0.20$). When estimates of the variances of two populations were not equal, pooled estimates were used for statistical comparisons.

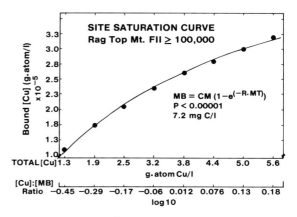

Fig. 1. Copper (Cu^{+2}) bound to ultrafilterable material (MB) as a function of the total Cu added to a solution containing 7.2 mg Cu/l of ultrafilter fraction II from Rag Top Mountain, Wyoming, USA. The total Cu added is also represented as the $\log_{10}$ of the ratio of total [Cu] to total concentration of binding sites. Closed circles represent points of titration curve determined by selective ion electrode. The curve represents the predicted function from the hyperbolic, site-saturation model. CM = the asymptotic, maximum binding capacity; R = a curve fitting parameter, which describes the rate of approach to CM and MT = the total Cu concentration.

Geochemical simulation

The thermodynamic geochemical simulation model GEOCHEM (Sposito & Mattigod, 1979) was parameterized for each of the six ponds studied or for the titration conditions for a particular fraction. The stability constants used for the simulations were those included in the thermodynamic data set supplied by the authors of GEOCHEM (Sposito & Mattigod, 1979) except for the interactions between Ca^{+2} and Fe^{+3} with humic materials which were determined previously (Alberts and Giesy, 1983; Giesy et al., 1983 and 1986a), and Cu^{+2}, which was determined for the systems on which we report here. Where necessary, such as with multiple ligand models, additional metal-humic interactions were added to the model. Simulations were conducted with and without particulates present, depending on the water sample. Sensitivity analyses were conducted by varying the magnitude of different constants or variables either singly or in combination.

Models

In a heterogeneous mixture of binding sites, such as humic and fulvic materials, an overall conditional stability constant ($\overline{K}'$) can be defined (equation 1).

$$\overline{K}' = \frac{[M_t] - [M_f]}{[M_f][L_f]} = \frac{[ML]}{[M_f][L_f]} \qquad (1)$$

where:

M_t = total molar concentration of metal
M_f = molar concentration of free metal
L_f = molar concentration of free binding sites
ML = molar concentration of metal bound to ligand

A number of methods have been used to estimate $\overline{K}'$ (van den Berg & Kramer, 1979; Scatchard, 1949; Mantoura & Riley, 1975; Giesy *et al.*, 1986a). In this study four methods to estimate the conditional stability constants and mole fractions were used.

The one-component Scatchard function which is a discrete model to give the maximum apparent conditional stability constant at small ratios of copper to binding sites was examined. The Scatchard function, which was first introduced to

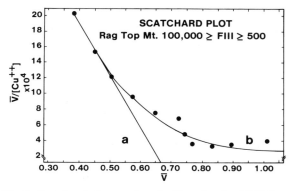

Fig. 2. Scatchard diagram for ultrafilter fraction F III from Rag Top Mountain, Wyoming, USA. 7.2 mg DOC/L titrated, pH = 6.5, ionic strength = 0.1 M. Solid circles represent observed values. Line 'b' represents values predicted from a two-component Scatchard function. Line 'a' represents the 'initial' or maximum one-component Scatchard approximation to the titration data.

determine the stability constants for metal binding by proteins (Scatchard, 1949), has been adapted for determining the overall conditional stability constants for binding of metals by more heterogeneous materials, such as humic and fulvic acids (Mantoura & Riley, 1975; Guy & Chakrabarti, 1976; Mantoura *et al.*, 1978; Giesy, 1980; Saar & Weber, 1979; Kribek & Podlaha, 1980; Li *et al.*, 1980; Shuman *et al.*, 1983; Perdue & Lytle, 1983; Perdue *et al.*, 1984; Guy *et al.*, 1975; Scheinberg, 1982). The Scatchard analysis can be interpreted graphically by plotting $\overline{V}/M_f$ as a function of $\overline{V}$ (Fig. 2) to estimate $\overline{K}'$ and N_i. In this approach, $\overline{K}'$ is the conditional stability constant for some range of metal to ligand ratio and N_i is the mole fraction of the total number of binding sites, which are estimated to have an overall stability constant represented by $\overline{K}'$. The function $\overline{V}$ is the proportion of the total binding sites which are filled (equation 2).

$$\overline{V} = \frac{[ML_t]}{[L_t]} \qquad (2)$$

where:

ML_t = number of binding sites of type L_t which are bound by the metal of interest
L_t = total number of binding sites available for the metal of interest.

If the resulting plot is curvilinear, more than one type of binding site is indicated. This is the case for all of the surface waters and fractions studied here. The Scatchard plot for the titration of F III from Rag Top Mountain Pond is representative of the type of plot observed for all of the waters studied (Fig. 2). The one-component Scatchard model cannot be used to give an overall conditional stability constant, which is valid for the entire range of metal ligand ratios, but can be used to estimate the maximum value for $\overline{K}'$ which would be observed at small MLR (line a in Fig. 2). When more than one type of binding site is indicated by a nonlinear Scatchard plot, one can use the two-component Scatchard model (equation 3) to represent the strength of binding by two average conditional stability constants ($\overline{K}'_1$ and $\overline{K}'_2$). The relative proportions of the sites represented by these two average values are given by N_1 and N_2.

$$\overline{V} = \frac{\overline{K}_1'[M_f]}{1 + \overline{K}_1[M_f]} \cdot \frac{[ML_1]}{[L_t]} +$$

$$+ \frac{\overline{K}_2'[M_f]}{1 + \overline{K}_2[M_f]} \cdot \frac{[ML_2]}{[L_t]} \qquad (3)$$

where:

$$N_1 = \frac{[ML_1]}{[L_t]}$$

$$N_2 = \frac{[ML_2]}{[L_t]}$$

When the entire titration range is used, this method produces weighted, curve-fitted parameters that are quasiparticle models useful in predicting the relative distribution of metal among binding sites. The two-component Scatchard function generally gives statistically significant fits to the observed titration data (line b in Fig. 2).

The Scatchard equation has been criticized for its lack of rigor in multi-ligand systems (Perdue et al., 1984). As indicated above, the Scatchard function can be used as a multi-component function. However, while the Scatchard function for multi-ligand binding adequately describes the observed distribution of free and bound metal, the values derived for the conditional stability constants are only number-average values which describe a number of similar but nonidentical metal binding sites. While the observed relationship between the conditional stability constant $\overline{K}'$ and the ratio of total metal to total binding capacity results in an apparent continuous distribution, this does not mean that there is necessarily a continuum of strengths of binding sites. In fact, a thermodynamic model, which includes two discrete types of sites, will generate a relative distribution of free and bound metal, such that when the apparent $\overline{K}^1$ values are calculated, it will appear that there are a large number of closely spaced types of sites due to the fact that each

apparent $\overline{K}_1'$ value is a function of both of the stability constants. It may be that the two-component Scatchard function has good predictive power for many fractions of organic carbon extracted from surface water because there are primarily two types of sites, as has been suggested by many authors (Schnitzer & Khan, 1972; Bresnahan et al., 1978; Alberts & Giesy, 1983). However, the goodness of fit of the two-component Scatchard function can neither support nor refute the existence of either the discrete sites or a continuum of sites. It can, however, be concluded that the two-component Scatchard function can be used in thermodynamic, geochemical simulation models to accurately predict the relative concentrations of free and bound metals over a wide range of MLR in systems in which metal speciation is dominated by the binding to organic carbon (Giesy et al., 1986a).

An alternative to discrete models, such as those described above, is continuous multi-ligand models, which describe the entire range of strengths of sites. Such a frequency distribution can be described by an average or mean binding constant, which is a measure of the central tendency and a standard deviation, which is a measure of the dispersion of the binding constants about the mean (Fig. 3).

The Log_{10}-Gaussian distribution has been suggested as an appropriate model to describe the frequency distribution of the stability constants for binding of metals by humic materials (Perdue et al., 1983; Giesy et al., 1986a). When the Log_{10}-Gaussian distribution is combined with the Scatchard function, the resulting symmetrical Log_{10}-Gaussian-Scatchard function makes it possible to define the shape of the frequency distribution at greater MLR and extrapolate to smaller ratios (Fig. 3). The Log_{10}-Gaussian-Scatchard function is less constrained than the two-component Scatchard discrete model (Perdue et al., 1983). Once the distribution of sites has been described, one can determine the mole-average proportion of the number of sites with a particular binding strength for a given metal. This would be the area under the curve for some speci-

GAUSSIAN FREQUENCY DISTRIBUTION OF LOG$_{10}$ K′

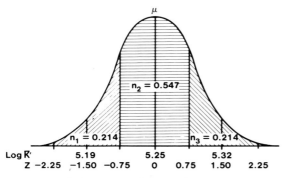

RAGGED TOP MOUNTAIN POOLED FII AND FIII

Fig. 3. Gaussian approximation of the normal probability function for Log$_{10}$ K'. μ = parametric mean. n_1, n_2 and n_3 represent the area under the curve and are relative portions of the total binding capacity assigned to each of 3 strengths of conditional stability constants.

fied regions. We have found that proportioning the binding sites into three regions allows for an adequate description of the distribution of sites across the ecologically relevant range of MLR (Fig. 3). The parametric mean (μ) and standard deviation (σ) are related by equation (4).

$$\frac{[ML]}{[L_t]} = \overline{V}_{\text{calc}} = \frac{1}{\sigma 2\pi} \int_{-\infty}^{\infty} \frac{[M_f]10^{\log K'}}{1 + [M_f]10^{\log K'}}$$

$$\times \left[e^{-\frac{1}{2} \frac{(\mu - \log K'^2)}{\sigma}} \cdot d \log K' \right] \quad (4)$$

where:

V_{calc} = empirical ratio $[ML]/[L_t]$ at each point of titration.

$[M_f]$ = molar concentration of free metal

$[ML]$ = molar concentration of metal bound to ligand

$[L_t]$ = molar concentration of total ligand sites for binding

K' = conditional stability constant observed for any ratio of total metal to total ligand concentrations

σ = parametric standard deviation of distribution of K' as a function of metal ligand ratio

μ = parametric mean conditional stability constant ($\overline{K}'$)

This function can be fit to the empirical titration data by non-linear, least squares techniques to provide estimates of the mean and standard deviation of the overall conditional stability constant ($\overline{K}'$). A single estimate or one-component Log$_{10}$-Gaussian estimate is derived by integrating the entire function from $-\infty$, which can be approximated by $\mu - 4\sigma$, and $+\infty$, which is approximated by $\mu + 4\sigma$. The least squares are evaluated numerically by substituting values of μ and σ such that the residual sums of squares are minimized (equation 5).

$$\text{ess} = \Sigma \frac{(\overline{V}_{\text{calc}} - \overline{V}_{\text{exp}})^2}{V_{\text{exp}}} \quad (5)$$

where:

ess = error sum of squares

$\overline{V}_{\text{calc}}$ = calculated $\overline{V}$ from observed titration data

$\overline{V}_{\text{exp}}$ = $\overline{V}$ expected based on the entire distribution

Better resolution can be obtained by partitioning the distribution into two or three regions. This is done by calculating the area under the frequency distribution (Fig. 3; Giesy et al., 1986a; 1986b); for instance, by centering the conditional stability constants at 0, +1.5 and −1.5 standard deviation units from the mean and calculating the area under the normal probability function, which represents the relative proportion of binding sites assigned to each of the three stability constants. In the work presented here, both the single-component and the three-component Gaussian-Scatchard continuous distributions have been used to describe the strength of binding and, where appropriate, the relative number of binding sites.

Results and discussion

% Ash content

The surface waters studied were selected to give a range of concentrations of organic and inorganic constituents (Tables 2 and 3). The concentrations of organic carbon ranged from 3.9 to 9.5 mg/l. The greatest proportion of the organic carbon was found to have a nominal molecular diameter, which was smaller than that retained by the smallest ultrafilter (Table 4).

The % ash content of the organic carbon retained by the ultrafilters ranged from as little as 21.9% to as great as 80.3% (Table 5). The mean % ash contents of F II and F III were not statistically different within systems. We did, however, observe statistically significant dif-ferences among the ash contents of DOC of both F II and F III from different locations (Table 5).

The % ash content of material in ultrafilter fractions F II and F III were positively correlated with the alkalinity, hardness, chloride concentration and concentration of particles greater than 0.15 μm but not correlated with the SO_4^{-2} concentration of the waters from which they were collected (Table 6). The % ash content of ultra-filtered material isolated from MSU pond was the smallest in both fractions F II and F III. This location also had the smallest hardness and alkalinity values and one of the smallest concentrations of suspended particulates (Table 2). Alsop Pond, which had the greatest hardness and alkalinity values and greatest concentration of suspended particulates (Table 2), also had the greatest % ash content in both fractions (Table 5).

Table 2. Chemical characteristics of six Wyoming surface waters.

Location	Hardness	Alkalinity	Cl⁻	$SO_4^=$	TOC	pH	Redox (mv)	Particulates >0.15 μM (mg/l)
	(——————— mg/l ———————)							
MSU	13	17	10	60.0	7.7	7.0	400	1.3
Lodgepole	138	128	53	8.1	6.4	6.5	345	7.0
Hunting Camp	110	113	18	0.8	9.5	7.4	300	3.5
Alsop	330	232	80	546.0	3.9	8.8	230	16.3
Rag Top Mt.	54	52	23	5.4	7.0	7.8	370	0.4
N. McKechnie	48	51	38	10.7	7.0	9.2	255	1.9

Table 3. Elemental constituents of six Wyoming surface waters.

Location	Na	K	Ca	PO_4*	Cu	Pb	Cd	Fe	Cu particulate** >0.15 μm (%)
	(——————— mg/l ———————)			(——————————— (μg/l) ———————————)					
MSU	1.2	0.3	2.9	5.9	2.0	22.1	>0.3	397	<1
Lodgepole	5.0	2.1	36.5	1.3	7.4	24.5	0.16	494	22
Hunting Camp	3.5	1.5	39.5	32.1	2.9	23.1	0.29	106	2
Alsop	128.0	5.1	35.3	11.8	10.7	30.6	0.23	35	5.5
Rag Top Mt.	3.8	1.5	16.3	5.5	3.0	14.3	0.15	49	19
N. McKechnie	4.5	1.9	16.9	20.5	5.1	13.6	0.12	115	12

* Soluble reactive $P \leq 0.45$ m.

** Determined by centrifugation.

Table 4. Concentrations (mgC/L) and relative distribution of organic carbon in three ultrafilter fractions from six Wyoming surface waters.

Location	Fraction		
	F II	F III	F IV
MSU	0.11	0.49	7.1
	(1.5)*	(6.3)	(92.2)
Rag Top Mt.	0.09	0.09	5.1
	(1.3)	(1.3)	(97.4)
N. McKechnie	0.05	0.15	7.0
	(<1.0)	(<1.0)	(99.0)
Lodgepole	0.11	0.17	6.5
	(<1.0)	(<1.0)	(99.0)
Alsop	0.14	0.14	3.9
	(<1.0)	(<1.0)	(99.0)
Hunting Camp	0.14	0.20	9.5
	(1.5)	(6.5)	(93.0)

* Percent of total in parentheses.

Acidity

Statistically significant differences between the average acidity of ash-free OC of F II and F III as well as among the locations (Table 5) were observed. However, there was no significant correlation between the acidity of F II and that of F III (Table 6). The acidity of the materials isolated by ultrafiltration, as measured by the Ba(OH)$_2$ technique, was not correlated with the alkalinity or hardness or concentrations of SO_4^{-2}, Cl^- or particulate material $> 0.15 \mu m$ nominal diameter (Table 6). There was no statistically significant correlation between the % ash content of F II, but a significant positive correlation between % ash content and acidity was observed for the smaller nominal size fraction, F III (Table 6). This indicates that, for the larger nominal size fraction, inorganic constituents were not contributing significantly to the number of equivalents measured, but that there may be a significant inorganic contribution to the overall acidity of the smaller nominal diameter ultrafilter fraction.

Cu-binding capacity of ultrafilterable material

The maximum binding capacities for Cu by DOC, extracted from each location, were determined by fitting a site saturation model to the titrations (Fig. 1). All of the predictions were highly significant ($P < 0.001$). There were significant differences among locations for both F II and F III (Table 7). The binding capacity of F III was significantly greater than that of F II and there was a significant positive correlation between the Cu-binding capacities of F II and those of F III (Table 6).

The locations for which ultrafiltered material

Table 5. Ash content and acidity of two ultrafilter fractions, isolated from six Wyoming surface waters.

Location	Fraction F II		Fraction F III	
	Ash (%)	Acidity* (meg/g Ash Free OC)	Ash (%)	Acidity* (meg/g Ash Free OC)
MSU	34.6[a]	63.7 ± 24.0[b]	21.9[a]	13.3 ± 4.9[a]
Lodgepole	68.9[b]	67.4 ± 9.6[b]	64.2[b]	48.9 ± 8.3[b]
Hunting Camp	69.8[b]	50.2 ± 10.5[a, b]	57.3[b, c]	22.0 ± 7.6[a]
Alsop	80.3[c]	52.8 ± 3.2[a, b]	79.5[b]	43.1 ± 3.5[b]
Rag Top Mt.	63.8[b]	40.7 ± 17.5[a]	63.8[b]	40.7 ± 17.5[b]
N. McKechnie	65.0[b]	39.4 ± 18.1[a]	39.4[a, c]	9.2 ± 8.0[a]
	$\bar{X}$ = 63.7 SD = 15.4	$\bar{X}$ = 52.4 SD = 11.5	$\bar{X}$ = 54.4 SD = 20.5	$\bar{X}$ = 29.5 SD = 16.8

* $n = 3$, ± SD.
[a, b, c] Values denoted by the same letters within a parameter are not significantly different from one another.

Table 6. Pearson, pairwise, product-moment correlations (R^2) among copper binding capacity (BCf-II and BCf-III) and chemical characteristics of surface waters and materials isolated by ultrafiltration.

	% Ash-fIII	% Ash-fII	Acidity-fII	Acidity-fIII	BCf-III	BCf-II	[Cl$^-$]	[SO$_4^=$]	[Part > 0.15 μm]	Alkalinity
Hardness	0.636***	0.549***	0.018 NS	0.018 NS	0.969***	0.366***	0.791***	0.787***	0.961***	0.972***
Alkalinity	0.692***	0.637***	0.025 NS	0.025 NS	0.932***	0.313***	0.752***	0.641***	0.913***	
[Part > 0.15 μm]	0.475***	0.402***	0.069 NS	0.069 NS	0.880***	0.259*	0.828***	0.818***		
[SO$_4^=$]	0.279***	0.190 NS	0.0045 NS	0.005 NS	0.782***	0.408***	0.600***			
[Cl$^-$]	0.512***	0.534***	0.009 NS	0.009 NS	0.681***	0.321**				
BCf-II	0.305**	0.492***	0.006 NS	−0.474***	0.412***					
BCf-III	0.705***	0.548***	0.394**	0.0032 NS						
Acidity-fII	−0.012 NS	−0.096 NS	−0.029 NS							
Acidity-fIII	0.701***									
% Ash-fII	0.798***	0.314***								

* $P < 0.05$.
** $P < 0.01$.
*** $P < 0.001$.
NS $P > 0.1$.

Table 7. Maximum Cu(II) binding capacities and percent of total equivalents available for divalent binding of Cu for organic carbon isolated from six Wyoming surface waters.

Location	Cu(II) binding capacities g – atom Cu		Equiv. available %	
	F II	F III	F II	F III
MSU	1.10×10^{-6a}	2.89×10^{-7a}	3.45^a	4.35^a
Lodgepole	1.45×10^{-6a}	2.11×10^{-6a}	2.26^a	8.63^a
Hunting Camp	$4.64 \times 10^{-6b, d}$	2.18×10^{-6a}	18.50^a	19.82^b
Alsop	8.26×10^{-6c}	5.79×10^{-6b}	31.29^b	26.87^b
Rag Top Mt.	$4.94 \times 10^{-6b, d}$	1.60×10^{-6a}	24.28^b	7.86^a
N. McKechnie	$6.10 \times 10^{-6c, d}$	5.95×10^{-7b}	30.96^b	12.98^b
	$\overline{X} = 4.42 \times 10^{-6}$ SD $= 2.75 \times 10^{-6}$	$\overline{X} = 2.1 \times 10^{-6}$ SD $= 2.0 \times 10^{-6}$	$\overline{X} = 18.46$ SD $= 12.98$	$\overline{X} = 13.41$ SD $= 8.46$

[a, b, c, d] Values within a fraction denoted by the same letters are not statistically different from one another.

exhibited the greatest binding capacities in F II were generally also greater in F III (Table 7). The material isolated from the waters with lesser hardness and alkalinity, such as MSU Pond, had lesser Cu binding capacities, while those from waters with greater hardness and alkalinity exhibited greater binding capacities (Table 7). The binding capacities of both F II and F III were significantly positively correlated with hardness, alkalinity and particulate matter $> 0.15 \mu m$ as well as the % ash content of the ultrafilterable material (Table 6). These results indicate that inorganic constituents, especially in F III, may be responsible for at least a part of the Cu-binding capacity of the ultrafiltered material. This is further supported by the correlation between % ash content and acidity for F III. Also, when the proportion of equivalents available for binding Cu is calculated, one finds that MSU Pond, which has the smallest Cu-binding capacity, has some of the smallest apparent proportions of the total $Ba(OH)_2$ equivalents available for binding Cu, and Alsop Lake has the greatest proportion (Table 7). If the ash content were binding only to organic binding sites and not contributing to the binding capacity, one would expect to see a relatively constant percentage.

Some researchers have chosen to try to eliminate all of the ash content from isolated fractions so that the 'pure' organic matter can be studied. It is our opinion that the ultrafilterable material is a heterogeneous but definable functional unit in nature and can and should be studied as such. Severe acid digestions and base extractions probably alter the material to such an extent that the results of studies on this material would probably not be very useful for making predictions of the metal-binding in surface waters.

Distribution of conditional stability constants

The conditional stability constants for the binding of Cu(II) by ultrafiltered materials ($\overline{K}$, equation 1) decreased with increasing MLR (Fig. 4). This is indicative of thermodynamically more stable sites filling before less stable sites, when Cu was added to a fixed concentration of those binding sites. Furthermore, this distribution would be expected if there were a large number of similar but non-identical copper-binding sites. This may be due to polyelectric effects induced by the sequential filling of sites and not necessarily indicate a large number of discrete types of sites (Giesy *et al.*, 1986a & 1986b). As discussed in the methods section, the continuous distribution of apparent conditional stability constants could be caused by the interaction of two or more discrete sites. The

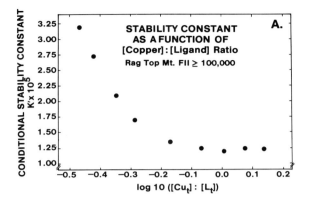

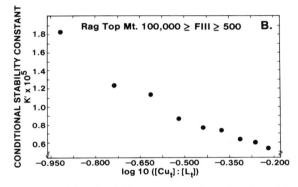

Fig. 4. Conditional stability constants K_i' as a function of the ratio of total Cu^{+2} concentration to total concentration of Cu^{+2} binding sites for two ultrafilter fractions from Rag Top Mountain Pond, Wyoming, USA.

frequency distributions of the observed apparent conditional stability constants, which were calculated from data points in the titration of ultrafilter fractionated material with Cu(II) were examined. The amount of Cu bound, at any given MLR, is a function of all of the individual 'true' stability constants. Since the 'true' values are unknown, what are reported and discussed are the average 'apparent' conditional stability constants (K_i').

The distributions of K_i' as a function of MLR were observed to be a continuous distribution, which was slightly skewed to the right (increasing MLR) with proportionately more weaker than stronger binding sites (Giesy *et al.*, 1986a and 1986b). Similarly, proton binding by humic acid has been observed to exhibit K_a values, which could be effectively described by a continuous

multi-ligand model (Posner, 1966; Perdue *et al.*, 1984). The conditional stability constants for binding of copper (Giesy *et al.*, 1986a) or uranium (Giesy *et al.*, 1986b) to humic matter have been shown to be log-normally distributed. That is, the frequency distribution of the $\log_{10}$ of the constant K_i' is symmetrical and can be described by the Gaussian or 'normal' probability function (equation 4).

The degree to which a function adequately described an observed distribution can be determined by a number of statistical tests of the 'goodness of the fit' by determining the residual deviation of the empirical data from some theoretical model of the function. Alternatively, one can rapidly evaluate the appropriateness of the normal distribution to describe a population of points graphically. This is done by a normal probability plot, in which the cumulative proportion of the strengths of sites filled is plotted as a function of K_i'. If the relative proportion of binding sites is normally distributed, such a plot will result in a straight line (Sokal & Rohlf, 1973). If, however, the resulting function is not a straight line, the normal distribution does not adequately describe the relative frequency of occurrence of sites with a particular strength.

In the case of Cu binding by material isolated by ultrafiltration from all six Wyoming surface waters, a plot of the cumulative probability as a function of the $\log_{10} K_i'$ results in a straight line (see example, Fig. 5). Thus, the distribution of binding sites is '$\log_{10}$-normally' distributed and allows us to use the $\log_{10}$ transformation of K_i' in our continuous distribution adaptation of the Scatchard function (equation 4) for all locations.

One means of quickly determining the reasonableness of the estimates of $\overline{K}'$ from the Gaussian-Scatchard model is to examine the probability plots. The median ($\overline{K}'$), which is the mean in a symmetrical distribution such as that used in this study, can be estimated by dropping a perpendicular from the point where the 50 percentile intersects the frequency function to the $\log_{10} K'$ axis. A value of 0 standard deviation units represents the parametric median (μ) value for $\overline{K}'$ (Fig. 5). When this was done for F III from Rag

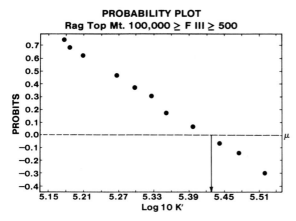

PROBABILITY PLOT
Rag Top Mt. 100,000 ≥ F III ≥ 500

Fig. 5. Normal probability plot of the frequency distribution of $\log_{10} K_i'$ for Cu^{+2} binding by F III from Rag Top Mountain Pond, Wyoming, USA. Dashed line represents median. Arrow represents the K_i' value associated with the median of the $\log_{10}$-transformed distribution.

Table 8. Estimates of the initial (maximum) overall conditional stability constant ($\overline{K}'$) for two organic carbon fractions isolated from six Wyoming surface waters. Estimated from a one-component Scatchard model of the initial portion of the titration.

Location	$\log_{10} \overline{K}'$	
	F II	F III
MSU	5.78 [a]	6.23 [a]
Lodgepole	5.59 [a, b]	5.68 [b]
Hunting Camp	5.78 [a]	5.79 [a]
Alsop	5.55 [b]	6.06 [a]
Rag Top Mt.	5.94 [a, c]	5.79 [a, b]
N. McKechnie	6.04 [c]	5.94 [a, b]
	$\overline{X} = 5.78$	$\overline{X} = 5.91$
	SD = 0.19	SD = 0.24

[a, b, c] Values, within a fraction, denoted by the same letter are not significantly different from one another.

Top Mountain, we estimated the $\log_{10} \overline{K}'$ value to be 5.42, which is very similar to the value of 5.5 and well within one standard deviation of the estimate of $\overline{K}'$ ($\sigma = 1.1$), which was determined for the same fraction by a non-linear, least square fit to the Gaussian-Scatchard function (equation 4). This rate of deviation (approximately 17% or 1.45% on the $\log_{10}$-Transformed scale) seems acceptable, when one is comparing an estimate made by eye to one which uses numerical estimation techniques to integrate areas under a function.

Estimation of conditional stability constants

One-Component Scatchard

The one-component Scatchard model has been used in a number of studies to estimate the conditional stability constants of materials isolated from surface waters (Mantoura & Riley, 1975; Guy & Chakrabarti, 1976; Li *et al.*, 1980). For this reason and because the ecologically-relevant region of the MLR ratio is often in the range -0.5 to -2.5 ($\log_{10}$), the one-component, or initial, maximum estimate of $\overline{K}'$ is often useful. This value is estimated from the initial, quasi-linear

region of the Scatchard function (line a, Fig. 2). The estimates of the maximum conditional stability constant ($\overline{K}'_{max}$) determined by the one-component Scatchard function were similar among locations (Table 8). None of the observed values for either fraction or location were statistically different from the overall mean for that fraction, and there was no trend of one fraction exhibiting greater values for $\log_{10} \overline{K}'$. The overall average $\log_{10} \overline{K}'$ across all six locations and both fractions was 5.84 (SD = 0.20). We concluded that the maximum overall conditional stability constants from the six locations and two fractions were very similar and that the overall average value could be used as an estimate for all locations.

Two-component Scatchard

The two-component Scatchard function proportions the total number of binding sites into two components and assigns an overall average conditional stability constant to each of these ($\overline{K}'_1$ and $\overline{K}'_2$) and assigns a mole fraction portion of the total number of sites to each of the average conditional stability constants (equation 3). Signifi-

Table 9. Conditional stability constants ($\overline{K}'_1$ and $\overline{K}'_2$) and mole fractions (n_1 and n_2), as determined by a two-component Scatchard function for each location.

Location	Fraction F II				Fraction F III			
	$\log_{10}\overline{K}'_1$	n_1	$\log_{10}\overline{K}'_2$	n_2	$\log_{10}\overline{K}'_1$	n_1	$\log_{10}\overline{K}'_2$	n_2
MSU	7.15 [b, c]	0.08 [a]	4.96 [b]	0.87 [a, b, c]	7.33 [b]	0.24	4.90 [a]	0.79 [a]
Lodgepole	6.59 [a]	0.12 [a]	4.52 [a]	0.85 [a, b, c]	6.30 [a]	0.15 [a]	4.69 [a]	0.99 [b]
Hunting Camp	7.26 [b, c]	0.07 [a]	4.92 [b]	0.83 [a, c]	6.63 [a, c]	0.17 [b]	4.78 [a]	0.98 [b]
Alsop	5.86	0.01 [a]	4.71 [a]	0.79 [a, d]	6.47 [a]	0.12 [a, b]	4.96 [a]	0.58
Rag Top Mt.	6.81 [a, c]	0.28 [b]	4.60 [a]	0.88 [b]	7.02 [b, d]	0.10 [a]	5.67 [b]	0.83 [a, b]
N. McKechnie	6.98 [a, c]	0.29 [b]	4.60 [a]	0.78 [d]	6.92 [c, d]	0.14 [a, b]	5.78 [b]	0.93 [a, b]
	$\overline{X}$ = 6.78	0.14	4.72	0.83	$\overline{X}$ = 6.78	0.15	5.13	0.85
	SD = 0.51	0.11	0.18	0.04	SD = 0.38	0.05	0.47	0.16

[a, b, c] Values within a fraction and parameter denoted by the same letter are not significantly different from one another.

cant differences in estimates of $\overline{K}'_1$ and $\overline{K}'_2$ among locations for both F II and F III (Table 9) were observed. However, there was no significant difference for the mean values of either $\overline{K}'_1$ or $\overline{K}'_2$ for F II and F III. Therefore we calculated averages of $\overline{K}'_1$ and $\overline{K}'_2$ across F II and F III, which are 6.78 and 4.93, respectively. Similarly, estimates of the relative proportion of binding sites to be assigned to each of the $\overline{K}'$ values varied more among locations than did the estimates of $\overline{K}'_i$ themselves (Table 9). However, again there were no statistical differences for F II and F III for the mean values of either n_1 or n_2. The overall average across F II and F III for n_1 and n_2 are 0.15 and 0.84, respectively.

The fact that we were able to obtain fits of our titration data with the two-component Scatchard model which were statistically significant ($P < 0.001$) does not indicate that there were only two types of binding sites present or even that the binding was predominated by two types of sites. As discussed above, the apparent conditional stability constants observed could be due to an actual continuum of closely spaced binding sites or due to the interaction of as few as two types of sites.

One-component Log₁₀-Gaussian-Scatchard model

Some statistical differences were observed among the $\overline{K}'$ values estimated for different locations for both F II and F III (Table 10). The mean $\overline{K}'$ values averaged across all six locations for F II and F III were not statistically different (estimate of standard deviation, pooled across F II and F III: $\overline{K}'$ = 0.142). Therefore, we calculated an overall average value of $\overline{K}'_1$ across all six locations and across both F II and F III to be 5.05. The parametric standard deviation of the distribution of K' was statistically smaller for F III than for F II (pooled estimate of the population standard deviation (SD) for F II and F III = 0.2613).

Three-component Log₁₀-Gaussian-Scatchard model

Instead of describing the conditional stability constant by one overall mean value from the one-component Log₁₀-Gaussian distribution, we used the mean and parametric standard deviation, determined by the one-component model with our knowledge of the normal distribution, to partition the total number of binding sites (1.0) into three regions, each of which was centered on a $\log_{10}\overline{K}'$ (Fig. 3 and Table 11). We chose the median (μ) value for $\log_{10}\overline{K}'$ as $\log_{10}\overline{K}'_2$, and chose $\log_{10}\overline{K}'_1$ and $\overline{K}'_3$ as values which were 1.5 parametric standard deviations (σ) below and above this value, respectively. The relative proportions (n_i) assigned to each of the three conditional stability constants represent the area under the probability function which is bounded by the

Table 10. Conditional stability constants ($\overline{K}'$) for binding of copper(II) estimated from the one-component Gaussian-Scatchard continuous distribution model (see Figure 3).

Location	Fraction F II		Fraction F III	
	$\log\overline{K}'$ (SD)	$\sigma\log\overline{K}'$ (SD)	$\log\overline{K}'$ (SD estimate)	$\sigma\log\overline{K}'$ (SD estimate)
MSU	4.9 (0.08)[b]	0.550 (0.09)[a]	5.20 (1.5)[b]	0.025 (0.08)[a]
Lodgepole	4.8 (0.85)[a, b]	1.200 (0.6)[b]	4.96 (0.020)[a]	0.020 (0.017)[a]
Hunting Camp	4.9 (0.015)[b]	0.530 (0.1)[a]	5.14 (0.113)[b]	0.017 (0.005)[a]
Alsop	4.7 (0.72)[a]	1.300 (0.72)[b]	4.69 (0.022)[a]	0.850 (0.13)
Rag Top Mt.	5.1 (0.054)[c]	0.019 (0.026)[a]	5.50 (1.1)[c]	0.069 (0.11)[a]
N. McKechnie	5.1 (0.074)[c]	0.037 (0.017)[a]	5.50 (1.0)[c]	0.120 (0.03)[a]
	$\overline{X}$ = 4.92	0.606	$\overline{X}$ = 5.17	0.184
	SD = 0.16	0.549	SD = 0.31	0.329

[a, b, c] Values denoted by the same letter are not significantly different from one another.

$\log_{10}\overline{K}'_i \pm 0.75\sigma$. This results in assigning values to n_1, n_2, and n_3 of 0.215, 0.547, and 0.215, respectively (Table 11; Fig. 3). Because the distribution is symmetrical, the values chosen are arbitrary; however, from experience we have found that the use of the three-component model adequately describes the relative distribution of sites (Giesy *et al.*, 1986a and b). Also, one could divide the distribution into two portions, such as the area associated with $\overline{K}'$ values centered at

$\mu \pm 2\sigma$. In this case, we would integrate the area under the cumulative frequency distribution, which is bounded by $\mu + 2$ and $\mu - 2\sigma$ ($\pm 2\sigma$), which would essentially be the right and left halves of the distribution. We have found that the use of this configuration does not give as good a representation as the three-component model. One could subdivide the area under the cumulative frequency distribution into any number of components, but we have found that the use of a

Table 11. Estimates of the average conditional stability constant ($\overline{K}'$) for three regions of the Gaussian distribution for fractions F II and F III. n_1 represents the fraction of the sites which are centered on each of the stability constants (see Figure 1).

Location	Fraction FII			Fraction FIII		
	$\overline{K}'_1$ $\overline{X}$ − 1.5 SD*	$\overline{K}'_2$ $\overline{X}$	$\overline{K}'_3$ $\overline{X}$ + 1.5 SD	$\overline{K}'_1$ $\overline{X}$ − 1.5 SD	$\overline{K}'_2$ $\overline{X}$	$\overline{K}'_3$ $\overline{X}$ + 1.5 SD
MSU	4.1	4.9	5.7	4.98	5.20	5.06
Lodgepole	3.0	4.8	6.6	4.93	5.96	4.99
Hunting Camp	4.1	4.9	5.7	5.11	5.14	5.16
Alsop	2.8	4.7	6.7	3.42	4.69	5.97
Rag Top Mt.	5.1	5.1	5.1	5.40	5.50	5.60
N. McKechnie	5.1	5.1	5.2	5.32	5.51	5.68
	$\overline{X}$ = 4.02	4.9	5.83	$\overline{X}$ = 4.86	5.33	5.41
	SD** = 0.97	0.16	0.68	SD = 0.73	0.43	0.340
	n_1 = 0.215	n_2 = 0.547	n_3 = 0.215	n_1 = 0.215	n_2 = 0.547	n_3 = 0.215

* SD = parametric standard deviation.
** SD = estimator of population variance.

Table 12. Conditional three-component stability constants extracted from the Gaussian-Scatchard model with F II and F III combined. N_i represents the proportion of the total number of binding sites associated with each component.

	$\overline{K}_1'$	$\overline{K}_2'$	$\overline{K}_3'$
	$\overline{X} - 1.5\,SD$	$\overline{X}$	$\overline{X} + 1.5\,SD$
MSU	4.62	5.05	5.48
Lodgepole	3.98	4.90	5.82
Hunting Camp	4.59	5.00	5.41
Alsop	3.09	4.70	6.32
Rag Top Mt.	5.19	5.25	5.32
N. McKechnie	5.17	5.25	5.33
	$n_1 = 0.215$	$n_2 = 0.547$	$n_3 = 0.215$

larger number of components does not increase the accuracy of the model.

Because there was no statistical difference between the means pooled across location of F II and F III, we calculated average conditional stability constants for the three-component model (Table 12). Overall conditional stability constants for the three-component, Log_{10}-Gaussian-Scatchard model, pooled across both location and fraction for $\overline{K}_1'$, $\overline{K}_2'$, and $\overline{K}_3'$ were calculated to be 4.44, 5.12, and 5.62, respectively.

Comparison of estimation techniques

To evaluate the predictive power of the four estimation techniques and the similarity among locations, we used the thermodynamic, geochemical simulation model GEOCHEM and parameterized it for the titration conditions of both fractions at each location. This resulted in 12 simulations for each fraction and each location, or a total of 144 simulations. The results of this analysis were similar for the six locations. Depending on the location or fraction, some average values, pooled across fraction or location, overestimated or underestimated the relative concentration of free Cu^{+2}.

Here, we report the results of the predictions with the actual titration curve for F II from Rag Top Mountain Pond (Fig. 6). The best prediction

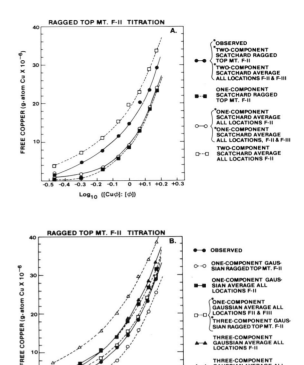

Fig. 6. Concentrations of free Cu^{+2} in a solution of 7.2 mg C/L, pH = 6.5, ionic strength = 0.1 M, as a function of the Log_{10} of the ratio of bound Cu to total available copper binding sites. When results of functions are plotted with the same symbol it indicates that the results were indistinguishable, values determined by selective ion electrode or predicted by GEOCHEM, based on different estimates of $\overline{K}'$ for F II of Rag Top Mountain Pond, Wyoming, USA.

of the Cu^{+2} concentration was given by the two-component Scatchard function, which had been fitted to the titration data for F II at Rag Top Mountain. The two-component Scatchard function, with $\overline{K}'$ values pooled across fractions, F II and F III and across all six locations, also predicted the same concentrations of free Cu^{+2} as measured by selective ion electrode. The two-component Scatchard function for F III, pooled across all six locations, predicted a free Cu^{+2} concentration which was greater than that measured by selective ion electrode. The overestimate was approximately 100% at the smallest ratio of $[Cu_t]:[L_t]$, ($\text{log}_{10} = -0.5$) and 16% at the greatest ratio ($\log_{10}$ = 1.12), which we studied for Rag Top Mountain.

$\overline{K}'$ values derived from the one-component Scatchard models all gave accurate predictions of the observed concentration of free Cu^{+2}, at the smallest MLR; however, the one-component Scatchard estimates all resulted in underestimates at the greater MLR. This observation results from an overestimate of the strengths of the binding sites, which are filled at the very greatest MLR. For this reason, the initial (maximum) one-component Scatchard estimates give accurate predictions only under conditions of a great excess of binding sites, which is generally the case under natural conditions, when the $\log_{10}$ of MLR may be less than -3.0, and under conditions of contamination, when the ratio may be greater than $+1.0$, in many circumstances. However, the predictions from the one-component Scatchard model rapidly deviate from the observed values, as the number of sites filled increases (Fig. 6A) and would underestimate the amount of Cu^{+2} present. The underestimation of the relative amount of free Cu^{+2} at MLR between $+0.2$ and -0.3 were approximately 13% and 42%, respectively.

The one-component Gaussian-Scatchard model for the fractions studied gave predictions which were similar to the measured values of Cu^{+2} (Fig. 6B). Depending on the location and fraction and whether an estimate pooled by fraction or location was used, we observed both over- and underestimates. The one-component Gaussian-Scatchard model averaged across both fractions and all locations closely predicted the observed concentrations of Cu. The greatest deviation was at the smallest MLR.

In general, the three-component Gaussian-Scatchard model gave the poorest predictions of Cu speciation. This was surprising, because in other studies, we have found this model to have good predictive power (Giesy, 1986a; 1986b). The three-component Gaussian-Scatchard model derived for each location most closely predicted the observed speciation of Cu in these titrations. However, the pooled estimates either over- or underestimated the proportion of free Cu^{+2} in

solution. For instance, in the example presented here, we observed an overestimation error of 255% at the lowest MLR.

Comparison of simulated to observed speciation in whole waters

We again used the thermodynamic, geochemical speciation model, GEOCHEM, to predict the speciation of Cu in whole waters from the six locations studied (Fig. 7). In this investigation, the model was parameterized to values measured in the waters. We compared the observed concentrations of Cu^{+2} as measured by selective ion electrode of two concentrations of copper added: 1.26×10^{-5} or 5.66×10^{-5} g · atom Cu/l (0.8 or 3.6 mg/l). Here we report the results of these simulations for F II from Rag Top Mountain Pond. At each concentration, we compared the observed concentration of the % Cu bound for the two total concentrations of copper added to these waters. We compared the results of all four models with estimates derived from the Rag Top Mountain F II data (RTM-II) or values for F II pooled across all six locations ($\overline{X}$-6 F II) or pooled across both F II and F III from all six locations ($\overline{X}^{-6}$, F II, F III).

The results of the simulations demonstrate that, for the two concentrations studied, the one-component Scatchard model consistently gives the best agreement with the observed speciation of copper. The one-component Gaussian model gave good predictions of Cu speciation at both concentrations when the Rag Top Mountain F II data was used, but underestimated the amount of Cu bound when pooled estimates were used. The two-component Scatchard model and three-component Log_{10}-Gaussian-Scatchard model both underestimated the degree of Cu binding depending on the data set and degree of pooling used. The maximum relative error, based on the proportion of bound Cu observed, was 20%. Therefore, we concluded that any one of the four models gives acceptable predictive power but some give better predictive power under different conditions. Furthermore, at least for the area of

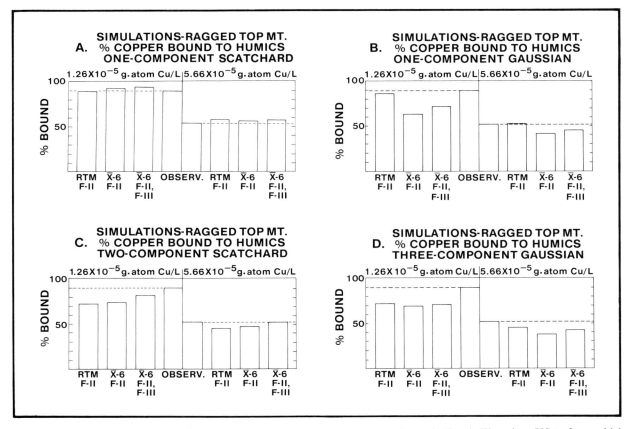

Fig. 7. Observed bound Cu (organic + inorganic) in water from Rag Top Mountain Pond, Wyoming, USA, from which particulates $\geq 0.15\ \mu m$ have been removed by centrifugation. Free copper was measured by selective ion electrode and bound Cu calculated by mass balance at two concentrations of total Cu. Speciation was predicted by GEOCHEM, parameterized with concentrations of constituents (Tables 4, 5, and 9) and the appropriate conditional stability constants. RTM F-II is the result of the simulation, using the value of $\overline{K}'$ determined by each of the four methods for that ultrafilter fraction binding of Cu: $\overline{X}6$ F-II is the mean value for $\overline{K}'$ averaged across all locations and used to predict the speciation of Cu in Rag Top Mountain; $\overline{X}6$-F II, F III is the result of the simulation where the overall conditional stability constant was the mean value pooled across fractions F II and F III for all six locations.

Wyoming studied, the Cu binding properties of ultrafilterable material from six surface waters of different inorganic compositions were fairly similar. While the best prediction was obtained by using estimates derived for a particular ultrafilter fraction and location, pooled estimates, based on both fractions and all six locations, gave adequate predictions of the speciation observed for other locations.

Conclusions

The binding of Cu by ultrafilterable material $\leq 0.15\ \mu m$ nominal diameter, which was isolated from six surface waters in the vicinity of Laramie, Wyoming, USA was examined. The material isolated from these six ponds had ash contents which ranged from 21 to 80% and was positively correlated with the dissolved inorganic constituents of the water from which it was isolated. The binding capacity for Cu(II) varied from location to location and was positively correlated with the ash content of the ultrafilterable material, which indicates that inorganic as well as organic matter in the ultrafilterable material may have been responsible for binding of Cu.

The binding capacity of the larger, nominal diameter material was approximately twice as great as that of the smaller diameter material, but this difference was not statistically significant. This suggests that while there seem to be differences in binding capacity by different nominal diameter fractions and from different locations within a region, that these differences, relative to the range of values, reported in the literature, were rather small. Therefore, extensive fractionation of ultrafilterable material does not seem warranted to obtain reasonable estimates for use in geochemical simulation models. Furthermore, the values from one location in a geographical region seem adequate to estimate the speciation of metals in adjacent waters.

Two discrete and two continuous models to estimate overall conditional stability constants for the binding of Cu by ultrafilterable materials were investigated. We found that each of the methods provided adequate estimates but that under the conditions studied here the two-component Scatchard model gave the best prediction of the speciation of Cu in both titrations and whole waters. Under conditions, where the ratio of metal to ligand concentrations is small the one-component Scatchard model is also appropriate. Furthermore, a sensitivity analysis demonstrated that average overall conditional stability constants, calculated for the region, could be used to predict speciation in adjacent but inorganically different surface waters.

The conditional stability constants, when used in the thermodynamic geochemical simulation model GEOCHEM, accurately predicted the speciation of Cu as measured by selective ion electrodes. The overall conditional stability constants which we measured do not represent perfect theoretical descriptions of the individual binding sites of ultrafilterable material for Cu and do not provide insight into the exact nature of these sites. The methods described here do, however, allow one to make predictions of the relative speciation of copper in surface waters over a range of metal to ligand ratios.

Acknowledgements

We gratefully acknowledge the assistance of L. Briese, J. Alberts, and L. Alberts in sample collection and preparation. Dr. H. Bergman, Department of Zoology and Physiology, University of Wyoming, Laramie, Wyoming and his staff provided laboratory space and technical assistance. Manuscript preparation was supported by the Michigan Agricultural Experiment Station, from which this is contribution no. 12980, and the University of Georgia Marine Institute, contribution no. 628, as well as by a grant from the Fulbright Commission to J. P. Giesy and Prof. O. Hutzinger, chair of Ecological Chemistry at the University of Bayreuth, Germany. This research was supported in part by Contract EX-76-C-09-0819 between the University of Georgia's Savannah River Ecology Laboratory and the U.S. Department of Energy.

References

Alberts, J. J. & J. P. Giesy, 1983. Conditional stability constants of trace metals and naturally occurring humic materials: Application in equilibrium models and verification with field data. In: R. F. Christman & E. T. Gjessing (eds.), Aquatic and Humic Substances. pp. 333–348. Ann Arbor Science Publishers, Ann Arbor, MI.

Alberts, J. J., J. P. Giesy & D. W. Evans, 1984. Distribution of dissolved organic carbon and metal-binding capacity among ultrafilterable fractions isolated from selected surface waters of the southeastern United States. Envir. Geol. Wat. Sci. 6: 91–101.

American Public Health Association. 1976. Standard Methods for the Examination of Water and Waste Water. 14 ed., 1193 pp.

Baumann, E. W., 1976. Nephelometric determination of microgram quantities of sulfate with 3-aminoperimidine. Savannah River Laboratory Report No. ERDA, DP-1437, 17 pp.

Bloom, P. R. & M. B. McBride, 1979. Metal ion binding and exchange with hydrogen ions in acid-washed peat. Soil Sci. Soc. Am. J. 43: 687–692.

Bresnahan, W. T., C. L. Grant & S. H. Weber, 1978. Stability constants for the complexation of copper(II) ions with

L. I. H. E.
THE MARKLAND LIBRARY
STAND PARK RD., LIVERPOOL, L16 9JD

water and soil fulvic acids measured by an ion selective electrode. Analyt. Chem. 50: 1675–1679.

Buffle, J., 1980. A critical comparison of studies of complex formation between copper(II) and fulvic substance of natural waters. Analyt. Chim. Acta 118: 29–44.

Buffle, J. & P. Deladoey, 1982. Analysis and characterization of natural organic matters in freshwaters, II. Comparison of the properties of waters of various origins and their annual trend. Schweiz. Z. Hydrol. 44: 363–391.

Buffle, J., F. Greter & W. Haerdi, 1977. Measurement of complexation properties of humic and fulvic acids in natural waters with lead and copper ion-selective electrodes. Analyt. Chem. 49: 216–222.

Bunzl, K., A. Wolf & B. Sansoni, 1976. Kinetics of ion exchange in soil organic matter. V. Differential ion exchange reactions of Cu^{+2}, Cd^{+2}, Zn^{+2} and Ca^{+2} ions in humic acid. Z. Pf/Ernahr. Dung. Bodenk. 4: 415–485.

Choppin, G. R. & L. Kullberg, 1978. Protonation thermodynamics of humic acid. J. Inorg. Nucl. Chem. 40: 651–654.

Christman, R. F. & E. T. Gjessing (eds.), 1983. Aquatic and Terrestrial Humic Materials. Ann Arbor Science Publishers, Ann Arbor, MI. 538 pp.

Dye, J. L. & V. A. Nicely, 1971. A general purpose curve-fitting program for class and research use. J. Chem. Education. 48: 443–448.

Gamble, D. S., A. W. Underdown & C. H. Langford, 1980. Copper(II) titration of fulvic acid ligand sites with theoretical potentiometric and spectrophotometric analysis. Analyt. Chem., 52: 1901–1908.

Giesy, J. P., 1980. Cadmium interactions with naturally occurring organic ligands. In: J. O. Nriagu (ed.), Cadmium in the Environment, Part I. John Wiley & Sons, New York, NY, pp. 237–256.

Giesy, J. P., 1983a. Biological control of trace metal equilibria in surface waters. In: G. G. Leppard (ed.), Trace Element Speciation in Surface Waters and its Ecological Implications. pp. 195–210. North Atlantic Treaty Organisation, Plenum Press.

Giesy, J. P.,, 1983b. Metal binding capacity of soft, acid, organic-rich waters. Toxicol. Envir. Chem. 6: 203–224.

Giesy, J. P., J. J. Alberts & D. W. Evans, 1986a. Conditional stability constants and binding capacities for copper(II) by dissolved organic carbon isolated from surface waters of the southeastern United States. Envir. Toxicol. Chem. 5: 139–154.

Giesy, J. P. & L. A. Briese, 1977. Metals associated with organic carbon extracted from Okefenokee Swamp water. Chem. Geol. 20: 109–120.

Giesy, J. P., R. A. Geiger, N. R. Kevern & J. J. Alberts, 1986b. UO_2^{+2}-Humate interactions in soft, acid, humate-rich waters. J. Envir. Radioactivity 4: 39–64.

Giesy, J. P., A. Newell & G. J. Leversee, 1983. Copper speciation in soft, acid, humic waters: Effects on copper bioaccumulation by and copper toxicity to Simocephalus serrulatus (Daphnidae). Sci. Total Envir. 28: 23–26.

Grim, R. E., 1968. Clay Mineralogy. McGraw-Hill Book Co., New York, 596 15 pp.

Guy, R. P. & C. L. Chakrabarti, 1976. Studies of metal-organic interactions in model systems pertaining to natural waters. Can. J. Chem. 54: 2600–2611.

Guy, R. D., C. L. Chakrabarti & L. L. Schrumm, 1975. The application of a simple chemical model of natural waters to metal fixation in particulate matter. Can. J. Chem. 53: 661–669.

Helwig, J. T. & K. A. Council, 1979. SAS User's Guide. SAS Institute, Raleigh, NC. 494 pp.

Kribek, B. & J. Podlaha, 1980. Transport of iron by natural and synthetic humic acids in oxidation environment. Org. Geochim. 2: 93–97.

Li, W. C., D. M. Victor & C. L. Chakrabarti, 1980. Effect of pH and uranium concentration on interaction of uranium (VI) and uranium (IV) with organic ligands in aqueous solutions. Analyt. Chem. 52: 520–523.

Mantoura, R. F. C. & J. P. Riley, 1975. The use of gel filtration in the study of metal binding by humic acids and related compounds. Analyt. Chim. Acta 78: 193–200.

Mantoura, R. F. C. & A. Dickson & J. P. Riley, 1978. The complexation of metals with humic materials in natural waters. Estuar. Coast. Mar. Sci. 6: 387–409.

McKnight, D. M., G. L. Feder, E. M. Thurman, R. L. Weshaw & J. C. Westall, 1983. Empirical estimation of complexation of copper by aquatic humic substances. Sci. Total Envir. 28: 65–76.

Perdue, E. M. & C. R. Lytle, 1983. A critical examination of metal-ligand complexation models: Application to defined multiligand mixtures. In: R. F. Christman & E. T. Gjessing (eds.), Aquatic and Terrestrial Humic Materials. Ann Arbor Science Publishers, Ann Arbor, MI. pp. 295–314.

Perdue, E. M., J. H. Reuter & R. S. Parrish, 1984. A statistical model of proton binding by humus. Geochem. Cosmochim. Acta 48: 1257–1263.

Perhac, R. M., 1972. Distribution of Cd, Co, Cu, Fe, Mn, Ni, Pb and Zn in dissolved and particulate solids from two streams in Tennessee. J. Hydrobiol. 15: 177–186.

Posner, A. M., 1966. Humic acids extracted by various reagents from a soil. I. Yield, inorganic components and titration curves. J. Soil Sci. 17: 65–78.

Sanders, J. R. & C. Bloomfield, 1980. The influence of pH, ionic strength and reactant concentration on copper complexing by humified organic matter. J. Soil Sci. 31: 53–63.

Saar, R. A. & J. H. Weber, 1979. Complexation of cadmium(II) with water- and soil-derived fulvic acids: Effect of pH and fulvic acid concentration. Can. J. Chem. 57: 1263–1268.

Scatchard, G., 1949. The attractions of proteins for small molecules and ions. Ann. N.Y. Acad. Sci. 51: 660–672.

Scheinberg, I. H., 1982. Scatchard plots. Science 215: 312–313.

Schnitzer, M. & S. U. Khan, 1972. Humic Substances in the Environment. Marcel Dekker, Inc., New York. p. 203.

Shuman, M. S., B. J. Collins, P. J. Fitzgerald & D. C. Olson, 1983. Spectrum analysis of ion-selective and potentiometric data. In: R. F. Christman & E. T. Gjessing (eds.), Aquatic and Terrestrial Humic Materials. Ann Arbor Science Publishers, Ann Arbor, MI. pp. 349–370.

Sokal, R. R. & F. J. Rolf, 1973. Introduction to iostatistics. W. H. Freeman, San Francisco, CA, 368 pp.

Sposito, G. 1981. Trace metals in contaminated waters. Envir. Sci. Technol. 15: 396–403.

Sposito, G. & S. V. Mattigod, 1979. GEOCHEM: A computer program for the calculation of chemical equilibria in soil solutions and other natural water systems. Kearney Foundation of Soil Science, University of California at Riverside. 179 pp.

Vandenberg, C. M. G. & J. R. Kramer, 1979. Determination of complexing capacities of ligands in natural waters and conditional stability constants of copper complexes by means of manganese dioxide. Analyt. Chim. Acta 106: 113–120.

Zunino, H. & J. P. Martin, 1977. Metal-binding organic macromolecules in soil: 1. Hypothesis interpreting the role of soil organic matter in the translocation of metal ions from racks to biological systems. Soil Sci. 123: 65–76.

Zunino, H., M. Aguilera, M. Caiozzi, P. Peirano, F. Borie & J. P. Martin, 1977. Metal binding organic molecules in soil: 2. Characterization of the maximum binding ability of the macromolecules. Soil Sci. 128: 257–265.

279178